Solid State Physics

Solid State Physics

GIUSEPPE GROSSO

Professor of Solid State Physics, Department of Physics,
University of Pisa, and INFM

GIUSEPPE PASTORI PARRAVICINI

Professor of Solid State Physics, Department of Physics,
University of Pavia, and INFM

ACADEMIC PRESS
A Harcourt Science and Technology Company

San Diego • San Francisco • New York • Boston
London • Sydney • Tokyo • Toronto

Academic Press
A Harcourt Science and Technology Company
525 B Street, Suite 1900, San Diego, California 92101-4495, USA
http://www.academicpress.com

Academic Press
32 Jamestown Road, London NW1 7BY, UK
http://www.academicpress.com

ISBN 0-12-304460-X

Library of Congress Catalog Card Number: 99-64032

A catalogue record for this book is available from the British Library

Printed in Great Britain by Cambridge University Press, Cambridge

00 01 02 03 04 05 CUP 9 8 7 6 5 4 3 2 1

Contents

Chapter XII   Optical properties of semiconductors and insulators   425

Chapter XIII   Transport in intrinsic and homogeneously doped semiconductors   473

Chapter XIV   Transport in inhomogeneous semiconductors   506

CONTENTS SYNOPSIS

I, II, III	Introductory information and propaedeutic subjects
IV, V VI, VII	Electronic structure of crystals
VIII, IX	Adiabatic principle and lattice vibrations
X, XI, XII XIII, XIV	Scattering; optical and transport properties
XV, XVI XVII	Magnetic field effects and magnetism
XVIII	Superconductivity

Preface

This textbook has developed from the experience of the authors in teaching the course of Solid State Physics to students in physics at the Universities of Pavia and Pisa. The book is addressed to students at the graduate and advanced level, both oriented toward theoretical and experimental activity. No particular prerequisite is required, except for the ordinary working knowledge of wave mechanics. The degree of difficulty increases somewhat as the book progresses; however, the contents develop always in very gradual steps.

The material presented in the book has been assembled to make as economical as possible the didactical task of teaching, or learning, the various subjects. The general organization in chapters, and groups of chapters, is summarized in the synoptic table of contents. The first three chapters have a propaedeutic nature to the main entries of the book. Chapter IV starts with the analysis of the electronic structure of crystals, one of the most traditional subjects in solid state physics; chapters V and VI concern the band theory of solids and a number of specific applications; the concepts of excitons and plasmons are given in chapter VII. Then, in chapters VIII and IX, the adiabatic principle and the interdependence of electronic states and lattice dynamics are studied. Having established the electronic and vibrational structure of crystals, the successive chapters from X to XIV describe several investigative techniques of crystalline properties; these include scattering of particles, optical spectroscopy, and transport measurements. Chapters XV, XVI and XVII concern the electronic magnetism of tendentially delocalized or localized electronic systems, and cooperative magnetic effects. The final chapter is an introduction to the world of superconductivity.

From a didactical point of view, an effort is made to remain as rigorous as possible, while keeping the presentation at an accessible level. In this book, on one hand we aim to give a clear presentation of the basic physical facts, on the other hand we wish to describe them by rigorous theoretical and mathematical tools. The technical side is given the due attention and is never considered optional; in fact, a clear supporting theoretical formalism (without being pedantic) is essential to establish the limits of the physical models, and is basic enough to allow the reader eventually to move on his or her own legs, this being the ultimate purpose of a useful book.

The various chapters are organized in a self-contained way for the contents, appendixes (if any), references, and have their own progressive numeration for tables, figures and formulae (the chapter number is added when addressing items in chapters differing from the running one). With regards to the references, these are intended only as orientative, since it is impossible to mention, let alone to comment on, all the relevant contributions of the wide literature. Since the chapters are presented in a (reasonably) self-contained way, the lecturers and readers are not compelled to follow the order in which the various subjects are discussed; the chapters, or group of chapters, can be taken up with great flexibility, selecting those topics that best fit personal tastes or needs. We will be very interested and very pleased to receive (either directly or by correspondence) comments and suggestions from lecturers and readers.

The preparation of a textbook, although general in nature, requires also a great deal of specialized information. We consider ourselves fortunate for the generous help of the colleagues and friends at the Physics Departments of the University of Pavia, University of Pisa, and Scuola Normale Superiore of Pisa; they have contributed to making this textbook far better by sharing their expertise with us. Very special thanks are due to Emilio Doni, who survived the task of reading and commenting on the whole preview-manuscript. Several other colleagues helped with their critical reading of specific chapters; we are particularly grateful to Lucio Claudio Andreani, Antonio Barone, Pietro Carretta, Alberto Di Lieto, Giorgio Guizzetti, Franco Marabelli, Liana Martinelli, Attilio Rigamonti. Many heartfelt thanks are due to Saverio Moroni for his right-on-target comments.

Before closing, we wish to thank all contributors and publishers, who gave us permission to reproduce their illustrations; their names and references are indicated adjacent each figure. We wish also to express our gratitude to Gioia Ghezzi for her encouragement, to Serena Bureau, Manjula Goonawardena, Cordelia Sealy and Bridget Shine for their assistance in the preparation of the manuscript. Last, but always first, we thank our families for their never ending trust that the manuscript would eventually be completed and would be useful to somebody.

Pavia and Pisa, May 1999

Giuseppe Grosso
Giuseppe Pastori Parravicini

I

Electrons in one-dimensional periodic potentials

Low dimensionality offers a unique opportunity to introduce some relevant concepts of solid state physics, keeping the treatment at a reasonably simple level. In the early days of solid state physics, just this desire for technical simplicity was the motivation for the studies on one-dimensional periodic potentials. More recently, with the development of semiconductor microstructures, low-dimensional models have had a renaissance for the understanding of a number of realistic situations. At the same time a note of warning is necessary: the features of one-dimensional systems that have any relevance beyond dimensionality must be assessed situation by situation; cavalier extensions of one-dimensional results to actual three-dimensional crystals may be misleading or even completely unreliable.

The material of this chapter is organized and presented so that it can also be embodied in standard courses of "Structure of Matter" or "Quantum Mechanics"; in fact no prior knowledge is needed other than elementary ideas about wave mechanics.

We begin with the presentation of the Bloch theorem for one-dimensional periodic lattices. A peculiar aspect of the energy spectrum of an electron in a periodic potential is the presence of allowed and forbidden energy regions. One-dimensional approaches are particularly suited to show from different points of view (weak binding, tight-binding, quantum tunneling) the mechanism of formation of energy bands in solids.

The semiclassical dynamics of electrons in energy bands is then considered; together with the Pauli exclusion principle for occupation of states, it gives an orientative distinction between metals, semiconductors and insulators.

1   The Bloch theorem for one-dimensional periodicity

Consider an electron in a one-dimensional potential $V(x)$ and the corresponding Schrödinger equation

$$-\frac{\hbar^2}{2m}\frac{d^2\psi(x)}{dx^2} + V(x)\,\psi(x) = E\,\psi(x)\;. \tag{1}$$

The solutions of Eq. (1) for several typical forms of the potential $V(x)$ are well known; familiar models include the free-electron case $V(x)=0$, the harmonic oscillator $V(x) = (1/2)Kx^2$, the potential $V(x)=eFx$ due to a uniform electric field F, quantum wells, and others. We focus here on the general properties of Eq. (1) in the case where $V(x)$ is the periodic potential of a one-dimensional crystal of lattice constant a.

A potential $V(x)$, of period a, satisfies the relation

$$V(x) = V(x + ma)\;, \tag{2}$$

with m arbitrary integer. The Fourier transform of a periodic potential $V(x)$ includes only plane waves of wavenumbers $h_n{=}n\,2\pi/a$, and $V(x)$ can be expressed in the form

$$V(x) = \sum_{n=-\infty}^{+\infty} V_n\,e^{ih_n x}\;. \tag{3}$$

In general, if $V(x)$ is not periodic, it can still have a continuous Fourier transform $V(q)$ so that

$$V(x) = \int_{-\infty}^{+\infty} V(q)\,e^{iqx}\,dq\;. \tag{4}$$

We wish to analyse the implications on the eigenfunctions and eigenvalues of Eq. (1) brought about by the fact that the potential $V(x)$ is periodic, and hence its Fourier spectrum is discrete, according to Eq. (3).

Let us start considering Eq. (1) in the particular case that the periodic potential $V(x)$ vanishes. In the free-electron case, the wavefunctions are simply plane waves and can be written in the form

$$W_k(x) = \frac{1}{\sqrt{L}}e^{ikx}\;. \tag{5}$$

The normalization constant has been chosen so that $W_k(x)$ is normalized to 1 in the interval $0 \leq x \leq L$ (and the length L of the crystal is understood in the limit $L \to \infty$ whenever necessary). The wavenumbers k are real and the eigenvalues are $E(k){=}\hbar^2k^2/2m$. The plane waves (5) constitute a complete set of orthonormal functions, that can be conveniently used as an expansion set.

Let us now consider the eigenvalue problem (1), when the potential $V(x)$ is periodic and thus satisfies Eq. (3). If we apply the operator $H = (p^2/2m) + V(x)$ to the plane wave $W_k(x)$, we see that $H|W_k(x)\rangle$ belongs to the subspace $\mathbf{S}_k$ of plane waves of wavenumbers $k + h_n$:

$$\mathbf{S}_k \equiv \{W_k(x), W_{k+h_1}(x), W_{k-h_1}(x), W_{k+h_2}(x), W_{k-h_2}(x)\ldots\} \ .$$

We also notice that the subspace $\mathbf{S}_k$ is *closed* under the application of the operator H to any of its elements; thus the diagonalization of the Hamiltonian operator within the given subspace $\mathbf{S}_k$ provides rigorous eigenfunctions of H, that can be labelled as $\psi_k(x)$. Notice that two subspaces $\mathbf{S}_k$ and $\mathbf{S}_{k'}$ are different if k and k' are *not* related by integer multiples (positive, negative or zero) of $2\pi/a$; on the contrary, if $k \equiv k' + n2\pi/a$, then the two subspaces $\mathbf{S}_k$ and $\mathbf{S}_{k'}$ coincide. This allows us to define a fundamental region of k-space, limited by $-\pi/a < k \leq \pi/a$, which includes all the different k labels giving independent $\mathbf{S}_k$ subspaces; this fundamental region, of length $2\pi/a$, is named *first Brillouin zone* (or simply *Brillouin zone*).

Any generic wavefunction $\psi_k(x)$, obtained by diagonalization of H within the subspace $\mathbf{S}_k$, can be expressed as an appropriate linear combination of the type

$$\psi_k(x) = \sum_n c_n(k) \frac{1}{\sqrt{L}} e^{i(k+h_n)x} \ . \tag{6}$$

It is convenient to denote by $u_k(x)$ the function

$$u_k(x) = \sum_n c_n(k) \frac{1}{\sqrt{L}} e^{ih_n x} = \sum_n c_n(k) \frac{1}{\sqrt{L}} e^{in(2\pi/a)x} \ .$$

It is evident that $u_k(x)$ is a function with the same periodicity as $V(x)$; Eq. (6) then takes the form

$$\boxed{\psi_k(x) = e^{ikx} u_k(x)} \tag{7}$$

where $u_k(x + a) = u_k(x)$ is a *periodic function*. This expresses the Bloch theorem: *any (physically acceptable) solution of the Schrödinger equation in a periodic potential takes the form of a travelling plane wave modulated on the microscopic scale by an appropriate function with the lattice periodicity.*

The Bloch theorem, summarized by Eq. (7), can also be written in the alternative form

$$\boxed{\psi_k(x + t_n) = e^{ikt_n} \psi_k(x)} \tag{8}$$

where $t_n = na$ is any translation in the direct lattice. It is easy to verify that Eq. (7) implies Eq. (8), and vice versa [to demonstrate the opposite case, multiply both members of Eq. (8) by $\exp(-ikx - ikt_n)$ and denote by $u_k(x)$ the resulting periodic function]. The Bloch theorem, in the form of Eq. (8), shows that the values of the wavefunction $\psi_k(x)$ in any two points differing by a translation t_n are related by the phase $\exp(ikt_n)$.

We can finally notice that, in the case of a general potential of type (4), the discretized expansion of Eq. (6) does not hold any more; in general the expansion of $\psi(x)$

includes all plane waves in the form

$$\psi(x) = \int_{-\infty}^{+\infty} c(q)\, e^{iqx}\, dq \ . \tag{9}$$

Nothing specific and general can be inferred from Eq. (9) about the properties of the wavefunctions and the energy spectrum of H: in fact, for aperiodic potentials, it is possible to find localized wavefunctions for the whole spectrum, itinerant ones, both types (separated by mobility edges), or even fractal solutions.

The Bloch theorem plays a central role in the physics of periodic systems; not only it characterizes the *itinerant form of the wavefunctions* (summarized by Eq. 7 and Eq. 8), but also entails the fact that the energy spectrum consists, in general, of *allowed energy regions separated by energy gaps* (as discussed in the forthcoming sections). The eigenvalues $E = E(k)$ of the Schrödinger equation (1), when plotted as a function of k within the first Brillouin zone, describe the band structure of the crystal; notice that $E(k) = E(-k)$, as can be seen from direct inspection of Eq. (1) under complex conjugate operation (or in a more formal way by use of group theory analysis of time-reversal symmetry).

A feature of one-dimensional periodic potentials is that the allowed energy bands *cannot cross* each other; in fact for any allowed energy E there are only two linearly independent solutions of the differential equations (1), and the degeneracy $E(k) = E(-k)$ rules out any degeneracy at a given k. In any allowed energy band the dispersion relation $E(k)$ is monotonic function of k for $0 \le k \le \pi/a$; the extremal energies occur only at $k = 0$ and π/a, where dE/dk in general vanishes. [The non-crossing of one-dimensional bands is confirmed also by group theory analysis; in one-dimension the group of symmetry operations (neglecting spin) is too small to imply degeneracy of bands; on the contrary, in three-dimensional crystals crossing of energy bands is possible at high symmetry points or lines in the Brillouin zone.]

So far we have considered the equation (1) in the infinite interval $-\infty \le x \le \infty$; it is essentially equivalent from a physical point of view to consider Eq. (1) in the macroscopic region $0 \le x \le L \equiv Na$ where N is a very large but finite number (N is the number of unit cells of the crystal and is of order 10^8 for $L = 1$ cm). The reason to consider a very large macroscopic region, rather than an infinite one, is simply a matter of convenience, mainly for counting states and distributing electrons in the energy bands. In order not to affect the physics by boundary effects we use cyclic or Born–von Karman boundary conditions for the wavefunctions. This consists in the requirement

$$\psi(x + Na) \equiv \psi(x) \ , \tag{10a}$$

i.e. the points x and $x + Na$ are considered as physically equivalent.

The wavefunction $\psi(x)$ must be a Bloch function of wavenumber k; then the boundary condition (10a) restricts the acceptable values of k to the ones that satisfy

$$e^{ikNa} = 1 \ ;$$

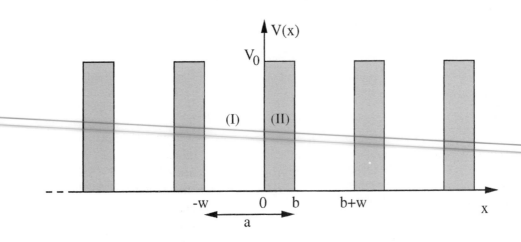

Fig. 1 One-dimensional potential formed by a periodic array of quantum wells.

the allowed values of k are discrete and are given by

$$k = \frac{2\pi}{Na} n \qquad n = 0, \pm 1, \pm 2 \ldots \tag{10b}$$

The density-of-states in k space is $L/2\pi$, which is proportional to the length of the crystal. When the macroscopic length $L = Na$ is very large, the variable k must be thought of as a dense (although discrete) variable; notice that the first Brillouin zone contains *a number of uniformly distributed k points, equal to the number N of cells of the lattice.*

2   Energy levels in a periodic array of quantum wells

One of the most elementary problems of quantum mechanics is the study of the energy levels of a particle in a single quantum well. Similarly, one of the most elementary applications of the Bloch theorem is the study of the energy bands of a particle moving in a periodic array of quantum wells. The periodically repeated quantum well model was introduced by Kronig and Penney in 1931 to replace the actual crystal potential with a much more manageable piecewise constant potential; in this way, in each well (or in each barrier) the linear independent solutions of the Schrödinger equation are simple trigonometric (or exponential) functions. Standard boundary conditions of continuity of wavefunctions and currents, combined with Bloch conditions required by periodicity, easily lead to an analytic compatibility equation for the eigenvalues of the crystal Hamiltonian. In the last decade, the Kronig–Penney model, with appropriate generalizations, has found a revival in the very active field of research of the electronic states of superlattices (superlattices are artificial materials, obtained by growing on a substrate a controlled number of layers of two or more chemically similar crystals, in an appropriate periodic sequence).

Consider an infinite periodic sequence of rectangular wells, as shown in Fig. 1; the lattice constant of the periodic array is $a = w + b$. Within the unit cell $-w < x < b$, the well region $-w < x < 0$ and the barrier region $0 < x < b$ are denoted as I and II, respectively. In regions I and II, the general solution of the Schrödinger equation (1) for energies $0 < E < V_0$ has the form

$$\begin{cases} \psi_I(x) = A\,e^{iqx} + B\,e^{-iqx} & -w < x < 0 \qquad q(E) = \sqrt{2mE/\hbar^2} \\ \psi_{II}(x) = C\,e^{\beta x} + D\,e^{-\beta x} & 0 < x < b \qquad \beta(E) = \sqrt{2m(V_0 - E)/\hbar^2} \end{cases} \qquad (11)$$

where $q(E)$ is the propagation wavenumber in the well, $\beta(E)$ refers to the barrier, and A, B, C, D are arbitrary constants (we consider here positive energies E smaller than V_0; energies E larger than V_0 could be dealt with in a similar way).

The four arbitrary constants A, B, C, D must be chosen in such a way to satisfy the following four boundary conditions:

$$\psi_I(0) = \psi_{II}(0)\,, \qquad \left(\frac{d\psi_I}{dx}\right)_{x=0} = \left(\frac{d\psi_{II}}{dx}\right)_{x=0}\,, \qquad (12a)$$

$$\psi_{II}(b) = e^{ika}\,\psi_I(-w)\,, \qquad \left(\frac{d\psi_{II}}{dx}\right)_{x=b} = e^{ika}\left(\frac{d\psi_I}{dx}\right)_{x=-w}\,. \qquad (12b)$$

The two conditions (12a) impose the continuity of the wavefunction and its derivative at $x = 0$. The two conditions (12b) connect the wavefunction and its derivative at $x = -w$ and $x = b$ via the phase factor $\exp(ika)$, as required by the Bloch theorem.

The conditions (12) lead to the following linear homogeneous equations for the coefficients A, B, C, D:

$$\begin{cases} A + B = C + D \\ A\,iq - B\,iq = C\,\beta - D\,\beta \\ C\,e^{\beta b} + D\,e^{-\beta b} = e^{ika}\left[A\,e^{-iqw} + B\,e^{+iqw}\right] \\ C\,\beta e^{\beta b} - D\,\beta e^{-\beta b} = e^{ika}\left[A\,iqe^{-iqw} - B\,iqe^{+iqw}\right] \end{cases}\,.$$

The above four equations have a solution only if the determinant of the coefficients A, B, C, D vanishes:

$$\begin{vmatrix} 1 & 1 & -1 & -1 \\ iq & -iq & -\beta & \beta \\ -e^{ika-iqw} & -e^{ika+iqw} & e^{\beta b} & e^{-\beta b} \\ -iqe^{ika-iqw} & iqe^{ika+iqw} & \beta e^{\beta b} & -\beta e^{-\beta b} \end{vmatrix} = 0\,.$$

The determinant can be evaluated, for instance, by expanding it with respect to the first row and by direct evaluation of the four 3×3 minors. With easy calculations and collection of terms we obtain

$$\boxed{\frac{\beta^2 - q^2}{2q\beta}\,\sinh\beta b\,\sin qw + \cosh\beta b\,\cos qw = \cos ka}\,. \qquad (13)$$

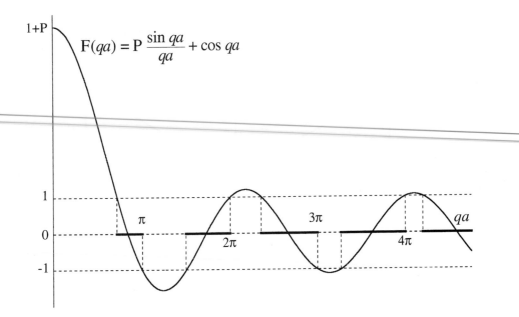

Fig. 2 Graphical solution of the compatibility equation $P(qa)^{-1}\sin qa + \cos qa = \cos ka$ of the Kronig–Penney model. The dimensionless parameter P has been set equal to $3\pi/2$; the qa regions where the compatibility equation is satisfied have been enhanced for convenience.

It is possible to solve graphically the compatibility equation (13); although not necessary, it is however standard practice to adopt the further simplification to let the width of the potential barriers to approach zero and simultaneously the height of the barriers to approach infinity so to conserve finite the area underneath. We thus arrive at a model potential consisting of a periodic sequence of δ-like potential barriers. In Eq. (13) we consider $b \to 0$ under the constraint $V_0 b$ constant; we obtain the simplified compatibility equation

$$\boxed{P\frac{\sin qa}{qa} + \cos qa = \cos ka} \tag{14}$$

where $P = mV_0 ba/\hbar^2$ is a dimensionless parameter, proportional to the area $V_0 b$ of the barrier.

The graphical solution of Eq. (14) is obtained plotting the function $F(qa)$, given by the first member of Eq. (14), versus the dimensionless variable qa (see Fig. 2); the qa regions, where $|F(qa)| \leq 1$, provide the energy levels $E=\hbar^2 q^2/2m$ which are allowed. From Fig. 2, it can be seen by inspection that the energy levels are grouped into allowed energy bands separated by forbidden energy regions.

It is instructive to consider the compatibility equation (14) in the limiting cases of P very small or very large. In the case $P = 0$ we obviously recover the free-electron result. In the case $P \to \infty$ the allowed energy bands become extremely narrow and the energy spectrum is composed by lines at energies E such that $q(E)a = n\pi$ $(n = 1, 2, \dots)$.

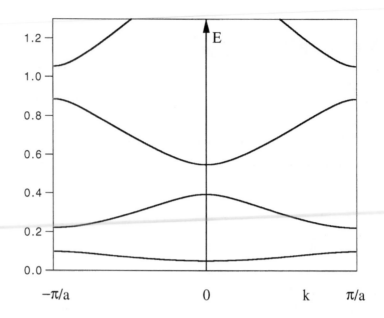

Fig. 3 Energy bands of lower energy in the Kronig–Penney model for $P = 3\pi/2$. The energy E (in Rydberg) is plotted as a function of k in the first Brillouin zone; the value of the lattice parameter considered is $a = 10\,a_B$ (a_B Bohr radius).

The energy levels are

$$E_n = n^2 \frac{\hbar^2}{2m} \frac{\pi^2}{a^2} \qquad (n = 1, 2, \dots) \,, \tag{15}$$

and coincide (as expected) with the energy levels of a quantum well, of width a, and infinitely high potential barriers; Eq. (15) can also be written in the form

$$E_n = n^2 \frac{\hbar^2}{2m\,a_B^2} \frac{\pi^2}{(a/a_B)^2} \quad \text{or equivalently} \quad E_n = n^2 \frac{9.87}{(a/a_B)^2} \quad \text{in Rydberg}$$

where $a_B = 0.529\,\text{Å}$ is the Bohr radius, and $\hbar^2/2m\,a_B^2 = 1\,\text{Ryd} = 13.606\,\text{eV}$.

The allowed energy bands in the Kronig–Penney model can be calculated for any chosen value of P, and in Fig. 3 we give the typical energy band structure. We can see in particular that the extrema of E versus k appear at the center and at the border of the Brillouin zone (as required for any one-dimensional potential).

3   Electron tunneling and energy bands

3.1   Transmission and reflection of electrons through an arbitrary potential

Generalities on the transfer matrix method

In this section we consider the electron propagation in one-dimensional crystals from the point of view of the *electron optics in solid state*, via the determination of the

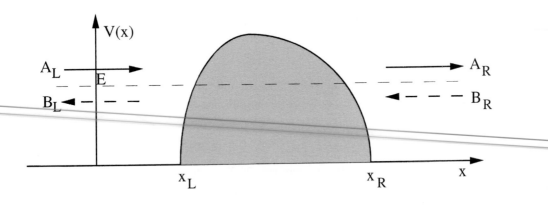

Fig. 4 Schematic representation of a potential (of finite range and arbitrary shape) connecting two leads at zero potential. Arrows indicate the wavevectors of the propagating plane waves (of energy E), and A_L, B_L, A_R, B_R the amplitudes.

reflected and transmitted components of a wave impinging on a given potential. We present first some general features of the elastic tunneling through a potential of arbitrary shape; then, in the case of periodic potentials, we illustrate the formation of the energy bands from the point of view of quantum tunneling.

Consider a potential of arbitrary profile, extending in the region $x_L \leq x \leq x_R$, and connecting two semi-infinite regions (called "leads") at the same constant potential, taken to be zero for simplicity (see Fig. 4). The general solution of the Schrödinger equation (1) for a positive energy E, in the left and right leads, can be written as

$$\begin{cases} \psi_L(x) = A_L\, e^{iqx} + B_L\, e^{-iqx} & x \leq x_L \\ \psi_R(x) = A_R\, e^{iqx} + B_R\, e^{-iqx} & x \geq x_R \end{cases} \tag{16}$$

where $q(E) = \sqrt{2mE/\hbar^2}$. Since the Schrödinger equation is linear and of second order, the amplitudes (A_R, B_R) on the right-side lead depend linearly from the amplitudes (A_L, B_L) on the left-side lead; we can thus write

$$\begin{pmatrix} A_R \\ B_R \end{pmatrix} = S(E) \begin{pmatrix} A_L \\ B_L \end{pmatrix} = \begin{pmatrix} s_{11}(E) & s_{12}(E) \\ s_{21}(E) & s_{22}(E) \end{pmatrix} \begin{pmatrix} A_L \\ B_L \end{pmatrix} \tag{17}$$

where $S(E)$ is called *transfer matrix*.

From Eq. (17) we can write

$$\begin{cases} A_R = s_{11}\, A_L + s_{12}\, B_L \\ B_R = s_{21}\, A_L + s_{22}\, B_L \end{cases} \tag{18}$$

(for simplicity of notations, at times the energy dependence of variables is not indicated explicitly). The transfer matrix S, corresponding to a given potential $V(x)$ and chosen energy E, can be obtained by (numerical or analytic) integration of the Schrödinger equation; before considering specific examples, we discuss some general properties.

First of all we notice that the complex conjugate of any solution of the Schrödinger equation (1) is on its own right a solution at the same energy. It follows

$$s_{11} = s_{22}^* \quad \text{and} \quad s_{12} = s_{21}^* \ . \tag{19}$$

We can also show that the matrix S is unimodular, i.e. its determinant equals unity. In fact, from the physical meaning of the amplitudes A_L, B_L, A_R, B_R in Eq. (16), and conservation of the electronic current density, we have

$$|A_L|^2 - |B_L|^2 \equiv |A_R|^2 - |B_R|^2 \ . \tag{20}$$

Using Eqs. (18) and Eqs. (19), we can rewrite Eq. (20) in the form

$$|A_L|^2 - |B_L|^2 = |s_{11} A_L + s_{12} B_L|^2 - |s_{21} A_L + s_{22} B_L|^2$$

$$= [\,|s_{11}|^2 - |s_{21}|^2\,][\,|A_L|^2 - |B_L|^2\,] \ .$$

We have thus $|s_{11}|^2 - |s_{21}|^2 = \det S = 1$: *the transfer matrix between two leads at the same potential is thus unimodular.*

Reflection and transmission amplitudes and coefficients

The transfer matrix provides an intuitive description of the electron tunneling through the given potential region. Consider in fact the stationary solution of the Schrödinger equation in the form of an impinging wave partially reflected and partially transmitted through the potential region. In this specific case we have $B_R \equiv 0$, and Eqs. (18) read

$$\begin{cases} A_R = s_{11} A_L + s_{12} B_L \\ 0 \ = s_{21} A_L + s_{22} B_L \end{cases} ; \tag{21}$$

from these two equations, we can easily obtain the reflection and transmission amplitudes and coefficients as follows.

The *reflection amplitude* r, defined as $r = B_L/A_L$, is given by

$$\boxed{r = -\frac{s_{21}}{s_{22}}} \ ; \tag{22a}$$

the *reflection coefficient* is

$$R = r\,r^* = |r|^2 = \left|\frac{s_{21}}{s_{22}}\right|^2 \ . \tag{22b}$$

The *transmission amplitude* $t = A_R/A_L$ is given (remember the unimodularity of the transfer matrix S)

$$\boxed{t = \frac{1}{s_{22}}} \ ; \tag{23a}$$

the *transmission coefficient* is

$$T = t\,t^* = |t|^2 = \left|\frac{1}{s_{22}}\right|^2 \ . \tag{23b}$$

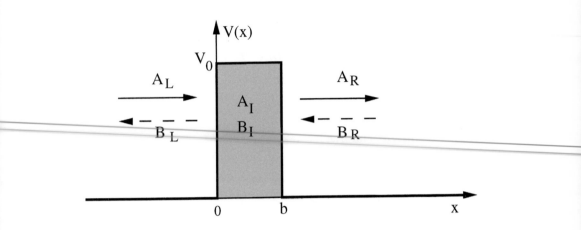

Fig. 5 Potential barrier of height V_0 and width b connecting two leads at zero potential.

We can easily verify the general property $R + T = 1$. The transfer matrix can be conveniently written in the form

$$
S = \begin{pmatrix} 1/t^* & -r^*/t^* \\ -r/t & 1/t \end{pmatrix},
$$

with a clear physical meaning of its matrix elements.

In the following we need to connect the transfer matrix S, corresponding to a given potential $V(x)$, with the transfer matrix $S(d)$ corresponding to the rigidly displaced potential $V(x-d)$. It is seen by simple elaboration of the defining equations (16) that

$$
S(d) = \begin{pmatrix} s_{11} & s_{12}\, e^{-2iqd} \\ s_{21}\, e^{2iqd} & s_{22} \end{pmatrix} ; \tag{24}
$$

in shifting the potential along the x-axis by the displacement d, the diagonal matrix elements of the transfer matrix do not change, while the off-diagonal elements acquire the phase $\exp(\pm 2iqd)$.

Transfer matrix for a rectangular barrier and for a piecewise potential

As an application of the concepts explained so far, we determine explicitly the transfer matrix for a rectangular barrier; such a matrix also constitutes the basic ingredient to build up the transfer matrix of any piecewise potential.

Consider the potential barrier of height V_0 in the interval $0 \le x \le b$, shown in Fig. 5. In the whole x axis, we can distinguish three regions: the left lead, the intermediate barrier region, and the right lead. In the three regions, the general solution of the

Schrödinger equation (1) for energies $0 < E < V_0$ has the form

$$\psi_L(x) = A_L\, e^{iqx} + B_L\, e^{-iqx} \qquad x < 0$$

$$\psi_I(x) = A_I\, e^{\beta x} + B_I\, e^{-\beta x} \qquad 0 < x < b$$

$$\psi_R(x) = A_R\, e^{iqx} + B_R\, e^{-iqx} \qquad x > b$$

where $q(E) = \sqrt{2mE/\hbar^2}$ and $\beta(E) = \sqrt{2m(V_0 - E)/\hbar^2}$.

The standard boundary conditions of continuity of wavefunction and its derivative at $x = 0$ give

$$\begin{cases} A_L + B_L &= A_I + B_I \\ A_L\, iq - B_L\, iq &= A_I\, \beta - B_I\, \beta \end{cases} . \qquad (25)$$

From Eqs. (25) we can express (A_I, B_I) in terms of (A_L, B_L) in the matrix form

$$\begin{pmatrix} A_I \\ B_I \end{pmatrix} = \frac{1}{2\beta} \begin{pmatrix} (iq + \beta) & (-iq + \beta) \\ (-iq + \beta) & (iq + \beta) \end{pmatrix} \begin{pmatrix} A_L \\ B_L \end{pmatrix} . \qquad (26a)$$

Similarly, we can consider the boundary conditions at $x = b$ and we can express (A_R, B_R) in terms of (A_I, B_I) in the form

$$\begin{pmatrix} A_R \\ B_R \end{pmatrix} = \frac{1}{2iq} \begin{pmatrix} (iq + \beta)\, e^{(-iq+\beta)b} & (iq - \beta)\, e^{(-iq-\beta)b} \\ (iq - \beta)\, e^{(iq+\beta)b} & (iq + \beta)\, e^{(iq-\beta)b} \end{pmatrix} \begin{pmatrix} A_I \\ B_I \end{pmatrix} . \qquad (26b)$$

The direct multiplication of the transfer matrices in Eq. (26a) and Eq. (26b) provides the transfer matrix S for the rectangular barrier; with straightforward calculations we obtain

$$\boxed{\begin{array}{ll} s_{11} = e^{-iqb}\left[\cosh \beta b + i\dfrac{q^2 - \beta^2}{2q\beta} \sinh \beta b\right] & s_{22} = s_{11}^* \\[2ex] s_{12} = e^{-iqb}\,(-i)\dfrac{q^2 + \beta^2}{2q\beta} \sinh \beta b & s_{21} = s_{12}^* \end{array}} \qquad (27)$$

Expressions (27) provide the transfer matrix of a barrier of height V_0 in the interval $[0, b]$; for barriers shifted by a displacement d, appropriate phase factors (see Eq. 24) are to be included. The transfer matrix for any arbitrary piecewise potential is obtained simply multiplying in the appropriate order the matrices corresponding to each component barrier.

With the transfer matrix given in Eq. (27), we can analyse explicitly the electron tunneling through a rectangular potential barrier; from the expression of the matrix element s_{11} we obtain

$$|s_{11}|^2 = 1 + \frac{(q^2 + \beta^2)^2}{4q^2\beta^2} \sinh^2 \beta b .$$

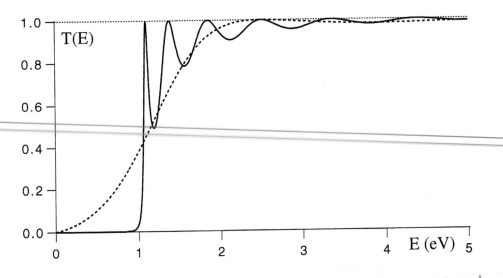

Fig. 6 Transmission coefficient through a barrier of height $V_0 = 1\,\text{eV}$ and length $b = 5\,\text{Å}$ (dashed line), and through a barrier of height $V_0 = 1\,\text{eV}$ and length $b = 20\,\text{Å}$ (solid line).

The transmission coefficient of the barrier $T(E) = 1/|s_{11}|^2$ is thus

$$T(E) = \frac{1}{1 + \dfrac{V_0^2}{4E(V_0 - E)} \sinh^2 \sqrt{\dfrac{2m(V_0 - E)b^2}{\hbar^2}}} \qquad 0 \le E \le V_0 \,. \qquad (28a)$$

For $E \ge V_0$ similar calculations can be performed and the transmission coefficient becomes

$$T(E) = \frac{1}{1 + \dfrac{V_0^2}{4E(E - V_0)} \sin^2 \sqrt{\dfrac{2m(E - V_0)b^2}{\hbar^2}}} \qquad E \ge V_0 \,. \qquad (28b)$$

For $E = V_0$ we have

$$T(V_0) = \frac{1}{1 + \dfrac{1}{4}\beta_0^2 b^2} \qquad \text{with} \qquad \hbar^2 \beta_0^2 / 2m = V_0 \,. \qquad (28c)$$

The typical behaviour of the transmission coefficient (28) of a rectangular barrier is reported in Fig. 6. For $0 \le E \le V_0$, $T(E)$ varies monotonically from zero to $T(V_0)$; for a reasonable thick barrier characterized by $\beta_0 b \gg 1$, the transmission is near to zero. For $E > V_0$ the transmission coefficient has an oscillatory behaviour and approaches asymptotically to 1. In particular $T(E)$ is exactly equal to 1 at energies such that the argument of the trigonometric function in Eq. (28b) equals an integer multiple of π.

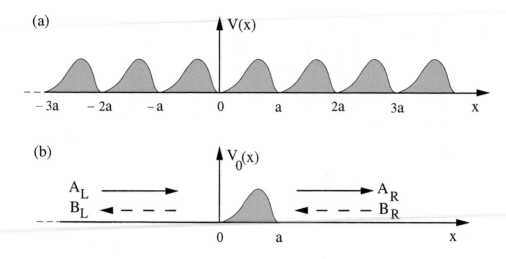

Fig. 7 Schematic representation of the periodic potential $V(x)$ of a one-dimensional crystal (a), and of the potential $V_0(x)$ within the unit cell (b).

3.2   Electron tunneling through a periodic potential

Consider a one-dimensional potential $V(x)$, of periodicity a, and let $V_0(x)$ denote the potential within the unit cell $0 \leq x \leq a$ (see Fig. 7). For simplicity, the minima of $V(x)$ are assumed to be zero and located at the positions $t_n = na$; this is always possible by an appropriate choice of origin in the energy axis and in the x axis. With the help of the Bloch theorem and the knowledge of the transfer matrix S corresponding to the potential $V_0(x)$ in the unit cell, we can express the condition of propagation of Bloch-type wavefunctions throughout the system.

As indicated in Fig. 7b, we consider the electron tunneling through the potential $V_0(x)$, which separates the two leads at zero potential extending in the regions $x \leq 0$ and $x \geq a$. We specify Eqs. (16) to the present geometry and we have

$$\begin{cases} \psi_L(x) = A_L\, e^{iqx} + B_L\, e^{-iqx} & x \leq 0 \\ \psi_R(x) = (s_{11}A_L + s_{12}B_L)\, e^{iqx} + (s_{21}A_L + s_{22}B_L)\, e^{-iqx} & x \geq a \end{cases} \qquad (29)$$

where $q(E) = \sqrt{2mE/\hbar^2}$ is the propagation wavenumber in the leads, the energy E is positive, and $s_{ij}(E)$ are the matrix elements of the transfer matrix through the potential $V_0(x)$.

We look for solutions of Eq. (29) which satisfy, at the points $x = 0$ and $x = a$, the boundary conditions required by the Bloch theorem

$$\psi_R(a) = e^{ika}\,\psi_L(0) \quad \text{and} \quad \left(\frac{d\psi_R}{dx}\right)_{x=a} = e^{ika}\left(\frac{d\psi_L}{dx}\right)_{x=0}. \qquad (30)$$

Inserting Eqs. (29) into Eqs. (30), we obtain two linear homogeneous equations for the arbitrary coefficients A_L and B_L; the two equations can be written in the compact matrix form

$$\begin{pmatrix} s_{11}\,e^{iqa} & s_{12}\,e^{iqa} \\ s_{21}\,e^{-iqa} & s_{22}\,e^{-iqa} \end{pmatrix} \begin{pmatrix} A_L \\ B_L \end{pmatrix} = e^{ika} \begin{pmatrix} A_L \\ B_L \end{pmatrix} . \tag{31}$$

The above equation shows that the 2×2 matrix in the first member of Eq. (31), has eigenvalue $\exp(ika)$; since the matrix is unimodular, it has also the eigenvalue $\exp(-ika)$ and trace $2\cos ka$. The condition of Bloch propagation and existence of allowed energy levels is thus

$$\boxed{s_{11}\,e^{iqa} + s_{22}\,e^{-iqa} = 2\cos ka} . \tag{32}$$

It is convenient to express the above equation in terms of the transmission amplitude $t = |t|\,\exp(i\phi)$; we have

$$s_{11}(E) = \frac{1}{t^*(E)} = \frac{1}{|t(E)|}\,e^{i\phi(E)} .$$

The compatibility condition (32) for the tunneling of Bloch-type wavefunctions takes the compact and significant form

$$\boxed{\frac{1}{|t(E)|}\cos\left[\phi(E) + q(E)\,a\right] = \cos ka} , \tag{33}$$

where $\phi(E)$ and $|t(E)|$ are the phase and the modulus of the transmission amplitude $t(E)$ through the potential $V_0(x)$ of the unit cell. Notice that Eq. (33) can be satisfied only for those values of E for which the function $(1/|t|)\cos(\phi + qa)$ is in magnitude less than (or equal to) unity; since $|t(E)| \leq 1$, we can infer that the compatibility equation (33) leads quite generally to a sequence of allowed and forbidden energy regions.

It is instructive to analyse the band structure, produced by the periodic array of barriers of Fig. 1, from the point of view of the electron tunneling framework. The transfer matrix S for a single barrier is given by Eq. (27). The energy bands are determined by Eq. (32), which reads in the present case

$$\mathrm{Re}\left\{\left[\cosh\beta b + i\frac{q^2 - \beta^2}{2q\beta}\sinh\beta b\right]e^{iq(a-b)}\right\} = \cos ka$$

where Re denotes the real part of its argument. With $a - b = w$ (see Fig. 1), the above equation becomes

$$\cosh\beta b\,\cos qw - \frac{q^2 - \beta^2}{2q\beta}\sinh\beta b\,\sin qw = \cos ka . \tag{34}$$

We thus see that the compatibility equation (34) for the tunneling of Bloch-type wavefunctions through periodically repeated barriers, exactly coincides with Eq. (13), previously obtained using the ordinary boundary conditions procedure.

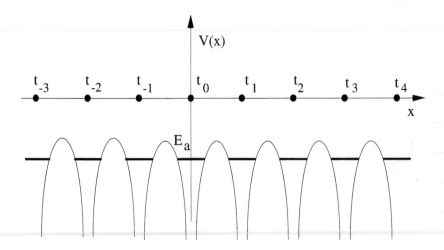

Fig. 8 Schematic representation of the crystal potential as a superposition of atomic-like potentials, centered at the lattice sites t_n. In the tight-binding approximation, the interaction between nearby atomic-like orbitals ϕ_a of energy E_a leads to the formation of energy bands.

4   The tight-binding approximation

4.1   Expansion in localized orbitals

In the previous section we have considered the origin of the energy bands within the framework of electron tunneling through a periodic potential. In this section we adopt another point of view; we imagine construction of a crystal from a hypothetical periodic one-dimensional sequence of N equal atoms. In the case of negligible interaction among atoms, the same atomic orbitals centered in the different lattice sites would have the same energy; in the presence of interaction this N fold degeneracy is removed and evolves into an energy band. As usual in this chapter we keep our considerations at a modelistic level, referring to Chapter V for a more realistic analysis.

Let us consider a one-dimensional crystal, made up by equal atoms centered in the lattice positions $t_n = na$. For each atom we focus our attention on a given local orbital ϕ_a of energy E_a (for simplicity, the atomic orbital ϕ_a is assumed to be non-degenerate and already in real form). We try to obtain crystal wavefunctions using as basis set the N orbital functions $\phi_a(x - t_n)$ centered in the N atomic sites t_n. A schematic picture of the model is given in Fig. 8.

It is convenient to represent the crystal Hamiltonian H on the localized functions $\{\phi_a(x - t_n)\}$. We do not need here to be very specific on the crystal Hamiltonian, but only exploit its translational symmetry. Because of translational symmetry, the diagonal matrix elements of H on atomic orbitals are all equal; similarly the hopping integrals between nearest neighbour orbitals are equal. We have

$$\langle \phi_a(x - t_n)|H|\phi_a(x - t_n)\rangle = E_0 \quad \text{and} \quad \langle \phi_a(x - t_n)|H|\phi_a(x - t_{n\pm 1})\rangle = \gamma \ . \quad (35)$$

For simplicity, due to the localized nature of atomic orbitals, the hopping integrals

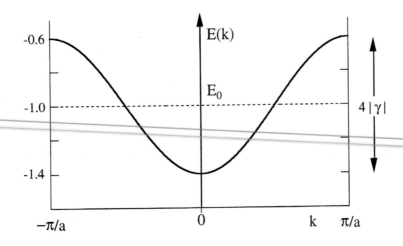

Fig. 9 Energy band $E(k) = E_0 + 2\gamma \cos ka$ for a tight-binding model with a single orbital per site and nearest neighbour interactions; in the figure $E_0 = -1\,\text{eV}$ and $\gamma = -0.2\,\text{eV}$.

involving second or further apart neighbours are assumed negligible; the value of the interaction energy γ is taken to be negative (to simulate the case of s-like orbitals and attractive atomic potentials).

The localized functions $\{\phi_a(x - t_n)\}$ do not satisfy the Bloch theorem; but we can easily remedy this by considering the linear combinations of atomic orbitals of the type

$$\Phi(k, x) = \frac{1}{\sqrt{N}} \sum_n e^{ik\,t_n}\, \phi_a(x - t_n) \,, \tag{36}$$

where N is the number of unit cells of the crystal. The function defined by Eq. (36) is named *Bloch sum*, it has itinerant character, and satisfies the Bloch theorem; in fact

$$\Phi(k, x + t_m) = \frac{1}{\sqrt{N}} \sum_n e^{ik t_n}\, \phi_a(x + t_m - t_n)$$

$$= e^{ik t_m} \frac{1}{\sqrt{N}} \sum_n e^{ik(t_n - t_m)}\, \phi_a(x - t_n + t_m) = e^{ik t_m} \Phi(k, x) \,.$$

For simplicity we assume orthonormality of orbitals centered on different atoms; in this case, the Bloch sums (36) are also orthonormal.

The N itinerant Bloch sums $\{\Phi(k, x)\}$ span the same Hilbert space as the N localized functions $\{\phi_a(x - t_n)\}$, but the great advantage is that Bloch sums of different k values cannot mix under the influence of a periodic potential. The energy dispersion of the energy band originated from the atomic orbitals $\{\phi_a(x - t_n)\}$ is thus given by

$$\boxed{E(k) = \langle \Phi(k, x) | H | \Phi(k, x) \rangle} \,.$$

In the particular case that the matrix elements of H between atomic orbitals are given

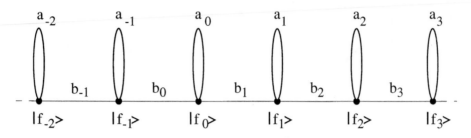

Fig. 10 Graphical representation of the most general tridiagonal Hamiltonian. The diagonal matrix elements are denoted by a_n; the off-diagonal hopping integrals are denoted by b_n.

by Eqs. (35), the above dispersion relation becomes

$$E(k) = E_0 + 2\gamma \cos ka \; ; \qquad (37)$$

this expression clearly shows at the most elementary level that the N-fold degenerate states of the non-interacting atoms are smeared into a continuous band of width $4|\gamma|$ (see Fig. 9).

We can expand the second member of Eq. (37) in powers of k, for small k, and retain terms up to the second order; we have

$$E(k) \approx E_0 + 2\gamma - \gamma a^2 k^2 \; .$$

The quadratic term $-\gamma a^2 k^2$ can be written in the form $\hbar^2 k^2 / 2m^*$, similar to the one pertaining to the free-electron case, provided the effective mass m^* is defined as

$$m^* = \frac{\hbar^2}{2|\gamma| a^2} \; . \qquad (38)$$

The effective mass is small if the hopping parameter γ is large, and vice versa. Notice also that the effective mass, at the border of the Brillouin zone, is negative and is the opposite of expression (38); this can be seen performing a series development of the second member of Eq. (37) around the border point $k = \pi/a$.

4.2   Tridiagonal matrices and continued fractions

A matrix, whose non-vanishing elements occur only on the principal diagonal and the next upper and lower ones, is said to be tridiagonal. The crystal Hamiltonian H, with matrix elements given by Eqs. (35), can be conveniently represented on a basis of localized orbitals $|f_n\rangle = |\phi_a(x - t_n)\rangle$ in the form

$$H = E_0 \sum_n |f_n\rangle\langle f_n| + \gamma \sum_n [\, |f_n\rangle\langle f_{n+1}| + |f_{n+1}\rangle\langle f_n| \,] \; . \qquad (39)$$

The one-dimensional tight-binding Hamiltonian (39), with one orbital per site and nearest neighbour interactions, is the simplest example of problems described by a tridiagonal matrix.

The most general Hamiltonian in tridiagonal form is

$$H = \sum_n a_n |f_n\rangle\langle f_n| + \sum_n b_{n+1} [\, |f_n\rangle\langle f_{n+1}| + |f_{n+1}\rangle\langle f_n| \,] \tag{40}$$

with appropriate values for diagonal and off-diagonal matrix elements a_n and b_n (since H is hermitian, all a_n are real; also b_n can be brought to real form, without loss of generality, just embodying appropriate phases in the basis functions). The tridiagonal operator (40) can be represented graphically as indicated in Fig. 10.

Numerous physical problems can be described by a tridiagonal Hamiltonian of the type (40); among them we mention the Harper model for incommensurate systems and the Anderson model for disordered systems. In the Harper model the diagonal energies are taken as $a_n = V_0 \cos(2\pi n\alpha)$ with α irrational number, while the hopping integrals are constant (taken equal to unity); it can be shown that if $0 \le V_0 < 2$ all the states are delocalized, but when $V_0 > 2$ all the states become localized. In the Anderson model the diagonal energies a_n are randomly distributed in an energy interval, while the hopping integrals are constant; in the Anderson model all states are localized, for any degree of disorder. We cannot pursue here the interesting problems posed by localization, and the wide and restless research on Hamiltonians of type (40).

Much more generally, *the importance of tridiagonal matrices is due to the fact that, by the Lanczos method* (see Chapter V), *it is possible to map any quantum mechanical problem into a semi-infinite chain represented by a tridiagonal matrix.* This is the reason why tridiagonal operators have a special relevance in quantum mechanics [see for instance D. W. Bullett, R. Haydock, V. Heine and M. J. Kelly, Solid State Physics Vol.**35** (Academic Press, New York 1980)].

Tridiagonal matrices have simple and elegant mathematical properties, that allow focus on physical aspects. From a physical point of view, the most remarkable aspect is that the inversion of tridiagonal matrices is straightforward and thus the Green's function of tridiagonal operators is easily accessible.

A note on the inversion of tridiagonal matrices

Consider a tridiagonal matrix M of the form

$$M = \begin{pmatrix} \alpha_0 & \beta_1 & 0 & 0 & \cdot \\ \beta_1 & \alpha_1 & \beta_2 & 0 & \cdot \\ 0 & \beta_2 & \alpha_2 & \beta_3 & \cdot \\ 0 & 0 & \beta_3 & \alpha_3 & \cdot \\ \cdot & \cdot & \cdot & \cdot & \cdot \end{pmatrix} \tag{41}$$

(for simplicity the rank of the matrix M is supposed to be finite, although arbitrary large). We consider the inverse matrix M^{-1}, and we are here primarily interested in obtaining the explicit expression of its upper left corner element $(M^{-1})_{00}$.

Let us indicate with D_0 the determinant of the matrix M; we denote by D_1 the determinant of the matrix obtained by M suppressing the first column and the first row; similarly D_2 denotes the determinant of the matrix obtained from M suppressing also the successive row and column, and so on. We now develop the determinant D_0 along the elements of the first row; we have

$$D_0 = \alpha_0\, D_1 - \beta_1^2\, D_2 \ .$$

From the above result and considering that $(1/M)_{00} = D_1/D_0$ we obtain

$$\left(\frac{1}{M}\right)_{00} \equiv \frac{1}{D_0/D_1} = \frac{1}{\alpha_0 - \beta_1^2\dfrac{1}{D_1/D_2}} \ .$$

Iterating the procedure for D_1/D_2 we obtain the continued fraction expansion

$$\left(\frac{1}{M}\right)_{00} = \cfrac{1}{\alpha_0 - \cfrac{\beta_1^2}{\alpha_1 - \cfrac{\beta_2^2}{\cdots}}} \ . \tag{42}$$

The continued fraction expansion (42) is truncated at the nth step if β_n vanishes.

In the case the tridiagonal matrix M extends both to positive and negative n (as for instance in Fig. 10), we obtain

$$\left(\frac{1}{M}\right)_{00} = \cfrac{1}{\alpha_0 - \cfrac{\beta_1^2}{\alpha_1 - \cfrac{\beta_2^2}{\cdots}} - \cfrac{\beta_0^2}{\alpha_{-1} - \cfrac{\beta_{-1}^2}{\cdots}}} \ . \tag{43}$$

In the case $\beta_0 = 0$ we recover Eq. (42). It is also evident that we can obtain the expression of any diagonal matrix element $(1/M)_{nn}$ (this simply requires an appropriate relabelling of states). Also off-diagonal matrix elements of $(1/M)$ could be expressed, with appropriate elaboration, in terms of continued fractions.

Density-of-states and Green's function

Consider a system described by a Hamiltonian H, and let us indicate by ψ_m and E_m its normalized eigenfunctions and eigenvalues (supposed to be countable, for simplicity). The *total density-of-states* of the system is defined as

$$D(E) = \sum_m \delta(E - E_m) \ . \tag{44}$$

It is evident that the integral $\int D(E)\, dE$ in any energy interval $[E_1, E_2]$ provides the number of states of the system therein.

The *density-of-states projected* on any arbitrarily chosen state of interest $|f_0\rangle$ (normalized to unity) is defined as

$$n_0(E) = \sum_m |\langle f_0|\psi_m\rangle|^2\, \delta(E - E_m) \ . \tag{45}$$

Differently from $D(E)$, the projected density-of-states $n_0(E)$ (also called *local density-of-states*) gives information uniquely on the spectral region investigated by the orbital

$|f_0\rangle$; we also see by inspection that $\int n_0(E) \, dE \equiv 1$ (due to the normalization of $|f_0\rangle$). Furthermore, the total density-of-states is the sum of the projected density-of-states on any complete orthonormal set $\{f_n\}$.

The (retarded) Green's function (or resolvent) of the operator H is defined as

$$G(E + i\varepsilon) = \frac{1}{E + i\varepsilon - H} , \qquad (46)$$

where the real energy E is accompanied by an (infinitesimal) positive imaginary part (when a scalar quantity such as $E + i\varepsilon$ is added to an operator such as H in matrix form, it is implicitly understood that the scalar quantity is preliminarily multiplied by the identity operator).

A most useful property of the Green's function is its connection with the density-of-states of the system. Consider in fact a diagonal matrix element of the Green's function, for instance

$$G_{00}(E + i\varepsilon) = \langle f_0 | \frac{1}{E + i\varepsilon - H} | f_0 \rangle . \qquad (47a)$$

It is possible to put in evidence some properties of the resolvent inserting into Eq. (47a) the unit operator $1 = \sum |\psi_m\rangle\langle\psi_m|$, where ψ_m are the eigenfunctions of H. We have

$$G_{00}(E + i\varepsilon) = \langle f_0 | \sum_m |\psi_m\rangle\langle\psi_m| \frac{1}{E + i\varepsilon - H} | f_0 \rangle$$

$$= \sum_m |\langle f_0|\psi_m\rangle|^2 \frac{1}{E + i\varepsilon - E_m} = \sum_m |\langle f_0|\psi_m\rangle|^2 \frac{E - E_m - i\varepsilon}{(E - E_m)^2 + \varepsilon^2} . \qquad (47b)$$

From Eq. (47b), we see that for any $\varepsilon > 0$, $G_{00}(E + i\varepsilon)$ is analytic and its imaginary part is negative (Herglotz property). On the real energy axis, the real part of $G_{00}(E)$ exhibits poles in correspondence of the *discrete* eigenvalues of H, while the imaginary part exhibits δ-like singularities; this can be seen keeping the limit $\varepsilon \to 0^+$ in Eq. (47b) and using the result

$$\lim_{\varepsilon \to 0^+} \frac{1}{\pi} \frac{\varepsilon}{(E - E_m)^2 + \varepsilon^2} = \delta(E - E_m) .$$

From Eq. (45) and Eq. (47b), we obtain the standard spectral theorem

$$n_0(E) = -\frac{1}{\pi} \lim_{\varepsilon \to 0^+} \text{Im} \, G_{00}(E + i\varepsilon) . \qquad (48a)$$

The total density-of-states of the system can be expressed as the trace of the Green's function on any chosen complete orthonormal set

$$D(E) = -\frac{1}{\pi} \lim_{\varepsilon \to 0^+} \text{Im} \, \text{Tr} \, G(E + i\varepsilon) . \qquad (48b)$$

Relations (48) hold regardless of the fact that the energy spectrum of H is discrete or continuous.

In general the calculation of the Green's function of an operator H requires the preliminary diagonalization of H. However for tridiagonal operators the calculation of the diagonal matrix elements of the resolvent is straightforward. In fact, if the operator H has the tridiagonal form of type (40), the operator $E - H$ is also in tridiagonal form. Then from the inversion properties of tridiagonal matrices (see Eq. 43) we have for the Green's function diagonal matrix element

$$G_{00}(E) = \cfrac{1}{E - a_0 - \cfrac{b_1^2}{E - a_1 - \cfrac{b_2^2}{E - a_2 - \cfrac{b_3^2}{\cdots}}} - \cfrac{b_0^2}{E - a_{-1} - \cfrac{b_{-1}^2}{E - a_{-2} - \cfrac{b_{-2}^2}{\cdots}}}} . \tag{49}$$

Application to the tight-binding case

Let us consider the Hamiltonian (39) for the one-dimensional tight-binding crystal with one orbital per site. In this case the diagonal elements a_n are constant and equal to E_0 (without loss of generality we take $E_0 = 0$); the off-diagonal elements b_n are constant and equal to γ. The Green's function (49) for the infinite linear chain becomes

$$G_{00}(E) = \langle f_0 | \frac{1}{E - H} | f_0 \rangle = \cfrac{1}{E - \cfrac{2\gamma^2}{E - \cfrac{\gamma^2}{E - \cfrac{\gamma^2}{\cdots}}}} . \tag{50a}$$

To perform the sum of the continued fraction let us indicate with $t(E)$ the quantity

$$t(E) = \cfrac{\gamma^2}{E - \cfrac{\gamma^2}{E - \cfrac{\gamma^2}{\cdots}}} = \frac{\gamma^2}{E - t(E)} ;$$

the solution of this equation gives $t(E) = (1/2)(E \pm \sqrt{E^2 - 4\gamma^2})$. The Green's function (50a) can be summed exactly to the value

$$G_{00}(E) = \frac{1}{E - 2t(E)} = \frac{1}{\pm\sqrt{E^2 - 4\gamma^2}} ; \tag{50b}$$

the sign of the square root must be chosen so that $\operatorname{Im} G_{00}(E + i\varepsilon) < 0$ for $\varepsilon \to 0^+$. In particular from the spectral theorem the density-of-states becomes

$$n(E) = \frac{1}{\pi} \frac{1}{\sqrt{4\gamma^2 - E^2}} \qquad |E| < 2|\gamma| . \tag{50c}$$

The result (50c) can also be obtained starting directly from the dispersion relation $E(k) = 2\gamma \cos ka$; in fact, with simple elaborations, we have for the total density-of-states

$$D(E) = \sum_k \delta(E - 2\gamma \cos ka) = \frac{L}{2\pi} \int_{-\pi/a}^{+\pi/a} \delta(E - 2\gamma \cos ka)\, dk$$

$$= \frac{L}{2\pi}\, 2\, \frac{1}{|2\gamma a \sin k_0 a|} = N \frac{1}{\pi} \frac{1}{\sqrt{4\gamma^2 - E^2}} \qquad |E| < 2|\gamma|\,.$$

In the above expression $L = Na$ denotes the length of the crystal; k_0 is one of the two zeroes of the expression $E - 2\gamma \cos ka$ when $|E| < 2|\gamma|$; furthermore, use has been made of the property $\delta[f(x)] \equiv \sum \delta(x - x_0)/|f'(x_0)|$, where the sum is over any simple zero x_0 of the function $f(x)$.

It is instructive to calculate the Green's function just on the initial site (surface site) of a semi-infinite linear chain. The Green's function at the surface orbital is

$$G_{00}^{(s)}(E) = \cfrac{1}{E - \cfrac{\gamma^2}{E - \cfrac{\gamma^2}{\cdots}}} = \frac{E \pm \sqrt{E^2 - 4\gamma^2}}{2\gamma^2}\,; \qquad (51a)$$

the sign of the square root must be chosen so that $\operatorname{Im} G_{00}^{(s)}(E + i\varepsilon) < 0$ for $\varepsilon \to 0^+$. The projected density-of-states on the initial orbital is thus

$$n_0^{(s)}(E) = -\frac{1}{\pi} \lim_{\varepsilon \to 0^+} \operatorname{Im} G_{00}^{(s)}(E + i\varepsilon) = \frac{\sqrt{4\gamma^2 - E^2}}{2\pi\gamma^2} \qquad |E| < 2\gamma\,. \qquad (51b)$$

Similarly one could calculate the Green's function $G_{11}^{(s)}(E)$, $G_{22}^{(s)}(E)$ etc. on the next sites of the semi-infinite constant chain, and obtain the corresponding local density-of-states $n_1^{(s)}(E)$, $n_2^{(s)}(E)$, etc.; a simple calculation gives

$$n_1^{(s)}(E) = n_0^{(s)}(E) \frac{E^2}{\gamma^2}\,, \qquad n_2^{(s)}(E) = n_0^{(s)}(E) \left[\frac{E^2}{\gamma^2} - 1\right]^2\,. \qquad (51c)$$

The local density-of-states in the bulk of a constant infinite chain, and at the surface orbital and at the next two ones of a semi-infinite chain is reported in Fig. 11.

We do not pursue further the properties of the Green's function of general tight-binding Hamiltonians of type (40); we only mention that several significant properties such as localization effects in incommensurate systems, dimerized systems, disordered systems etc. can be inferred with a systematic study of the resolvents of the corresponding Hamiltonians [see for instance J. B. Sokoloff, Phys. Rep. **126**, 190 (1985); R. Farchioni, G. Grosso and G. Pastori Parravicini, Phys. Rev. B**45**, 6383 (1992); B**53**, 4294 (1996) and references quoted therein].

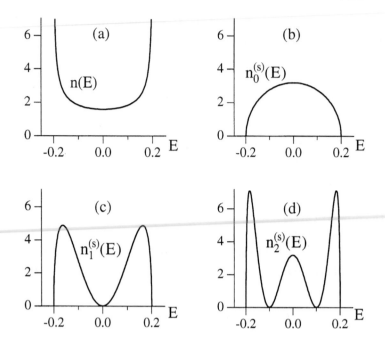

Fig. 11 Local density-of-states on the bulk orbital of a constant infinite linear chain (a), at the surface orbital of a semi-infinite chain (b), and on the next two orbitals (c) and (d). The chosen value of γ is $\gamma = -0.1\,\mathrm{eV}$. Energies are in eV, and the density-of-states is in electron/eV.

5 Plane waves and nearly free-electron approximation

5.1 Expansion in plane waves

In this section we consider the origin of the energy bands starting from the "empty lattice" situation and then switching on a weak periodic potential. For this purpose, we consider first the case of vanishing periodic potential, where the energy dispersion $E(k) = \hbar^2 k^2 / 2m$ is a continuous function of k, without gaps; then we study how a periodic potential modifies the free electron behaviour, introducing energy gaps. In Chapter V we will see that plane wave expansion of crystal wavefunctions, together with the concept of pseudopotential, is indeed one of the most effective methods of band structure calculation.

In Fig. 12 we report the free-electron energy dispersion $E(k) = \hbar^2 k^2 / 2m$ as k varies in the reciprocal lattice; we also show the free-electron parabola folded within the first Brillouin zone. Folding is easily performed by means of appropriate reciprocal lattice translations by $n\,2\pi/a$ of arches of parabola; the convenience of folding stands in the fact that, according to the Bloch theorem, *only states vertical in the first Brillouin zone may interact* under the influence of a potential with the lattice periodicity.

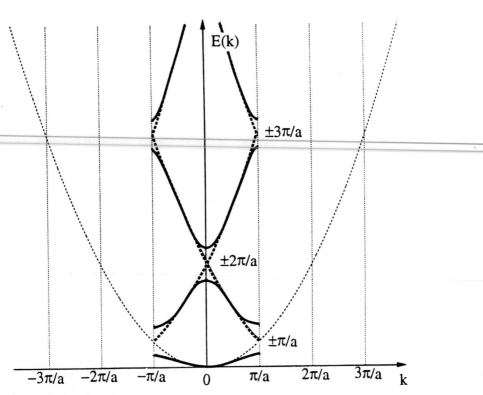

Fig. 12 Empty lattice states (free-electron parabola) and folding in the first Brillouin zone. At the center and at the border of the first Brillouin zone we have also indicated the wavenumbers of the degenerate plane waves. In the presence of a periodic potential, gaps open at the center and at the border of the Brillouin zone, as shown qualitatively by continuous lines.

We wish now to obtain the eigenvalues and eigenvectors of the crystal Hamiltonian

$$H = -\frac{\hbar^2}{2m}\frac{d^2}{dx^2} + V(x) \tag{52a}$$

starting from the empty lattice eigenvalues and eigenfunctions. At a given k value, the crystal wavefunctions $\psi_k(x)$ can be expanded as a linear combination of plane waves of the type

$$W_{k_n}(x) = \frac{1}{\sqrt{L}}e^{i(k+h_n)x} , \tag{52b}$$

where L is the length of the crystal and $h_n = n\,2\pi/a$. The matrix elements of H between the basis functions (52b) take the form

$$\langle W_{k_m}|H|W_{k_n}\rangle = \frac{\hbar^2(k+h_n)^2}{2m}\delta_{mn} + \frac{1}{L}\int_0^L e^{-i(h_m-h_n)x}\,V(x)\,dx$$

$$= \frac{\hbar^2(k+h_n)^2}{2m}\delta_{mn} + V(h_m - h_n) , \tag{52c}$$

where $V(h)$ denotes the Fourier transform of $V(x)$.

Diagonalization of H on the basis set of plane waves (52b) then leads to the secular equation for the energy eigenvalues

$$\left\| \left[\frac{\hbar^2 (k + h_n)^2}{2m} - E \right] \delta_{mn} + V(h_m - h_n) \right\| = 0 . \tag{53}$$

We consider now some significant particular cases of Eq. (53) in order to make more transparent the origin of allowed and forbidden energy bands in periodic potentials.

Nearly free-electron approximation

Let us consider the general secular equation (53) at a given k value, for instance at $k = \pi/a$. From Fig. 12 we see that the interacting plane waves are characterized by the wavenumbers $\pm\pi/a$ with energy $E_0 = (\hbar^2/2m)(\pi^2/a^2)$, by the wavenumbers $\pm 3\pi/a$ with energy $9 E_0$, etc. For small strength of the periodic potential, we may confine our attention to the two basis functions (degenerate in the empty lattice analysis)

$$\psi_1(x) = \frac{1}{\sqrt{L}} \exp(i\frac{\pi}{a}x) \qquad \psi_2(x) = \frac{1}{\sqrt{L}} \exp(-i\frac{\pi}{a}x) . \tag{54a}$$

The diagonalization of the crystal Hamiltonian (52a) on the two wavefunctions (54a) leads to the 2×2 secular equation

$$\left\| \begin{matrix} E_0 - E & V_1 \\ V_1^* & E_0 - E \end{matrix} \right\| = 0 \tag{54b}$$

where $V_1 = V(2\pi/a)$ is the Fourier transform of the crystal potential corresponding to the lowest reciprocal lattice wavenumber $h_1 = 2\pi/a$. We thus see that the twofold degenerate empty lattice states of energy E_0 are split by the periodic potential in the form

$$E = E_0 \pm |V_1| . \tag{54c}$$

The same reasoning holds for the other empty lattice degenerate states at the points $k = 0$ and $k = \pm\pi/a$, and we thus can understand qualitatively, in the nearly free-electron approximation, the origin of the energy gaps of the one-dimensional crystal as due to the splitting of the twofold degeneracy of the empty lattice produced by the periodic potential.

We can also determine analytically the behaviour of the energy band near the boundary of the first Brillouin zone. Let us suppose that the wavenumber k is very near to the boundary, but not exactly at $\pm\pi/a$. Instead of the basis functions (54a), we consider now the basis functions

$$\begin{cases} \psi_1(x) = \frac{1}{\sqrt{L}} e^{i(\pi/a - \Delta k)x} & \text{with energy} \quad E_1 = \frac{\hbar^2}{2m}(\frac{\pi}{a} - \Delta k)^2 \\[2ex] \psi_2(x) = \frac{1}{\sqrt{L}} e^{i(-\pi/a - \Delta k)x} & \text{with energy} \quad E_2 = \frac{\hbar^2}{2m}(\frac{\pi}{a} + \Delta k)^2 \end{cases} . \tag{55a}$$

We have the 2×2 determinantal equation

$$\left\| \begin{matrix} E_1 - E & V_1 \\ V_1^* & E_2 - E \end{matrix} \right\| = 0 . \tag{55b}$$

The two eigenvalues of (55b) are given by

$$E = \frac{1}{2}(E_1 + E_2 \pm \sqrt{(E_1 - E_2)^2 + 4|V_1|^2}) . \tag{55c}$$

Inserting in the above expression the values of E_1 and E_2 given by Eqs. (55a), we obtain the energy dispersion curves near the boundary of the Brillouin zone

$$E(\Delta k) = E_0 + \frac{\hbar^2 (\Delta k)^2}{2m} \pm \frac{1}{2} \sqrt{16 E_0 \frac{\hbar^2 (\Delta k)^2}{2m} + 4|V_1|^2}$$

where $E_0 = (\hbar^2/2m)(\pi/a)^2$. When Δk is small, a series development of the square root term gives

$$E(\Delta k) = E_0 + \frac{\hbar^2 (\Delta k)^2}{2m} \pm |V_1| \left[1 + \frac{2 E_0}{|V_1|^2} \frac{\hbar^2 (\Delta k)^2}{2m} \right] + \cdots ;$$

the electronic effective masses for the upper and lower energy bands are

$$\frac{1}{m^*} = \frac{1}{m} \left(1 \pm \frac{2 E_0}{|V_1|} \right) . \tag{56}$$

In the case of small energy gap ($|V_1| \ll E_0$) also the effective masses are expected to be small (at parity of other conditions). Qualitatively this is the trend actually observed in some materials; semiconductors with small energy gaps are often characterized by small effective masses (and high mobility) of carriers.

5.2   The Mathieu potential and the continued fraction solution

We consider now the general secular equation (53) in the case the periodic potential is a simple cosine potential $V(x) = 2 V_1 [1 - \cos(2\pi x/a)]$ (the energy origin is chosen so that $V(x) \geq 0$). The Hamiltonian operator becomes

$$H = \frac{p^2}{2m} + 2 V_1 - V_1 \left[e^{i(2\pi/a)x} + e^{-i(2\pi/a)x} \right] . \tag{57}$$

The matrix elements of the Mathieu Hamiltonian (57) between plane waves have the form

$$\langle W_{k_m} | H | W_{k_n} \rangle = \left[\frac{\hbar^2 (k + h_n)^2}{2m} + 2V_1 \right] \delta_{mn} - V_1 \delta_{m,n+1} - V_1 \delta_{m,n-1} . \tag{58}$$

The Mathieu Hamiltonian is another important case of problems with tridiagonal form, and it is thus candidate to be solved by means of the continued fraction apparatus.

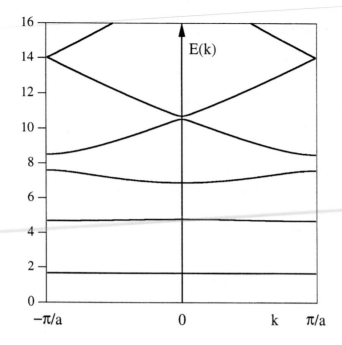

Fig. 13 Energy bands for the periodic Mathieu potential $V(x) = 2V_1[1 - \cos(2\pi x/a)]$, with $V_1 = 5(\hbar^2/2m)(\pi/a)^2$. Energies are in Rydberg and $a = 5a_B$.

The tridiagonal form (58) is well suited for the evaluation of the Green's function. The diagonal matrix element $G_{00}(E)$, in particular, is given by the expression

$$G_{00}(E) = \cfrac{1}{E - a_0 - \cfrac{V_1^2}{E - a_1 - \cfrac{V_1^2}{E - a_2 - \cfrac{V_1^2}{\cdots}}} - \cfrac{V_1^2}{E - a_{-1} - \cfrac{V_1^2}{E - a_{-2} - \cfrac{V_1^2}{\cdots}}}} \tag{59}$$

where $a_n(k) = 2V_1 + (\hbar^2/2m)(k + n\,2\pi/a)^2$.

It is now easy to plot the expression (59) as a function of E and from the poles of $G_{00}(E)$ we can obtain the eigenvalues of the Mathieu problem. In general the numbers of steps to be effectively included in the continued fraction before truncation depends on the relative strength of V_1 with respect to $E_0 = (\hbar^2/2m)(2\pi/a)^2$; if this ratio is of order of unity (or smaller) a few tens of steps in the continued fraction are more than sufficient.

In Fig. 13 we report the energy bands corresponding to the Mathieu potential in the case $V_1 = 5E_0$. Notice that all the degeneracies of the empty lattice are fully removed (even if the direct coupling between degenerate wavefunctions of the empty lattice occurs only for $k = \pm\pi/a$) [for further aspects of the Mathieu problem see J. C. Slater, Phys. Rev. **87**, 807 (1952)].

6 Some dynamical aspects of electrons in band theory

Velocity, quasi-momentum and effective mass of an electron in a band

It is useful to introduce the concept of velocity, quasi-momentum and effective mass for an electron in a band, within a semiclassical picture. In order to introduce this argument we consider first the free electron situation, with eigenfunctions and eigenvalues given by $W(k, x) = (1/\sqrt{L}) \exp(ikx)$ and $E(k) = \hbar^2 k^2 / 2m$. The plane waves are eigenfunctions of the operator $p = -i\hbar d/dx$ with eigenvalue $\hbar k$, which represents the momentum of the free electron.

We consider now an electron in a periodic potential. We focus on a given energy band and we indicate with $E(k)$ and $\psi(k, x)$ energies and wavefunctions as k varies in the Brillouin zone. We notice, first of all, that a Bloch function $\psi(k, x) = \exp(ikx) u(k, x)$ is not (in general) an eigenfunction of the momentum operator; in fact we have

$$p\psi(k, x) = -i\hbar \frac{d}{dx}[e^{ikx} u(k, x)] = \hbar k \psi(k, x) + e^{ikx}(-i)\hbar \frac{d}{dx} u(k, x) ;$$

this relation clearly shows that the eigenfunctions of the crystal Hamiltonian are not eigenfunctions of the p operator (except for the trivial empty lattice situation), and thus $\hbar k$ cannot be considered (rigorously speaking) the momentum of the electron in the state $\psi(k, x)$.

A Bloch wavefunction $\psi(k, x)$ is not, in general, an eigenfunction of the momentum operator $p = -i\hbar d/dx$. However, $\psi(k, x)$ can always be expressed as a linear combination of plane waves of wavenumbers k, $k \pm 2\pi/a$, $k \pm 4\pi/a, \ldots$ and the *only* possible values of a measure of the observable p on the Bloch function $\psi(k, x)$ are thus $\hbar k$, $\hbar k \pm \hbar(2\pi/a)$, $\hbar k \pm \hbar(4\pi/a), \ldots$. For this reason, the quantity $\hbar k$ (with k usually but not necessarily, taken within the first Brillouin zone) is called *quasi-momentum* of the electron in the crystal, or also *crystal momentum*. For brevity, and with due caution, $\hbar k$ is often addressed simply as the *momentum of the electron*, keeping in mind that the wavenumber k and the momentum $\hbar k$ of an electron in a crystal are defined within arbitrary integer values of $2\pi/a$ and $\hbar 2\pi/a$, respectively.

It is of interest to consider the expectation value of the momentum operator in the state $\psi(k, x)$, and the closely related semiclassical electron velocity $v(k)$ defined as

$$v(k) = \langle \psi(k, x) | \frac{p}{m} | \psi(k, x) \rangle . \tag{60}$$

As shown below, the electron velocity is connected to the energy gradient by the relation

$$\boxed{v(k) = \frac{1}{\hbar} \frac{dE(k)}{dk}} ; \tag{61}$$

in particular, the electron velocity $v(k)$ vanishes at the extrema of the band dispersion curve $E(k)$.

To demonstrate Eq. (61), we start from the relation

$$\langle \psi(k, x) | \frac{p^2}{2m} + V(x) | \psi(k, x) \rangle = E(k) \; ,$$

and express the crystal wavefunction in the Bloch form $\psi(k, x) = \exp(ikx) \, u(k, x)$; it follows

$$\langle u(k, x) | \frac{1}{2m} (p + \hbar k)^2 + V(x) | u(k, x) \rangle = E(k) \; . \tag{62}$$

We now derive both members of Eq. (62) with respect to the parameter k. From the derivation of the first member of Eq. (62) we obtain three contributions, two of them coming from the derivation of $u^*(k, x)$ and $u(k, x)$, and the other from the derivation of the operator; the two terms coming from the derivation of u^* and u equal $E(k)(d/dk)\langle u(k, x) | u(k, x) \rangle$ and give zero because the wavefunctions are normalized to one. Thus we have

$$\langle u(k, x) | \frac{\hbar}{m} (p + \hbar k) | u(k, x) \rangle = \frac{dE(k)}{dk} \; ;$$

by expressing $u(k, x)$ as $\exp(-ikx)\psi(k, x)$, equation (61) follows.

We examine now the effect of an external electric field on the dynamics of an electron in a periodic potential $V(x)$. In the presence of a uniform electric field F (in the positive x-direction) the total Hamiltonian is

$$H = \frac{p^2}{2m} + V(x) + eFx \; ,$$

where e is the absolute value of the electronic charge. Suppose that at the time $t = t_0 \equiv 0$ an electron is prepared in the Bloch state $\psi(k_0, x)$ of wavenumber k_0; the time evolved state at time t is given by

$$\psi(x, t; F) = e^{(-i/\hbar)[p^2/2m + V(x) + eFx] \, t} \, \psi(k_0, x) \; .$$

We replace the x variable with the translated variable $x + a$ in both members of the above relation, and we obtain

$$\psi(x + a, t; F) = e^{(-i/\hbar)[p^2/2m + V(x+a) + eF(x+a)] \, t} \, \psi(k_0, x + a)$$

$$= e^{(-i/\hbar)[p^2/2m + V(x) + eFx] \, t} \, e^{-(i/\hbar)eFat} e^{ik_0 a} \psi(k_0, x) \; .$$

We have thus

$$\psi(x + a, t; F) = e^{ik(t)a} \, \psi(x, t; F) \; ,$$

where

$$k(t) = -\frac{1}{\hbar} eFt + k_0 \; . \tag{63}$$

The time evolved wavefunction $\psi(x, t; F)$ is a Bloch-type wavefunction, whose wavenumber $k(t)$ changes linearly in time according to Eq. (63). The time derivative of

Eq. (63) gives

$$\boxed{\frac{d(\hbar k)}{dt} = -eF} \; ; \tag{64}$$

thus, for what concerns the effect of external forces (at least for applied electric field effects), it is indeed justified to identify $\hbar k$ as the quasi-momentum of the electron in the crystal.

In general the wavefunction $\psi(x, t; F)$, in the presence of the field F, can be expressed as linear combination of the crystal eigenstates (without field) of wavenumber $k(t)$ *with contributions from all the bands of the crystal.* However, in ordinary situations of electric field strength and energy gap, the interband mixing among different bands induced by the electric field F can be neglected. For instance for a gap $E_G = 1\,\mathrm{eV}$ and $F = 10^4$ volt/cm interband mixing would require tunneling of the electron through a triangular barrier of height $1\,\mathrm{eV}$ and width $d = 10^{-4}\,\mathrm{cm} = 10^4\,\text{Å}$ $(eFd = E_G)$, and it is thus negligible. *Thus in the absence of interband tunneling and scattering processes, an electron belonging to a given band, under the influence of an applied electric field explores the whole band to which it belongs.*

When the electron explores a unique band, it is instructive to obtain a semiclassical expression for its acceleration, defined as the time derivative of the velocity (61); using Eq. (64) we obtain

$$\frac{dv(k)}{dt} = \frac{d}{dt}\frac{1}{\hbar}\frac{dE(k)}{dk} = \frac{1}{\hbar}\frac{d^2 E(k)}{dk^2}\frac{dk}{dt} = \frac{1}{\hbar^2}\frac{d^2 E(k)}{dk^2}(-e)\,F \; . \tag{65}$$

Thus we have reduced the equation of motion for the electron in a Newton-like form, where only *external forces* appear, provided we define the electron effective mass in the form

$$\frac{1}{m^*} = \frac{1}{\hbar^2}\frac{d^2 E(k)}{dk^2} \; . \tag{66}$$

We can also say that the lattice periodic potential has modified the inertia of the electron according to Eq. (66).

Eq. (66) relates the effective mass of an electron directly to the *local curvature* of the energy band at the specified k vector. The concept of effective mass is particularly useful in the neighbourhood of regions in k space, where the energy dispersion curve $E(k)$ has a parabolic energy-momentum relationship. In these regions the "effective mass" is constant, and this simplifies the treatment of transport effects of carriers. Since the effective mass concept is related to the curvature of the energy band dispersion curve $E(k)$, one can encounter carriers whose effective mass is much different from the bare electron mass, and also with negative effective mass.

A digression on the Bloch oscillator and Stark–Wannier ladder

Consider an electron moving in an energy band $E(k)$ under the influence of a steady electric field F. In the presence of ordinary scattering processes, the electron wavenumber would remain near some initial value. On the contrary, in the ideal situation of the

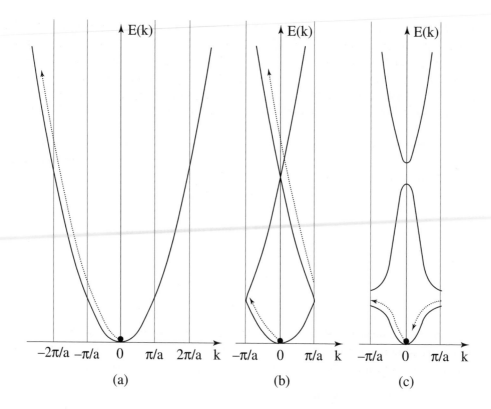

-2π/a -π/a   0   π/a 2π/a k   -π/a   0   π/a   k   -π/a   0   π/a   k

(a)   (b)   (c)

Fig. 14 Schematic representation in k space of the motion of an electron under the influence of a constant electric field F in x direction. In the free-electron case (Fig. 14a), the momentum, velocity and energy increase indefinitely in time. Fig. 14b is the same as Fig. 14a, with the free-electron parabola folded in the first Brillouin zone for convenience. In the presence of the electric field and of a periodic potential, the electron velocity oscillates and the electron undertakes a very fast oscillating motion (Fig. 14c).

absence of scattering processes, the semiclassical motion of the electron is described by the equations

$$k(t) = k_0 - \frac{1}{\hbar}eFt \quad \text{and} \quad v(t) = \frac{1}{\hbar}\left[\frac{dE(k)}{dk}\right]_{k=k(t)}, \tag{67}$$

where k_0 is the electron wavenumber at the initial time $t = 0$. It is interesting to analyse closely the implications of Eqs. (67) in the free-electron case and for an electron belonging to an actual energy band.

In the free-electron case, the momentum, the velocity, and the kinetic energy of the electron increase indefinitely in time, as schematically shown in Fig. 14a.

The situation becomes drastically different for an electron belonging to an energy band, whenever interband transitions can be neglected. With reference to Fig. 14c, let us start for instance with an electron at $k = 0$ in some energy band. Under the influence of an electric field the electron is first accelerated, and acquires energy and velocity. At the top of the energy band at $k = -\pi/a$ the velocity vanishes; in the absence of interband tunneling, the electron continues its path on the same band from $k = +\pi/a$ and it begins to lose energy, until it reaches the initial state at $k = 0$. Differently from the free-electron evolution, the motion is now periodic also in real space, where the electron oscillates between its initial position and an end point.

The time T_B to complete a cycle can be obtained from Eq. (67) and is given by

$$\frac{2\pi}{a} = \frac{1}{\hbar} eF T_B \; ;$$

the angular frequency ω_B of the "Bloch oscillator" is

$$\omega_B = \frac{2\pi}{T_B} = \frac{eFa}{\hbar} \; . \tag{68a}$$

For instance, with a field of intensity $F = 10^4 \, \text{volt/cm}$, we have $eF = 10^{-4} \, \text{eV/Å}$; taking $\hbar = 6.5822 \cdot 10^{-16} \, \text{eV} \cdot \text{sec}$, we obtain

$$\omega_B = \begin{cases} 1.5192 \times 10^{11} \, \text{rad/sec} & \text{for } a = 1\text{Å} \\ 1.5192 \times 10^{13} \, \text{rad/sec} & \text{for } a = 100\text{Å} \end{cases} . \tag{68b}$$

As an exemplification, let us consider the case that the energy band $E(k)$ is given by the tight-binding expression $E(k) = E_0 + 2\gamma \cos ka$ (see Eq. 37). In this case, from Eqs. (67), we obtain for the semiclassical electron velocity the expression

$$v(t) = -\frac{2\gamma a}{\hbar} \sin \left[(k_0 - \frac{eFt}{\hbar}) a \right] \; ; \tag{69a}$$

the semiclassical position $x(t)$ of the oscillator is then

$$x(t) = x_0 - \frac{2\gamma}{eF} \cos \left[(k_0 - \frac{eFt}{\hbar}) a \right] . \tag{69b}$$

Thus the electron performs a harmonic motion of angular frequency ω_B and amplitude $A = 2|\gamma|/eF$ around some equilibrium position x_0; notice that $2eFA$ represents the maximum energy change of the semiclassical oscillator at $x = x_0 \pm A$; as expected, $2eFA$ equals the band width $4|\gamma|$ of the energy band in consideration.

Until now we have considered the semiclassical motion of the electron in the highly idealized situation of a collisionless regime. Unavoidable deviations from ideal period- icity of any realistic system must be taken into account; phenomenologically we assume that the collisionless motion is limited by a time τ, that somehow represents the av- erage time between two successive collisions. In order to achieve the actual realization of a Bloch oscillator of frequency ω_B, we must require $T_B \ll \tau$, or equivalently

$$\omega_B \tau \gg 1 \; , \tag{70}$$

so that the electron may complete several Bloch oscillations before scattering events take place. In standard lattices, where the lattice parameter is of the order of the Angstrom and $\omega_B \approx 10^{11} \, \text{rad/sec}$, the condition (70) is hardly satisfied. However in man-made superlattices, both the lattice parameter ($a \approx 50-100 \, \text{Å}$) and the relaxation time are significantly higher than in ordinary materials; then the realization of Bloch oscillators becomes possible and the effects of the quantization become observable.

The periodic semiclassical motion is an hint that quantization of the whole band occurs when a static electric field F is present. Indeed, strictly speaking, the electric field breaks translational symmetry, and the band structure scheme can at most be a pictorial initial approximation. To try to understand more on the subtle nature of states in periodic potentials and constant electric fields, let us consider the stationary eigenvalue equation

$$\left[\frac{p^2}{2m} + V(x) + eFx \right] \psi(x) = E\,\psi(x) \; , \tag{71}$$

and the following heuristic considerations. Suppose that $\psi(x)$ is an eigenfunction of energy E of the eigenvalue equation (71); by making the translation $x \to x - t_n$ we have that $\psi(x - t_n)$ is also an eigenfunction of Eq. (71) with eigenvalue $E + eFt_n$; thus the solutions of the Schrödinger equation (71) are organized in what are called the Stark–Wannier ladders, with rungs separated by $\Delta W = eFa$. For a review of the theory of a Bloch electron moving in the presence of a homogeneous electric field see for instance G. J. Iafrate, J. P. Reynolds, J. He and J. B. Krieger, International Journal of High Speed Electronics **9**, 223 (1998) and references quoted therein.

Velocity, quasi-momentum and effective mass of a hole in a band

Until now we have considered *a single electron in an empty band*; here we wish to consider the dynamical aspects of a *missing electron in a completely filled band* (the missing electron is called *hole*). We notice that the current associated to a band fully filled with electrons vanishes; in fact we have

$$I = 2 \sum_k (-e) \frac{v(k)}{L} = 2 \frac{-e}{L} \frac{1}{\hbar} \sum_k \frac{dE(k)}{dk} \equiv 0 \; ,$$

where $L = Na$ is the length of the crystal, $L/v(k)$ is the time required by the electron to cross the crystal, and the factor 2 takes into account spin degeneracy; the current is zero because of the general property that $E(k) = E(-k)$ is an even function of k. If an electron (of given spin) is missing in a state of wavenumber k_h, we have for the current

$$I_h = 2 \sum_k (-e) \frac{v(k)}{L} - (-e) \frac{v(k_h)}{L} \equiv +e \frac{v(k_h)}{L} \; . \tag{72}$$

From Eq. (72) we see that the effective current due to the presence of a hole is the opposite to the current carried by an electron in that state; thus "holes" and "electrons" have opposite charges. A similar reasoning can be done for the variation of the crystal momentum; in fact we have for a hole

$$\frac{d(\hbar k)}{dt} = +eF \; .$$

Finally, if we are interested on holes at the top of a fully occupied band, the negative local curvature of the energy band can be interpreted as a positive effective mass of the hole.

Conductors, semiconductors and insulators

A proper distinction between conductors, semiconductors and insulators, can be done only in the three-dimensional case. However it is of interest to show how the band structure theory can explain the huge difference of conductivity in materials. We first of all notice that a fully occupied band is completely ineffective for electron conductivity. In fact the electrons subjected to electric fields cannot absorb energy, unless interband transitions are involved.

The energy bands of a crystal are occupied by the available electrons according to the Pauli principle and Fermi–Dirac statistics. An insulator is composed by fully occupied bands and fully unoccupied bands and is thus ineffective to the conduction. A typical model of a conductor is constituted by a partially filled band. In this case the energy gap between occupied and unoccupied states is zero, and it is possible to have a response to a steady electric field; the conductivity of a metal usually decreases with increasing temperature, due to the reduction of the relaxation time elapsing between two successive collisions of the free carriers (with the lattice vibrations, for instance). In semiconductors we have fully occupied and fully empty bands at $T = 0$, but the energy gap between occupied and unoccupied states is small (typically less than $1 - 2\,\text{eV}$). Varying the temperature a number of electrons occupy the conduction band and a number of holes are left in the valence bands. These thermally excited carriers depend strongly on temperature, and give rise to a conductivity, that highly increases with temperature.

Further reading

G. Bastard "Wave Mechanics Applied to Semiconductor Heterostructures" (Halstead, New York 1988)

S. Datta "Electron Transport in Mesoscopic Systems" (Cambridge University Press 1995)

E. N. Economou "Green's Function in Quantum Physics" (Springer, Berlin 1990)

D. K. Ferry "Quantum Mechanics. An Introduction for Device Physicists and Electrical Engineers" (Institute of Physics Publishing, Bristol 1995)

M. J. Kelly "Low-Dimensional Semiconductors" (Clarendon Press, Oxford 1995)

E. Merzbacher "Quantum Mechanics" (Wiley, New York 1970). A discussion of transfer matrix is given in Chapter 6.

R. E. Peierls "Quantum Theory of Solids" (Oxford University Press, London 1955)

J. S. Walker and J. Gathright "Exploring One-Dimensional Quantum Mechanics with Transfer Matrix" Am. J. Phys. **62**, 408 (1994)

G. H. Wannier "Elements of Solid State Theory" (Cambridge University Press 1959)

C. Weisbuch and B. Vinter "Quantum Semiconductor Structures" (Academic Press, London 1991)

II

Geometrical description of crystals: direct and reciprocal lattices

Crystalline solids are constituted by a regular array of identical units, periodically repeated in space. In this chapter we consider some geometrical aspects of periodic systems, and describe a few crystal structures of wide interest. The description of translational symmetry in the ordinary direct space would not be complete without the complementary notion of the reciprocal space; the Brillouin zone concept in reciprocal space, together with the Bloch theorem, is the fundamental tool for the classification of states in quantum mechanical problems with translational symmetry. Other subjects, best studied with the help of direct and reciprocal spaces, include indicization of lattice planes and directions, choice of special points in the Brillouin zones for averaging of physical quantities, some general properties of itinerant wavefunctions, energy dispersion curves and density-of-states.

1   Simple lattices and composite lattices

A crystal is characterized by a regular array of atoms, which repeat periodically in the space. A crystalline solid is the most familiar example of a system with long-range order; the knowledge of physical properties (for instance electron charge density, crystalline field, etc.) within any arbitrary unit cell in ideal crystals implies their knowledge within any other unit cell, even macroscopically far removed from the reference one. In amorphous solids, or in liquids, the long-range order is destroyed, although some sort of short-range order may survive on a local microscopic scale. In this section we describe some general properties entailed by periodicity, and focus on the basic distinction between simple lattices and composite lattices.

1.1   Periodicity and Bravais lattices

A *Bravais lattice* is defined as a regular periodic arrangement of points in space, all of them connected by translation vectors

$$\mathbf{t}_n = n_1\mathbf{t}_1 + n_2\mathbf{t}_2 + n_3\mathbf{t}_3 \; ; \tag{1}$$

the non coplanar vectors $\mathbf{t}_1, \mathbf{t}_2, \mathbf{t}_3$ are called *primitive or fundamental translational vectors* and n_1, n_2, n_3 are any tern of *integer* numbers (negative, positive or zero). The parallalepiped formed by $\mathbf{t}_1, \mathbf{t}_2, \mathbf{t}_3$ is called *primitive unit cell* (or primitive cell); its volume is

$$\Omega = \mathbf{t}_1 \cdot (\mathbf{t}_2 \times \mathbf{t}_3) \tag{2}$$

(it is understood that the order of the primitive translation vectors is chosen in such a way to form a right handed system). The primitive cell contains just one lattice point. The parallelepiped primitive cell is usually specified providing the lengths a, b, c of the edges and the angles α, β, γ between each pair of them.

There are several properties of the Bravais lattices, implicit in the definition of Eq. (1), worth to be noticed. First of all, *the choice of the primitive translation vectors is not unique* (and hence also the shape of the primitive cell is not unique). This means that it is possible to describe the *same infinite set of points* of a Bravais lattice with (infinite) different choices of primitive vectors. The origin (and irrelevance) of this arbitrariness is easily understood. Instead of the triad of translation vectors $\mathbf{t}_1, \mathbf{t}_2, \mathbf{t}_3$, consider the triad of vectors $\tilde{\mathbf{t}}_1, \tilde{\mathbf{t}}_2, \tilde{\mathbf{t}}_3$ related to the previous one by the transformation

$$\tilde{\mathbf{t}}_1 = m_{11}\mathbf{t}_1 + m_{12}\mathbf{t}_2 + m_{13}\mathbf{t}_3$$

$$\tilde{\mathbf{t}}_2 = m_{21}\mathbf{t}_1 + m_{22}\mathbf{t}_2 + m_{23}\mathbf{t}_3 \tag{3}$$

$$\tilde{\mathbf{t}}_3 = m_{31}\mathbf{t}_1 + m_{32}\mathbf{t}_2 + m_{33}\mathbf{t}_3$$

where $m_{ij}(i, j = 1, 2, 3)$ are *integer numbers and the 3×3 transformation matrix M has unit determinant*. Then M^{-1} is also a matrix of integer numbers and unit determinant; inversion of Eq. (3) shows that the set of points $\{\mathbf{t}_n\}$ generated by $\mathbf{t}_1, \mathbf{t}_2, \mathbf{t}_3$ coincides with the set of points $\{\tilde{\mathbf{t}}_n\}$ generated by $\tilde{\mathbf{t}}_1, \tilde{\mathbf{t}}_2, \tilde{\mathbf{t}}_3$. Furthermore, the volume $\widetilde{\Omega} = \widetilde{\mathbf{t}}_1 \cdot (\widetilde{\mathbf{t}}_2 \times \widetilde{\mathbf{t}}_3)$ and $\Omega = \mathbf{t}_1 \cdot (\mathbf{t}_2 \times \mathbf{t}_3)$ are equal, as can be established from Eq. (3) and the fact that $\det M = 1$; thus the *volume of the primitive cell is independent* from the choices of the primitive translation vectors. Since triads related by Eq. (3) are mathematically equivalent and describe the same Bravais lattice, it is customary to choose the parallelepiped primitive cell, whose shape has the highest possible symmetry.

An elementary example of the mentioned arbitrariness in the choice of the primitive translation vectors is illustrated in Fig. 1. We consider the two-dimensional rectangular Bravais lattice generated by the fundamental vectors $\mathbf{t}_1$ and $\mathbf{t}_2$, given by

$$\mathbf{t}_1 = (a, 0, 0) \qquad \mathbf{t}_2 = (0, b, 0) \ . \tag{4a}$$

The same Bravais lattice can be described, for instance, by the primitive vectors

$$\widetilde{\mathbf{t}}_1 = (2a, -b, 0) \qquad \widetilde{\mathbf{t}}_2 = (-a, b, 0) \ . \tag{4b}$$

In fact the vectors $\widetilde{\mathbf{t}}_1, \widetilde{\mathbf{t}}_2$ are connected to $\mathbf{t}_1, \mathbf{t}_2$ by the relation

$$\begin{pmatrix} \widetilde{\mathbf{t}}_1 \\ \widetilde{\mathbf{t}}_2 \end{pmatrix} = \begin{pmatrix} 2 & -1 \\ -1 & 1 \end{pmatrix} \begin{pmatrix} \mathbf{t}_1 \\ \mathbf{t}_2 \end{pmatrix} , \tag{4c}$$

and the transformation matrix has integer elements and unit determinant. As shown in Fig. 1, the two choices are perfectly equivalent and the area of the unit cell is invariant; it is also evident that the choice (4a) of primitive vectors and rectangular unit cell appears preferable (although equivalent) to the choice (4b), where the shape of the unit cell is an oblique parallelogram.

From the definition of Bravais lattices, it is evidently possible to describe the whole set of lattice points (1) using non-primitive cells (called *conventional unit cells*), containing an integer number of primitive cells and a corresponding number of extra lattice points. An elementary example of the opportunity to describe some Bravais lattices with a conventional unit cell, is illustrated in Fig. 2, that reports the two-dimensional Bravais lattice, known as the centered rectangular lattice.

In Fig. 2, we choose as primitive lattice vectors

$$\mathbf{t}_1 = (a, 0, 0) \quad \text{and} \quad \mathbf{t}_2 = (\frac{a}{2}, \frac{b}{2}, 0) \ . \tag{5a}$$

The primitive unit cell is a parallelogram of area $ab/2$. Instead of the primitive unit cell (which is a particular case of an oblique parallelogram), it is sometimes preferred (although equivalent) to describe the lattice by a *conventional unit cell of rectangular shape, double area and an extra lattice point at the center of the rectangle*. The fundamental vectors of the conventional unit cell are

$$\mathbf{t}_1^{(c)} = (a, 0, 0) \quad \text{and} \quad \mathbf{t}_2^{(c)} = (0, b, 0) \ . \tag{5b}$$

The matrix M connecting $\mathbf{t}_1^{(c)}, \mathbf{t}_2^{(c)}$ with $\mathbf{t}_1, \mathbf{t}_2$ is given by

$$\begin{pmatrix} \mathbf{t}_1^{(c)} \\ \mathbf{t}_2^{(c)} \end{pmatrix} = \begin{pmatrix} 1 & 0 \\ -1 & 2 \end{pmatrix} \begin{pmatrix} \mathbf{t}_1 \\ \mathbf{t}_2 \end{pmatrix} \tag{5c}$$

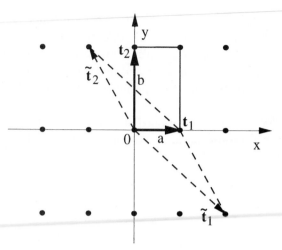

Fig. 1 Illustration of a two-dimensional rectangular Bravais lattice. The lattice can be described by the primitive translation vectors $\mathbf{t}_1 = a(1,0,0)$ and $\mathbf{t}_2 = b(0,1,0)$. Although this is the most natural choice, infinite other choices are possible; the figure illustrates the choice $\tilde{\mathbf{t}}_1 = (2a, -b, 0)$ and $\tilde{\mathbf{t}}_2 = (-a, b, 0)$. Whatever choice is done, the area of the unit cell is invariant and equals ab.

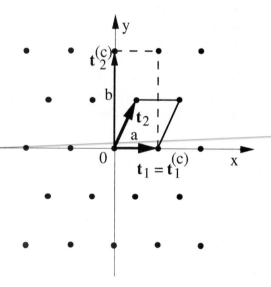

Fig. 2 Illustration of the primitive unit cell and conventional unit cell for a two-dimensional centered rectangular Bravais lattice.

and has determinant 2; the conventional cell defined by $\mathbf{t}_1^{(c)}$ and $\mathbf{t}_2^{(c)}$ contains two primitive cells.

Notice that the conventional cell is not the repeating block of smallest volume; thus in the study of the effects related to the translational symmetry (for instance classification of states with wavevector quantum number, Brillouin zone concept, and in general in any problem directly or indirectly related to the *full* translational symmetry) we must *consider only the primitive translational vectors and the primitive unit cell*. Conventional cells are however of help for a pictorial description of a number of crystals.

The number of different Bravais lattices is determined by symmetry considerations. In two dimensions, the different Bravais lattices are five: oblique, rectangular (primitive and centered), square, hexagonal.

In three dimensions, symmetry considerations lead to fourteen different Bravais lattices grouped into seven crystal systems: triclinic, monoclinic (primitive and base-centered), orthorhombic (primitive, base-centered, body-centered and face-centered), trigonal, tetragonal (primitive and base-centered), hexagonal, cubic (primitive, body-centered and face-centered). This classification is based on the equality or inequality of the lengths a, b, c of the edges of the conventional cell, on the angles α, β, γ between each pair of them, and on possible occurrence of additional lattice points at the centre of opposite faces or at the center of the cell. The seven crystal systems, in order of increasing symmetry (triclinic, monoclinic, orthorhombic, trigonal, tetragonal, hexagonal, cubic) and the fourteen Bravais lattices are reported in Fig. 3.

Before closing, we remark that the symmetry properties of a crystal are not confined to the translational invariance, but include in general also appropriate point group operations. Throughout this chapter, we shall focus essentially on the translational symmetry and its entailed consequences; for an account of the full crystal symmetry, we refer to textbooks on group theory.

1.2 Simple and composite crystal structures

The geometrical description of a crystal requires the specification of the *primitive translation vectors* $\mathbf{t}_1, \mathbf{t}_2, \mathbf{t}_3$ of the underlying Bravais lattice, as well as the specification of the atoms (or ions) in the primitive cell. The contents of the unit cell is described by means of an appropriate set of *basis vectors* $\mathbf{d}_1, \mathbf{d}_2, \ldots, \mathbf{d}_\nu$, which individuate the equilibrium positions of the nuclei of all the atoms (or ions) in the unit cell; for consistency, the positions of the atoms in the basis *cannot* be related by translational vectors. The above considerations can be summarized as follows

$$
\text{Crystal structure} \Rightarrow
\begin{cases}
\mathbf{t}_1, \mathbf{t}_2, \mathbf{t}_3 & \text{(primitive translation vectors)} \\
\mathbf{d}_1, \mathbf{d}_2, \ldots, \mathbf{d}_\nu & \text{(basis of equal or different atoms)}
\end{cases}
.
$$

The geometrical description of the crystal structure is thus specified by the *triad of primitive translation vectors* of the Bravais lattice and by *the vectors forming the basis*.

A crystal with a single atom in the primitive unit cell is called *simple crystal* (or *simple lattice*); in this case, the basis contains a single vector, which can be taken to be zero (with appropriate choice of the origin in the crystal). In a simple lattice the (equilibrium) atomic positions $\mathbf{R}_n$ coincide with the Bravais translation vectors $\mathbf{t}_n$, and we have

$$\mathbf{R}_n = n_1\mathbf{t}_1 + n_2\mathbf{t}_2 + n_3\mathbf{t}_3 \ . \tag{6a}$$

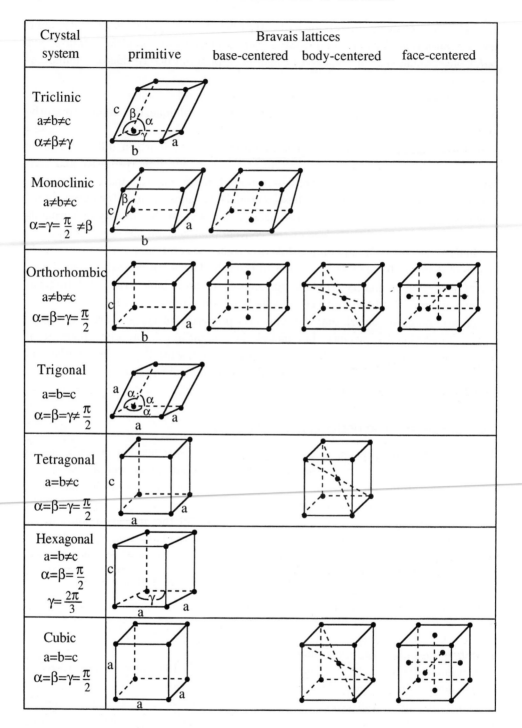

Fig. 3 Classification of the fourteen Bravais lattices into the seven crystal systems. The conventional parallelepiped cells are shown.

A crystal with two or more atoms (or ions) in the primitive unit cell is called a *composite crystal* (or a *composite lattice*). The equilibrium atomic positions are at the points

$$\mathbf{R}_n^{(1)} = \mathbf{d}_1 + n_1\mathbf{t}_1 + n_2\mathbf{t}_2 + n_3\mathbf{t}_3$$

$$\mathbf{R}_n^{(2)} = \mathbf{d}_2 + n_1\mathbf{t}_1 + n_2\mathbf{t}_2 + n_3\mathbf{t}_3 \tag{6b}$$

$$\cdots\cdots\cdots\cdots\cdots$$

$$\mathbf{R}_n^{(\nu)} = \mathbf{d}_\nu + n_1\mathbf{t}_1 + n_2\mathbf{t}_2 + n_3\mathbf{t}_3 \ .$$

A composite lattice can be thought of as composed by a number of interpenetrating simple lattices (called *sublattices*) equal to the number of vectors of the basis. All the points of a given sublattice are related by translation vectors and must thus be occupied by atoms of the same type; different sublattices are occupied by atoms of the same or different type.

2 Geometrical description of some crystal structures

Crystal structure of rare-gas solids (face-centered cubic lattice)

We begin our brief description of some crystals of interest with the case of rare-gas solids (Ne, Ar, Kr, Xe). This crystal structure (illustrated in Fig. 4) is obtained repeating periodically in space a *face-centered cube* (fcc), i.e. a conventional cubic cell (of edge a) with atoms at the corners and at the center of the faces. The primitive translation vectors of the fcc Bravais lattice are

$$\boxed{\mathbf{t}_1 = \frac{a}{2}(0,1,1) \quad \mathbf{t}_2 = \frac{a}{2}(1,0,1) \quad \mathbf{t}_3 = \frac{a}{2}(1,1,0)} \ . \tag{7}$$

Notice that $-\mathbf{t}_1+\mathbf{t}_2+\mathbf{t}_3 = a(1,0,0), \mathbf{t}_1-\mathbf{t}_2+\mathbf{t}_3 = a(0,1,0)$ and $\mathbf{t}_1+\mathbf{t}_2-\mathbf{t}_3 = a(0,0,1)$. The volume of the primitive cell is $\Omega = \mathbf{t}_1 \cdot (\mathbf{t}_2 \times \mathbf{t}_3) = a^3/4$; thus the primitive cell, which expresses the building block of minimum volume of the crystal, is four times smaller than the volume a^3 of the conventional cubic cell.

In the fcc structure, each atom has 12 nearest neighbours (the number of nearest neighbours is called *coordination number*). The atom at the origin has 12 nearest neighbours in the positions $(a/2)(0,\pm1,\pm1)$ (and cyclic permutations) at distance $(a/2)\sqrt{2}$. It has 6 second nearest neighbours in the positions $(a/2)(\pm2,0,0)$ at distance a; 24 third nearest neighbours $(a/2)(\pm2,\pm1,\pm1)$ at distance $(a/2)\sqrt{6}$, etc..

The lattice constants of rare-gas solids are: $a = 4.43\,\text{Å}$ for Ne; $a = 5.26\,\text{Å}$ for Ar; $a = 5.72\,\text{Å}$ for Kr; $a = 6.20\,\text{Å}$ for Xe. Besides rare-gas solids, there is a number of elements with monoatomic face-centered cubic structure; among them, we may mention several metals (Ag, Al, Au, Cu, Pd, Pt and others) and some rare earth elements.

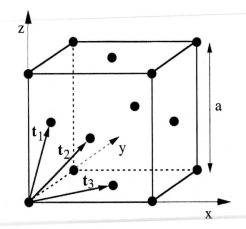

Fig. 4 Face-centered cubic lattice and primitive translation vectors t_1, t_2, t_3 given by Eqs. (7) in the text.

Crystal structure of alkali metals (body-centered cubic lattice)

The crystal structure of alkali metals (see Fig. 5) is obtained repeating periodically in space a body-centered cube (bcc), i.e. a conventional cubic cell with atoms at the corners and at the center of a cube of edge a. The primitive translation vectors of the bcc Bravais lattice are

$$t_1 = \frac{a}{2}(-1,1,1) \quad t_2 = \frac{a}{2}(1,-1,1) \quad t_3 = \frac{a}{2}(1,1,-1) \ . \tag{8}$$

Notice that $t_1 + t_2 + t_3 = (a/2)(1,1,1)$, and also $t_2 + t_3 = a(1,0,0)$, $t_1 + t_3 = a(0,1,0)$, $t_1 + t_2 = a(0,0,1)$. The atom at the origin has 8 nearest neighbours in the positions $(a/2)(\pm1, \pm1, \pm1)$ at distance $(a/2)\sqrt{3}$; there are 6 second nearest neighbours in the positions $(a/2)(\pm2, 0, 0)$ (and cyclic permutations) at distance a, etc. The volume of the primitive unit cell is $\Omega = t_1 \cdot (t_2 \times t_3) = a^3/2$; thus the conventional cubic cell is twice as big as the primitive unit cell.

The lattice constants of alkali metals are: $a = 3.49\,\text{Å}$ for Li; $a = 4.23\,\text{Å}$ for Na; $a = 5.23\,\text{Å}$ for K; $a = 5.59\,\text{Å}$ for Rb; $a = 6.05\,\text{Å}$ for Cs. Other elements which crystallize in the monoatomic body-centered cubic structure include Cr ($a = 2.88\,\text{Å}$), Mo ($a = 3.15\,\text{Å}$), W ($a = 3.16\,\text{Å}$), Fe ($a = 2.87\,\text{Å}$), Ba($a = 5.03\,\text{Å}$).

Sodium chloride structure

The sodium chloride structure is shown in Fig. 6. The crystal structure can be described as two interpenetrating fcc lattices displaced by $(a/2)(1,1,1)$ along the body diagonal of the conventional cube; one of the two fcc sublattices is composed by cations (Na$^+$) and the other by anions (Cl$^-$). The primitive translation vectors and the two

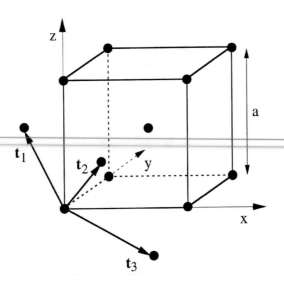

Fig. 5 Body-centered cubic lattice and primitive translation vectors $\mathbf{t}_1, \mathbf{t}_2, \mathbf{t}_3$ given by Eqs. (8) in the text.

vectors of the basis of the NaCl crystal structure are given by

$$\mathbf{t}_1 = \tfrac{a}{2}(0,1,1) \quad \mathbf{t}_2 = \tfrac{a}{2}(1,0,1) \quad \mathbf{t}_3 = \tfrac{a}{2}(1,1,0)$$
$$\mathbf{d}_1 = 0 \quad \mathbf{d}_2 = \tfrac{a}{2}(1,1,1)$$

(9)

In the NaCl structure, each anion in a given sublattice has 6 nearest neighbouring cations (on the other sublattice) at distance $a/2$, 12 second nearest neighbours of the same type at distance $(a/2)\sqrt{2}$ etc. Some lattice constants of crystals with the NaCl arrangement are: $a = 5.63\,\text{Å}$ for NaCl; $a = 4.02\,\text{Å}$ for LiF; $a = 5.77\,\text{Å}$ for AgBr; $a = 6.99\,\text{Å}$ for BaTe; $a = 6.02\,\text{Å}$ for SnSe; $a = 4.61\,\text{Å}$ for ZrN.

Cesium chloride structure

The cesium chloride structure is shown in Fig. 7. The crystal structure of CsCl can be described as two interpenetrating simple cubic lattices displaced by $(a/2)(1,1,1)$ along the body diagonal of the cubic cell. The underlying Bravais lattice is simple cubic, and there are two atoms Cs and Cl in the primitive unit cell; the primitive translation vectors and the two vectors of the basis are

$$\mathbf{t}_1 = a(1,0,0) \quad \mathbf{t}_2 = a(0,1,0) \quad \mathbf{t}_3 = a(0,0,1)$$
$$\mathbf{d}_1 = 0 \quad \mathbf{d}_2 = \tfrac{a}{2}(1,1,1)$$

(10)

In the CsCl structure, each anion in a given sublattice has 8 nearest neighbouring cations (on the other sublattice) at distance $(a/2)\sqrt{3}$; there are 6 second nearest

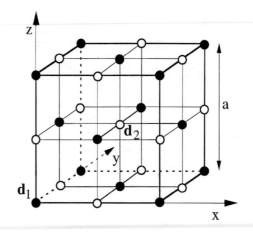

Fig. 6 Crystal structure of sodium chloride; the end-points of the basis vectors $\mathbf{d}_1$ and $\mathbf{d}_2$, given in Eqs. (9) of the text, are also indicated.

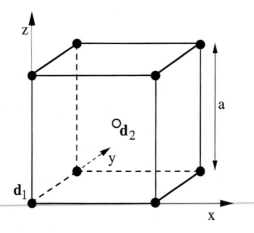

Fig. 7 The cesium chloride structure; the end-points of the basis vectors $\mathbf{d}_1$ and $\mathbf{d}_2$, given in Eqs. (10) of the text, are also indicated.

neighbours of the same type at distance a, etc. Some lattice constants of crystals with the CsCl arrangement are: $a = 4.12\,\text{Å}$ for CsCl; $a = 3.97\,\text{Å}$ for TlBr; $a = 3.33\,\text{Å}$ for AgCd; $a = 2.95\,\text{Å}$ for CuZn.

Cubic perovskite structure

The cubic perovskite structure is shown in Fig. 8 for $BaTiO_3$; this crystal remains cubic in the temperature interval from 201°C (lattice constant $a = 4.01\,\text{Å}$) up to 1372°C ($a = 4.08\,\text{Å}$). The crystal structure of barium titanide presents one molecule per unit cell, and can be described as five interpenetrating simple cubic lattices. The

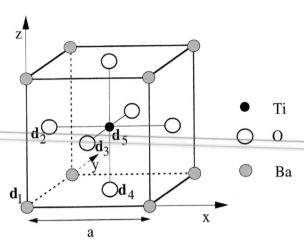

Fig. 8 Cubic structure of barium titanide; the end-points of the basis vectors $\mathbf{d}_1, \mathbf{d}_2, \mathbf{d}_3, \mathbf{d}_4$ and $\mathbf{d}_5$ are also indicated.

primitive translation vectors and the five vectors of the basis are

$$
\begin{array}{c}
\mathbf{t}_1 = a(1,0,0) \quad \mathbf{t}_2 = a(0,1,0) \quad \mathbf{t}_3 = a(0,0,1) \\
\mathbf{d}_1 = 0 \quad \mathbf{d}_2 = \tfrac{a}{2}(0,1,1) \quad \mathbf{d}_3 = \tfrac{a}{2}(1,0,1) \quad \mathbf{d}_4 = \tfrac{a}{2}(1,1,0) \quad \mathbf{d}_5 = \tfrac{a}{2}(1,1,1)
\end{array}
\tag{11}
$$

In Fig. 8 barium is at the origin $\mathbf{d}_1$, oxygens are in $\mathbf{d}_2, \mathbf{d}_3, \mathbf{d}_4$, and titanium at $\mathbf{d}_5$. Each titanium is octrahedrally coordinated to six oxygens; each oxygen is coordinated with two titanium and four barium sites; each barium is surrounded by twelve oxygen sites.

Diamond structure and zincblende structure

The diamond structure can be described as two interpenetrating fcc lattices displaced by $(a/4)(1,1,1)$ along the body diagonal of the conventional cube. The underlying Bravais lattice is fcc with two carbon atoms forming the basis; the primitive translation vectors and the two vectors of the basis are

$$
\begin{array}{c}
\mathbf{t}_1 = \tfrac{a}{2}(0,1,1) \quad \mathbf{t}_2 = \tfrac{a}{2}(1,0,1) \quad \mathbf{t}_3 = \tfrac{a}{2}(1,1,0) \\
\mathbf{d}_1 = 0 \quad \mathbf{d}_2 = \tfrac{a}{4}(1,1,1)
\end{array}
\tag{12}
$$

The coordination number is 4; each atom in a given sublattice is surrounded by four atoms of the other sublattice, at distance $(a/4)\sqrt{3}$, in a tetrahedral configuration;

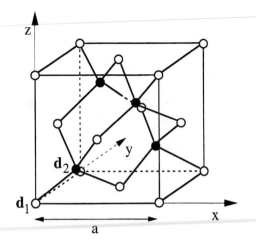

Fig. 9 Crystal structure of diamond and zincblende; the end-points of the vectors $\mathbf{d}_1$ and $\mathbf{d}_2$ of the basis are also indicated. In the diamond structure the two sublattices are occupied by atoms of the same type; in the zincblende structure the two sublattices are occupied by different types of atoms.

there are 12 second neighbours at distance $(a/2)\sqrt{2}$. For instance, the carbon atom at the origin $(\mathbf{t}_n = 0, \mathbf{d}_1 = 0)$ is surrounded by four atoms in the positions

$$\mathbf{d}_2 = \tfrac{a}{4}(1,1,1) \qquad \mathbf{d}_2 - \mathbf{t}_1 = \tfrac{a}{4}(1,-1,-1)$$
$$\mathbf{d}_2 - \mathbf{t}_2 = \tfrac{a}{4}(-1,1,-1) \quad \mathbf{d}_2 - \mathbf{t}_3 = \tfrac{a}{4}(-1,-1,1) \ .$$

The diamond structure is also typical of silicon, germanium and grey tin. The lattice constants are: $a = 3.57\,\text{Å}$ for C (diamond); $a = 5.43\,\text{Å}$ for Si; $a = 5.65\,\text{Å}$ for Ge; $a = 6.49\,\text{Å}$ for α-Sn (grey tin).

The zincblende (ZnS) structure is similar to that of diamond; now the two sublattices are occupied by two different types of atoms or ions. Among crystals with this structure we mention: cubic ZnS $(a = 5.41\,\text{Å})$, AgI $(a = 6.47\,\text{Å})$, cubic BN $(a = 3.62\,\text{Å})$, CuCl $(a = 5.41\,\text{Å})$. This structure is also typical of several III–V and II–VI groups semiconducting compounds; for instance: GaAs $(a = 5.65\,\text{Å})$, AlAs $(a = 5.66\,\text{Å})$, InAs $(a = 6.04\,\text{Å})$, GaSb $(a = 6.12\,\text{Å})$, CdTe $(a = 6.48\,\text{Å})$, HgTe $(a = 6.48\,\text{Å})$, ZnTe $(a = 6.09\,\text{Å})$.

Crystal structure of two- and three-dimensional graphite

The two-dimensional honeycomb crystal structure of graphite is indicated in Fig. 10. With the choice of axes as indicated in Fig. 10, the primitive translation vectors and the two vectors of the basis are

$$\begin{array}{ll} \mathbf{t}_1 = a(\tfrac{1}{2}, \tfrac{\sqrt{3}}{2}, 0) & \mathbf{t}_2 = a(-\tfrac{1}{2}, \tfrac{\sqrt{3}}{2}, 0) \\[2mm] \mathbf{d}_1 = 0 & \mathbf{d}_2 = (0, \tfrac{a}{\sqrt{3}}, 0) \end{array} \qquad (13)$$

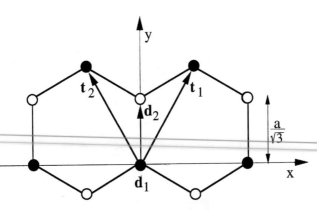

Fig. 10 Layer structure of two-dimensional graphite. The primitive translation vectors $\mathbf{t}_1$ and $\mathbf{t}_2$, and the basis vectors $\mathbf{d}_1$ and $\mathbf{d}_2$, given by Eqs. (13) of the text, are also indicated.

with $a = 2.46\,\text{Å}$. The crystal structure of graphite is obtained occupying each vertex of the hexagons by carbon atoms.

It can be easily noticed, by inspection, that it would be impossible to define a lattice translational vector joining an atom with any of its first neighbours; in fact, the opposite of such a vector points at the center of the hexagons, where no atom exists. The two-dimensional graphite structure is composite and consists of two sublattices of carbon atoms, corresponding to the basis vectors $\mathbf{d}_1 = 0$ and $\mathbf{d}_2 = (0, a/\sqrt{3}, 0)$. If we replace one sublattice with boron atoms and the other with nitrogen atoms, we obtain the two-dimensional structure of hexagonal boron nitride.

The three-dimensional structure of graphite is indicated in Fig. 11. The value of the lattice parameter in the z-direction is $c = 6.71\,\text{Å}$. The fundamental unit cell of graphite contains four atoms. The primitive translation vectors and the basis for the three-dimensional graphite are given by

$$
\begin{array}{llll}
\mathbf{t}_1 = a(\tfrac{1}{2}, \tfrac{\sqrt{3}}{2}, 0) & \mathbf{t}_2 = a(-\tfrac{1}{2}, \tfrac{\sqrt{3}}{2}, 0) & \mathbf{t}_3 = c(0, 0, 1) & \\
\mathbf{d}_1 = 0 & \mathbf{d}_2 = (0, \tfrac{a}{\sqrt{3}}, 0) & \mathbf{d}_3 = (0, 0, \tfrac{c}{2}) & \mathbf{d}_4 = (0, \tfrac{2a}{\sqrt{3}}, \tfrac{c}{2})
\end{array}
\tag{14}
$$

In the case of graphite two consecutive layers, stacked along the z-axis, are rotated by $2\pi/6$. As a consequence the atoms $\mathbf{d}_1$ and $\mathbf{d}_3$ have two nearest neighbours on different adjacent planes at distance $c/2$.

We wish also to mention a third interesting crystalline form of carbon, besides the graphite and diamond. It is constituted by molecules C_{60} with the structure of a soccerball, i.e. a truncated icosahedron, originally proposed by H. W. Kroto, J. R. Heath, S. C. O'Brien, R. F. Curl and R. E. Smalley, Nature **318**, 162 (1985). The structure of C_{60} molecule is shown in Fig. 12; the centers of the molecules are arranged in a fcc lattice, and the primitive cell of the solid contains a soccerball molecule of

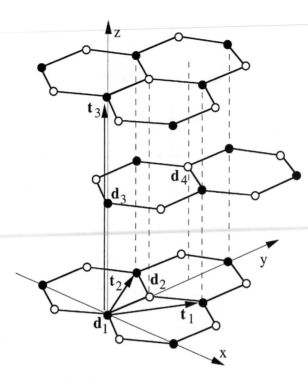

Fig. 11 Three-dimensional crystal structure of graphite. The primitive translation vectors t_1, t_2, t_3 and the end-points of the basis vectors d_1, d_2, d_3 and d_4, given in Eqs. (14) of the text, are also indicated.

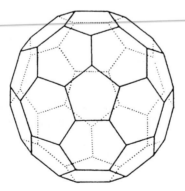

Fig. 12 The structure of C_{60} molecule in its regular truncated icosahedron geometry. The polygon has 60 vertices and 32 faces, 12 of which are pentagonal and 20 hexagonal. The bond lengths forming pentagons are 1.47 Å; the bond lengths common to two hexagons are 1.41 Å [from P. Milani, Rivista del Nuovo Cimento **19**, N.11 (1996); copyright 1996 by Società Italiana di Fisica].

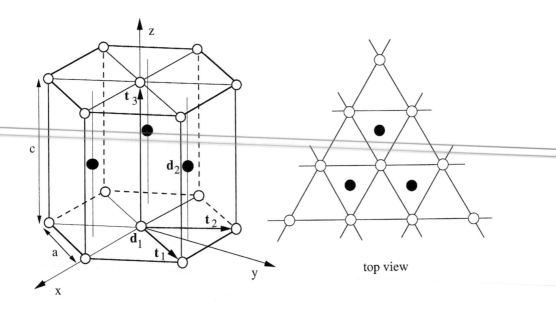

Fig. 13 Hexagonal closed-packed structure. The primitive translation vectors $\mathbf{t}_1, \mathbf{t}_2, \mathbf{t}_3$ and the end points of the basis vectors $\mathbf{d}_1$ and $\mathbf{d}_2$, given in Eqs. (15) of the text, are also indicated. The top view of the structure is also shown for convenience.

sixty carbon atoms. This form of carbon has been given the name of *fullerene*, as a tribute to the geodesic studies of the architect R. Buckminster Fuller.

Hexagonal close-packed structure

Several elements crystallize in the hexagonal close-packed structure (hcp). To describe this structure consider a two-dimensional array of equilater triangles (or regular hexagons *including* their centers) of edge a, as indicated in Fig. 13. Then one stacks along the z axis a second plane of equilateral triangles, above the centers of every other triangle of the first plane, as indicated in Fig. 13. The sequence of planes ABAB AB ... is then repeated: the third plane lies directly over the first one, the fourth over the second, and so on.

The hexagonal close-packed structure can be pictured as formed by two interpenetrating simple hexagonal Bravais lattices; the primitive translation vectors and the two basis vectors of the hcp structure are

$$
\begin{aligned}
\mathbf{t}_1 = a(\tfrac{1}{2}, \tfrac{\sqrt{3}}{2}, 0) \qquad \mathbf{t}_2 = a(-\tfrac{1}{2}, \tfrac{\sqrt{3}}{2}, 0) \qquad \mathbf{t}_3 = c(0,0,1) \\
\mathbf{d}_1 = 0 \qquad \mathbf{d}_2 = (0, \tfrac{a}{\sqrt{3}}, \tfrac{c}{2})
\end{aligned}
\quad ; \qquad (15)
$$

the two sublattices specified by $\mathbf{d}_1$ and $\mathbf{d}_2$ are occupied by atoms of the same type. In the case the atoms can be thought of as hard spheres (of radius r_0) touching each other, we have $2r_0 = a$ (contact condition within a hexagonal plane) and also

$|\mathbf{d}_2| = \sqrt{(a^2/3) + (c^2/4)} = 2r_0$ (contact condition between two adjacent hexagonal planes). The *ideal ratio* c/a becomes

$$\frac{c}{a} = \sqrt{\frac{8}{3}} = 1.633 \ ;$$

this ratio is only approximately verified in actual hcp crystals. Several metals have the hcp structure, among them Be ($a = 2.29$ Å; $c = 3.58$ Å; $c/a = 1.56$), Cd ($a = 2.98$ Å; $c = 5.62$ Å; $c/a = 1.88$), Mg ($a = 3.21$ Å; $c = 5.21$ Å; $c/a = 1.62$), Ti ($a = 2.95$ Å; $c = 4.69$ Å; $c/a = 1.59$), Zn ($a = 2.66$ Å; $c = 4.95$ Å; $c/a = 1.86$).

Finally it is worth to mention that another close-packing arrangement is provided by the fcc structure; the sequence of hexagonal planes is in this case ABC, ABC, etc. where C is rotated by $2\pi/6$ with respect to the B plane. These closed packed hexagonal planes are arranged orthogonally to the body diagonal of the conventional fcc cube. To see this, we remember that in the fcc structure the atom at the origin has twelve neighbours in the positions $(a/2)(0, \pm 1, \pm 1)$ and cyclic permutations. The three points $P_1 = (a/2)(0, -1, -1)$, $P_2 = (a/2)(-1, 0, -1)$ and $P_3 = (a/2)(-1, -1, 0)$, describe a regular triangle (of edge $a\sqrt{2}/2$) orthogonal to the body diagonal versor $\mathbf{n} = (1/\sqrt{3})(1, 1, 1)$ and belong to the plane A, distant $a/\sqrt{3}$ from the origin. The six points given by $(a/2)(0, 1, -1)$ and $(a/2)(0, -1, 1)$ and cyclic permutations, describe a regular hexagon (orthogonal to $\mathbf{n}$) around the atom at the origin, and belong to plane B; finally, the three points given by $(a/2)(0, 1, 1)$ and cyclic permutations, belong to plane C. The ratio between $2a/\sqrt{3}$ and $a\sqrt{2}/2$ equals the ideal value $\sqrt{8/3}$.

Hexagonal wurtzite structure

The wurtzite structure can be considered as formed by two interpenetrating hexagonal closed-packed lattices; in the unit cell there are four atoms of two different types, forming two molecules (see Fig. 14). The primitive translation vectors and the basis vectors are

$$
\boxed{
\begin{array}{lll}
\mathbf{t}_1 = a(\tfrac{1}{2}, \tfrac{\sqrt{3}}{2}, 0) & \mathbf{t}_2 = a(-\tfrac{1}{2}, \tfrac{\sqrt{3}}{2}, 0) & \mathbf{t}_3 = c(0, 0, 1) \\[6pt]
\mathbf{d}_1 = 0 \quad \mathbf{d}_2 = (0, 0, uc) & \mathbf{d}_3 = (0, \tfrac{a}{\sqrt{3}}, \tfrac{c}{2}) & \mathbf{d}_4 = (0, \tfrac{a}{\sqrt{3}}, \tfrac{c}{2} + uc)
\end{array}
}
\tag{16}
$$

where a and c are the lattice constants, and u is dimensionless; $\mathbf{d}_1$ and $\mathbf{d}_3$ are occupied by the same type of atom, $\mathbf{d}_2$ and $\mathbf{d}_4$ are occupied by the other type of atom. Each atom is surrounded by an (almost regular) tetrahedron of atoms of opposite sort; the tetrahedral blocks are regular if the axis ratio has the ideal value $c/a = 1.633$ and the parameter $u = 3/8 = 0.375$. The adjacent tetrahedral blocks are stacked along the c-axis with the relative orientation shown in Fig. 14 as an example. [Notice that in the cubic zincblende structure, the adjacent tetrahedral blocks are stacked along the body diagonal of the conventional cubic cell, and are rotated by $2\pi/6$ with respect to each other]. The lattice parameters for ZnO (or zincite) are: $a = 3.25$ Å, $c = 5.21$ Å, $u = 0.345$. For the the hexagonal form of CdS we have: $a = 4.14$ Å, $c = 6.75$ Å, $c/a = 1.63$; and for ZnS, or wurtzite, we have: $a = 3.81$ Å, $c = 6.23$ Å, $c/a = 1.64$.

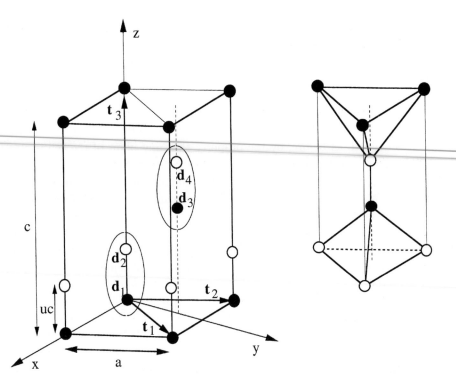

Fig. 14 Hexagonal wurtzite structure. The primitive translation vectors $\mathbf{t}_1, \mathbf{t}_2, \mathbf{t}_3$ and the end points of the basis vectors $\mathbf{d}_1, \mathbf{d}_2, \mathbf{d}_3$ and $\mathbf{d}_4$, given by Eq. (16) of the text, are indicated; the two molecules of the unit cell are encircled for convenience. The orientations of adjacent tetrahedra surrounding $\mathbf{d}_3$ and $\mathbf{d}_4$ are also shown.

We close here our brief description of crystal structures, and refer to the literature for further information [see for instance R. W. G. Wyckoff "Crystal Structures" vols.1-5 (Wiley, New York, 1963-1968)]. Beyond specific descriptions, it should be clear the main message of this section: a crystal structure is specified by the knowledge of the *primitive translation vectors* and the *basis vectors* individuating the atomic positions in the unit cell; from this knowledge, any geometrical property of interest (neighbours, distances and bond lengths, angles, planes, directions, and also symmetry operations) can be easily worked out.

3   Wigner–Seitz primitive cells

In describing crystal structures, we have found convenient to take the fundamental primitive cell as a parallelepiped defined by the vectors $\mathbf{t}_1, \mathbf{t}_2, \mathbf{t}_3$. There are infinite possible choices for the fundamental cell (one can vary the origin, or the primitive vectors keeping the volume constant, or both). An important and very useful choice for the primitive cell has been suggested by Wigner and Seitz.

Consider first a *simple lattice*; the translation vectors individuate all the atoms of the crystal. *The Wigner–Seitz cell about a (reference) lattice point is defined by the property that any point of the cell is closer to that lattice point than to any other.* The Wigner–Seitz cell can be operatively obtained by bisecting with perpendicular planes the vectors joining one atom with the nearest neighbours, second nearest neighbours, and so on, and considering the smallest volume enclosed.

If we have *composite lattices*, we focus on the underlying Bravais lattice; the Wigner–Seitz cell is defined, as before, by the property that any point of the cell is closer to that lattice point than any other. However, in this case, we can obtain *subcells* by bisecting with perpendicular planes the vectors joining one atom with nearest neighbours, second nearest neighbours etc. and considering the smallest volume enclosed. For composite lattices, the Wigner–Seitz cell is constituted by a number of subcells equal to the number of atoms of the basis. Each subcell has an atom at its center. The Wigner–Seitz cells are particularly convenient in those methods of electronic state calculations that exploit appropriately the "local spherical symmetry" of the periodic crystal potential.

4   Reciprocal lattices

4.1   Definitions and basic properties

For the study of crystals, besides *the direct lattice in the ordinary space*, it is important to consider also the *reciprocal lattice in the dual (or reciprocal) space*.

Given a crystal with primitive translation vectors $\mathbf{t}_1, \mathbf{t}_2, \mathbf{t}_3$ in the direct space, we consider the three primitive vectors $\mathbf{g}_1, \mathbf{g}_2, \mathbf{g}_3$ in the reciprocal space, defined by the relations

$$\boxed{\mathbf{t}_i \cdot \mathbf{g}_j = 2\pi \delta_{ij}} \tag{17a}$$

(the numerical factor 2π is introduced as a matter of convenience to simplify some expressions later on). If $\mathbf{t}_1, \mathbf{t}_2, \mathbf{t}_3$ are non co-planar vectors and form a right-handed system, also $\mathbf{g}_1, \mathbf{g}_2, \mathbf{g}_3$ are non co-planar vectors and form a right-handed system. If a *crystal rotation of* $\mathbf{t}_1, \mathbf{t}_2, \mathbf{t}_3$ *is performed in the direct space, the same rotation of* $\mathbf{g}_1, \mathbf{g}_2, \mathbf{g}_3$ *occurs in the reciprocal space*. Notice that the propagation wavevector $\mathbf{k}$ of a general plane wave $\exp(i\mathbf{k} \cdot \mathbf{r})$ has "reciprocal length" dimension, and can be conveniently represented in the reciprocal space.

All the points defined by the vectors of the type

$$\mathbf{g}_m = m_1\mathbf{g}_1 + m_2\mathbf{g}_2 + m_3\mathbf{g}_3 \tag{17b}$$

(with m_1, m_2, m_3 integer numbers, negative, zero, or positive) constitute the *reciprocal lattice*. Notice that the reciprocal lattice is related only to the translational properties of the crystal and not to the basis. Crystals with the same translational symmetry, but completely different basis, have the same reciprocal lattice. In Chapter X, in the study of elastic diffraction, we will see that the geometry of the diffracted beams determines *reciprocal lattice vectors*, while the *intensity* of the diffracted beams provides information on *the basis*.

We can solve Eq. (17a) explicitly. For instance, notice that $\mathbf{g}_1$ must be orthogonal both to $\mathbf{t}_2$ and $\mathbf{t}_3$ and is thus parallel to $\mathbf{t}_2 \times \mathbf{t}_3$. The condition $\mathbf{g}_1 \cdot \mathbf{t}_1 = 2\pi$ fully determines $\mathbf{g}_1$. The other vectors are similarly obtained with cyclic permutations. We have

$$\mathbf{g}_1 = \frac{2\pi}{\Omega} \mathbf{t}_2 \times \mathbf{t}_3 , \qquad \mathbf{g}_2 = \frac{2\pi}{\Omega} \mathbf{t}_3 \times \mathbf{t}_1 , \qquad \mathbf{g}_3 = \frac{2\pi}{\Omega} \mathbf{t}_1 \times \mathbf{t}_2 , \qquad (18)$$

where $\Omega = \mathbf{t}_1 \cdot (\mathbf{t}_2 \times \mathbf{t}_3)$ is the volume of the primitive cell in the direct lattice. From Eq. (17a) we see that the reciprocal of the reciprocal lattice is the original direct lattice.

The direct and reciprocal lattices obey some simple useful properties. First of all, we begin to observe that *the volume Ω_k of the unit cell in the reciprocal space is $(2\pi)^3$ times the reciprocal of the volume of the unit cell in the direct lattice*. In fact

$$\Omega_k = \mathbf{g}_1 \cdot (\mathbf{g}_2 \times \mathbf{g}_3) = \frac{(2\pi)^3}{\Omega^3}(\mathbf{t}_2 \times \mathbf{t}_3) \cdot [(\mathbf{t}_3 \times \mathbf{t}_1) \times (\mathbf{t}_1 \times \mathbf{t}_2)] = \frac{(2\pi)^3}{\Omega} .$$

To perform the calculation of the quantity within square brackets, we have used the following relation for the vector product among any three vectors

$$\mathbf{v}_1 \times (\mathbf{v}_2 \times \mathbf{v}_3) \equiv \mathbf{v}_2(\mathbf{v}_1 \cdot \mathbf{v}_3) - \mathbf{v}_3(\mathbf{v}_1 \cdot \mathbf{v}_2) .$$

From equations (17), it is evident that the scalar product of any reciprocal lattice vector $\mathbf{g}_m$ with any translation vector $\mathbf{t}_n$ is an integer number of 2π. We can thus write

$$\mathbf{g}_m \cdot \mathbf{t}_n = \text{integer} \cdot 2\pi \qquad (19a)$$

for any translation vector $\mathbf{t}_n$. Furthermore, if a vector $\mathbf{q}$ satisfies the relation

$$\mathbf{q} \cdot \mathbf{t}_n = \text{integer} \cdot 2\pi \qquad (19b)$$

for any $\mathbf{t}_n$, then $\mathbf{q}$ must be a reciprocal lattice vector. To show this, we notice that it is always possible to express $\mathbf{q}$ in terms of the non-collinear vectors $\mathbf{g}_1, \mathbf{g}_2, \mathbf{g}_3$; we thus can write $\mathbf{q} = c_1\mathbf{g}_1 + c_2\mathbf{g}_2 + c_3\mathbf{g}_3$ with c_i real numbers. To satisfy Eq. (19b) for any $\mathbf{t}_n$ and in particular for $\mathbf{t}_1, \mathbf{t}_2, \mathbf{t}_3$ it is necessary that c_1, c_2, c_3 are *integer* numbers, and this means that $\mathbf{q}$ is a reciprocal lattice vector. Eqs. (19) say that all and only the reciprocal lattice vectors have scalar product with any lattice translation vector equal to an integer number of 2π.

An immediate consequence of Eqs. (19) is that a *plane wave* $\exp(i\mathbf{k} \cdot \mathbf{r})$ *has the lattice periodicity if and only if the wavevector $\mathbf{k}$ equals a reciprocal lattice vector*. In fact the function

$$W(\mathbf{r}) = e^{i\mathbf{g}_m \cdot \mathbf{r}} \qquad (20a)$$

remains unchanged if we replace $\mathbf{r} \to \mathbf{r} + \mathbf{t}_n$ (the opposite follows from Eq. 19b). Thus a *function $f(\mathbf{r})$ periodic in the direct lattice*, can always be expanded in the form

$$f(\mathbf{r}) = \sum_{\mathbf{g}_m} f_m \, e^{i\mathbf{g}_m \cdot \mathbf{r}} , \tag{20b}$$

where the sum is over the reciprocal lattice vectors. Similarly, a function $F(\mathbf{k})$ *periodic in the reciprocal lattice*, can always be expanded in the form

$$F(\mathbf{k}) = \sum_{\mathbf{t}_m} F_m \, e^{i\mathbf{k} \cdot \mathbf{t}_m} , \tag{20c}$$

where the sum is over the translation lattice vectors.

4.2   Planes and directions in Bravais lattices

Consider a Bravais lattice, its dual space, and a given vector $\mathbf{g}_m = m_1 \mathbf{g}_1 + m_2 \mathbf{g}_2 + m_3 \mathbf{g}_3$ of the reciprocal lattice. From Eqs. (17), it is seen by inspection that the possible values of the scalar products $\mathbf{g}_m \cdot \mathbf{t}_n$ of a given *fixed* vector $\mathbf{g}_m$ with *all* translation vectors $\mathbf{t}_n$ are by necessity organized in constant steps of the type $0, \pm 2\pi\nu, \pm 4\pi\nu, \pm 6\pi\nu, \ldots$ (where ν is an integer number, not necessarily equal to 1). If $\nu \neq 1$ then the vector $\mathbf{g}_m/\nu$ is also a reciprocal lattice vector (in fact the scalar products of $\mathbf{g}_m/\nu$ with any translation vector are integer multiples of 2π); this means that the integer values m_1, m_2, m_3 of $\mathbf{g}_m$ have the integer ν as common divisor.

   In this section, we confine our attention to reciprocal lattice vectors $\mathbf{g}_m$ with the *restriction that the tern of integers m_1, m_2, m_3 has no common divisor* (otherwise we divide by it); this means that we are considering the reciprocal vector of minimum length among all its possible multiples (in the direction of $\mathbf{g}_m$). The scalar product of $\mathbf{g}_m$ with any translation vector $\mathbf{t}_n$ gives any *integer multiples of* 2π; namely, we have

$$\mathbf{g}_m \cdot \mathbf{t}_n = 0, \ \pm 2\pi, \ \pm 4\pi, \ \pm 6\pi, \ldots \tag{21a}$$

where $\mathbf{t}_n$ is any translation vector, and m_1, m_2, m_3 are integers without common integer divisor.

   We can interpret the result of Eq. (21a) in a geometrical form. Consider the family of planes in the direct space defined by the equations

$$\mathbf{g}_m \cdot \mathbf{r} = 0, \ \pm 2\pi, \ \pm 4\pi, \ \pm 6\pi, \ \ldots \tag{21b}$$

The family of planes is represented in Fig. 15. Because of equation (21a) we see that *all translation vectors* belong to the family of planes described by Eq. (21b); it is also apparent that the distance between two consecutive planes is

$$d = \frac{2\pi}{g_m} . \tag{21c}$$

In conclusion: *Every reciprocal lattice vector $\mathbf{g}_m$ is normal to a family of parallel and equidistant planes containing all the direct lattice points*; the distance between two successive planes is $d = 2\pi/g_m$. This property is the basis for the interpretation of the Bragg diffraction (presented in Section X-2).

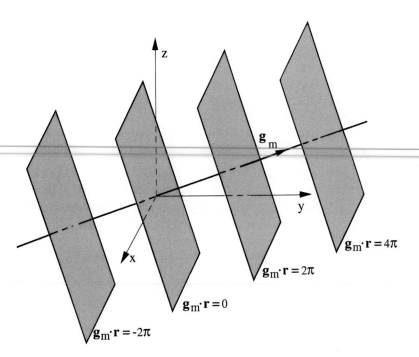

Fig. 15 Family of planes in direct space defined by the equations $\mathbf{g}_m \cdot \mathbf{r} = 0, \pm 2\pi, \pm 4\pi \ldots$; all the translation vectors of the Bravais lattice belong to the family of planes.

From the above properties it turns out natural to label a family of parallel lattice planes, in which we can resolve a Bravais lattice, by means of the tern of integers m_1, m_2, m_3 (with no common integer divisor) of the $\mathbf{g}_m$ vector normal to them. The three indices, conventionally enclosed by round brackets (m_1, m_2, m_3), are used to denote the orientation of the family of planes. The family of all the planes in direct space, equivalent to (m_1, m_2, m_3) by point symmetry, is indicated by curly brackets $\{m_1, m_2, m_3\}$.

It is interesting to analyse the equivalence of the present indicization scheme, based on the properties of the dual spaces, with the more traditional one, introduced by Miller in crystallography.

According to the prescription introduced by Miller, a lattice plane (i.e. a plane through three non-collinear lattice points) can be described starting from the intercepts with the primitive axes (expressed in units of $\mathbf{t}_1, \mathbf{t}_2, \mathbf{t}_3$). The intercepts themselves are not used (to avoid the occurrence of ∞ if the plane is parallel to a primitive translation vector). The reciprocal of the intercepts, multiplied by the smallest factor to convert them into integer numbers, are called *Miller indices*.

As a working illustration of the Miller indicization, consider for example in Fig. 16 the lattice plane that intersects the axes of the primitive translation vectors at the points

$$\mathbf{R}_1 = 2\mathbf{t}_1 \qquad \mathbf{R}_2 = 3\mathbf{t}_2 \qquad \mathbf{R}_3 = \mathbf{t}_3 \ . \tag{22}$$

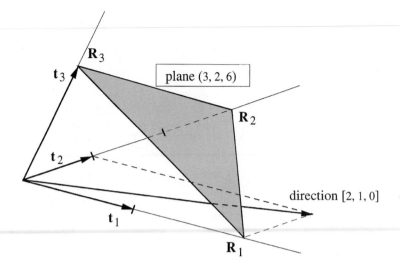

Fig. 16 Miller indices of planes and directions in a Bravais lattice.

The intercepts are 2, 3, 1; the corresponding reciprocals are $1/2$, $1/3$, 1 and the Miller indices $(3, 2, 6)$. We can immediately observe that the reciprocal lattice vector

$$\mathbf{g}_m = 3\mathbf{g}_1 + 2\mathbf{g}_2 + 6\mathbf{g}_3$$

is perpendicular to the plane passing through $\mathbf{R}_1, \mathbf{R}_2, \mathbf{R}_3$. In fact we have (by construction of the Miller indices)

$$\mathbf{g}_m \cdot \mathbf{R}_1 \equiv \mathbf{g}_m \cdot \mathbf{R}_2 \equiv \mathbf{g}_m \cdot \mathbf{R}_3 \ .$$

Thus $\mathbf{g}_m$ is perpendicular to $\mathbf{R}_1 - \mathbf{R}_2$ and to $\mathbf{R}_2 - \mathbf{R}_3$ and thus to their plane. It is evident, in general, the full equivalence of the Miller indicization scheme and the reciprocal lattice indicization of families of planes.

Before concluding, we give also the generally adopted convention to label a crystallographic direction joining two lattice points of a Bravais lattice. The relative vector, connecting the two lattice points, can always be expressed in the form

$$\mathbf{t}_n = n_1\mathbf{t}_1 + n_2\mathbf{t}_2 + n_3\mathbf{t}_3 \ ,$$

where n_1, n_2, n_3 are integer numbers. Dividing the numbers n_1, n_2, n_3 by their highest common factor, we can reduce them to a non-divisible tern l_1, l_2 and l_3. The direction is specified by enclosing this tern in square brackets $[l_1, l_2, l_3]$. All the directions equivalent by symmetry to $[l_1, l_2, l_3]$ are indicated by angular brackets $\langle l_1, l_2, l_3 \rangle$.

In the particular case of the primitive cubic Bravais lattice, where the axes have the same length and are orthogonal, we see that the $[m_1, m_2, m_3]$ direction is always orthogonal to the (m_1, m_2, m_3) family of planes. This is not generally true in any other crystal system. For fcc and bcc lattices, it is customary to indicate directions and planes with respect to the underlying non-primitive simple cubic lattice.

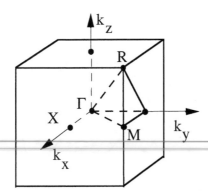

Fig. 17 Brillouin zone for the simple cubic lattice. Some high symmetry points are indicated: $\Gamma = 0$; $X = (2\pi/a)(1/2, 0, 0)$; $M = (2\pi/a)(1/2, 1/2, 0)$; $R = (2\pi/a)(1/2, 1/2, 1/2)$.

5 Brillouin zones

The *first Brillouin zone* (or simply the *Brillouin zone*) of the reciprocal lattice has the same definition as the Wigner–Seitz cell in the direct lattice: it has *the property that any point of the cell is closer to the chosen lattice point* (say $\mathbf{g} \equiv 0$) *than to any other.* The first Brillouin zone can be obtained by bisecting with perpendicular planes nearest neighbours reciprocal lattice vectors, second nearest neighbours (and other orders of neighbours if necessary) and considering the smallest volume enclosed. Similarly, the second Brillouin zone is obtained continuing the bisecting operations and delimiting the second volume enclosed (with exclusion of the first zone), etc. The shape of the Brillouin zone is connected to the geometry of the direct Bravais lattice, irrespectively of the content of the basis. We illustrate the construction of the first Brillouin zone with some examples.

Brillouin zone for the simple cubic lattice

The fundamental vectors in direct space of a simple cubic lattice are

$$\mathbf{t}_1 = a(1, 0, 0) \quad \mathbf{t}_2 = a(0, 1, 0) \quad \mathbf{t}_3 = a(0, 0, 1) \ . \tag{23a}$$

From Eqs. (18) we have

$$\mathbf{g}_1 = (2\pi/a)(1, 0, 0) \quad \mathbf{g}_2 = (2\pi/a)(0, 1, 0) \quad \mathbf{g}_3 = (2\pi/a)(0, 0, 1) \ . \tag{23b}$$

Thus the reciprocal lattice is still a simple cube, with edge $2\pi/a$.

The first Brillouin zone is indicated in Fig. 17. Points of high symmetry in the cubic Brillouin zone are indicated by conventional letters: Γ denotes the origin of the Brillouin zone; X is the center of a square face at the boundaries; M is the center of a cube edge, and R is the vertex of the cube.

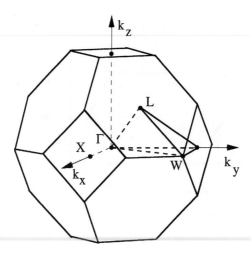

Fig. 18 Brillouin zone for the face-centered cubic lattice (truncated octahedron). Some high symmetry points are: $\Gamma = 0$; $X=(2\pi/a)(1,0,0)$; $L=(2\pi/a)(1/2,1/2,1/2)$; $W=(2\pi/a)(1/2,1,0)$.

Brillouin zone for the face-centered cubic lattice

The fundamental vectors for an fcc Bravais lattice are

$$\mathbf{t}_1 = \frac{a}{2}(0,1,1) \quad \mathbf{t}_2 = \frac{a}{2}(1,0,1) \quad \mathbf{t}_3 = \frac{a}{2}(1,1,0) . \tag{24a}$$

By applying Eqs. (18), we have for the fundamental vectors of the reciprocal lattice

$$\mathbf{g}_1 = \frac{2\pi}{a}(-1,1,1) \quad \mathbf{g}_2 = \frac{2\pi}{a}(1,-1,1) \quad \mathbf{g}_3 = \frac{2\pi}{a}(1,1,-1) . \tag{24b}$$

Thus the reciprocal lattice of an fcc lattice is a bcc lattice. The Brillouin zone is the truncated octahedron shown in Fig. 18.

Brillouin zone for body-centered cubic lattice

The fundamental vectors for a bcc lattice in direct and reciprocal space are

$$\mathbf{t}_1 = \frac{a}{2}(-1,1,1) \quad \mathbf{t}_2 = \frac{a}{2}(1,-1,1) \quad \mathbf{t}_3 = \frac{a}{2}(1,1,-1) \tag{25a}$$

and

$$\mathbf{g}_1 = \frac{2\pi}{a}(0,1,1) \quad \mathbf{g}_2 = \frac{2\pi}{a}(1,0,1) \quad \mathbf{g}_3 = \frac{2\pi}{a}(1,1,0) . \tag{25b}$$

The Brillouin zone is the regular rhombic dodecahedron shown in Fig. 19.

Brillouin zone for the hexagonal lattice

The fundamental translation vectors of the Bravais hexagonal lattice are

$$\mathbf{t}_1 = a(\frac{1}{2}, \frac{\sqrt{3}}{2}, 0) \quad \mathbf{t}_2 = a(-\frac{1}{2}, \frac{\sqrt{3}}{2}, 0) \quad \mathbf{t}_3 = c(0,0,1) , \tag{26a}$$

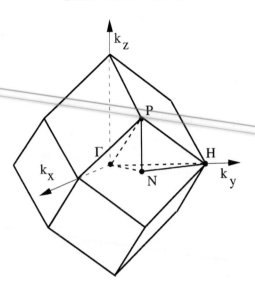

Fig. 19 Brillouin zone for the body-centered cubic lattice (rhombic dodecahedron). Some high symmetry points are also indicated: $\Gamma = 0$; $N = (2\pi/a)(1/2, 1/2, 0)$; $P = (2\pi/a)(1/2, 1/2, 1/2)$; $H = (2\pi/a)(0, 1, 0)$.

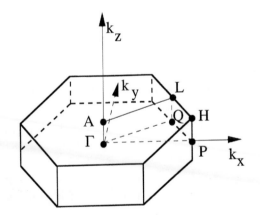

Fig. 20 Brillouin zone for the hexagonal Bravais lattice. Some high symmetry points are also indicated: $\Gamma = 0$; $P = (2\pi/a)(2/3, 0, 0)$; $Q = (\pi/a)(1, 1/\sqrt{3}, 0)$; $A = (\pi/c)(0, 0, 1)$.

and the fundamental vectors of the reciprocal lattice are

$$\mathbf{g}_1 = \frac{2\pi}{a}(1, \frac{1}{\sqrt{3}}, 0) \quad \mathbf{g}_2 = \frac{2\pi}{a}(-1, \frac{1}{\sqrt{3}}, 0) \quad \mathbf{g}_3 = \frac{2\pi}{c}(0, 0, 1) . \tag{26b}$$

Thus the reciprocal of the hexagonal lattice is still an hexagonal lattice. The Brillouin zone is shown in Fig. 20.

6 Translational symmetry and quantum mechanical aspects

6.1 Translational symmetry and Bloch wavefunctions

In Section I-1 we have analysed the important consequences of the periodicity on the quantum mechanical propagation of electrons in one-dimensional crystals. In this section, we consider the effect of periodicity in actual three-dimensional crystals; for this purpose we have already prepared the needed tools, in particular the notion of Brillouin zone and reciprocal lattice. Similarly to the one-dimensional case, also for three-dimensional crystals we can demonstrate the Bloch theorem: *the wavefunctions of the crystal Hamiltonian can be written as the product of a plane wave of wavevector* **k** *within the first Brillouin zone, times an appropriate periodic function.* The Bloch theorem is a general consequence of translation symmetry of the crystals; additional symmetry properties of the crystals may add further information on the wavefunctions, but cannot remove their Bloch form.

Consider the Schrödinger equation

$$\left[\frac{\mathbf{p}^2}{2m} + V(\mathbf{r}) \right] \psi(\mathbf{r}) = E\,\psi(\mathbf{r}) \ , \tag{27}$$

where $V(\mathbf{r}) = V(\mathbf{r} + \mathbf{t}_n)$ is a periodic potential. Since $V(\mathbf{r})$ has the lattice periodicity, we can expand it in plane waves of vectors $\mathbf{g}_m$, where $\mathbf{g}_m$ are vectors of the reciprocal lattice

$$V(\mathbf{r}) = \sum_{\mathbf{g}_m} V(\mathbf{g}_m)\, e^{i\mathbf{g}_m \cdot \mathbf{r}} \ . \tag{28a}$$

From expansion (28a), it follows that the matrix elements of $V(\mathbf{r})$ between (normalized) plane waves of vectors **k** and **k**′ are different from zero only if $\mathbf{k}' = \mathbf{k} + \mathbf{g}_m$. We have in fact

$$\langle \frac{1}{\sqrt{V}} e^{i\mathbf{k}' \cdot \mathbf{r}} | V(\mathbf{r}) | \frac{1}{\sqrt{V}} e^{i\mathbf{k} \cdot \mathbf{r}} \rangle = \sum_{\mathbf{g}_m} V(\mathbf{g}_m) \int \frac{1}{V} e^{i(-\mathbf{k}' + \mathbf{k} + \mathbf{g}_m) \cdot \mathbf{r}}\, d\mathbf{r}$$

$$= \begin{cases} 0 & \text{if } \mathbf{k}' \neq \mathbf{k} + \mathbf{g}_m \\ V(\mathbf{g}_m) & \text{if } \mathbf{k}' = \mathbf{k} + \mathbf{g}_m \end{cases} \tag{28b}$$

($d\mathbf{r}$ denotes the volume element in real space).

Let us denote by **k** and **k**′ two different wavevectors chosen within the first Brillouin zone (thus $\mathbf{k}' \neq \mathbf{k} + \mathbf{g}_m$ for whatever $\mathbf{g}_m$). According to Eq. (28b), we see that the periodic potential $V(\mathbf{r})$ can never mix plane waves of vectors $\mathbf{k} + \mathbf{g}_n$ with plane waves of vectors $\mathbf{k}' + \mathbf{g}'_n$. Thus the eigenfunctions of the crystal Hamiltonian (27) can be labelled as $\psi(\mathbf{k}, \mathbf{r})$, with the wavevector **k** defined within the first Brillouin zone.

A crystal wavefunction $\psi(\mathbf{k}, \mathbf{r})$, of vector **k**, must be an appropriate linear combination of plane waves of vectors $\mathbf{k} + \mathbf{g}_n$ (this being the only kind of mixing allowed by a periodic potential, and hence by the crystal Hamiltonian); we have in general

$$\psi(\mathbf{k}, \mathbf{r}) = \sum_{\mathbf{g}_n} a_n(\mathbf{k})\, e^{i(\mathbf{k} + \mathbf{g}_n) \cdot \mathbf{r}} \ . \tag{29}$$

The functions $\exp(i\mathbf{g}_n \cdot \mathbf{r})$, as well as any linear combination, have the lattice periodicity; we can thus write (29) in the form

$$\psi(\mathbf{k}, \mathbf{r}) = e^{i\mathbf{k}\cdot\mathbf{r}} u(\mathbf{k}, \mathbf{r}) \, , \tag{30a}$$

where $u(\mathbf{k}, \mathbf{r})$ denotes a *periodic function*. Thus the eigenfunctions of the Schrödinger equation (27) must take the form of the product of a plane wave, modulated on atomic scale by a periodic function (*Bloch wavefunctions*). The Bloch wavefunctions extend on the whole crystals and have thus a "delocalized" (or "itinerant") nature.

There is another equivalent way of expressing the Bloch theorem, summarized by Eq. (30a). We notice that

$$\psi(\mathbf{k}, \mathbf{r} + \mathbf{t}_n) = e^{i\mathbf{k}\cdot(\mathbf{r}+\mathbf{t}_n)} u(\mathbf{k}, \mathbf{r} + \mathbf{t}_n) = e^{i\mathbf{k}\cdot\mathbf{t}_n} \psi(\mathbf{k}, \mathbf{r}) \; ; \tag{30b}$$

thus, the values of the wavefunction of vector $\mathbf{k}$ in any two points connected by a lattice translation $\mathbf{t}_n$ are related by the phase factor $\exp(i\mathbf{k}\cdot\mathbf{t}_n)$. Notice that Eq. (30b) implies (30a); in fact if Eq. (30b) is multiplied by $\exp[-i\mathbf{k} \cdot (\mathbf{r} + \mathbf{t}_n)]$, we see that the function $\exp(-i\mathbf{k} \cdot \mathbf{r})\psi(\mathbf{k}, \mathbf{r})$ is periodic, which is the same information given by Eq. (30a).

From the present discussion on translational symmetry, we see that the eigenfunctions of a crystal Hamiltonian with periodic potential, are characterized by a wavevector $\mathbf{k}$ defined within the first Brillouin zone. For any given $\mathbf{k}$ vector, we have in general a countably infinite set of solutions; different solutions can be labelled with an index, say n, following for instance the increasing value of energy. Suppose to vary continuously $\mathbf{k}$ in the first Brillouin zone and to follow the crystal eigenvalues $E_n(\mathbf{k})$ and the crystal wavefunctions $\psi_n(\mathbf{k}, \mathbf{r})$; we arrive thus at the concept of the *energy band structure* of electrons in crystals.

Similarly to what already discussed in Section I-6, the quantity $\hbar\mathbf{k}$ can be interpreted as *quasi-momentum* or *crystal momentum* of the electron in the crystal. For brevity, $\hbar\mathbf{k}$ is also addressed as the *momentum of the electron*, keeping always in mind that the *momentum of an electron in the crystal is to be understood within reciprocal lattice vectors times $\hbar$*.

The methods and technique for the calculation of the band structure of specific materials are presented in Chapter V; here we wish to discuss some general properties of the band structure of materials, that are essentially connected once again to the periodicity (rather than the actual details of the crystal potential).

6.2   The parametric $k \cdot p$ Hamiltonian

The general form of the Bloch wavefunctions, as the product of a plane wave and a periodic part, allows further general elaborations. We insert the Bloch form $\psi_n(\mathbf{k}, \mathbf{r}) = \exp(i\mathbf{k} \cdot \mathbf{r}) u_n(\mathbf{k}, \mathbf{r})$ into the Schrödinger equation (27) and obtain

$$\left[\frac{\mathbf{p}^2}{2m} + V(\mathbf{r}) \right] e^{i\mathbf{k}\cdot\mathbf{r}} u_n(\mathbf{k}, \mathbf{r}) = E_n(\mathbf{k}) \, e^{i\mathbf{k}\cdot\mathbf{r}} u_n(\mathbf{k}, \mathbf{r}) \; ;$$

thus, the periodic parts $u_n(\mathbf{k},\mathbf{r})$ of the electronic wavefunctions satisfy the modified Schrödinger equation

$$\left[\frac{1}{2m}(\mathbf{p}+\hbar\mathbf{k})^2 + V(\mathbf{r})\right] u_n(\mathbf{k},\mathbf{r}) = E_n(\mathbf{k})\, u_n(\mathbf{k},\mathbf{r}) \ . \tag{31}$$

The eigenvalue equation (31) can also be written in the form

$$H(\mathbf{k})\, u_n(\mathbf{k},\mathbf{r}) = E_n(\mathbf{k})\, u_n(\mathbf{k},\mathbf{r}) \ ,$$

where

$$H(\mathbf{k}) = \frac{1}{2m}(\mathbf{p}+\hbar\mathbf{k})^2 + V(\mathbf{r}) = \frac{\mathbf{p}^2}{2m} + V(\mathbf{r}) + \frac{\hbar}{m}\mathbf{k}\cdot\mathbf{p} + \frac{\hbar^2 k^2}{2m} \ .$$

The operator $H(\mathbf{k})$, besides the crystal Hamiltonian and the scalar quantity $\hbar^2 k^2/2m$, contains the operator $(\hbar/m)\,\mathbf{k}\cdot\mathbf{p}$, which lends the name of $\mathbf{k}\cdot\mathbf{p}$ to the whole procedure.

The operator $H(\mathbf{k})$ contains $\mathbf{k}$ as a parameter, and is one of the simplest examples of parameter dependent operators (see Section VIII-4 for other examples and applications). Following a Feynman procedure, we derive both members of Eq. (31) with respect to the parameter $\mathbf{k}$ and obtain

$$\frac{\hbar}{m}(\mathbf{p}+\hbar\mathbf{k})\, u_n(\mathbf{k},\mathbf{r}) + H(\mathbf{k})\frac{\partial u_n(\mathbf{k},\mathbf{r})}{\partial \mathbf{k}} = \frac{\partial E_n(\mathbf{k})}{\partial \mathbf{k}} u_n(\mathbf{k},\mathbf{r}) + E_n(\mathbf{k})\frac{\partial u_n(\mathbf{k},\mathbf{r})}{\partial \mathbf{k}} \ . \tag{32}$$

The projection of Eq. (32) on $\langle u_m(\mathbf{k},\mathbf{r})|$ with $m \equiv n$ gives

$$\langle u_n(\mathbf{k},\mathbf{r})|\frac{\hbar}{m}(\mathbf{p}+\hbar\mathbf{k})|u_n(\mathbf{k},\mathbf{r})\rangle = \frac{\partial E_n(\mathbf{k})}{\partial \mathbf{k}} \ ;$$

expressing $u_n(\mathbf{k},\mathbf{r}) = \exp(-i\mathbf{k}\cdot\mathbf{r})\,\psi_n(\mathbf{k},\mathbf{r})$ we obtain

$$\boxed{\langle\psi_n(\mathbf{k},\mathbf{r})|\frac{\mathbf{p}}{m}|\psi_n(\mathbf{k},\mathbf{r})\rangle = \frac{1}{\hbar}\frac{\partial E_n(\mathbf{k})}{\partial \mathbf{k}}} \ , \tag{33}$$

which is the trivial generalization of Eq. (I-61) to the three-dimensional crystal. The projection of Eq. (32) on $\langle u_m(\mathbf{k},\mathbf{r})|$ with $m \neq n$ gives

$$\boxed{\langle\psi_m(\mathbf{k};\mathbf{r})|\frac{\hbar}{m}\mathbf{p}|\psi_n(\mathbf{k};\mathbf{r})\rangle = -[E_m(\mathbf{k}) - E_n(\mathbf{k})]\langle u_m(\mathbf{k};\mathbf{r})|\frac{\partial}{\partial \mathbf{k}}u_n(\mathbf{k};\mathbf{r})\rangle} \ . \tag{34}$$

The expressions (33) and (34) of the diagonal and off-diagonal matrix elements of the operator $\mathbf{p}$ among band wavefunctions are quite general and extremely useful. In particular, Eq. (33) allows to link the band structure $E_n(\mathbf{k})$ with the electron velocity $\mathbf{v}_n(\mathbf{k}) = (1/\hbar)\,\partial E_n(\mathbf{k})/\partial\mathbf{k}$ in semi-classical transport theories.

In many situations it is convenient, instead of the momentum operator $\mathbf{p}$, to evaluate the matrix elements of the coordinate operator $\mathbf{r}$. From standard commutation rules, we have

$$[H,\mathbf{r}] = \left[\frac{\mathbf{p}^2}{2m} + V(\mathbf{r}), \mathbf{r}\right] = -i\frac{\hbar}{m}\mathbf{p} \ .$$

Taking the matrix elements of both members of the above equation between two crystal eigenfunctions we obtain

$$\langle \psi_m(\mathbf{k}, \mathbf{r})| - i\frac{\hbar}{m}\mathbf{p}|\psi_n(\mathbf{k}, \mathbf{r})\rangle = \langle \psi_m(\mathbf{k}, \mathbf{r})|[H, \mathbf{r}]|\psi_n(\mathbf{k}, \mathbf{r})\rangle$$

$$= [E_m(\mathbf{k}) - E_n(\mathbf{k})]\langle \psi_m(\mathbf{k}, \mathbf{r})| \mathbf{r} |\psi_n(\mathbf{k}, \mathbf{r})\rangle . \tag{35}$$

Comparison of Eq. (35) and Eq. (34) gives the relation

$$\boxed{\langle u_m(\mathbf{k}, \mathbf{r})| \mathbf{r} |u_n(\mathbf{k}, \mathbf{r})\rangle = i\langle u_m(\mathbf{k}, \mathbf{r})| \frac{\partial}{\partial \mathbf{k}} u_n(\mathbf{k}, \mathbf{r})\rangle} \tag{36}$$

(which holds provided $m \neq n$). This is a useful property of the off-diagonal matrix elements of the operator $\mathbf{r}$ and the parameter operator $i\partial/\partial\mathbf{k}$, on the set of periodic functions $\{u_j(\mathbf{k}, \mathbf{r})\}$ at a given selected $\mathbf{k}$ point.

In periodic materials, thanks to the Bloch theorem, the knowledge of a wavefunction $\psi_n(\mathbf{k}, \mathbf{r})$ within the unit cell entails its knowledge in the whole real space. We now show that the knowledge of the set of wavefunctions $\{\psi_n(\mathbf{k}_0, \mathbf{r})\}$ and band energies $E_n(\mathbf{k}_0)$ just at a *single* $\mathbf{k}_0$ *vector*, implies the knowledge (at least in principle) of the whole band structure (wavefunctions and energies) on the entire Brillouin zone.

From the Schrödinger equation (31), and the general properties of an eigenvalue equation, we have that the periodic wavefunctions $\{u_n(\mathbf{k}_0, \mathbf{r})\}$, at a specific vector $\mathbf{k}_0$, constitute a complete set for the expansion of any other periodic wavefunction $u(\mathbf{k}, \mathbf{r})$ of wavevector $\mathbf{k}$. It follows that a crystal wavefunction $\psi(\mathbf{k}, \mathbf{r})$ of vector $\mathbf{k}$ can always be expanded in the form

$$\psi(\mathbf{k}, \mathbf{r}) = \sum_n c_n(\mathbf{k}) \left[e^{i(\mathbf{k}-\mathbf{k}_0)\cdot\mathbf{r}}\psi_n(\mathbf{k}_0, \mathbf{r})\right] . \tag{37a}$$

In the above expansion, the wavevector $\mathbf{k}_0$ specifies the chosen reference point in the Brillouin zone where eigenfunctions and eigenvalues are supposed to be known, while $\mathbf{k}$ specifies the point where eigenfunctions and eigenvalues are to be found.

The matrix elements of the crystal Hamiltonian among the basis functions in the square brackets of Eq. (37a) are

$$M_{nn'}(\mathbf{k}_0, \mathbf{k}) = \langle e^{i(\mathbf{k}-\mathbf{k}_0)\cdot\mathbf{r}}\psi_n(\mathbf{k}_0, \mathbf{r})|\frac{\mathbf{p}^2}{2m} + V(\mathbf{r})|e^{i(\mathbf{k}-\mathbf{k}_0)\cdot\mathbf{r}}\psi_{n'}(\mathbf{k}_0, \mathbf{r})\rangle$$

$$= \langle \psi_n(\mathbf{k}_0, \mathbf{r})|\frac{\mathbf{p}^2}{2m} + V(\mathbf{r}) + \frac{\hbar}{m}(\mathbf{k}-\mathbf{k}_0)\cdot\mathbf{p} + \frac{\hbar^2(\mathbf{k}-\mathbf{k}_0)^2}{2m}|\psi_{n'}(\mathbf{k}_0, \mathbf{r})\rangle$$

$$= \left[E_n(\mathbf{k}_0) + \frac{\hbar^2(\mathbf{k}-\mathbf{k}_0)^2}{2m}\right]\delta_{nn'} + \frac{\hbar}{m}\langle \psi_n(\mathbf{k}_0, \mathbf{r})|(\mathbf{k}-\mathbf{k}_0)\cdot\mathbf{p}|\psi_{n'}(\mathbf{k}_0, \mathbf{r})\rangle. \tag{37b}$$

The matrix elements $M_{nn'}$ essentially require the knowledge of the band energies $E_n(\mathbf{k}_0)$ and the matrix elements of the momentum operator $\mathbf{p}$ between the band wavefunctions at the selected $\mathbf{k}_0$ vector.

From expansion (37a), matrix elements (37b), and standard variational methods in the expansion coefficients, we have that band energies and band wavefunctions at any wavevector $\mathbf{k}$ are obtained by the eigenvalues and eigenvectors of the secular equation

$$\boxed{\|M_{nn'}(\mathbf{k}_0, \mathbf{k}) - E\,\delta_{nn'}\| = 0}\;. \tag{38}$$

The secular equation (38) is exact and can be used in principle (and at times it is convenient also in practice) to extend the band structure, known at a particular $\mathbf{k}_0$ point, to the *whole* Brillouin zone (most often, for convenience, the point $\mathbf{k}_0$ is the center of the Brillouin zone or some other high symmetry point).

A most significant use of Eq. (38) is to confine $\mathbf{k}$ to a small region of the Brillouin zone, around the point $\mathbf{k}_0$. In this case $\mathbf{k} - \mathbf{k}_0 \approx 0$, the off-diagonal matrix elements in Eq. (37b) are small, and thus can be safely treated with standard perturbation theory. In the case the band of interest (say the nth one) is non-degenerate at $\mathbf{k} = \mathbf{k}_0$, the perturbation theory for non-degenerate states, up to second order in $\mathbf{k} - \mathbf{k}_0$, gives for the band energy $E_n(\mathbf{k})$ the expression

$$E_n(\mathbf{k}) = E_n(\mathbf{k}_0) + \frac{\hbar^2(\mathbf{k} - \mathbf{k}_0)^2}{2m} + \frac{\hbar}{m}\langle\psi_n(\mathbf{k}_0, \mathbf{r})|(\mathbf{k} - \mathbf{k}_0)\cdot\mathbf{p}|\psi_n(\mathbf{k}_0, \mathbf{r})\rangle$$

$$+ \frac{\hbar^2}{m^2}\sum_{n'(\neq n)}\frac{|\langle\psi_{n'}(\mathbf{k}_0, \mathbf{r})|(\mathbf{k} - \mathbf{k}_0)\cdot\mathbf{p}|\psi_n(\mathbf{k}_0, \mathbf{r})\rangle|^2}{E_n(\mathbf{k}_0) - E_{n'}(\mathbf{k}_0)}\;. \tag{39}$$

Suppose that $\mathbf{k}_0$ is an extremal of $E_n(\mathbf{k})$, and thus $\langle\psi_n(\mathbf{k}_0, \mathbf{r})|\mathbf{p}|\psi_n(\mathbf{k}_0, \mathbf{r})\rangle$ vanishes because of Eq. (33). The right-hand side of Eq. (39) is now a quadratic function of $\mathbf{k} - \mathbf{k}_0$, and the expression of $E_n(\mathbf{k})$ becomes

$$E_n(\mathbf{k}) = E_n(\mathbf{k}_0) + \frac{\hbar^2(\mathbf{k} - \mathbf{k}_0)^2}{2m}$$

$$+ \frac{\hbar^2}{m^2}\sum_{n'(\neq n)}\frac{\langle\psi_{n\,\mathbf{k}_0}|(\mathbf{k} - \mathbf{k}_0)\cdot\mathbf{p}|\psi_{n'\,\mathbf{k}_0}\rangle\langle\psi_{n'\,\mathbf{k}_0}|(\mathbf{k} - \mathbf{k}_0)\cdot\mathbf{p}|\psi_{n\,\mathbf{k}_0}\rangle}{E_n(\mathbf{k}_0) - E_{n'}(\mathbf{k}_0)}\;.$$

This can be written in the form

$$E_n(\mathbf{k}) = E_n(\mathbf{k}_0) + \sum_{\alpha\beta}\frac{\hbar^2}{2m}\left(\frac{m}{m^*}\right)_{\alpha\beta}(\mathbf{k} - \mathbf{k}_0)_\alpha(\mathbf{k} - \mathbf{k}_0)_\beta\;,$$

where the effective mass tensor is given by

$$\left(\frac{m}{m^*}\right)_{\alpha\beta} = \delta_{\alpha\beta} + \frac{2}{m}\sum_{n'(\neq n)}\frac{\langle\psi_{n\,\mathbf{k}_0}|\mathbf{p}_\alpha|\psi_{n'\,\mathbf{k}_0}\rangle\langle\psi_{n'\,\mathbf{k}_0}|\mathbf{p}_\beta|\psi_{n\,\mathbf{k}_0}\rangle}{E_n(\mathbf{k}_0) - E_{n'}(\mathbf{k}_0)} \tag{40}$$

with $\alpha, \beta = x, y, z$.

Expression (40) has been often used to evaluate effective masses in crystals. For an isotropic two-band model semiconductor, with energy gap E_G and matrix element $\mathbf{p}_{cv}$ between the valence and the conduction wavefunctions at the bandedges, expression (40) gives for the effective masses of the two bands

$$\frac{m}{m^*} \approx 1 \pm \frac{2}{m}\frac{|\mathbf{p}_{cv}|^2}{E_G}\;;$$

this relation shows that small effective masses for electrons and holes are to be expected in small energy gap semiconductors, a qualitative trend confirmed by experience. The $k \cdot p$ approach has been profitably generalized and extended to the study of degenerate bands, also in the presence of spin-orbit coupling.

6.3 Cyclic boundary conditions

In the previous sections we have considered ideal infinite crystals. Suppose now to have a macroscopic (but finite) crystal, of volume V, in the form of a parallelepiped with edges $N_1 \mathbf{t}_1, N_2 \mathbf{t}_2$ and $N_3 \mathbf{t}_3$, where N_1, N_2, N_3 are large (but finite) integer numbers. The Born–von Karman cyclic boundary conditions on the electronic wavefunctions require

$$\psi(\mathbf{r}) = \psi(\mathbf{r} + N_1 \mathbf{t}_1) = \psi(\mathbf{r} + N_2 \mathbf{t}_2) = \psi(\mathbf{r} + N_3 \mathbf{t}_3) \ .$$

Since $\psi(\mathbf{r})$ must be a Bloch function of vector $\mathbf{k}$, we have

$$e^{i\mathbf{k} \cdot N_1 \mathbf{t}_1} = e^{i\mathbf{k} \cdot N_2 \mathbf{t}_2} = e^{i\mathbf{k} \cdot N_3 \mathbf{t}_3} = 1 \ . \tag{41}$$

The possible $\mathbf{k}$ vectors, compatible with cyclic boundary conditions (41), are thus

$$\mathbf{k} = \frac{m_1}{N_1} \mathbf{g}_1 + \frac{m_2}{N_2} \mathbf{g}_2 + \frac{m_3}{N_3} \mathbf{g}_3 \tag{42}$$

with m_i integer numbers.

The allowed $\mathbf{k}$ vectors (42) within the primitive unit cell of edges $\mathbf{g}_1, \mathbf{g}_2, \mathbf{g}_3$ are obtained choosing $0 \le m_i < N_i$ ($i = 1, 2, 3$); their number is $N = N_1 \cdot N_2 \cdot N_3$. The first Brillouin zone has the same volume as the primitive unit cell in the reciprocal lattice. *Thus the number of allowed $\mathbf{k}$ vectors in the first Brillouin zone equals the number of primitive unit cells of the crystal.*

The density of allowed $\mathbf{k}$ vectors (42) in the reciprocal space is uniform and is given by

$$W(\mathbf{k}) = \frac{1}{\frac{1}{N_1}\mathbf{g}_1 \cdot \left(\frac{1}{N_2}\mathbf{g}_2 \times \frac{1}{N_3}\mathbf{g}_3\right)} = \frac{N_1 N_2 N_3}{\Omega_k} = \frac{V}{(2\pi)^3} \ , \tag{43}$$

where V is the volume of the crystal. In some problems, we have to perform sums in reciprocal space of a given function of $\mathbf{k}$. According to Eq. (43), the discrete sum can be replaced by an integral as follows

$$\boxed{\sum_{\mathbf{k}} f(\mathbf{k}) \Rightarrow \frac{V}{(2\pi)^3} \int f(\mathbf{k}) \, d\mathbf{k}} \ , \tag{44}$$

where $d\mathbf{k}$ is the volume element in reciprocal space.

Notice also that the $\mathbf{k}$ vectors (42) within the first Brillouin zone satisfy the relation

$$\sum_{\mathbf{k}}^{\text{B.Z.}} e^{i\mathbf{k} \cdot \mathbf{t}_n} = \begin{cases} N_1 N_2 N_3 & \text{if} \quad \mathbf{t}_n = 0 \\ 0 & \text{if} \quad \mathbf{t}_n \neq 0 \end{cases} \ , \tag{45a}$$

where $\mathbf{t}_n$ is any translation of the crystal such that $0 \le n_i < N_i$ ($i = 1, 2, 3$). The proof is simply based on the elementary algebraic property that the sum of the N_i-th roots of the unity equals zero. Similarly we have

$$\sum_{\mathbf{t}_n}^{N_1 N_2 N_3} e^{i\mathbf{k}\cdot\mathbf{t}_n} = \begin{cases} N_1 N_2 N_3 & \text{if} \quad \mathbf{k} = \mathbf{g}_m \\ 0 & \text{if} \quad \mathbf{k} \neq \mathbf{g}_m \end{cases} , \tag{45b}$$

where $\mathbf{k}$ is any vector (42) of the reciprocal space.

6.4   Special k points for averaging over the Brillouin zone

In several studies of crystals (for instance in the calculation of charge density, total energies and other properties) we have to perform the average of appropriate $\mathbf{k}$-dependent functions throughout the Brillouin zone. The average value of a function $F(\mathbf{k})$ is given by

$$F_0 = \frac{1}{N} \sum_{\mathbf{k}}^{\text{B.Z.}} F(\mathbf{k}) , \tag{46}$$

where N is the (large) number of allowed $\mathbf{k}$ vectors in the first Brillouin zone. The calculation of $F(\mathbf{k})$ at a given $\mathbf{k}$ vector may require laborious numerical work; then it would be highly desirable to find sets of "special $\mathbf{k}$ points" (composed by a reasonably modest number of $\mathbf{k}$ vectors) to determine efficiently the average F_0.

To see the basic principle for this achievement we proceed as follows. Consider a function $F(\mathbf{k})$ with $\mathbf{k}$ defined in the first Brillouin zone, or, equivalently, consider a function $F(\mathbf{k})$ with the periodicity of the reciprocal lattice; its most general expansion, according to Eq. (20c) is

$$F(\mathbf{k}) = F_0 + \sum_{\mathbf{t}_n \neq 0} F_n e^{i\mathbf{k}\cdot\mathbf{t}_n} , \tag{47}$$

where $\mathbf{t}_n$ are translation vectors of the direct lattice, F_n are the appropriate Fourier coefficients, and for convenience the average value F_0 has been separated from all the other terms. If we sum both members of Eq. (47) over all the allowed $\mathbf{k}$ vectors in the first Brillouin zone, use Eq. (45a) and divide by N, we recover expression (46). Thus, in general, only the sum over a very large number of $\mathbf{k}$ vectors allows to obtain the exact average value of the function $F(\mathbf{k})$.

Consider now a function $G(\mathbf{k})$ represented in the form

$$G(\mathbf{k}) = G_0 + \sum_{\mathbf{t}_i^{(I)}} G_i^{(I)} e^{i\mathbf{k}\cdot\mathbf{t}_i^{(I)}} , \tag{48a}$$

where $\mathbf{t}_i^{(I)}$ denotes translation vectors belonging to the first shell of neighbours (all other Fourier coefficients are assumed to be zero). Let $\{\mathbf{k}_s\}$ indicate an appropriately selected set of n_s points of the Brillouin zone, chosen in such a way that

$$\sum_{\mathbf{k}\in\{\mathbf{k}_s\}} e^{i\mathbf{k}\cdot\mathbf{t}_i^{(I)}} = 0 \tag{48b}$$

for any translation vector of the first shell. Then the average value G_0 of the function $G(\mathbf{k})$ is rigorously expressed by

$$G_0 = \frac{1}{n_s} \sum_{\mathbf{k} \in \{\mathbf{k}_s\}} G(\mathbf{k}) . \qquad (48c)$$

As an example, consider a simple cubic lattice; the six translation vectors $\mathbf{t}_i^{(I)}$ of the first shell are: $(\pm a, 0, 0)$, $(0, \pm a, 0)$, $(0, 0, \pm a)$. It is seen by inspection that the set of just two vectors $\mathbf{k}_1$ and $\mathbf{k}_2$, given by $\mathbf{k}_1 = -\mathbf{k}_2 = (\pi/a)(1/2, 1/2, 1/2)$, satisfies Eq. (48b) for any of the six vectors $\mathbf{t}_i^{(I)}$. The procedure can be extended to find systematically sets of wavevectors $\{\mathbf{k}_s\}$ for which $\sum_s \exp(i\mathbf{k}_s \cdot \mathbf{t}_n) = 0$ for all the translational vectors $\mathbf{t}_n$ belonging to the first shell, second shell, and so on, up to the outermost shell of interest, and we refer to the literature for details. When the number of shells to account for becomes very large, the set of special points approaches the uniform mesh of points (42) confined within the primitive reciprocal cell.

In several situations of physical interest, the Fourier coefficients in expression (47) depend only on the absolute value $|\mathbf{t}_n|$; in these situations, expansion (47) can be recast to the form

$$F(\mathbf{k}) = F_0 + F_1 \sum_{\mathbf{t}_i^{(I)}} e^{i\mathbf{k} \cdot \mathbf{t}_i^{(I)}} + F_2 \sum_{\mathbf{t}_i^{(II)}} e^{i\mathbf{k} \cdot \mathbf{t}_i^{(II)}} + \dots \qquad (49)$$

where $\mathbf{t}_i^{(I)}$ denotes the first shell of neighbour translation vectors, $\mathbf{t}_i^{(II)}$ denotes the second shell, etc. In this case, the determination of sets of special points is particularly instructive.

As an example let us consider a simple cubic lattice. The six translation vectors in the first shell are given by $(\pm a, 0, 0)$, $(0, \pm a, 0)$, $(0, 0, \pm a)$. The twelve vectors in the second shell are given by $(\pm a, \pm a, 0)$, $(\pm a, 0, \pm a)$, $(0, \pm a, \pm a)$. The eight vectors in the third shell are given by $(\pm a, \pm a, \pm a)$. For the simple cubic lattice, the expansion (49) reads

$$F(\mathbf{k}) = F_0 + 2F_1 \left(\cos k_x a + \cos k_y a + \cos k_z a \right)$$

$$+ 4F_2 \left(\cos k_x a \cos k_y a + \cos k_x a \cos k_z a + \cos k_y a \cos k_z a \right)$$

$$+ 8F_3 \cos k_x a \cos k_y a \cos k_z a + \dots$$

In the above expression, it is seen by inspection that a single point (the Baldereschi point) reduces to zero the contributions from the first, second and third shells; this mean value point is just $\mathbf{k} = (\pi/a)(1/2, 1/2, 1/2)$. Similar procedures can be performed with the other Bravais lattices. The mean value procedure, and the generated extensions, are particularly useful for instance in performing self-consistent calculations, keeping a relative manageable number of k points into account.

Finally we mention that another appealing feature of dual spaces is the possibility to use fast Fourier transforms for passing from one space to the other; we cannot enter here in details, and we refer to the literature for further aspects [see for instance W. H.

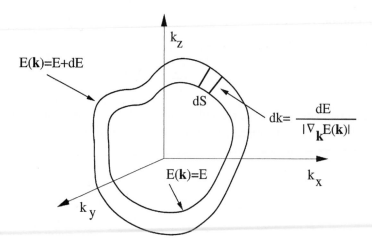

Fig. 21 Schematic representation of two isoenergetic surfaces in the **k** space.

Press, B. P. Flannery, S. A. Teukolsky and W. T. Vetterling "Numerical Recipes" (Cambridge University Press 1986); J. L. Martins and M. L. Cohen, Phys. Rev. B**37**, 6134 (1988) and references quoted therein].

7   Density-of-states and critical points

In several problems the primary interest is not in the detailed wavevector dependence of the crystal band structure but only on the density-of-states in a given energy range. Consider for simplicity a non-degenerate band $E(\mathbf{k})$; the corresponding density-of-states for this band is given by

$$D(E) = 2 \sum_{\mathbf{k}} \delta(E(\mathbf{k}) - E) = 2 \int_{B.Z.} \frac{V}{(2\pi)^3} \delta(E(\mathbf{k}) - E) \, d\mathbf{k} , \qquad (50)$$

where $d\mathbf{k}$ is the volume element of the reciprocal space and the factor 2 accounts for the spin degeneracy (we suppose this degeneracy is not removed). Equation (50) shows that contributions to the density-of-states $D(E)$ at energy E occur from the band states (if any) such that $E(\mathbf{k}) = E$; the factor $V/(2\pi)^3$ gives the uniform density of allowed **k** vectors in **k**-space, as discussed in Eq. (43).

In Fig. 21 we indicate schematically the two isoenergetic surfaces $E(\mathbf{k}) = E$ and $E(\mathbf{k}) = E + dE$. The distance dk between the two isoenergetic surfaces is obtained by observing that $dE = \nabla_{\mathbf{k}} E(\mathbf{k}) \cdot d\mathbf{k} = |\nabla_{\mathbf{k}} E(\mathbf{k})| \, dk$. Counting the volume in **k**-space enclosed between the two surfaces, the expression (50) for the density-of-states can be transformed into the integral on the constant energy surface $E(\mathbf{k}) = E$ in the form

$$D(E) = 2 \int_{E(\mathbf{k}) \equiv E} \frac{V}{(2\pi)^3} \frac{dS}{|\nabla_{\mathbf{k}} E(\mathbf{k})|} . \qquad (51)$$

Expression (51) clearly shows that singularities in the density-of-states are expected at the *critical points*, defined as those points in **k**-space for which

$$\nabla_{\mathbf{k}} E(\mathbf{k}) = 0 ; \tag{52}$$

for these points the density-of-states is expected to exhibit anomalies as a function of energy.

At the critical points, Eq. (50) or Eq. (51) can be integrated analytically. In more general situations the density-of-states can be obtained numerically by appropriate sampling of the **k** vectors in the Brillouin zone.

Near a critical point, where Eq. (52) holds, we can expand the energy as a function of the wavevector, in quadratic form. Indicating with k_x, k_y, k_z the principal axes of the quadratic form and taking the origin at the critical point itself, we have

$$E(\mathbf{k}) = E_c \pm \frac{\hbar^2}{2m_x} k_x^2 \pm \frac{\hbar^2}{2m_y} k_y^2 \pm \frac{\hbar^2}{2m_z} k_z^2 \tag{53}$$

(where we choose $m_x, m_y, m_z > 0$, while the occurrence of plus or minus sign specifies the kind of critical point). The critical point M_0 denotes zero negative signs in expression (53) and is thus a minimum of $E(\mathbf{k})$; the critical point M_3 denotes three negative signs in expression (53) and is thus a maximum; M_1 and M_2 denote one and two negative signs, respectively, and are thus two saddle points. Thus in three dimensions we have four types of critical points. Similarly, in two-dimensional crystals we have three types of critical points M_0, M_1 and M_2: a minimum, a saddle point and a maximum, respectively. In one-dimensional crystals we have two types of critical points: a minimum M_0 and a maximum M_1.

Near a three-dimensional critical point, the expression (50) for the density-of-states becomes

$$D(E) = \frac{2V}{(2\pi)^3} \int \delta(E_c \pm \frac{\hbar^2 k_x^2}{2m_x} \pm \frac{\hbar^2 k_y^2}{2m_y} \pm \frac{\hbar^2 k_z^2}{2m_z} - E) \, dk_x \, dk_y \, dk_z . \tag{54a}$$

For the two-dimensional case we have

$$D(E) = \frac{2L_x L_y}{(2\pi)^2} \int \delta(E_c \pm \frac{\hbar^2 k_x^2}{2m_x} \pm \frac{\hbar^2 k_y^2}{2m_y} - E) \, dk_x \, dk_y . \tag{54b}$$

For the one-dimensional case we have

$$D(E) = \frac{2L_x}{2\pi} \int \delta(E_c \pm \frac{\hbar^2 k_x^2}{2m_x} - E) \, dk_x . \tag{54c}$$

All types of integrals (54) can be easily calculated analytically with the help of the following property of the delta function

$$\delta[f(x)] = \sum_n \frac{\delta(x - x_n)}{|f'(x_n)|} , \tag{55}$$

where x_n are the simple zeroes of the function $f(x)$.

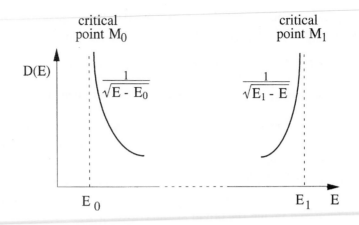

Fig. 22 Density-of-states at the critical points M_0 (minimum) and M_1 (maximum) of a one-dimensional crystal.

Critical points in one-dimensional crystals

In one-dimensional crystals, for the critical point of type M_0 at energy E_0, we consider the integral

$$D(E) = \frac{2L_x}{2\pi} \int \delta\left(E_0 + \frac{\hbar^2 k_x^2}{2m_x} - E\right) dk_x = L_x \frac{\sqrt{2m_x}}{\pi\hbar} \int \delta(E_0 + q_x^2 - E)\, dq_x \qquad (56)$$

where the change of variable $q_x = (\hbar/\sqrt{2m_x})k_x$ has been performed. Consider the function $f(q_x) = E_0 + q_x^2 - E$; the zeroes of this function occur for $q_{x_0} = \pm\sqrt{E - E_0}$ and $f'(q_{x_0}) = 2q_{x_0} = \pm 2\sqrt{E - E_0}$. The integral (56) then becomes

$$D(E) = L_x \frac{\sqrt{2m_x}}{\pi\hbar} \frac{1}{\sqrt{E - E_0}} \qquad E > E_0 . \qquad (57)$$

Similarly for a critical point of type M_1 at energy E_1 we have

$$D(E) = L_x \frac{\sqrt{2m_x}}{\pi\hbar} \frac{1}{\sqrt{E_1 - E}} \qquad E < E_1 . \qquad (58)$$

In Fig. 22 we give the behaviour of the density-of-states for the two possible critical points in one-dimensional crystals. It can be noticed that the singularities appearing in Fig. I-11a, which gives the bulk density-of-states for a linear chain evaluated with the Green's function technique, are (as expected) of the type illustrated in Fig. 22.

Critical points in two-dimensional crystals

In two-dimensional crystals, for a critical point of type M_0 at energy E_0, we have to calculate the integral

$$D(E) = \frac{2L_x L_y}{(2\pi)^2} \int \delta\left(E_0 + \frac{\hbar^2 k_x^2}{2m_x} + \frac{\hbar^2 k_y^2}{2m_y} - E\right) dk_x\, dk_y . \qquad (59)$$

It is convenient to introduce the variable $q_x = (\hbar/\sqrt{2m_x})k_x$ and $q_y = (\hbar/\sqrt{2m_y})k_y$ in Eq. (59), and denote by $S=L_x L_y$ the surface of the two-dimensional crystal; we have

$$D(E) = S\frac{\sqrt{m_x m_y}}{\pi^2 \hbar^2} \int \delta(E_0 + q_x^2 + q_y^2 - E)\, dq_x\, dq_y \ .$$

We pass to cylindrical coordinates and easily obtain

$$D(E) = S\frac{\sqrt{m_x m_y}}{\pi \hbar^2} \quad \text{for} \quad E > E_0 \ . \tag{60}$$

A similar expression holds for the critical point which is a maximum, for $E < E_2$. Thus the density-of-states is step-like at the points M_0 and M_2 of a two-dimensional crystal.

For the saddle critical point M_1 we have to evaluate the expression

$$D(E) = S\frac{\sqrt{m_x m_y}}{\pi^2 \hbar^2} \int \delta(E_1 + q_x^2 - q_y^2 - E)\, dq_x\, dq_y \ . \tag{61}$$

We suppose momentarily $E > E_1$. Consider the function $f(q_x) = E_1 + q_x^2 - q_y^2 - E$; the zeroes of this function occur for $q_{x_0} = \pm\sqrt{E - E_1 + q_y^2}$ and $f'(q_{x_0}) = 2q_{x_0}$. The integral (61) thus becomes

$$D(E) = S\frac{\sqrt{m_x m_y}}{\pi^2 \hbar^2} \int_{-q_c}^{q_c} \frac{1}{\sqrt{E - E_1 + q_y^2}}\, dq_y \ ,$$

where the integral has been confined to a cutoff where the series expansion is supposed to hold. With the help of the indefinite integral

$$\int \frac{dx}{\sqrt{a^2 + x^2}} = \ln(x + \sqrt{a^2 + x^2}) \ ,$$

we obtain

$$D(E) = S\frac{\sqrt{m_x m_y}}{\pi^2 \hbar^2} \ln \frac{q_c + \sqrt{E - E_1 + q_c^2}}{-q_c + \sqrt{E - E_1 + q_c^2}} \ . \tag{62}$$

For $E \approx E_1$, with appropriate series development in the second member of Eq. (62), we obtain

$$D(E) = S\frac{\sqrt{m_x m_y}}{\pi^2 \hbar^2} \ln \frac{4q_c^2}{|E - E_1|} \ , \tag{63}$$

and we see that a logarithmic divergence occurs at the critical point (with the absolute value $|E - E_1|$, Eq. (63) holds also for $E < E_1$). In Fig. 23 we show the behaviour of the density-of-states at the two-dimensional critical points.

Critical points in three-dimensional crystals

For three-dimensional crystals we have four types of critical points: M_0 (minimum), M_1 and M_2 (saddle points), M_3 (maximum). The analytic behaviour at the three-dimensional critical points can be obtained with procedures similar to the previous

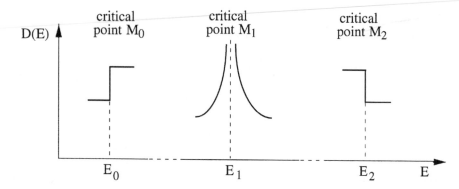

Fig. 23 Density-of-states at the critical points of type M_0 (minimum), M_1 (saddle point), and M_2 (maximum) of a two-dimensional crystal.

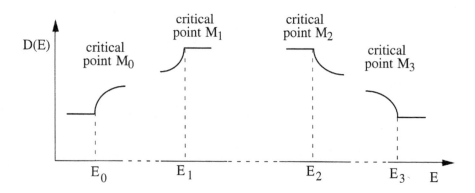

Fig. 24 Density-of-states at the critical points of type M_0 (minimum), M_1 and M_2 (saddle points), and M_3 (maximum) of a three-dimensional crystal.

reported calculations; we have:

$$\text{critical point } M_0 \quad D(E) = C_0 + V \frac{\sqrt{2m_x m_y m_z}}{\pi^2 \hbar^3} \sqrt{E - E_0} \quad \text{for } E > E_0 \qquad (64a)$$

$$\text{critical point } M_1 \quad D(E) = C_1 - V \frac{\sqrt{2m_x m_y m_z}}{\pi^2 \hbar^3} \sqrt{E_1 - E} \quad \text{for } E < E_1 \qquad (64b)$$

$$\text{critical point } M_2 \quad D(E) = C_2 - V \frac{\sqrt{2m_x m_y m_z}}{\pi^2 \hbar^3} \sqrt{E - E_2} \quad \text{for } E > E_2 \qquad (64c)$$

$$\text{critical point } M_3 \quad D(E) = C_3 + V \frac{\sqrt{2m_x m_y m_z}}{\pi^2 \hbar^3} \sqrt{E_3 - E} \quad \text{for } E < E_3 \, . \qquad (64d)$$

In the above expressions C_i indicate either a constant (including zero) or a smoothly energy dependent quantity, while the terms with the square root are present only when the argument is positive. The results (64) are schematically indicated in Fig. 24.

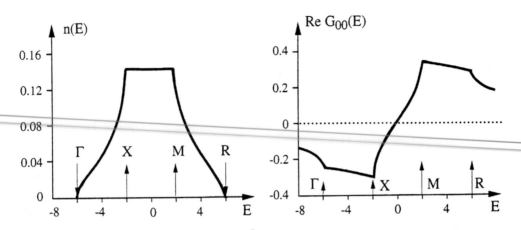

Fig. 25 Density-of-states of the cubium, to illustrate the three-dimensional critical points. The energy E is in units of $|\gamma|$, and $n(E)$ is normalized to one. Besides $n(E) = -(1/\pi)\mathrm{Im}\, G_{00}(E)$, also the real part of the Green's function diagonal matrix element $G_{00}(E)$, on a localized orbital of the cubium, is reported for convenience [from D. J. Lohrmann, L. Resca, G. Pastori Parravicini and R. D. Graft, Phys. Rev. B**40**, 8404 (1989); copyright 1989 by the American Physical Society].

As an example of three-dimensional density-of-states consider the case of the cubium, a simple cubic crystal with a single s-like orbital per site, and nearest neighbour interactions only. With a straightforward generalization of the procedures leading to Eq. (I-37), we have that the band energy $E(\mathbf{k})$ of the cubium is given by the expression

$$E(\mathbf{k}) = \alpha + 2\gamma \left(\cos k_x a + \cos k_y a + \cos k_z a\right) \tag{65}$$

(for simplicity we take $\alpha = 0$; the hopping parameter γ is supposed to be negative). The Brillouin zone of the cubium is indicated in Fig. 17, together with the symmetry points Γ, X, M and R. It is seen by inspection of Eq. (65) that Γ is a critical point of type M_0 and energy $E_0 = -6|\gamma|$. Similarly, X and M are saddle points of type M_1 and M_2, and energy $E_1 = -2|\gamma|$ and $E_2 = 2|\gamma|$, respectively. The point R is a critical point of type M_3 and energy $E_3 = 6|\gamma|$. The density-of-states corresponding to the energy band (65) can be computed numerically, for instance via the general definition (50) and appropriate sampling of the Brillouin zone, or by means of the Green's function technique and Lanczos procedure (see Section V-8.2). The computed density-of-states of cubium is reported in Fig. 25, and the presence of the critical points of the type and energy discussed above can be clearly noticed.

Further reading

S. L. Altmann "Band Theory of Solids: an Introduction from the Point of View of Symmetry" (Clarendon Press, Oxford 1994)

A. Baldereschi "Mean-value point in the Brillouin zone" Phys. Rev. B**7**, 5212 (1973). For further aspects on special points see for instance: D. J. Chadi and M. L. Cohen "Special points in the Brillouin zone" Phys. Rev. B**8**, 5747 (1973); J. Hama and M. Watanabe "General formulae for the special points and their weighting factors in k-space integration" J. Phys. C**4**, 4583 (1992)

F. Bassani and G. Pastori Parravicini "Electronic States and Optical Transitions in Solids" (Pergamon Press, Oxford 1975)

C. Giacovazzo, H. L. Monaco, D. Viterbo, F. Scordari, G. Gilli, G. Zanotti and M. Catti "Fundamentals of Crystallography " edited by C. Giacovazzo (Oxford University Press, Oxford 1992)

G. Gilat "Methods of Brillouin zone integration" in Methods in Computational Physics **15**, 317 (Academic Press, New York 1976). This article contains a survey of k-space integrations. For further aspects see for instance: G. Lehmann and M. Taut "On the numerical calculation of the density of states and related properties" Phys. Stat. Sol. (b) **54**, 469 (1972); P. Lambin and J. P. Vigneron "Computation of crystal Green's function in the complex energy plane with the use of the analytical tetrahedron method" Phys. Rev. B**29**, 3430 (1984); C. J. Pickard and M. C. Payne "Extrapolative approaches to Brillouin–zone integration" Phys. Rev. B**59**, 4685 (1999).

H. Jones "The Theory of Brillouin Zones and Electronic States in Crystals" (North-Holland, Amsterdam 1975)

G. F. Koster, J. O. Dimmock, R. G. Wheeler and H. Statz "Properties of the Thirty-Two Points Groups" (MIT Press, Cambridge, Massachusetts 1963). This book provides the conventionally used notations for symmetry points, symmetry lines, and irreducible representations.

D. Schwarzenbach "Crystallography" (Wiley, Chichester 1996)

S. Sternberg "Group Theory and Physics" (Cambridge University Press 1994)

R. W. G. Wyckoff "Crystal Structures" vols. 1-5 (Wiley, New York 1963-1968). These volumes contain a most comprehensive description of crystal structures. See also G. Burns and A. M. Glazer "Space Groups for Solid State Scientists" (New York, Academic Press 1978); B. K. Vainshtein "Modern Crystallography" Springer Series in Solid State Science, vol. 15 (Springer, Berlin 1981)

The Sommerfeld free-electron theory of metals

In this chapter we discuss the free-electron theory of metals, originally developed by Sommerfeld and others. The free-electron model, with its parabolic energy-wavevector dispersion curve, provides a reasonable description for conduction electrons in simple metals; it also may give useful guidelines for other metals with more complicated conduction bands. Because of the simplicity of the model and its density-of-states, we can work out explicitly thermodynamic properties, and in particular the specific heat and the thermionic emission. In the Appendices we summarize for a general quantum system the thermodynamic functions of more frequent use in statistical physics. The Fermi–Dirac and Bose–Einstein distribution functions for independent fermions and bosons are discussed; finally we obtain the modified Fermi–Dirac distribution function, in a model case of correlation among electrons in localized states.

1   Quantum theory of the free-electron gas

An electron in a crystal feels the potential energy due to all the nuclei and all the other electrons. The determination of the crystalline potential in specific materials is a rather demanding problem (see Chapters IV and V). However in several metals, it turns out reasonable to assume that the conduction electrons feel a potential which is constant throughout the sample; this model, suggested by Sommerfeld in 1928, is

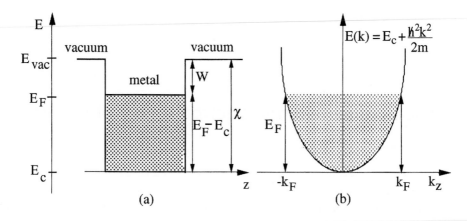

Fig. 1 (a) The Sommerfeld model for a metal. The energy E_c denotes the bottom of the conduction band; E_{vac} denotes the energy of an electron at rest in the vacuum; the electron affinity is $\chi = E_{vac} - E_c$. The Fermi energy is denoted by E_F, and the work function W equals $\chi - E_F$. (b) Free-electron energy dispersion curve along a direction, say k_z, in the reciprocal space. At $T = 0$ all the states with $k < k_F$ are occupied by two electrons of either spin.

still of value for an orientative understanding of a number of properties of simple metals.

In the Sommerfeld model, the Schrödinger equation for conduction electrons inside the metal takes the simple form

$$\left[\frac{\mathbf{p}^2}{2m} + E_c \right] \psi(\mathbf{r}) = E \, \psi(\mathbf{r}) \, , \tag{1}$$

where E_c denotes an appropriate constant specific of the metal under investigation (see Fig. 1). The number of "free electrons" of the metal is assumed to correspond to the number of electrons in the most external shell of the composing atoms. For instance in alkali metals, we expect one free electron per atom, since an alkali atom has just one electron in the most external shell. In aluminum (atomic configuration $1s^2, 2s^2 2p^6, 3s^2 3p^1$) we expect three free electrons per atom.

At first sight it could appear surprising that the conduction electrons in actual metals (or in any type of solid) feel an approximately constant potential, as strong spatial variations of the crystalline potential occur near the nuclei. However, the conduction electrons are hardly sensitive to the region close to the nuclei, because of the orthogonalization effects due to the core electron states; thus the "effective potential" or "pseudopotential" (see Section V-4) for conduction electrons may indeed become a smoothly varying quantity, eventually approximated with a constant; these are the underlying reasons for the success of the free-electron model (or of the nearly-free-electron implementations) in a number of actual metals.

The eigenfunctions of the Schrödinger equation (1), normalized to one in the volume

V of the metal, are the plane waves

$$\psi(\mathbf{k}, \mathbf{r}) = \frac{1}{\sqrt{V}} e^{i\mathbf{k}\cdot\mathbf{r}} ; \qquad (2)$$

the plane waves are just a particular case of Bloch functions, when the periodic modulating part is constant. The eigenvalues of Eq. (1) are

$$E(\mathbf{k}) = E_c + \frac{\hbar^2 k^2}{2m} . \qquad (3)$$

For simplicity, throughout this chapter, we set the zero of energy at the bottom E_c of the conduction band, and thus take $E_c = 0$; whenever necessary, we can reinstate the actual value of E_c by an appropriate shift in the energy scale. [In some situations, for instance in the discussion of photoelectron emission, it can be more convenient to set the zero of energy at the vacuum level, i.e. $E_{\text{vac}} = 0$; in other problems, for instance in the discussion of many-body effects, the zero of energy is generally taken at the Fermi level, i.e. $E_F = 0$; of course any choice is lawful, provided one keeps in mind which choice has been done].

Consider now an electron gas with N free electrons in a volume V, and electron density $n = N/V$. To specify the electron density of a metal it is customary to consider the dimensionless parameter r_s connected to n by the relation

$$\frac{4}{3}\pi r_s^3 a_B^3 = \frac{1}{n} = \frac{V}{N} , \qquad (4)$$

where $a_B = 0.529 \,\text{Å}$ is the Bohr radius; $r_s a_B$ represents the radius of the sphere that contains in average one electron.

We evaluate r_s for alkali metals (for instance); Li, Na, K, Rb have bcc structure with cube edge a equal to 3.49 Å, 4.23 Å, 5.23 Å and 5.59 Å, respectively. In the volume a^3 we have two atoms and two conduction electrons. From Eq. (4) we obtain $(4/3)\pi r_s^3 a_B^3 = a^3/2$ and thus

$$r_s = \frac{a}{a_B} \frac{1}{2} \left(\frac{3}{\pi}\right)^{1/3} = 0.492 \frac{a}{a_B} ;$$

the values for r_s are 3.25, 3.94, 4.87 and 5.20 for Li, Na, K and Rb, respectively. A similar reasoning can be done for other crystals. For instance, the crystal structure of Al is fcc, with lattice constant $a = 4.05$ Å. In the volume a^3 we have four atoms and twelve conduction electrons; from Eq. (4) we have $(4/3)\pi r_s^3 a_B^3 = a^3/12$ and $r_s = 2.07$. Metallic densities of conduction electrons occur mostly in the range $2 < r_s < 6$.

The ground state of the electron system at $T = 0$ is obtained accommodating the N available electrons into the N lowest available energy levels up to the energy E_F, called the *Fermi energy*; each state of wavevector $\mathbf{k}$ and energy $E(\mathbf{k}) < E_F$ accommodates two electrons of either spin. In the $\mathbf{k}$-space, the *Fermi surface* $E(\mathbf{k}) = E_F$ separates the occupied band states of energy $E(\mathbf{k}) < E_F$ from the unoccupied band states of energy $E(\mathbf{k}) > E_F$. For the free-electron system at $T = 0$, the occupied states fill a Fermi sphere of radius k_F, as schematically indicated in Fig. 2.

To determine k_F, we require that the total number of electrons within the Fermi

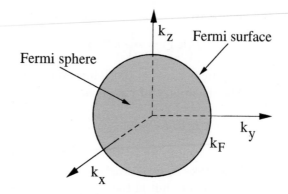

Fig. 2 Schematic representation in the **k**-space of the Fermi sphere of occupied states and of the Fermi surface of the free-electron gas. At $T = 0$ each state of wavevector **k**, with $k < k_F$, is occupied by two electrons of either spin.

sphere of wavevector k_F equals the number N of electrons available in the crystal; we must have

$$N = \sum_{\mathbf{k}}^{k<k_F} 2 = 2\frac{V}{(2\pi)^3}\frac{4}{3}\pi k_F^3 \ , \tag{5}$$

where we have used the standard prescription of Eq. (II-44) for converting the *sum* over **k** into the corresponding *integral* times $V/(2\pi)^3$. From Eq. (5), the Fermi wavevector k_F becomes

$$k_F^3 = 3\pi^2 n \ ; \tag{6a}$$

using Eq. (4), we have

$$k_F = \left(\frac{9\pi}{4}\right)^{1/3}\frac{1}{r_s a_B} = \frac{1.92}{r_s a_B} \ . \tag{6b}$$

The *Fermi velocity* is given by

$$v_F = \frac{\hbar}{m}k_F \ . \tag{7a}$$

It is convenient to consider the dimensionless ratio v_F/c, where c is the velocity of the light. We divide both members of Eq. (7a) by c, and remember that the Bohr radius is $a_B = \hbar^2/m e^2$ and that the fine structure constant is $1/\alpha \equiv \hbar c/e^2 = 137.036$; we obtain

$$\frac{v_F}{c} = \frac{\hbar}{m c}k_F = \alpha\, a_B\, k_F = \frac{1}{137.036}\frac{1.92}{r_s} \ . \tag{7b}$$

We thus see that in the ordinary range of metallic densities, the Fermi velocity is about one hundredth of the velocity of light.

The *Fermi energy* at $T = 0$ is given by

$$E_F = \frac{\hbar^2 k_F^2}{2m} = \frac{\hbar^2}{2m\, a_B^2}\frac{1.92^2}{r_s^2} \ .$$

We notice that $\hbar^2/2ma_B^2 = 1\,\mathrm{Rydberg} = 13.606\,\mathrm{eV}$; we can express the Fermi energy as

$$E_F = \frac{3.683}{r_s^2} \text{ (in Rydberg) .} \tag{8}$$

Typical values of E_F for metals are in the range $1 - 10$ eV. Since $1\,\mathrm{eV}/k_B = 11605$ K, the *Fermi temperature* T_F, defined as $k_B T_F = E_F$ is of the order of $10^4 - 10^5$ K.

The ground-state energy E_0 at $T = 0$ of the free-electron gas is given by

$$E_0 = 2 \sum_{\mathbf{k}}^{k < k_F} \frac{\hbar^2 k^2}{2m} .$$

Converting as usual the sum over $\mathbf{k}$ into an integral times $V/(2\pi)^3$, one has

$$E_0 = 2 \frac{V}{(2\pi)^3} \int_0^{k_F} 4\pi k^2 \frac{\hbar^2 k^2}{2m} \, dk = 8\pi \frac{V}{(2\pi)^3} \frac{\hbar^2}{2m} \frac{k_F^5}{5} .$$

Using Eq. (5) and Eq. (8) we obtain

$$\frac{E_0}{N} = \frac{3}{5} E_F = \frac{2.21}{r_s^2} \text{ (in Rydberg) .} \tag{9}$$

The average energy of each electron at zero temperature is $(3/5)\,E_F$.

We can here anticipate that the ground-state energy, given by Eq. (9), is only the leading term in the high density limit ($r_s \ll 1$) of the exact expression of the ground-state energy of the homogeneous electron gas (see for further aspects and discussion Sections IV-7 and IV-8). In spite of this and other limitations, the Sommerfeld picture is nevertheless of help to describe orientatively and preliminarly a number of properties of simple metals.

From the parabolic dispersion relation $E(\mathbf{k}) = \hbar^2 k^2/2\,m$ of the free-electron gas, we can calculate the density-of-states $D(E)$ from the relationship

$$D(E) = 2 \frac{V}{(2\pi)^3} \int \delta\left(E - \frac{\hbar^2 k^2}{2m}\right) d\mathbf{k} ,$$

where $d\mathbf{k}$ is the volume element in the reciprocal space. Performing the integral with the same procedures adopted in Section II-7, we obtain

$$D(E) = V \frac{(2m)^{3/2}}{2\pi^2 \hbar^3} E^{1/2} \qquad E > 0 . \tag{10a}$$

Since the integrated density-of-states $\int D(E)\,dE$ for $0 \le E \le E_F$ equals N, we can write the density-of-states (10a) in the form

$$D(E) = \frac{3}{2} \frac{N}{E_F} \left(\frac{E}{E_F}\right)^{1/2} \qquad E > 0 . \tag{10b}$$

In particular for $E = E_F$ we have

$$D(E_F) = \frac{3}{2} \frac{N}{E_F} . \tag{11}$$

The plot of $D(E)$ is shown in Fig. 3.

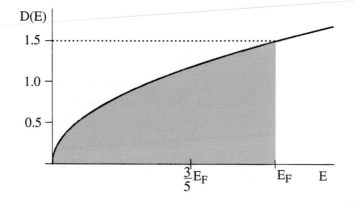

Fig. 3 Density-of-states, in units of N/E_F, for a free-electron gas (the conduction band region occupied by electrons has been shadowed); the average electron energy $(3/5)\,E_F$ is also indicated.

2   Fermi–Dirac distribution function and chemical potential

Fermi–Dirac distribution function

Consider an assembly of identical particles at thermal equilibrium moving independently in a given volume. If the particles obey the Pauli exclusion principle, the occupation probability of a one-particle quantum state of energy E is given by the Fermi–Dirac function

$$f(E) = \frac{1}{e^{(E-\mu)/k_B T} + 1} \, , \tag{12}$$

where μ is the chemical potential, k_B is the Boltzmann constant and T the absolute temperature (see Appendix B). The chemical potential μ for the Fermi–Dirac distribution is also commonly addressed as the Fermi energy or Fermi level E_F; following a common use, the terms "chemical potential" and "Fermi energy", as well as the symbols μ and E_F, are used by us as synonymous.

The behaviour of the Fermi–Dirac function is shown in Fig. 4. If $\mu - E \gg k_B T$ the distribution function approaches unity; if $E - \mu \gg k_B T$, $f(E)$ falls exponentially to zero with a Boltzmann tail. At $T = 0$ the Fermi–Dirac distribution function becomes the step function $\Theta(\mu_0 - E)$, where μ_0 denotes the Fermi energy at $T = 0$; at zero temperature, all the states with energy lower than μ_0 are occupied and all the states with energy higher than μ_0 are empty. At finite temperature T, $f(E)$ deviates from the step function only in the thermal energy range of order $k_B T$ around $\mu(T)$.

We consider now the properties of the function $(-\partial f/\partial E)$. At $T = 0$ we have $(-\partial f/\partial E) = \delta(E - \mu_0)$. At finite temperature $T \ll T_F$, it holds approximately that $(-\partial f/\partial E) \approx \delta(E - \mu)$ (see Fig. 4). In fact, at finite temperature the function $(-\partial f/\partial E)$ is very steep with its maximum at $E = \mu$ and differs significantly from zero in the energy range of the order of $k_B T$ around μ. It is easy to verify that $(-\partial f/\partial E)$ is an even function of E around μ and vanishes exponentially for $|E - \mu| \gg k_B T$; we

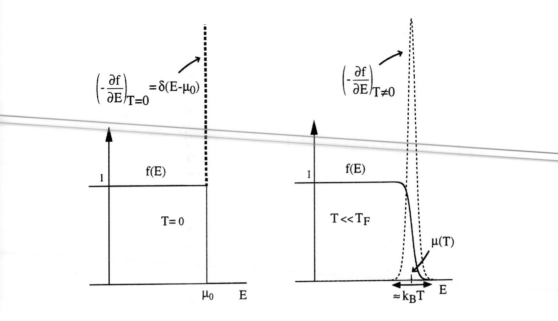

Fig. 4 The Fermi–Dirac distribution function $f(E)$ and energy derivative $(-\partial f/\partial E)$ at $T = 0$ and at a finite temperature T, with $T \ll T_F$; μ_0 denotes the chemical potential $\mu(T)$ at $T = 0$.

also have

$$\int_{-\infty}^{+\infty}\left(-\frac{\partial f}{\partial E}\right) dE = -[f(\infty) - f(-\infty)] = 1 \ . \tag{13}$$

Sommerfeld expansion

Let us consider an integral of the type

$$I = \int_{-\infty}^{+\infty} G(E)\left(-\frac{\partial f}{\partial E}\right) dE \ , \tag{14}$$

where $G(E)$ is a function regular and (infinitely) differentiable around μ. We can demonstrate the following relation

$$\boxed{\int_{-\infty}^{+\infty} G(E)\left(-\frac{\partial f}{\partial E}\right) dE = G(\mu) + \frac{\pi^2}{6}(k_B T)^2 G''(\mu) + \frac{7\pi^4}{360}(k_B T)^4 G''''(\mu) + \ldots} \tag{15}$$

where energy differentiation order is indicated by the corresponding number of apices.

To prove Eq. (15), consider the series expansion of $G(E)$ in powers of E around the chemical potential

$$G(E) = G(\mu) + (E - \mu)\left(\frac{dG}{dE}\right)_{E=\mu} + \frac{1}{2}(E - \mu)^2 \left(\frac{d^2 G}{dE^2}\right)_{E=\mu} + \ldots \ ;$$

inserting this expression into Eq. (14), and using Eq. (13), we obtain

$$\int_{-\infty}^{+\infty} G(E)\left(-\frac{\partial f}{\partial E}\right) dE = G(\mu) + \frac{1}{2} G''(\mu) \int_{-\infty}^{+\infty} (E-\mu)^2 \left(-\frac{\partial f}{\partial E}\right) dE + \dots .$$

Odd powers of $(E - \mu)$ in the expansion have been omitted because they give zero contribution. The coefficient of $G''(\mu)$ can be written as

$$\frac{1}{2} \int_{-\infty}^{+\infty} (E-\mu)^2 \left(-\frac{\partial f}{\partial E}\right) dE = \frac{1}{2} \int_{-\infty}^{+\infty} (E-\mu)^2 \frac{e^{(E-\mu)/k_BT}}{\left[e^{(E-\mu)/k_BT}+1\right]^2} \frac{1}{k_BT} dE$$

$$= k_B^2 T^2 \int_0^{+\infty} x^2 \frac{e^x}{(e^x+1)^2} dx$$

(in the last passage we have denoted by x the dimensionless variable $x = (E-\mu)/k_BT$). To perform the integral in the x variable, we notice that

$$\int_0^{+\infty} x^2 e^{-x}[1+e^{-x}]^{-2} dx = \int_0^{+\infty} x^2 e^{-x}[1 - 2e^{-x} + 3e^{-2x} - \dots] dx$$

$$= 2\left(1 - \frac{1}{2^2} + \frac{1}{3^2} - \frac{1}{4^2} + \dots\right) = \frac{\pi^2}{6} .$$

This explains the numerical coefficient in front of $G''(\mu)$ in expression (15). Successive terms in expansion (15) can be calculated in a similar way.

Equation (15) is known as the *Sommerfeld expansion*. To appreciate the fact that it is a rapidly convergent expansion for $T \ll T_F$, consider the case in which $G(E)$ has a power dependence on E of the type $G(E) \approx E^p$; then it is easy to verify that the terms of the expansion (15) are of the type of $(k_BT/\mu)^2$, $(k_BT/\mu)^4$ or equivalently $(T/T_F)^2$, $(T/T_F)^4$ etc..

The Sommerfeld expansion (15) can be rewritten also in a slightly different form. Most often the function $G(E)$ vanishes for energies E below some threshold energy and, furthermore, $G(E)$ behaves reasonably well at infinity so that $G(E) f(E) \to 0$ for $E \to \infty$. When these conditions are verified, it is convenient to perform an integration by parts in the first member of Eq. (15); we have

$$\int_{-\infty}^{+\infty} \frac{dG(E)}{dE} f(E) dE \equiv G(\mu) + \frac{\pi^2}{6} (k_BT)^2 G''(\mu) + \dots . \qquad (16)$$

Let us indicate with $\Gamma(E)$ the derivative $G'(E)$, or equivalently

$$G(E) = \int_{-\infty}^{E} \Gamma(E') dE' .$$

Eq. (16) becomes

$$\boxed{\int_{-\infty}^{+\infty} \Gamma(E) f(E) dE = \int_{-\infty}^{\mu} \Gamma(E) dE + \frac{\pi^2}{6} (k_BT)^2 \left(\frac{d\Gamma(E)}{dE}\right)_{E=\mu} + O(T^4)} \qquad (17)$$

which is a useful expression for the given function $\Gamma(E)$.

Temperature dependence of the chemical potential

The chemical potential μ depends (slightly) on the temperature; in the study of several transport properties, the temperature dependence of the chemical potential (whatever small it might appear at first sight) has quite important consequences, and now we work out explicitly the behaviour of $\mu(T)$.

Let $D(E)$ be the (single-particle) density-of-states for both spin directions for the metallic sample of volume V; no particular assumption on $D(E)$ or on the conduction band energy $E(\mathbf{k})$ is done at this stage. Let N be the total number of conduction electrons of the sample. At any temperature T, we have that $\mu(T)$ is determined (implicitly) enforcing the equality

$$N = \int_{-\infty}^{+\infty} D(E)\, f(E)\, dE \ .$$

Such an integral has in general a rather complex structure. However if $T \ll T_F$ and if $D(E)$ is reasonably smooth at $E \approx \mu$ (both conditions are in practice quite well satisfied), we can perform the integral using the Sommerfeld expansion (17), where the function $\Gamma(E)$ stands now for $D(E)$. We have

$$N = \int_{-\infty}^{\mu} D(E)\, dE + \frac{\pi^2}{6} k_B^2 T^2 D'(\mu) + O(T^4) \ , \qquad (18)$$

where the prime indicates differentiation with respect to the energy, and terms of the order of $(T/T_F)^4$ can be safely omitted since $T \ll T_F$ at the ordinary temperatures of interest. We now differentiate both members of Eq. (18) with respect to the temperature, taking into account that $\mu = \mu(T)$. We obtain

$$0 = D(\mu)\frac{d\mu}{dT} + \frac{\pi^2}{3} k_B^2 T\, D'(\mu) \qquad (19)$$

[we have neglected the term $(\pi^2/6)k_B^2 T^2 D''(\mu)\,(d\mu/dT)$ because of order $(T/T_F)^2$ with respect to $D(\mu)\,(d\mu/dT)$]. From Eq. (19) it follows

$$\frac{d\mu}{dT} = -\frac{\pi^2}{3} k_B^2 T \frac{D'(\mu)}{D(\mu)} \ . \qquad (20)$$

This equation, together with the obvious condition $\mu(T) = \mu_0$ for $T = 0$ allows us to obtain $\mu(T)$ at any temperature T, provided $T \ll T_F$. We notice that:

(a) If the density-of-states $D(E)$ for $E \approx \mu_0$ increases with energy, then the chemical potential μ decreases with temperature (and vice versa). In the case $D(E)$ is constant for $E \approx \mu_0$, the chemical potential μ is temperature independent.

(b) For a free-electron gas $D(E) \propto E^{1/2}$. Then $D'(\mu)/D(\mu) = 1/2\mu$. The integration of Eq. (20) is straightforward and gives $\mu^2 - \mu_0^2 = -(\pi^2/6)k_B^2 T^2$. Then

$$\mu(T) = \mu_0 \left[1 - \frac{\pi^2}{12}\left(\frac{k_B T}{\mu_0}\right)^2 \right] = \mu_0 \left[1 - \frac{\pi^2}{12}\left(\frac{T}{T_F}\right)^2 \right] \ . \qquad (21)$$

This relation gives μ as a function of T and shows that $\mu(T)$ decreases slowly with increasing temperature for the free-electron gas.

3   Electronic specific heat in metals and thermodynamic functions

The *heat capacity at constant volume* of a sample is defined by

$$C_V = \left(\frac{\delta Q}{dT} \right)_V , \tag{22}$$

where δQ is the amount of heat transferred from the external world to the system, and dT is the corresponding change in temperature of the system, kept at constant volume V. The heat capacity is an extensive quantity, i.e. a quantity proportional to the volume of the sample; for this reason, depending on the nature of the system under investigation, it may become preferable to introduce the *specific heat* per mole, or the specific heat per unit volume, or per unit cell, or per composing atoms or electrons.

We can express the heat capacity (22) in convenient alternative forms. From the first law of thermodynamics, we know that the change dU of the internal energy of a system, in a transformation in which an infinitesimal quantity of heat δQ is received by the system and δL is the work done on the system by external forces, is given by

$$dU = \delta Q + \delta L .$$

In the case the external forces are only mechanical forces, exerting a pressure p on the system, we have $\delta L = -p\,dV$; if the volume V of the system is kept constant during the transformation then $\delta L = 0$; it follows $dU = \delta Q$ and

$$C_V = \left(\frac{dU}{dT} \right)_V . \tag{23a}$$

In the case of reversible transformations, the second law of thermodynamics states that $\delta Q = T\,dS$, where S is the entropy of the system. We have thus

$$C_V = T \left(\frac{dS}{dT} \right)_V . \tag{23b}$$

We now calculate the heat capacity of an electron gas, using Eq. (23a). The internal energy of the Fermi gas is

$$U(T) = \int_{-\infty}^{+\infty} E\,D(E)\,f(E)\,dE$$

$$= \int_{-\infty}^{\mu} E\,D(E)\,dE + \frac{\pi^2}{6} k_B^2 T^2 \left[D(\mu) + \mu\,D'(\mu) \right] + O(T^4) , \tag{24}$$

where the standard Sommerfeld expansion (17) has been used. By differentiating with respect to the temperature both members of Eq. (24), and keeping only the most relevant among the terms containing $(d\mu/dT)$, we have

$$C_V = \left(\frac{dU}{dT} \right)_V = \mu\,D(\mu) \frac{d\mu}{dT} + \frac{\pi^2}{3} k_B^2 T \left[D(\mu) + \mu\,D'(\mu) \right] .$$

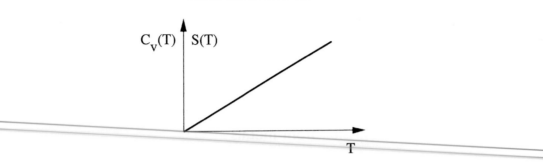

Fig. 5 Electronic contribution $C_V(T)$ to the heat capacity of a metal at constant volume, as a function of temperature; the same expression holds for the electronic contribution to the entropy.

Using Eq. (19), and replacing $D(\mu)$ with $D(\mu_0)$, we obtain

$$\boxed{C_V(T) = \frac{\pi^2}{3} k_B^2 T \, D(\mu_0)} \; . \tag{25}$$

From Eq. (23b), it can be noticed that expression (25) represents also the entropy of the electron system.

The specific heat per unit volume $c_V = C_V/V$ becomes

$$c_V(T) = \frac{\pi^2}{3} k_B^2 T \frac{D(\mu_0)}{V} = \gamma T \; , \tag{26a}$$

where

$$\gamma = \frac{\pi^2}{3} k_B^2 \frac{D(\mu_0)}{V} \; . \tag{26b}$$

The contribution to the specific heat from conduction electrons is proportional to T for all temperatures of interest, since T is always much less than T_F. The electronic contribution becomes the leading one at sufficiently low temperatures, where it prevails over the T^3 Debye contribution originated by the lattice vibrations (see Section IX-5). We also notice that the knowledge of the density-of-states $D(\mu_0)$ at the Fermi energy μ_0 completely determines C_V; this gives a first insight on the importance of the electronic states at or near the Fermi surface in metals. In discussing transport phenomena, impurity screening etc. we will further see the basic role played by the electronic states lying in the thermal interval of the order $k_B T$ around the Fermi energy.

It is interesting to specify Eq. (25) for the case of the free-electron gas, where $D(\mu_0) = (3/2)\, N/\mu_0$ according to Eq. (11). We have

$$C_V(T) = \frac{\pi^2}{2} k_B N \frac{T}{T_F} \; . \tag{27}$$

For comparison, we recall that the classical statistical mechanics would give the expression $C_V = (3/2) k_B N$ for the heat capacity of a gas of N non-interacting particles.

The correct result (27) can be interpreted noticing that only the electrons in the thermal interval $k_B T$ around the Fermi energy $\mu_0 = k_B T_F$ can vary their energy, and thus the effective number of electrons with "classical behaviour" is not N but rather the fraction T/T_F of N.

4   Thermionic emission from metals

We can apply the Fermi–Dirac statistics to study under very simplified conditions the thermionic emission from metals, i.e. the emission of electrons from a metal in the vacuum because of the effect of a finite temperature. For our semi-quantitative considerations, we do not consider in detail possible reflection of electrons impinging at the surface; we simply assume that all the electrons that arrive at the surface with an energy sufficient to overcome the surface barrier are transferred to the vacuum (and swept away by some small applied electric field without accumulation of space charge). The model electronic structure of the metal, with electron affinity χ and work function W, is illustrated in Fig. 1; we wish to obtain the number of electrons which escape from the metal, kept at temperature T.

In the metal, the electrons are distributed in energy according to the Fermi–Dirac statistics. Let us indicate with z the direction normal to the surface; the current density of escaping electrons is given by

$$J_s = (-e) \int_{\sqrt{2m\chi/\hbar^2}}^{+\infty} dk_z \int_{-\infty}^{+\infty} dk_x \int_{-\infty}^{+\infty} dk_y \frac{2}{(2\pi)^3} \frac{1}{e^{(E(\mathbf{k})-\mu)/k_B T}+1} v_z \, , \qquad (28)$$

where $v_z = \hbar k_z/m$. Notice that expression (28), in the case v_z is just replaced by a drift velocity v independent on $\mathbf{k}$ (and the integration over k_z extends from $-\infty$ to $+\infty$), would give the standard density $J_s = (-e)n\,v$. The limitation in Eq. (28) for the integration on k_z is just to make sure that the escaping electrons have enough kinetic energy $\hbar^2 k_z^2/2m \geq \chi$ in the z direction to leave the metal.

In order to perform the integral (28), let us notice that

$$E(\mathbf{k}) - \mu = \frac{\hbar^2 k_x^2}{2m} + \frac{\hbar^2 k_y^2}{2m} + \frac{\hbar^2 k_z^2}{2m} - \mu \geq \chi - \mu \equiv W \, .$$

Since in general the work function $W \gg k_B T$, we can safely neglect the unity in the Fermi distribution function in Eq. (28). We thus obtain

$$J_s = \frac{(-e)}{4\pi^3} \frac{\hbar}{m} \int_{\sqrt{2m\chi/\hbar^2}}^{+\infty} k_z \, dk_z \int_{-\infty}^{+\infty} dk_x \int_{-\infty}^{+\infty} dk_y \exp\left[\frac{\mu}{k_B T} - \frac{\hbar^2(k_x^2 + k_y^2 + k_z^2)}{2mk_B T} \right] \, .$$

We now use the standard result for Gaussian functions

$$\int_{-\infty}^{+\infty} dk_x \exp\left[-\frac{\hbar^2 k_x^2}{2mk_B T} \right] = \frac{\sqrt{2\pi m k_B T}}{\hbar}$$

Table 1 Experimental values of the work function W for some metals.

metal	W(eV)	metal	W(eV)
Li	2.49	Al	4.2
Na	2.28	Ga	4.1
K	2.24	Sn	4.4
Cs	1.81	Pb	4.0
Ag	4.3	Pt	5.6
Au	5.2	W	4.5

and obtain

$$J_s = \frac{(-e)\,k_B T}{2\pi^2}\frac{1}{\hbar} \int_{\sqrt{2m\chi/\hbar^2}}^{+\infty} k_z\,dk_z \exp\left[\frac{\mu}{k_B T} - \frac{\hbar^2 k_z^2}{2m k_B T}\right] .$$

The integral can be easily performed with the change of variable $(\hbar^2 k_z^2/2m) - \mu = x$, and we arrive at the Richardson expression

$$J_s = \frac{(-e)m\,k_B^2}{2\pi^2\hbar^3}\,T^2\,e^{-W/k_B T} . \tag{29}$$

The absolute value of the numerical factor in the Richardson law (29) is

$$\frac{e\,m\,k_B^2}{2\pi^2\hbar^3} = 120.4 \text{ amp}\cdot\text{cm}^{-2}\cdot\text{K}^{-2} , \tag{30}$$

in many metals, measured values are indeed in the range $50 - 120$ amp $\cdot$ cm$^{-2}\cdot$ K^{-2}.

A quantitative treatment of the electron emission from metals requires a number of refinements of the simplified model here considered. In the model of Fig. 1, the metal–vacuum boundary is represented by an abrupt discontinuity in the potential. In reality an electron outside the metal feels an attractive image potential (to be determined in principle quantum mechanically) with consequences on the reflection coefficient of escaping electrons. The image potential also leads to a reduction of the apparent work function in the presence of applied electric fields (Schottky effect). The work function is also sensitive to various effects, for instance, surface impurities and charge modifications at the surface. Observed values of the work function for some metals are reported in Table 1; for a more complete list see for instance H. B. Michaelson "Handbook of Chemistry and Physics" edited by R. C. Weast (CRC Press, Cleveland 1962); D. E. Eastman, Phys. Rev. B**2**, 1 (1970) and references quoted therein.

APPENDIX A. Outline of statistical physics and thermodynamic relations

A1. Microcanonical ensemble and thermodynamic quantities

The basic principles of classical and quantum statistics can be found in standard textbooks on statistical physics; the purpose of this appendix is simply to summarize the "recipes" for the connection between statistics and thermodynamics.

Let us consider a physical system composed by N identical particles confined within a volume V. Quantum mechanics provides for this confined system discretized energy levels; we label with an integer number m all the distinct eigenstates of the system in increasing energy order E_m ($\ldots \le E_m \le E_{m+1} \le E_{m+2} \le \ldots$). The eigenstates of the system are also called *microstates* or *accessible states* of the system.

For systems with a large number of particles and volume, the spacings of the different levels becomes extremely small (if not zero), and the total energy E of the system can be regarded as an almost continuous variable. Instead of focusing on each individual microstate, it is convenient to consider the quantity $W(E, V, N)$ which gives the total number of distinct microstates lying in the interval $(E - \Delta, E + \Delta)$, where $\Delta \ll E$.

Let us indicate with (E, V, N) the parameters that define a particular macrostate of the given system. In general, for a given macrostate (E, V, N), there is a large number $W(E, V, N)$ of corresponding microstates; according to the standard concepts of the *microcanonical ensemble*, it is assumed that the macrostate (E, V, N) is equally likely to be in any of the distinct microstates $W(E, V, N)$ lying in the energy interval $(E - \Delta, E + \Delta)$ around E.

The thermodynamics of the physical system is completely determined by the expression of the entropy, which is given by

$$\boxed{S = k_B \ln W(E, V, N)} \; ; \tag{A1}$$

from the basic expression $(A1)$ of the entropy, the rest of the thermodymics follows.

Let us consider quasistatic transformations with (infinitesimally) slow varations of the macroscopic parameters E, V, N so as to guarantee that the transformations involve only thermodynamic equilibrium states. The change of entropy $(A1)$ for an infinitesimal transformation can be written as

$$dS = \left(\frac{\partial S}{\partial E} \right)_{V,N} dE + \left(\frac{\partial S}{\partial V} \right)_{E,N} dV + \left(\frac{\partial S}{\partial N} \right)_{E,V} dN \; . \tag{A2}$$

From the first and second principle of thermodynamics, we can also write for any infinitesimal reversible transformation

$$dE = T \, dS - p \, dV + \mu \, dN \; . \tag{A3}$$

From equations $(A2)$ and $(A3)$ we obtain

$$\frac{1}{T} dE + \frac{p}{T} dV - \frac{\mu}{T} dN = \left(\frac{\partial S}{\partial E} \right)_{V,N} dE + \left(\frac{\partial S}{\partial V} \right)_{E,N} dV + \left(\frac{\partial S}{\partial N} \right)_{E,V} dN \; .$$

We have thus the three basic relations

$$\frac{1}{T} = \left(\frac{\partial S}{\partial E} \right)_{V,N} \tag{A4}$$

$$\frac{p}{T} = \left(\frac{\partial S}{\partial V} \right)_{E,N} \tag{A5}$$

$$-\frac{\mu}{T} = \left(\frac{\partial S}{\partial N}\right)_{E,V} . \tag{A6}$$

Relation $(A4)$ provides the *absolute temperature* T; relation (A5) provides the *equation of state* (i.e. a relation among p, V and T); relation $(A6)$ provides the *chemical potential* μ.

Formulae $(A4)$, $(A5)$ and $(A6)$ are just the ones that follow from the basic expression $(A1)$ of entropy and from the basic laws of thermodynamics; from them any other desired thermodynamic function can be obtained. For instance, the *Helmholtz free energy* is given by $F = E - TS$; the *Gibbs free energy* is $G = E - TS + pV$; the *enthalpy* is given by $H = E + pV = G + TS$.

A2. Canonical ensemble and thermodynamic quantities

Consider a physical system containing N identical particles confined within the volume V; as before we label with an integer number m all the distinct eigenstates of the system in increasing order of energy E_m.

We are now interested in the statistical description of the system in equilibrium with a surrounding thermal bath at temperature T, and thus free to exchange energy with it. The equilibrium properties of the system are calculated by averaging over all its accessible states of energy E_m, assigning to each state a weight P_m proportional to the Boltzmann factor $\exp(-\beta E_m)$ with $\beta = 1/k_B T$. By requiring the normalization of the weights to unity, we obtain the *Gibbs* or *canonical distribution probability*

$$P_m = \frac{e^{-\beta E_m}}{\left(\sum e^{-\beta E_m}\right)} . \tag{A7}$$

The denominator in Eq. $(A7)$ is known as *canonical partition function*

$$\boxed{Z(T,V,N) = \sum_m e^{-\beta E_m}} . \tag{A8}$$

We now show how the thermodynamic functions of the system can be evaluated from the canonical partition function.

To establish the connection between statistics and thermodynamics it is convenient to define the *canonical potential* $\Omega = \Omega(T,V,N)$ so that

$$e^{-\beta\Omega} = \sum_m e^{-\beta E_m} \equiv Z(T,V,N) . \tag{A9}$$

The probability distribution $(A7)$ then takes the form

$$P_m = e^{-\beta(E_m - \Omega)} . \tag{A10}$$

We now show that the Helmholtz free energy F is just given by the canonical thermodynamic potential Ω defined in Eq. $(A9)$.

Consider in fact an infinitesimal reversible transformation in which the parameter

T is changed (while V and N are kept constant). Taking the differential of Eq. ($A9$) we have

$$e^{-\beta\Omega}\, d(\beta\Omega) = \sum_m e^{-\beta E_m} E_m\, d\beta.$$

Multiplying by $\exp(\beta\Omega)$ both members of the above equation, we obtain

$$d(\beta\Omega) = \sum_m e^{-\beta(E_m - \Omega)} E_m\, d\beta = \sum_m P_m E_m\, d\beta = U\, d\beta\;, \qquad (A11)$$

where U denotes the mean internal energy of the system; Eq. ($A11$) can be re-written as

$$d(\beta\Omega) = d(\beta U) - \beta\, dU\;. \qquad (A12)$$

From the first and second law of thermodynamics, for reversible transformations with constant volume and constant number of particles, we have $dU = \delta Q = T\, dS$. Eq. ($A12$) thus becomes

$$d(\beta\Omega) = d(\beta U) - \beta T\, dS\;,$$

and we have

$$d\,\frac{\Omega - U}{T} = -dS\;.$$

Integrating the above expression from zero temperature to an arbitrary temperature T, and using the third principle of thermodynamics to set to zero the arbitrary integration constant, we have $(\Omega - U)/T = -S$ and $\Omega = U - TS$; hence the *canonical potential* Ω equals the *Helmholtz free energy* $F = U - TS$. Using Eq. ($A9$), we obtain that the Helmholtz free energy is related to the partition function by the expression

$$\boxed{F = -k_B T \ln Z(T, V, N) = -k_B T \ln \sum_m e^{-\beta E_m}}\;. \qquad (A13)$$

This is the basic result of the canonical ensemble apparatus.

Other thermodynamic relationships can now be easily established. For instance, by differentiating the free energy $F = U - TS$, we have

$$dF = dU - T\, dS - S\, dT\;.$$

On the other hand, from the standard laws of thermodynamics, we also have

$$dU = T\, dS - p\, dV + \mu\, dN\;.$$

Hence

$$dF = -S\, dT - p\, dV + \mu\, dN\;,$$

and we obtain the relations:

$$S = -\left(\frac{\partial F}{\partial T}\right)_{V,N} \qquad (A14)$$

$$p = -\left(\frac{\partial F}{\partial V}\right)_{T,N} \tag{A15}$$

$$\mu = \left(\frac{\partial F}{\partial N}\right)_{T,V} . \tag{A16}$$

Let us consider, for instance, expression $(A14)$ for the entropy. With the help of expression $(A13)$ we have

$$S = -\left(\frac{\partial F}{\partial T}\right)_{V,N} = k_B \ln \sum_m e^{-\beta E_m} + k_B T \sum_m e^{-\beta E_m} \frac{E_m}{k_B T^2} \Big/ \sum_m e^{-\beta E_m} .$$

It is now easy to verify that the above expression becomes

$$S = -k_B \sum_m P_m \ln P_m , \tag{A17}$$

where P_m are defined in Eq. $(A7)$. This shows that the entropy is determined by the probability values P_m.

Similarly for the pressure we have

$$p = -\left(\frac{\partial F}{\partial V}\right)_{T,N} = k_B T \left(\frac{\partial \ln Z}{\partial V}\right)_{T,N} = -\sum_m \left(\frac{\partial E_m}{\partial V}\right)_{T,N} e^{-\beta(E_m - \Omega)} . \tag{A18}$$

We notice that $-\partial E_m/\partial V$ can be identified as the pressure acting on the physical system in the pure quantum mechanical state of energy E_m. Then the weighted average of pressures in the sum in Eq. $(A18)$ can be identified with the pressure p acting on the physical system in thermodynamic equilibrium with a thermal bath of temperature T. In particular, at zero temperature we have

$$p = -\frac{\partial E_0}{\partial V} ; \tag{A19}$$

at $T = 0$, the pressure is just the derivative of the ground-state energy of the system with respect to the volume.

A3. Grand canonical ensemble and thermodynamic quantities

Consider a physical system containing N_s identical particles confined within the volume V. The number N_s of identical particles of the system can now fluctuate. For any chosen value of N_s, we label with an integer number m all the distinct eigenstates of the system and we indicate with $E_{m,s}$ the eigenvalues of the system with N_s particles.

We are now interested in the description of the system in thermodynamic equilibrium with a thermal bath at temperature T and with a particle reservoir with chemical potential μ. The equilibrium properties of the system are calculated by averaging over all its accessible states of energy $E_{m,s}$ assigning to each state a probability $P_{m,s}$ proportional to

$$P_{m,s} \propto e^{-\beta(E_{m,s} - \mu N_s)} .$$

With normalization of the weights to unity, we obtain the *grand canonical distribution* probability

$$P_{m,s} = \frac{e^{-\beta(E_{m,s}-\mu N_s)}}{\sum\limits_{m,s} e^{-\beta(E_{m,s}-\mu N_s)}} \quad . \tag{A20}$$

The denominator in ($A20$) is known as the *grand canonical partition function*

$$\boxed{Z_{\text{grand}}(T,V,\mu) = \sum_{m,s} e^{-\beta(E_{m,s}-\mu N_s)}} \quad . \tag{A21}$$

In analogy with the Appendix A2, we now show that all the thermodynamic functions of the system can be calculated from the knowledge of the grand canonical partition function.

To establish the connection between statistics and thermodynamics, we define the *grand canonical potential* $\Omega_{\text{grand}} = \Omega_{\text{grand}}(T,V,\mu)$ so that

$$e^{-\beta\Omega_{\text{grand}}} \equiv \sum_{m,s} e^{-\beta(E_{m,s}-\mu N_s)} = Z_{\text{grand}}(T,V,\mu) \tag{A22}$$

or equivalently

$$\Omega_{\text{grand}} = -k_B T \ln Z_{\text{grand}}(T,V,\mu) = -k_B T \ln \sum_{m,s} e^{-\beta(E_{m,s}-\mu N_s)} \quad . \tag{A23}$$

Following "mutatis mutandis" the reasoning done in the Appendix A2, we can easily establish that the grand canonical potential equals

$$\boxed{\Omega_{\text{grand}} = U - TS - \overline{N}\mu = -k_B T \ln Z_{\text{grand}}(T,V,\mu)} \tag{A24}$$

where U is the mean internal energy and $\overline{N}$ is the mean particle number in the grand canonical distribution. Relation ($A24$) is the basic result of the grand canonical apparatus; all the other thermodynamic relationships follow from it.

From the first and second principles of thermodynamics we have

$$dU = T\,dS - p\,dV + \mu\,d\overline{N} \quad .$$

By differentiating Ω_{grand} and using the above expression, we obtain

$$d\Omega_{\text{grand}} = dU - T\,dS - S\,dT - \overline{N}d\mu - \mu d\overline{N} = -S\,dT - p\,dV - \overline{N}d\mu \quad .$$

It follows:

$$S = -\left(\frac{\partial\Omega_{\text{grand}}}{\partial T}\right)_{V,\mu} \tag{A25}$$

$$p = -\left(\frac{\partial\Omega_{\text{grand}}}{\partial V}\right)_{T,\mu} \tag{A26}$$

$$\overline{N} = -\left(\frac{\partial\Omega_{\text{grand}}}{\partial\mu}\right)_{T,V} \quad . \tag{A27}$$

Relation $(A25)$ provides the entropy of the system; relation $(A26)$ provides the equation of state; relation $(A27)$ provides the mean number of particles; all other thermodynamic functions can be easily obtained from the equations so far established.

APPENDIX B. Fermi–Dirac and Bose–Einstein statistics for independent particles

Fermi–Dirac statistics

We consider a physical system composed by N identical particles confined within a volume V. The particles of the system are regarded as *non-interacting* (except for an arbitrary small interaction, to ensure thermodynamic equilibrium). We thus discuss the energy levels of the many-body system in terms of independent *one-particle states*.

Quantum mechanics provides for a single particle confined within the volume V discretized energy levels; we label with an integer number i all the distinct eigenstates, of energies ε_i, of the single-particle quantum problem. Since the particles are non-interacting, the total energy of the many-body system is the sum of the energies of the individual particles

$$E = \sum_i \varepsilon_i n_i \,, \tag{B1}$$

where n_i denotes the number of particles with energy ε_i; the total number of particles is

$$N = \sum_i n_i \,. \tag{B2}$$

We now consider specifically a system of identical Fermi particles; as a consequence of the Pauli principle, the possible values of the occupation numbers n_i are either 0 or 1. The most general accessible state for the system of indistinguishable particles is defined by a set of numbers $\{n_i\}$ (i.e. any sequence of integer numbers equal to 0 or 1). The grand partition function $(A21)$ for a quantum system of non-interacting fermions becomes

$$Z_{\text{grand}}(T, V, \mu) = \sum_{\{n_i\}} e^{-\beta \sum n_i(\varepsilon_i - \mu)} \,. \tag{B3}$$

This sum can be carried out exactly and gives

$$Z_{\text{grand}}(T, V, \mu) = \prod_i \left[1 + e^{-\beta(\varepsilon_i - \mu)} \right] \,. \tag{B4}$$

The passage from Eq. $(B3)$ to Eq. $(B4)$ can be done, for instance, with the following reasoning. Consider first the particular situation of a system with a *single* one-electron level, say ε_1; in this case the sum in Eq. $(B3)$ provides $Z_1 = 1 + \exp[-\beta(\varepsilon_1 - \mu)]$. Consider then the particular situation of a system with only two levels, say ε_1 and ε_2; the sum defined in Eq. $(B3)$ provides $Z = Z_1 Z_2$. Similarly, for a system with n levels

$\varepsilon_1, \varepsilon_2, \dots, \varepsilon_n$, the sum defined in Eq. $(B3)$ provides $Z = Z_1 Z_2 \dots Z_n$ and for a system with *any* number of levels we obtain Eq. $(B4)$.

Now that the grand canonical partition function is known, we can calculate whatever desired thermodynamic quantity. For instance we can calculate the average occupation number $f(\varepsilon_i)$ of a given state ε_i. We have

$$f(\varepsilon_i) = \langle n_i \rangle = \frac{1}{Z_{\text{grand}}} \sum_{\{n_j\}} n_i e^{-\beta \sum n_j (\varepsilon_j - \mu)} \,.$$

The sum over the configurations $\{n_j\}$ can be easily carried out following, mutatis mutandis, the procedure done for the calculation of Z_{grand}. We have

$$\langle n_i \rangle = \frac{1}{Z_{\text{grand}}} e^{-\beta(\varepsilon_i - \mu)} \prod_{j(\neq i)} \left[1 + e^{-\beta(\varepsilon_j - \mu)} \right] = \frac{e^{-\beta(\varepsilon_i - \mu)}}{1 + e^{-\beta(\varepsilon_i - \mu)}} \,.$$

We thus recover immediately the Fermi–Dirac statistics

$$\boxed{ f(\varepsilon_i) = \frac{1}{e^{\beta(\varepsilon_i - \mu)} + 1} } \,. \tag{B5}$$

We can obtain the expression of the free energy and of the entropy of a system of non-interacting fermions. We start from the expression of the grand canonical potential

$$\Omega_{\text{grand}} = -k_B T \ln Z_{\text{grand}} = -k_B T \sum_i \ln \left[1 + e^{-\beta(\varepsilon_i - \mu)} \right] \,. \tag{B6}$$

From Eq. $(A24)$ we have for the free energy

$$F = \overline{N} \mu - k_B T \sum_i \ln \left[1 + e^{-\beta(\varepsilon_i - \mu)} \right] \,. \tag{B7}$$

From Eq. $(A25)$, we have for the entropy

$$S = -\left(\frac{\partial \Omega_{\text{grand}}}{\partial T} \right)_{V,\mu}$$

$$= k_B \sum_i \ln[1 + e^{-\beta(\varepsilon_i - \mu)}] + k_B T \sum_i \frac{e^{-\beta(\varepsilon_i - \mu)}}{1 + e^{-\beta(\varepsilon_i - \mu)}} \frac{\varepsilon_i - \mu}{k_B T^2}$$

$$= k_B \sum_i \ln \frac{1}{1 - f(\varepsilon_i)} + k_B \sum_i f(\varepsilon_i) \beta (\varepsilon_i - \mu) \,.$$

We use the identity

$$\beta(\varepsilon_i - \mu) \equiv \ln \frac{1 - f_i}{f_i} \,,$$

where f_i denotes by brevity $f(\varepsilon_i)$. We obtain

$$S = k_B \sum_i \left[-\ln(1 - f_i) + f_i \ln \frac{1 - f_i}{f_i} \right] \,,$$

and finally

$$S = -k_B \sum_i [f_i \ln f_i + (1 - f_i) \ln(1 - f_i)] \tag{B8}$$

which is the desired expression for the entropy of non-interacting fermions.

Another important expression can be proved for the total number of particles. From Eq. (A27), we have

$$\overline{N} = -\left(\frac{\partial \Omega_{\text{grand}}}{\partial \mu}\right)_{T,V} = k_B T \sum_i \frac{\partial}{\partial \mu} \ln[1 + e^{-\beta(\varepsilon_i - \mu)}]$$

$$= \sum_i \frac{e^{-\beta(\varepsilon_i - \mu)}}{1 + e^{-\beta(\varepsilon_i - \mu)}} = \sum_i f_i .$$

Finally for the equation of state (A26) we have

$$p = -\left(\frac{\partial \Omega_{\text{grand}}}{\partial V}\right)_{T,\mu} = k_B T \sum_i \frac{e^{-\beta(\varepsilon_i - \mu)}}{1 + e^{-\beta(\varepsilon_i - \mu)}}(-\beta)\frac{\partial \varepsilon_i}{\partial V} = -\sum_i f_i \frac{\partial \varepsilon_i}{\partial V} . \tag{B9}$$

The energy eigenvalues of a particle confined in a cubic box of volume $V = L^3$ are given by

$$\varepsilon = \frac{\hbar^2}{2m}\left(\frac{\pi}{L}\right)^2 (n_x^2 + n_y^2 + n_z^2) \qquad n_x, n_y, n_z = 1, 2, 3, \ldots$$

We have thus $\partial \varepsilon / \partial L = -2\varepsilon / L$ and also

$$\frac{\partial \varepsilon}{\partial V} = \frac{\partial \varepsilon}{\partial L}\frac{\partial L}{\partial V} = -\frac{2}{3}\frac{\varepsilon}{V} . \tag{B10}$$

Expression (B9) thus becomes

$$p = \frac{2}{3}\frac{1}{V}\sum_i f_i \varepsilon_i$$

and then

$$pV = \frac{2}{3}U . \tag{B11}$$

Bose–Einstein statistics

In this case, differently from the previous treatment, the sequence $\{n_i\}$ can contain any integer from zero to infinity. The grand partition function for a system of non-interacting bosons is still given by expression (B3), keeping in mind that n_i can now take any integer value from zero to infinity. The sum (B3) can be carried out exactly and gives for bosons

$$Z_{\text{grand}}(T, V, \mu) = \prod_i \frac{1}{1 - e^{-\beta(\varepsilon_i - \mu)}} . \tag{B12}$$

The passage from Eq. $(B3)$ to Eq. $(B12)$ can be done (as before) considering first the particular case of a single one-particle level, say ε_1. In this case the sum in Eq. $(B3)$ becomes

$$Z_1 = \sum_{n_1=0}^{\infty} \left[e^{-\beta(\varepsilon_1 - \mu)} \right]^{n_1} = \frac{1}{1 - e^{-\beta(\varepsilon_1 - \mu)}} \;.$$

For a system with arbitrary number of one-particle levels we thus obtain Eq. $(B12)$.

The grand canonical potential for a system of independent bosons is

$$\Omega_{\mathrm{grand}} = -k_B T \ln Z_{\mathrm{grand}} = k_B T \sum_i \ln \left[1 - e^{-\beta(\varepsilon_i - \mu)} \right] \;. \qquad (B13)$$

We can now obtain all the thermodynamic quantities of interest for a system of non interacting bosons. The average occupation number of a given state is given by

$$\boxed{f(\varepsilon_i) = \frac{1}{e^{\beta(\varepsilon_i - \mu)} - 1}}\;. \qquad (B14)$$

Similarly, the entropy of a system of non-interacting bosons is given by

$$\boxed{S = -k_B \sum_i \left[f_i \ln f_i - (1 + f_i) \ln(1 + f_i) \right]}\;. \qquad (B15)$$

APPENDIX C. Modified Fermi–Dirac statistics in a model of correlation effects

In the previous Appendix we have considered a physical system composed by N indistinguishable non-interacting fermions, confined within a volume V. Suppose that the one-electron Hamiltonian does not remove the spin degeneracy; in the independent particle approximation, the occupation probability of the *space orbital* (of energy ε_i), *regardless of the spin degeneracy*, is then given by

$$f(\varepsilon_i) = 2 \frac{1}{e^{\beta(\varepsilon_i - \mu)} + 1} \;, \qquad (C1)$$

where the factor 2 accounts for the spin degeneracy of the orbital level.

In Chapter XIII, in the study of doped semiconductors, we need to know not only the occupation probability of valence and conduction states (described by standard delocalized Bloch wavefunctions), but also the occupation probability of impurity localized states in the energy gap. In several situations (for instance for donor levels) the Coulomb repulsion between electrons may *prevent double occupation of a given localized orbital*. We now discuss how this effect (which is the simplest example of correlation beyond the one-electron approximation) modifies the Fermi–Dirac distribution function.

For sake of clarity, consider a band state in an allowed energy region of the crystal and described by a Bloch wavefunction; a band level can be empty, or occupied by

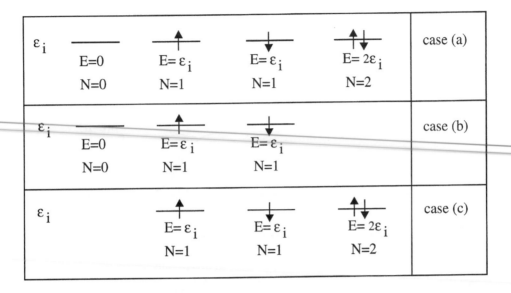

Fig. 6 Schematic illustration of possible occupation of a given spatial orbital of energy ε_i. (a) The given level can accept *two* electrons of either spin (this is the common situation for band states) (b) The given level can accept only *one* electron of either spin (this situation is common for donor impurity levels in semiconductors) (c) The given level can accommodate one or two electrons, *but not zero* (this situation is common for acceptor levels in semiconductors).

one electron of either spin, or by two electrons of opposite spin; the four possibilities are illustrated in Fig. 6a. Consider now an impurity state within the energy gap and described by a localized wavefunction; a donor level, for instance, can be empty, or occupied by one electron of either spin, but not by two electrons of opposite spin, because of the penalty in the electrostatic repulsion energy; the situation is illustrated in Fig. 6b.

We calculate for both situations the average number of electrons in the state ε_i (regardless of the spin direction); the average number is given by

$$\langle n_i \rangle = \frac{\sum\limits_{m,N} N\, e^{-\beta(E_{m,N}-\mu N)}}{\sum\limits_{m,N} e^{-\beta(E_{m,N}-\mu N)}} \ . \tag{C2}$$

In the case of Fig. 6a, Eq. $(C2)$ gives

$$\langle n_i \rangle = \frac{e^{-\beta(\varepsilon_i-\mu)} + e^{-\beta(\varepsilon_i-\mu)} + 2e^{-\beta(2\varepsilon_i-2\mu)}}{1 + e^{-\beta(\varepsilon_i-\mu)} + e^{-\beta(\varepsilon_i-\mu)} + e^{-\beta(2\varepsilon_i-2\mu)}}$$

$$= \frac{2e^{-\beta(\varepsilon_i-\mu)}\left[1 + e^{-\beta(\varepsilon_i-\mu)}\right]}{\left[1 + e^{-\beta(\varepsilon_i-\mu)}\right]\left[1 + e^{-\beta(\varepsilon_i-\mu)}\right]} = 2\frac{1}{e^{\beta(\varepsilon_i-\mu)} + 1} \ ; \tag{C3}$$

as expected, the result $(C1)$ is recovered.

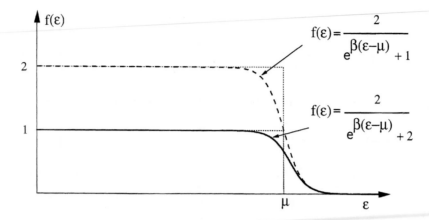

Fig. 7 Average occupation number (dashed line) for an orbital of energy ε that can accept up to two electrons of either spin (standard Fermi–Dirac statistics). The average occupation number (solid line) for an orbital of energy ε that can accept only one electron of either spin is also reported. For $\varepsilon - \mu \gg k_B T$ the two curves coincide.

In the case of Fig. 6b, we have instead

$$\langle n_i \rangle = \frac{e^{-\beta(\varepsilon_i - \mu)} + e^{-\beta(\varepsilon_i - \mu)}}{1 + e^{-\beta(\varepsilon_i - \mu)} + e^{-\beta(\varepsilon_i - \mu)}} = \frac{1}{\frac{1}{2} e^{\beta(\varepsilon_i - \mu)} + 1} \ . \tag{C4}$$

The occurrence of the factor $1/2$ in Eq. ($C4$), can be easily understood qualitatively in the limiting case of a Boltzmann tail (see Fig. 7).

For sake of completeness, we consider also the statistics of the acceptor levels, whose electronic structure is studied in Chapter XIII. An acceptor level can be occupied by two paired electrons, or one of either spin, but cannot be empty because of the penalty in electrostatic repulsion energy between the two holes; the situation is schematically indicated in Fig. 6c. The application of Eq. ($C2$) gives

$$\langle n_i \rangle = \frac{e^{-\beta(\varepsilon_i - \mu)} + e^{-\beta(\varepsilon_i - \mu)} + 2 e^{-\beta(2\varepsilon_i - 2\mu)}}{e^{-\beta(\varepsilon_i - \mu)} + e^{-\beta(\varepsilon_i - \mu)} + e^{-\beta(2\varepsilon_i - 2\mu)}} = 2 - \frac{1}{\frac{1}{2} e^{-\beta(\varepsilon_i - \mu)} + 1} \ . \tag{C5}$$

If we indicate by $\langle p_i \rangle = 2 - \langle n_i \rangle$ the mean number of holes, we obtain

$$\langle p_i \rangle = \frac{1}{\frac{1}{2} e^{\beta(\mu - \varepsilon_i)} + 1} \ . \tag{C6}$$

Further reading

N. W. Ashcroft and N. D. Mermin "Solid State Physics" (Holt, Rinehart and Winston, New York 1976)

H. B. Callen "Thermodynamics and an Introduction to Thermostatics" (Wiley, New York 1985, second edition)

K. Huang "Statistical Mechanics" (Wiley, New York 1987, second edition)

C. Kittel "Elementary Statistical Physics" (Wiley, New York 1958)

R. Kubo "Statistical Mechanics" (North-Holland, Amsterdam 1988, seventh edition)

A. Münster "Statistical Thermodynamics" Vols. I and II (Springer, Berlin 1969)

R. K. Pathria "Statistical Mechanics" (Pergamon Press, Oxford 1972)

L. E. Reichl "A Modern Course in Statistical Physics" (Arnold, London 1980)

A. H. Wilson "The Theory of Metals" (Cambridge University Press 1954)

IV

The one-electron approximation and beyond

In this chapter on the quantum mechanics of electrons in solids, we strike at the heart of one of the most traditional subjects of solid state physics. A solid is constituted by a large number of electrons and nuclei in mutual interaction, and the dynamics of these particles, in general, cannot be considered separately. Throughout this chapter, however, we assume that the nuclei are *fixed in a given configuration* (usually the equilibrium configuration), so that we can focus on the electronic problem. Among the approaches to the many-electron system, we consider the historic (but always actual) Hartree–Fock theory, because of its central role in the general framework of the many-body theory. We will then discuss aspects beyond the one-electron approximation, in particular the density functional theory, which has been so successful in the description of the ground-state properties of solids.

From the point of view of the organization of this book, Chapter IV is the backbone and paves the way for the sequence of logically linked subjects, presented in Chapters from V through X. In Chapter V we consider in detail the methods of electronic band structure calculations within the one-electron approximation. In Chapter VI some applications to selected classes of solids are provided. In Chapter VII we consider aspects beyond the one-electron approximation, with the concepts of excitons and plasmons quasiparticles. In all the mentioned chapters, the focus is on the electronic problem (within the one-electron approximation or beyond it), with the nuclei pictured as a fixed source of a static potential. Later, in Chapters VIII, IX and X we discuss the interdependence of electronic and nuclear dynamics, the elastic field and the vibrational structure of solids, the electronic and lattice effects on the scattering of particles from crystals.

1 Introductory remarks on the many-electron problem

Atoms, molecules, clusters or solids, are systems composed by mutually interacting electrons and nuclei. The total non-relativistic Hamiltonian of a system of electrons (of coordinates $\mathbf{r}_i$, momenta $\mathbf{p}_i$, and charge $-e$) and nuclei (of coordinates $\mathbf{R}_I$, momenta $\mathbf{P}_I$, and charge $+z_I e$) in mutual interaction via Coulomb forces, can be written as

$$H_{\text{tot}} = \sum_i \frac{\mathbf{p}_i^2}{2m} + \sum_I \frac{\mathbf{P}_I^2}{2M_I} + \sum_i V_{\text{nucl}}(\mathbf{r}_i) + \frac{1}{2}\sum_{i\neq j} \frac{e^2}{|\mathbf{r}_i - \mathbf{r}_j|} + \frac{1}{2}\sum_{I\neq J} \frac{z_I z_J e^2}{|\mathbf{R}_I - \mathbf{R}_J|} \quad (1)$$

with

$$V_{\text{nucl}}(\mathbf{r}) = -\sum_I \frac{z_I e^2}{|\mathbf{r} - \mathbf{R}_I|} . \quad (2)$$

The terms appearing in Eq. (1) represent the kinetic energy of the electrons, the kinetic energy of the nuclei, the electron-nucleus attractive potential energy, the electron-electron and nucleus-nucleus repulsive potential energies. The expression containing the electron-electron interaction involves a sum on the indices i and j, with the obvious exclusion of the self-interaction terms $i = j$; similarly in the nucleus-nucleus interaction the sum runs over I and J, with the exclusion of the self-interaction terms $I = J$. We notice that, for simplicity, the Hamiltonian (1) omits any energy term (of relativistic or classical origin) related to the spin and the magnetic moment of the particles in interaction.

The first and obvious hint to attack the Hamiltonian (1) is suggested by the large difference between electron and nuclear masses. The simplest approximation (although a drastic one) is just to drop the kinetic energy of the nuclei in Eq. (1) [a more gentle and fruitful approach within the adiabatic principle is the subject of Chapter VIII]. When the drastic approximation is anyway adopted, the nuclear positions $\mathbf{R}_I$ become classical variables (or parameters) and the nuclei can be considered as fixed in a given selected configuration; most often the fixed configuration is the equilibrium configuration.

Consistently with our neglecting the nuclear kinetic energy, we gain the further bonus that the nuclear repulsion (represented by the last term in Eq. 1) is a constant for any fixed configuration; its value increases with the sample volume (and diverges if not combined with appropriate attractive terms due to the overall charge neutrality of the system) and is important to determine the total energy of the system or its cohesive energy. However for what concerns the *electronic states in the rigid lattice approximation* (i.e. *without relaxation of the nuclear positions*), we can disregard the last term of Eq. (1), and take it into account only when needed.

The many-body Hamiltonian for a system of N interacting electrons in the presence of nuclei fixed in some selected configuration can thus be written in the form

$$H_e = \sum_i^N h(\mathbf{r}_i) + \frac{1}{2} \sum_{i \neq j}^N \frac{e^2}{r_{ij}} \ , \tag{3}$$

where

$$h(\mathbf{r}) = \frac{\mathbf{p}^2}{2m} + V_{\text{nucl}}(\mathbf{r}) \tag{4}$$

denotes for each electron the kinetic energy operator plus the potential energy due to the nuclei (see Eq. 2). The purpose of this chapter is to present various approximations towards the solution of the many-body eigenvalue problem

$$H_e \, \Psi(\mathbf{r}_1\sigma_1, \mathbf{r}_2\sigma_2, \dots, \mathbf{r}_N\sigma_N) = E \, \Psi(\mathbf{r}_1\sigma_1, \mathbf{r}_2\sigma_2, \dots, \mathbf{r}_N\sigma_N) \ , \tag{5}$$

where $\mathbf{r}_i\sigma_i$ are the space and spin variables of the i-th electron. Although the Hamiltonian H_e does not include spin-dependent energy terms, the electron spin plays a fundamental role since the correct many-electron wavefunctions $\Psi(\mathbf{r}_1\sigma_1, \mathbf{r}_2\sigma_2, \dots, \mathbf{r}_N\sigma_N)$ must be antisymmetric for interchange of the spatial and spin coordinates of any two electrons.

The pioneering approach to solve Eq. (5) is the *Hartree theory*, where the total N-electron ground-state wavefunction is represented by the best *simple product of N one-electron spin-orbitals*. The next major improvement, that correctly embodies the Pauli exclusion principle for identical fermions, is the *Hartree–Fock theory*, where the total N-electron ground-state wavefunction is represented by the best *antisymmetrized product of N one-electron spin-orbitals*. We also consider some aspects beyond the one-electron approximation; in this wide field, we focus on the particular role of the *density functional theory* and Kohn–Sham orbitals for the ground-state electronic density.

2   The Hartree equations

To represent many-electron wavefunctions, it is convenient to consider preliminarily a complete set of orthonormal one-electron orbitals $\{\phi_i(\mathbf{r})\}$. From this, we can form a complete set of orthonormal *spin-orbitals* $\{\psi_i(\mathbf{r}\sigma)\}$, where $\psi_i(\mathbf{r}\sigma) = \phi_i(\mathbf{r})\chi_i(\sigma)$ denotes the product of the *spatial orbital* $\phi_i(\mathbf{r})$ and the *spin function* $\chi_i(\sigma)$ (either spin-up function α or spin-down function β). In short-hand notations, space and spin variables are (frequently) omitted. In order to avoid possible source of confusion, throughout this chapter, we use the notation ψ_i exclusively to denote spin-orbitals; similarly we use the notation ϕ_i and χ_i exclusively to denote the orbital part and the spin part of the spin-orbital ψ_i. Thus the arguments of ψ_i, ϕ_i and χ_i are understood to be $\psi_i(\mathbf{r}\sigma)$, $\phi_i(\mathbf{r})$ and $\chi_i(\sigma)$, respectively.

In the Hartree theory the ground-state wavefunction of the many-body system is expressed as simple product of orthonormalized one-electron spin-orbitals in the form

$$\Psi_0(\mathbf{r}_1\sigma_1, \mathbf{r}_2\sigma_2, \ldots, \mathbf{r}_N\sigma_N) = \psi_1(\mathbf{r}_1\sigma_1)\psi_2(\mathbf{r}_2\sigma_2)\ldots\psi_N(\mathbf{r}_N\sigma_N) . \tag{6}$$

In Eq. (6), the many-body wavefunction Ψ_0 is constructed by assigning any given electron to some given spin-orbital (for intance the electron of coordinates $\mathbf{r}_1\sigma_1$ is assigned to the spin-orbital ψ_1, the electron $\mathbf{r}_2\sigma_2$ is assigned to the spin-orbital ψ_2, and so on). It is apparent that the simple Hartree product (6) does not have the correct antisymmetry character for interchange of space and spin coordinates of any two electrons; the Pauli principle is taken into account ad hoc, avoiding multiple occupancy of any given spin-orbital. Furthermore, a simple product of type (6) completely neglects any correlation in the position of the electrons. Because of these serious drawbacks, we mention the Hartree theory mostly for its historical role, and confine our discussion to some heuristic considerations [later, for the conceptually more important Hartree–Fock equations and Kohn–Sham equations, we use rigorous variational principles].

The electronic charge density $\rho(\mathbf{r})$ corresponding to the Hartree wavefunction (6) is given by

$$\rho(\mathbf{r}) = (-e) \sum_{j}^{(\text{occ})} \phi_j^*(\mathbf{r})\phi_j(\mathbf{r}) , \tag{7}$$

where the sum runs over all occupied spin-orbitals, entering the ground state Ψ_0. The Coulomb potential (also called the *Hartree potential*) corresponding to the electronic charge density (7) is

$$V_{\text{coul}}(\mathbf{r}) = \sum_{j}^{(\text{occ})} \int \phi_j^*(\mathbf{r}')\frac{e^2}{|\mathbf{r} - \mathbf{r}'|}\phi_j(\mathbf{r}')\, d\mathbf{r}' . \tag{8}$$

In the Hartree self-consistent approximation, it is assumed that each electron moves in the effective field corresponding to the Coulomb potential generated by the charge distribution of all the other $N - 1$ electrons; for simplicity, to avoid the determination of as many (slightly) different effective fields as the number of orbitals, it is assumed that the effective field for any electron is given by the Hartree potential (8). We thus obtain that the spin-orbitals entering in the product wavefunction (6) satisfy the Hartree equations

$$\left[\frac{\mathbf{p}^2}{2m} + V_{\text{nucl}}(\mathbf{r}) + V_{\text{coul}}(\mathbf{r})\right] \psi_i = \varepsilon_i \psi_i . \tag{9}$$

The Hartree potential is defined in terms of the occupied orbitals ψ_i and must thus be determined in a self-consistent way (usually with appropriate iterations). From an initial reasonable guess of the functions $\psi_1, \psi_2, \ldots, \psi_N$ one evaluates the space charge distribution (7) and the corresponding Hartree potential (8). A new set of improved wavefunctions is then obtained from the solution of the Hartree equations (9). The corresponding Hartree potential (or, more often, an appropriate weighted average of input and output ones) is used to start a new cycle. Cycles are repeated up to self-consistency of input and output functions and potentials.

3   Identical particles and determinantal wavefunctions

In this section we summarize some basic results on the antisymmetry principle for fermions, and in particular the possibility to write the basis functions of a many-body system in the form of determinantal states. It is well known that a many-electron wavefunction must be antisymmetric for interchange of the coordinates of any two electrons. Consider an N-electron system and a set $\{\psi_i\}$ $(i = 1, 2, \ldots, N)$ of *orthonormal one-particle spin-orbitals*. A proper antisymmetric N-electron wavefunction takes the form

$$\Psi_0(\mathbf{r}_1\sigma_1, \mathbf{r}_2\sigma_2, \ldots, \mathbf{r}_N\sigma_N) = A\{\psi_1(\mathbf{r}_1\sigma_1)\psi_2(\mathbf{r}_2\sigma_2)\ldots\psi_N(\mathbf{r}_N\sigma_N)\}, \qquad (10a)$$

where A denotes the antisymmetrization operator

$$A = \frac{1}{\sqrt{N!}}\sum_{i=1}^{N!}(-1)^{p_i}P_i \qquad (10b)$$

(the sum extends to all the $N!$ permutations P_i of the electronic coordinates, and $(-1)^{p_i}$ equals $+1$ or -1 for permutations of even or odd class with respect to the fundamental one). Notice that Ψ_0 is normalized to one, if the composing spin-orbitals ψ_i are orthonormal.

Expression (10) can be conveniently written in the determinantal form suggested by Slater:

$$\Psi_0(\mathbf{r}_1\sigma_1, \mathbf{r}_2\sigma_2, \ldots, \mathbf{r}_N\sigma_N) = \frac{1}{\sqrt{N!}}\begin{vmatrix} \psi_1(\mathbf{r}_1\sigma_1) & \psi_1(\mathbf{r}_2\sigma_2) & \ldots & \psi_1(\mathbf{r}_N\sigma_N) \\ \psi_2(\mathbf{r}_1\sigma_1) & \psi_2(\mathbf{r}_2\sigma_2) & \ldots & \psi_2(\mathbf{r}_N\sigma_N) \\ \ldots & \ldots & \ldots & \ldots \\ \psi_N(\mathbf{r}_1\sigma_1) & \psi_N(\mathbf{r}_2\sigma_2) & \ldots & \psi_N(\mathbf{r}_N\sigma_N) \end{vmatrix}. \qquad (11)$$

It is evident that the interchange of two columns changes the sign of the determinant consistently with the antisymmetry property of the wavefunction; moreover, occupancy of the same spin-orbital by two electrons gives two equal rows and thus the determinant equals zero.

An important property automatically embodied in determinantal wavefunctions is that electrons with parallel spin are (correctly) kept apart. To better realize this point, consider for simplicity the determinantal state (11) in the particular case in which the spin-orbitals have *all spin parallel* (for instance spin up). The determinantal state (11) then keeps the form:

$$\Psi_0(\mathbf{r}_1\sigma_1, \mathbf{r}_2\sigma_2, \dots, \mathbf{r}_N\sigma_N) = \frac{1}{\sqrt{N!}} \begin{vmatrix} \phi_1(\mathbf{r}_1) & \phi_1(\mathbf{r}_2) & \dots & \phi_1(\mathbf{r}_N) \\ \phi_2(\mathbf{r}_1) & \phi_2(\mathbf{r}_2) & \dots & \phi_2(\mathbf{r}_N) \\ \dots & \dots & \dots & \dots \\ \phi_N(\mathbf{r}_1) & \phi_N(\mathbf{r}_2) & \dots & \phi_N(\mathbf{r}_N) \end{vmatrix} \alpha(1)\alpha(2)\dots\alpha(N) .$$

(12)

It is evident that *nodes* of Ψ_0 occur whenever $\mathbf{r}_i \equiv \mathbf{r}_j$; thus any two electrons cannot be in the same spatial position (with the same spin).

The many-body wavefunctions can be written in a more compact form, leaving implicit the space and spin coordinates of the electrons. Two frequently used shorthand notations for the wavefunction Ψ_0 are

$$\Psi_0 = A\{\psi_1\psi_2\dots\psi_N\} \quad \text{or} \quad \Psi_0 = \frac{1}{\sqrt{N!}}\det\{\psi_1\psi_2\dots\psi_N\} ;$$

both notations clearly list the spin-orbitals entering the antisymmetrization operator or forming the Slater determinant.

4   Matrix elements between determinantal states

The purpose of this section is to provide the explicit expression of the matrix elements of one-electron and two-electron operators between determinantal states. These expressions constitute a technical (but nevertheless essential) tool for handling the Hartree–Fock theory, as well as the second quantization formalism of the many-body problem. The readers more interested in the structure and physical meaning of the Hartree–Fock equations can skip this section, using their own knowledge when meeting matrix elements between determinantal states.

Expectation values of one-particle and two-particle operators on determinantal states

The many-body electron Hamiltonian (3) contains two types of operators. One is the sum of one-particle operators of the form

$$G_1 = \sum_i^N h(\mathbf{r}_i) ;$$

(13a)

the other type is the sum of two-particle operators of the form

$$G_2 = \frac{1}{2} \sum_{i \neq j}^N \frac{e^2}{|\mathbf{r}_i - \mathbf{r}_j|} ,$$

(13b)

which describes electron-electron interactions. We wish now to express the matrix elements of one- and two-particle operators between Slater determinants in terms of monoelectronic and bielectronic integrals, respectively. The elaboration is based on the elementary properties of determinants and the invariance of G_1 and G_2 under permutations of the electronic coordinates.

We begin with the evaluation of the expectation value of G_1 on a given determinantal state Ψ_0; we have

$$\langle \Psi_0 | G_1 | \Psi_0 \rangle = \frac{1}{N!} \langle \det \{\psi_1 \psi_2 \ldots \psi_N\} | G_1 | \det \{\psi_1 \psi_2 \ldots \psi_N\} \rangle . \tag{14a}$$

In the above expression, we can just select the identical permutation in the determinant in the left part; in fact, all the other $N! - 1$ terms contributing to the determinant in the left side are equal to the selected one. We can thus cancel $N!$ at the denominator of Eq. (14a), and write

$$\langle \Psi_0 | G_1 | \Psi_0 \rangle = \langle \psi_1 \psi_2 \ldots \psi_N | G_1 | \det \{\psi_1 \psi_2 \ldots \psi_N\} \rangle . \tag{14b}$$

Because of the orthonormality of spin-orbitals it is also evident that only the identical permutation survives in the determinant on the right part of the above matrix element. We have thus

$$\langle \Psi_0 | G_1 | \Psi_0 \rangle = \langle \psi_1 \psi_2 \ldots \psi_N | G_1 | \psi_1 \psi_2 \ldots \psi_N \rangle . \tag{14c}$$

This relation shows that, for what concerns the matrix elements of the operator G_1, it is irrelevant to use antisymmetrized products of spin-orbitals or simple (Hartree) products of the same spin-orbitals.

Inserting the expression (13a) of G_1 into expression (14c), we obtain

$$\boxed{\langle \Psi_0 | G_1 | \Psi_0 \rangle = \sum_i \langle \psi_i | h | \psi_i \rangle} . \tag{15}$$

Thus, we see that the expectation value on a determinantal state of a sum of one-electron operators equals the sum of the expectation values of the one-electron operator on the spin-orbitals, which enter in the determinantal state.

We evaluate now the expectation value of G_2 on a given determinantal state; we have

$$\langle \Psi_0 | G_2 | \Psi_0 \rangle = \frac{1}{N!} \langle \det \{\psi_1 \psi_2 \ldots \psi_N\} | G_2 | \det \{\psi_1 \psi_2 \ldots \psi_N\} \rangle$$

$$= \langle \psi_1 \psi_2 \ldots \psi_N | G_2 | \det \{\psi_1 \psi_2 \ldots \psi_N\} \rangle . \tag{16a}$$

Consider for instance the contribution of the term e^2/r_{12}. We restore for clarity the space and spin electronic coordinates, and we easily verify that

$$\langle \psi_1(\mathbf{r}_1 \sigma_1) \psi_2(\mathbf{r}_2 \sigma_2) \ldots \psi_N(\mathbf{r}_N \sigma_N) | \frac{e^2}{r_{12}} | \det \{\psi_1(\mathbf{r}_1 \sigma_1) \psi_2(\mathbf{r}_2 \sigma_2) \ldots \psi_N(\mathbf{r}_N \sigma_N)\} \rangle$$

$$= \langle \psi_1 \psi_2 | \frac{e^2}{r_{12}} | \psi_1 \psi_2 \rangle - \langle \psi_1 \psi_2 | \frac{e^2}{r_{12}} | \psi_2 \psi_1 \rangle ; \tag{16b}$$

in fact only the identical permutation and the permutation (of odd class) that interchanges coordinates $\mathbf{r}_2 \sigma_2$ with coordinates $\mathbf{r}_1 \sigma_1$ survive. The definition and elementary properties of the Coulomb and exchange bielectronic integrals appearing in the right-hand side of Eq. (16b) are discussed in Appendix A; in particular we remind the reader that bielectronic exchange integrals are definite positive quantities, and are different from zero only for parallel spins of the spin-orbitals entering in their definition.

The expectation value of G_2 on a given determinantal state thus becomes

$$\langle \Psi_0 | G_2 | \Psi_0 \rangle = \frac{1}{2} \sum_{ij} \left[\langle \psi_i \psi_j | \frac{e^2}{r_{12}} | \psi_i \psi_j \rangle - \langle \psi_i \psi_j | \frac{e^2}{r_{12}} | \psi_j \psi_i \rangle \right] . \tag{17}$$

In Eq. (17) we can write indifferently $\sum_{ij}$ or $\sum_{i \neq j}$, since Coulomb and exchange contributions exactly cancel each other for $i \equiv j$. We can thus recast Eq. (17) in the form

$$\langle \Psi_0 | G_2 | \Psi_0 \rangle = \frac{1}{2} \sum_{i \neq j} \langle \psi_i \psi_j | \frac{e^2}{r_{12}} | \psi_i \psi_j \rangle - \frac{1}{2} \sum_{i \neq j} \langle \psi_i \psi_j | \frac{e^2}{r_{12}} | \psi_j \psi_i \rangle . \tag{18}$$

It is important to notice explicitly that the *electron-electron repulsive energy* $\langle \Psi_0 | G_2 | \Psi_0 \rangle$ *on the antisymmetrized product of spin-orbitals, is always lower than the electron-electron repulsive energy on the simple Hartree product of the same spin-orbitals*; the difference is just represented by the second term in the right-hand side of Eq. (18). Since exchange bielectronic integrals are positive quantities, different from zero only for spin-orbitals with parallel spins, we link the decrease in energy with the physical fact that electrons with parallel spin are kept apart in real space in Slater determinantal states (as already discussed in Eq. 12).

In the following we report other matrix elements of interest of one-electron and two-electron operators among determinantal states; however we omit from now on any proof, as this could be easily worked out following "mutatis mutandis" the reasoning leading to Eq. (15) and Eq. (17).

Determinantal states with replacement of one spin-orbital

Consider two N-electron determinantal states which differ by one of the one-particle functions, say

$$\Psi_0 = A \{ \psi_1 \psi_2 \dots \psi_m \dots \psi_N \}$$

and

$$\Psi_{\mu,m} = A \{ \psi_1 \psi_2 \dots \psi_\mu \dots \psi_N \} .$$

The notation $\Psi_{\mu,m}$ helps to memorize that the spin-orbital ψ_m has been replaced by the spin-orbital ψ_μ, all other spin-orbitals being equal and in the same sequence.

It can be easily verified that

$$\langle \Psi_{\mu,m} | G_1 | \Psi_0 \rangle = \langle \psi_\mu | h | \psi_m \rangle . \tag{19}$$

Thus, we see that the matrix elements of G_1 connecting two determinantal states, which differ by one of the one-particle functions, equal the matrix elements of the one-particle operator between the two (replacing and replaced) spin-orbitals. For the operator G_2 we have

$$\langle \Psi_{\mu,m} | G_2 | \Psi_0 \rangle = \sum_j \left[\langle \psi_\mu \psi_j | \frac{e^2}{r_{12}} | \psi_m \psi_j \rangle - \langle \psi_\mu \psi_j | \frac{e^2}{r_{12}} | \psi_j \psi_m \rangle \right] . \tag{20}$$

Determinantal states with replacement of two spin-orbitals

Consider two determinantal states which differ by two of the one-particle functions, say

$$\Psi_0 = A \{ \psi_1 \psi_2 \ldots \psi_m \ldots \psi_n \ldots \psi_N \}$$

and

$$\Psi_{\mu\nu,mn} = A \{ \psi_1 \psi_2 \ldots \psi_\mu \ldots \psi_\nu \ldots \psi_N \} .$$

A sum of one-electron operators, as G_1, has vanishing matrix elements between determinantal states which differ by two spin-orbitals. The matrix elements of G_2 are given by

$$\langle \Psi_{\mu\nu,mn} | G_2 | \Psi_0 \rangle = \langle \psi_\mu \psi_\nu | \frac{e^2}{r_{12}} | \psi_m \psi_n \rangle - \langle \psi_\mu \psi_\nu | \frac{e^2}{r_{12}} | \psi_n \psi_m \rangle . \tag{21}$$

Finally, the matrix elements of the operators G_1 and G_2 between determinantal states which differ by *more than two spin-orbitals are zero*.

In this section we have provided useful expressions of the matrix elements of the operator G_1 and G_2 between determinantal states. With respect to the matrix elements involving simple product wavefunctions, it can be observed that the antisimmetrization does not alter the matrix elements of the one-body operator G_1, while, for the two-body operator G_2, exchange bielectronic integrals are to be appropriately included. A more convenient formal and conceptual point of view to treat Slater determinants and related matrix elements is the use of the second quantization formalism, which is outlined in Appendix B.

5   The Hartree–Fock equations

5.1   Variational approach and Hartree–Fock equations

Let us consider a system of N interacting electrons described by the Hamiltonian H_e given in Eq. (3). We attempt to describe *the ground state of the many-electron system with a single determinantal state*; the best possible choice of spin-orbitals entering this determinantal state is obtained minimizing the electron energy, using the variational principle. We find that the "optimized" orbitals are solutions of the self-consistent integro-differential equations, known as Hartree–Fock equations.

We approximate the ground state of the system with a single determinantal state in the form

$$\Psi_0 = A \{ \psi_1 \psi_2 \ldots \psi_N \} .$$

A determinantal state Ψ_0, constituted by doubly occupied spatial orbitals, is called *spin-restricted Slater determinant*; this spin configuration is the appropriate choice for *closed-shell systems*, where electrons with spin-up and spin-down balance for each spatial orbital. However, to discuss also *open-shell systems*, we do not put at this stage any restriction on the spin part of the spin-orbitals entering the ground state Ψ_0, which is thus in general a *spin-unrestricted Slater determinant*. The variational procedure concerns the optimized determination of the orbital parts of the spin-orbitals (for a given chosen spin configuration).

The expectation value $E_0 = \langle\Psi_0|H_e|\Psi_0\rangle$ of the electronic Hamiltonian H_e on the (normalized) state Ψ_0 can be calculated by means of Eq. (15) and Eq. (17); it is given by

$$E_0 = \sum_i \langle\psi_i|h|\psi_i\rangle + \frac{1}{2}\sum_{ij}\left[\langle\psi_i\psi_j|\frac{e^2}{r_{12}}|\psi_i\psi_j\rangle - \langle\psi_i\psi_j|\frac{e^2}{r_{12}}|\psi_j\psi_i\rangle\right]. \qquad (22)$$

According to the variational principle we vary the N contributing spin-orbitals $\{\psi_i\}$ ($i=1,2,\ldots,N$) until the energy $E_0 \equiv E_0(\{\psi_i\})$ achieves its minimum value. This minimum value is higher (or at most equal) than the correct ground-state energy. The variation of the functional $E_0(\{\psi_i\})$ must be done under the constraint that $\{\psi_i\}$ are orthonormal

$$\langle\psi_i|\psi_j\rangle = \delta_{ij} \ (i,j = 1,2,\ldots,N) \ .$$

With the standard technique of undetermined Lagrange multipliers, we minimize without constraints the functional

$$G(\{\psi_i\}) = \sum_i \langle\psi_i|h|\psi_i\rangle + \frac{1}{2}\sum_{ij}\left[\langle\psi_i\psi_j|\frac{e^2}{r_{12}}|\psi_i\psi_j\rangle - \langle\psi_i\psi_j|\frac{e^2}{r_{12}}|\psi_j\psi_i\rangle\right] - \sum_{ij}\varepsilon_{ij}\langle\psi_i|\psi_j\rangle$$

where ε_{ij} are at this stage N^2 undetermined Lagrange multipliers.

We can now calculate the variation δG when the orbital part of the spin-orbitals ψ_i are changed by an infinitesimal amount $\psi_i + \delta\psi_i$. Consider, for instance, the variation of the term $\langle\psi_i|h|\psi_i\rangle$; we have to first order in the variations

$$\delta\langle\psi_i|h|\psi_i\rangle = \langle\delta\psi_i|h|\psi_i\rangle + \langle\psi_i|h|\delta\psi_i\rangle \ .$$

Similar considerations can be performed for any term appearing in the expression of $G(\{\psi_i\})$, and the variation δG can be calculated.

We can consider the variations $\delta\psi_i$ and $\delta\psi_i^*$ completely independent; for instance taking the variations $\delta\psi_i^*$ arbitrary and the variations $\delta\psi_i \equiv 0$ we obtain for δG:

$$\delta G = \sum_i \langle\delta\psi_i|h|\psi_i\rangle + \sum_{ij}\left[\langle\delta\psi_i\psi_j|\frac{e^2}{r_{12}}|\psi_i\psi_j\rangle - \langle\delta\psi_i\psi_j|\frac{e^2}{r_{12}}|\psi_j\psi_i\rangle\right] - \sum_{ij}\varepsilon_{ij}\langle\delta\psi_i|\psi_j\rangle \ .$$

$$(23)$$

In stationary conditions $\delta G = 0$; since the variations $\delta\psi_i^*$ are arbitrary, we have that the quantity multiplying any chosen $\delta\psi_i^*$ in Eq. (23) must be zero. We obtain that the best spin-orbitals composing the ground determinantal state satisfy the set of non-linear integro-differential equations

$$\left[\frac{\mathbf{p}^2}{2m} + V_{\text{nucl}}(\mathbf{r}) + V_{\text{coul}}(\mathbf{r}) + V_{\text{exch}}\right]\psi_i = \sum_j^{(\text{occ})} \varepsilon_{ij}\psi_j \;, \tag{24a}$$

with the Coulomb and exchange operators defined as

$$V_{\text{coul}}\psi_i(\mathbf{r}\sigma) = \sum_j^{(\text{occ})} \psi_i(\mathbf{r}\sigma) \int \psi_j^*(\mathbf{r}_2\sigma_2)\frac{e^2}{|\mathbf{r}-\mathbf{r}_2|}\psi_j(\mathbf{r}_2\sigma_2)\,d(\mathbf{r}_2\sigma_2) \tag{24b}$$

$$V_{\text{exch}}\psi_i(\mathbf{r}\sigma) = -\sum_j^{(\text{occ})} \psi_j(\mathbf{r}\sigma) \int \psi_j^*(\mathbf{r}_2\sigma_2)\frac{e^2}{|\mathbf{r}-\mathbf{r}_2|}\psi_i(\mathbf{r}_2\sigma_2)\,d(\mathbf{r}_2\sigma_2) \tag{24c}$$

(the sum over $j = 1, 2, \ldots, N$ has been relabelled as sum over the occupied spin-orbitals).

If we perform explicitly the integration over the spin variables in Eq. (24b), we see that the Coulomb operator is the standard classical Hartree potential $V_{\text{coul}}(\mathbf{r})$, defined by Eq. (8). The exchange operator of Eq. (24c) (also improperly called "exchange potential") is an integral operator, since the wavefunction on which it operates enters under integral sign. If the integration over the spin variables is performed in Eq. (24c), it is seen that the occupied spin-orbitals contributing to $V_{\text{exch}}\psi_i(\mathbf{r}\sigma)$ are only the *spin-orbitals with spin parallel to* $\psi_i(\mathbf{r}\sigma)$.

We can always perform a unitary transformation within the basis of occupied spin-orbitals ψ_i to put the matrix ε_{ij} in diagonal form, $\varepsilon_{ij} = \varepsilon_i\delta_{ij}$; this is possible because both the Coulomb and exchange operators (24) keep the same form under any unitary transformation. We arrive thus at the *canonical Hartree–Fock equations*

$$\left[\frac{\mathbf{p}^2}{2m} + V_{\text{nucl}}(\mathbf{r}) + V_{\text{coul}}(\mathbf{r}) + V_{\text{exch}}\right]\psi_i = \varepsilon_i\psi_i \;; \tag{25a}$$

the electronic Coulomb and exchange potentials, again expressed by Eq. (24b) and Eq. (24c), can be written in the form

$$V_{\text{coul}}(\mathbf{r}) = \sum_j^{(\text{occ})} \langle\psi_j(\mathbf{r}_2\sigma)\,|\,\frac{e^2}{|\mathbf{r}-\mathbf{r}_2|}\,|\,\psi_j(\mathbf{r}_2\sigma)\rangle \tag{25b}$$

and

$$V_{\text{exch}}\psi_i(\mathbf{r}\sigma) = -\sum_j^{(\text{occ})} \langle\psi_j(\mathbf{r}_2\sigma)\,|\,\frac{e^2}{|\mathbf{r}-\mathbf{r}_2|}\,|\,\psi_i(\mathbf{r}_2\sigma)\rangle\,\psi_j(\mathbf{r}\sigma) \;. \tag{25c}$$

The Hartree–Fock equations can be written in a more compact form by introducing the Fock operator

$$F = h(\mathbf{r}) + V_{\text{coul}}(\mathbf{r}) + V_{\text{exch}} \;, \tag{26a}$$

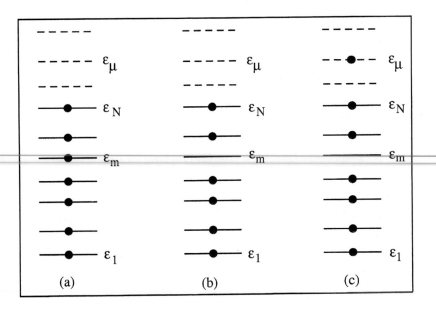

Fig. 1 Schematic representation (a) of occupied spin-orbitals (solid lines), in increasing order of energy, for the ground state of an N-electron system in the Hartree–Fock approximation; virtual spin-orbitals are represented with broken lines. (b) represents an ionized state of the N-electron system, with an electron removed from the occupied spin-orbital ψ_m (and transferred to infinity). (c) represents an excited state of the N-electron system, with an electron removed from the occupied spin-orbital ψ_m and transferred to the virtual spin-orbital ψ_μ.

where $h(\mathbf{r}) = \mathbf{p}^2/2m + V_{\text{nucl}}(\mathbf{r})$; we have

$$F\psi_i = \varepsilon_i \psi_i . \tag{26b}$$

The Fock operator is defined in terms of the occupied spin-orbitals ψ_i and must thus be determined in a self-consistent way (usually with appropriate iterations). The Fock operator reduces the many-electron problem to a one-particle problem (within the stated approximations); although this is a major formal and practical achievement, the highest caution must be applied for a correct interpretation of the meaning of eigenfunctions and eigenvalues of the Fock operator in connection with the original many-electron problem under investigation; we thus analyse further some general properties of the Hartree–Fock equations.

5.2 Ground-state energy, ionization energies and transition energies

Consider the Hartree–Fock approximation for a system of N interacting electrons and suppose we know (at least in principle) the exact self-consistent Fock operator of the system, its eigenfunctions ψ_i and the spin-orbital energies ε_i. The N eigenfunctions of lower energy of the Fock operator represent the occupied spin-orbitals, used to construct the Fock operator itself; all the other eigenfunctions of the operator F are unoccupied, do not enter in the construction of F itself, and are called *virtual spin-orbitals*. In Fig. 1 we represent schematically occupied and virtual spin-orbitals, in increasing order of energy, for an interacting system of N electrons; ionized states and excited states of the N-electron system are also schematically represented.

Ground-state electronic energy

The ground-state electronic energy E_0 is given by expression (22), where the ψ_i satisfy the Hartree–Fock equations (26); from Eqs. (26) we have for the energies $\varepsilon_i = \langle \psi_i | F | \psi_i \rangle$ the following expression

$$\varepsilon_i = \langle \psi_i | h | \psi_i \rangle + \sum_{j}^{\text{(occ)}} \left[\langle \psi_i \psi_j | \frac{e^2}{r_{12}} | \psi_i \psi_j \rangle - \langle \psi_i \psi_j | \frac{e^2}{r_{12}} | \psi_j \psi_i \rangle \right] . \qquad (27)$$

From Eq. (22) and Eq. (27), we see that the ground-state electronic energy in the Hartree–Fock approximation is given by

$$E_0^{\text{(HF)}} = \sum_{i}^{\text{(occ)}} \varepsilon_i - \frac{1}{2} \sum_{ij}^{\text{(occ)}} \left[\langle \psi_i \psi_j | \frac{e^2}{r_{12}} | \psi_i \psi_j \rangle - \langle \psi_i \psi_j | \frac{e^2}{r_{12}} | \psi_j \psi_i \rangle \right] . \qquad (28)$$

Equation (28) shows that the ground-state electronic energy *is not equal to the sum of the Hartree–Fock spin-orbital energies of the occupied states*; this can be understood intuitively from the fact that otherwise the Coulomb and exchange interactions would be counted twice. The total energy of the electronic-nuclear system in the Hartree–Fock approximation is obtained adding to $E_0^{\text{(HF)}}$ the nucleus-nucleus Coulomb repulsion $(1/2) \sum_{I \neq J} z_I z_J e^2 / R_{IJ}$ which constitutes the last term of the Hamiltonian of Eq. (1).

It is to be noticed that, in agreement with the variational principle, the Hartree–Fock energy $E_0^{\text{(HF)}}$ is *higher* than the exact ground-state energy $E_0^{\text{(exact)}}$ of the many-body system (the difference $E_0^{\text{(HF)}} - E_0^{\text{(exact)}}$ is called *correlation energy*). In spite of the importance and achievements of the Hartree–Fock approximation, corrections beyond it are often to be considered; this is due to the fact that a single determinantal state, even when built up with the best possible orbitals, remains in general a rather poor representation of the complicated ground-state wavefunction of a many-particle system.

Ionization energy and Koopmans' theorem

Consider an (ideal) ionization process in which an electron in the occupied state ψ_m is removed from the system and transferred to infinity (i.e. far enough from the electronic system) with zero kinetic energy; it is assumed that no modification (or "relaxation") of the other electronic orbitals occurs in the ionization process.

In the Hartree–Fock approximation, the energy $E_0^{\text{(HF)}}(N)$ of the ground state is given by Eq. (22); the energy $E_0^{\text{(HF)}}(N-1)$ of the ionized states given by the same expression (22), provided the spin-orbital ψ_m is excluded from the sums. We have for the energy difference

$$E_0^{(\text{HF})}(N) - E_0^{(\text{HF})}(N-1)$$

$$= \langle \psi_m | h | \psi_m \rangle + \sum_j^{(\text{occ})} \left[\langle \psi_m \psi_j | \frac{e^2}{r_{12}} | \psi_m \psi_j \rangle - \langle \psi_m \psi_j | \frac{e^2}{r_{12}} | \psi_j \psi_m \rangle \right] \equiv \varepsilon_m \ .$$

We thus obtain the following important result, known as *Koopmans' theorem: the energy required to remove (without relaxation) an electron from the spin-orbital ψ_m is simply the opposite of the Hartree–Fock eigenvalue ε_m.* The parameters ε_m appearing in the Hartree–Fock equations (25) as purely formal Lagrange multipliers are thus rescued to physical interpretation by the Koopmans' theorem.

Excited states and transition energies

We consider now the ground state $\Psi_0 = A\{\psi_1 \psi_2 \ldots \psi_m \ldots \psi_N\}$, and the excited state $\Psi_{\mu,m} = A\{\psi_1 \psi_2 \ldots \psi_\mu \ldots \psi_N\}$, where an electron has been removed from the occupied spin-orbital ψ_m and transferred in the previously unoccupied spin-orbital ψ_μ (we assume again that all the other occupied orbitals are unchanged); we want to determine the transition energy

$$\Delta E = \langle \Psi_{\mu,m} | H_e | \Psi_{\mu,m} \rangle - \langle \Psi_0 | H_e | \Psi_0 \rangle \ .$$

The quantity $\langle \Psi_0 | H_e | \Psi_0 \rangle$ is given by expression (22); with straightforward manipulations we can obtain also $\langle \Psi_{\mu,m} | H_e | \Psi_{\mu,m} \rangle$; the transition energy ΔE then becomes

$$\Delta E = \langle \psi_\mu | h | \psi_\mu \rangle - \langle \psi_m | h | \psi_m \rangle$$

$$+ \sum_j^{(\text{occ})} \left[\langle \psi_\mu \psi_j | \frac{e^2}{r_{12}} | \psi_\mu \psi_j \rangle - \langle \psi_\mu \psi_j | \frac{e^2}{r_{12}} | \psi_j \psi_\mu \rangle \right]$$

$$- \left[\langle \psi_\mu \psi_m | \frac{e^2}{r_{12}} | \psi_\mu \psi_m \rangle - \langle \psi_\mu \psi_m | \frac{e^2}{r_{12}} | \psi_m \psi_\mu \rangle \right]$$

$$- \sum_j^{(\text{occ})} \left[\langle \psi_j \psi_m | \frac{e^2}{r_{12}} | \psi_j \psi_m \rangle - \langle \psi_j \psi_m | \frac{e^2}{r_{12}} | \psi_m \psi_j \rangle \right] \ .$$

Using Eq. (27) for the spin-orbital energies, we obtain

$$\Delta E = \varepsilon_\mu - \varepsilon_m - \left[\langle \psi_\mu \psi_m | \frac{e^2}{r_{12}} | \psi_\mu \psi_m \rangle - \langle \psi_\mu \psi_m | \frac{e^2}{r_{12}} | \psi_m \psi_\mu \rangle \right] \ . \tag{29}$$

The physical meaning of Eq. (29) is transparent indeed: to promote an electron from the occupied spin-orbital ψ_m to the unoccupied spin-orbital ψ_μ we need an energy equal to the difference $\varepsilon_\mu - \varepsilon_m$ of the spin-orbital energies, corrected by the *Coulomb* and *exchange* energy between the "extra electron" in the spin-orbital ψ_μ and the "missing electron" (hole) left in the initial spin-orbital ψ_m.

In crystals, ψ_μ and ψ_m are normalized Bloch functions, and thus the bielectronic integrals are of the order of O(1/N) (where N is the number of unit cells of the crystal). This appears to provide a good justification for neglecting the "electron-hole interaction", expressed by the bielectronic integrals in square brackets in Eq. (29); thus when the wavefunctions are itinerant, we expect $\Delta E \approx \varepsilon_\mu - \varepsilon_m$ for the transition energies. Although for Bloch functions, bielectronic integrals are of the order of O(1/N), we must remember that also the allowed $\mathbf{k}$ vectors (and hence wavefunctions for each band) in the Brillouin zone equal the number N of unit cells of the crystal; thus it is not a priori justified to neglect N contributions of order O(1/N), and the exciton theory will discuss how to include electron-hole corrections, whenever necessary (see Section VII-1).

As a final property we wish to highlight the matrix elements $\langle \Psi_{\mu,m} | H_e | \Psi_0 \rangle$ of H_e between the Hartree–Fock *ground state and any singly excited state* are zero; this is known as *Brillouin theorem*.

5.3   Hartree–Fock equations and transition energies in closed-shell systems

In the Hartree–Fock theory developped until now, no restriction has been considered on the spin part of the spin-orbitals entering the single Slater determinant, adopted to approximate the ground state wavefunction of the electronic system. For closed-shell systems, the trial Slater determinant can be written in the more specific form

$$\Psi_0 = A\{\phi_1\alpha\ \phi_1\beta\ \ldots\phi_m\alpha\ \phi_m\beta\ldots\phi_{N/2}\alpha\ \phi_{N/2}\beta\}\ , \tag{30}$$

where $\phi_i(\mathbf{r})$ $(i = 1, 2, \ldots N/2)$ indicate orthonormal orbitals. Notice that the state Ψ_0 has total spin component S_z equal to zero and total spin S equal to zero.

The general Hartree–Fock equations (25), carrying out explicitly the summation over spin variables, take the form

$$\left[\frac{\mathbf{p}^2}{2m} + V_{\text{nucl}}(\mathbf{r}) + V_{\text{coul}}(\mathbf{r}) + V_{\text{exch}} \right] \phi_i(\mathbf{r}) = \varepsilon_i \phi_i(\mathbf{r}) \tag{31a}$$

$(i = 1, 2, \ldots, N/2)$, with

$$V_{\text{coul}}(\mathbf{r}) = 2 \sum_j^{N/2} \langle \phi_j(\mathbf{r}_2)\,|\,\frac{e^2}{|\mathbf{r} - \mathbf{r}_2|}\,|\,\phi_j(\mathbf{r}_2)\rangle \tag{31b}$$

$$V_{\text{exch}}\phi_i(\mathbf{r}) = - \sum_j^{N/2} \langle \phi_j(\mathbf{r}_2)\,|\,\frac{e^2}{|\mathbf{r} - \mathbf{r}_2|}\,|\,\phi_i(\mathbf{r}_2)\rangle\,\phi_j(\mathbf{r})\ . \tag{31c}$$

We have already seen that ε_i can be interpreted as ionization energy in the Koopmans' approximation.

We wish to discuss now the transition energies in closed-shell systems. When in the determinantal state (30), an occupied orbital $\phi_m(\mathbf{r})$ is replaced by a virtual orbital

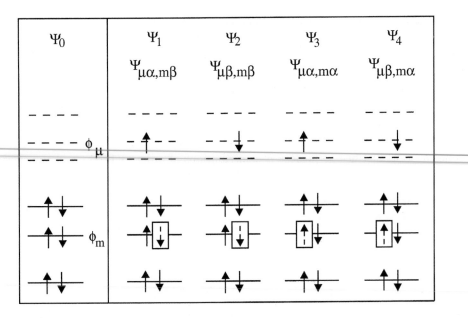

Fig. 2 Schematic representation of Hartree–Fock occupied and virtual states (here supposed to be discrete) of a closed-shell electronic system. In the four trial excited states $\Psi_1, \Psi_2, \Psi_3, \Psi_4$, an electron is promoted from the occupied orbital $\phi_m(\mathbf{r})$ to the virtual orbital $\phi_\mu(\mathbf{r})$.

$\phi_\mu(\mathbf{r})$, we have the following four possible wavefunctions for excited states depending on the initial or final spin functions:

$$\Psi_1 = \Psi_{\mu\alpha,m\beta} \qquad \Psi_2 = \Psi_{\mu\beta,m\beta} \qquad \Psi_3 = \Psi_{\mu\alpha,m\alpha} \qquad \Psi_4 = \Psi_{\mu\beta,m\alpha} \ . \qquad (32)$$

The ground state Ψ_0 and the four trial excited states $\Psi_1, \Psi_2, \Psi_3, \Psi_4$, where an electron is created in ϕ_μ and a hole is left in ϕ_m, are shown schematically in Fig. 2.

We can now diagonalize the many-electron Hamiltonian H_e on the four determinantal states (32) and construct the linear combinations with definite values of the total spin component S_z and total spin multiplicity S. For this purpose, we calculate the matrix M, whose elements are

$$M_{ij} = \langle \Psi_i | H_e | \Psi_j \rangle \ (i,j = 1,2,3,4) \ . \qquad (33a)$$

Let us begin for instance with M_{11}; for its evaluation we can use Eq. (29) (with appropriate relabelling of orbitals). If we indicate with E_0 the ground state energy, we obtain

$$M_{11} = E_0 + \varepsilon_\mu - \varepsilon_m - \left[\langle \phi_{\mu\alpha}\phi_{m\beta} | \frac{e^2}{r_{12}} | \phi_{\mu\alpha}\phi_{m\beta} \rangle - \langle \phi_{\mu\alpha}\phi_{m\beta} | \frac{e^2}{r_{12}} | \phi_{\mu\beta}\phi_{m\alpha} \rangle \right]$$

$$= E_0 + \varepsilon_\mu - \varepsilon_m - Q_{\mu m} \ ,$$

where $Q_{\mu m}$ is the Coulomb bielectronic integral

$$Q_{\mu m} = \langle \phi_\mu(\mathbf{r}_1)\phi_m(\mathbf{r}_2)| \frac{e^2}{r_{12}} |\phi_\mu(\mathbf{r}_1)\phi_m(\mathbf{r}_2)\rangle \ .$$

For the matrix element M_{22} we have

$$M_{22} = E_0 + \varepsilon_\mu - \varepsilon_m - Q_{\mu m} + J_{\mu m} \ ,$$

where $J_{\mu m}$ is the exchange bielectronic integral

$$J_{\mu m} = \langle \phi_\mu(\mathbf{r}_1)\phi_m(\mathbf{r}_2)| \frac{e^2}{r_{12}} |\phi_m(\mathbf{r}_1)\phi_\mu(\mathbf{r}_2)\rangle \ .$$

With similar methods we can elaborate all the matrix elements (33a) and we obtain for the matrix M the expression

$$M = E_0 + \varepsilon_\mu - \varepsilon_m + \begin{pmatrix} -Q_{\mu m} & 0 & 0 & 0 \\ 0 & -Q_{\mu m} + J_{\mu m} & J_{\mu m} & 0 \\ 0 & J_{\mu m} & -Q_{\mu m} + J_{\mu m} & 0 \\ 0 & 0 & 0 & -Q_{\mu m} \end{pmatrix} \tag{33b}$$

(it is implicitly understood that the scalar quantity $E_0 + \varepsilon_\mu - \varepsilon_m$ is multiplied by the 4×4 identity matrix before addition is performed).

It is immediately seen that the diagonalization of the matrix (33b) gives one triplet state with energy

$$E_{\text{triplet}} = E_0 + \varepsilon_\mu - \varepsilon_m - Q_{\mu m} \ , \tag{34a}$$

and one singlet state with energy

$$E_{\text{singlet}} = E_0 + \varepsilon_\mu - \varepsilon_m - Q_{\mu m} + 2J_{\mu m} \ . \tag{34b}$$

For *triplet states* ($S = 1$; $S_z = 1, 0, -1$), the appropriate combinations of the basis functions (32) with definite spin component are

$$\Psi_{\mu\alpha,m\beta} \ , \qquad \frac{1}{\sqrt{2}}(\Psi_{\mu\alpha,m\alpha} - \Psi_{\mu\beta,m\beta}) \ , \qquad \Psi_{\mu\beta,m\alpha} \ .$$

For the *singlet state* ($S = 0$; $S_z = 0$) the appropriate combination is

$$\frac{1}{\sqrt{2}}(\Psi_{\mu\alpha,m\alpha} + \Psi_{\mu\beta,m\beta}) \ .$$

For brevity, we indicate trial excited states of a given spin multiplicity by $\Psi^{(S)}_{\mu,m}$ with $S = 0$ for the singlet state and $S = 1$ for the triplet state (for the triplet, we select arbitrary one of the three possible partners); equations (34) can be summarized in the compact form

$$\langle \Psi^{(S)}_{\mu,m}|H_e|\Psi^{(S)}_{\mu,m}\rangle = E_0 + \varepsilon_\mu - \varepsilon_m - \langle\phi_\mu\phi_m|\frac{e^2}{r_{12}}|\phi_\mu\phi_m\rangle + 2\delta_{S,0}\langle\phi_\mu\phi_m|\frac{e^2}{r_{12}}|\phi_m\phi_\mu\rangle \ . \tag{35}$$

Since exchange integrals are definite positive quantities, we see that the triplet (magnetic) state is lower in energy than the singlet (non-magnetic) state.

Before concluding we wish also to mention other matrix elements of the operator H_e among single excited states for closed-shell systems. With methods completely similar to those used to prove Eq. (35), we obtain

$$\langle\Psi_{\mu,m}^{(S)}|H_e|\Psi_{\nu,n}^{(S)}\rangle = -\langle\phi_\mu\phi_n|\frac{e^2}{r_{12}}|\phi_\nu\phi_m\rangle + 2\delta_{S,0}\langle\phi_\mu\phi_n|\frac{e^2}{r_{12}}|\phi_m\phi_\nu\rangle . \tag{36}$$

We will need these results in the theory of excitons in solids of Section VII-1.

5.4 Hartree–Fock–Slater and Hartree–Fock–Roothaan approximations

The major obstacle to the solution of the Hartree–Fock equations is the integro-differential nature of the Fock operator. Various physical and technical elaborations have been devised for it in the literature; it would be impossible to describe all of them, but we wish here to mention, in their general lines, the ideas of the Slater approximation and the Roothaan approximation, because of their basic role in solid state physics and in atomic and molecular physics, respectively.

Hartree–Fock–Slater approximation

The Slater approximation starts from the fact that the exchange integral operator in the homogeneous free-electron gas can be well approximated with a local potential (as shown in Section 7). Then Slater proposes to replace, also for non-homogeneous systems, the exchange integral operator with the ordinary local potential, corresponding to the free-electron gas of the same local density $n(\mathbf{r})$; in this way the exchange operator takes the form

$$V_{\text{exch}}^{(\text{Slater})}(\mathbf{r}) = -\frac{3}{2}\frac{e^2}{\pi}[3\pi^2 n(\mathbf{r})]^{1/3} .$$

With the above approximation, the Fock operator (26) becomes an ordinary differential operator with a local potential to be determined self-consistently. The method (with appropriate implementations) has been of wide use mainly in solid state physics.

The Hartree–Fock–Slater approximation becomes particularly agile when applied to atoms, because of the spherical symmetry of the (average) atomic electron density $n(r)$. The effective atomic potential in the Hartree–Fock–Slater approximation is related to $n(r)$ in the form

$$V^{(\text{HFS})}(r) = -\frac{Ze^2}{r}+\frac{e^2}{r}\int_0^r 4\pi r'^2 n(r')\,dr' + e^2\int_r^\infty 4\pi r' n(r')\,dr' - \frac{3}{2}\frac{e^2}{\pi}[3\pi^2 n(r)]^{1/3} . \tag{37}$$

The first term in the right-hand side of Eq. (37) is the nuclear potential, the second and the third are the Hartree potential of the spherically symmetric charge distribution $(-e)n(r)$; the last is the Slater local exchange potential [to be precise the potential $V^{(\text{HFS})}(r)$ should be appropriately corrected to preserve the asymptotic behaviour $-e^2/r$, when acting on occupied wavefunctions]. Usually the Numerov method is adopted to solve the radial Schrödinger equation; self-consistency of wavefunctions and potential can be achieved with reasonable modest computational labour. Complete calculations (programme code included) for all atoms is reported, for instance, by F. Herman and S. Skillman "Atomic Stucture Calculations" (Prentice Hall, Englewood Cliffs, New Jersey 1963).

The Hartree–Fock–Roothaan method

Because of its importance in theoretical chemical physics, it is worthwhile to mention the method of Roothaan for the solution of the Hartree–Fock equations. This method introduces a finite number of localized basis spin-orbitals, on which to expand each Hartree–Fock orbital; also the integro-differential Hartree–Fock operator is represented on the chosen basis set, and the Hartree–Fock equations are thus converted into a set of self-consistent algebraic equations in the expansion coefficients, to be solved by matrix techniques. A nice feature of the Hartree–Fock–Roothaan procedure is that the quality of the results can be tested by enlarging the basis set; a limitation of the method is that numerical labour rapidly increases with the number of electrons; in practice, besides atoms, only molecules or clusters with small to moderate number of electrons (up to a few tens or so) can be handled (unless further simplifying assumptions are inserted).

As a final point we should remember the basis functions used in most calculations. The localized atomic functions are often expressed in terms of Slater-type functions; the normalized Slater-type functions are defined as

$$\phi_{nlm}(\mathbf{r}) = \frac{(2\alpha)^{n+(1/2)}}{\sqrt{(2n)!}} \, r^n e^{-\alpha r} Y_{lm}(\mathbf{r}) \, ,$$

where $Y_{lm}(\mathbf{r})$ denote normalized harmonic functions. Optimized coefficients for atoms are reported for instance by E. Clementi and C. Roetti "Roothaan–Hartree–Fock Atomic Wavefunctions" Atomic Data Nuclear Data Tables **14**, 177 (1974).

Another convenient set is constituted by Gaussian-type functions; normalized Gaussian orbitals are defined as

$$g_{nlm}(\mathbf{r}) = \left(\frac{2}{\pi}\right)^{1/4} \frac{2^{n+(1/2)}}{\sqrt{(2n-1)!!}} \, \alpha^{(2n+1)/4} \, r^{n-1} e^{-\alpha r^2} Y_{lm}(\mathbf{r}) \, .$$

This choice makes it possible to perform many integrals analytically [see for instance I. Shavitt in "Methods in Computational Physics" (eds. B. Adler, S. Fernbach and M. Rotenberg, Academic Press, New York 1963)]. On the other hand a large number of Gaussians are in general needed to represent atomic-like wavefunctions [for further aspects see for instance S. Huzinaga, J. Andzelm, M. Klobukowski, E. Radzio-Andzelm, Y. Sakai and H. Takewaki (eds) "Gaussian Basis Sets for Molecular Calculation" (Elsevier, Amsterdam 1984); R. Poirier, R. Kari and I. Csizmadia "Handbook of Gaussian Basis Sets" (Elsevier, Oxford 1985)].

6   Overview of approaches beyond the one-electron approximation

In the one-electron approximation, electrons with parallel spin are correctly taken apart by the use of antisymmetrized wavefunctions (see the discussion on Eq. 12); thus, the basic inadequacy of the one-electron approximation is that electrons with opposite spin remain uncorrelated. The depletion of parallel spin particles around a reference electron occurs in real space as a consequence of Fermi statistics (this fact is what is essentially referred in the literature as the *exchange hole or Fermi hole*). For antiparallel spin, correlation is not included, and attempts beyond the Hartree–Fock approximation are mainly motivated to overcome this limitation.

Despite its limitations, the Hartree–Fock approximation is a milestone in the many-body electron problem from several points of view. From a *conceptual point of view*, it allows to separate exchange effects from correlation effects; from a *technical point of view*, it introduces the technicities to face the demanding self-consistent equations that have found applications also in many other fields; from a *practical point of view*, a number of problems can be understood (at least qualitatively) within this approximation.

In most many-body systems, however, the Hartree–Fock method is insufficient for the desired accuracy; in these cases the Hartree–Fock approximation is usually considered as a starting point before more sophisticated approaches are implemented. It is outside the scope of this book to present a thorough discussion of the main lines of improvements and their physical or heuristic motivations; thus we refrain from any quantitative discussion and simply mention some developments.

One of the most important ways to go beyond the Hartree–Fock theory is offered by the quantum chemistry approaches, which consider the Hartree–Fock method as a starting point to generate a convenient set of one-electron spin-orbitals, and then insert excited configurations. These methods have different names such as configuration interaction approach, multi-configuration self-consistent field, many-body perturbation theory, depending on the adopted implementations. In essence, besides the ground Hartree–Fock state, one considers single excited states (in which one of the occupied spin-orbitals is replaced by a virtual spin-orbital), double excited states, etc. and diagonalizes the many-body Hamiltonian among a judiciously chosen set of (trial) excited states. This procedure has been widely used in theoretical quantum chemistry for atoms, molecules and clusters, to obtain a better description of the ground-state and the lowest lying excitations of the many-electron system. In solid state physics, *excitons* in insulators and semiconductors are often studied in this spirit (see Chapter VII).

For open-shell systems, the situation may become rather complicated, when several spin-configurations are expected to produce levels within narrow energy intervals. In some cases, it is possible to describe the low-lying states by approximate *equivalent Hamiltonians* (for instance of the type of the Heisenberg spin Hamiltonian in magnetism, see Chapter XVII).

A very important line of progress, is given by theoretical developments of the Green's function many-body formalism. In essence the main novelties with respect to the one-electron formalism, is the screening of the exchange potential and the introduction of the Coulomb hole (COHSEX: Coulomb hole and screened exchange approximation). This line (though not routine at all) has lead a long way towards a better quantitative description of correlation effects in solids [see for instance L. Hedin and S. Lundqvist, Solid State Physics **23**, 1 (1969) and references quoted therein].

A special importance has been acquired by *the density functional theory*, in the study of the ground-state properties of a many-body system. This theory focuses on the ground-state electron density (rather than eigenfunctions) and provides a number of highly accurate results for the ground-state properties. We discuss the main aspects of the density functional theory later in this chapter. Finally we mention the important role assumed by the quantum Monte Carlo method in many-body calculations: it gives practically "exact" results for small systems and for the homogeneous electron gas. The Monte Carlo method has been used also as a precious test of the approximations introduced within the density functional theory in atoms or molecules with small number of electrons.

7   Electronic properties and phase diagram of the homogeneous electron gas

In this section we discuss some relevant properties of the *homogeneous electron gas or fluid*; this many-body system is constituted by *interacting electrons embedded in a uniform neutralizing background of positive charges (jellium model)*. The homogeneous electron gas is of major interest from several points of view. At the most elementary level, it provides an orientative description of the conduction electrons in simple metals (Sommerfeld model of Chapter III). It is the prototype system, on which many-body theory has been developed and applied. With lowering density, the electron gas displays transitions to magnetic ordering and to Wigner crystallization. Furthermore, the results of the electron gas are a basic ingredient also for the treatment of real materials (as in the local density approximation to the density functional theory, discussed in Section 8).

Normal ground-state of the homogeneous electron gas in the Hartree–Fock approximation

It is instructive to analyse the jellium model within the Hartree–Fock theory, even if it is well known that this theory is completely inadequate for the description of the electron gas. Consider a system of N electrons, in the volume V, embedded in a uniform neutralizing background of positive charges. We indicate with $n = N/V$ the electron density and with r_s the dimensionless parameter related to the electronic density n by the equality $(4\pi/3)r_s^3 a_B^3 = 1/n$, where a_B is the Bohr radius; $r_s a_B$ represents the radius of the sphere containing (in average) one electron. The case of the non-interacting electron gas, i.e. the case that the electron–electron interaction can be neglected, constitutes the free-electron Sommerfeld model, characterized by a parabolic energy dispersion curve $E(\mathbf{k}) = \hbar^2 k^2/2m$ (schematically indicated in Fig. 3). We discuss now within the Hartree–Fock framework the interacting electron gas.

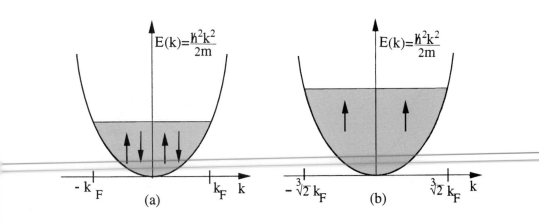

Fig. 3 (a) Schematic representation of the *normal state* of the free-electron gas. (b) Schematic representation of the *fully polarized state* of the free-electron gas.

The Hartree–Fock equations of the homogeneous (interacting) electron gas can be solved exactly using plane waves $W(\mathbf{k}_i, \mathbf{r}) = (1/\sqrt{V}) \exp(i\mathbf{k}_i \cdot \mathbf{r})$. The normal ground-state of the interacting gas is approximated with a single Slater determinant, formed by doubly occupied plane waves with wavevectors filling the Fermi sphere of radius k_F; we have

$$\Psi_0 = A\{W_{\mathbf{k}_1}\alpha\, W_{\mathbf{k}_1}\beta\, \ldots W_{\mathbf{k}_{N/2}}\alpha\, W_{\mathbf{k}_{N/2}}\beta\} \, . \tag{38}$$

In the jellium model $V_{\text{nucl}}(\mathbf{r})$ and $V_{\text{coul}}(\mathbf{r})$ exactly cancel, and the Fock operator becomes

$$F = \frac{\mathbf{p}^2}{2m} + V_{\text{exch}} \, , \tag{39}$$

where V_{exch} is the exchange integral operator of type (31c) built with the occupied plane waves. In the case the Hartree approximation is considered, the exchange operator V_{exch} is omitted and one recovers the free-electron Sommerfeld model.

It is well known that the kinetic energy operator $\mathbf{p}^2/2m$ is diagonal in a plane wave representation. We now show that also the exchange integral operator, built up with plane waves, is diagonal in a plane wave representation. In fact we have

$$V_{\text{exch}} e^{i\mathbf{k}\cdot\mathbf{r}} = -\sum_{\mathbf{q}}^{(\text{occ})} \frac{1}{\sqrt{V}} e^{i\mathbf{q}\cdot\mathbf{r}} \int \frac{1}{\sqrt{V}} e^{-i\mathbf{q}\cdot\mathbf{r}'} \frac{e^2}{|\mathbf{r}-\mathbf{r}'|} e^{i\mathbf{k}\cdot\mathbf{r}'}\, d\mathbf{r}'$$

$$= -e^{i\mathbf{k}\cdot\mathbf{r}} \frac{1}{V} \sum_{\mathbf{q}}^{(\text{occ})} \int e^{-i(\mathbf{k}-\mathbf{q})\cdot(\mathbf{r}-\mathbf{r}')} \frac{e^2}{|\mathbf{r}-\mathbf{r}'|}\, d\mathbf{r}'$$

$$= -e^{i\mathbf{k}\cdot\mathbf{r}} \frac{1}{V} \sum_{q<k_F} \frac{4\pi e^2}{|\mathbf{k}-\mathbf{q}|^2} \, . \tag{40}$$

The explicit evaluation of the sum in Eq. (40) is given in Appendix C.

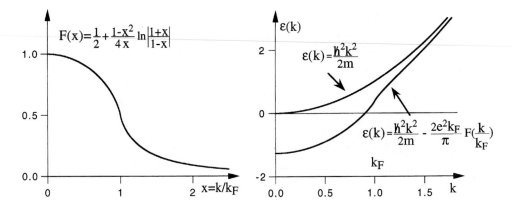

Fig. 4 (a) Schematic plot of the function $F(x)$. (b) Kinetic energy and Hartree–Fock orbital energy as a function of the wavevector k for the homogeneous electron gas. Energies are in Rydbergs, k is in units of a_B^{-1} (inverse Bohr radius), and we have taken $k_F = 1/a_B$.

From Eq. (40) and Eq. (C3), we see that plane waves are eigenfunctions of V_{exch} and satisfy the eigenvalue equation

$$V_{\text{exch}} e^{i\mathbf{k}\cdot\mathbf{r}} = -\frac{2e^2 k_F}{\pi} F\left(\frac{k}{k_F}\right) e^{i\mathbf{k}\cdot\mathbf{r}} \ , \tag{41a}$$

where k_F is the Fermi wavevector, and

$$F(x) = \frac{1}{2} + \frac{1 - x^2}{4x} \ln\left|\frac{1 + x}{1 - x}\right| \ . \tag{41b}$$

The plane waves are thus also eigenfunctions of the Fock operator (39) with eigenvalues

$$\varepsilon(k) = \frac{\hbar^2 k^2}{2m} - \frac{2e^2 k_F}{\pi} F\left(\frac{k}{k_F}\right) \ . \tag{42}$$

Notice that the exchange contribution gradually vanishes as the electron wavevector k overcomes the Fermi wavevector k_F. The behaviour of $F(x)$ and $\varepsilon(k)$ are given in Fig. 4.

From Eq. (42), we can see that the Hartree–Fock treatment of the homogeneous electron gas presents quite serious drawbacks. For instance, we notice that $d\varepsilon(k)/dk$ presents an (unphysical) logarithmic divergence at $k = k_F$; this divergence would make the density-of-states vanish for $\varepsilon \approx \varepsilon_F$; on the contrary, experimental evidence shows that the density-of-states varies rather regularly throughout the whole conduction band, Fermi energy region included. The Hartree–Fock results also produce unphysical large widths of conduction bands in metals (and also unphysical large band gaps in insulators up to a factor 2 or so).

In spite of these and other inadequacies of the Hartree–Fock theory, it is interesting to evaluate the Hartree–Fock ground-state energy, as this constitutes an *upper bound* of the exact ground-state energy of the many-body system, independently on how good or how bad the Hartree–Fock approximation actually might be. The total Hartree–Fock ground-state energy can be obtained summing up the orbital energies given by Eq. (42) and introducing the factor 1/2 in the exchange electron-electron contribution (to avoid its double counting); we have

$$E_0^{(\mathrm{HF})} = 2 \sum_{k<k_F} \frac{\hbar^2 k^2}{2m} - \sum_{k<k_F} \frac{2e^2 k_F}{\pi} F\left(\frac{k}{k_F}\right) . \tag{43}$$

In Eq. (43), we can perform explicitly the sum over k (see Appendix C) and obtain

$$E_0^{(\mathrm{HF})} = N \left[\frac{3}{5} \frac{\hbar^2 k_F^2}{2m} - \frac{3}{4} \frac{e^2 k_F}{\pi} \right] . \tag{44a}$$

Using the dimensionless parameter r_s and the relations $\hbar^2/2ma_B^2 = 1\,\mathrm{Rydberg} = e^2/2a_B$, $k_F a_B = 1.92/r_s$, we have for the ground-state energy per particle

$$\frac{E_0^{(\mathrm{HF})}}{N} = \frac{2.21}{r_s^2} - \frac{0.916}{r_s} \quad (\text{in Rydberg}) . \tag{44b}$$

Since $2 < r_s < 6$ in ordinary metals, we see that the exchange energy is an important contribution to the binding energy of metals. In the high density limit (i.e. $r_s < 1$) the exact leading terms in the ground-state energy are given by Eq. (44b).

Another interesting remark can be made on Eqs. (41). From Eq. (41b) we see that $F(x)$ varies smoothly (from one to one half) as k goes from zero to k_F. In the hypothetical case that we replace the function $F(x)$ by its average value 3/4 (see Eq. C5), we would have that the operator V_{exch} becomes a constant equal to

$$V_{\mathrm{exch}}^{(\mathrm{Slater})} = -\frac{3}{2} \frac{e^2}{\pi} k_F = -\frac{3}{2} \frac{e^2}{\pi} (3\pi^2 n)^{1/3} . \tag{45a}$$

Slater suggested that expression (45a) could be adopted also for non-homogeneous electronic systems of *local density* $n(\mathbf{r})$, so obtaining the famous "$n^{1/3}$ local approximation" to the exchange operator

$$V_{\mathrm{exch}}^{(\mathrm{Slater})}(\mathbf{r}) = -\frac{3}{2} \frac{e^2}{\pi} [3\pi^2 n(\mathbf{r})]^{1/3} . \tag{45b}$$

It is worthwhile to mention that the Slater local exchange approximation has had historically an important role in making many properties of actual materials accessible to the theory and to the interpretation; the reason is that the Hartree–Fock equations become ordinary differential equations, when the exchange operator is approximated with a local potential. In the literature, many attempts have been done to correct and improve the Slater expression (45b); among them we mention the once popular "X_α local approximation", where a semi-empirical parameter α multiplies the original Slater expression of Eq. (45b). All these attempts have come to an end with the advances of the density functional theory, mainly in the local density approximation (see Section 8).

Fully spin-polarized ground-state in the Hartree–Fock approximation

In the Hartree–Fock approximation, the normal ground-state of the electron gas has been pictured as a Slater determinant constituted by doubly occupied plane waves of wavevectors confined within the Fermi sphere. Besides the normal ground-state, it is of interest to consider the extreme situation in which all electrons have parallel spin; we wish to compare the ground-state energy of the "normal phase" (also called the "paramagnetic phase') with the energy of the "fully polarized phase" (also called "ferromagnetic phase").

In the free-electron model, it is evident that the normal phase is *always* more stable than the polarized phase (see Fig. 3); this is no more true in the Hartree–Fock approach. Consider in fact the Hartree–Fock ground-state energy of the normal phase given by Eq. (44a); the ground-state energy for the fully magnetized state, is obtained by Eq. (44a) with the replacement $k_F \rightarrow 2^{1/3} k_F$ (to account for the new occupied volume in reciprocal space). Thus the "fully polarized phase" is lower in energy than the "normal phase" when the condition

$$\frac{3}{5} \frac{\hbar^2}{2m} (2^{1/3} k_F)^2 - \frac{3e^2}{4\pi} (2^{1/3} k_F) < \frac{3}{5} \frac{\hbar^2}{2m} k_F^2 - \frac{3e^2}{4\pi} k_F$$

is verified. The above inequality is satisfied for

$$k_F < \frac{5}{2\pi} \frac{1}{2^{1/3} + 1} \frac{1}{a_B} \quad \text{and thus} \quad r_s > 5.45 \tag{46}$$

$(a_B = \hbar^2/me^2 = \text{Bohr radius}; k_F a_B = 1.92/r_s)$.

The Hartree–Fock approximation predicts that, at sufficiently low density, the spin-aligned state of the electrons is more stable than the normal unpolarized state. However, the critical value for the transition, estimated at $r_s \approx 5.45$ (a value still in the range of ordinary metallic densities), is completely unreliable; this is another example of the inadequacy of the Hartree–Fock method for treating quantitatively the electron gas. When electron correlation effects are taken into account in the jellium model, it is seen that the region of stability of the normal phase extends up to $r_s \approx 75$, as discussed below with Monte Carlo methods.

Considerations on Wigner crystallization

At high density (small r_s), the electron jellium becomes an (almost) ideal Fermi gas with kinetic energy dominant with respect to electrostatic repulsion of electrons; the ground-state is represented by double occupied plane waves filling the Fermi sphere, in order to minimize the electron kinetic energy. At intermediate densities, a fully polarized fluid with all the spins aligned has energy lower than the normal fluid (the presently accredited range of stability is $75 < r_s < 100$). With further increase of the mean interparticle distance $r_s a_B$, each electron becomes trapped in the potential cage created by the Coulomb repulsion of its neighbors, and we have then a Mott insulator. It is generally accepted that electrons localize in a body-centered cubic lattice. The crystallization into a regular lattice of a low density electron gas was predicted by E. P. Wigner, Phys. Rev. **46**, 1002 (1934); Trans. Faraday Soc. **34**, 678 (1938).

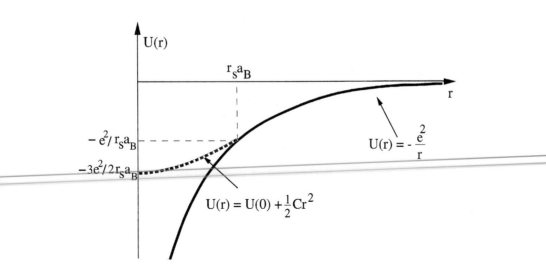

Fig. 5 Potential energy of an electron in the Coulomb field due to a uniformely charged sphere of radius $r_s a_B$ and total charge $+e$. For $0 \le r \le r_s a_B$ the potential is quadratic, with "elastic constant" $C = e^2/r_s^3 a_B^3$.

The origin of the low density Wigner crystallization can be intuitively argued as follows. In the ground state of the Wigner lattice, electrostatic repulsion of electrons is dominant on kinetic energy and the electrons perform zero-point oscillations around their lattice sites. The region in which each electron vibrates can be estimated by constructing a sphere of radius $r_s a_B$ surrounding the lattice site. Within such a sphere, the electron feels the restoring potential generated by the positive background charge within the sphere (see Fig. 5); contributions from other spheres are higher order multipole terms and are here neglected for qualitative remarks. From classical electrostatic, the potential energy of an electron, at position r from the center of a sphere of radius $r_s a_B$ with a uniform charge $+e$ is

$$U(r) = U(0) + \frac{1}{2}\frac{e^2}{r_s^3 a_B^3}r^2 \qquad 0 < r < r_s a_B \qquad (47a)$$

with $U(0) = -3e^2/2r_s a_B$. This potential is just that of an harmonic oscillator.

From the elementary properties of the harmonic oscillator, we know that the ground-state wavefunction of a particle of mass m bound to a spring of elastic constant $C = e^2/r_s^3 a_B^3$ is a Gaussian function given by

$$\phi(r) = \left(\frac{\alpha}{\pi}\right)^{3/4} \exp(-\frac{1}{2}\alpha r^2) , \qquad (47b)$$

where

$$\alpha^2 = \frac{mC}{\hbar^2} = \frac{me^2}{\hbar^2 r_s^3 a_B^3} = \frac{1}{a_B^4 r_s^3} , \qquad (47c)$$

and a_B is the Bohr radius. The description of each electron in each cell with the localized wavefunction $\phi(r)$ is valid provided $\phi(r_s a_B) \ll \phi(0)$ (say at least one order of magnitude or so). To be conservative, we require $\alpha r_s^2 a_B^2 > 10$ and hence $\alpha^2 r_s^4 a_B^4 > 100$; using Eq. (47c) the condition becomes $r_s > 100$. While this numerical estimation of the critical density for Wigner crystallization is only orientative, it nevertheless hints that as r_s increases, the electrons tend to localize around appropriate lattice sites.

The phase diagram of the homogeneous electron gas by Monte Carlo methods

Among the numerous attempts to put on a quantitative basis the electron gas embedded in a uniform neutralizing background of positive charges, of particular interest are the Monte Carlo studies of D. Ceperley and B. J. Alder, Phys. Rev. Lett. **45**, 566 (1980). In their approach the normal unpolarized fluid, the fully polarized fluid and the Wigner crystal, are described starting from trial many-body wavefunctions of the Slater–Jastrow form

$$\Psi_T(\{\mathbf{r}_i \sigma_i\}) = D(N) \exp\left[-\sum_{i<j} u(r_{ij}) \right] . \tag{48}$$

where $r_{ij} = |\mathbf{r}_i - \mathbf{r}_j|$. The trial wavefunction Ψ_T is the product of two-body correlation functions (whose tendency is to keep apart two electrons approaching each other) times a Slater determinant $D(N)$ of single-particle wavefunctions. For the normal fluid, $D(N)$ is a Slater determinant of plane waves with paired electrons; for the fully polarized fluid it is a determinant of plane waves with only one spin for each spatial state. For the fully polarized crystal phase, $D(N)$ is a Slater determinant of Gaussian orbitals centered around the sites of a body-centered cubic lattice. The function $u(r_{ij})$, which correlates the motion of pairs of electrons in the Jastrow function, is most often parametrized along the lines given by D. Ceperley, Phys. Rev. B**18**, 3126 (1978).

The basic strategy of the Monte Carlo method consists in the direct evaluation of the multi-dimensional integrals involved in the definition of the total energy

$$E_T = \frac{\langle \Psi_T | H_e | \Psi_T \rangle}{\langle \Psi_T | \Psi_T \rangle} ;$$

this is obtained expressing the integrand on a representative random sampling of points. The variational Monte Carlo method is used to optimize the correlated many-body trial function Ψ_T. For high-accuracy energy calculations, the diffusion Monte Carlo method is used to remove the variational bias; for this purpose, the trial wavefunction is allowed to evolve in imaginary time, so to filter out from Ψ_T the ground state (with the same symmetry).

In Fig. 6 the energy of several phases of interest of the electron gas are reported. It can be seen that the normal fluid is stable for $r_s < 75$; the fully polarized fluid is stable in the region approximately given by $75 < r_s < 100$ and the Wigner crystal is stable for higher values of r_s. For sake of comparison, Fig. 6 also considers a fictitious system of "charged bosons", with the same charge, mass and density as the electrons. For the boson fluid the trial wavefunction is the Jastrow function $\Psi_T = \exp\left[-\sum_{i<j} u(r_{ij}) \right]$;

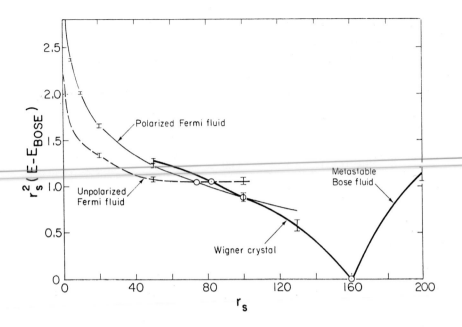

Fig. 6 Energy (times r_s^2) in Rydberg versus r_s for several phases of the homogeneous electron gas. The figure shows that the normal Fermi fluid is stable below $r_s \approx 75$, the polarized (ferromagnetic) Fermi fluid is stable between $r_s \approx 75$ and $r_s \approx 100$, the Wigner crystal above $r_s \approx 100$. In the figure the reference energy E_{BOSE} is that of the lowest boson phase, i.e. the Bose fluid for $r_s < 160$, and the Wigner crystal for $r_s > 160$ [from D. Ceperley and B. J. Alder, Phys. Rev. Lett. **45**, 566 (1980); copyright 1980 by the American Physical Society].

for the Wigner crystal of bosons, the trial wavefunction is constructed by the product of Gaussian orbitals centered at bcc sites and the Jastrow factor. Notice that the Wigner crystals of bosons and fermions have practically the same energy for the large values of r_s (>50) of interest in the Fig. 6 (actually, the product of Gaussian orbitals needs not to be symmetrized for large r_s).

Studies of electron gas by quantal simulation techniques have also been extended to the exact determination of static response functions, thus providing invaluable checks of theories and approximations [see for instance S. Moroni, D. M. Ceperley and G. Senatore, Phys. Rev. Lett. **69**, 1837 (1992); Phys. Rev. Lett. **75**, 689 (1995) and references quoted therein]. The Monte Carlo calculations have been used also to provide precious tests of the approximations inherent the Hartree–Fock and the density functional theory in atoms, molecules and clusters with small number of electrons. As an example, in Table 1, we report the binding energies of small silicon clusters calculated with different theoretical methods, and compared with the experimental results. The binding energies with the diffusion Monte Carlo method are within a few percent from experimental data. The Hartree–Fock calculations understimate the binding energy by about 50 percent or so. The calculations within the local density approximation overestimate somewhat the binding energies, but are very accurate in the prediction of geometry and energetics of clusters with the same number of atoms.

Table 1 Binding energies (eV/atom) of clusters of silicon calculated by the Hartree–Fock (HF), local density approximation (LDA) and diffusion Monte Carlo (DMC) methods compared with experimental data [from J. C. Grossman and L. Mitáš, Phys. Rev. Lett. **74**, 1323 (1995); copyright 1995 by the American Physical Society].

	HF	LDA	DMC	Exp.
Si_2 (D_{2h})	0.85	1.98	1.580(7)	1.61(4)
Si_3 (C_{2v})	1.12	2.92	2.374(8)	2.45(6)
Si_4 (D_{2h})	1.61	3.50	2.86(2)	3.01(6)
Si_6 (C_{2v})	1.82	4.00	3.26(1)	3.42(4)
Si_7 (D_{5h})	1.91	4.14	3.43(2)	3.60(4)
Si_9 (C_s)	1.74	4.06	3.28(2)	......
Si_9 (D_{3h})	1.77	4.14	3.39(2)	......
Si_{10} (T_d)	1.94	4.25	3.44(2)	......
Si_{10} (C_{3v})	1.89	4.32	3.48(2)	......
Si_{13} (I_h)	1.41	3.98	3.12(2)	......
Si_{13} (C_{3v})	1.80	4.28	3.41(1)	......
Si_{13}^- (C_{3v})	1.88	4.43	3.56(1)	......
Si_{20} (I_h)	1.61	4.10	3.23(3)	......
Si_{20} (C_{3v})	1.55	4.28	3.43(3)	......

8   The density functional theory and the Kohn–Sham equations

In the previous sections, the many-electron problem has been attacked by approximating as better as possible the exact ground-state many-electron wavefunction $\Psi_G(\mathbf{r}_1\sigma_1,\dots,\mathbf{r}_N\sigma_N)$. In the density functional theory the emphasis shifts from the ground-state wavefunction to the much more manageable ground-state *one-body electron density* $n(\mathbf{r})$. The density functional theory shows that the ground-state energy of a many-particle system can be expressed as a functional of the one-body density; minimization of this functional allows in principle the determination of the actual ground-state density. The success of the theory is also to provide a reasonable approximation of the functional to be minimized. The peculiarity of the density functional approach to the many-body theory, is to attain rigorously a one-electron Schrödinger equation with a local effective potential in the study of the ground-state properties of the many-electron systems.

The Hohenberg–Kohn theorem

Consider a system of N electrons, described by the standard many-electron Hamiltonian H_e (see Eq. 3). For our reasoning, it is convenient to decompose H_e into the sum of an "internal" part (kinetic energy of the electrons plus electron-electron Coulomb interactions) and "external" part (in our specific case the external part is given by the electronic-nuclear interactions). With obvious rearrangement of notations, we have:

$$H_e = H_{\text{int}} + V_{\text{ext}} \tag{49a}$$

$$H_{\text{int}} = T + V_{\text{ee}} = \sum_i \frac{\mathbf{p}_i^2}{2m} + \frac{1}{2} \sum_{i \neq j} \frac{e^2}{|\mathbf{r}_i - \mathbf{r}_j|} \qquad (49\text{b})$$

$$V_{\text{ext}} = \sum_i v_{\text{ext}}(\mathbf{r}_i) \quad \text{with} \quad v_{\text{ext}}(\mathbf{r}) \equiv V_{\text{nucl}}(\mathbf{r}) = - \sum_I \frac{z_I e^2}{|\mathbf{r} - \mathbf{R}_I|} \ . \qquad (49\text{c})$$

For simplicity, we suppose that the ground-state $|\Psi_G\rangle$ is non-degenerate (in principle, any degeneracy can be removed by an arbitrary small perturbation that appropriately lowers the symmetry of the system).

Let us consider as only variable of the many-electron problem the external potential $v_{\text{ext}}(\mathbf{r})$; the mass of the electrons, their charge, their number N, the form of the internal interactions are vice versa supposed to be fixed. The Hohenberg–Hohn theorem states that *there is a one-to-one correspondence between the ground-state density of a N electron system and the external potential acting on it*; in this sense the ground-state electron density becomes the variable of interest.

The first part of the theorem is almost trivial. Suppose, in fact, that we know $v_{\text{ext}}(\mathbf{r})$ and hence the total Hamiltonian H_e of the system. Thus we know exactly (at least in principle) all its eigenfunctions and eigenvalues; in particular we know the ground-state $\Psi_G(\mathbf{r}_1, \mathbf{r}_2, \dots, \mathbf{r}_N)$. [In the following, for simplicity, we confine our attention to systems without spin polarization, and do not indicate explicitly the spin-variables, since this is not essential to the "abstract" reasoning we are going to present]. The ground-state Ψ_G depends on the chosen external potential $v_{\text{ext}}(\mathbf{r})$ and is also denoted with the short-hand functional notation $\Psi_G[v_{\text{ext}}]$. From the knowledge of Ψ_G we obtain the one-body ground-state density $n(\mathbf{r})$ defined as

$$n(\mathbf{r}) = \langle \Psi_G(\mathbf{r}_1, \mathbf{r}_2, \dots, \mathbf{r}_N) | \sum_i \delta(\mathbf{r} - \mathbf{r}_i) | \Psi_G(\mathbf{r}_1, \mathbf{r}_2, \dots, \mathbf{r}_N) \rangle \ . \qquad (50)$$

Incidentally, we notice the property

$$\langle \Psi_G | V_{\text{ext}} | \Psi_G \rangle = \int n(\mathbf{r}) \, v_{\text{ext}}(\mathbf{r}) \, d\mathbf{r} \ , \qquad (51)$$

which follows from definition (50) and from the fact that the potential V_{ext} can be written as

$$V_{\text{ext}} = \sum_i \int v_{\text{ext}}(\mathbf{r}) \delta(\mathbf{r} - \mathbf{r}_i) \, d\mathbf{r} \ .$$

We can summarize the contents of the previous reasoning by the following scheme:

$$v_{\text{ext}}(\mathbf{r}) \Rightarrow \Psi_G[v_{\text{ext}}] \Rightarrow n(\mathbf{r}) \ .$$

Thus the knowledge of $v_{\text{ext}}(\mathbf{r})$ entails the knowledge of $\Psi_G[v_{\text{ext}}]$ and hence the knowledge of $n(\mathbf{r})$. In other words there exists a functional that links $n(\mathbf{r})$ and $v_{\text{ext}}(\mathbf{r})$, and we write

$$n(\mathbf{r}) = F[v_{\text{ext}}] \ . \qquad (52)$$

It is also evident that two external potentials, which differ by a constant in the whole

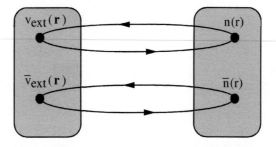

Fig. 7 Schematic representation of the Hohenberg–Kohn theorem. *Two different* ground-state density functions must correspond to *two different* potentials, and *vice versa*.

space, lead to the same $n(\mathbf{r})$ (and can be considered essentially the same, upon appropriate shift by an additive constant). The conceptual scheme of this reasoning is indicated in Fig. 7, focusing on $v_{\mathrm{ext}}(\mathbf{r})$ and following the arrows towards right. The real novelty of the theorem is that the functional relation (52) can be inverted in the form

$$\boxed{v_{\mathrm{ext}}(\mathbf{r}) = G[n(\mathbf{r})]}\,, \tag{53}$$

which means that from the knowledge of the ground-state density $n(\mathbf{r})$ we can determine uniquely the external potential (to within a non-essential additive constant) and thus the Hamiltonian of the system.

In order to prove Eq. (53), we should establish that for any given pair of external potentials $v_{\mathrm{ext}}(\mathbf{r})$ and $\bar{v}_{\mathrm{ext}}(\mathbf{r})$ such that $v_{\mathrm{ext}}(\mathbf{r}) \neq \bar{v}_{\mathrm{ext}}(\mathbf{r})$ we have $n(\mathbf{r}) \neq \bar{n}(\mathbf{r})$. This follows from the minimum property of the ground-state energy. Consider in fact the two Hamiltonians

$$H = H_{\mathrm{int}} + V_{\mathrm{ext}} \tag{54a}$$

$$\overline{H} = H_{\mathrm{int}} + \overline{V}_{\mathrm{ext}}\,, \tag{54b}$$

and their ground-states $|\Psi_G\rangle$ and $|\overline{\Psi}_G\rangle$ with eigenvalues E_G and $\overline{E}_G$ respectively. We have

$$\langle\overline{\Psi}_G|H|\overline{\Psi}_G\rangle = \langle\overline{\Psi}_G|H_{\mathrm{int}} + V_{\mathrm{ext}} + \overline{V}_{\mathrm{ext}} - \overline{V}_{\mathrm{ext}}|\overline{\Psi}_G\rangle = \overline{E}_G + \langle\overline{\Psi}_G|V_{\mathrm{ext}} - \overline{V}_{\mathrm{ext}}|\overline{\Psi}_G\rangle$$

$$= \overline{E}_G + \int \bar{n}(\mathbf{r})[v_{\mathrm{ext}}(\mathbf{r}) - \bar{v}_{\mathrm{ext}}(\mathbf{r})]\,d\mathbf{r}\,,$$

where the last passage follows from Eq. (51). Since the ground eigenvalue of H is strictly lower than the mean value of the Hamiltonian on any other state, we have

$$E_G < \overline{E}_G + \int \bar{n}(\mathbf{r})[v_{\mathrm{ext}}(\mathbf{r}) - \bar{v}_{\mathrm{ext}}(\mathbf{r})]\,d\mathbf{r}\,. \tag{55a}$$

From a similar reasoning using $\langle\Psi_G|\overline{H}|\Psi_G\rangle$, one obtains the above expression with barred and unbarred quantities exchanged

$$\overline{E}_G < E_G + \int n(\mathbf{r})[\overline{v}_{\text{ext}}(\mathbf{r}) - v_{\text{ext}}(\mathbf{r})]\, d\mathbf{r} \ . \tag{55b}$$

Eq. (55a) and Eq. (55b) are obviously incompatible in the hypothesis $n(\mathbf{r}) = \overline{n}(\mathbf{r})$. Therefore for $v_{\text{ext}}(\mathbf{r}) \neq \overline{v}_{\text{ext}}(\mathbf{r})$ it must be $n(\mathbf{r}) \neq \overline{n}(\mathbf{r})$, which completes the proof of the Hohenberg–Kohn theorem (schematically represented in Fig. 7).

The Hohenberg–Kohn theorem has far reaching consequences, that require simple (although subtle) elaborations. First of all, it is convenient to define other functionals of the ground-state density of an electron system. We start from the following (obvious) functionals of the external potential $v_{\text{ext}}(\mathbf{r})$:

$$v_{\text{ext}}(\mathbf{r}) \Rightarrow \Psi_G[v_{\text{ext}}] \begin{cases} \Rightarrow & E[v_{\text{ext}}] \\ \Rightarrow & T[v_{\text{ext}}] \\ \Rightarrow & V_{\text{ee}}[v_{\text{ext}}] \end{cases} . \tag{56}$$

The above logical sequence allows one to state that the ground-state energy is a functional $E[v_{\text{ext}}]$ of the external potential $v_{\text{ext}}(\mathbf{r})$; the expectation value of the electron kinetic energy on the ground-state wavefunction is a functional $T[v_{\text{ext}}]$, and the expectation value of the electron–electron interaction on the ground-state wavefunction is a functional $V_{\text{ee}}[v_{\text{ext}}]$.

Because of the Hohenberg–Kohn theorem, $v_{\text{ext}}(\mathbf{r})$ and $n(\mathbf{r})$ are in one-to-one correspondence. We thus have that the ground-state energy is a functional $E[n]$ of the ground-state density, that the expectation value of the electronic kinetic energy is a functional $T[n]$ of the ground-state density and so is also the expectation value $V_{\text{ee}}[n]$ of the electron-electron interaction.

A most important consequence of the Hohenberg–Kohn theorem is the formulation of a variational principle concerning the ground-state density of a system. Consider a system of interacting electrons in a given nuclear external potential, denoted by $v_{\text{ext}}(\mathbf{r})$. Following Hohenberg and Kohn, we consider the following functional

$$E^{(\text{HK})}[n(\mathbf{r}); v_{\text{ext}}(\mathbf{r})] \equiv \langle \Psi_G[n] | \, T + V_{\text{ee}} + V_{\text{ext}} \, | \Psi_G[n] \rangle \ , \tag{57}$$

where $v_{\text{ext}}(\mathbf{r})$ is taken fixed, $n(\mathbf{r})$ is allowed to vary, and $\Psi_G[n]$ is the ground-state of the system with ground-state density $n(\mathbf{r})$. It is evident that the absolute minimum of the energy functional (57) occurs when $\Psi_G[n]$ is the ground-state of the operator $H_e = T + V_{\text{ee}} + V_{\text{ext}}$, i.e. when $n(\mathbf{r})$ is the exact electron density of the system.

The properties of the Hohenberg–Kohn energy functional (57), here rewritten in the form

$$\boxed{E^{(\text{HK})}[n(\mathbf{r}); v_{\text{ext}}(\mathbf{r})] = T[n(\mathbf{r})] + V_{\text{ee}}[n(\mathbf{r})] + \int v_{\text{ext}}(\mathbf{r})n(\mathbf{r})\, d\mathbf{r}} \ , \tag{58}$$

can be summarized as follows. The energy functional $E^{(\text{HK})}[n(\mathbf{r}); v_{\text{ext}}(\mathbf{r})]$, which exists and is unique, is *minimal at the exact ground-state density*, and its minimum gives the *exact ground-state energy* of the many-body electron system. We also notice that the functional $F[n]=T[n]+V_{ee}[n]$ is *universal*, i.e. it does not depend on v_{ext}; this good news, however, is no compensation for the fact that the functional $F[n]$ is not known explicitly, and must be appropriately approximated.

The Kohn–Sham equations

The Kohn–Sham equations are obtained by minimizing the functional (58) with respect to $n(\mathbf{r})$. To carry out explicitly the variational procedure, we decompose the ground-state density $n(\mathbf{r})$ of an interacting electron system into the sum of N independent orbital contributions of the form

$$n(\mathbf{r}) = \sum_i \phi_i^*(\mathbf{r})\phi_i(\mathbf{r}) \; , \qquad (59)$$

where $\{\phi_i(\mathbf{r})\}(i=1,2,\dots,N)$ are orthonormal orbitals.

We can argue that the decomposition (59) is exact and unique for any chosen (well behaved) density $n(\mathbf{r})$. In fact we can always imagine a fictitious *system of non-interacting electrons, whose ground-state density $n_0(\mathbf{r})$ equals $n(\mathbf{r})$*. The Hohenberg–Kohn theorem applies to any electron system, whether interacting or not. For the system of non-interacting electrons of density $n_0(\mathbf{r})$, the Hohenberg–Kohn theorem guarantees the existence and unicity of an external potential $v_{ext}^0(\mathbf{r})$, producing the ground-state density $n_0(\mathbf{r})$. If $\{\phi_i(\mathbf{r})\}(i=1,2,\dots,N)$ denote the N orbitals of lowest energy of the system of non-interacting electrons, the ground-state wavefunction is just the Slater determinant $A\{\phi_1 \phi_2 \dots \phi_N\}$ and the electron density is given by $n_0(\mathbf{r}) = \sum_i \phi_i^*(\mathbf{r})\phi_i(\mathbf{r})$. Since $n_0(\mathbf{r})$ and $n(\mathbf{r})$ have been taken equal, the decomposition (59) is fully justified also for the interacting electrons, regardless of the fact that *the exact ground-state of an interacting electron system is not the Slater determinant formed with the orbitals contributing to $n(\mathbf{r})$.*

Before adopting the variational procedure on the functional (58), it is convenient to extract from it the inter-electron Coulomb interaction $V_H[n]$ (called Hartree potential)

$$V_H[n] = \frac{1}{2}\int n(\mathbf{r})\frac{e^2}{|\mathbf{r}-\mathbf{r}'|}n(\mathbf{r}')\,d\mathbf{r}\,d\mathbf{r}' \equiv \frac{1}{2}\sum_{ij}\langle\phi_i\phi_j|\frac{e^2}{r_{12}}|\phi_i\phi_j\rangle \; ,$$

and the kinetic energy $T_0[n]$ (of a system of non-interacting electrons with the same density) defined as

$$T_0[n] = \sum_i \langle\phi_i(\mathbf{r})| - \frac{\hbar^2\nabla^2}{2m}|\phi_i(\mathbf{r})\rangle \; .$$

The Hohenberg–Kohn functional (58) can be written as

$$E^{(HK)}[n(\mathbf{r}); v_{ext}(\mathbf{r})] = T_0[n] + V_H[n] + \int v_{ext}(\mathbf{r})n(\mathbf{r})\,d\mathbf{r} + E_{xc}[n] \; , \qquad (60a)$$

where the *exchange-correlation functional* $E_{xc}[n]$ is defined as

$$E_{xc}[n] = T[n] - T_0[n] + V_{ee}[n] - V_H[n] \; . \qquad (60b)$$

The functional (60a) can be recast in the form

$$E^{(\mathrm{HK})}[n(\mathbf{r}); v_{\mathrm{ext}}(\mathbf{r})] = \sum_i \langle \phi_i | -\frac{\hbar^2 \nabla^2}{2m} + v_{\mathrm{ext}} | \phi_i \rangle + \frac{1}{2} \sum_{ij} \langle \phi_i \phi_j | \frac{e^2}{r_{12}} | \phi_i \phi_j \rangle + E_{\mathrm{xc}}[n] .$$

(61)

According to the standard variational procedure, we vary the N contributing orbitals $\phi_1, \phi_2, \ldots, \phi_N$ so to make stationary the energy functional (61), under the constraint of orthonormalization of the wavefunctions $\{\phi_i\}$. The variational expression for the first two terms in the right-hand side of Eq. (61) has been considered in Section 5.1, and need not to be repeated here. The only novelty from a technical point of view is thus the calculation of the variation $\delta E_{\mathrm{xc}}[n]$; the variation of the functional is defined as

$$\delta E_{\mathrm{xc}}[n] = \int V_{\mathrm{xc}}(\mathbf{r}) \delta n(\mathbf{r}) \, d\mathbf{r} = \int V_{\mathrm{xc}}(\mathbf{r}) \, \delta \sum_i \phi_i^*(\mathbf{r}) \phi_i(\mathbf{r}) \, d\mathbf{r}$$

(62)

where

$$V_{\mathrm{xc}}(\mathbf{r}) \equiv \frac{\delta E_{\mathrm{xc}}[n]}{\delta n(\mathbf{r})}$$

(63)

is the functional derivative of $E_{\mathrm{xc}}[n]$.

A straightforward variational calculation of the functional (61), keeping in mind Eq. (62) and Eq. (63), leads to the Kohn–Sham equations

$$\left[-\frac{\hbar^2 \nabla^2}{2m} + V_{\mathrm{nucl}}(\mathbf{r}) + V_{\mathrm{coul}}(\mathbf{r}) + V_{\mathrm{xc}}(\mathbf{r}) \right] \phi_i(\mathbf{r}) = \varepsilon_i \phi_i(\mathbf{r}) ,$$

(64)

where $V_{\mathrm{coul}}(\mathbf{r})$ denotes the Hartree potential, $V_{\mathrm{xc}}(\mathbf{r})$ is the functional derivative of $E_{\mathrm{xc}}[n]$, and $V_{\mathrm{nucl}}(\mathbf{r})$ is the external potential in consideration. Once the Kohn–Sham orbitals and energies have been determined, the exact total ground-state energy (61) of the electronic system can be expressed as:

$$E_0 = \sum_i \varepsilon_i - \frac{1}{2} \sum_{ij} \langle \phi_i \phi_j | \frac{e^2}{r_{12}} | \phi_i \phi_j \rangle + E_{\mathrm{xc}}[n] - \int V_{\mathrm{xc}}(\mathbf{r}) n(\mathbf{r}) \, d\mathbf{r} .$$

(65)

We notice that the Kohn–Sham equations are standard differential equations with a rigorously local effective potential $V_{\mathrm{eff}}(\mathbf{r}) = V_{\mathrm{nucl}}(\mathbf{r}) + V_{\mathrm{coul}}(\mathbf{r}) + V_{\mathrm{xc}}(\mathbf{r})$; any difficulty in the procedure has been confined to a reasonable guess of the exchange-correlation functional $E_{\mathrm{xc}}[n]$ (which is known only in principle). Conceptually, the Kohn–Sham equations determine exactly the electron density and the electronic energy of the ground-state; however, the orbital energies ε_i appearing in Eq. (64) are and remain purely formal Lagrange multipliers (not rescued for physical interpretation by any Koopmans' theorem); any identification of ε_i with (occupied or non-occupied) one-particle energies is to be justified (and often heuristically corrected) situation by situation. Experience shows that the calculations, performed within one version or another of the density functional theory, tend to understimate the energy band gap in semiconductor and insulators; however, the general trend of the dispersion curves of the valence and the conduction bands is often represented to reasonably accuracy.

Local density approximation for the exchange-correlation functional

The formal definition of the exchange-correlation functional $E_{xc}[n]$, provided by Eq. (60b), would be of no practical use, unless workable approximations can be given. Among them, the most popular is the *local density approximation*, which is particularly justified in systems with reasonably slowly varying spatial density $n(\mathbf{r})$. In the local density approximation (LDA), the exchange-correlation functional (60b) is approximated in the form

$$E_{xc}^{(LDA)}[n(\mathbf{r})] = \int \varepsilon_{xc}(n(\mathbf{r}))n(\mathbf{r}) \, d\mathbf{r} \, , \tag{66a}$$

where $\varepsilon_{xc}(n(\mathbf{r}))$ is the many-body exchange-correlation energy per electron *of a uniform gas of interacting electrons of density* $n(\mathbf{r})$. The exchange-correlation potential produced by the functional (66a) becomes

$$V_{xc}^{(LDA)}(\mathbf{r}) \equiv \frac{\delta E_{xc}^{(LDA)}[n]}{\delta n(\mathbf{r})} = \varepsilon_{xc}(n(\mathbf{r})) + n(\mathbf{r}) \frac{d\varepsilon_{xc}(n(\mathbf{r}))}{dn(\mathbf{r})} \, . \tag{66b}$$

In the local density approximation, the total ground-state energy $E_0^{(LDA)}$ takes the expression

$$E_0^{(LDA)} = \sum_i \varepsilon_i - \frac{1}{2} \int n(\mathbf{r}) \frac{e^2}{|\mathbf{r} - \mathbf{r}'|} n(\mathbf{r}') \, d\mathbf{r} \, d\mathbf{r}' - \int n(\mathbf{r}) \frac{d\varepsilon_{xc}(n(\mathbf{r}))}{dn(\mathbf{r})} n(\mathbf{r}) \, d\mathbf{r} \, , \tag{66c}$$

where $d\varepsilon_{xc}(n(\mathbf{r}))/dn(\mathbf{r})$ denotes $d\varepsilon_{xc}/dn$ calculated for $n = n(\mathbf{r})$.

Among the various forms, adopted for ε_{xc}, we report here the expression given by J. P. Perdew and A. Zunger, Phys. Rev. **B23**, 5048 (1981), who have parametrized the numerical results obtained by Ceperley and Alder with Monte Carlo calculations. For the unpolarized homogeneous electron gas, we have:

$$\varepsilon_{xc}(r_s) = \varepsilon_x(r_s) + \varepsilon_c(r_s)$$

$$\varepsilon_x(r_s) = -\frac{0.4582}{r_s}$$

$$\varepsilon_c(r_s) = \begin{cases} -0.1423/(1 + 1.0529\sqrt{r_s} + 0.3334\,r_s) & \text{for } r_s \geq 1 \\ -0.0480 + 0.0311 \ln r_s - 0.0116\,r_s + 0.0020\,r_s \ln r_s & \text{for } r_s \leq 1 \end{cases} \, .$$

In the above expressions, energies are in Hartrees (1 Hartree = 2 Rydberg); r_s is the usual dimensionless parameter defined by $(4\pi/3)(r_s a_B)^3 = 1/n$; the distinction between the exchange term and the correlation term is just for comparison with the Hartree–Fock approach. The density functional theory, especially in conjunction with the local density approximation, has become a most successful tool for the investigation of the chemical and physical ground state properties of electronic systems.

APPENDIX A. Bielectronic integrals among spin-orbitals

Let us consider a complete set of orthonormal *spin-orbitals* of the form

$$\psi_i(\mathbf{r}\sigma) = \phi_i(\mathbf{r})\chi_i(\sigma) \; ,$$

where $\mathbf{r}$ and σ denote space and spin coordinates, $\phi_i(\mathbf{r})$ are spatial orbitals and $\chi_i(\sigma)$ equals α or β for spin-up or spin-down electrons, respectively.

Bielectronic integrals among any four spin-orbitals are defined as

$$\langle\psi_i\psi_j|\frac{e^2}{r_{12}}|\psi_m\psi_n\rangle = \int \psi_i^*(\mathbf{r}_1\sigma_1)\psi_j^*(\mathbf{r}_2\sigma_2)\frac{e^2}{|\mathbf{r}_1 - \mathbf{r}_2|}\psi_m(\mathbf{r}_1\sigma_1)\psi_n(\mathbf{r}_2\sigma_2)\,d(\mathbf{r}_1\sigma_1)\,d(\mathbf{r}_2\sigma_2)$$

$$(A1)$$

where integration on the space variables also includes summation on spin components. *Notice the order by which the arguments* $\mathbf{r}_1\sigma_1$ *and* $\mathbf{r}_2\sigma_2$ *appear in the definition* $(A1)$. Performing the integration over the spin variables in Eq. $(A1)$, we have

$$\langle\psi_i\psi_j|\frac{e^2}{r_{12}}|\psi_m\psi_n\rangle = \delta_{\chi_i,\chi_m}\delta_{\chi_j,\chi_n}\int \phi_i^*(\mathbf{r}_1)\phi_j^*(\mathbf{r}_2)\frac{e^2}{r_{12}}\phi_m(\mathbf{r}_1)\phi_n(\mathbf{r}_2)\,d\mathbf{r}_1\,d\mathbf{r}_2 \; . \quad (A2)$$

An obvious property of the definition $(A1)$ is

$$\langle\psi_i\psi_j|\frac{e^2}{r_{12}}|\psi_m\psi_n\rangle = \langle\psi_j\psi_i|\frac{e^2}{r_{12}}|\psi_n\psi_m\rangle \; .$$

In several situations, we encounter bielectronic integrals involving two different spin-orbitals. In this case, we can have *Coulomb bielectronic integrals* (also called *direct* bielectronic integrals) defined as

$$Q_{mn} \equiv \langle\psi_m\psi_n|\frac{e^2}{r_{12}}|\psi_m\psi_n\rangle \; , \quad (A3)$$

and the *exchange bielectronic integrals*, defined as

$$J_{mn} \equiv \langle\psi_m\psi_n|\frac{e^2}{r_{12}}|\psi_n\psi_m\rangle \; . \quad (A4)$$

The physical meaning and the nature of Coulomb and exchange bielectronic integrals is almost obvious. Consider in fact the Coulomb term defined by Eq. $(A3)$ and let us perform explicitly the integration on the spin variables; we see that this term *does not depend on the spin part of the spin-orbitals* and we have

$$Q_{mn} \equiv \langle\psi_m\psi_n|\frac{e^2}{r_{12}}|\psi_m\psi_n\rangle = \int \phi_m^*(\mathbf{r}_1)\phi_n^*(\mathbf{r}_2)\frac{e^2}{r_{12}}\phi_m(\mathbf{r}_1)\phi_n(\mathbf{r}_2)\,d\mathbf{r}_1\,d\mathbf{r}_2 \; . \quad (A5)$$

The Coulomb bielectronic integral Q_{mn} represents the interaction between two charge distributions given by $(-e)|\phi_m(\mathbf{r}_1)|^2$ and $(-e)|\phi_n(\mathbf{r}_2)|^2$, respectively; this term is of long-range nature and behaves as $1/R$ for orbitals localized at distance R.

For what concerns the exchange term, we have

$$J_{mn} \equiv \langle\psi_m\psi_n|\frac{e^2}{r_{12}}|\psi_n\psi_m\rangle = \delta_{\chi_m,\chi_n}\int \phi_m^*(\mathbf{r}_1)\phi_n^*(\mathbf{r}_2)\frac{e^2}{r_{12}}\phi_n(\mathbf{r}_1)\phi_m(\mathbf{r}_2)\,d\mathbf{r}_1\,d\mathbf{r}_2 \; . \quad (A6)$$

Thus exchange bielectronic integrals may be different from zero only for spin-orbitals with parallel spin. Furthermore exchange bielectronic integrals are of short-range nature; in fact they go rapidly to zero if ϕ_m and ϕ_n are localized far from each other.

The exchange bielectronic integral of Eq. $(A6)$ can be formally interpreted as the classical electrostatic interaction between the exchange charge $(-e)\phi_m^*(\mathbf{r})\phi_n(\mathbf{r})$ at $\mathbf{r}_1$ and the complex conjugate charge $(-e)\phi_m(\mathbf{r})\phi_n^*(\mathbf{r})$ at $\mathbf{r}_2$. Such "exchange charge" is not necessarily real; even when the orbital wavefunctions are real, we see that the exchange charge is somewhere positive in space and somewhere negative; in any case, *the total exchange charge* (if so we decide to address the quantity $(-e)\int\phi_m^*(\mathbf{r})\phi_n(\mathbf{r})d\mathbf{r}$) is *exactly zero*, because of orthogonality of different spatial orbitals.

It is important to note that the *exchange bielectronic integrals, as well as the Coulomb ones, are definite-positive quantities*. That the Coulomb integrals are definite positive quantities is obvious. It is easy to show that this is true also for any exchange integral of type $(A6)$ (unless identically zero when the spins of the spin-orbitals are antiparallel). In fact, expanding e^2/r_{12} in plane waves

$$\frac{1}{|\mathbf{r}_1 - \mathbf{r}_2|} = \frac{1}{(2\pi)^3}\int\frac{4\pi}{k^2}e^{i\mathbf{k}\cdot(\mathbf{r}_1-\mathbf{r}_2)}\,d\mathbf{k}\ , \qquad (A7)$$

and replacing expression $(A7)$ into $(A6)$, it is evident that the integral in Eq. $(A6)$ is a definite-positive quantity.

We finally notice that $Q_{mn} \geq J_{mn} \geq 0$; this means that the exchange bielectronic integrals not only are positive but also are smaller that the corresponding Coulomb integrals. This can be proved noticing that $\int |\Psi(\mathbf{r}_1,\mathbf{r}_2)|^2(e^2/r_{12})d\mathbf{r}_1\,d\mathbf{r}_2 > 0$ and keeping $\Psi(\mathbf{r}_1,\mathbf{r}_2) = \phi_m(\mathbf{r}_1)\phi_n(\mathbf{r}_2) - \phi_m(\mathbf{r}_2)\phi_n(\mathbf{r}_1)$.

APPENDIX B. Outline of second quantization formalism for identical fermions

In this Appendix we briefly summarize some elements of second quantization for the description of a system of identical fermions. The method of second quantization is closely linked to the Slater determinants, introduced in Section 3, but exhibits several advantages. In fact, second quantization allows to rewrite in a more compact notation the Slater determinantal states; the Slater rules for matrix elements of operators between determinantal states are automatically embodied within the anticommutation rules of creation and annihilation operators. The second quantization formalism is of particular value for the treatment of systems where the number of particles can vary, or to discuss and handle contributions beyond the one-electron approximation.

Let us consider a complete set of orthonormal spin-orbitals $\{\psi_i\}$. The set of all different (normalized) Slater determinants formed selecting N (different) spin-orbitals from $\{\psi_i\}$, constitutes a complete set for the description of the N-particle system. Let us indicate with $|\Psi_0(N)\rangle$ a given Slater determinantal state with contributing spin-orbitals $\psi_{i_1}, \psi_{i_2}, \ldots, \psi_{i_N}$; we have

$$|\Psi_0(N)\rangle = A\{\psi_{i_1} \psi_{i_2} \ldots \psi_{i_N}\} . \qquad (B1)$$

Next we consider a Slater determinant for $N+1$ particles, obtained by adding an extra spin-orbital ψ_m to the contributing spin-orbitals $\psi_{i_1}, \psi_{i_2}, \ldots, \psi_{i_N}$; the change can be represented by defining a *creation operator* $c_m^\dagger$ that applied to the state $|\Psi_0(N)\rangle$ gives

$$c_m^\dagger |\Psi_0(N)\rangle = A\{\psi_m \psi_{i_1} \psi_{i_2} \ldots \psi_{i_N}\} . \qquad (B2)$$

In the case ψ_m belongs to the occupied set $\psi_{i_1}, \psi_{i_2}, \ldots, \psi_{i_N}$, the new determinant $(B2)$ has two identical rows and thus vanishes. In a similar way, we can denote by $c_n^\dagger c_m^\dagger |\Psi_0(N)\rangle$ the operation of adding two extra spin-orbitals ψ_n and ψ_m to $|\Psi_0(N)\rangle$ and define

$$c_n^\dagger c_m^\dagger |\Psi_0(N)\rangle = A\{\psi_n \psi_m \psi_{i_1} \psi_{i_2} \ldots \psi_{i_N}\} . \qquad (B3)$$

In the case ψ_n and ψ_m are equal (or belong to the occupied set $\psi_{i_1}, \psi_{i_2}, \ldots, \psi_{i_N}$) the determinant $(B3)$ vanishes; this means $c_n^{\dagger 2} = c_m^{\dagger 2} = 0$.

It is evident, from the fact that Slater determinants change sign for exchange of the order of any two spin-orbitals, that we have

$$c_n^\dagger c_m^\dagger |\Psi_0(N)\rangle = -c_m^\dagger c_n^\dagger |\Psi_0(N)\rangle . \qquad (B4)$$

Eq. $(B4)$ can be written in the form

$$(c_n^\dagger c_m^\dagger + c_m^\dagger c_n^\dagger)|\Psi_0(N)\rangle \equiv 0 . \qquad (B5)$$

What is important to notice is that Eq. $(B5)$ holds *whatever is the state* $|\Psi_0(N)\rangle$ (i.e. for any choice of the number N and for any choice of the contributing spin-orbitals); thus equation $(B5)$ is satisfied if and only if the anticommutation of creation operators equals zero, i.e.

$$\{c_n^\dagger, c_m^\dagger\} \equiv c_n^\dagger c_m^\dagger + c_m^\dagger c_n^\dagger = 0 . \qquad (B6)$$

A determinantal state $|\Psi_0(N)\rangle$ of type $(B1)$ is denoted listing the contributing spin-orbitals $\psi_{i_1}, \psi_{i_2}, \ldots, \psi_{i_N}$; from the properties of the creation operators we can equally well denote the state $|\Psi_0(N)\rangle$ listing the corresponding creation operators applied to the vacuum state $|0\rangle$ (defined as the state where no electron is present). We have

$$|\Psi_0(N)\rangle = c_{i_1}^\dagger c_{i_2}^\dagger \ldots c_{i_N}^\dagger |0\rangle .$$

We can now define the *annihilation operator* c_m as the operator that removes a spin-orbital from any given Slater deteminant as follows

$$c_m A\{\psi_m \psi_{i_1} \psi_{i_2} \ldots \psi_{i_N}\} = A\{\psi_{i_1} \psi_{i_2} \ldots \psi_{i_N}\} . \qquad (B7)$$

If there is no orbital ψ_m in the given Slater determinant, the action of the operator c_m is then defined to give zero. If the spin-orbital ψ_m is not in the first row of the determinant (as defined in Eq. $B7$), it must be brought in first position; according to the number of necessary transpositions this may generate a minus sign in front of the state. In a rather similar way, as done for creation operators, we can show the anticommutation property of annihilation operators

$$\{c_n, c_m\} \equiv c_n c_m + c_m c_n = 0 \ . \tag{B8}$$

Finally if the anticommutation between creation and annihilation operators are examined, we can verify that creation and annihilation operators corresponding to different spin-orbitals anticommute; in the case of the same spin-orbital the anticommutation gives the identity operator. In formulae we have

$$\{c_n, c_m^\dagger\} \equiv c_n c_m^\dagger + c_m^\dagger c_n = \delta_{mn} \ . \tag{B9}$$

It is also easy to verify that the annihilation operator c_m is the adjoint of the creation operator $c_m^\dagger$, i.e. $c_m = \left(c_m^\dagger\right)^\dagger$.

Operators in second quantization form

The Hamiltonian of a many-electron system (see for instance Eq. 3) usually contains one-electron operators and two-electron operators. It is convenient to express not only determinantal states, but also one- and two-electron operators in second quantization form.

Consider first a one-electron operator of the type

$$G_1 = \sum_i h(\mathbf{r}_i) \ . \tag{B10}$$

Let us indicate with $\{\psi_i\}$ a complete set of orthonormal spin-orbitals, and with $h_{mn} = \langle \psi_m | h | \psi_n \rangle$ the matrix element of the monoelectronic operator $h(\mathbf{r})$ between any two spin-orbitals. The expression of G_1 in second quantization form is given by

$$G_1 = \sum_i h(\mathbf{r}_i) \equiv \sum_{mn} \langle \psi_m | h | \psi_n \rangle \, c_m^\dagger c_n \ . \tag{B11}$$

The above equivalence means that the matrix elements of the operator G_1 calculated among Slater determinantal states with the Slater rules, and the matrix elements of the operator on the right-hand side, calculated using the anticommutation rules of creation and annihilation operators are perfectly equal on any basis set of determinantal states.

As an example, consider the determinantal state $|\Psi_0\rangle = A\{\psi_1, \psi_2, \dots, \psi_N\}$. From the Slater rules we have seen that

$$\langle \Psi_0 | G_1 | \Psi_0 \rangle = \sum_i \langle \psi_i | h | \psi_i \rangle \ .$$

Operating with creation and annihilation operators we observe that

$$\langle \Psi_0 | c_m^\dagger c_n | \Psi_0 \rangle = \begin{cases} 1 & \text{if } m = n = \text{occupied spin-orbital} \\ 0 & \text{otherwise} \end{cases}$$

and the equivalence is thus proved for the diagonal matrix elements of G_1; similar considerations show that the equivalence holds also for off-diagonal matrix elements.

Notice that in the particular case in which the operator h is the unit operator, we obtain just the operator that counts the number of particles

$$N_{\text{op}} = \sum_m c_m^\dagger c_m .$$

In a rather similar way, we can see that a two-body operator can be expressed in second quantization form as follows

$$G_2 = \frac{1}{2} \sum_{i \neq j} \frac{e^2}{r_{ij}} \equiv \frac{1}{2} \sum_{klmn} \langle \psi_k \psi_l | \frac{e^2}{r_{ij}} | \psi_m \psi_n \rangle c_k^\dagger c_l^\dagger c_n c_m . \qquad (B12)$$

APPENDIX C. An integral on the Fermi sphere

Consider the integral of the type

$$I(k) = \frac{1}{V} \sum_{q < k_F} \frac{4\pi e^2}{|\mathbf{k} - \mathbf{q}|^2} = \frac{4\pi e^2}{(2\pi)^3} \int_{q < k_F} \frac{1}{|\mathbf{k} - \mathbf{q}|^2} \, d\mathbf{q} . \qquad (C1)$$

The fixed vector $\mathbf{k}$ can be taken to lie on the z direction, without loss of generality (we also suppose momentarily that k is larger than k_F, although later we will also consider the case k smaller than k_F).

To calculate the integral, let us introduce for $\mathbf{q}$ polar coordinates with polar axis along z; the integral $(C1)$ becomes

$$I(k) = \frac{4\pi e^2}{(2\pi)^3} \int_{q < k_F} \frac{1}{q^2 - 2kq \cos\theta + k^2} q^2 \sin\theta \, d\theta \, d\phi \, dq . \qquad (C2)$$

The integration in $d\phi$ gives 2π. Also the angular integration in $d\theta$ is easily performed with the change of variable $t = \cos\theta$ and $dt = -\sin\theta \, d\theta$. The integral $I(k)$ becomes

$$I(k) = \frac{e^2}{\pi} \frac{1}{k} \int_0^{k_F} q \ln \frac{k+q}{k-q} \, dq = \frac{e^2}{\pi} \frac{1}{k} \left[kq - \frac{1}{2}(k^2 - q^2) \ln \frac{k+q}{k-q} \right]_0^{k_F} .$$

We thus obtain

$$\boxed{I(k) = \frac{2e^2 k_F}{\pi} F\left(\frac{k}{k_F}\right)} , \qquad (C3)$$

where the function $F(x)$ is given by

$$F(x) = \frac{1}{2} + \frac{1 - x^2}{4x} \ln \left| \frac{1+x}{1-x} \right| \qquad (C4)$$

(the absolute value is introduced because this same expression holds also for $k < k_F$).

It is of interest to calculate the average of the function $F(k/k_F)$ when k runs uniformly within the Fermi sphere. Let us indicate by x the dimensionless quantity

$x = k/k_F$. In the interval $0 \le x \le 1$ the function $F(x)$ varies from 1 to $1/2$; it is easily seen that its average value is just $3/4$; in fact

$$F_{av} = \int_0^1 x^2 F(x)\, dx \Big/ \int_0^1 x^2\, dx = 3 \int_0^1 x^2 F(x)\, dx = \frac{3}{4}\ ; \qquad (C5)$$

the integral in Eq. $(C5)$ has been performed with the help of the indefinite integral

$$\int x(1 - x^2) \ln \frac{1 + x}{1 - x}\, dx = \frac{1}{2}x - \frac{1}{6}x^3 - \frac{1}{4}(1 - x^2)^2 \ln \frac{1 + x}{1 - x}\ .$$

Further reading

P. W. Atkins and R. S. Friedman "Molecular Quantum Mechanics" (Oxford University Press, third edition 1997)

E. Clementi (ed.) "Modern Techniques in Computational Chemistry: MOTECC-91" (Escom, Leiden 1991); "Methods and Techniques in Computational Chemistry: METECC-94" (Stef, Cagliari 1993)

R. M. Dreizler and E. K. U. Gross "Density Functional Theory" (Springer, Berlin 1990)

P. Hohenberg and W. Kohn "Inhomogeneous Electron Gas" Phys. Rev. **136**, B 864 (1964); W. Kohn and L. J. Sham "Self Consistent Equations Including Exchange and Correlation Effects" Phys. Rev. **140**, A 1133 (1965)

S. Lundqvist and N. H. March (eds) "Theory of the Inhomogeneous Electron Gas" (Plenum Press, New York 1983)

R. McWeeny and B. T. Sutcliffe "Methods of Molecular Quantum Mechanics" (Academic Press, New York 1969)

C. Pisani, R. Dovesi and C. Roetti "Hartree–Fock Ab Initio Treatment of Crystalline Systems" (Springer, New York 1988); R. Dovesi, V. R. Saunders and C. Roetti "An Ab Initio Hartree–Fock LCAO Program for Periodic Systems CRYSTAL92"

F. Seitz "The Modern Theory of Solids" (McGraw-Hill, New York 1940)

J. C. Slater "Quantum Theory of Atomic Structure" vols I and II (McGraw Hill, New York 1960); "The Self-Consistent Field for Molecules and Solids" vol IV (McGraw Hill, New York 1974)

A. Szabo and N. S. Ostlund "Modern Quantum Chemistry" (McGraw Hill, New York 1989)

V

Band theory of crystals

1 Basic assumptions of the band theory

In this chapter we describe the basic methods for calculating electronic states in crystals; we focus mainly on physical motivations, without entering into all the subtleties of the art of band structure calculations. Our purpose is to illustrate to a reader, already familiar with the Schrödinger equation for one-dimensional periodic potentials (Chapter I), the physical concepts and the mathematical techniques which have been developed for solving the Schrödinger equation in actual three-dimensional crystals.

In the band theory of crystals, one considers the one-electron Schrödinger equation

$$\left[\frac{\mathbf{p}^2}{2m} + V(\mathbf{r}) \right] \psi(\mathbf{r}) = E\psi(\mathbf{r}) ,$$

(1a)

where

$$V(\mathbf{r} + \mathbf{t}_n) = V(\mathbf{r})$$

(1b)

is the periodic crystalline potential, $\mathbf{t}_n$ are translation vectors, and the eigenfunctions $\psi(\mathbf{r})$ must be of Bloch type. The eigenvalue equation (1) tacitly contains a number of assumptions, which have been discussed in Chapter IV, and are here summarized for convenience.

(i) *Rigid lattice approximation*. The nuclei are taken as fixed at their equilibrium positions; the large difference between the masses of the electrons and the nuclei is the basic justification for this (nuclear vibrations and related effects are considered in Chapter VIII and following).

(ii) *One-electron approximation in local form*. In essence, the complicated many-body electron problem is simplified to a single one-electron problem with an appropriate local potential. A conceptual scheme for this is provided by the Hartree–Fock theory, with a local approximation to the exchange potential, or more rigorously by the density functional method.

(iii) *Relativistic effects are neglected*. Whenever necessary, one should replace the Schrödinger equation (1) with the Dirac equation, all the techniques developed for the former equation being generalizable to the latter. Often one includes relativistic terms of interest (such as spin-orbit coupling) by perturbation theory.

Treatments of the electronic problem beyond the above approximations are numerous in the literature; however to maintain technicities at a reasonable simple level, we prefer in general to describe the methods of band structure calculations with reference to the one-electron Schrödinger equation (1), with a periodic local potential. The various procedures are normally illustrated for *simple lattices* with one atom per unit cell; the appropriate generalizations necessary for *composite lattices*, with a basis of two or more atoms in the unit cell, are usually obvious (or almost obvious). Furthermore we do not enter in the details of the construction of the self-consistent crystal potential; conceptually, this can be done by solving the Schrödinger equation (1) with a trial periodic potential and then iteratively updating the potential itself and the wavefunctions up to self-consistency; Bloch wavefunctions, and corresponding electronic charge density and potential, are usually calculated at a restricted set of "special $\mathbf{k}$ vectors" (notion discussed in Section II-6.4). For simple semiqualitative considerations, it may be reasonable to approximate the crystalline potential as sum of spherically symmetric atomic-like potentials centred at the atomic positions.

To solve the Schrödinger equation with a periodic potential we can distinguish two main points of view. A first line of approach consists in expanding the crystal states in a convenient set of *energy independent Bloch functions*: Bloch sums made with atomic orbitals in the tight-binding method, plane waves orthogonalized to core states in the orthogonalized plane wave method, plane waves in the pseudopotential method. A

Table 1 Schematic summary of some features of methods of band structure calculations.

POINT OF VIEW	METHODS	ENERGY INDEPENDENT BASIS FUNCTIONS
Whole crystal formulation	Tight binding method	Bloch sums
	Orthogonalized plane wave method	Bloch sums and plane waves
	Pseudopotential method	Plane waves

POINT OF VIEW	METHODS	ENERGY DEPENDENT BASIS FUNCTIONS
Single cell formulation	Cellular method	Spherical waves
	Augmented plane wave method	Spherical and plane waves
	Green's function method	Spherical waves

second line of approach (which includes the cellular method, the augmented plane wave method and the Green's function method) is based on the single cell formulation and adopts *energy dependent basis functions*. In the cellular formulation one considers the Wigner–Seitz cell (or subcells), an appropriate spherically symmetric potential within it, and the energy dependent solutions of the radial Schrödinger equation regular at the origin; these constitute the basis set for the expansion of the crystal wavefunctions, which must satisfy appropriate boundary conditions required by the Bloch theorem; the modality of imposing the boundary conditions characterizes the different cellular methods.

In Table 1 we summarize in a schematic form the basic methods of band structure calculations and the basis functions used to describe the crystal states. Of course this scheme has to be taken only as orientative, being impossible to summarize in a concise way all the interesting variations, refinements or interplay among the various lines of approaches, and the wealth of this traditional subject of solid state physics.

2 The tight-binding method (LCAO method)

2.1 Description of the method for simple lattices

The tight-binding method, suggested by Bloch in 1928, consists in expanding the crystal states in linear combinations of atomic orbitals (LCAO) of the composing atoms. This method (when not applied in oversimplified forms) provides a reasonable description of occupied states in any type of crystal (metals, semiconductors and insulators) and often also of the lowest lying conduction states.

For sake of simplicity and for keeping notations at the minimum, *we begin to consider a simple crystal, with one atom per unit cell.* We indicate by $\phi_i(\mathbf{r})$ an atomic orbital of quantum numbers i and energy E_i of the atom centred in the reference unit cell; similarly, we indicate by $\phi_i(\mathbf{r}-\mathbf{t}_m)$ the same orbital for the atom in the unit cell $\mathbf{t}_m$. In correspondence to the atomic orbital ϕ_i, we construct the *Bloch sum* of $\mathbf{k}$ wavevector

$$\Phi_i(\mathbf{k},\mathbf{r}) = \frac{1}{\sqrt{N}} \sum_{\mathbf{t}_m} e^{i\mathbf{k}\cdot\mathbf{t}_m} \phi_i(\mathbf{r}-\mathbf{t}_m) \ , \tag{2}$$

where N is the number of unit cells of the crystal. In the tight-binding method, a (judicious) number of Bloch sums of vector $\mathbf{k}$ are used for expanding the crystal wave functions of vector $\mathbf{k}$ in the form

$$\psi(\mathbf{k},\mathbf{r}) = \sum_i c_i(\mathbf{k})\Phi_i(\mathbf{k},\mathbf{r}) \ , \tag{3}$$

where the coefficients $c_i(\mathbf{k})$ are to be determined with standard variational methods.

From the eigenvalue equation (1), expansion (3), and application of the variational principle, we have that crystal eigenvalues and eigenfunctions are obtained from the determinantal compatibility equation

$$\boxed{\|M_{ij}(\mathbf{k}) - E\,S_{ij}(\mathbf{k})\| = 0} \tag{4}$$

where $M_{ij}(\mathbf{k})$ are the matrix elements of the crystal Hamiltonian $H = \mathbf{p}^2/(2m)+V(\mathbf{r})$ between Bloch sums, and $S_{ij}(\mathbf{k})$ are the overlap matrix elements; namely

$$M_{ij}(\mathbf{k}) = \langle \Phi_i(\mathbf{k},\mathbf{r})|H|\Phi_j(\mathbf{k},\mathbf{r})\rangle \tag{5a}$$

$$S_{ij}(\mathbf{k}) = \langle \Phi_i(\mathbf{k},\mathbf{r})|\Phi_j(\mathbf{k},\mathbf{r})\rangle \ . \tag{5b}$$

The matrix elements (5) can be evaluated numerically, performing the appropriate integrals within any chosen unit cell in real space. Frequently, however, the tight-binding method is used in a semi-empirical way; here we wish to illustrate this aspect, because of its capability to give a vivid picture of some features of the band structure of several solids, with a modest amount of computational labour.

Semi-empirical tight-binding framework

In semi-empirical tight-binding descriptions of crystals, some rather drastic (and yet meaningful) assumptions are often performed on the overlap and Hamiltonian matrix elements (5). We begin to notice that, in the case of extremely localized atomic orbitals, the overlap between atomic-like functions centred on different atoms becomes negligible. This justifies the assumption that the localized atomic orbitals are orthonormal, and so are the corresponding Bloch sums; in this approximation, the overlap matrix $S_{ij}(\mathbf{k})$ in Eq. (5b) is taken as the unit matrix δ_{ij}.

We are thus left to estimate the Hamiltonian matrix elements of Eq. (5a). We have

$$M_{ij}(\mathbf{k}) = \frac{1}{N} \sum_{\mathbf{t}_m \mathbf{t}_n} e^{i\mathbf{k}\cdot(\mathbf{t}_n-\mathbf{t}_m)} \langle \phi_i(\mathbf{r}-\mathbf{t}_m)|H|\phi_j(\mathbf{r}-\mathbf{t}_n)\rangle \ ;$$

because of the translational invariance of H we can choose $\mathbf{t}_m = 0$ in the above expression, drop the sum over $\mathbf{t}_m$ and the factor $1/N$, and write

$$M_{ij}(\mathbf{k}) = \sum_{\mathbf{t}_n} e^{i\mathbf{k}\cdot\mathbf{t}_n} \langle \phi_i(\mathbf{r}) | H | \phi_j(\mathbf{r} - \mathbf{t}_n) \rangle . \tag{6}$$

For semi-empirical evaluations, it is convenient to express the crystal potential as sum of spherically symmetric atomic-like potentials $V_a(\mathbf{r} - \mathbf{t}_n)$, centred at the lattice positions. We thus approximate the crystal Hamiltonian H in the form

$$H = \frac{\mathbf{p}^2}{2m} + \sum_{\mathbf{t}_n} V_a(\mathbf{r} - \mathbf{t}_n) . \tag{7}$$

From Eq. (6) and Eq. (7), we obtain

$$M_{ij}(\mathbf{k}) = \sum_{\mathbf{t}_n} e^{i\mathbf{k}\cdot\mathbf{t}_n} \int \phi_i^*(\mathbf{r}) \left[\frac{\mathbf{p}^2}{2m} + V_a(\mathbf{r}) + V'(\mathbf{r}) \right] \phi_j(\mathbf{r} - \mathbf{t}_n) \, d\mathbf{r} , \tag{8}$$

where $d\mathbf{r}$ is the volume element in direct space, and $V'(\mathbf{r})$ denotes the sum of all the atomic potentials of the crystal, except the contribution $V_a(\mathbf{r})$ of the atom at the origin.

We use the property that the atomic orbital $\phi_i(\mathbf{r})$ is eigenfunction of the atomic Hamiltonian with energy E_i; this fact, together with the assumption of orthonormality of localized atomic orbitals, allows us to write Eq. (8) in the form

$$M_{ij}(\mathbf{k}) = E_i \delta_{ij} + \sum_{\mathbf{t}_n} e^{i\mathbf{k}\cdot\mathbf{t}_n} \int \phi_i^*(\mathbf{r}) \, V'(\mathbf{r}) \, \phi_j(\mathbf{r} - \mathbf{t}_n) \, d\mathbf{r} . \tag{9}$$

In Eq. (9) the term $\mathbf{t}_n = 0$ gives the so called *crystal field* integrals

$$I_{ij} = \int \phi_i^*(\mathbf{r}) \, V'(\mathbf{r}) \, \phi_j(\mathbf{r}) \, d\mathbf{r} .$$

If the tails of the neighbouring atomic-like potentials are almost constant in the region where the wave functions $\phi_i(\mathbf{r})$ extend, the matrix with elements I_{ij} becomes a constant diagonal matrix (which produces a rigid shift of the whole band structure, but does not influence the dispersion curves); this justifies the fact that the term $\mathbf{t}_n = 0$ in Eq. (9) is often neglected.

For what concerns the terms with $\mathbf{t}_n \neq 0$ in Eq. (9), we invoke again the localized nature of the atomic orbitals, so that we can limit the sum in Eq. (9) to a small number of neighbours. For instance, we can include first neighbour contributions, second neighbour contributions etc. For easy qualitative "do it yourself" band structure calculations, besides assuming orthonormality of local orbitals, one disregards crystal field terms, and also considers nearest neighbour interaction adopting the *two-center approximation* (integrals involving *three* different centers are considered negligible). The Hamiltonian matrix elements (9) thus simplify in the form

$$M_{ij}(\mathbf{k}) = E_i \delta_{ij} + \sum_{\mathbf{t}_I} e^{i\mathbf{k}\cdot\mathbf{t}_I} \int \phi_i^*(\mathbf{r}) V_a(\mathbf{r} - \mathbf{t}_I) \phi_j(\mathbf{r} - \mathbf{t}_I) \, d\mathbf{r} , \tag{10}$$

where the sum over $\mathbf{t}_I$ indicates the sum over first neighbours.

Table 2 Expression of two-center interaction integrals involving atomic orbitals $\phi_i(\mathbf{r})$ and $\psi_j(\mathbf{r} - \mathbf{R})$ of the type s, p_x, p_y, p_z; $V_a(\mathbf{r} - \mathbf{R}) = V_a(|\mathbf{r} - \mathbf{R}|)$ denotes a spherically symmetric potential. The director cosines of the two-center distance $\mathbf{R}$ are indicated by l_x, l_y, l_z. The other expressions of interest are obtained by cyclic permutations of indices x, y, z.

$$\int \phi_s^*(\mathbf{r}) V_a(\mathbf{r} - \mathbf{R}) \psi_s(\mathbf{r} - \mathbf{R}) \, d\mathbf{r} = V(ss\sigma)$$

$$\int \phi_s^*(\mathbf{r}) V_a(\mathbf{r} - \mathbf{R}) \psi_x(\mathbf{r} - \mathbf{R}) \, d\mathbf{r} = l_x \, V(sp\sigma)$$

$$\int \phi_x^*(\mathbf{r}) V_a(\mathbf{r} - \mathbf{R}) \psi_x(\mathbf{r} - \mathbf{R}) \, d\mathbf{r} = l_x^2 \, V(pp\sigma) + (1 - l_x^2) \, V(pp\pi)$$

$$\int \phi_x^*(\mathbf{r}) V_a(\mathbf{r} - \mathbf{R}) \psi_y(\mathbf{r} - \mathbf{R}) \, d\mathbf{r} = l_x l_y [V(pp\sigma) - V(pp\pi)]$$

$$\int \phi_x^*(\mathbf{r}) V_a(\mathbf{r} - \mathbf{R}) \psi_z(\mathbf{r} - \mathbf{R}) \, d\mathbf{r} = l_x l_z [V(pp\sigma) - V(pp\pi)]$$

The two-center integrals can be expressed in terms of a small number of independent parameters, which are evaluated either analytically, or numerically, or semi-empirically. In Table 2, we give the expression of the two-center integrals involving atomic orbitals of type s and p; we see that, for a given distance R between the two centers, the independent integrals are four and have been labelled $V(ss\sigma), V(sp\sigma), V(pp\sigma)$ and $V(pp\pi)$; the convention for notations is that s or p specify the angular momentum of the orbitals, while σ, π, δ, etc. denote that the angular part with respect to the axis of quantization (along the two centres) is characterized by $\exp(im\phi)$ with $m = 0, \pm 1, \pm 2, \ldots$ A pictorial representation of the space arrangement of the orbitals involved in $V(ss\sigma), V(sp\sigma), V(pp\sigma)$ and $V(pp\pi)$ is given as an exemplification in Fig. 1. More complete tables concerning also d orbitals and f orbitals are reported in the literature [J. C. Slater and G. F. Koster, Phys. Rev. **94**, 1498 (1954); see also K. Lendi, Phys. Rev. B9, 2433 (1974); B10, 1768 (1974); A. K. McMahan, Phys. Rev. B58, 4293 (1998)].

2.2   Description of the tight-binding method for composite lattices

The extension of the tight-binding method to composite lattices formally needs only some obvious generalization, that we briefly outline. Consider a crystal with a basis of atoms in the positions $\mathbf{d}_1, \mathbf{d}_2, \ldots, \mathbf{d}_{\nu_b}$ in the unit cell. Let $\phi_{\mu i}(\mathbf{r} - \mathbf{d}_\mu)$ indicate an atomic orbital of quantum numbers i and energy $E_{\mu i}$ for the atom in position $\mathbf{d}_\mu$. In correspondence to the atomic orbitals $\phi_{\mu i}(\mathbf{r} - \mathbf{d}_\mu - \mathbf{t}_m)$ we construct the Bloch sum

$$\Phi_{\mu i}(\mathbf{k}, \mathbf{r}) = \frac{1}{\sqrt{N}} \sum_{\mathbf{t}_m} e^{i\mathbf{k} \cdot \mathbf{t}_m} \phi_{\mu i}(\mathbf{r} - \mathbf{d}_\mu - \mathbf{t}_m) \ . \tag{11a}$$

A number of Bloch sums (corresponding to atomic orbitals with the same energy or reasonably near in energy) are used as basis functions to describe the crystal wave-functions in the chosen energy region of interest.

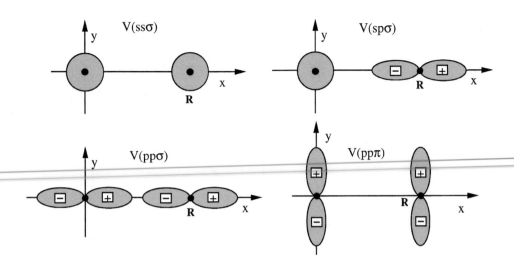

Fig. 1 Pictorial representation of the independent two-center integrals concerning s and p atomic wavefunctions. One center is taken at the origin and the other center $\mathbf{R}$ is taken in the x-direction.

For semi-empirical evaluation, it is convenient to approximate the general one-electron crystalline Hamiltonian in the form

$$H = \frac{\mathbf{p}^2}{2m} + \sum_{\mathbf{t}_n \mathbf{d}_\nu} V_{a\nu}(\mathbf{r} - \mathbf{d}_\nu - \mathbf{t}_n) \;, \qquad (11b)$$

where $V_{a\nu}(\mathbf{r} - \mathbf{d}_\nu - \mathbf{t}_n)$ is the atomic-like spherically symmetric potential for the atom centred in the position $\mathbf{d}_\nu$ of the lattice cell $\mathbf{t}_n$. The matrix elements of the Hamiltonian (11b) between the Bloch sums (11a) can be evaluated with the same techniques presented in the previous section, and the Hamiltonian matrix is then diagonalized. The mixing of Bloch sums, with different local symmetry or belonging to different sublattices, describes often intuitively hybridization effects among atomic orbitals; in particular we mention here the $sp^3 s^*$ hybridization scheme for semiconductors with diamond or zincblende structures [P. Vogl, H. P. Hjalmarson and J. D. Dow, J. Phys. Chem. Solids **44**, 365 (1983)].

The tight-binding approach allows to describe, on a rather intuitive basis, trends in energy bands and bonds in a variety of periodic systems. However, one of the greatest merits of the formalism is the possibility to describe electronic states also in non-periodic systems; within the tight-binding framework, significant models have been developed to describe electronic states in incommensurate systems, in disordered systems (in this context, we just mention the Anderson model for diagonal disorder), or even to embody effects beyond the independent electron approximation (as for instance the Hubbard model of one-site correlation effects). Further bonuses are the convenience to describe electron transport properties on a localized basis, and the possibility to carry out molecular dynamics simulations with reasonable computational

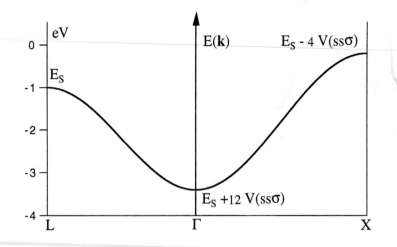

Fig. 2 Schematic behaviour of an s-like band in a fcc crystal lattice with the tight-binding method; E_s is the energy of the s-like orbital; $V(ss\sigma) < 0$ is the two-center potential integral between nearest neighbour s-like atomic orbitals. The values of the wavevector $\mathbf{k}$ at the points Γ, X and L are $\mathbf{k} = 0$, $\mathbf{k} = (2\pi/a)(1, 0, 0)$ and $\mathbf{k} = (2\pi/a)(1/2, 1/2, 1/2)$, respectively. In the figure, we have taken $E_s = -1$ eV and $V(ss\sigma) = -0.2$ eV.

effort. These are among the reasons of the continuous renaissance of the tight-binding method, too often declared in the literature as obsolete. We do not dwell on these and other aspects of the tight-binding formalism, and confine our next considerations to some elementary illustrative examples.

2.3   Illustrative applications of the tight-binding scheme

Energy dispersion of an s-like band in face-centered cubic crystals

As a first illustrative example we consider a face-centered cubic crystal with one atom per unit cell, and a single s-like atomic orbital ϕ_s. From the localized orbitals $\phi_s(\mathbf{r} - \mathbf{t}_m)$ we form the itinerant Bloch sum of $\mathbf{k}$ wavevector

$$\Phi_s(\mathbf{k}, \mathbf{r}) = \frac{1}{\sqrt{N}} \sum_{\mathbf{t}_m} e^{i\mathbf{k}\cdot\mathbf{t}_m} \phi_s(\mathbf{r} - \mathbf{t}_m) \ . \tag{12a}$$

The energy of the s-like band is (neglecting interaction with Bloch sums formed with other atomic states)

$$E(\mathbf{k}) = \frac{\langle \Phi_s(\mathbf{k}, \mathbf{r})|H|\Phi_s(\mathbf{k}, \mathbf{r})\rangle}{\langle \Phi_s(\mathbf{k}, \mathbf{r})|\Phi_s(\mathbf{k}, \mathbf{r})\rangle} \ . \tag{12b}$$

In the empirical tight-binding method, we assume orthonormality of localized orbitals and thus the denominator in Eq. (12b) is unit; the numerator is evaluated in the two-center approximation, retaining only nearest neighbour interactions (see Eq. 10).

We obtain

$$E(\mathbf{k}) = E_s + \sum_{\mathbf{t}_I} e^{i\mathbf{k}\cdot\mathbf{t}_I} \int \phi_s^*(\mathbf{r}) V_a(\mathbf{r} - \mathbf{t}_I) \phi_s(\mathbf{r} - \mathbf{t}_I) d\mathbf{r} , \qquad (12c)$$

where E_s is the atomic eigenvalue, $V_a(\mathbf{r})$ is the atomic-like potential, and the twelve nearest neighbour vectors $\mathbf{t}_I$ are $(a/2)(0, \pm 1, \pm 1)$; $(a/2)(\pm 1, 0, \pm 1)$; $(a/2)(\pm 1, \pm 1, 0)$.

The value $V(ss\sigma)$ of the two-center potential integral appearing in Eq. (12c) is independent from $\mathbf{t}_I$, and can thus be factorized out of the sum. We denote by $F(\mathbf{k})$ the sum

$$F(\mathbf{k}) = \sum_{\mathbf{t}_I} e^{i\mathbf{k}\cdot\mathbf{t}_I} = 4(\cos\frac{k_x a}{2}\cos\frac{k_y a}{2} + \cos\frac{k_y a}{2}\cos\frac{k_z a}{2} + \cos\frac{k_z a}{2}\cos\frac{k_x a}{2}) .$$

The energy dispersion curve (12c) thus becomes

$$E(\mathbf{k}) = E_s + V(ss\sigma)F(\mathbf{k}) . \qquad (12d)$$

In Fig. 2 we report the behaviour of the dispersion relation (12d). It is evident that s-like bands *bend upward*, as $\mathbf{k}$ increases from the center of the Brillouin zone towards the border, since $V(ss\sigma)$ is negative and $F(\mathbf{k})$ has its maximum value at $\mathbf{k} = 0$.

Energy dispersion of p-like bands in face-centered cubic lattices

As a second illustrative example we consider the band structure corresponding to p-like atomic states in face-centered cubic crystals. We have now the three Bloch sums

$$\Phi_i(\mathbf{k}, \mathbf{r}) = \frac{1}{\sqrt{N}} \sum_{\mathbf{t}_m} e^{i\mathbf{k}\cdot\mathbf{t}_m} \phi_i(\mathbf{r} - \mathbf{t}_m) \quad (i = x, y, z) \qquad (13a)$$

(we neglect interactions with Bloch sums originated from other atomic orbitals). The matrix elements of the crystal Hamiltonian on the basis functions (13a) are obtained in the two-center nearest neighbour approximation and using Table 2 for the expression of independent integrals; we have

$$M_{xx}(\mathbf{k}) = E_p + 2\cos\frac{k_x a}{2}(\cos\frac{k_y a}{2} + \cos\frac{k_z a}{2})[V(pp\sigma) + V(pp\pi)]$$
$$+4\cos\frac{k_y a}{2}\cos\frac{k_z a}{2}V(pp\pi)$$
$$M_{xy}(\mathbf{k}) = -2\sin\frac{k_x a}{2}\sin\frac{k_y a}{2}[V(pp\sigma) - V(pp\pi)]$$

and cyclic permutations. The 3×3 secular determinant has the form

$$\begin{Vmatrix} M_{xx}(\mathbf{k}) - E & M_{xy}(\mathbf{k}) & M_{xz}(\mathbf{k}) \\ M_{xy}^*(\mathbf{k}) & M_{yy}(\mathbf{k}) - E & M_{yz}(\mathbf{k}) \\ M_{xz}^*(\mathbf{k}) & M_{yz}^*(\mathbf{k}) & M_{zz}(\mathbf{k}) - E \end{Vmatrix} = 0 . \qquad (13b)$$

The diagonalization of the matrix (13b) is straightforward at the symmetry points

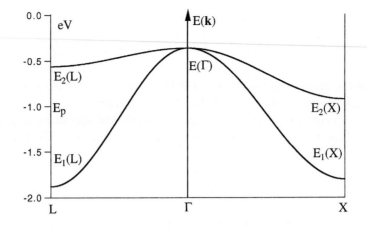

Fig. 3 Schematic behaviour of p-like bands in a fcc crystal lattice with the tight-binding method; E_p is the energy of the p-like orbitals. The values of $E(\Gamma)$, $E_1(X)$, $E_2(X)$, $E_1(L)$, $E_2(L)$ are given by Eqs. (14) of the text. The figure refers to the case $E_p = -1$ eV, $V(pp\sigma) = 0.2$ eV and $V(pp\pi) = -0.02$ eV.

and lines of the Brillouin zone. At the point $\Gamma(\mathbf{k} = 0)$ we have the three-fold degenerate eigenvalue

$$E(\Gamma) = E_p + 4V(pp\sigma) + 8V(pp\pi) . \tag{14a}$$

At the point X, $\mathbf{k} = (2\pi/a)(1,0,0)$, we have a non-degenerate level $E_1(X)$ and a two-fold degenerate level $E_2(X)$ given by

$$E_1(X) = E_p - 4V(pp\sigma), \qquad E_2(X) = E_p - 4V(pp\pi) . \tag{14b}$$

At the point L, $\mathbf{k} = (2\pi/a)(1/2, 1/2, 1/2)$, we have a non-degenerate level $E_1(L)$ and a two-fold degenerate level $E_2(L)$ given by

$$E_1(L) = E_p - 4V(pp\sigma) + 4V(pp\pi), \qquad E_2(L) = E_p + 2V(pp\sigma) - 2V(pp\pi) . \tag{14c}$$

At a general $\mathbf{k}$ point of the Brillouin zone, the energy bands are given by solution of the secular equation (13b).

In Fig. 3 we report the behaviour of the p-like energy bands. In general, from the graphical representation of the independent integrals $V(pp\sigma)$ and $V(pp\pi)$ of Fig. 1, we expect that the former is positive and the latter is negative, and furthermore $|V(pp\sigma)| > |V(pp\pi)|$ (in the strong binding limit). Thus we see that p-bands *bend downward* as k varies from the center of the Brillouin zone towards its boundaries.

Energy dispersion of π and σ bands in two-dimensional graphite

As a final example we consider the two-dimensional model of graphite (see Fig. II-10), which presents two carbon atoms per unit cell in the positions $\mathbf{d}_1 = 0$ and $\mathbf{d}_2 = (a/\sqrt{3})(0, 1, 0)$. The translation vectors for the two-dimensional hexagonal lattice are $\mathbf{t}_1 = (a/2)(1, \sqrt{3}, 0)$ and $\mathbf{t}_2 = (a/2)(-1, \sqrt{3}, 0)$.

The electronic configuration of the carbon atom is $1s^2, 2s^2 2p^2$, and the orbital energies are approximately $E_{1s} = -21.37$, $E_{2s} = -1.29$ and $E_{2p} = -0.66$ (Rydberg). The $1s$ core atomic orbitals are strongly localized near the nuclei, and give rise to dispersionless crystal core bands. In order to study the valence and the (lowest) conduction bands, we diagonalize the crystal Hamiltonian on the basis set of the eight Bloch sums, formed with $2s, 2p_x, 2p_y$ and $2p_z$ orbitals for each of the two carbon atoms in the unit cell. Since there are eight valence electrons per unit cell, we expect four completely occupied bands (if the four lowest lying bands do not overlap in energy with upper lying four energy bands).

The band wavefunctions orginated from s, p_x and p_y orbitals (σ bands) are even under reflection in the plane of graphite; they do not mix with band wavefunctions originated from p_z orbitals (π bands), which are odd under reflection in the plane of graphite. Thus σ-bands and π-bands can be studied separately. For π-bands, we start from the two basis Bloch sums built from the p_z orbital of the two carbon atoms in the unit cell

$$\Phi_1(\mathbf{k}, \mathbf{r}) = \frac{1}{\sqrt{N}} \sum_{\mathbf{t}_m} e^{i\mathbf{k}\cdot\mathbf{t}_m} \phi_z(\mathbf{r} - \mathbf{d}_1 - \mathbf{t}_m)$$

$$\Phi_2(\mathbf{k}, \mathbf{r}) = \frac{1}{\sqrt{N}} \sum_{\mathbf{t}_m} e^{i\mathbf{k}\cdot\mathbf{t}_m} \phi_z(\mathbf{r} - \mathbf{d}_2 - \mathbf{t}_m) .$$

(15a)

We express now the matrix elements of the crystal Hamiltonian between the basis functions (15a), by means of the semi-empirical scheme previously discussed; we obtain the 2×2 determinantal equation

$$\left\| \begin{array}{cc} E_p - E & V(pp\pi)F(\mathbf{k}) \\ V(pp\pi)F(\mathbf{k})^* & E_p - E \end{array} \right\| = 0 .$$

(15b)

In Eq. (15b), $V(pp\pi)$ denotes the two-center potential integral between nearest neighbour p_z orbitals; $F(\mathbf{k})$ denotes the sum

$$F(\mathbf{k}) = \sum_{\mathbf{t}_I} e^{i\mathbf{k}\cdot\mathbf{t}_I} = 1 + 2\cos\frac{k_x a}{2} \exp(-i\frac{\sqrt{3}k_y a}{2}) ;$$

(15c)

the atom at $\mathbf{d}_1$ has three nearest neighbours located at $\mathbf{d}_2, \mathbf{d}_2 - \mathbf{t}_1$ and $\mathbf{d}_2 - \mathbf{t}_2$; thus the sum in Eq. (15c) runs over the vectors $0, -\mathbf{t}_1, -\mathbf{t}_2$.

From the secular equation (15b) and expression (15c), we can easily understand the overall behaviour of the π-bands of graphite (illustrated in Fig. 4). We notice that $F(\mathbf{k})$ takes its maximum value at $\Gamma(\mathbf{k} = 0)$; here the separation between π-bands is maximum. The value of $F(\mathbf{k})$ decreases to one at $Q[\mathbf{k} = (2\pi/a)(1/2, 1/2\sqrt{3}, 0)]$, which turns out to be a saddle point (see Section II-7). The value of $F(\mathbf{k})$ vanishes at the point $P[\mathbf{k} = (2\pi/a)(2/3, 0, 0)]$ of the two-dimensional Brillouin zone; the π-bands are degenerate at the point P, and the material is said to be a semi-metal. All these features are confirmed by the detailed calculations on the band structure of graphite, performed within the framework of the tight-binding scheme, and reported in

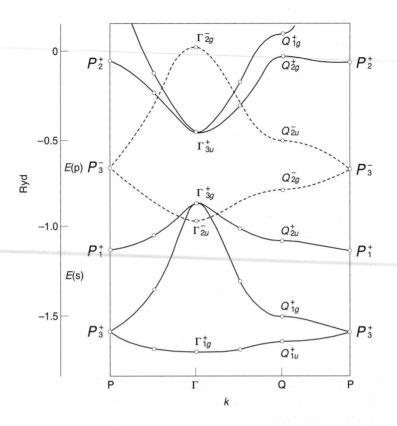

Fig. 4 Band structure of two-dimensional graphite, obtained with the tight-binding method. Bands which are even and odd under reflection in the plane of graphite are indicated with continuous and broken lines respectively. The energies of the valence states of the carbon atom are also indicated. The values of $\mathbf{k}$ at the points Γ, P and Q, are $\mathbf{k} = 0$, $\mathbf{k} = (2\pi/a)(2/3, 0, 0)$ and $\mathbf{k} = (2\pi/a)(1/2, 1/2\sqrt{3}, 0)$, respectively [from F. Bassani and G. Pastori Parravicini, Nuovo Cimento **50**B, 95 (1967); copyright 1967 by Società Italiana di Fisica].

Fig. 4. In this figure we also report the σ-bands obtained by setting up the appropriate 6×6 secular determinant, and we refer to the literature for further experimental and theoretical aspects.

3   The orthogonalized plane wave (OPW) method

Considerations on plane waves expansion

The choice of plane waves as basis functions to expand crystal wavefunctions (Sommerfeld and Bethe 1933) appears to be very attractive, because of formal simplicity of matrix elements of the Hamiltonian. In fact the kinetic energy operator is diagonal in the plane wave representation and matrix elements of the potential simply involve the Fourier transforms of the atomic potential (and appropriate phase factors in the case of several atoms in the unit cell).

Table 3 Expression of some fundamental constants (Bohr radius, Rydberg energy, Hartree energy, and fine structure constant). The Hartree atomic units and the Rydberg atomic units are summarized.

Expression of some fundamental constants
$a_B = \dfrac{\hbar^2}{me^2} =$ Bohr radius $= 0.529$ Å $\qquad$ 1 Rydberg $= \dfrac{\hbar^2}{2ma_B^2} = \dfrac{e^2}{2a_B} = 13.606$ eV
$\alpha^{-1} = \dfrac{\hbar c}{e^2} = 137.036 \qquad$ 1 Hartree $= \dfrac{e^2}{a_B} = 27.21$ eV
Hartree atomic units: $\quad \hbar = 1 \quad m = 1 \quad e = 1$
unit of length $= a_B \qquad$ unit of energy $= 1$ Hartree $\qquad$ velocity of light $= \alpha^{-1}$
Rydberg atomic units: $\quad \hbar = 1 \quad m = 1/2 \quad e^2 = 2$
unit of length $= a_B \qquad$ unit of energy $= 1$ Rydberg $\qquad$ velocity of light $= 2 \cdot \alpha^{-1}$

Let us consider a crystal, simple or composite, with one or more atoms in the positions $\mathbf{d}_1, \mathbf{d}_2, \ldots, \mathbf{d}_{\nu_b}$ in the unit cell. For sake of simplicity, we assume that the crystal potential is approximated as sum of atomic-like potentials centred at the atomic positions; the crystal Hamiltonian thus becomes

$$H = -\nabla^2 + V(\mathbf{r}) = -\nabla^2 + \sum_{\mathbf{t}_n \mathbf{d}_\nu} V_{a\nu}(\mathbf{r} - \mathbf{d}_\nu - \mathbf{t}_n) \ . \tag{16}$$

For convenience, Rydberg atomic units are being used from now on (see Table 3), so that the kinetic energy operator $-(\hbar^2/2m)\,\nabla^2 \equiv -(\hbar^2/2m\,a_B^2)\,a_B^2\,\nabla^2$ simply becomes $-\nabla^2$ in Rydberg atomic units.

Chosen a vector $\mathbf{k}$ in the Brillouin zone, consider the set of plane waves $W(\mathbf{k}_i, \mathbf{r})$, of vector $\mathbf{k}_i = \mathbf{k} + \mathbf{h}_i$, and normalized to one, given by

$$W(\mathbf{k}_i, \mathbf{r}) = \frac{1}{\sqrt{N\Omega}} e^{i(\mathbf{k}+\mathbf{h}_i)\cdot\mathbf{r}} \ ,$$

where Ω is the volume of the primitive unit cell, and $\mathbf{h}_i$ are reciprocal lattice vectors [we suppose that the plane waves of vector $\mathbf{k}_i = \mathbf{k} + \mathbf{h}_i$ have been ordered according to increasing values of the moduli $|\mathbf{k}_i|$ and hence of the kinetic energy k_i^2]. The plane waves $\{W(\mathbf{k}+\mathbf{h}_i, \mathbf{r})\}$ constitute a complete orthonormal set for the expansion of Bloch functions of vector $\mathbf{k}$.

The matrix elements of the crystal Hamiltonian (16) between plane waves are very simple; in fact we have

$$\langle W_{\mathbf{k}_i}| - \nabla^2 + V(\mathbf{r})|W_{\mathbf{k}_j}\rangle$$

$$= k_i^2 \delta_{ij} + \frac{1}{N\Omega} \sum_{\mathbf{t}_n \mathbf{d}_\nu} \int e^{-i(\mathbf{k}+\mathbf{h}_i)\cdot\mathbf{r}} V_{a\nu}(\mathbf{r} - \mathbf{d}_\nu - \mathbf{t}_n) e^{i(\mathbf{k}+\mathbf{h}_j)\cdot\mathbf{r}} \, d\mathbf{r}$$

$$= k_i^2 \delta_{ij} + \sum_{\mathbf{d}_\nu} e^{-i(\mathbf{h}_i-\mathbf{h}_j)\cdot\mathbf{d}_\nu} V_{a\nu}(\mathbf{h}_i - \mathbf{h}_j) \tag{17a}$$

where

$$V_{a\nu}(\mathbf{h}_i - \mathbf{h}_j) = V_{a\nu}(|\mathbf{h}_i - \mathbf{h}_j|) = \frac{1}{\Omega} \int e^{-i(\mathbf{h}_i - \mathbf{h}_j)\cdot \mathbf{r}} V_{a\nu}(r)\, d\mathbf{r} \ . \qquad (17b)$$

The phase $\exp[-i(\mathbf{h}_i - \mathbf{h}_j) \cdot \mathbf{d}_\nu]$ is called *structure factor* for the atom in the position $\mathbf{d}_\nu$ in the unit cell, while $V_{a\nu}(\mathbf{h}_i - \mathbf{h}_j)$ is called *form factor* for the atom. We can now expand a crystal wavefunction of vector $\mathbf{k}$ in the form

$$\psi(\mathbf{k}, \mathbf{r}) = \sum_i c_i(\mathbf{k})\, W(\mathbf{k}_i, \mathbf{r}) \ ;$$

standard variational methods on the expansion coefficients $c_i(\mathbf{k})$ lead to the determinantal compatibility equation

$$\left\| (k_i^2 - E)\delta_{ij} + \sum_{\mathbf{d}_\nu} e^{-i(\mathbf{h}_i - \mathbf{h}_j)\cdot \mathbf{d}_\nu} V_{a\nu}(\mathbf{h}_i - \mathbf{h}_j) \right\| = 0 \ . \qquad (18)$$

The secular equation (18) in the plane wave representation is built with elementary ingredients, which are the *empty lattice* kinetic energies k_i^2, and the *form and structure factors* of the atoms in the primitive unit cell.

Following a simple consideration due to Herring, it is easily seen that a *pure expansion of crystal states into plane waves is seriously flawed by the so called "variational collapse" problem*. Consider in fact the secular equation (18) and imagine to truncate and solve it including a finite number n of plane waves (with modulus $|\mathbf{k}_i|$ up to some cut-off value $k_{\max}$). As $n \to \infty$, the lowest roots of the secular equation (18) precipitate towards the lowest core states of the crystal. Since core states are strongly localized in real space, their accurate description with plane waves requires an impossible large number of them. Thus the secular equation (18) is completely unreliable for the crystal states of lowest energy, and is thus unreliable for any other higher energy state too.

To understand better the nature of the "variational collapse" problem, consider for instance the case of silicon (Z=14), whose structure is fcc with lattice constant a=5.43 Å=10.26 a_B, and unit cell volume $\Omega = a^3/4$. The radius a_{1s} of the core state $1s$ of silicon atom can be estimated to be $a_{1s} \approx a_B/Z$, and the cut-off value $k_{\max} \approx 2\pi/a_{1s} = 2\pi Z/a_B$. The number n of reciprocal lattice vectors inside a sphere of radius $k_{\max}$ is given by $(4/3)\pi k_{\max}^3 = n\Omega_k = n(2\pi)^3/\Omega$; we get $n \approx Z^3(a/a_B)^3 \approx 10^6$. Thus, the secular matrix (18) should be of the order of $10^6 \times 10^6$ to reproduce the $1s$ state of silicon, and even larger for heavier elements.

The principle of orthogonalization

The variational collapse problem can be cured with the orthogonalization procedure, introduced by C. Herring, Phys. Rev. **57**, 1169 (1940). We illustrate first the essential idea, and later the development of the orthogonalized plane wave method. Consider an operator H and a complete orthonormal set $\{\phi_i\}$. The eigenvalues and eigenfunctions of H are obtained diagonalizing H within this set, i.e. solving the standard secular equation

$$\|\langle\phi_i|H|\phi_j\rangle - E\delta_{ij}\| = 0 \ .\tag{19}$$

Suppose that the n_c *lowest energy eigenvalues and eigenfunctions of the operator H are already known* (this is the situation of practical interest). We denote these states and eigenvalues by ψ_c and E_c; they satisfy the eigenvalue equation

$$H\psi_c = E_c\psi_c \qquad c = 1, 2, \ldots, n_c \ .\tag{20}$$

We wish to exploit this knowledge to simplify the problem of obtaining all the other eigenvalues and eigenfunctions of the operator H.

A possible way to proceed is to consider for any basis state $|\phi_i\rangle$ the modified state $|\widetilde{\phi}_i\rangle$ orthogonalized to all the known ψ_c eigenstates and given by

$$|\widetilde{\phi}_i\rangle = |\phi_i\rangle - \sum_c |\psi_c\rangle\langle\psi_c|\phi_i\rangle \ .\tag{21}$$

An eigenfunction $|\psi_i\rangle$ of H, orthogonal to the states ψ_c, can be expanded on the set of states $\{\widetilde{\phi}_i\}$ in the form

$$|\psi_i\rangle = \sum_j c_{ij}|\widetilde{\phi}_j\rangle \ ;\tag{22}$$

with standard variational methods, the energies E_i and the expansion coefficients c_{ij} are determined by solving the secular equation

$$\|\langle\widetilde{\phi}_i|H|\widetilde{\phi}_j\rangle - E\langle\widetilde{\phi}_i|\widetilde{\phi}_j\rangle\| = 0 \ .\tag{23}$$

The secular equation (23) *produces only the eigenstates of H orthogonal to the already known states* $\{\psi_c\}$, while the original secular equation (19) obviously produces all the eigenstates of H.

We discuss now in more detail some features of the modified secular equation (23). We begin to notice that the modified states (21) are not orthonormal; in fact the overlap can be expressed in the form

$$\widetilde{S}_{ij} = \langle\widetilde{\phi}_i|\widetilde{\phi}_j\rangle = \delta_{ij} - \langle\phi_i|\left[\sum_c |\psi_c\rangle\langle\psi_c|\right]|\phi_j\rangle \ .\tag{24a}$$

Thus, a preliminary check is to verify that the overlap matrix $\widetilde{S}$ (truncated at a given finite rank of interest) is not singular and that its determinant is actually positive, i.e. $\det\widetilde{S} > 0$; this assures that the orthogonalized basis functions $\{\widetilde{\phi}_i\}$ under attention are still linearly independent and the set is not "overcomplete" [experience shows that the matrix $\widetilde{S}$ tends to become singular as its rank increases; in this case the cure consists in diagonalizing the matrix $\widetilde{S}$ and dropping the redundant linear combination(s)].

Using Eq. (20) and Eq. (21), it is immediately seen that the matrix elements $\widetilde{M}_{ij}$ of the secular equation (23) can be expressed as follows

$$\widetilde{M}_{ij} = \langle\widetilde{\phi}_i|H - E|\widetilde{\phi}_j\rangle = \langle\phi_i|\left[H - E + \sum_c(E - E_c)|\psi_c\rangle\langle\psi_c|\right]|\phi_j\rangle \ .\tag{24b}$$

Thus the secular equation (23) can be written in the alternative and perfectly equivalent form

$$\left\| \langle \phi_i | H + V^{(\mathrm{rep})} | \phi_j \rangle - E \delta_{ij} \right\| = 0 \, , \qquad (25\mathrm{a})$$

where the operator $V^{(\mathrm{rep})}$ is defined as

$$V^{(\mathrm{rep})} = \sum_c (E - E_c) | \psi_c \rangle \langle \psi_c | \, . \qquad (25\mathrm{b})$$

In the secular equation (25a), the effect of orthogonalization is exactly accounted for by adding to H the energy dependent operator (25b), and then considering formally the matrix elements of the *modified operator* $H + V^{(\mathrm{rep})}$ between the set of *unmodified basis functions* ϕ_i. We have supposed that $\{\psi_c\}$ are the lowest eigenfunctions of H; this fact gives a very simple interpretation to the additional operator $V^{(\mathrm{rep})}$ defined by Eq. (25b). For any $E > E_c$ $(c = 1, 2, \ldots, n_c)$, *the expectation value of* $V^{(\mathrm{rep})}$ *on any state can never become negative*; this justifies the interpretation of $V^{(\mathrm{rep})}$ as a kind of repulsive potential, even if its actual formal definition (25b) is quite different from an ordinary local potential.

Before proceeding, it is worthwhile to stress that the secular equation (25) is identical to the secular equation (23), with the same matrix elements $\widetilde{M}_{ij}$ and the same basis functions $|\tilde{\phi}_j\rangle$. In particular it is evident that, once the eigenvalues E_i and corresponding eigenvectors c_{ij} are obtained from Eq. (25), the crystal wavefunctions are $\psi_i = \sum_j c_{ij} \tilde{\phi}_j$ and not $\psi_i = \sum_j c_{ij} \phi_j$; furthermore the normalization $\langle \psi_i | \psi_i \rangle = 1$ implies $\sum_{jj'} c_{ij}^* c_{ij'} \widetilde{S}_{jj'} = 1$ and not $\sum_j c_{ij}^* c_{ij} = 1$. Thus the naive interpretation of the secular equation (25) as the diagonalization of the "non-orthodox" operator $H + V^{(\mathrm{rep})}$ on the orthonormal wavefunctions $\{\phi_i\}$ is not possible; although not possible in a rigorous way, the naive interpretation is neverthless appealing for heuristic elaborations, and is at the basis of the pseudopotential concepts of the next section.

Expansion in orthogonalized plane waves

The basic difficulty of simple plane waves expansion consists in its impossibility to describe (with a reasonable number of plane waves) the strongly localized core states; on the other hand these states are accurately described by the tight-binding method and Herring suggested how to use from the very beginning this information. There is a large number of crystals whose electronic states can be sharply separated into two classes: (i) *inner states*, which are very localized spatially and very deep in energy (core states); (ii) *outer states* (valence and/or conduction states) which are spread out spatially and at higher energy. Herring proposed to describe the former by Bloch sums built from localized orbitals and the latter by plane waves orthogonalized to the core states. Orthogonalized plane waves appear a convenient set to describe itinerant states, since they are atomic-like near the nuclei where the crystal potential is atomic-like and plane waves in the interstitial regions, where the crystal potential is smooth.

Consider a plane wave $W(\mathbf{k}_j, \mathbf{r})$ of vector $\mathbf{k}_j = \mathbf{k} + \mathbf{h}_j$; we orthogonalize it to all the core Bloch sums $\Phi_c(\mathbf{k}, \mathbf{r})$ and obtain the OPW function in the form

$$|\widetilde{W}(\mathbf{k}_j, \mathbf{r})\rangle = |W(\mathbf{k}_j, \mathbf{r})\rangle - \sum_{\text{core}} |\Phi_c(\mathbf{k}, \mathbf{r})\rangle\langle\Phi_c(\mathbf{k}, \mathbf{r})|W(\mathbf{k}_j, \mathbf{r})\rangle . \tag{26}$$

We now expand *outer crystal states* of vector $\mathbf{k}$ into orthogonalized plane waves of vectors $\mathbf{k}_j = \mathbf{k} + \mathbf{h}_j$; standard variational procedures on the coefficients of the expansion lead to the determinantal compatibility equation

$$\|\langle\widetilde{W}_{\mathbf{k}_i}|H|\widetilde{W}_{\mathbf{k}_j}\rangle - E\langle\widetilde{W}_{\mathbf{k}_i}|\widetilde{W}_{\mathbf{k}_j}\rangle\| = 0 . \tag{27a}$$

The lowest eigenvalue of (27a) corresponds to the lowest (valence or conduction) crystal eigenstate. The secular equation (27a), using the expression (26), can be written in the alternative and perfectly equivalent form

$$\boxed{\|\langle W_{\mathbf{k}_i}| -\nabla^2 + V(\mathbf{r}) + V^{(\text{rep})}|W_{\mathbf{k}_j}\rangle - E\delta_{ij}\| = 0} , \tag{27b}$$

where $H = -\nabla^2 + V(\mathbf{r})$ is the crystal Hamiltonian and the repulsive operator $V^{(\text{rep})}$ is defined as

$$\boxed{V^{(\text{rep})} = \sum_{\text{core}} (E - E_c)|\Phi_{c\mathbf{k}}\rangle\langle\Phi_{c\mathbf{k}}|} .$$

The operator $V^{(\text{rep})}$, defined above, is evidently a "non-orthodox" operator, which is energy-dependent and non-local; *qualitatively*, it can be interpreted as a repulsive potential produced by the presence of core states.

In the orthogonalized plane wave method, one sets up the secular equation (27a), or the equivalent secular equation (27b), using a finite number n of OPWs (up to a given $k_{\max}$); then the stability of the eigenvalues in the energy range of interest is checked as the order of the secular equation is increased. In general a reasonable rapid convergence is obtained, a number of ten to hundred OPWs being sufficient in most cases. In the OPW method, a check must be always performed to assure that the overlap matrix $\widetilde{S}_{ij} = \langle\widetilde{W}_{\mathbf{k}_i}|\widetilde{W}_{\mathbf{k}_j}\rangle$ is not singular (otherwise the corresponding redundant linear combinations must be suppressed). From a mathematical point of view, the rapid convergence of the method is related to the *cancellation effect* between orthogonalization terms and Fourier transforms of the crystal potential when the wavevector transfers $|\mathbf{h}_i - \mathbf{h}_j|$ are large.

Evaluation of the matrix elements of the OPW method

The ingredients to set up the secular equation (27) of the OPW method are the matrix elements of the crystal Hamiltonian and of the repulsive operator between (ordinary) plane waves. The former have already been discussed in Eqs. (17); the latter are now discussed beginning with the case of a simple lattice. Consider the Bloch sum

$$\Phi_c(\mathbf{k}, \mathbf{r}) = \frac{1}{\sqrt{N}} \sum_{\mathbf{t}_n} e^{i\mathbf{k}\cdot\mathbf{t}_n} \phi_c(\mathbf{r} - \mathbf{t}_n)$$

corresponding to the atomic core function $\phi_c(\mathbf{r})$; the overlap between core functions and plane waves is

$$\langle \Phi_{c\mathbf{k}} | W_{\mathbf{k}_j} \rangle = \frac{1}{\sqrt{N}} \sum_{\mathbf{t}_n} \int e^{-i\mathbf{k}\cdot\mathbf{t}_n} \phi_c^*(\mathbf{r} - \mathbf{t}_n) \frac{1}{\sqrt{N\Omega}} e^{i(\mathbf{k}+\mathbf{h}_j)\cdot\mathbf{r}} \, d\mathbf{r}$$

$$= \frac{1}{N\sqrt{\Omega}} \sum_{\mathbf{t}_n} \int e^{i(\mathbf{k}+\mathbf{h}_j)\cdot(\mathbf{r}-\mathbf{t}_n)} \phi_c^*(\mathbf{r} - \mathbf{t}_n) \, d\mathbf{r}$$

$$= \frac{1}{\sqrt{\Omega}} \int e^{i(\mathbf{k}+\mathbf{h}_j)\cdot\mathbf{r}} \phi_c^*(\mathbf{r}) \, d\mathbf{r} \ . \tag{28}$$

We use the familiar expansion of plane waves in spherical harmonics

$$e^{i\mathbf{k}_j\cdot\mathbf{r}} = 4\pi \sum_{lm} i^l j_l(k_j r) Y_{lm}^*(\mathbf{k}_j) Y_{lm}(\mathbf{r}) \ , \tag{29}$$

where $Y_{lm}(\mathbf{r})$ stands for $Y_{lm}(\theta_\mathbf{r}, \phi_\mathbf{r})$, $Y_{lm}(\mathbf{k})$ stands for $Y_{lm}(\theta_\mathbf{k}, \phi_\mathbf{k})$ and $j_l(x)$ is the spherical Bessel function of order l. We write the atomic core state as product of a radial part and an angular part in the usual form $\phi_c(\mathbf{r}) = R_{nl}(r) Y_{lm}(\mathbf{r})$. The orthogonalization coefficient (28), re-labelled as $\langle \Phi_{nlm\mathbf{k}} | W_{\mathbf{k}_j} \rangle$, becomes

$$\langle \Phi_{nlm\mathbf{k}} | W_{\mathbf{k}_j} \rangle = \frac{4\pi}{\sqrt{\Omega}} i^l Y_{lm}^*(\mathbf{k}_j) \int_0^\infty R_{nl}(r) j_l(k_j r) r^2 \, dr \ .$$

The matrix elements of the operator $V^{(\mathrm{rep})}$ between plane waves is

$$\langle W_{\mathbf{k}_i} | V^{(\mathrm{rep})} | W_{\mathbf{k}_j} \rangle = \sum_{nlm}^{\mathrm{core}} (E - E_{nl}) \langle W_{\mathbf{k}_i} | \Phi_{nlm\mathbf{k}} \rangle \langle \Phi_{nlm\mathbf{k}} | W_{\mathbf{k}_j} \rangle \ .$$

The sum over m ($m = -l, -l+1, \dots, +l$) can be performed using the addition theorem for spherical harmonics

$$P_l(\widehat{\mathbf{k}}_i \cdot \widehat{\mathbf{k}}_j) = \frac{4\pi}{2l+1} \sum_{m=-l}^{+l} Y_{lm}^*(\mathbf{k}_i) Y_{lm}(\mathbf{k}_j) \ , \tag{30}$$

where $\widehat{\mathbf{k}}_i$ and $\widehat{\mathbf{k}}_j$ denote the unit vectors of $\mathbf{k}_i$ and $\mathbf{k}_j$, and P_l is the Legendre polynomial of order l. After defining for every core state the *orthogonalization coefficients*

$$A_{nl}(k_i) = \left[\frac{4\pi(2l+1)}{\Omega} \right]^{1/2} \int_0^\infty R_{nl}(r) j_l(k_i r) r^2 \, dr \ ,$$

we obtain the expression

$$\langle W_{\mathbf{k}_i} | V^{(\mathrm{rep})} | W_{\mathbf{k}_j} \rangle = \sum_{nl}^{\mathrm{core}} (E - E_{nl}) P_l(\widehat{\mathbf{k}}_i \cdot \widehat{\mathbf{k}}_j) A_{nl}(k_i) A_{nl}(k_j) \ . \tag{31}$$

The ingredients for the matrix elements of the OPW method are thus the Fourier transforms of the atomic potential at reciprocal lattice vectors, and the orthogonalization coefficients.

The matrix elements of the OPW method for *composite lattices*, with atoms in the

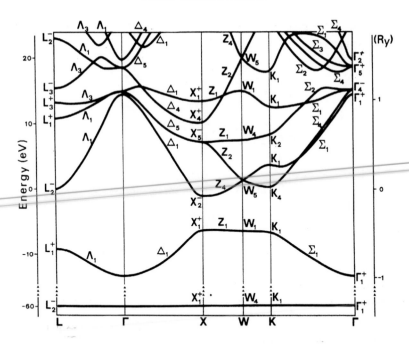

Fig. 5 Band structure of solid lithium hydride with the orthogonalized plane wave method [from S. Baroni, G. Pastori Parravicini and G. Pezzica, Phys. Rev. B**32**, 4077 (1985); copyright 1985 by the American Physical Society].

positions $\mathbf{d}_1, \mathbf{d}_2, \ldots, \mathbf{d}_{\nu_b}$ in the primitive unit cell, are obtained in a rather similar procedure; each atom contributes with an appropriate number of core states, and the corresponding orthogonalization terms include the appropriate structure factor, due to the atomic position in the unit cell.

We conclude this section recalling some properties of the functions used in the OPW method. The Legendre polynomials are obtained from the recursion relation

$$(l+1)P_{l+1}(x) - (2l+1)x\,P_l(x) + l\,P_{l-1}(x) = 0$$

with starting conditions $P_0(x) = 1$ and $P_1(x) = x$. The Bessel functions of interest can be obtained from the definitions

$$j_0(x) = \frac{\sin x}{x}, \qquad j_1(x) = \frac{\sin x - x\cos x}{x^2}, \qquad (32a)$$

and the recurrence relation

$$j_{l+1}(x) = \frac{2l+1}{x} j_l(x) - j_{l-1}(x) ; \qquad (32b)$$

for small values of x it holds

$$j_l(x) \rightarrow \frac{x^l}{1 \cdot 3 \cdot \ldots \cdot (2l+1)} \qquad \text{as } x \rightarrow 0 . \qquad (32c)$$

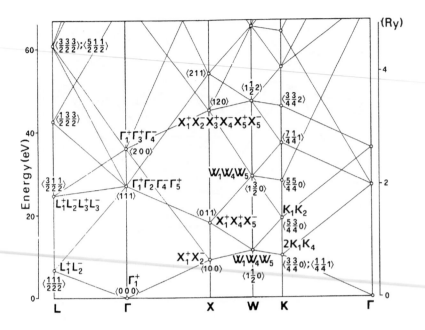

Fig. 6 Empty lattice energy states for face-centered cubic lattices at the symmetry points L, Γ, X, W, K; the lattice parameter is taken to be $a = 7.72\, a_B$, as in LiH. The straight lines are simply a guide to the eye to link empty lattice states at different symmetry points. The symmetry notations are those of G. F. Koster, J. O. Dimmock, R. C. Wheeler and H. Statz "Properties of the Thirty-Two Point Groups" (MIT, Cambridge, Massachusetts 1963).

Illustrative example: valence and conduction bands of lithium hydride

As an application of the orthogonalized plane wave method, we consider the band structure of lithium hydride. The crystal structure of LiH is fcc, with lattice parameter $a = 7.72$ a.u.; there are two ions in the unit cell: H^- in the position $\mathbf{d}_1 = 0$ and Li^+ in the position $\mathbf{d}_2 = (a/2)(1,0,0)$. There are only four electrons in the unit cell, and this simplicity adds to the interest in this prototype material.

The Li^+ ion has the closed shell electronic configuration $1s^2$. We have only a core state and thus the application of the OPW method is particularly simple and significant. To describe valence and conduction bands, we orthogonalize the plane waves to the Bloch sum formed with the $1s$ wavefunctions of Li^+. The matrix elements appearing in the OPW method, including the exchange potential in its non-local form, can be evaluated analytically. The band structure of lithiun hydride is reported in Fig. 5. The core band at ≈ -60 eV originates from the $1s$ core orbitals of the Li^+ ions. The valence band is s-like, with its minimum at Γ and its maximum at the point X of the Brillouin zone; the valence band is mostly related to the $1s$ orbitals of the H^- ions. The conduction band has its minimum at X; the material is an insulator with a direct energy gap of about 5 eV.

It is of interest to report the empty lattice energy states in face-centered cubic lattices at some relevant symmetry points (Fig. 6). The general trend of valence and conduction bands of Fig. 5 can be orientatively understood from the empty lattice states of Fig. 6. In fact the accidental degeneracies of the empty lattice are removed by the introduction of the crystal potential and repulsive potential (due to orthogonalization) in agreement with the qualitative rules of perturbation theory; this suggests that the "effective potential" $V(\mathbf{r}) + V^{(\mathrm{rep})}$ is reasonably smooth, so that valence and conduction band wavefunctions can be well approximated by a relatively small number of orthogonalized plane waves.

4 The pseudopotential method

Introduction and empirical pseudopotentials for crystals

In the study of the electronic states in solids, we can often separate the chemically inert core states of the atoms from the chemically active valence states. The former do not change significantly from free atoms to solids, and do not influence the chemical properties of materials; the latter, with self-consistent readjustment of wavefunctions and charge densities to different environment situations, determine the actual properties of materials.

We have seen that *the basic strategy of the* OPW *method is to exploit the presence of the core states* for an efficient determination of the valence and conduction energy bands, as well as of the crystalline wavefunctions in the whole sample, i.e. *both in the core regions and outside them.* As we shall see, *the basic strategy of the pseudopotential method is to get rid of the core states, by replacing the strong crystalline potential with a weak pseudopotential,* capable of an efficient determination of valence and conduction energy bands; for what concerns the corresponding pseudowavefunctions, these are required to represent the genuine crystalline wavefunctions only *outside the core regions,* without worrying of the smoothing occurring in the chemically inert core regions. The link and evolution between the OPW method and the pseudopotential method can be understood with the following arguments.

In the orthogonalized plane wave method, the core states are taken into account by enforcing the orthogonalization requirement of valence and conduction states to them. According to the OPW method, the valence and conduction states are obtained solving the secular equation corresponding to the effective operator

$$H^{(\mathrm{eff})} = -\nabla^2 + V(\mathbf{r}) + \sum_{\mathrm{core}} (E - E_c) |\Phi_{c\mathbf{k}}\rangle \langle \Phi_{c\mathbf{k}}| \,, \tag{33}$$

where $-\nabla^2 + V(\mathbf{r})$ is the ordinary crystal Hamiltonian; the additional operator is an energy-dependent non-local operator, whose origin is related to the presence of core states. At first sight, the *formal aspect* of the orthogonalization-related terms appearing in Eq. (33) (with their non-local and energy dependent nature) does not seem particularly encouraging for simplifications (in the OPW these terms are evaluated just as they stand); nevertheless the physical aspect is rather appealing. In essence, the effect of the orthogonalization terms is to cancel (to a large extent) the true all-electron crystal potential $V(\mathbf{r})$ just in the core region, where $V(\mathbf{r})$ is particularly strong. From

this, the hope arises that the "effective potential" (true potential plus orthogonalization terms) appearing in Eq. (33), can be somehow mimicked by an appropriate
weak *pseudopotential* $V^{(pseudo)}$, smoother than the true potential $V(\mathbf{r})$ *within* the core
region, and equal to it *outside*.

To corroborate this point of view with a simple example, consider for instance the
case of Na atom (electronic configuration: $1s^2; 2s^2 2p^6; 3s^1$). The atomic potential $V_a(r)$
behaves as $-11\, e^2/r$ for small r and as $-e^2/r$ for large r. We consider $1s, 2s$ and $2p$
shells as atomic core states, and we try to infer the atomic pseudopotential acting
on the $3s$ optical electron; from the fact that the orbital energy of the $3s$ state is
≈ -5.14 eV, we can safely conclude that the effective potential in the core region is
much softer than the hydrogen potential $-e^2/r$ (whose ground energy is -13.6 eV);
thus the true potential $V_a(r)$ has been largely cancelled in the core region. Much more
general plausibility and formal arguments are available in the literature, to corroborate
this point of view; in particular, pseudopotentials have appeared in solid state physics
following the pioneering works of J. C. Phillips and L. Kleinman, Phys. Rev. **116**, 287
(1959) and F. Bassani and V. Celli, J. Phys. Chem. Solids **20**, 64 (1961).

The above guidelines have been combined with physical ingenuity to mimic as better
as possible the exact operator $V(\mathbf{r})+\sum_c (E-E_c)|\Phi_{c\mathbf{k}}\rangle\langle\Phi_{c\mathbf{k}}|$ of Eq. (33) with a smooth
pseudopotential (possibly local and energy independent). The simplest *empirical* way
to introduce a crystal pseudopotential is to express it as sum of local spherically
symmetric atomic pseudopotentials centred in the atomic positions. In this case the
secular equation (19) for *valence and conduction* states becomes

$$\left\| (k_i^2 - E)\delta_{ij} + \sum_{\mathbf{d}_\nu} e^{-i(\mathbf{h}_i - \mathbf{h}_j)\cdot \mathbf{d}_\nu} V_{a\nu}^{(pseudo)}(\mathbf{h}_i - \mathbf{h}_j) \right\| = 0 \quad . \tag{34}$$

It is worthwhile to stress that Eq. (34) is formally similar to Eq. (18), but strongly
different from it in the meaning. Eq. (18) is built with the true atomic potentials;
it thus describes all crystal states (core, valence and conduction) and is subject to
the "variational collapse" problem, previously discussed. Eq. (34) is built with atomic
pseudopotentials; it thus describes only valence and conduction states, and is mathematically free of the variational collapse problem.

The matrix elements of the secular equation (34) can be expressed in terms of a few
Fourier components of the atomic pseudopotentials at the reciprocal lattice vectors;
often these components are considered as disposable parameters. Their number can
be significantly limited by requiring that these parameters are zero when the transfer
wavevector $q = |\mathbf{h}_i - \mathbf{h}_j|$ is sufficiently large (so that the cancellation, expected for
small r is at work). The small number of disposable parameters are determined in
such a way to fit basic properties of the crystal (for instance energy gap, some relevant
transition energies, etc.). With a little amount of computational labour and with a
limited knowledge of the experimental data of the solid, one can derive the whole band
structure and explain a number of properties (such as optical constants, photoemission

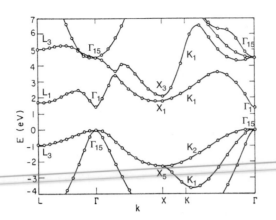

Fig. 7 Energy bands of GaAs by the empirical pseudopotential method [from M. L. Cohen and T. K. Bergstresser, Phys. Rev. **141**, 789 (1966); copyright 1966 by the American Physical Society].

spectra, structural properties); this kind of approach is referred in the literature as *empirical pseudopotential method*.

As an illustrative application, we briefly consider the semiconductors with the diamond or zincblende structure; this structure is fcc with two atoms in the unit cell in the positions $\mathbf{d}_1 = 0$ and $\mathbf{d}_2 = (a/4)(1,1,1)$. The secular equation (34) of the empirical pseudopotential method becomes:

$$\left\| (k_i^2 - E)\delta_{ij} + V_1^{(\text{pseudo})}(\mathbf{h}_i - \mathbf{h}_j) + e^{-i(\mathbf{h}_i - \mathbf{h}_j)\cdot\mathbf{d}_2} V_2^{(\text{pseudo})}(\mathbf{h}_i - \mathbf{h}_j) \right\| = 0 \qquad (35)$$

[in actual calculations, it is generally preferred to choose the origin of coordinates halfway between the two atoms of the unit cell, and break up the crystal pseudopotential into a symmetric and an antisymmetric part; for details we refer to the classic work of M. L. Cohen and T. K. Bergstresser, Phys. Rev. **141**, 789 (1966)].

The first five shells of fcc reciprocal lattice vectors are: $(2\pi/a)(0,0,0)$ (1 vector); $(2\pi/a)(\pm1,\pm1,\pm1)$ [8 vectors]; $(2\pi/a)$ $(\pm2,0,0)$ and permutations [6 vectors]; $(2\pi/a)$ $(\pm2,\pm2,0)$ and permutations [12 vectors]; $(2\pi/a)(\pm3,\pm1,\pm1)$ and permutations [24 vectors]. The magnitude of the vectors of the first five shells are 0, $\sqrt{3}$, $\sqrt{4}$, $\sqrt{8}$, $\sqrt{11}$ times $(2\pi/a)$ (for the lowest ones, see Fig. 6 at Γ), and only these magnitudes are allowed to have a non-zero pseudopotential in the paper of Cohen and Bergstresser. The value of the crystal pseudopotential for $\mathbf{h} = 0$ is irrelevant, since it merely adds a constant to all energy levels. We notice that $\exp(-i\mathbf{h} \cdot \mathbf{d}_2) = -1$ for any of the six $\mathbf{h}$ vectors of the type $(2\pi/a)(\pm2,0,0)$; and $\exp(-i\mathbf{h} \cdot \mathbf{d}_2) = +1$ for any of the twelve $\mathbf{h}$ vectors of the type $(2\pi/a)(\pm2,\pm2,0)$. Thus, the secular Eq. (35) for homopolar substances contains only three independent parameters: $V(\sqrt{3})$, $V(\sqrt{8})$ and $V(\sqrt{11})$. For instance, for silicon, the fitted pseudopotential form factors are found to be: $V(\sqrt{3}) = -0.21$ Ry, $V(\sqrt{8}) = 0.04$ Ry, $V(\sqrt{11}) = 0.08$ Ry, while for germanium: $V(\sqrt{3}) = -0.23$ Ry, $V(\sqrt{8}) = 0.01$ Ry, and $V(\sqrt{11}) = 0.06$ Ry. For heteropolar compounds, the secular equation (35) contains six independent parameters, namely: $V_1(\sqrt{3})$, $V_2(\sqrt{3})$, $V_1(\sqrt{4}) - V_2(\sqrt{4})$, $V_1(\sqrt{8}) + V_2(\sqrt{8})$, $V_1(\sqrt{11})$, $V_2(\sqrt{11})$. A so reasonably small number of independent parameters is appealing, and explains why the empirical pseudopotential method has been extremely useful for understanding band structure and trends in several crystals.

As an example, we report in Fig. 7 the energy bands of GaAs, a direct gap material of great importance in optoelectronics. The valence bands of GaAs have the maximum at $\mathbf{k} = 0$; the symmetry of the top valence state is Γ_{15} (a three-fold degenerate level). The minimum of the conduction bands, which also occurs at $\mathbf{k} = 0$, is the total-symmetric state Γ_1; the energy bandgap is $E_G = E(\Gamma_1) - E(\Gamma_{15}) \approx 1.4\,\text{eV}$. Notice that the minima of the conduction bands at L_1 and X_1 are not far above Γ_1, and this is at the origin of the Gunn effect in the transport properties of GaAs (see Section XIII-5.1).

Atomic pseudopotentials

An appealing evolution of the pseudopotential framework appears in the literature in the 1970's, and consists in a skillful use of atomic properties to fix the pseudopotentials of the atoms composing the solid. There is not a unique recipe for the elaboration of atomic pseudopotentials, and this has sometimes cast (unjustified) doubts on the reliability of the procedure. In spite of some degree of arbitrariness, the essential guidelines apply on whatever pseudopotential model and assure, in general, a reasonable standard of quality.

The first step is an *all-electron* self-consistent atomic calculation, usually in the local density approximation. The atomic potential $V_a(r)$ and the atomic wavefunctions are solutions of the radial Schrödinger equation

$$\left[-\frac{d^2}{dr^2} + \frac{l(l+1)}{r^2} + V_a(r) \right] r\, R_{nl}(r) = E_{nl}\, r\, R_{nl}(r) \ . \tag{36}$$

Once the all-electron potential $V_a(r)$ is known, one proceeds towards the determination of a *pseudopotential* $V_a^{(\text{pseudo})}(r)$ *for the pseudoatom free from the core states*, and yet capable to describe the physics and chemistry of the external valence electrons. Most often, the pseudopotential $V_a^{(\text{pseudo})}(r)$ is taken in the form of the type

$$V_a^{(\text{pseudo})}(r) = \begin{cases} V(r; \{\lambda_i\}) & \text{for} \quad r < r_c \\ V_a(r) & \text{for} \quad r > r_c \end{cases} . \tag{37}$$

Inside the core region, the pseudopotential (37) is approximated by some reasonable functional form, containing one or more adjustable parameters $\{\lambda_i\}$; outside the core region, the pseudopotential (37) coincides with $V_a(r)$. The radius r_c (although it carries some degree of arbitrariness from model to model) is however indicative of the extension of the core states of the atom. The one or more adjustable parameters $\{\lambda_i\}$ are then determined by fitting to a number of experimental inputs (for instance atomic spectroscopic data for outer electrons, electron-atom cross section etc.).

In considering the needed steps to elaborate an all-electron atomic calculation into a useful atomic pseudopotential, we illustrate the specific case of the Na atom. We suppose to have performed the all-electron calculation and to know the atomic potential $V_{Na}(r)$, the eigenvalues and the radial wavefunctions of the occupied states. We consider $1s, 2s$ and $2p$ states as core states, and we outline possible procedures to construct an atomic pseudopotential free from the core states, and yet capable to describe the physics and chemistry of the external $3s$ optical electron.

The simplest form of parametrization of type (37), as suggested by Ashcroft, and Heine and Abarenkov, is

$$V_{Na}^{(pseudo)}(r) = \begin{cases} A & \text{for} \quad r < r_c \\ -\dfrac{e^2}{r} & \text{for} \quad r > r_c \end{cases} \tag{38}$$

[for other forms see for instance F. Nogueira, C. Fiolhais and J. P. Perdew, Phys. Rev. B**59**, 2570 (1999) and references quoted therein]. In Eq. (38), the pseudopotential is constant for $r < r_c$, and correctly equals $-e^2/r$ for $r > r_c$. The adjustable parameters A and r_c are chosen so that the ground state energy of the pseudopotential Hamiltonian $-\nabla^2 + V_{Na}^{(pseudo)}(r)$ equals the orbital energy $E_{3s} = -5.14$ eV of the Na atom (this quantity is taken either from theoretical calculations or from experimental spectroscopic measurements). If the flexibility of the pseudopotential allows it, other properties (for instance higher eigenvalues) of the optical electron are enforced.

Ab initio norm-conserving pseudopotentials

A major breakthrough in the theory of pseudopotentials has occurred in 1979 with the concept of norm-conserving pseudopotentials of Hamann, Schlüter and Chiang. This has further improved the quality of the last generation pseudopotentials, allowing major developments in the field of electronic state calculations. The concept starts from the technical observation *that in the region outside the core, where the true potential and the pseudopotential coincide, the atomic radial wavefunction $R_a(r)$ and the corresponding atomic pseudowavefunction $R_a^{(pseudo)}(r)$ are proportional to each other, but in general are not rigorously equal* (as they should be). In fact integration of the radial Schrödinger equation (at the same energy) from infinity obviously produces the same wavefunctions within a normalization factor as far as the pseudopotential and true potential coincide (the two normalizations are in general different, as they depend on the integration up to the origin, the very region where the true potential and the pseudopotential are different). The basic principle of the *norm-conserving pseudopotential is just to enforce the condition* $R_a^{(pseudo)}(r) \equiv R_a(r)$ for $r > r_c$; this assures that pseudo-charge density and true charge density outside the core region are perfectly equal.

We present now some considerations for the generation of norm-conserving pseudopotentials for atoms; for simplicity, we make reference to some aspects of the procedure introduced by G. P. Kerker, J. Phys. C**13**, L189 (1980) (the wealth of implementations and variants could not be summarized here; in the "art" of pseudopotentials there is wide space for special requirements and ingenuity). In agreement to the general concepts of norm-conserving pseudopotentials, the Kerker procedure *does not parametrize the pseudopotential in the core region* (as done for instance in Eq. 38); *rather it parametrizes the ground-state pseudowavefunction* in the form

$$R_a^{(\text{pseudo})}(r) = \begin{cases} r^l e^{p(r)} & \text{for } r < r_c \\ R_a(r) & \text{for } r > r_c \end{cases} . \tag{39a}$$

In Eq. (39a), l is the angular momentum of the radial wavefunction $R_a(r)$; $p(r)$ is a polynomial of degree 4 of the type

$$p(r) = \lambda_0 + \lambda_2\, r^2 + \lambda_3\, r^3 + \lambda_4\, r^4 \tag{39b}$$

and $\lambda_0, \lambda_2, \lambda_3, \lambda_4$ are four disposable parameters; the coefficient λ_1 is taken as zero so that $p'(r) \div r$ for $r \to 0$, and $V_a^{(\text{pseudo})}(r)$ at $r = 0$ is not singular, as can be seen from Eq. (39c) below. It is evident that also for $r < r_c$ the radial pseudowavefunction $R_a^{(\text{pseudo})}(r)$ is nodeless (this requirement *excludes* that undesired pseudowavefunctions of lower energy might exist). The four coefficients of the polynomial $p(r)$ are determined by the following four conditions: normalization to 1 of the wavefunction (this means that charge conservation within the core region is guaranteed), continuity of function, first and second derivative at r_c.

Once the coefficients $\lambda_0, \lambda_2, \lambda_3, \lambda_4$ and hence $p(r)$ are determined, $R_a^{(\text{pseudo})}(r)$ is replaced into the radial Schrödinger equation (36), and one obtains for the norm-conserving pseudopotential the analytic expression

$$V_a^{(\text{pseudo})}(r) = \begin{cases} E_a + \dfrac{l+1}{r} p'(r) + p''(r) + [p'(r)]^2 & \text{for} \quad r < r_c \\ V_a(r) & \text{for} \quad r > r_c \end{cases} , \tag{39c}$$

where E_a is the energy of the atomic radial wavefunction $R_a(r)$ under consideration.

In the construction of pseudopotentials one arrives in general to a *pseudopotential different for each angular momentum*. The matrix elements of such angular momentum dependent pseudopotential between plane waves are still manageable enough to be explicitly calculated, using the standard spherical waves expansion of plane waves. Another significant aspect of the art of generating norm-conserving pseudopotentials concerns their transferability to different environments, this being a prerequisite for accurate first principle calculations of properties of solids. As an example, we consider the silicon atom and in Fig. 8a we report the all-electron wavefunctions $3s$, $3p$ and $3d$, and the corresponding nodeless pseudowavefunctions; the latter coincide with the true wavefunctions for large r, and appropriately extrapolate to zero for small r. The non-local pseudopotentials of Si for $l=0, 1, 2$ are reported in Fig. 8b, and are shown to be rather smooth compared with the Coulomb potential of a (fictitious) point-like atomic core.

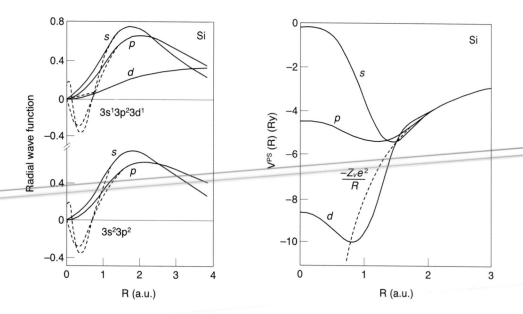

Fig. 8 (a) Comparison of the pseudowavefunction (solid lines) and the corresponding all-electron wavefunctions (dashed lines) for the configurations $3s^2 3p^2$ and $3s^1 3p^2 3d^1$ of silicon. (b) Non-local pseudopotential of Si for angular momentum $l = 0, 1, 2$. The dashed line denotes the Coulomb potential of a point-like atomic core [from M. T. Yin and M. L. Cohen, Phys. Rev. B**26**, 5668 (1982); copyright 1982 by the American Physical Society].

5 The cellular method

The cellular method was originally introduced by Wigner and Seitz (1933-1934) in the study of the s-like conduction band in alkali metals, and explicitly formulated in a more general context by Slater (1934). In the cellular method attention is focused to the Wigner–Seitz unit cell; within it, *the crystal potential is assumed to be spherically symmetric*, the radial Schrödinger equation is solved, and appropriate *boundary conditions at the surface of the unit cell* are applied. The cellular method has been widely used for about three decades since its original formulation; then the advent of large computers made it possible to perform calculations with the more satisfactory augmented plane wave method and Green's function method, and the cellular method has thus lost most of its importance for practical calculations. Despite this, the cellular method well deserves illustration, because of its important concepts and its historical relevance.

We consider first crystals with one atom per unit cell. As unit cell we choose the Wigner–Seitz polyhedron, and we make the basic assumption that the actual crystal potential within the polyhedron has spherical symmetry; in many practical cases this assumption is well justified because of the dominant contribution of the atomic potential at the centre of the Wigner–Seitz cell.

A crystal state $\psi(\mathbf{k}, \mathbf{r})$ of energy $E(\mathbf{k})$ can be expanded within the Wigner–Seitz cell in the form

$$\psi(\mathbf{k}, \mathbf{r}) = \sum_{lm} c_{lm}(\mathbf{k}) \, Y_{lm}(\mathbf{r}) \, R_l(E, r) \qquad \mathbf{r} \in \Omega_{WS} , \tag{40}$$

where $l = 0, 1, 2, \ldots$ and $m = -l, -l+1, \ldots, +l$; $Y_{lm}(\mathbf{r})$ are spherical harmonic functions [$Y_{lm}(\mathbf{r})$ stands for $Y_{lm}(\theta_{\mathbf{r}}, \phi_{\mathbf{r}})$ where $\theta_{\mathbf{r}}, \phi_{\mathbf{r}}$ are the polar coordinates of $\mathbf{r}$ and the origin is at the center of the Wigner–Seitz cell Ω_{WS}]; c_{lm} are appropriate coefficients not yet specified. The function $R_l(E, r)$ is the solution regular at the origin of the radial wave equation

$$\frac{d^2(r\,R_l)}{dr^2} = \left[V(r) + \frac{l(l+1)}{r^2} - E \right] (r\,R_l)$$

and $V(r)$ is the spherically symmetric cellular potential; $R_l(E, r)$ can be obtained by integrating numerically, outward from the origin, the above radial wave equation.

At the surface of the Wigner–Seitz polyhedron, a crystal state of vector $\mathbf{k}$ must satisfy the boundary conditions required by the Bloch theorem

$$\psi(\mathbf{k}, \mathbf{R}_2) = e^{i\mathbf{k}\cdot\mathbf{t}_{12}} \, \psi(\mathbf{k}, \mathbf{R}_1) \tag{41a}$$

$$\mathbf{n} \cdot \nabla \psi(\mathbf{k}, \mathbf{R}_2) = e^{i\mathbf{k}\cdot\mathbf{t}_{12}} \, \mathbf{n} \cdot \nabla \psi(\mathbf{k}, \mathbf{R}_1) \tag{41b}$$

for every pair of points $\mathbf{R}_1$ and $\mathbf{R}_2$ on opposite faces of the Wigner–Seitz cell, connected by a lattice translational vector $\mathbf{t}_{12}$; $\mathbf{n}$ is a unit vector normal to the pair of faces to which $\mathbf{R}_1$ and $\mathbf{R}_2$ belong (see Fig. 9).

In principle the arbitrary coefficients $c_{lm}(\mathbf{k})$ and the crystal energies $E(\mathbf{k})$ are determined by inserting expansion (40) into Eqs. (41), considering the resulting homogeneous equations in the coefficients c_{lm} and requiring that the corresponding determinant vanishes. In practice one restricts the sum in Eq. (40) to a finite number of l values (for instance if one wishes to reproduce s, p and d character of crystal states, $l_{max} = 2$ and there are 9 disposable coefficients c_{lm}) and selects a corresponding finite number of boundary conditions. Practical recipes for judicious selections of surface points have been provided, and often tested for the case $V(r) = 0$, for which the exact eigenfunctions are plane waves and the exact eigenvalues are $(\mathbf{k} + \mathbf{h}_n)^2$ (Shockley empty lattice test). However the unavoidable arbitrariness in the choice of surface points constitutes the basic limitation of the cellular method.

At this point it is interesting to mention the original calculation of Wigner and Seitz for the bottom of the conduction band (which occurs at $\mathbf{k} = 0$) for alkali metals. Wigner and Seitz limited the expansion (40) to the term $R_0(E, r)$ and replaced the Wigner–Seitz polyhedron with a sphere of radius r_s of equal volume. For $\mathbf{k} = 0$, the boundary condition (41a) is automatically satisfied, while (41b) implies $dR_0(E, r)/dr = 0$ for $r = r_s$. This condition for the solid should be contrasted with the standard condition of regularity at infinity for the radial wavefunctions of the isolated atoms; this difference in boundary conditions explains in part the cohesive energy of alkali metals.

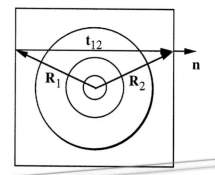

Fig. 9 Schematic visualization of the Wigner–Seitz cell, of the spherically symmetric cellular potential, and of a pair of surface conjugate points $\mathbf{R}_1$ and $\mathbf{R}_2$ for imposing boundary conditions.

The cellular method has been also extended to composite crystals. In this case the unit cell is divided into a number of subcells equal to the number of atoms in the unit cell. Within each Wigner–Seitz subcell, the crystal potential is assumed to have spherical symmetry, polar coordinates with respect to the center are introduced, and expansion in spherical waves considered. Besides boundary conditions at the surface of the cell, one has now to consider standard continuity conditions at the subcell surfaces.

6   The augmented plane wave (APW) method

6.1   Description of the method

The augmented plane wave (APW) method was originally proposed by Slater in 1937 in order to overcome the problem of boundary conditions of the cellular method. The APW method finds its motivation in the fact that, in many solids, the crystal potential energy can be well approximated by a *muffin-tin potential*: so is called a potential spherically symmetric within non-overlapping spheres centered about each atom, and constant elsewhere (see Fig. 10); the shape of the potential is reminiscent of the shape of a popular kitchen mold for muffins. The essential idea of the APW method consists in representing the crystal states by cellular type functions within the muffin-tin spheres of radius r_{MT} and plane waves outside them; a much easier matching problem at the sphere surface provides the band states more satisfactorily than in the cellular method.

We begin to illustrate the method for the case of simple crystals with one atom per unit cell. We assume the crystal potential in the muffin-tin form; the constant value of the muffin-tin potential outside the non-overlapping spheres of radius r_s can be taken as zero, upon appropriate shift of the energy scale.

Within the unit cell we consider the Schrödinger equation (in atomic units)

$$[-\nabla^2 + V(r)]\psi(\mathbf{k}, \mathbf{r}) = E\psi(\mathbf{k}, \mathbf{r}) \, ,$$

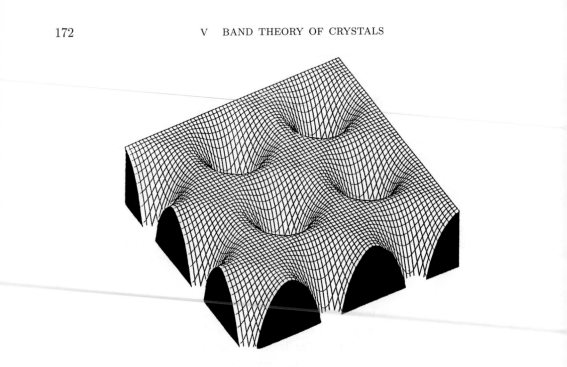

Fig. 10 Schematic representation of the muffin-tin potential.

where $V(r)$ is the spherically symmetric muffin-tin potential ($V(r) = 0$ for $r > r_s$). As in the cellular method, we indicate by $R_l(E, r)$ the radial wavefunction which is regular at the origin and satisfies the radial wave equation

$$\frac{d^2(r R_l)}{dr^2} = \left[V(r) + \frac{l(l+1)}{r^2} - E \right] (r R_l) . \tag{42}$$

An augmented plane wave $A(\mathbf{k} + \mathbf{h}_i, \mathbf{r}, E)$ is defined as a plane wave of vector $\mathbf{k} + \mathbf{h}_i$ outside the sphere r_s, which joins continuously with a cellular type function inside it. By using the familiar expansion (29) of a plane wave in spherical harmonics, we easily find the following expression for an augmented plane wave:

$$A(\mathbf{k}_i, \mathbf{r}, E)$$

$$= \begin{cases} A_{\text{in}}(\mathbf{k}_i, \mathbf{r}, E) = \dfrac{4\pi}{\sqrt{\Omega}} \displaystyle\sum_{lm} i^l \dfrac{j_l(k_i r_s)}{R_l(E, r_s)} Y_{lm}^*(\mathbf{k}_i) Y_{lm}(\mathbf{r}) R_l(E, r) & \mathbf{r} \in \Omega_s \\[2ex] A_{\text{out}}(\mathbf{k}_i, \mathbf{r}, E) = \dfrac{e^{i\mathbf{k}_i \cdot \mathbf{r}}}{\sqrt{\Omega}} \equiv \dfrac{4\pi}{\sqrt{\Omega}} \displaystyle\sum_{lm} i^l j_l(k_i r) Y_{lm}^*(\mathbf{k}_i) Y_{lm}(\mathbf{r}) & \mathbf{r} \in \Omega_{WS} - \Omega_s \end{cases}$$

$$\tag{43}$$

where Ω is the volume of the unit cell, j_l is the spherical Bessel functions of order l, Ω_{WS} and Ω_s indicate the Wigner–Seitz cell and the inscribed sphere of radius r_s, respectively. Notice that an augmented plane wave is a continuous function, but in general a discontinuity in its slope at $r = r_s$ remains.

A crystal eigenfunction $\psi(\mathbf{k}, \mathbf{r})$ with energy $E(\mathbf{k})$ can be expanded within the

Wigner–Seitz cell, in the form

$$\psi(\mathbf{k}, \mathbf{r}) = \sum_j a_j(\mathbf{k}) A(\mathbf{k} + \mathbf{h}_j, \mathbf{r}, E) \ ,$$

where $\mathbf{h}_j$ are the vectors of the reciprocal space. The function $\psi(\mathbf{k}, \mathbf{r})$ and its gradient are well behaved everywhere in the unit cell, though the expansion functions have discontinuous first derivative at the surface of the sphere of radius r_s. By substituting the above expansion in the Schrödinger equation and using the variational procedure with respect to the coefficients a_j, the following secular equation is obtained

$$\|\langle A(\mathbf{k}_i, \mathbf{r}, E) | H - E | A(\mathbf{k}_j, \mathbf{r}, E) \rangle\| = 0 \ . \tag{44}$$

The explicit expression of matrix elements $M_{ij}(\mathbf{k}, E)$ of the secular equation (44) is derived in the next Section 6.2; we quote here the final form

$$M_{ij}(\mathbf{k}, E) = (k_j^2 - E)\delta_{ij} - \frac{4\pi r_s^2}{\Omega}(\mathbf{k}_i \cdot \mathbf{k}_j - E)\frac{j_1(|\mathbf{k}_i - \mathbf{k}_j|r)}{|\mathbf{k}_i - \mathbf{k}_j|}$$

$$+ \frac{4\pi r_s^2}{\Omega}\sum_{l=0}^{\infty}(2l+1)P_l(\widehat{\mathbf{k}}_i \cdot \widehat{\mathbf{k}}_j)j_l(k_i r_s)j_l(k_j r_s)\frac{R_l'(E, r_s)}{R_l(E, r_s)} \tag{45}$$

where $\mathbf{k}_i$ is a short hand notation for $\mathbf{k} + \mathbf{h}_i$, $\widehat{\mathbf{k}}_i$ is the versor of $\mathbf{k}_i$ and $P_l(z)$ is the Legendre polynomial of order l; we have also indicated

$$R_l'(E, r_s) = \left[\frac{dR_l(E, r)}{dr}\right]_{r=r_s} \ .$$

The generalization to composite lattice is straightforward. Instead of a single atom per unit cell we have a number ν_b of atoms, centred at $\mathbf{d}_1, \mathbf{d}_2, \dots, \mathbf{d}_{\nu_b}$, each of them sorrounded by non-overlapping spheres of radii $r_{s1}, r_{s2}, \dots, r_{s\nu_b}$. The matrix elements for composite lattices are obtained putting an appropriate label on the second and third terms in the right-hand side of Eq. (45), multiplying by the structure factor $\exp[-i(\mathbf{h}_i - \mathbf{h}_j) \cdot \mathbf{d}_\nu]$ ($\nu = 1, 2, \dots, \nu_b$) and summing over ν. The matrix elements $M_{ij}(\mathbf{k}, E)$, given by Eq. (45), contain E both explicitly and implicitly through the logarithmic derivative R_l'/R_l. For practical calculations one may proceed in this way: (i) an appropriate truncation in $|\mathbf{k}_i|_{max}$ (i.e. in the order of the determinant) and l_{max} is chosen; (ii) the matrix elements $M_{ij}(\mathbf{k}, E)$ are computed at regular intervals of E (say 0.001 Rydberg); the determinant $\|M_{ij}(\mathbf{k}, E)\|$ is plotted as a function of E and eigenvalues are obtained as the zeroes of the curve (when needed, the corresponding eigenvectors can be determined and crystal wavefunctions explicitly found); (iii) the convergence and stability of the method can be tested by increasing both $|\mathbf{k}_i|_{max}$ and l_{max}. The favourable aspect of the method lies in its very rapid convergence, a number of 10–30 augmented plane waves (and $l_{max} \approx 5$) being sufficient in most practical cases; this well rewards the amount of computational labour involved.

Another aspect which makes the APW method such a powerful one is the fact that the degree of localization of crystal states is irrelevant. Remember that the tight-binding method is applicable for reasonably localized states; on the other hand, the

OPW method or its pseudo-potential version requires an arbitrary (and sometimes questionable) separation of crystal states in well localized core states and spread-out valence or conduction states. The APW method describes in a natural way strong localization, weak localization, or any possible intermediate situation. The limitation of the APW method is the requirement that the crystal potential must be well approximated by a muffin-tin type potential; we notice however that the APW method is indeed rather flexible and (moderate) corrections from muffin-tin form can be taken into account with perturbative schemes.

6.2   Expression and evaluation of the matrix elements of the APW method

This section provides some technical aspects concerning the APW method, and can be skipped if the reader is more interested in the physical aspects. The explicit expression of the matrix elements in APW method requires some care because of the discontinuity in the slope at $r = r_s$ of the basis wavefunctions; thus, we begin to show how to handle the discontinuity.

Let $u(\mathbf{r})$ and $v(\mathbf{r})$ be *continuous functions with a kink, i.e. with a finite discontinuity in the gradient*, at $r = r_s$. Let $u_{\text{in}}(\mathbf{r})$ and $u_{\text{out}}(\mathbf{r})$ indicate the function $u(\mathbf{r})$ inside and outside the sphere of radius r_s. We consider the integral of the kinetic energy operator $-\nabla^2$ between $v(\mathbf{r})$ and $u(\mathbf{r})$ through an arbitrary small volume Ω_ε containing the spherical surface S of radius $r = r_s$ (the region of integration is shown in Fig. 11). We have

$$\int_{\Omega_\varepsilon} v(\mathbf{r})[-\nabla^2 u(\mathbf{r})]\, d\mathbf{r} = -\int_{\Omega_\varepsilon} \nabla \cdot [v(\mathbf{r})\,\nabla u(\mathbf{r})]\, d\mathbf{r} + \int_{\Omega_\varepsilon} \nabla v(\mathbf{r}) \cdot \nabla u(\mathbf{r})\, d\mathbf{r} \qquad (46a)$$

(as can be seen by inspection of the integrands). The first integral in the right-hand side of Eq. (46a) can be performed using the divergence theorem, while the second integral in the right-hand can be neglected for $\Omega_\varepsilon \to 0$. It follows

$$\int_{\Omega_\varepsilon} v(\mathbf{r})[-\nabla^2 u(\mathbf{r})]\, d\mathbf{r} = \int_S v(\mathbf{r}) \left[\frac{\partial}{\partial r} u_{\text{in}}(\mathbf{r}) - \frac{\partial}{\partial r} u_{\text{out}}(\mathbf{r}) \right] dS ; \qquad (46b)$$

the two surface contributions in the right-hand side of Eq. (46b) do not cancel because of the discontinuity of the gradient of the wavefunction $u(\mathbf{r})$ at $r = r_s$.

We have now all the ingredients for the calculation of the matrix elements $M_{ij}(\mathbf{k}, E)$ of the crystal Hamiltonian operator $H - E$ between any two augmented plane waves; their expression reads

$$M_{ij}(\mathbf{k}, E) = \int_{\Omega_{WS}} A^*(\mathbf{k}_i, \mathbf{r}, E)[-\nabla^2 + V(r) - E]\, A(\mathbf{k}_j, \mathbf{r}, E)\, d\mathbf{r} . \qquad (47a)$$

The matrix elements $M_{ij}(\mathbf{k}, E)$ can be split into four contributions

$$M_{ij} = M_{ij}^{(1)} + M_{ij}^{(2)} + M_{ij}^{(3)} + M_{ij}^{(4)} ,$$

i.e. two volume integrals corresponding to the two regions in which the unit cell is partitioned, and two surface integrals at the spheres of radius $r_s \pm \varepsilon$ (see Fig. 11). The surface contributions appear due to the kink of the augmented plane waves at $r = r_s$.

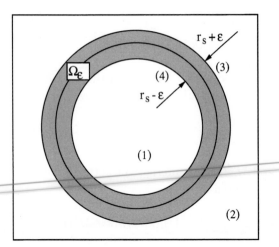

Fig. 11 Schematic representation of the four contributions (two volume contributions and two surface contributions) for the calculation of the matrix elements of the crystal Hamiltonian between augmented plane waves.

Let us consider separately these four contributions. For the first volume integral we have

$$M_{ij}^{(1)} = \int_{\Omega_s} A_{\mathrm{in}}^*(\mathbf{k}_i, \mathbf{r}, E)[-\nabla^2 + V(r) - E]\, A_{\mathrm{in}}(\mathbf{k}_j, \mathbf{r}, E)\, d\mathbf{r} \equiv 0 \;.$$

This term is exactly zero because an APW satisfies the Schrödinger equation inside the sphere Ω_s. The second volume integral is

$$M_{ij}^{(2)} = \int_{\Omega_{WS}-\Omega_s} A_{\mathrm{out}}^*(\mathbf{k}_i, \mathbf{r}, E)[-\nabla^2 - E]\, A_{\mathrm{out}}(\mathbf{k}_j, \mathbf{r}, E)\, d\mathbf{r} \;.$$

In the interstitial region $\Omega_{WS} - \Omega_s$, the APW is simply a plane wave; then we have

$$M_{ij}^{(2)} = \frac{k_j^2 - E}{\Omega} \left[\int_{\Omega_{WS}} e^{i(\mathbf{k}_j - \mathbf{k}_i)\cdot \mathbf{r}}\, d\mathbf{r} - \int_{\Omega_s} e^{i(\mathbf{k}_j - \mathbf{k}_i)\cdot \mathbf{r}}\, d\mathbf{r} \right]$$

$$= (k_j^2 - E)\delta_{ij} - (k_j^2 - E)\frac{4\pi r_s^2}{\Omega}\frac{j_1(|\mathbf{k}_i - \mathbf{k}_j|r_s)}{|\mathbf{k}_i - \mathbf{k}_j|} \;.$$

Let us consider now the surface integrals. The surface integral $M_{ij}^{(3)}$ on the external side of the spherical surface, using Eq. (43), becomes

$$M_{ij}^{(3)} = -\int_S A_{\mathrm{out}}^*(\mathbf{k}_i, \mathbf{r}, E)\frac{\partial}{\partial r} A_{\mathrm{out}}(\mathbf{k}_j, \mathbf{r}, E)\, dS$$

$$= -\frac{16\pi^2}{\Omega} \sum_{lml'm'} \int_S (-i)^l j_l(k_i r_s) Y_{lm}(\mathbf{k}_i) Y_{lm}^*(\mathbf{r}) i^{l'} j'_{l'}(k_j r_s) Y_{l'm'}^*(\mathbf{k}_j) Y_{l'm'}(\mathbf{r})\, dS \;,$$

where

$$j'_l(k_j r) = \left[\frac{d}{dr} j_l(k_j r) \right]_{r=r_s} .$$

Using the standard orthonormality relations of the spherical harmonics, and the addition theorem (30) for spherical harmonics, it follows

$$M_{ij}^{(3)} = -\frac{4\pi r_s^2}{\Omega} \sum_l (2l+1) j_l(k_i r_s) j_l(k_j r_s) P_l(\widehat{\mathbf{k}}_i \cdot \widehat{\mathbf{k}}_j) \frac{j'_l(k_j r_s)}{j_l(k_j r_s)} .$$

The surface integral $M_{ij}^{(4)}$ on the internal side of the spherical surface can be obtained in a similar way and is given by

$$M_{ij}^{(4)} = \frac{4\pi r_s^2}{\Omega} \sum_l (2l+1) j_l(k_i r_s) j_l(k_j r_s) P_l(\widehat{\mathbf{k}}_i \cdot \widehat{\mathbf{k}}_j) \frac{R'_l(E, r_s)}{R_l(E, r_s)} .$$

Summing up the previous results we have

$$M_{ij}(\mathbf{k}, E) = (k_j^2 - E)\delta_{ij} - \frac{4\pi r_s^2}{\Omega}(k_j^2 - E)\frac{j_1(|\mathbf{k}_i - \mathbf{k}_j| r_s)}{|\mathbf{k}_i - \mathbf{k}_j|}$$
$$+ \frac{4\pi r_s^2}{\Omega} \sum_l (2l+1) j_l(k_i r_s) j_l(k_j r_s) P_l(\widehat{\mathbf{k}}_i \cdot \widehat{\mathbf{k}}_j) \left[\frac{R'_l(E, r_s)}{R_l(E, r_s)} - \frac{j'_l(k_j r_s)}{j_l(k_j r_s)} \right]. \tag{47b}$$

Eq. (47b) does not show, at first sight, the condition $M_{ij} = M_{ji}$; to put in evidence explicitly the hermiticity of the (real) matrix M, we perform a last elaboration on Eq. (47b). For this purpose we exploit the identity

$$j_0(|\mathbf{k}_i - \mathbf{k}_j| r_s) = \sum_l (2l+1) P_l(\widehat{\mathbf{k}}_i \cdot \widehat{\mathbf{k}}_j) j_l(k_i r_s) j_l(k_j r_s) ,$$

and derive both members with respect to k_j. In performing the derivation, we remember the standard relation $j'_0(x) = -j_1(x)$ and also

$$\frac{\partial}{\partial k_j} |\mathbf{k}_i - \mathbf{k}_j| = \frac{\partial}{\partial k_j} \sqrt{k_i^2 + k_j^2 - 2k_i k_j \cos \Theta_{ij}} = \frac{1}{|\mathbf{k}_i - \mathbf{k}_j|} \frac{1}{k_j} (k_j^2 - \mathbf{k}_i \cdot \mathbf{k}_j) .$$

We obtain

$$(\mathbf{k}_i \cdot \mathbf{k}_j - k_j^2) \frac{j_1(|\mathbf{k}_i - \mathbf{k}_j| r_s)}{|\mathbf{k}_i - \mathbf{k}_j|} \equiv \sum_l (2l+1) P_l(\widehat{\mathbf{k}}_i \cdot \widehat{\mathbf{k}}_j) j_l(k_i r_s) j'_l(k_j r_s) .$$

Using this identity we can transform the matrix elements (47b) into the form given previously in Eq. (45).

Final remarks on the APW matrix elements

From a computational point of view, the most demanding part of the APW matrix elements (45) is constituted by the evaluation of the logarithmic derivatives $R'_l(E, r)/R_l(E, r)$ at the sphere radius r_s. The wavefunction $R_l(E, r)$ is obtained by integrating numerically (usually with the very simple Numerov method) the radial wave equation (42), from the origin outward. The functions $R_l(E, r)$ are thus determined, except for an inessential multiplicative constant, which is irrelevant for the logarithmic derivatives $R_l'(E, r)/R_l(E, r)$.

We notice that the differential equation (42) implies the identity

$$\int_0^{r_s} R_l^2(E, r)\, r^2\, dr \equiv -R_l^2(E, r_s)\, r_s^2 \frac{\partial}{\partial E} \frac{R_l'(E, r_s)}{R_l(E, r_s)} \,, \tag{48}$$

where $R_l(E, r)$ is the solution of the radial Schrödinger equation regular at the origin. [Eq. (48) can be proved as follows: write Eq. (42) for $U_l(E_1, r) \equiv r\, R_l(E_1, r)$ and for $U_l(E_2, r) \equiv r\, R_l(E_2, r)$, multiply the radial equation for $U_l(E_1, r)$ by $U_l(E_2, r)$ and vice versa, subtract, integrate over the radial coordinate and perform an integration by parts, and finally let $E_1 \to E_2$]. From Eq. (48) we notice that the first member is always positive; it follows that the logarithmic derivatives $R_l'(E, r)/R_l(E, r)$ *are always monotonically decreasing functions of energy* and exhibit poles at the resonance energies, which are defined as the energies for which $R_l(E, r_s) = 0$.

As a final comment, we observe that the matrix elements of the APW method given by Eq. (45) depend obviously on r_s. Notice that r_s is to some extent arbitrary and must satisfy the relation $r_{\mathrm{MT}} \leq r_s \leq r_{\max}$, where r_{MT} is the radius of the assumed muffin-tin potential and $r_{\max}$ is the maximum radius consistent with non-overlapping spheres. In practical situations one usually chooses $r_{\mathrm{MT}} \equiv r_s \equiv r_{\max}$. However, in principle, one can have $r_{\mathrm{MT}} < r_{\max}$; in this situation, r_s can be chosen arbitrarily in the interval $[r_{\mathrm{MT}}, r_{\max}]$, the matrix elements M_{ij} do depend on r_s, but the eigenvalues of Eq. (45) do not.

It would be more satisfactory, from a pure formal point of view, a method whose matrix elements are independent from r_s. This would occur if matrix elements do not contain the *logarithmic derivatives of the radial wavefunctions, but rather the phase shifts of the muffin-tin potential*; for $r > r_{\mathrm{MT}}$ the logarithmic derivatives depend on r, while the phase shifts are independent from it. This last step is achieved in the Green's function method, which is the most far reaching development of the cellular and APW methods.

7   The Green's function method (KKR method)

The Green's function method was proposed by Korringa (1947), Kohn and Rostoker (1954) (and therefore called also KKR method) and Morse (1956) in different though equivalent forms. In the KKR method the Schrödinger equation is transformed into an equivalent integral equation and, once a muffin-tin potential is assumed, the problem of boundary conditions is elegantly solved through the fulfilment of appropriate surface integrals, related to the Green's theorem. A nice feature of the method is the sharp separation of the procedure into two parts: (i) structural aspects of the lattice; (ii) phase shifts of the spherical muffin-tin potential.

From the point of view of electronic state calculation for periodic crystals, it is generally agreed that KKR is even more rapidly convergent than the APW method; but this advantage is to some extent counterbalanced by the somewhat more complicated expressions of the KKR structural coefficients. However the real peculiar novelty of the KKR methodology (i.e. sharp separation of *geometry* from *potential*) has made the method generalizable to the treatment of impurities, clusters, homogeneously disordered alloys, and also photon energy bands in metallic ceramics. Probably these are the most important modern aspects of the KKR method; however, we keep in line with the purpose of this chapter and we limit ourselves to the study of electrons in fully periodic materials.

7.1   Scattering integral equation for a generic potential

In this section we briefly summarize the Green's function technique to transform into integral form the Schrödinger equation, here rewritten for convenience as

$$(E + \nabla^2)\,\psi(\mathbf{r}) = V(\mathbf{r})\,\psi(\mathbf{r}) \ . \tag{49}$$

The free-particle Green's function $g(\mathbf{r}, \mathbf{r}_0, E)$ for a δ-like source at $\mathbf{r}_0$ is defined by means of the equation

$$(E + \nabla^2)\,g(\mathbf{r}, \mathbf{r}_0, E) = \delta(\mathbf{r} - \mathbf{r}_0) \ , \tag{50}$$

and the requirement that $g(\mathbf{r}, \mathbf{r}_0, E)$ obeys the same boundary conditions as $\psi(\mathbf{r})$. It is straightforward to verify that the differential eigenvalue equation (49) is equivalent to the integral eigenvalue equation

$$\psi(\mathbf{r}) = \phi(\mathbf{r}) + \int g(\mathbf{r}, \mathbf{r}_0, E)\,V(\mathbf{r}_0)\,\psi(\mathbf{r}_0)\,d\mathbf{r}_0 \ , \tag{51}$$

where $\phi(\mathbf{r})$ denotes a solution at energy E (if any) of the homogeneous differential equation $(E + \nabla^2)\,\phi(\mathbf{r}) = 0$. The equivalence of Eq. (51) with Eq. (49) follows by direct application of the operator $(E + \nabla^2)$ to both members of Eq. (51).

It is convenient to express the Green's function $g(\mathbf{r}, \mathbf{r}_0, E)$ of the free-particle Hamiltonian $H_0 = -\nabla^2$ without sticking to the $\mathbf{r}$-representation. The free-particle retarded Green's function $g(E)$ can be expressed as

$$g(E) = \frac{1}{E + i\varepsilon - H_0} = \frac{1}{E + i\varepsilon + \nabla^2} \qquad \varepsilon \to 0^+ \ ; \tag{52}$$

it is understood that the real energy E is always accompanied by an infinitesimal positive imaginary part (sometimes, for brevity of notations, the imaginary part is not explicitly indicated; however, the imaginary part $i\varepsilon$ must never be overlooked, and must be reinstated whenever necessary). The integral eigenvalue equation (51) can now be written in the form

$$|\psi\rangle = |\phi\rangle + \frac{1}{E + i\varepsilon - H_0}\,V|\psi\rangle \ . \tag{53}$$

It is convenient to obtain a spectral representation of the Green's function $g(E)$ in terms of the eigenvalues and eigenfunctions of the operator H_0 itself; these are plane

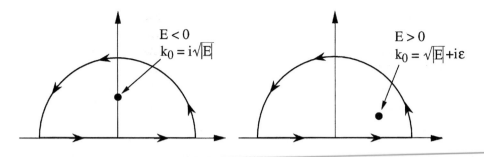

Fig. 12 Integration contour for evaluating the Green's function of the free-electron problem.

waves $W(\mathbf{k}, \mathbf{r}) = (1/\sqrt{V}) \exp(i\mathbf{k} \cdot \mathbf{r})$ of energy k^2. The Green's function (52) can be evidently written as

$$g(E) = \frac{1}{E + i\varepsilon - H_0} \sum_{\mathbf{k}} |W_{\mathbf{k}}\rangle\langle W_{\mathbf{k}}| = \sum_{\mathbf{k}} \frac{1}{E + i\varepsilon - k^2} |W_{\mathbf{k}}\rangle\langle W_{\mathbf{k}}| . \qquad (54a)$$

In the $\mathbf{r}$-representation we have

$$g(\mathbf{r}, \mathbf{r}_0, E) = \frac{1}{V} \sum_{\mathbf{k}} \frac{e^{i\mathbf{k} \cdot \mathbf{r}} \, e^{-i\mathbf{k} \cdot \mathbf{r}_0}}{E + i\varepsilon - k^2} . \qquad (54b)$$

Notice that the free-particle Green's function depends on the difference $\mathbf{r} - \mathbf{r}_0$, i.e. $g(\mathbf{r}, \mathbf{r}_0, E)$ has the form $g(\mathbf{r} - \mathbf{r}_0, E)$.

The discrete sum over $\mathbf{k}$ in Eq. (54b) can be replaced (as usual) with $V/(2\pi)^3$ times the integral on the $\mathbf{k}$ variable; we have

$$g(\mathbf{r} - \mathbf{r}_0, E) = \frac{1}{(2\pi)^3} \int \frac{1}{E + i\varepsilon - k^2} e^{i\mathbf{k} \cdot (\mathbf{r} - \mathbf{r}_0)} \, d\mathbf{k} , \qquad (55a)$$

where $d\mathbf{k}$ denotes the volume element in the reciprocal space. Performing the integration we obtain

$$g(\mathbf{r} - \mathbf{r}_0, E) = -\frac{1}{4\pi} \frac{e^{i\,\alpha|\mathbf{r} - \mathbf{r}_0|}}{|\mathbf{r} - \mathbf{r}_0|} , \qquad (55b)$$

where $\alpha^2 = E$ or, more specifically,

$$\alpha(E) = \begin{cases} \sqrt{E} & \text{for} \quad E > 0 \\ i\sqrt{|E|} & \text{for} \quad E < 0 \end{cases} . \qquad (55c)$$

The integration in Eq. (55a) has been carried out with the following procedure. We introduce polar coordinates in $\mathbf{k}$ space with $\mathbf{r} - \mathbf{r}_0$ as k_z direction, and obtain

$$g(\mathbf{r} - \mathbf{r}_0, E) = \frac{1}{(2\pi)^3} 2\pi \int_0^\infty \frac{k^2}{E + i\varepsilon - k^2} \, dk \int_0^\pi e^{ik|\mathbf{r} - \mathbf{r}_0| \cos\theta} \sin\theta \, d\theta .$$

The integral in the angular variable θ can be performed and, with straightforward manipulations, we have

$$g(\mathbf{r} - \mathbf{r}_0, E) = \frac{1}{4\pi^2 i |\mathbf{r} - \mathbf{r}_0|} \int_{-\infty}^{\infty} \frac{k}{E + i\varepsilon - k^2} e^{ik|\mathbf{r}-\mathbf{r}_0|} \, dk \ . \tag{56}$$

Consider first the integration in Eq. (56) in the case $E < 0$. The integral can be easily evaluated closing the integration contour in the upper part of the complex plane, as indicated in Fig. 12. We have to consider one simple pole at $k = k_0 \equiv i\sqrt{|E|}$ and the residue of the integrand function is $(-1/2) \exp(ik_0|\mathbf{r} - \mathbf{r}_0|)$. We thus obtain Eq. (55b) for negative energies. A similar procedure proves that Eq. (55b) holds also for positive energies.

7.2   Scattering integral equation for a periodic muffin-tin potential

In the case of a periodic crystal potential, we can specify the integral equation (51) taking into account that the wavefunctions must be of Bloch type; we have

$$\psi(\mathbf{k}, \mathbf{r}) = \int_{\text{crystal}} g(\mathbf{r} - \mathbf{r}_0, E) \, V(\mathbf{r}_0) \, \psi(\mathbf{k}, \mathbf{r}_0) \, d\mathbf{r}_0 \ , \tag{57a}$$

where the integration extends on the crystal volume; the solution $\phi(\mathbf{r})$ of the homogeneous equation is set equal to zero, since in general there is no incident plane wave in the infinite crystal.

The wavefunction $\psi(\mathbf{k}, \mathbf{r})$, at any two points $\mathbf{r}$ and $\mathbf{r} + \mathbf{t}_n$ related by a translation vector, takes values related by the Bloch phase factor $\exp(i\mathbf{k} \cdot \mathbf{t}_n)$. This allows to restrict the integration in Eq. (57a) to a single unit cell (for convenience the Wigner–Seitz cell centred at $\mathbf{t}_n = 0$). We obtain the following integral equation

$$\psi(\mathbf{k}, \mathbf{r}) = \int_{\Omega_{\text{WS}}} g(\mathbf{k}, \mathbf{r} - \mathbf{r}_0, E) \, V(\mathbf{r}_0) \, \psi(\mathbf{k}, \mathbf{r}_0) \, d\mathbf{r}_0 \ , \tag{57b}$$

where the kernel of the integral equation has the expression

$$g(\mathbf{k}, \mathbf{r} - \mathbf{r}_0, E) = \sum_{\mathbf{t}_n} e^{i\mathbf{k} \cdot \mathbf{t}_n} g(\mathbf{r} - \mathbf{r}_0 - \mathbf{t}_n, E) \tag{57c}$$

and is called *the greenian of vector* $\mathbf{k}$. Using Eq. (55b), we can express the greenian of vector $\mathbf{k}$ in the form

$$g(\mathbf{k}, \mathbf{r} - \mathbf{r}_0, E) = -\frac{1}{4\pi} \sum_{\mathbf{t}_n} e^{i\mathbf{k} \cdot \mathbf{t}_n} \frac{e^{i\alpha|\mathbf{r}-\mathbf{r}_0-\mathbf{t}_n|}}{|\mathbf{r} - \mathbf{r}_0 - \mathbf{t}_n|} \ , \tag{58}$$

where the sum is over the translation vectors of the direct lattice.

We now assume the muffin-tin form of the potential; the integration region in Eq. (57b) can then be reduced to the sphere Ω_s of radius r_s ($\geq r_{\text{MT}}$) inscribed in the Wigner–Seitz polyhedron (we limit our attention to simple crystals, since the extension to composite crystals only requires some straightforward generalizations). We have thus that the crystal eigenfunctions of vector $\mathbf{k}$ satisfy the integral equation

$$\psi(\mathbf{k}, \mathbf{r}) = \int_{\Omega_s} g(\mathbf{k}, \mathbf{r} - \mathbf{r}_0, E)\, V(\mathbf{r}_0)\, \psi(\mathbf{k}, \mathbf{r}_0)\, d\mathbf{r}_0 \ .$$

Because of Eq. (49), we have also

$$\int_{\Omega_s} g(\mathbf{k}, \mathbf{r} - \mathbf{r}_0, E)\, (E + \nabla_0^2)\, \psi(\mathbf{k}, \mathbf{r}_0)\, d\mathbf{r}_0 = \psi(\mathbf{k}, \mathbf{r}) \ . \tag{59}$$

We use the identity

$$g\, \nabla^2 \psi = (\nabla^2 g)\psi + \nabla \cdot [g\, \nabla\psi - \psi\, \nabla g] \ ,$$

where g and ψ are any two continuous and differentiable functions. We insert this identity into Eq. (59) and use the Green's theorem to transform the volume integral of $\operatorname{div}(g\, \nabla\psi - \psi\, \nabla g)$ into a more convenient surface integral; we arrive at the basic "boundary condition" equation

$$\boxed{\int_{S_0} \left[g(\mathbf{k}, \mathbf{r} - \mathbf{r}_0, E) \frac{\partial \psi(\mathbf{k}, \mathbf{r}_0)}{\partial r_0} - \psi(\mathbf{k}, \mathbf{r}_0) \frac{\partial g(\mathbf{k}, \mathbf{r} - \mathbf{r}_0, E)}{\partial r_0} \right]_{r_0 = r_s} dS_0 = 0} \ , \tag{60}$$

where S_0 is the surface of the sphere Ω_s and $\mathbf{r}$ is any point within it. In a crystal, with muffin-tin form of the potential, any crystal wavefunction of vector $\mathbf{k}$ and energy E is such that the surface integral, defined in the left-hand side of Eq. (60), identically vanishes for any $\mathbf{r} \in \Omega_s$. In a crystal, with *no* muffin-tin form of the potential, a similar result holds, but the great complication occurs that the integral involves the surface of the primitive cell rather than the quite convenient spherical surface S_0.

In order to perform explicitly the surface integrals in Eq. (60) we need an expression of the greenian in terms of appropriate (structure) coefficients, multiplied by factorized functions of $\mathbf{r}$ and $\mathbf{r}_0$ separately. The desired expression of the greenian (discussed in the next Section 7.3) is

$$g(\mathbf{k}, \mathbf{r} - \mathbf{r}_0, E) = \alpha \sum_{lm} j_l(\alpha r)\, Y_{lm}(\mathbf{r})\, n_l(\alpha r_0)\, Y_{lm}^*(\mathbf{r}_0)$$

$$+ \sum_{lml'm'} j_l(\alpha r)\, Y_{lm}(\mathbf{r})\, \Gamma_{lm,l'm'}(\mathbf{k}, E)\, j_{l'}(\alpha r_0)\, Y_{l'm'}^*(\mathbf{r}_0) \ , \tag{61}$$

where $\alpha = \sqrt{E}$, $\mathbf{r}$ and $\mathbf{r}_0$ are within the Wigner–Seitz cell, and $r < r_0$; in the case $r > r_0$ we have to exchange r and r_0 in the first term in the right-hand side of Eq. (61); the coefficients $\Gamma_{lm,l'm'}(\mathbf{k}, E)$ are called *structure coefficients* and their explicit expression is reported in Section 7.3.

As in the cellular method, also in the Green's function method the crystal wavefunctions are expanded in spherical waves

$$\psi(\mathbf{k}, \mathbf{r}_0) = \sum_{l''m''} c_{l''m''}(\mathbf{k})\, R_{l''}(E, r_0)\, Y_{l''m''}(\mathbf{r}_0) \ , \tag{62}$$

but now the unknown coefficients c_{lm} and the crystal eigenvalues are determined in a natural way through the fulfilment of the boundary conditions summarized by Eq. (60). As usual $R_l(E, r)$ indicates the solution, regular at the origin, of the radial Schrödinger equation. For convenience, the normalization of $R_l(E, r)$ is chosen so as to satisfy the following asymptotic behaviour

$$R_l(E, r) = \cos \eta_l(E) \cdot j_l(\alpha r) - \sin \eta_l(E) \cdot n_l(\alpha r) \quad \text{for } r \geq r_{\text{MT}} , \tag{63}$$

where j_l and n_l are the spherical Bessel and Neumann functions, and $\eta_l(E)$ are the *phase shifts* due to the muffin-tin potential at the energy E and for the wave of angular momentum l. It is useful to recall the Wronskian relations for Bessel and Neumann functions

$$[j_l(x), n_l(x)] \equiv j_l(x) \, n_l'(x) - j_l'(x) \, n_l(x) = \frac{1}{x^2} ;$$

from it and Eq. (63), the following two Wronskian relations hold for $r = r_s \geq r_{\text{MT}}$

$$[j_l, R_l] = -\sin \eta_l(E) \frac{1}{\alpha r_s^2} , \qquad [n_l, R_l] = -\cos \eta_l(E) \frac{1}{\alpha r_s^2} .$$

We now insert expression (61) for the greenian and expansion (62) for the trial wavefunction into Eq. (60); we obtain

$$\alpha \sum_{lm} j_l(\alpha r) Y_{lm}(\mathbf{r}) [n_l, R_l] c_{lm} + \sum_{lml'm'} j_l(\alpha r) Y_{lm}(\mathbf{r}) \Gamma_{lm,l'm'} [j_{l'}, R_{l'}] c_{l'm'} = 0$$

for any $\mathbf{r} \in \Omega_s$. Because of this arbitrariness, it follows

$$\alpha \cos \eta_l(E) c_{lm} + \sum_{l'm'} \Gamma_{lm,l'm'}(\mathbf{k}, E) \sin \eta_{l'}(E) c_{l'm'} = 0 .$$

The above system of linear homogeneous equations for the arbitrary coefficients c_{lm} leads to the compatibility determinantal equation

$$\boxed{\left\| \Gamma_{lm,l'm'}(\mathbf{k}, E) + \sqrt{E} \cot \eta_l(E) \, \delta_{ll'} \, \delta_{mm'} \right\| = 0} . \tag{64}$$

The determinant in Eq. (64) is plotted versus the energy E and its zeroes provide the crystal eigenvalues at the chosen $\mathbf{k}$ vector.

A remarkable feature of the basic secular equation (64) is the net separation of the information concerning the crystal structure, completely embodied in the structure coefficients $\Gamma_{lm,l'm'}(\mathbf{k}, E)$ and the effect of the potential summarized by the phase shifts $\eta_l(E)$. Furthermore, the dimension of the secular equation needed in actual calculations is expected to be reasonably small; for instance, to reproduce the s, p and d character of the bands of interest, it is sufficient to consider the angular momentum values $l = 0, 1, 2$ and the secular equation becomes a 9×9 matrix. From a formal point of view the KKR method is the most satisfactory tool to solve the problem of boundary conditions, when the muffin-tin form for the potential is assumed.

7.3   Expression and evaluation of the structure coefficients

We summarize briefly some mathematical aspects of interest for the expression of the structure coefficients of the KKR method. We first remember the Neumann expansion

$$-\frac{1}{4\pi}\frac{e^{i\alpha|\mathbf{r}-\mathbf{r}'|}}{|\mathbf{r}-\mathbf{r}'|} = \alpha \sum_{lm} j_l(\alpha r)\, Y_{lm}(\mathbf{r})[n_l(\alpha r') - i\, j_l(\alpha r')]\, Y_{lm}^*(\mathbf{r}') , \tag{65}$$

where $r < r'$ and j_l and n_l are spherical Bessel and Neumann functions, respectively. In the case $r > r'$ we have to exchange $\mathbf{r}$ and $\mathbf{r}'$ in the right-hand side of Eq. (65).

We now consider the expression (58) for the greenian, separate the term with $\mathbf{t}_n = 0$ from all the others, and use the The Neumann expression (65) with $\mathbf{r}' = \mathbf{r}_0$ for the term $\mathbf{t}_n = 0$, and $\mathbf{r}' = \mathbf{t}_n$ for all the others. With straightforward manipulations, for $|\mathbf{r} - \mathbf{r}_0| < t_n\,(\neq 0)$, one obtains

$$g(\mathbf{k}, \mathbf{r}-\mathbf{r}_0, E) = -\frac{1}{4\pi}\frac{\cos \alpha|\mathbf{r}-\mathbf{r}_0|}{|\mathbf{r}-\mathbf{r}_0|} + \sum_{lm} D_{lm}(\mathbf{k}, E)\, j_l(\alpha|\mathbf{r}-\mathbf{r}_0|)\, Y_{lm}(\mathbf{r}-\mathbf{r}_0) \tag{66}$$

where

$$D_{lm}(\mathbf{k}, E) = \alpha \sum_{\mathbf{t}_n(\neq 0)} e^{i\mathbf{k}\cdot\mathbf{t}_n}\, [n_l(\alpha t_n) - i\, j_l(\alpha t_n)] - i\frac{\alpha}{\sqrt{4\pi}}\,\delta_{l0}\,\delta_{m0} \tag{67}$$

are the so called *reduced structure coefficients* of the particular lattice in consideration.

We now use the identity

$$i^L j_L(\alpha R)\, Y_{LM}(\mathbf{R}) = 4\pi \sum_{lml'm'} i^{l-l'}\, C_{LM;lm,l'm'}\, j_l(\alpha r)\, j_{l'}(\alpha r_0)\, Y_{lm}(\mathbf{r})\, Y_{l'm'}^*(\mathbf{r}_0) \tag{68}$$

where $R = |\mathbf{r} - \mathbf{r}_0|$, and

$$C_{LM;lm,l'm'} = \int Y_{LM}(\mathbf{k})\, Y_{lm}^*(\mathbf{k})\, Y_{l'm'}(\mathbf{k})\, dS_{\mathbf{k}}$$

are the Gaunt coefficients. The identity (68) is proved by expanding $\exp[i\mathbf{k}\cdot(\mathbf{r}-\mathbf{r}_0)]$ in spherical harmonics and then comparing with the product of the expansion in spherical harmonics of $\exp(i\mathbf{k}\cdot\mathbf{r})$ and $\exp(-i\mathbf{k}\cdot\mathbf{r}_0)$ separately, and putting $|\mathbf{k}| = \alpha$.

With the help of the identity (68) and the Neumann expansion (65), Eq. (66) takes the form

$$g(\mathbf{k}, \mathbf{r} - \mathbf{r}_0, E) = \alpha \sum_{lm} j_l(\alpha r)\, Y_{lm}(\mathbf{r})\, n_l(\alpha r_0)\, Y_{lm}^*(\mathbf{r}_0)$$

$$+ \sum_{lm\,l'm'} j_l(\alpha r)\, Y_{lm}(\mathbf{r})\, \Gamma_{lm,l'm'}(\mathbf{k}, E)\, j_{l'}(\alpha r_0)\, Y_{l'm'}^*(\mathbf{r}_0) \tag{69}$$

where $r < r_0 < t_n\,(\neq 0)$, and

$$\Gamma_{lm,l'm'}(\mathbf{k}, E) = 4\pi i^{l-l'} \sum_{LM} i^{-L}\, D_{LM}(\mathbf{k}, E)\, C_{LM;lm,l'm'} \tag{70}$$

are the *structure coefficients* entering the secular equation of the KKR method.

The last step of our procedure is to obtain a workable evaluation of the reduced structure coefficients $D_{lm}(\mathbf{k}, E)$, expressed in real space by Eq. (67). For negative values of the energy, the Green's function $g(\mathbf{r} - \mathbf{r}_0, E)$ is exponentially damped since $\alpha = i\sqrt{|E|}$, and the direct-lattice sum in Eq. (67) converges exponentially fast. However, for positive values of the energy, the Green's function $g(\mathbf{r} - \mathbf{r}_0, E)$ has a long-range oscillatory behaviour, and the direct-lattice sum in Eq. (67) becomes slowly conditionally convergent.

The standard technique for achieving a satisfactory accuracy in the evaluation of the structure constants $D_{lm}(\mathbf{k}, E)$ at any energy, is the application of the (rather laborious) Ewald algorithm, which includes and appropriately balances summations both in direct and reciprocal space (for a worked out application of the Ewald algorithm in a different context, see Section VI-2). For convenience, we quote here the final expression in the case of one atom per unit cell. The reduced structure coefficients are given by the following expression

$$D_{lm}(\mathbf{k}, E) = D_{lm}^{(1)}(\mathbf{k}, E) + D_{lm}^{(2)}(\mathbf{k}, E) + D_{00}^{(3)}(E)\,\delta_{l0}\,\delta_{m0}$$

where

$$D_{lm}^{(1)}(\mathbf{k}, E) = \frac{4\pi}{\Omega}\,\frac{i^l}{\sqrt{E}^l}\sum_{\mathbf{k}_n}\frac{k_n^l}{E - k_n^2}Y_{lm}^*(\mathbf{k}_n)\exp[(E - k_n^2)/\eta]$$

$$D_{lm}^{(2)}(\mathbf{k}, E) = \frac{-1}{\sqrt{\pi}}\,\frac{2^{l+1}}{\sqrt{E}^l}\sum_{\mathbf{t}_n \neq 0}e^{i\mathbf{k}\cdot\mathbf{t}_n}\,|\mathbf{t}_n|\,Y_{lm}^*(\mathbf{t}_n)\int_{\sqrt{\eta}/2}^{\infty}x^{2l}\exp\left[-x^2 t_n^2 + \frac{E}{4x^2}\right]dx$$

$$D_{00}^{(3)}(E) = -\frac{\sqrt{\eta}}{2\pi}\sum_{n=0}^{\infty}\frac{(E/\eta)^n}{n!\,(2n - 1)}\,.$$

The arbitrary parameter η is chosen to optimize simultaneously the convergence of the series in configuration and reciprocal spaces.

8 Other methods and developments in electronic structure calculations

8.1 The linearized cellular methods

We have seen that in the cellular methods the crystal wavefunctions are expanded in *energy dependent* spherical waves $R_l(E, r)\,Y_{lm}(\mathbf{r})$ within the Wigner–Seitz cell or the inscribed muffin-tin sphere. Boundary conditions are then considered either at conjugate points at the Wigner–Seitz surfaces, or at the surface of the muffin-tin sphere. The matrix elements of the secular equations have thus a complicate *non-linear* energy dependence. For repeated calculations of energy bands towards self-consistency, or optical properties and other integrated properties through the whole Brillouin zone, this may become a real disadvantage and a limitation.

The possibility and concepts of linearization were introduced by O. K. Andersen at the beginning of seventies. The basic idea of linearization consists in representing the crystal wavefunctions with the combined use both of the radial functions $R_l(E, r)$ and of their derivatives $\dot{R}_l(E, r)$ with respect to the energy, specified at a particular but arbitrary value E_0 of the energy. The key point that determines the success of the method lies in the fact that the radial wavefunctions within the atomic sphere vary smoothly with energy, in an appreciable energy range. Thus one can expand them into Taylor series near the energy of interest and keep only the zero order term and the term linear in energy. The combined use of the wavefunctions $R_l(E_0, r)$ and $\dot{R}_l(E_0, r)$ leads to secular equations which are *linear* in energy, and the search of eigenvalues and eigenvectors can be handled as a usual algebraic problem.

In the augmented plane wave method, for instance, we have seen that a plane wave in the interstitial region can be matched continuously with the spherical waves $R_l(E, r) Y_{lm}(\mathbf{r})$, but discontinuities in derivatives cannot be eliminated (see Eq. 43). If inside the muffin-tin sphere we use not only the spherical waves $R_l(E_0, r) Y_{lm}(\mathbf{r})$ but also $\dot{R}_l(E_0, r) Y_{lm}(\mathbf{r})$, it is possible to match a plane wave in value and first derivative. The use of these linearized augmented plane waves leads to the linearized augmented plane wave method (LAPW).

We do not dwell on the physical and mathematical aspects of linearization. We simply mention that harmonic functions of the Neumann type in the interstitial region can be "augmented" with spherical waves and their energy derivatives, inside non-overlapping muffin-tin spheres; their use in representing the crystal wavefunctions has lead to the linear muffin-tin orbital method (LMTO), a most precious method for self-consistent calculations.

8.2 The Lanczos or recursion method

The Lanczos method (1950), or the closely related recursion method (Haydock, Heine and Kelly, 1972), is a very convenient approach for the determination of the eigensolutions of matrices, expecially those of very large rank and sparse character (i.e. with many matrix elements equal to zero). In the field of electronic state calculations, the recursion method has been originally introduced in connection with a local basis representation of the electronic states in solids; successively it has been used also in connection with other basis sets (i.e. plane waves, muffin-tin orbitals). The method is very useful for periodic systems; it becomes really essential in aperiodic materials, when a very large number of orbitals must be taken into account and when translational symmetry is (partially or completely) absent, so that the concepts and simplifications implied by the Bloch theorem are not at work.

The essential principle of the Lanczos recursion method is very simple and very general at the same time. Consider a quantum system, an operator H, and a number N (arbitrary large) of orthonormal basis states $\{|\phi_i\rangle\}$ $(i = 1, 2, \ldots, N)$. Starting from any given state $|f_0\rangle$ belonging to the space spanned by $\{|\phi_i\rangle\}$, and operating with H, *the recursion method provides a one-dimensional chain representation of the original quantum system.*

To illustrate the procedure, consider an operator H (usually the Hamiltonian of the system) and an initial normalized state $|f_0\rangle$, arbitrarily chosen; the starting state

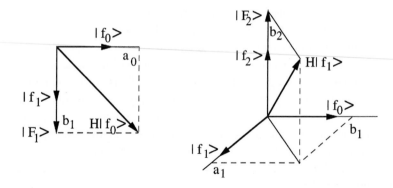

Fig. 13 Schematic representation of the Lanczos recursion procedure.

$|f_0\rangle$ is often addressed as the *seed state* of the procedure. We apply the operator H to the initial state and subtract from $H|f_0\rangle$ its projection on the initial state, so to obtain a new state $|F_1\rangle$, orthogonal to $|f_0\rangle$. The operation of orthogonalization can be conveniently performed by means of the projection operator $P_0 = |f_0\rangle\langle f_0|$, and the state $|F_1\rangle$ can be expressed as

$$|F_1\rangle = (1 - P_0)\, H\, |f_0\rangle = H\, |f_0\rangle - a_0|f_0\rangle \;, \qquad (71a)$$

where

$$a_0 = \langle f_0|H|f_0\rangle \;. \qquad (71b)$$

Let us indicate with b_1 the normalization of $|F_1\rangle$, and with $|f_1\rangle$ the corresponding normalized state, namely

$$b_1^2 = \langle F_1|F_1\rangle, \qquad |f_1\rangle = \frac{1}{b_1}|F_1\rangle \;.$$

In a rather similar way we proceed now by applying the operator H to the state $|f_1\rangle$; then $H|f_1\rangle$ is orthogonalized to both $|f_1\rangle$ and $|f_0\rangle$, obtaining the state $|F_2\rangle$. We have

$$|F_2\rangle = (1 - P_1)\,(1 - P_0)\, H\, |f_1\rangle = (1 - P_1 - P_0)\, H\, |f_1\rangle \;,$$

where $P_1 = |f_1\rangle\langle f_1|$ denotes the projection operator on the state $|f_1\rangle$. We thus obtain

$$|F_2\rangle = H\, |f_1\rangle - a_1|f_1\rangle - b_1|f_0\rangle \;, \qquad (72a)$$

where

$$a_1 = \langle f_1|H|f_1\rangle, \qquad b_2^2 = \langle F_2|F_2\rangle \quad \text{and} \quad |f_2\rangle = \frac{1}{b_2}|F_2\rangle \;. \qquad (72b)$$

The procedure is indicated schematically in Fig. 13.

Let us now proceed to the next logical step of the sequence, and here we arrive at the key point of the ingenious, and yet so simple, Lanczos procedure. Consider in fact

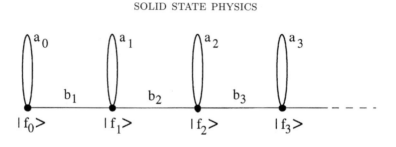

Fig. 14 Schematic representation of the linear chain generated by the Lanczos procedure.

the state $|F_3\rangle$ obtained by orthogonalization of $H|f_2\rangle$ to the previous states f_0, f_1, f_2; we have

$$|F_3\rangle = (1 - P_2)(1 - P_1)(1 - P_0)H|f_2\rangle = (1 - P_2 - P_1 - P_0)H|f_2\rangle .$$

We notice that $P_0 H|f_2\rangle \equiv 0$ since $\langle f_0|H|f_2\rangle \equiv 0$; *thus the iteration procedure leads to a three term relation.* The most remarkable feature of the method is that the iteration never includes more than three terms.

Iterating n times the procedure, orthogonalizing the new generated state $H|f_n\rangle$ only to the two predecessors, $|f_n\rangle$ and $|f_{n-1}\rangle$ we obtain

$$|F_{n+1}\rangle = H|f_n\rangle - a_n|f_n\rangle - b_n|f_{n-1}\rangle . \tag{73a}$$

The next pairs of coefficients are given by

$$b_{n+1}^2 = \langle F_{n+1}|F_{n+1}\rangle \quad \text{and} \quad a_{n+1} = \langle f_{n+1}|f_{n+1}\rangle . \tag{73b}$$

The three term relation (73a) is all we need to generate the orthonormal basis set $\{|f_n\rangle\}$ and the set of coefficients $\{a_n\}$ and $\{b_n\}$. This is a remarkable advantage, both conceptual and technical, in the specific calculations carried out by a computer code. On the basis set $\{|f_n\rangle\}$, the operator H is represented by the tridiagonal matrix

$$H = \begin{pmatrix} a_0 & b_1 & & & \\ b_1 & a_1 & b_2 & & \\ & b_2 & a_2 & b_3 & \\ & & \cdot & \cdot & \cdot \\ & & & \cdot & \cdot \end{pmatrix} . \tag{74a}$$

The operator (74a) can also be written in the form

$$H = \sum_{n=0}^{\infty} a_n |f_n\rangle\langle f_n| + \sum_{n=0}^{\infty} b_{n+1} \left[|f_n\rangle\langle f_{n+1}| + |f_{n+1}\rangle\langle f_n| \right] \tag{74b}$$

and can be represented diagrammatically as in Fig. 14.

A most important feature of the tridiagonal matrix (74a) is the possibility to obtain very easily the Green's function; in particular the matrix element $G_{00}(E)$ of the Green's function is given by the continued fraction expansion (see Section I-4.2)

$$G_{00}(E) = \langle f_0 | \frac{1}{E - H} | f_0 \rangle = \cfrac{1}{E - a_0 - \cfrac{b_1^2}{E - a_1 - \cfrac{b_2^2}{E - a_2 - \cdots}}} . \qquad (75a)$$

From $G_{00}(E)$ the local density-of-states projected on $|f_0\rangle$ of the Hamiltonian operator H can be obtained; according to Eq. (I-48a), the local density-of-states is given by

$$n_0(E) = -\frac{1}{\pi} \lim_{\varepsilon \to 0^+} \mathrm{Im}\, G_{00}(E + i\varepsilon) \qquad (75b)$$

where ε is an infinitesimal positive quantity.

The recursion method allows (at least in principle) the one-dimensional chain representation of any quantum mechanical system, with an arbitrary number N of states. No explicit diagonalization of the original matrix H is required for performing such a transformation. The one-dimensional chain representation allows to describe economically the system through the parameters a_n and b_n (i.e. $2N - 1$ parameters rather than $N(N+1)/2$ matrix elements $\langle \phi_i | H | \phi_j \rangle$; for $N = 1000$ for instance, we have to compare approximately two thousand with one million!).

A peculiar feature of the Lanczos procedure is that the number of steps to be carried out in specific problems may be reasonably small (say ten up to hundred or so). Once the original matrix H is put into tridiagonal form, successive diagonalizations of small order tridiagonal matrices allow to infer the eigenvalues of H, especially those whose eigenfunctions have large overlap with the seed state.

In some problems, it is even possible to reasonably infer the asymptotic behaviour of expansion (75a) and reproduce the analytic properties of the Green's function through the analytic theory of continued fractions. The asymptotic behaviour of the continued fraction coefficients a_n and b_n is connected with the compactness of the spectrum and with the presence of critical points. We cannot dwell on this and other interesting aspects of the recursion method, and on its interplay with other formalisms, such as the method of moments, the memory function approach, and the equation of motion method [see for instance the review articles by G. Grosso and G. Pastori Parravicini, Adv. Chem. Phys. **63**, 81 (1985); **63**, 133 (1985) and references quoted therein]; we rather prefer to clarify the procedure with an example.

Illustrative example of the recursion method

To illustrate the recursion method, we consider for instance the case of a "quadratum", which is a simple square lattice with a single orbital per site and nearest neighbour

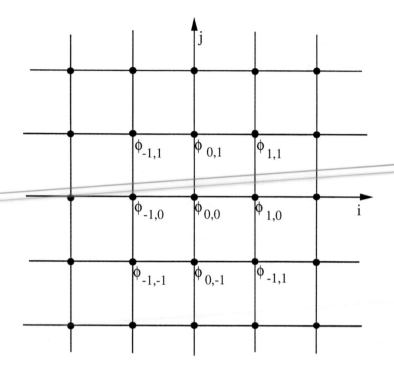

Fig. 15 Schematic graphical representation of a simple square lattice, with a single orbital per site, and nearest neighbour interactions; ϕ_{ij} indicates the orbital localized in the position (i, j) (the edge of the quadratum is taken as unit).

interactions. The Hamiltonian is given by

$$H = E_0 \sum_{ij} |\phi_{ij}\rangle\langle\phi_{ij}|$$

$$+t \sum_{ij} \left[|\phi_{ij}\rangle\langle\phi_{i+1,j}| + |\phi_{ij}\rangle\langle\phi_{i-1,j}| + |\phi_{i+1,j}\rangle\langle\phi_{ij}| + |\phi_{i-1,j}\rangle\langle\phi_{ij}| \right]$$

$$+t \sum_{ij} \left[|\phi_{ij}\rangle\langle\phi_{i,j+1}| + |\phi_{ij}\rangle\langle\phi_{i,j-1}| + |\phi_{i,j+1}\rangle\langle\phi_{ij}| + |\phi_{i,j-1}\rangle\langle\phi_{ij}| \right] . \quad (76)$$

We take for convenience $E_0 = 0$ and $t = 1$. The Hamiltonian operator (76) is represented by a sparse matrix (non-vanishing interactions are confined to nearest neighbour orbitals), and is indicated graphically in Fig. 15.

Before starting the Lanczos procedure, we need to establish explicitly the effect of the operator H on a generic state of the form

$$|u\rangle = \sum_{ij} c_{ij} |\phi_{ij}\rangle . \quad (77a)$$

From Eq. (76), we see that

$$H\,|u\rangle = \sum_{ij} \gamma_{ij}|\phi_{ij}\rangle \tag{77b}$$

with

$$\gamma_{ij} = c_{i-1,j} + c_{i+1,j} + c_{i,j-1} + c_{i,j+1} \ . \tag{77c}$$

With the help of Eqs. (77), one can easily carry out the Lanczos procedure; choosing as seed state the orbital at the origin, i.e. $|f_0\rangle = |\phi_{00}\rangle$, one obtains

$$|F_1\rangle = |\phi_{0,1}\rangle + |\phi_{0,-1}\rangle + |\phi_{1,0}\rangle + |\phi_{-1,0}\rangle \ ;$$

the normalization of $|F_1\rangle$ gives $b_1^2 = \langle F_1|F_1\rangle = 4$. At the second step, one obtains

$$|F_2\rangle = |\phi_{1,1}\rangle + |\phi_{-1,1}\rangle + |\phi_{-1,-1}\rangle + |\phi_{1,-1}\rangle + \frac{1}{2}\left[|\phi_{2,0}\rangle + |\phi_{0,2}\rangle + |\phi_{-2,0}\rangle + |\phi_{0,-2}\rangle\right] \ .$$

The normalization of $|F_2\rangle$ gives $b_2^2 = \langle F_2|F_2\rangle = 5$. The procedure can be continued so to obtain all the desired continued fraction parameters (with the help of a straightforward computer code). The density-of-states can then be computed from Eq. (75b).

As a second illustrative example, the reader could consider the case of the "cubium", which is a simple cubic lattice with a single orbital per site and nearest neighbour interactions. Choosing as seed state an atomic orbital, the continued fraction expression of the Green's function can be established; the real part of the Green's function, as well as the density-of-states of the cubium calculated by means of Eq. (75b), have been reported in Fig. II-25.

8.3   Modified Lanczos method for excited states

In the Lanczos method, the states $\{|f_n\rangle\}$ generated by the three-term relations (73) are in principle orthonormal. However, it is well known that finite precision arithmetic induces rounding errors in the procedure, which may give rise to loss of precision in the results (loss of orthogonality among the generated states and appearance of spurious "ghost" states). In a number of situations where high accuracy is needed, there is little benefit to pursue further steps of the Lanczos approach, and alternative procedures must be found.

There is a vast literature that concerns the analysis and solution of the numerical problems that can be met in the Lanczos method; in particular, for the ground state of quantum systems, we refer to the paper of E. Dagotto, Rev. Mod. Phys. **66**, 763 (1994) and references quoted therein. Here, we briefly mention an intuitive extension of the Lanczos procedure for the evaluation of the *excited states* of quantum systems in a desired energy range [G. Grosso, L. Martinelli and G. Pastori Parravicini, Nuovo Cimento D**15**, 269 (1993); Phys. Rev. B**51**, 13033 (1995)].

To obtain excited states of H, we consider a trial energy E_t and the auxiliary operator $A = (H - E_t)^2$ (other forms are suitable as well, but we focus on this form due to its simplicity). The ground state of the operator A is obviously the excited state of H nearest in energy to the chosen energy E_t. We can thus start the Lanczos iterations with the operator A and a seed state $|f_0\rangle$; however, to avoid any possible numerical problem, we perform only one Lanczos step and obtain the state $|f_1\rangle$. At this stage we diagonalize A on the two states $|f_0\rangle$, $|f_1\rangle$, and select the lowest eigenvalue of the 2×2 matrix and the corresponding eigenfunction $|f_g\rangle$. We can now use $|f_g\rangle$ as seed state and iterate the whole procedure; we obtain thus an iterative approach that systematically converges towards the exact eigenvalue and eigenfunction of the ground state of the operator A (which is also the excited state of H nearest in energy to the chosen energy E_t). The convergence can be conveniently accelerated with appropriate procedures, and we refer to the literature for a more detailed description of the required technical aspects and applications.

8.4   Renormalization method for electronic systems

The basic idea of the renormalization method is to infer the properties of systems, whose description requires a large (or infinite) number of basis functions, by reducing progressively the dimension of the space of preserved basis functions. In its ultimate motivation, the origin of this strategy can be traced back to the concept of renormalization introduced in the theory of phase transitions by Wilson (see Section XVII-6).

We begin our treatment of the renormalization procedure for electronic state calculations by considering the Dyson equation for the Green's function of a quantum system. Consider an operator H, arbitrarily split into two parts

$$H = H_0 + W \ . \tag{78a}$$

If we indicate with $g(E)$ the Green's function of the operator H_0, and with $G(E)$ the Green's function of the operator H, we have that the two Green's functions are related via the Dyson equation

$$G(E) = g(E) + g(E) \, W \, G(E) \ . \tag{78b}$$

The proof is simple indeed. We start from the identity

$$E - H_0 = (E - H_0 - W) + W \ ,$$

and multiply both members of the above equation by $1/(E - H_0)$ (multiplication is performed, say, on the left); then we multiply by $1/(E - H_0 - W)$ (on the right) and obtain

$$\frac{1}{(E - H_0 - W)} = \frac{1}{(E - H_0)} + \frac{1}{(E - H_0)} W \frac{1}{(E - H_0 - W)} \ ,$$

which proves the Dyson equation (78b).

Consider now an operator H represented in a given orthonormal set by a $N \times N$ matrix matrix (in specific applications N can be very large, say from thousand to a million). We separate the representative space S of dimension N into a subspace S_A of dimension N_A and a subspace S_B of dimension $N_B = N - N_A$ (in some situations the subspace S_A has dimension $N - 1$, and the subspace S_B is composed just by a single state). Whatever arbitrary separation has been performed, the operator H can be written, and then split, in the form

$$H = \begin{pmatrix} H_{AA} & H_{AB} \\ H_{BA} & H_{BB} \end{pmatrix} \quad \text{and} \quad H = H_0 + W = \begin{pmatrix} H_{AA} & 0 \\ 0 & H_{BB} \end{pmatrix} + \begin{pmatrix} 0 & H_{AB} \\ H_{BA} & 0 \end{pmatrix}.$$

Application of the Dyson equation (78b) gives

$$\begin{pmatrix} G_{AA} & G_{AB} \\ G_{BA} & G_{BB} \end{pmatrix} = \begin{pmatrix} g_{AA} & 0 \\ 0 & g_{BB} \end{pmatrix} + \begin{pmatrix} g_{AA} & 0 \\ 0 & g_{BB} \end{pmatrix} \begin{pmatrix} 0 & H_{AB} \\ H_{BA} & 0 \end{pmatrix} \begin{pmatrix} G_{AA} & G_{AB} \\ G_{BA} & G_{BB} \end{pmatrix}$$

where $g_{AA}(E) = (E - H_{AA})^{-1}$ and $g_{BB}(E) = (E - H_{BB})^{-1}$. Performing the algebraic matrix operations indicated in the above expression, it follows

$$G_{AA}(E) = g_{AA}(E) + g_{AA}(E) \, H_{AB} \, G_{BA}(E)$$

$$G_{BA}(E) = g_{BB}(E) \, H_{BA} \, G_{AA}(E)$$

(and similar expressions for exchange of the subscripts A and B). With straightforward properties of matrix multiplication we have

$$G_{AA}(E) = \cfrac{1}{E - H_{AA} - H_{AB} \cfrac{1}{E - H_{BB}} H_{BA}}. \tag{79}$$

Until now we have performed exact algebraic transformations. The interpretation of Eq. (79) is immediate. The Green's function in the subspace S_A is determined by a "renormalized" or "effective" Hamiltonian which is given by

$$H^{(\text{eff})}(E) = H_{AA} + H_{AB} \frac{1}{E - H_{BB}} H_{BA}.$$

The physical meaning of the renormalized Hamiltonian is self-explanatory: besides H_{AA}, it contains the effect of an excursion from A to B, a propagation within B, and an excursion back to A.

Suppose that the splitting of the original space spanned by H is such that in the subspace S_B the Green's function is known. We can *eliminate* (or as it is commonly said in the literature *decimate*) all the states of the subspace S_B, considering within the subspace S_A the effective Hamiltonian

$$H^{(\text{eff})}(E) = H_{AA} + \Sigma_{AA}(E) \tag{80a}$$

with

$$\Sigma_{AA}(E) = H_{AB} \frac{1}{E - H_{BB}} H_{BA}. \tag{80b}$$

The operator $\Sigma_{AA}(E)$, whose origin is linked to the elimination of the subspace S_B, is called *self-energy operator*; the self-energy operator $\Sigma_{AA}(E)$, added to H_{AA}, gives the effective Hamiltonian on the preserved subspace S_A, now formally decoupled from the subspace S_B.

The decimation procedure can always be performed, but it is of practical help when the advantages outnumber the disadvantages. Let us see in general when this can occur.

The advantage of the decimation procedure stands in having to handle a subspace S_A of dimension smaller than the original one. The price for this is the necessity to evaluate the appropriate self-energy operator, whose matrix elements are energy-dependent. In general situations it is not convenient to proceed towards a partial elimination of the states of the preserved subspace S_A, because of the complicated structure of the self-energy operator $\Sigma_{AA}(E)$. However in particular but very important cases (Bethe lattices, one-dimensional lattices, multilayer structures) it is possible to find peculiar decimations (the analogue of the "renormalization group transformations" of phase transitions), which can be applied "ad libitum" to the preserved subspace itself.

In particular, multilayer structures, superlattices and quantum wells, as well as very interesting incommensurate systems have been studied with the iterative renormalization procedure; most important, the high numerical stability of the procedure allows investigation of the asymptotic properties of systems with million of sites [for further details and references see for instance G. Grosso, S. Moroni and G. Pastori Parravicini, Physica Scripta T25, 316 (1989); R. Farchioni, G. Grosso and G. Pastori Parravicini, Phys. Rev. B47, 2394 (1993)]. Here we limit our attention to two particular significant examples, which are exactly soluble with the renormalization method, and are encountered very often in the forthcoming chapters.

Example 1. Discrete state interacting with a manifold (discrete or continuous) of states

The first example we consider is that of a discrete state interacting with a continuum of states; for simplicity, the continuum is mimicked with an (arbitrary) large number N of discrete states. The Hamiltonian can be written as

$$H = H_0 + W \tag{81a}$$

$$H_0 = E_0|\phi_0\rangle\langle\phi_0| + \sum_{i=1}^{N} E_i|\phi_i\rangle\langle\phi_i| \tag{81b}$$

$$W = \frac{1}{\sqrt{N}} \sum_{i=1}^{N} [\gamma_i |\phi_0\rangle\langle\phi_i| + \gamma_i^* |\phi_i\rangle\langle\phi_0|] . \tag{81c}$$

The energies E_i of the quasi-continuum manyfold of states $|\phi_i\rangle$ $(i = 1, 2, \ldots, N)$ extend from $E_1 = E_{\min}$ to $E_N = E_{\max}$; the energy E_0 of the discrete state can be either degenerate with the manifold (we shall consider this case in discussing the Fano model in Section XII-6) or at lower energy (we shall consider this case in discussing the Kondo effect in Section XVI-7). We discuss here explicitly this latter case, schematically indicated in Fig. 16.

Using the renormalization procedure, we can *eliminate* one-by-one all the states

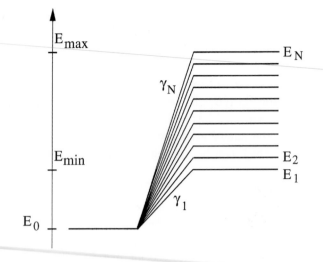

Fig. 16 Schematic representation of a system consisting of a discrete state of energy E_0, interacting with a manifold of states in the energy range $E_1 = E_{\min}$ and $E_N = E_{\max}$.

$|\phi_i\rangle \, (i = 1, 2, \ldots, N)$ and we obtain

$$G_{00}(E) = \cfrac{1}{E - E_0 - \cfrac{1}{N} \sum_i \cfrac{|\gamma_i|^2}{E - E_i}} \, . \tag{82a}$$

The poles of this Green's function are given by the equation

$$\boxed{E - E_0 = \frac{1}{N} \sum_i \frac{|\gamma_i|^2}{E - E_i}} \, . \tag{82b}$$

The above equation is easily solved graphically, as indicated in Fig. 17. The solutions are given by the intercepts of the straight line through E_0 and slope one, with the monotonically decreasing curve $(1/N) \sum |\gamma_i|^2/(E - E_i)$. Besides the solutions reminescent of the individual states of the quasi-continuum, Fig. 17 shows that there is always formation of a "collective state", with energy less than E_0; the adjective "collective" is used to indicate that this state is jointly determined by the discrete state and the states of the quasi-continuum (at least those of lower energy).

Example 2. Eigenvalues of an operator in the presence of a constant coupling between basis functions

Consider an Hamiltonian of the form

$$H = H_0 + W \tag{83a}$$

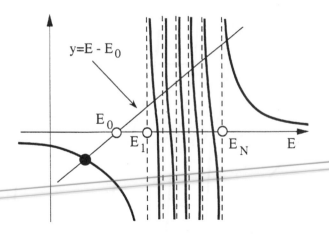

Fig. 17 Graphical solution of the eigenvalues of the Hamiltonian describing a discrete state E_0 interacting with a quasi-continuum of states extending from $E_{min} = E_1$ to $E_{max} = E_N$. We illustate the case $E_0 < E_{min}$. Notice that a "collective state" always appears with energy smaller than E_0.

with

$$H_0 = \sum_i E_i |\phi_i\rangle\langle\phi_i| \quad \text{and} \quad W = \frac{w_0}{N} \sum_{ij} |\phi_i\rangle\langle\phi_j| . \tag{83b}$$

The operator H_0 is diagonal with energies E_i $(i = 1, 2, \dots, N)$; the operator W has all the diagonal and off-diagonal matrix elements equal to a constant, denoted as w_0/N. We can obtain the eigenvalues of $H = H_0 + W$ with the following procedure.

Consider the (normalized) state

$$|u_0\rangle = \frac{1}{\sqrt{N}} \sum_i |\phi_i\rangle .$$

The diagonal matrix element on the state $|u_0\rangle$ of the Dyson equation (78b) reads

$$\langle u_0|G(E)|u_0\rangle = \langle u_0|g(E)|u_0\rangle + \langle u_0|g(E)\, W\, G(E)|u_0\rangle ;$$

since W can be expressed as $w_0 |u_0\rangle\langle u_0|$, we immediately obtain

$$\langle u_0|G(E)|u_0\rangle = \frac{\langle u_0|g(E)|u_0\rangle}{1 - w_0\langle u_0|g(E)|u_0\rangle} . \tag{84}$$

We notice that

$$g_{00}(E) = \langle u_0|g(E)|u_0\rangle = \frac{1}{N} \sum_i \frac{1}{E - E_i} ;$$

we thus have that the poles of $G(E)$, given by the zeroes of the denominator of Eq. (84), are determined by the condition

$$\boxed{\frac{1}{w_0} = \frac{1}{N} \sum_i \frac{1}{E - E_i}} . \tag{85}$$

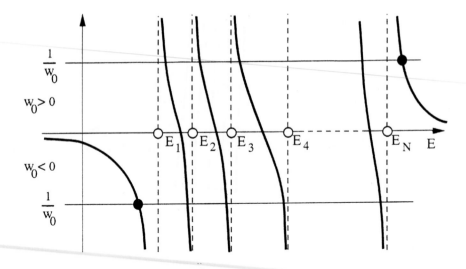

Fig. 18 Graphical solution of Eq. (85) of the text and formation of split-off states.

The graphical solution of Eq. (85) is indicated in Fig. 18. The graphical solution is obtained by the intercepts of the horizontal line $1/w_0$ with the curve $g_{00}(E)$. The curve $g_{00}(E)$ is regular and monotonically decreasing in the intervals $-\infty < E < E_1$ and $E_N < E < \infty$; furthermore we have:

$$\lim_{E \to E_1} g_{00}(E) = -\infty \quad \text{for } E < E_1 \;, \quad \lim_{E \to E_N} g_{00}(E) = +\infty \quad \text{for } E > E_N \;. \qquad (86)$$

In the case the N states E_i are discretized (case of Fig. 18), we have always the formation of a *split-off state* (or "collective state"), with energy outside the interval $E_1 \leq E \leq E_N$; the split-off state has energy smaller than E_1 if $w_0 < 0$, and energy larger than E_N if $w_0 > 0$.

It is interesting to consider the limiting case of a genuine continuum, i.e. the case $N \to \infty$. In this case $g_{00}(E)$ is again regular and monotonically decreasing in the intervals $-\infty < E < E_1$ and $E_N < E < \infty$, but the limits (86) can be either finite or infinite. In the case the limits (86) are finite, a critical strength $|w_0|$ is necessary to form split-off states. We shall encounter in other parts of this book examples of either type (for instance, in the case of Cooper pairs in metals, their formation occurs whatever small the interaction is; in the theory of deep impurity states, a threshold is in general required for their formation).

Further Reading

S. L. Altmann "Band Theory of Solids" (Clarendon Press, Oxford 1994)

F. Bassani and G. Pastori Parravicini "Electronic States and Optical Transitions in Solids" (Pergamon Press, Oxford 1975)

J. Callaway "Quantum Theory of the Solid State" (Academic Press, New York 1991, 2nd edition)

E. N. Economou "Green's Functions in Quantum Physics" (Springer, Berlin 1990)

H. Eschrig "Optimized LCAO method and the Electronic Structure of Extended Systems" (Springer Verlag, Berlin 1989)

G. C. Fletcher "Electron Band Theory of Solids" (North-Holland, Amsterdam 1971)

P. Giannozzi, G. Grosso and G. Pastori Parravicini "Theory of Electronic States in Lattices and Superlattices" Rivista del Nuovo Cimento **13**, N.3 (1990)

C. M. Goringe, D. R. Bowler and E. Hernández "Tight-binding modelling of materials" Rep. Prog. Phys. **60**, 1447 (1997)

D. R. Hamann, M. Schlüter and C. Chiang "Norm-Conserving Pseudopotential" Phys. Rev. Lett. **43**, 1494 (1979); G. B. Bachelet, D. R. Hamann and M. Schlüter "Pseudopotentials that work: from H to Pu" Phys. Rev. B**26**, 4199 (1982)

W. A. Harrison "Electronic Structure and the Properties of Solids" (Freeman, San Francisco 1980)

V. Heine "Electronic Structure from the Point of View of the Local Atomic Environment"; D. W. Bullett "The Renaissance and Quantitative Development of the Tight-Binding Method"; R. Haydock "The Recursive Solution of the Schrödinger Equation"; M. J. Kelly "Applications of the Recursion Method to the Electronic Structure from an Atomic Point of View" in Solid State Physics vol.35 (Academic Press, New York 1980)

V. Heine "The Pseudopotential Concept"; M. L. Cohen and V. Heine "The Fitting of Pseudopotentials to Experimenal Data and their Subsequent Application"; V. Heine and D. Weaire "Pseudopotential Theory of Cohesion and Structure" Solid State Physics vol.24 (1970)

C. Lanczos "Applied Analysis" (Prentice Hall, Englewood Cliffs, New York 1958)

T. L. Loucks "Augmented Plane Wave Method: a Guide to Perform Electronic Structure Calculations" (Benjamin, New York 1967)

V. V. Nemoshkalenko and V. N. Antonov "Computational Methods of Band Theory" (Gordon and Breach, Amsterdam 1998)

D. J. Singh "Planewaves, Pseudopotentials and the LAPW Method" (Kluwer Academic, Boston 1993)

H. L. Skriever "The LMTO Method" (Springer Verlag, Berlin 1984)

J. C. Slater "Quantum Theory of Molecules and Solids" Volumes 1-4 (McGraw Hill, New York 1974)

A. Sutton "Electronic Structure of Materials" (Clarendon Press, Oxford 1993)

VI

Electronic properties of selected crystals

In this chapter we give a brief survey of the electronic structure of some solids, of particular interest from a fundamental or technological point of view. The methods of electronic band structure calculations, in conjunction with the advances in the density functional formalism and many-body techniques, have made possible in principle (and often also in practice) to describe the band structure, charge density and cohesive energy of crystals, starting from the very knowledge of their chemical composition. However, simple heuristic schemes are still of great value, because of their capability in describing physical properties on a more intuitive basis.

In this chapter we discuss the electronic band structure and ground-state properties of some selected materials; the focus is not on the properties of single specific crystals, but rather on the trends in similar compounds. We begin with the description of rare-gas solids, which are large gap insulators formed by weakly interacting closed-shell neutral atoms. We then describe ionic crystals, constituted by strongly interacting closed-shell ions. In the discussion of crystals made up by open-shell units, we consider typical examples of covalent semiconductors and the metallic bond. In spite of the narrowness of the topics included, this chapter should nevertheless convey some of the ideas underlying the material science investigations in solids.

1   Band structure and cohesive energy of rare-gas solids

1.1   General features of band structure of rare-gas solids

Rare-gas solids (Ne, Ar, Kr, Xe) are the simplest crystals made up of closed-shell atoms (helium is a quantum crystal), and constitute a class of large gap insulators. Neon atom has the closed-shell electronic configuration $1s^2\, 2s^2\, 2p^6$. The electronic configuration of argon is [Ne] $3s^2\, 3p^6$; the configuration of krypton is [Ar] $3d^{10}\, 4s^2\, 4p^6$; that of xenon is [Kr] $4d^{10}\, 5s^2\, 5p^6$. Rare-gas solids crystallize in the face-centered cubic lattice (as described in Chapter II), with one atom per primitive unit cell.

The band structure of rare-gas solids is rather simple, at least in its essential lines. The inner core states give rise to (almost) dispersionless fully occupied bands. The occupied p-states of higher energy give rise to fully occupied valence bands of p-like nature; these bands present the maximum at the point Γ of the Brillouin zone, and bend downward as $\mathbf{k}$ increases out of the center. This general trend is expected qualitatively from a tight-binding description of localized p states in fcc Bravais lattices (as discussed in Section V-2.3). The conduction bands are almost free-electron like; in fact electrons can move almost freely in the crystal, with exclusion of the (small) fraction of volume, where core and valence orbitals extend. Solid neon with an energy gap of $\approx 21.5\,\mathrm{eV}$ is the crystal with the highest energy gap.

We consider, as an exemplification, the band structure of solid argon. In the ground state of the argon atom, the orbital energies are $E_{1s} = -3205.9\,\mathrm{eV}$, $E_{2s} = -326.3\,\mathrm{eV}$, $E_{2p} = -249.34\,\mathrm{eV}$, $E_{3s} = -29.24\,\mathrm{eV}$ and $E_{3p} = -15.82\,\mathrm{eV}$. The inner core states $1s, 2s, 2p$ of the argon atom give rise to (fully occupied) crystal core bands. The valence bands, originated from the $3s$ and $3p$ orbitals, and the conduction bands of solid argon can be described, for instance, by the orthogonalized plane wave method, where plane waves are orthogonalized to the five Bloch sums formed with $1s, 2s, 2p_x, 2p_y, 2p_z$ orbitals of argon. Using a Gaussian representation of atomic orbitals, the matrix elements appearing in the OPW method can be evaluated analytically, and accurate calculations can be performed. In the calculations reported in Fig. 1, a maximum of 259 OPWs were used; at Γ this corresponds to consider 16 shells of reciprocal lattice vectors, up to the waves of type $(2\pi/a)(4,4,2)$ and $(2\pi/a)(6,0,0)$.

The general trend of the valence bands of Fig. 1 can be understood within the tight-binding framework, because of the strong localization of the occupied orbitals on the scale of the lattice parameter. The general trend of the conduction bands can be understood starting from the empty lattice analysis for fcc structures (Fig. V-6); the removal of the empty lattice degeneracies is in qualitative agreement with what expected from the presence of a smooth repulsive pseudopotential.

From the correlated energy bands of Fig. 1, we see that the maximum of the valence bands of argon is the three-fold degenerate state $E(\Gamma_4^-) = -13.7\,\mathrm{eV}$. The width of the valence bands is $E(\Gamma_4^-) - E(L_2^-) = 1.7\,\mathrm{eV}$. The minimum of the conduction bands is the total symmetric state Γ_1^+ of energy $E(\Gamma_1^+) = 0.9\,\mathrm{eV}$. The calculated energy gap is $E_G = E(\Gamma_1^+) - E(\Gamma_4^-) = 14.6\,\mathrm{eV}$. All the above calculated values are in rather good agreement with the experimental band structure parameters of Ar reported in Table 1.

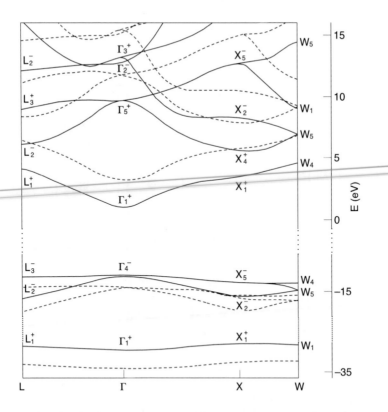

Fig. 1 Energy bands of solid argon with inclusion of correlation effects (solid line). For comparison the energy bands in the Hartree–Fock approximation are also reported (dashed lines); correlation effects shift (almost rigidly) the Hartree–Fock conduction bands to lower energies, and the Hartree–Fock valence bands to higher energies [from S. Baroni, G. Grosso and G. Pastori Parravicini, Phys. Rev. B**29**, 2891 (1984); copyright 1984 by the American Physical Society].

The table also reports the relevant band structure parameters of Ne, Kr and Xe; for completeness the spin-orbit splittings of the more external occupied atomic p-states and of the corresponding valence bands at $\mathbf{k} = 0$ are also given.

From the data of Table 1, it can be noticed that there is a systematic trend of the relevant band structure parameters passing from neon to xenon. The trend can be best noticed also from Fig. 2, which reports the experimental energy distribution curves for the photoelectrons emitted from the valence bands of solid Ne, Ar, Kr and Xe. Photoemission measurements basically consist in shining monochromatic photons of energy $\hbar\omega$ on a target crystal and measuring the kinetic energy of the electrons escaped from the surface.

The analysis of the energy distribution curves of the emitted photoelectrons permits us to infer relevant information on the band structure. In fact the energy distribution curves are expected to be reminescent of the major features and edges of the actual

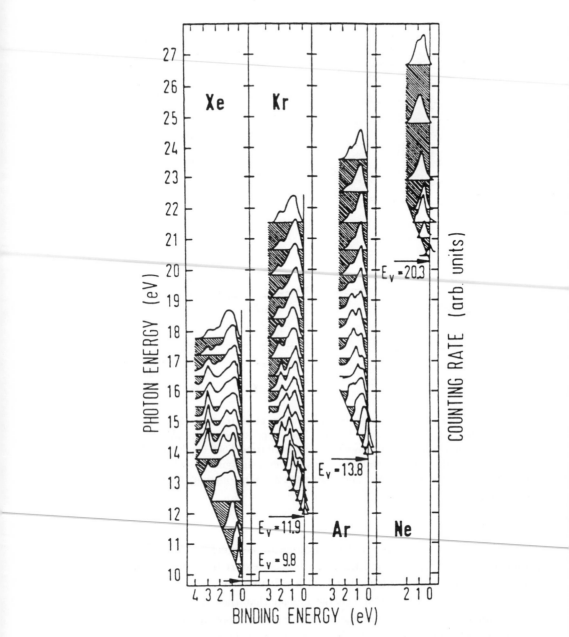

Fig. 2 Energy distribution curves of emitted photoelectrons for solid Ne, Ar, Kr, and Xe, at various photon energies. The zero-count line for each individual curve is shifted upwards proportional to the exciting photon energy. E_V represents the vacuum energy measured from the top of the valence band [from N. Schwentner, F.-J. Himpsel, V. Saile, M. Skibowski, W. Steinmann and E. E. Koch, Phys. Rev. Lett. **34**, 528 (1975); copyright 1975 by the American Physical Society].

Table 1 Relevant parameters of the band structure of rare-gas solids as obtained from experiments. The energies are in eV [from N. Schwentner, F.-J. Himpsel, V. Saile, M. Skibowski, W. Steinmann and E. E. Koch, Phys. Rev. Lett. **34**, 528 (1975); copyright 1975 by the American Physical Society].

	Ne	Ar	Kr	Xe
band-gap energy E_G	21.7	14.2	11.6	9.3
top valence band $E(\Gamma_4^-)$	-20.3	-13.8	-11.9	-9.8
bottom conduction band $E(\Gamma_1^+)$	1.4	0.4	-0.3	-0.5
valence band width	1.3	1.7	2.3	3.0
$\Delta_{SO}^{(gas)}$	0.14	0.22	0.67	1.31
$\Delta_{SO}^{(solid)}$	≈ 0.1	0.2	0.64	1.3

density-of-states of occupied bands, shifted by $\hbar\omega$. This is shown for rare-gas solids in Fig. 2 for different energies of the incident photons; when the energy of the incident photon is high enough to excite electrons from the bottom of the valence band, the widths of the energy distibution curves are independent from $\hbar\omega$ and provide the total width of the valence bands; furthermore, with appropriate extrapolation, the position of the vacuum level E_V with respect to the top of the valence band can be obtained with high accuracy.

1.2 Cohesive energy of rare-gas solids

The study of the ground-state energy of solids allows us to investigate several properties and in particular the equilibrium lattice constant, cohesive energy, compressibility and bulk modulus. In principle, the density functional approach permits us to obtain the ground-state charge density of a system and hence any ground-state property of interest; however, we wish here to confine ourselves to intuitive semi-quantitative considerations.

In rare-gas solids, the total energy can be expressed reasonably well in terms of pair interaction (or two-body interaction). For the present discussion we make the standard approach to consider only an effective two-body interaction (although the importance of three or multi-body interactions is well established, for instance in the description of the stability of the fcc phase with respect to the hcp phase).

The interaction potential $U(R)$ between two isolated atoms can be calculated, in principle, starting from the quantum mechanical approach. A widely used semiempirical description approximates the interatomic potential between two neutral closed-shell atoms with the *Lennard–Jones potential* of the type

$$U(R) = \frac{A}{R^{12}} - \frac{B}{R^6} \ . \tag{1}$$

The first term is a crude way to represent classically the fact that filled atomic shells constitute an almost impenetrable core. The inverse twelve power law has been chosen

Table 2 Parameters of the Lennard–Jones potential $U(R) = \varepsilon\left[(\sigma/R)^{12} - 2(\sigma/R)^6\right]$ for a pair of rare-gas atoms, as provided by S. Gonçalves and H. Bonadeo, Phys. Rev. B**46**, 10738 (1992). In the table, we also report the static properties of rare-gas solids (nearest neighbour distance, cohesive energy and bulk modulus) calculated from the given set of parameters ε and σ. The experimental values of the nearest neighbour distance are taken from R. W. G. Wyckoff "Crystal Structures" (Interscience, New York 1963). The experimental binding energies are quoted by E. R. Dobbs and G. O. Jones, Rep. Progr. Phys. **20**, 516 (1957). The experimental values of the bulk modulus are quoted by P. Korpium and E. Lüscher in "Rare Gas Solids" (edited by M. L. Klein and J. A. Venables) vol. II, p. 729 (Academic Press, London 1977).

Pair of rare-gas atoms		Rare-gas solids					
		Nearest neighbour distance (Å)		Binding energy (eV/atom)		Bulk modulus (kbar)	
σ (Å)	ε (eV)	calc.(Eq. 6)	exp.	calc.(Eq. 7)	exp.	calc.(Eq. 10)	exp.
Ne	3.25	0.0024	3.16	3.13	0.021	0.020	11.9	11.1
Ar	3.87	0.0098	3.76	3.72	0.084	0.080	28.8	28.6
Kr	4.11	0.0135	3.99	4.05	0.116	0.112	33.1	34.1
Xe	4.46	0.0185	4.33	4.38	0.159	0.166	35.5	37.9

for analytic simplicity (and for historical reasons); a high exponent in the repulsive term simulates its short-range nature. The second term in Eq. (1) corresponds to the fluctuating dipole–dipole interaction, which is attractive; the inverse sixth power law is the leading term for the interaction of a dipole with a polarizable neutral atom (van der Waals interaction).

It is interesting to notice that a pure Hartree–Fock calculation of *two* isolated rare-gas atoms at distance R predicts repulsion between the two atoms at any R. The attractive term is beyond the Hartree–Fock approximation. Thus rare-gas solids would not be bound in the Hartree–Fock approximation.

The Lennard–Jones potential (1) for a pair of rare-gas atoms at distance R can be re-written in the equivalent and more convenient form

$$U(R) = \varepsilon\left[\left(\frac{\sigma}{R}\right)^{12} - 2\left(\frac{\sigma}{R}\right)^6\right] . \tag{2}$$

The physical meaning of ε and σ is obvious: the potential $U(R)$, defined by Eq. (2), is minimal when the pair distance R equals σ, and its value there is $-\varepsilon$. The quantities ε and σ are often considered as two phenomenological parameters, to be determined via some appropriate elaboration of experimental data. As an example, in Table 2 we report a set of parameters for rare-gas atoms, inferred from molecular dynamics simulation of their vibrational and structural properties.

Once the pair interaction is known (either semi-empirically or from first principles), we can obtain the total energy of the solid summing up the interactions for all pairs of atoms composing the crystal (three-body corrections are assumed negligible; zero point vibrational energy is also neglected). We have for the total energy $U_s(R)$ of the solid

$$U_s(R) = \frac{1}{2} N\varepsilon \left[\sum_{j(\neq i)} \left(\frac{\sigma}{R_{ij}} \right)^{12} - 2 \sum_{j(\neq i)} \left(\frac{\sigma}{R_{ij}} \right)^{6} \right], \tag{3}$$

where N is the total number of atoms, R_{ij} is the distance between two atoms i and j, and the factor $1/2$ appears in order not to include twice each pair of atoms. Because of the translational symmetry, the sum over j in the right-hand side of Eq. (3) is independent from the chosen reference atom with label i; the ith atom can thus be taken at the origin.

Let us express R_{ij} in the product form $R p_{ij}$, where R is the distance between nearest neighbours and p_{ij} is dimensionless. For the fcc structure, we have 12 first neighbours at distance R ($p_{ij} = 1$), 6 second neighbours at distance $R\sqrt{2}$ ($p_{ij} = \sqrt{2}$), 24 third neighbours at distance $R\sqrt{3}$ ($p_{ij} = \sqrt{3}$) etc. We consider thus the lattice sums

$$A_{12} = \sum_{j(\neq i)} \left(\frac{1}{p_{ij}} \right)^{12} = 12 + \frac{6}{\sqrt{2}^{12}} + \frac{24}{\sqrt{3}^{12}} + \ldots = 12.13 \tag{4a}$$

$$A_{6} = \sum_{j(\neq i)} \left(\frac{1}{p_{ij}} \right)^{6} = 12 + \frac{6}{\sqrt{2}^{6}} + \frac{24}{\sqrt{3}^{6}} + \ldots = 14.45 . \tag{4b}$$

The lattice sums (4) are rapidly convergent and can be obtained by direct summation. Using the coefficients A_{12} and A_6, we can express the crystal energy (3) in the form

$$U_s(R) = \frac{1}{2} N\varepsilon \left[A_{12} \left(\frac{\sigma}{R} \right)^{12} - 2 A_6 \left(\frac{\sigma}{R} \right)^{6} \right] . \tag{5}$$

The behaviour of $U_s(R)$ is given in Fig. 3. From Eq. (5), we can now easily obtain the equilibrium lattice constant, the cohesive energy and the bulk modulus.

The *equilibrium nearest-neighbour constant* R_0 is determined by the condition that the derivative of $U_s(R)$ vanishes; we have

$$R_0 = \left(\frac{A_{12}}{A_6} \right)^{1/6} \cdot \sigma = 0.971\,\sigma . \tag{6}$$

The *equilibrium cohesive energy* (disregarding vibrational effects) is given by

$$U_s(R_0) = -\frac{1}{2} N\varepsilon \frac{A_6^2}{A_{12}} = -8.61\,N\varepsilon . \tag{7}$$

Equation (6) shows that the ratio R_0/σ is smaller than 1 and is independent from the material (at least within the assumed model). Equation (7) shows that the cohesive energy per atom is proportional to the binding energy ε, and again the proportionality constant is independent from the material.

We pass now to consider the *compressibility and bulk modulus* in our simple model.

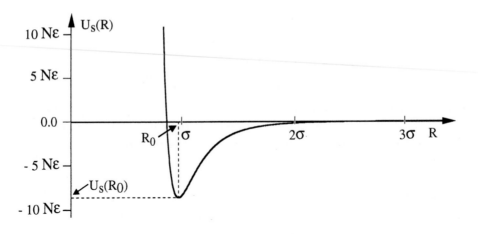

Fig. 3 Behaviour of the crystal energy $U_s(R)$ for a rare-gas solid as a function of the nearest neighbour distance R. The minimum of the curve provides the cohesive energy $U_s(R_0)$ and the equilibrium position R_0. The slope is related to the applied pressure, and the curvature near R_0 is related to the bulk modulus.

At $T = 0$ the pressure p is given by $p = -dU_s/dV$. The (isothermal) compressibility K is defined as $K = -(1/V)(dV/dp)$. The inverse of the compressibility is called bulk modulus B. From the definition of pressure and compressibility, we have for the bulk modulus at zero temperature

$$B_0 = V \frac{d^2 U_s}{dV^2} . \tag{8}$$

In fcc structures, the relation between the volume V and the nearest neighbour distance R is

$$V = \frac{1}{\sqrt{2}} N R^3$$

(in fact the volume V of a fcc structure with N atoms is $V = N a^3/4$, and the nearest neighbour distance is $R = a\sqrt{2}/2$). Thus

$$\frac{dU_s}{dV} = \frac{dU_s}{dR} \frac{dR}{dV} = \frac{\sqrt{2}}{3} \frac{1}{N R^2} \frac{dU_s}{dR} ;$$

we have also

$$\frac{d^2 U_s}{dV^2} = \frac{2}{9} \frac{1}{N^2 R^2} \frac{d}{dR} \left(\frac{1}{R^2} \frac{dU_s}{dR} \right).$$

At the equilibrium position, the bulk modulus (8) becomes

$$B_0 = \frac{\sqrt{2}}{9} \frac{1}{N R_0} \left(\frac{d^2 U_s}{dR^2} \right)_{R=R_0} . \tag{9}$$

Using Eq. (5) one has

$$B_0 = \frac{2\sqrt{2}}{3} \frac{\varepsilon}{\sigma^3} \frac{\sigma^9}{R_0^9} \left[13 A_{12} \frac{\sigma^6}{R_0^6} - 7 A_6 \right] ;$$

with the help of Eq. (6) it follows

$$B_0 = 4\sqrt{2}\frac{\varepsilon}{\sigma^3}\left(\frac{A_6}{A_{12}}\right)^{3/2} \qquad A_6 = 106.3\,\frac{\varepsilon}{\sigma^3} \ . \qquad (10)$$

For the evaluation of the bulk modulus with the data of Table 2, it is useful to re-member that $1\,\text{eV}/\text{Å}^3 = 1.602{\cdot}10^{12}\,\text{dyne}/\text{cm}^2 = 1.602\,\text{Mbar}$.

The values of the equilibrium distance, cohesive energy and bulk modulus for rare-gas solids, calculated using Eq. (6), Eq. (7) and Eq. (10), are also reported in Table 2; the reasonable description of these and other properties with a simple two-parameter model, explains the popularity of the Lennard–Jones scheme in the investigation of weakly bound van der Waals crystals.

2   Electronic properties of ionic crystals

2.1   Introductory remarks and Madelung constant

Crystals that can be pictured as formed by positive and negative ions are referred to as *ionic crystals*. Some properties of these solids can be understood starting from the extreme approximation of a lattice of pointwise charges (*point ion lattice*); the influence of the actual internal constitution of ions can then be introduced in the model, and overall trends in electronic properties can often be understood at a reasonable elementary level.

An important quantity in the study of ionic crystals is the Coulomb interaction energy (Madelung energy) between ions of the crystals in the point ion approximation. As we shall see, this quantity accounts for most of (i) cohesive energy of the crystal (ii) stability of the structure (iii) energy gap of the material. We thus begin the study of ionic crystals with the evaluation of the potential energy of a given reference ion (say a negative ion taken at the origin) due to the Coulomb interaction with all the other positive and negative ions of the crystal. It is usual to express this potential energy in the Madelung form

$$V(0) = -\alpha_M \frac{e^2}{R} \ , \qquad (11)$$

where R is the nearest neighbour distance, α_M is the Madelung constant discussed below, $\pm e$ are the point charges at the lattice sites. The Madelung constant is a dimensionless constant, characteristic of the geometrical crystal structure under consideration.

The evaluation of the Madelung constant requires a special care, because of the long-range character of the Coulomb interaction. To give an idea of the slow convergence of the series that are encountered in the direct evaluation of the Madelung constant, let us consider the one-dimensional model of ionic crystal indicated in Fig. 4. In this

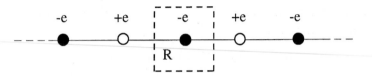

Fig. 4 Schematic representation of a one-dimensional ionic crystal in the point ion approximation, for the calculation of the Madelung constant. The nearest neighbour distance is denoted by R. The ion at the origin has been surrounded by a box.

Table 3 List of neighbours in NaCl structure with Cl^- at the origin; all ions up to the distance $R\sqrt{12}$ are indicated. In the last column, for each shell of neighbours, we indicate the charge *inside* the cube of edge $4R$ sorrounding the reference ion; including the ion at the origin, this cube contains 32 positive and 32 negative electronic charges.

	reference atom at the origin Cl^-		
number and type of ions	representative position	distance from from origin	charge within the cube of edge 4R
$6Na^+$	$R(1,0,0)$	$R\sqrt{1}$	$+6e$
$12Cl^-$	$R(1,1,0)$	$R\sqrt{2}$	$-12e$
$8Na^+$	$R(1,1,1)$	$R\sqrt{3}$	$+8e$
$6Cl^-$	$R(2,0,0)$	$R\sqrt{4}$	$-3e$
$24Na^+$	$R(2,1,0)$	$R\sqrt{5}$	$+12e$
$24Cl^-$	$R(2,1,1)$	$R\sqrt{6}$	$-12e$
$12Cl^-$	$R(2,2,0)$	$R\sqrt{8}$	$-3e$
$24Na^+$	$R(2,2,1)$	$R\sqrt{9}$	$+6e$
$6Na^+$	$R(3,0,0)$	$R\sqrt{9}$	0
$24Cl^-$	$R(3,1,0)$	$R\sqrt{10}$	0
$24Na^+$	$R(3,1,1)$	$R\sqrt{11}$	0
$8Cl^-$	$R(2,2,2)$	$R\sqrt{12}$	$-e$

case, we have for the potential energy of the ion at the origin

$$V(0) = -\frac{e^2}{R} 2 \left[\frac{1}{1} - \frac{1}{2} + \frac{1}{3} - \frac{1}{4} + \frac{1}{5} - \ldots \right] = -\frac{e^2}{R} 2\ln 2 .$$

The series in the square brackets is a slowly, conditionally convergent, series; in this case, however, the sum can be performed analytically and gives $\ln(1+x)$ with $x = 1$. The Madelung constant for a strictly one-dimensional model is thus $\alpha_M = 2\ln 2$.

In actual three-dimensional crystals, the evaluation of the Madelung constant is faced with the problem of summing up numerically series, which are slowly and conditionally convergent. Let us consider for instance the NaCl structure. Suppose we take a Cl^- anion at the origin; we find 6 ions Na^+ at distance R (where R denotes the nearest neighbour distance), 12 ions Cl^- at $R\sqrt{2}$, 8 ions Na^+ at $R\sqrt{3}$, 6 ions Cl^-

at distance $R\sqrt{4}$ etc.; the ions up to the distance $R\sqrt{12}$ are listed in Table 3. The Coulomb energy of this ion in the field of all the other ions is thus

$$V(0) = -\frac{e^2}{R}\left[\frac{6}{\sqrt{1}} - \frac{12}{\sqrt{2}} + \frac{8}{\sqrt{3}} - \frac{6}{\sqrt{4}} + \frac{24}{\sqrt{5}} - \frac{24}{\sqrt{6}} - \frac{12}{\sqrt{8}} + \dots\right]$$

$$= -\frac{e^2}{R}[6.000 - 8.486 + 4.619 - 3.000 + 10.733 - 9.798 - 4.243 + \dots]. \quad (12)$$

By inspection of the terms indicated in the square brackets in Eq. (12), it is not possible to infer when the direct summation converges, even approximately, to the value $\alpha_M = 1.7476$ of the NaCl structure (see Table 4). The most used procedures to evaluate lattice sums of the type shown in Eq. (12) are the Evjen and the Ewald methods.

The Evjen method for the evaluation of the Madelung constant

The Evjen method for the evaluation of the Madelung constant considers the crystal as formed by neutral (or approximately neutral) groups of ions (the Evjen cell). The electrostatic energy of a finite cell is evaluated by direct summation. As Evjen cells of increasing size are considered, one expects a rapidly convergent series, due to the short-range nature of the potential of groups of neutral atoms.

To give an exemplification of the Evjen method, we consider the NaCl structure, a reference ion (for instance Cl⁻ taken at the origin), and a cube of edge $4R$ surrounding the reference ion. In the Evjen method, all the ions *inside* the cube are considered as they stand and all the ions *outside* the cube are neglected. Ions at the *border* of the cube are assigned an appropriate fraction; this fraction is $1/2$ for an ion at the surface, $1/4$ for an ion at the edge and $1/8$ for an ion at the corner of the cube. In the last column of Table 3, the charge within the cube of edge $4R$, contributed by each shell of neighbours, is reported; from it, we see that the potential energy of the point ion at the origin is given by

$$V(0) = -\frac{e^2}{R}\left[\frac{6}{1} - \frac{12}{\sqrt{2}} + \frac{8}{\sqrt{3}} - \frac{3}{\sqrt{4}} + \frac{12}{\sqrt{5}} - \frac{12}{\sqrt{6}} - \frac{3}{\sqrt{8}} + \frac{6}{\sqrt{9}} - \frac{1}{\sqrt{12}}\right] = -1.751\frac{e^2}{R},$$

a result already very near to the exact one $-\alpha_M e^2/R$ with $\alpha_M = 1.7476$.

The Ewald method for the evaluation of the Madelung constant

The Ewald technique provides a very elegant procedure to transform one slow conditionally convergent lattice series into the sum of two fast absolutely convergent series in real and reciprocal space. The Ewald method essentially consists in using and balancing appropriately both real space and reciprocal space summations; this is achieved by splitting the function to be summed into a short-range part, to be computed as it stands in real space, and a long-range part to be represented and computed in reciprocal space. For the evaluation of the Madelung constant, the basic idea of Ewald is to replace point charges with Gaussian charge distributions, most conveniently handled in reciprocal space; the difference between the point charges and the Gaussian distributions is instead handled in real space. To implement the method, we need a few elementary facts concerning Gaussian functions, and the intimately related error function.

We remind the usual definition of the *error function*

$$\mathrm{erf}(x) = \frac{2}{\sqrt{\pi}} \int_0^x e^{-t^2}\, dt \ . \tag{13}$$

The error function increases monotonically from 0 to 1 as x varies from zero to infinity. We notice that

$$\mathrm{erf}(x) = \frac{2}{\sqrt{\pi}} \left(x - \frac{x^2}{3} + \dots \right) \quad \text{for} \quad x \ll 1 \ ;$$

the derivative of the error function is the Gaussian function

$$\frac{d}{dx}\mathrm{erf}(x) = \frac{2}{\sqrt{\pi}} e^{-x^2} \ .$$

It is also useful to consider the complementary error function $\mathrm{erfc}(x)$ defined from

$$\boxed{\mathrm{erf}(x) + \mathrm{erfc}(x) = 1} \ . \tag{14}$$

The reason for introducing the error function is the following. Let us consider a Gaussian charge distribution $\rho(r)$ (normalized to $+e$), centered at the origin in real space

$$\rho(r) \equiv (+e) \left(\frac{\eta}{\pi}\right)^{3/2} e^{-\eta r^2} \ . \tag{15a}$$

The electrostatic potential $\phi(\mathbf{r})$, corresponding to the Gaussian charge density of Eq. (15a) is simply

$$\phi(r) = +\frac{e}{r} \mathrm{erf}(\sqrt{\eta}\, r) \ . \tag{15b}$$

Let us in fact apply the operator $-\nabla^2$ to the electrostatic potential (15b); we obtain

$$-\nabla^2 \left[\frac{e}{r} \mathrm{erf}(\sqrt{\eta}\, r)\right] = -\frac{e}{r^2} \frac{\partial}{\partial r} \left[r^2 \frac{\partial}{\partial r} \frac{\mathrm{erf}(\sqrt{\eta}\, r)}{r}\right] = 4\pi e \left(\frac{\eta}{\pi}\right)^{3/2} e^{-\eta r^2} \ . \tag{15c}$$

The first passage in Eq. (15c) is just the action of the operator ∇^2 on a spherically symmetric function; the second passage is obtained performing explicitly the derivatives with respect to r. Eq. (15c) shows that the charge density (15a) and the electrostatic potential (15b) indeed satisfy the Poisson equation $-\nabla^2 \phi(\mathbf{r}) = 4\pi \rho(\mathbf{r})$.

Notice that in the particular case of $\eta \to \infty$, the Gaussian charge density (15a) represents the point charge $(+e)\, \delta(\mathbf{r})$, and the potential (15b) gives just the Coulomb potential e/r, which is singular at the origin. For a Gaussian charge density, the potential $\phi(r) = (e/r)\, \mathrm{erf}(\sqrt{\eta}\, r)$ is regular at the origin, where it takes the value

$$\phi(0) = 2e \left(\frac{\eta}{\pi}\right)^{1/2} \ . \tag{16}$$

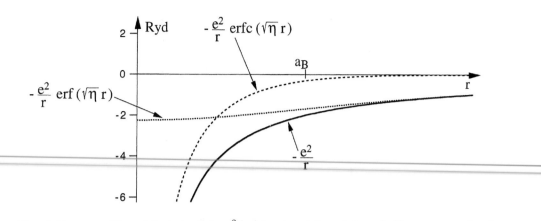

Fig. 5 Decomposition of the potential $-e^2/r$ (singular at the origin and of long-range nature) into a contribution $-(e^2/r)\,\mathrm{erf}(\sqrt{\eta}\,r)$ (regular at the origin and of long-range nature) and a contribution $-(e^2/r)\,\mathrm{erfc}(\sqrt{\eta}\,r)$ (singular at the origin and of short-range nature). Energies are expressed in Rydbergs, length in Bohr radius units; we have chosen $\sqrt{\eta} = 1\; a_B^{-1}$.

We now remember that the Fourier transform of a Gaussian function is still a Gaussian function. In particular we have the relation

$$\left(\frac{\eta}{\pi}\right)^{3/2} \int_{-\infty}^{+\infty} e^{-\eta r^2}\, e^{i\mathbf{k}\cdot\mathbf{r}}\, d\mathbf{r} = e^{-k^2/4\eta} \ . \tag{17a}$$

Another relation we need below is

$$\int_{-\infty}^{+\infty} \frac{e^2}{r}\,\mathrm{erf}(\sqrt{\eta}\,r)\, e^{i\mathbf{k}\cdot\mathbf{r}}\, d\mathbf{r} = \frac{4\pi e^2}{k^2}\, e^{-k^2/4\eta} \ . \tag{17b}$$

It is seen by inspection that the derivation with respect to η of both members of Eq. (17b) gives back Eq. (17a).

In the following we use the simple trick (suggested by Eq. 14) to write a point charge potential $-e^2/r$ (or similarly $+e^2/r$) in the form

$$-\frac{e^2}{r} \equiv -\frac{e^2}{r}\,\mathrm{erf}(\sqrt{\eta}\,r) - \frac{e^2}{r}\,\mathrm{erfc}(\sqrt{\eta}\,r) \tag{18}$$

(see Fig. 5). The first term on the right hand side of Eq. (18) is the potential generated by a Gaussian charge density; it is of *long-range nature and regular at the origin* (in lattice sums, it is conveniently handled in reciprocal space). The second term on the right hand side of Eq. (18) is the potential generated by a Gaussian charge density and a neutralizing point charge located at the origin; this contribution is of *short-range nature and singular at the origin* (in lattice sums with origin excluded, it is conveniently handled in real space).

Madelung constant with the Ewald method for a two-sublattice structure

We consider now a crystal constituted by two sublattices, occupied by anions and cations of point charges $(-e)$ and $(+e)$ respectively; the two sublattices are described

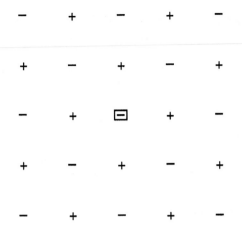

Fig. 6 Schematic representation of a two-sublattice ionic crystal in the point ion approximation, for the calculation of the Madelung constant. The ion at the origin has been surrounded
by a box.

by the positions $d_1 + t_n$ and $d_2 + t_n$ (with $d_1 = 0$, $d_2 \neq 0$ and t_n translational
vectors). We take, for instance, a negative ion position as a reference site, as in Fig. 6.
The potential energy $V(\mathbf{r})$ felt by a particle of charge $(-e)$ because of the field of all
the ions of the crystal (except the reference one) is

$$V(\mathbf{r}) = \sum_{t_n (\neq 0)} \frac{e^2}{|\mathbf{r} - \mathbf{t}_n|} - \sum_{t_n} \frac{e^2}{|\mathbf{r} - \mathbf{t}_n - \mathbf{d}_2|} \; . \tag{19}$$

The first term in the right-hand side of Eq. (19) is the repulsive contribution due to
the sublattice of negative ions, while the second term is the attractive contribution
due to the sublattice of positive ions.

Taking into account Eq. (18), we can write Eq. (19) in the form

$$V(\mathbf{r}) = V_1(\mathbf{r}) + V_2(\mathbf{r}) \; ,$$

where

$$V_1(\mathbf{r}) = \sum_{t_n (\neq 0)} \frac{e^2}{|\mathbf{r} - \mathbf{t}_n|} \, \mathrm{erfc}(\sqrt{\eta}|\mathbf{r} - \mathbf{t}_n|) - \sum_{t_n} \frac{e^2}{|\mathbf{r} - \mathbf{t}_n - \mathbf{d}_2|} \, \mathrm{erfc}(\sqrt{\eta}|\mathbf{r} - \mathbf{t}_n - \mathbf{d}_2|) \tag{20a}$$

and

$$V_2(\mathbf{r}) = \sum_{t_n (\neq 0)} \frac{e^2}{|\mathbf{r} - \mathbf{t}_n|} \, \mathrm{erf}(\sqrt{\eta}|\mathbf{r} - \mathbf{t}_n|) - \sum_{t_n} \frac{e^2}{|\mathbf{r} - \mathbf{t}_n - \mathbf{d}_2|} \, \mathrm{erf}(\sqrt{\eta}|\mathbf{r} - \mathbf{t}_n - \mathbf{d}_2|) \; . \tag{20b}$$

In particular, the term $V_1(0)$ is given by

$$V_1(0) = \sum_{t_n (\neq 0)} \frac{e^2}{t_n} \, \mathrm{erfc}(\sqrt{\eta} \, t_n) - \sum_{t_n} \frac{e^2}{|\mathbf{t}_n + \mathbf{d}_2|} \, \mathrm{erfc}(\sqrt{\eta} \, |\mathbf{t}_n + \mathbf{d}_2|) \; . \tag{21}$$

Because of the short-range nature of the complementary error function, the term $V_1(0)$ is conveniently calculated as it stands, i.e. by direct summation in real space.

We now elaborate the term $V_2(\mathbf{r})$ of Eq. (20b); this term is more conveniently transformed into a reciprocal space sum, with the following procedure. We write $V_2(\mathbf{r})$ in the form

$$V_2(\mathbf{r}) = U(\mathbf{r}) - \frac{e^2}{r}\,\mathrm{erf}(\sqrt{\eta}r)\ , \tag{22a}$$

where $U(\mathbf{r})$ is given by

$$U(\mathbf{r}) = \sum_{\mathbf{t}_n} \frac{e^2}{|\mathbf{r}-\mathbf{t}_n|}\,\mathrm{erf}(\sqrt{\eta}|\mathbf{r}-\mathbf{t}_n|) - \sum_{\mathbf{t}_n} \frac{e^2}{|\mathbf{r}-\mathbf{t}_n-\mathbf{d}_2|}\,\mathrm{erf}(\sqrt{\eta}|\mathbf{r}-\mathbf{t}_n-\mathbf{d}_2|)\ ; \tag{22b}$$

notice that $U(\mathbf{r})$ is a periodic function.

From the set of periodic functions $\left\{ (1/\sqrt{N\Omega})\,\exp(i\mathbf{h}_m\cdot\mathbf{r}) \right\}$, normalized to one in the volume $V = N\Omega$ of the crystal, we form the unit operator for projection of *periodic* functions

$$\frac{1}{N\Omega} \sum_{\mathbf{h}_m} |e^{i\mathbf{h}_m\cdot\mathbf{r}}\rangle\langle e^{i\mathbf{h}_m\cdot\mathbf{r}}| \equiv 1\ ,$$

where the sum extends to all the vectors $\mathbf{h}_m$ of the reciprocal lattice. By applying the above identity to the periodic function $U(\mathbf{r})$, we have

$$U(\mathbf{r}) = \frac{1}{N\Omega} \sum_{\mathbf{h}_m} e^{i\mathbf{h}_m\cdot\mathbf{r}} \langle e^{i\mathbf{h}_m\cdot\mathbf{r}}|U(\mathbf{r})\rangle\ . \tag{23a}$$

The scalar product appearing in Eq. (23a) can be evaluated explicitly, and one obtains

$$\langle e^{i\mathbf{h}_m\cdot\mathbf{r}}|U(\mathbf{r})\rangle = \int_V e^{-i\mathbf{h}_m\cdot\mathbf{r}}\,U(\mathbf{r})\,d\mathbf{r}$$

$$= N\left(1 - e^{-i\mathbf{h}_m\cdot\mathbf{d}_2}\right)\int_V e^{-i\mathbf{h}_m\cdot\mathbf{r}}\,\frac{e^2}{r}\,\mathrm{erf}(\sqrt{\eta}r)\,d\mathbf{r}$$

$$= N\left(1 - e^{-i\mathbf{h}_m\cdot\mathbf{d}_2}\right)\frac{4\pi e^2}{h_m^2}\,e^{-h_m^2/4\eta} \quad (\mathbf{h}_m \neq 0)\ , \tag{23b}$$

where use has been made of Eq. (22b) and Eq. (17b). When $\mathbf{h}_m \equiv 0$, the scalar product (23b) vanishes; in fact the average of $U(\mathbf{r})$ on the crystal volume is evidently equal to zero, because of the overall neutrality of the system of positive and negative (identical) charge distributions.

From Eqs. (23) we have

$$U(0) = \frac{4\pi e^2}{\Omega} \sum_{\mathbf{h}_m(\neq 0)} \left(1 - e^{-i\mathbf{h}_m\cdot\mathbf{d}_2}\right)\frac{1}{h_m^2}\,e^{-h_m^2/4\eta}\ .$$

From the above expression and from Eq. (22a) we obtain

$$V_2(0) = \frac{4\pi e^2}{\Omega} \sum_{\mathbf{h}_m(\neq 0)} \left(1 - e^{-i\mathbf{h}_m\cdot\mathbf{d}_2}\right)\frac{1}{h_m^2}\,e^{-h_m^2/4\eta} - 2\,e^2\sqrt{\frac{\eta}{\pi}}\ . \tag{24}$$

Table 4 Madelung constant for some cubic structures.

Crystal structure	Madelung constant
Cesium chloride	$\alpha_M = 1.7627$
Sodium chloride	$\alpha_M = 1.7476$
Zincblende	$\alpha_M = 1.6381$

Collecting Eq. (21) and Eq. (24), we obtain for the Madelung energy the final expression

$$V(0) = \sum_{\mathbf{t}_n (\neq 0)} \frac{e^2}{t_n} \operatorname{erfc}(\sqrt{\eta}\, t_n) - \sum_{\mathbf{t}_n} \frac{e^2}{|\mathbf{t}_n + \mathbf{d}_2|} \operatorname{erfc}(\sqrt{\eta}|\mathbf{t}_n + \mathbf{d}_2|)$$

$$+ \frac{4\pi e^2}{\Omega} \sum_{\mathbf{h}_m (\neq 0)} \left(1 - e^{-i\mathbf{h}_m \cdot \mathbf{d}_2}\right) \frac{1}{h_m^2} e^{-h_m^2/4\eta} - 2\,e^2 \sqrt{\frac{\eta}{\pi}}. \tag{25}$$

The arbitrary parameter η controls the convergence of the two summations, and is chosen to optimize simultaneously the convergence of the series in direct and reciprocal spaces.

Many ionic crystals occur in cubic structures: the cesium chloride structure (coordination number 8), sodium chloride structure (coordination number 6), zincblende structure (coordination number 4). The Madelung constant is (slightly) higher for higher coordination number, as can be seen from Table 4.

2.2   Considerations on bands and bonds in ionic crystals

In the class of ionic crystals, alkali halides (formed with one element of column I and one element of column VII of the periodic table) are the most typical compounds. Also II-VI crystals can be often pictured as formed by (doubly ionized) ions, although the partial covalent character of the atomic interactions may become significant. In III-V compounds the binding is mostly covalent in nature, while IV-IV materials are the typical covalent compounds. We confine here our attention to some aspects of the band structure of alkali halides crystals.

Alkali halides are in general large gap insulators; the electrons completely fill the core and valence bands, while the conduction bands are empty and well separated in energy from the valence states. To understand in essential lines the origin of the gap and the role of the Madelung energy, we consider as an exemplification the case of NaCl. In the ionic picture, the crystal is described as a system of interacting Cl^- and Na^+ ions; the ions have the closed-shell electronic configuration: Cl^- ($1s^2$, $2s^2\, 2p^6$, $3s^2\, 3p^6$) and Na^+ ($1s^2$, $2s^2\, 2p^6$).

We can orientatively envisage some features of the band structure of NaCl with the following considerations. The orbital energy $E_{3p}(Cl^-)$ is given by the (negative) of the ionization energy of the Cl^- ion; this in turns equals the electron affinity

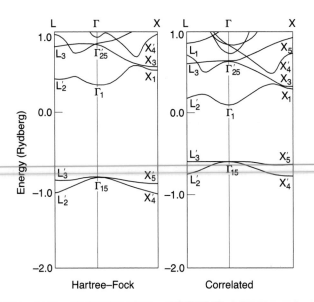

Fig. 7 Hartree–Fock and correlated energy bands of NaCl. The state Γ_{15} is the top of the valence bands and the state Γ_1 is the bottom of the conduction bands [from A. B. Kunz, Phys. Rev. B**26**, 2056 (1982); copyright 1982 by the American Physical Society].

of the Cl atom (≈ 3.7 eV); we thus estimate $E_{3p}(\mathrm{Cl}^-) = -3.7\,\mathrm{eV}$. In the crystal, an electron in the reasonably localized $3p$ orbital of Cl^- feels the Madelung energy $-\alpha_M e^2/R_0 \approx -9.0\,\mathrm{eV}$ ($R_0 = 5.30\,a_B$ in NaCl), due to all the surrounding ions in the point ion approximation. We thus estimate that the topmost valence bands (in the point ion approximation) are located around the energy $E_\mathrm{v} = E_{3p}(\mathrm{Cl}^-) - \alpha_M e^2/R_0 = -12.7\,\mathrm{eV}$. Rather in a similar way, we estimate the orbital energy $E_{3s}(\mathrm{Na}^+) = -5.1\,\mathrm{eV}$ as the (negative) of the ionization energy of the Na atom (≈ 5.1 eV). We can reasonably take the energy of the bottom of the conduction band as $E_\mathrm{c} = E_{3s}(\mathrm{Na}^+) + \alpha_M e^2/R_0 = +3.9\,\mathrm{eV}$. The estimated energy gap, in the point ion approximation, is $E_\mathrm{G} \approx E_{3s}(\mathrm{Na}^+) - E_{3p}(\mathrm{Cl}^-) + 2\alpha_M e^2/R_0 = 16.6\,\mathrm{eV}$, and is mostly determined by the Madelung energy.

From the above considerations, we expect that the topmost valence bands are p-like and originated from the $3p$ orbitals of the anion, while the bottom conduction band is s-like and originated from the $3s$ orbital of the cation. These speculations, as well as the role of the Madelung energy in making NaCl a large gap insulator, are corroborated by detailed band structure calculations (Fig. 7). The estimates done above within the point ion approximation for the band edges compare reasonably well with the Hartree–Fock calculations of NaCl. In Fig. 7, we also report the correlated energy bands, which properly take into account the polarizability of the Cl^- and Na^+ ions. With respect to the Hartree–Fock calculations, the correlation effects lower the conduction bands and increase the valence bands, similarly to the situation described in Fig. 1. [Incidentally, we notice that the correlation effects, of course, decrease the crystal total energy, which is *not* simply the sum of the orbital energies of the occupied bands, as discussed in Section IV-5.2]. In NaCl, the inclusion of correlation effects leads to a calculated band gap of ≈ 10 eV and a valence band width of ≈ 3 eV. Optical absorption measurements and photoemission measurements provide a fundamental band gap for NaCl of about 9 eV, and a valence band width of about 3 eV.

Similar considerations can be made for other ionic crystals; since the Madelung energy $\alpha_M e^2/R_0$ is ordinarily in the range 5–10 eV, this is also the expected order of magnitude of the energy gap. In particular, because of its small value of R_0, LiF is expected to have the largest gap among alkali halides; this is the origin of the utilization of LiF as a transparent window for near ultraviolet spectroscopy (up to the cut-off energy of ≈ 11.8 eV).

Considerations on cohesive energy in ionic crystals

We wish to mention here the semi-empirical Born model, to supply evidence for some features of bonds and cohesive energy in ionic crystals. Consider an ionic crystal, composed by N positive and N negative ions; we can write the total ground-state energy of the solid in the semi-empirical form

$$U_s(R) = N \left(\frac{\lambda}{R^n} - \alpha_M \frac{e^2}{R} \right) , \tag{26}$$

where R is the nearest neighbour distance, α_M is the Madelung constant, λ and n are semi-empirical parameters. The term $-\alpha_M e^2/R$ represents the attractive energy of an ion due to the presence of all the other ions, in the strictly point-ion approximation. The term λ/R^n represents the repulsive energy of an ion (of finite size) due to the presence of all the other ions (of finite size) of the lattice; it has been expressed by a power law, following an assumption by Born. Using Eq. (26), we can now proceed to the evaluation of the equilibrium lattice constant, the cohesive energy and the bulk modulus (similarly to what previously done in rare-gas solids in Section 1.2).

At the equilibrium distance, we have

$$\left(\frac{dU_s}{dR} \right)_{R=R_0} = 0 \quad \text{hence} \quad \lambda = \frac{\alpha_M e^2}{n} R_0^{n-1} .$$

The cohesive energy $U_s(R_0)$ becomes

$$U_s(R_0) = -N \alpha_M \frac{e^2}{R_0} \left(1 - \frac{1}{n} \right) . \tag{27}$$

Since n is of the order of ≈ 10, *we see that the cohesive energy of an ionic crystal is basically determined by the Madelung energy of the crystal.*

We can now evaluate the bulk modulus

$$B_0 = V_0 \left(\frac{d^2 U_s}{dV^2} \right)_{V=V_0} .$$

For the NaCl structure, the relation between volume and nearest-neighbour distance

R is $V = 2 N R^3$. With straightforward calculations (similar "mutatis mutandis" to those performed in Section 1.2) we have

$$B_0 = \frac{1}{18 N R_0} \left(\frac{d^2 U_s}{dR^2} \right)_{R=R_0} = (n-1) \frac{e^2 \alpha_M}{18 R_0^4} . \tag{28}$$

Equation (28) is often used to obtain n from the knowledge of B_0 and R_0; the values of n determined semi-empirically in this way are of the order of 10 and thus the relevance of the electrostatic contribution in the binding energy (27) is further confirmed.

Because of the relevance of the electrostatic interaction in the cohesive energy of ionic crystals, we can also obtain an orientative criterion to estimate the relative stability of the cesium chloride structure, compared with sodium chloride structure and zincblende stucture. Because of the higher Madelung constant (see Table 4), the CsCl structure should be more stable than the other two, at parity of nearest neighbour distance.

In ionic crystals, the ions have a closed-shell configuration, and this makes it possible to define (theoretically or experimentally) ionic radii, which are (approximately) transferable quantities. Thus we can analyse (at least at a preliminary level) the stability of different structures, requiring that positive and negative ions touch in the solid.

In the following reasoning, we assume ions as hard spheres of radii R_A and R_C (we suppose that R_A is the larger of the two radii). Let us consider the cesium chloride structure. In this structure, indicating by a the edge of the conventional cube, we have that the nearest neighbour distance and the second nearest neighbour distance are given respectively by

$$d_I = \frac{a\sqrt{3}}{2} \quad \text{and} \quad d_{II} = a \quad \text{(CsCl structure)} . \tag{29a}$$

The condition that positive and negative ions touch each other gives one equality and one inequality

$$\begin{cases} R_A + R_C = a\sqrt{3}/2 \\ 2 R_A \leq a \end{cases} .$$

Solving for a, we have

$$2 R_A \leq \frac{2}{\sqrt{3}} (R_A + R_C)$$

and

$$\frac{R_A}{R_C} \leq \frac{1}{\sqrt{3} - 1} = 1.37 \quad \text{(CsCl structure)} . \tag{29b}$$

Thus the cesium chloride structure allows touching of ions if the ratio of the ionic radii satisfies the condition (29b).

A similar reasoning can be done for the NaCl structure. In this structure, indicating by a the edge of the conventional cube, we have that the nearest neighbour distance and the second nearest neighbour distance are given by

$$d_I = \frac{a}{2} \quad \text{and} \quad d_{II} = a\frac{\sqrt{2}}{2} \quad \text{(NaCl structure)} . \tag{30a}$$

The condition that positive and negative ions can touch each other gives one equality and one inequality

$$\begin{cases} R_A + R_C = a/2 \\ 2\,R_A \leq a\sqrt{2}/2 \end{cases} .$$

Solving for a, we have

$$\frac{R_A}{R_C} \leq \frac{1}{\sqrt{2}-1} = 2.44 \quad \text{(NaCl structure)} . \tag{30b}$$

Thus the sodium chloride structure allows touching of ions if the ratio of the ionic radii satisfies the condition (30b).

Finally we can consider the zincblende structure. In this structure, indicating as usual by a the edge of the conventional cube, we have that nearest and second nearest neighbour distances are

$$d_I = \frac{a\sqrt{3}}{4} \quad \text{and} \quad d_{II} = a\frac{\sqrt{2}}{2} . \tag{31a}$$

The condition that positive and negative ions can touch each other leads to

$$R_A/R_C \leq 2 + \sqrt{6} = 4.45 \quad \text{(zincblende structure)} . \tag{31b}$$

The above argument leads to the orientative expectation that ions with almost equal radii tend to crystallize in CsCl structure, while ions with very different radii prefer sodium chloride or zincblende structure, a trend that is actually observed in ionic crystals.

3   Covalent crystals with diamond structure

Rare-gas crystals and ionic crystals are examples of crystals composed by closed-shell units; closed-shell units have no (or little) tendency to produce overlap of electron clouds. In crystals composed by open-shell units, there may be a tendency to a re-distribution of the electronic charge; the electronic charge, that accumulates midway two nearest neighbour nuclei, enjoys attractive interaction from both nuclei and is responsible of the covalent bond.

The group IV elements, diamond, silicon, germanium, and grey-tin, crystallize in the diamond structure, which is fcc with two atoms per unit cell. In the free atoms, there are four electrons in the most external shell: C $(2s^2, 2p^2)$, Si $(3s^2, 3p^2)$, Ge $(4s^2, 4p^2)$, Sn $(5s^2, 5p^2)$; inner shells are completely filled. The most external s and p_x, p_y, p_z atomic orbitals of the two atoms in the unit cell form bonding (and antibonding) combinations, which generate the highest valence bands (and the bottom conduction bands) of the crystal. For what concerns the energy gap, diamond is a strong insulator

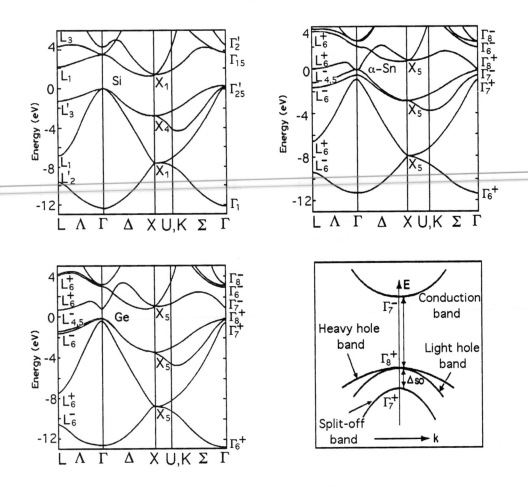

Fig. 8 Band structure for silicon, germanium and grey-tin [from J. R. Chelikowsky and M. L. Cohen, Phys. Rev. B**14**, 556 (1976); copyright 1976 by the American Physical Society]. The typical band structure of cubic semiconductors near the gap at **k** = 0 is also schematically reported; Δ_{so} denotes the spin-orbit splitting.

with energy gap 5.4 eV, silicon and germanium are well known semiconductors with energy gap 1.17 eV and 0.74 eV respectively (at $T = 0$), grey-tin is a semimetal and lead is a metal. In spite of this quite different behaviour, from band theory and symmetry considerations it is possible to understand the differences (and the similarities) in the electronic structure of these materials.

Among the numerous investigations in the literature on the group IV covalent solids, we briefly mention here some results obtained with the pseudopotential method; in this approach the effective Hamiltonian (sum of kinetic energy and a weak pseudopotential) is conveniently expressed and diagonalized using the basis set of plane waves. In Fig. 8 we give in sequence the band structure of silicon, germanium and grey-tin. For the heavier elements, the spin-orbit effects are included.

Of special interest are the bands of silicon and germanium around the fundamental

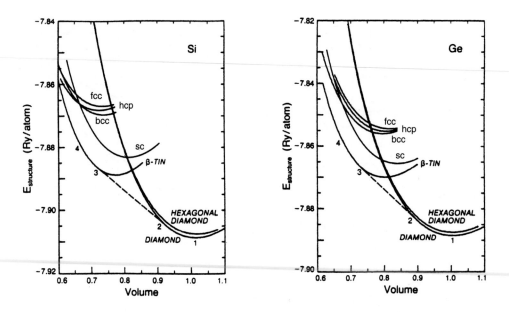

Fig. 9 Total energy curves for seven structures of Si and Ge, as a function of the atomic volume (normalized to the experimental volume). Dashed line is the common tangent of the energy curves for the diamond phase and the β-tin phase [from M. T. Yin and M. L. Cohen, Phys. Rev. B**26**, 5668 (1982); copyright 1982 by the American Physical Society].

gap. In silicon, in the ΓX direction, the minimum of the conduction band occurs at the point $\mathbf{k} \approx 0.85\,(2\pi/a)(1,0,0)$; there are thus six equivalent minima in the Brillouin zone. In germanium, the minimum of the conduction band occurs at the point $\mathbf{k} = (2\pi/a)(1/2,1/2,1/2)$ and there are thus four equivalent minima in the Brillouin zone (since $\mathbf{k}$ and $-\mathbf{k}$ are related by a reciprocal lattice vector). The top of the valence bands is at Γ'_{25}, both in germanium and in silicon. The state Γ'_{25} arises from the three p_x, p_y, p_z bonding orbitals, which are degenerate at the center of the Brillouin zone; considering the spin of the electrons, Γ'_{25} is sixfold degenerate. Spin-orbit interaction splits the topmost valence bands into two groups: $p_{3/2}$ bands (heavy-hole and light-hole bands) with degeneracy 4 at Γ point (Γ_8^+ state), and $p_{1/2}$ bands (splitt-off band) with degeneracy 2 (Γ_7^+ state). The spin-orbit separation of the valence bands at Γ is 0.044 eV for Si, 0.29 eV for Ge and 0.80 eV for grey-tin. A special feature occurs for grey-tin, where the energy of the s-like antibonding state decreases and becomes intermediate between the spin-orbit split Γ_8^+ and Γ_7^+ states; grey-tin becomes thus a semimetal with energy gap zero at the point Γ of the Brillouin zone.

We wish now to mention a classic investigation on the static structural properties, crystal stability and phase transformation of silicon and germanium, based on norm-conserving pseudopotential approach within the local-density-functional approximation. The total energy curves are reported in Fig. 9 for seven crystal structures: face-centered cubic lattice, body-centered cubic lattice, hexagonal close-packed, simple

Table 5 Calculated and measured equilibrium properties of silicon and germanium [from M. T. Yin and M. L. Cohen, Phys. Rev. B**26**, 5668 (1982); copyright 1982 by the American Physical Society].

	Lattice constant (Å)	Nearest neighbour distance (Å)	Binding energy (eV/atom)	Bulk modulus (Mbar)
Si				
Calculation	5.451	2.360	4.84	0.98
Experiment	5.429	2.351	4.63	0.99
Ge				
Calculation	5.655	2.449	4.26	0.73
Experiment	5.652	2.447	3.85	0.77

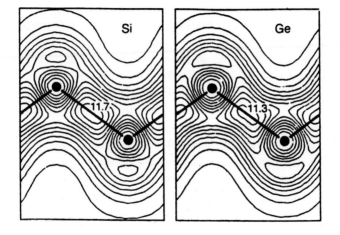

Fig. 10 Calculated pseudovalence charge density in the (110) plane for Si and Ge. The contours are in units of electrons per atomic volume, with a contour step of 1. Atomic positions are denoted by black dots; atomic chains are denoted by straight lines [from M. T. Yin and M. L. Cohen, Phys. Rev. B**26**, 5668 (1982); copyright 1982 by the American Physical Society].

cubic, cubic diamond, hexagonal diamond (wurtzite), and β-tin. From Fig. 9 it is seen that the diamond structure lies lowest in energy. The minimum of the total-energy curve, and its curvature, determine the lattice constant, the cohesive energy and the bulk modulus (see Table 5). It can also be noticed that the β-tin structure is lower in energy than the diamond structure at small volumes; hence a solid–solid phase transition should occur at the estimated transition pressure of ≈ 100 kbar, given by the slope of the dashed line in Fig. 9.

In Table 5 we report calculated and measured static properties of silicon and germanium, and notice the satisfactory agreement with experimental data. It is instructive to make a comparison of the order of magnitude of the static properties of rare-gas solids (Table 2) with those of the typical covalent crystals (Table 5); it can be seen that covalent crystals have smaller nearest neighbour distance, and stronger cohesive energy and bulk modulus.

Before concluding this section, it is of interest to show the pseudovalence charge densities for Si and Ge (see Fig. 10) [outside the core region pseudovalence charge distribution and true valence charge distribution coincide, because of the norm-conserving property of the pseudopotential used in the calculations]. The maximum value of the valence charge density occurs midway nearest neighbour atomic positions; the pile-up of the covalent bonding charge, elongated along the atomic chains, is clearly visualized in Fig. 10.

4   Band structures and Fermi surfaces of some metals

The metallic state is the most common situation for crystals of the elements of the periodic table: out of the more than hundred elements, about seventy are metals. A most important quantity of metals is the *Fermi energy* E_F, defined (at $T = 0$) by the requirement that the total number of band states with energy smaller than E_F exactly equals the total number of electrons available in the crystal. In metals, the Fermi level occurs within a band or intersects a number of bands; the surface in **k**-space that connects all the **k**-points, corresponding to crystal states of energy E_F, is called *Fermi surface*. The electron transport properties of metals are determined by the properties of the Fermi surface and of the density-of-states in the small thermal shell of energy $k_B T$ around it. The very rich phenomenology of metals is related to the variety of peculiar shapes of the Fermi surface, determined by the actual crystal potential.

In general, the study of metals requires the evaluation of the energy band structure $E_n(\mathbf{k})$ at a rather high number of **k**-points in the first Brillouin zone (say several thousand or so). This allows a reliable determination of the Fermi energy, of the topology of the Fermi surface, and the density-of-states curve. The points **k** belonging to the Fermi surface are given by the implicit equation

$$\boxed{E_n(\mathbf{k}) \equiv E_F} \,. \tag{32}$$

The equation can be solved graphically plotting any given branch $E = E_n(\mathbf{k})$ versus **k** in various directions in the first Brillouin zone; from the intersection of the curve with the line $E = E_F$ the points $\mathbf{k}_F$ on the Fermi surface can be obtained, and the Fermi surface drawn. We illustrate the above considerations with a few simple examples of band structures and Fermi surfaces.

The *alkali metals* (Li, Na, K, Rb, Cs) are the typical *simple metals*; so are addressed those metals with an almost free-electron picture for the conduction band. An alkali atom has one "optical electron" in the state ns^1, with $n = 2, 3, 4, 5, 6$ for lithium, sodium, potassium, rubidium and cesium, respectively, and fully occupied inner shells. The alkali metals crystallize in the bcc structure, with one atom per unit cell. The lowest conduction band is s-like in nature, corresponding to the partially occupied ns^1

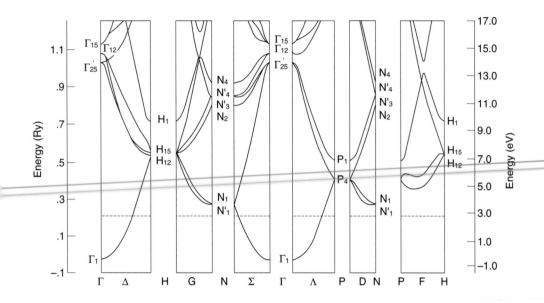

Fig. 11 Band structure of sodium. The Fermi energy is indicated by a dotted line [from D. A. Papaconstantopoulos "Handbook of the Band Structure of Elemental Solids" (Plenum, New York 1986)].

external atomic orbital; the Fermi surface lies within the Brillouin zone quite far from zone boundaries. As an example, the band structure and the Fermi level of sodium is reported in Fig. 11.

Although the Fermi surface in alkali metals is almost spherical and well inside the first Brillouin zone, there are nevertheless significant deviations, with influence on the conductivity, magnetoresistance, thermoelectric power, etc. These deviations are relatively high in lithium and almost absent in sodium, that can be considered the simplest of the simple metals. For instance, at the bottom of the conduction band, the effective mass m_c^* (in units of the free electron mass m) is $m_c^* = 1.33$, 0.97, 0.86, 0.78 and 0.73 for Li, Na, K, Rb and Cs, respectively. This trend of the effective mass in the sequence of alkali metals can be interpreted with the help of the $\mathbf{k} \cdot \mathbf{p}$ method (described in Section II-6.2). At the bottom of the conduction band, taking into account the cubic symmetry of the materials, we can rewrite Eq. (II-40) in the form

$$\frac{m}{m_c^*} = 1 + \frac{2}{m} \sum_{n(\neq c)} \frac{|\langle \psi_n(\mathbf{k}_0, \mathbf{r}) | p_x | \psi_c(\mathbf{k}_0, \mathbf{r}) \rangle|^2}{E_c(\mathbf{k}_0) - E_n(\mathbf{k}_0)} \,, \tag{33}$$

where $E_n(\mathbf{k}_0)$ and $\psi_n(\mathbf{k}_0, \mathbf{r})$ are eigenvalues and eigenfunctions at the center of the Brillouin zone, and the subscript c refers to the partially occupied conduction band. Eq. (33) easily accounts why Li has an effective mass larger than the free electron mass m (the only possible interactions are with higher energy bands); sodium has effective mass almost equal to m (there is an approximate balance in the interactions

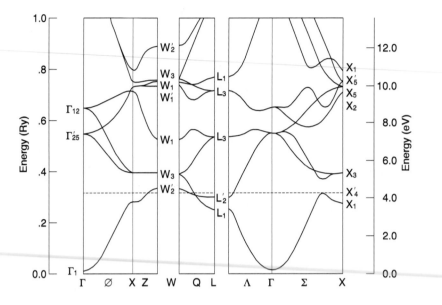

Fig. 12 Band structure of calcium. The Fermi energy is indicated by a dotted line [from D. A. Papaconstantopoulos "Handbook of the Band Structure of Elemental Solids" (Plenum, New York 1986)].

with lower and higher energy bands); finally the other alkali metals K, Rb and Cs have a conduction effective mass smaller than the free electron mass.

The *divalent metals* (Be, Mg , Ca, Sr, Ba and Zn, Cd, Hg) are formed by atoms with two electrons in the most external ns^2 orbital, and fully occupied inner shells. The lowest conduction bands are mostly s-like in nature, and closely related to the occupied ns^2 external atomic orbital. These crystals, having two optical electrons per atom, would give completely filled and completely empty bands, in the absence of energy overlap of different bands. This is not the case, as shown for instance in Fig. 12, where we report the conduction band structure of calcium (fcc Bravais lattice). The overlap of bands makes calcium a typical example of metal, with an equal number of electrons and holes: the carriers are constituted by the electrons occupying pockets in the second conduction band, and an equal number of empty levels (holes) to the full occupancy of the first conduction band.

We have seen that the conduction bands in alkali metals are essentially free-electron like (also referred as "itinerant"), with moderate modifications brought about by the crystalline potential. The situation can change dramatically in the presence of hybridization effects between tendentially itinerant bands (originating from orbitals of type s or p) and tendentially localized bands (originating from orbitals of type d or f). A very interesting case is constituted by noble metals. The *noble metals* (Cu, Ag, Au) are monovalent and crystallize in fcc Bravais lattice. The s and d mixing has a profound effect on the band structure and Fermi surface topology of these materials

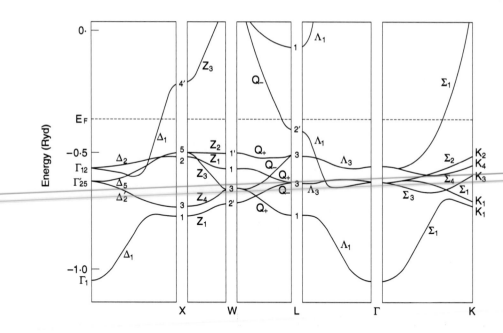

Fig. 13 Band structure of copper [from G. A. Burdick, Phys. Rev. **129**, 138 (1963); copyright 1963 by the American Physical Society].

(in the absence of s–d hybridization, these materials would be very much like the alkali metals).

The electronic configuration of Cu is [Ar] $3d^{10}\,4s^1$. Similarly for Ag we have [Kr] $4d^{10}\,5s^1$ and for Au the electronic configuration is [Xe] $5d^{10}\,6s^1$. We consider for instance the band structure of copper. The bands arising from $1s, 2s, 2p, 3s$ and $3p$ orbitals of Cu have very low energy and can be considered as flat core states, fully occupied. Thus we have to consider only the states arising from $3d$ and $4s$ atomic orbitals. In Fig. 13 we report the band structure of Cu; its overall aspect is just that of a rather wide s-like band, that hybridizes with rather narrow d-like bands.

From Fig. 13, we can now reconstruct the Fermi surface of copper, which is given schematically in Fig. 14. For this aim, we apply the procedure summarized in Eq. (32), along various directions in the Brillouin zone. Let us consider the direction ΓX, for instance; we see from Fig. 13 that there is one intersection of the band structure with the curve $E = E_F$; a similar situation occurs for instance along the ΓK direction. If we consider the ΓL direction, no intersection is found. If we consider the LW direction on the hexagonal face of the Brillouin zone, we find an intersection point. Putting these (and other similar) facts together, we can easily arrive at the Fermi surface schematically indicated in Fig. 14 and originally proposed by Pippard. The Fermi surface bulges out in $\langle 111 \rangle$ direction, makes contact with the hexagonal faces of the Brillouin zone, and is thus an *open Fermi surface*. As a typical consequence of this topology, profound effects on the transport properties (in magnetoresistance, for instance) may occur.

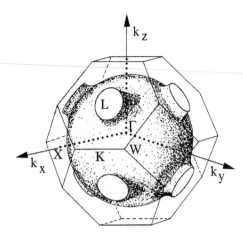

Fig. 14 Brillouin zone for the face-centered cubic lattice and Fermi surface of copper.

In noble metals, the mostly s-like band at the Fermi surface is separated in energy from the fully occupied d states, although their presence influences the energy dispersion of the s-like electrons, with the peculiar effects in the Fermi surface shape described above. In transition and rare-earth metals, the electronic states at and near the Fermi surface are determined by the mixing of localized inner d or f electrons with external s or p electrons; a variety of situations can occur in transition metals (partially filled $3d$, $4d$ or $5d$ orbitals), and rare-earth metals (partially filled $4f$ and $5f$ orbitals). An additional complication is that the ground state of several metals is magnetic, and that in general no simple model can include easily magnetic effects; we thus have to refer to the literature for investigations and details.

We wish to conclude these simple remarks on metals with a few considerations on the nature of the metallic bond. A preliminary, although *crude* estimation, can be inferred from the "idealized" jellium model of a metal, which assumes that the electrons of the Fermi sea are embedded in a uniform background of neutralizing positive charges. In the jellium model (see Section IV-7), the ground-state energy per electron in the Hartree–Fock approximation is given by

$$E_0 = \left[\frac{2.21}{r_s^2} - \frac{0.916}{r_s} \right] \quad \text{(in Rydberg)} . \tag{34}$$

The minimum of $E_0(r_s)$ occurs when the dimensionless parameter r_s equals $r_{s_0} = 4.825$; the corresponding cohesive energy is $E(r_{s_0}) = -1.29\,\text{eV}$.

In actual metals the correlation energy and the internal structure of ions are of major importance in determining the metallic bond. With the developments of self-consistent band structure calculations and implementation of the density functional formalism, much progress has been done for an accurate account for the total energy of metals [see for instance Y. Mishin, D. Farkas, M. J. Mehl and D. A. Papaconstantopoulos, Phys. Rev. B**59**, 3393 (1999) and references quoted therein].

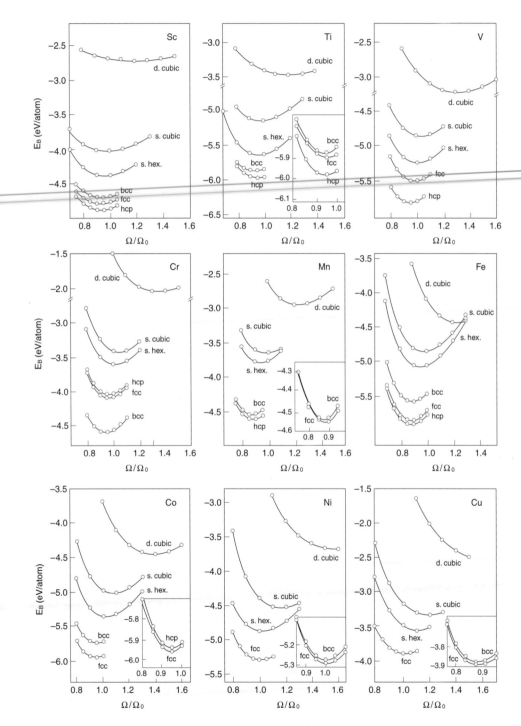

Fig. 15 Structural energy–volume curves in the 3d transition metals: cohesive energy (eV) versus atomic volume normalized to the experimental atomic volume Ω_0. Structures are fcc, bcc, hcp, simple hexagonal, simple cubic and diamond cubic; curves in insets are shown with expanded scales [from A. T. Paxton, M. Methfessel and H. M. Polatoglou, Phys. Rev. B**41**, 8127 (1990); copyright 1990 by the American Physical Society].

As an exemplification we report in Fig. 15 a study on the total energy of the first-row transition metals, for various structures; the study is based on the linearized muffin-tin orbital method and the local density functional approximation. The first-row transition elements from Sc to Cu have the core states $1s^2$, $2s^2\,2p^6$, $3s^2\,3p^6$ fully occupied; the electronic configuration of the external $3d$ and $4s$ shells is listed for convenience, together with the crystal structure: Sc $(3d^1\,4s^2$; hcp); Ti $(3d^2\,4s^2$; hcp); V $(3d^3\,4s^2$; bcc); Cr $(3d^5\,4s^1$; bcc); Mn $(3d^5\,4s^2$; nearly hcp); Fe $(3d^6\,4s^2$; bcc); Co $(3d^7\,4s^2$; hcp); Ni $(3d^8\,4s^2$; fcc); Cu $(3d^{10}\,4s^1$; fcc). In Fig. 15 we report for several structures the energy–volume curves. It must be noticed that the calculations do not include magnetic effects; thus the magnetic ground-state of transition metals near the end of the row are not expected to be determined very accurately; in principle, magnetic effects could be introduced by spin-polarized calculations. From Fig. 15, we see that the crystal structure of all non-magnetic crystals are correctly reproduced. We also see that the typical binding energy is of the order of 4–5 eV/atom. Also the value of the equilibrium lattice constant is predicted with good accuracy. The bulk modulus elaborated from the structural energy–volume curves is of the order of the Mbar for the various metals.

Further reading

C. A. Coulson "Valence" (Oxford University Press 1961)

B. H. Flowers and E. Mendoza "Properties of Matter" (Wiley, London 1970)

W. Harrison "Electronic Structure and Properties of Solids: the Physics of the Chemical Bond" (Freeman, San Francisco 1980)

K. H. Hellwege and O. Madelung (editors) "Semiconductors. Intrinsic Properties of Group IV Elements and III-V, II-VI, and I-VII Compounds" Landolt-Börnstein, New Series, Group III, vol. 22 (Springer Verlag, Berlin 1982)

M. L. Klein and J. A. Venables (editors) "Rare Gas Solids" vols. I and II (Academic Press, London 1976, 1977)

E. Mooser "Bonds and Bands in Semiconductors" in "Crystalline Semiconducting Materials and Devices" edited by P. N. Butcher, N. March and M. P. Tosi (Plenum Press, New York 1986)

V. V. Nemoshkalenko and V. N. Antonov "Computational Methods of Band Theory" (Gordon and Breach, Amsterdam 1998)

D. A. Papaconstantopoulos "Handbook of the Band Structure of Elemental Solids" (Plenum, New York 1986). This book contains useful discussions of trends of band structures in elemental crystals and tight-binding parametrization.

L. Pauling "The Nature of the Chemical Bond" (Cornell University Press, Ithaca, New York 1960)

J. C. Phillips "Bonds and Bands in Semiconductors" (Academic Press, New York 1973)

A. Sutton "Electronic Structure of Materials" (Clarendon Press, Oxford 1993)

M. P. Tosi "Cohesion of Ionic Solids in the Born Model" Solid State Physics **16**, 1 (1964) (edited by F. Seitz and D. Turnbull, Academic Press, New York)

VII

Excitons, plasmons and dielectric screening in crystals

In the previous two chapters we have considered the band theory of crystals and provided some specific applications. The whole treatment was based on the one-electron approximation; in spite of its great merits, it is often necessary to proceed beyond the independent particle approximation and investigate the many-body effects on the band structure of crystals.

Whithin the one-electron formalism, the elementary electronic excitations of a crystal are constituted by *individual electron–hole pairs*. In general, many-body effects not only influence the continuum of single-particle excitations (where resonances may be introduced), but more importantly may lead to new excitations in previously forbidden energy regions; the new features, referred to as *collective excitations*, are regarded as the fingerprint of many-body effects. In this chapter, we focus on *excitons* and *plasmons*, which are typical collective effects of a system of interacting electrons; they are present in any type of crystal (metals, semiconductors and insulators) and are commonly detected by optical measurements and electron energy-loss measurements.

In this chapter we also discuss the response of media to applied longitudinal perturbations; this is a quite complicated many-body problem, from which collective plasmon excitations can be inferred. We describe the dielectric screening in a number of relatively simple though significant models, that can be used as guidelines for more complicated situations.

1   Exciton states in crystals

Introductory remarks

Excitons are excited states of the crystals, whose description lies beyond the one-electron approximation and the band theory approach. We begin the discussion of many-body effects in crystals with the study of exciton states in semiconductors or insulators (excitonic effects on the optical properties are described in Chapter XII).

In the one-electron approximation, the electrons are considered to move independently in an appropriate self-consistent periodic potential. In semiconductors and insulators, the fully occupied valence bands are separated by the fully empty conduction bands by an energy gap E_G, which represents the threshold energy for band-to-band electronic transitions (in the independent particle approximation). Many-body effects modify the physical picture: the extra electron in the conduction band and the hole left behind in the valence band can be visualized as interacting via a Coulomb-like field. Thus, the possibility opens of bound "electron–hole" states (*excitons*) with excitation energies lower than the energy gap, as suggested in the original works of Frenkel and Peierls.

Exciton effects are particularly spectacular near the fundamental energy gap of insulators and semiconductor (*valence excitons*), and at the onset of transitions from deep energy bands (*core excitons*); they influence also the continuum of band-to-band transitions. Exciton binding energies vary drastically depending on the kind of material. For instance in neon, a large gap insulator, the exciton binding energy is about 4 eV (see Fig. 1). In the small gap semiconductor GaAs, the binding energy of valence excitons is a few meV (see Fig. 2). In metals, excitons from core states are believed to have vanishingly small binding energies, due to metallic screening of the electron–hole interaction; however, exciton effects may strongly enhance the optical absorption from core states to the continuum above the Fermi level, and actually produce the *exciton edge singularities* experimentally observed in many metals in the x-ray region.

Exciton states in the two-band model

We consider now the principles for the quantum description of the exciton states in solids. The basic Hamiltonian of the many-electron system is given by

$$H_e = \sum_i \frac{\mathbf{p}_i^2}{2m} - \sum_{i,I} \frac{z_I e^2}{|\mathbf{r}_i - \mathbf{R}_{I0}|} + \frac{1}{2} \sum_{i \neq j} \frac{e^2}{|\mathbf{r}_i - \mathbf{r}_j|} + \frac{1}{2} \sum_{I \neq J} \frac{e^2}{|\mathbf{R}_{I0} - \mathbf{R}_{J0}|} , \quad (1)$$

where the terms appearing in Eq. (1) represent the kinetic energy of the electrons, the electronic–nuclear, electronic–electronic and nuclear–nuclear interaction energies, respectively; the nuclei are supposed fixed at the equilibrium configuration $\{\mathbf{R}_{I0}\}$; spin-orbit terms and other relativistic corrections are omitted for simplicity. In Chapter IV, we have discussed at length how to approximate the many-electron Hamiltonian (1) with an appropriate one-electron operator (within the Hartree-Fock framework).

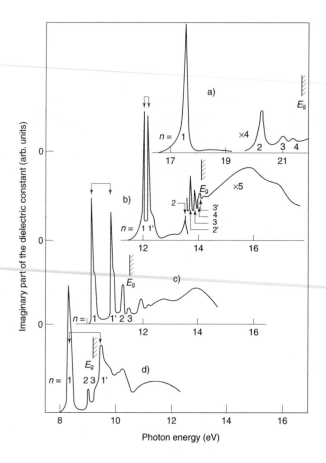

Fig. 1 Imaginary part of the dielectric constant of rare-gas solids [from G. Zimmerer, in "Excited-State Spectroscopy in Solids", Proc. Intern. School of Physics "Enrico Fermi", edited by U. M. Grassano and N. Terzi (North-Holland, Amsterdam 1987); copyright 1987 by Società Italiana di Fisica]. The pronounced structures in the optical absorption spectrum of Ne represent the exciton series $n = 1, 2, 3, \ldots$ converging to the energy gap $E_G = 21.69$ eV of the material. In the heavier rare-gas solids Ar, Kr, Xe the atomic spin-orbit splitting (and hence the splitting in the valence bands) increases; besides the exciton series $n = 1, 2, 3, \ldots$ converging to the energy gap, some spin-orbit partners $n' = 1', 2', 3', \ldots$ are resolved in the experiments.

Now, we face the task to overcome (in principle and, more importantly, in practice) the limitations inherent the independent particle picture.

To keep formalism at the essential, we consider here the simplest possible model of a semiconductor or insulator, i.e. a *two-band model*, with just one valence band (fully occupied with electrons of eiher spin) and a fully empty conduction band; a direct gap at the center of the Brillouin zone is assumed (see Fig. 3). All the other bands of the crystal are not accounted for in detail, but are assumed to influence some physical property (such as the dielectric constant ε of the medium), to be phenomenologically embodied in the two-band model.

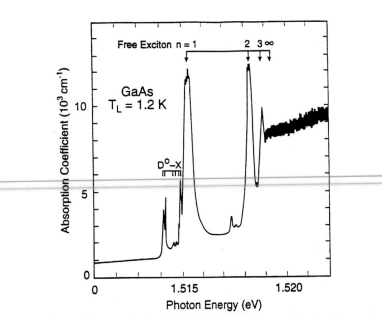

Fig. 2 Absorption spectrum at 1.2 K of a pure sample of GaAs containing residual donors. The $n = 1, 2, 3$ exciton peaks and the extrapolated energy gap are indicated by arrows. The peak $D^0 - X$ corresponds to the creation of an exciton (X) bound to a neutral donor (D^0) [from R. G. Ulbrich and C. Weisbuch, unpublished; C. Weisbuch, Thesis, Université Paris VII, 1977].

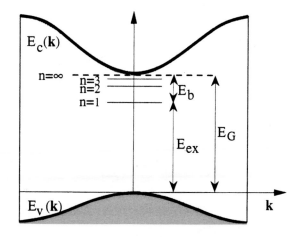

Fig. 3 Schematic energy band picture of a semiconductor in the two-band model representation. The lowest conduction band $E_c(\mathbf{k})$ and the highest valence band $E_v(\mathbf{k})$ are shown; the band extrema are supposed to occur both at $\mathbf{k} = 0$ (the valence band region has been shadowed to remind its full occupancy by electrons). The series of excitons (with exciton wavevector equal to zero), converging towards the fundamental energy gap E_G, are indicated by straight lines; E_{ex} and E_b are the energy and the binding energy of the $n = 1$ exciton state.

Within the one-electron approximation, the ground state of the two-band model is given by the Slater determinant, formed with double occupied valence wavefunctions $\psi_{v\mathbf{k}_i}$ of allowed wavevector $\mathbf{k}_i$ within the first Brillouin zone; we have

$$\Psi_0 = A\{\psi_{v\mathbf{k}_1}\alpha\ \psi_{v\mathbf{k}_1}\beta \ldots \psi_{v\mathbf{k}_N}\alpha\ \psi_{v\mathbf{k}_N}\beta\}\ , \tag{2}$$

where N is the number of unit cells of the crystal, of volume $V = N\Omega$; the valence (and conduction) wavefunctions are normalized to one in the volume V of the crystal.

The ground state Ψ_0 has total spin equal to zero and total wavevector equal to zero. In the one-electron approximation, the elementary excitations of given total wavevector $\mathbf{k}_{\mathrm{ex}}$ are constituted by electron–hole pairs in which an electron from the state $\psi_{v\mathbf{k}}$ is transferred to the state $\psi_{c\mathbf{k}+\mathbf{k}_{\mathrm{ex}}}$. The spin of the hole and the spin of the electron forming the pair can be either up or down, and it is thus possible to construct four different trial excited states; similarly to what discussed in Section IV-5.3, it is convenient to consider single-particle excitations $\Phi^{(\mathrm{S})}_{c\mathbf{k}+\mathbf{k}_{\mathrm{ex}},v\mathbf{k}}$ of definite spin multiplicity S (equal to 0 or to 1).

The orthonormal basis functions $\{\Phi^{(\mathrm{S})}_{c\mathbf{k}+\mathbf{k}_{\mathrm{ex}},v\mathbf{k}}\}$, when $\mathbf{k}$ varies throughout the first Brillouin zone, can be conveniently used to describe (at least approximately) the *excited states* (or *excitons*) of the many-electron Hamiltonian H_e. We thus expand the exciton wavefunctions of total wavevector $\mathbf{k}_{\mathrm{ex}}$ and total spin S in the form

$$\Psi_{\mathrm{ex}} = \sum_{\mathbf{k}'} A(\mathbf{k}')\Phi^{(\mathrm{S})}_{c\mathbf{k}'+\mathbf{k}_{\mathrm{ex}},v\mathbf{k}'}\ . \tag{3a}$$

We insert this expansion into the Schrödinger equation $H_e\Psi_{\mathrm{ex}} = E\Psi_{\mathrm{ex}}$, project on $\langle\Phi^{(\mathrm{S})}_{c\mathbf{k}+\mathbf{k}_{\mathrm{ex}},v\mathbf{k}}|$, and obtain for the expansion coefficients the set of linear homogeneous equations

$$\sum_{\mathbf{k}'}\langle\Phi^{(\mathrm{S})}_{c\mathbf{k}+\mathbf{k}_{\mathrm{ex}},v\mathbf{k}}|H_e|\Phi^{(\mathrm{S})}_{c\mathbf{k}'+\mathbf{k}_{\mathrm{ex}},v\mathbf{k}'}\rangle A(\mathbf{k}') = E\,A(\mathbf{k})\ . \tag{3b}$$

Using Eq. (IV-35) and Eq. (IV-36), the matrix elements of H_e in Eq. (3b) can be expressed in terms of bielectronic integrals involving the band wavefunctions. Thus the coefficients $A(\mathbf{k})$ satisfy the integral equation

$$\boxed{[E_c(\mathbf{k}+\mathbf{k}_{\mathrm{ex}}) - E_v(\mathbf{k}) - E]\,A(\mathbf{k}) + \sum_{\mathbf{k}'} U(\mathbf{k},\mathbf{k}')A(\mathbf{k}') = 0}\ , \tag{4}$$

where $\mathbf{k}$ and $\mathbf{k}'$ can be thought of as continuous variables; the kernel $U(\mathbf{k},\mathbf{k}')$ of the exciton integral equation (4) reads

$$U(\mathbf{k},\mathbf{k}') = -\ \langle\psi_{c\mathbf{k}+\mathbf{k}_{\mathrm{ex}}}\psi_{v\mathbf{k}'}|\frac{e^2}{r_{12}}|\psi_{c\mathbf{k}'+\mathbf{k}_{\mathrm{ex}}}\psi_{v\mathbf{k}}\rangle$$

$$+\ 2\delta_{\mathrm{S},0}\langle\psi_{c\mathbf{k}+\mathbf{k}_{\mathrm{ex}}}\psi_{v\mathbf{k}'}|\frac{e^2}{r_{12}}|\psi_{v\mathbf{k}}\psi_{c\mathbf{k}'+\mathbf{k}_{\mathrm{ex}}}\rangle\ . \tag{5}$$

The general properties of the bielectronic integrals are discussed in Appendix IV-A.

Approximate solution of the exciton integral equation (two-band model)

The integral equation (4), with the kernel (5), is the basic eigenvalue equation for excitons in the two-band model; although the accurate solution may present a formidable task, it is instructive to show that the integral equation (4) can be brought (approximately) to the form of the Schrödinger equation of a hydrogen-like system, with appropriate effective mass and screened Coulomb interaction.

In the kernel (5), we write the Bloch wavefunctions as the product of plane waves and periodic parts, and obtain

$$U(\mathbf{k}, \mathbf{k}')$$

$$= - \int u^*_{c\mathbf{k}+\mathbf{k}_{ex}}(\mathbf{r}_1)u^*_{v\mathbf{k}'}(\mathbf{r}_2)u_{c\mathbf{k}'+\mathbf{k}_{ex}}(\mathbf{r}_1)u_{v\mathbf{k}}(\mathbf{r}_2)e^{-i(\mathbf{k}-\mathbf{k}')\cdot(\mathbf{r}_1-\mathbf{r}_2)}\frac{e^2}{r_{12}}d\mathbf{r}_1\,d\mathbf{r}_2$$

$$+ 2\delta_{S,0} \int u^*_{c\mathbf{k}+\mathbf{k}_{ex}}(\mathbf{r}_1)u^*_{v\mathbf{k}'}(\mathbf{r}_2)u_{v\mathbf{k}}(\mathbf{r}_1)u_{c\mathbf{k}'+\mathbf{k}_{ex}}(\mathbf{r}_2)e^{-i\mathbf{k}_{ex}\cdot(\mathbf{r}_1-\mathbf{r}_2)}\frac{e^2}{r_{12}}d\mathbf{r}_1\,d\mathbf{r}_2 . \quad (6)$$

The above expression is so complicated that some simplification must be done; a reasonable one is just to replace the *product of any two periodic functions having the same argument, with the average value on the volume V of the crystal.* We have for instance

$$u^*_{v\mathbf{k}'}(\mathbf{r}_2)u_{v\mathbf{k}}(\mathbf{r}_2) \approx \frac{\displaystyle\int u^*_{v\mathbf{k}'}(\mathbf{r}_2)u_{v\mathbf{k}}(\mathbf{r}_2)\,d\mathbf{r}_2}{\displaystyle\int d\mathbf{r}_2} = \frac{1}{V}\langle u_{v\mathbf{k}'}|u_{v\mathbf{k}}\rangle \approx \frac{1}{V} , \quad (7a)$$

where the last passage is rigorous if $\mathbf{k} = \mathbf{k}'$, and is acceptable if $\mathbf{k}$ and $\mathbf{k}'$ are reasonably near in the Brillouin zone.

With similar procedures, we have

$$u^*_{c\mathbf{k}+\mathbf{k}_{ex}}(\mathbf{r}_1)u_{v\mathbf{k}}(\mathbf{r}_1) \approx \frac{1}{V}\langle u_{c\mathbf{k}+\mathbf{k}_{ex}}|u_{v\mathbf{k}}\rangle . \quad (7b)$$

We are particularly interested in excitons with $\mathbf{k}_{ex}$ small with respect to the Brillouin zone, because of their role in optical transitions (see Chapter XII). In the long wavelength limit ($\mathbf{k}_{ex} \to 0$) of the exciton wavevector, we can expand the conduction band wavefunction to first order in $\mathbf{k}_{ex}$, namely $u_{c\mathbf{k}+\mathbf{k}_{ex}} = u_{c\mathbf{k}} + \mathbf{k}_{ex} \cdot \partial u_{c\mathbf{k}}/\partial\mathbf{k}$. Then we obtain

$$\langle u_{c\mathbf{k}+\mathbf{k}_{ex}}|u_{v\mathbf{k}}\rangle = \mathbf{k}_{ex} \cdot \langle\frac{\partial}{\partial\mathbf{k}}u_{c\mathbf{k}}|u_{v\mathbf{k}}\rangle = i\mathbf{k}_{ex} \cdot \langle u_{c\mathbf{k}}|\,\mathbf{r}\,|u_{v\mathbf{k}}\rangle ,$$

where the last passage has been done using the complex conjugate of Eq. (II-36). For small $\mathbf{k}_{ex}$, Eq. (7b) can then be written as

$$u^*_{c\mathbf{k}+\mathbf{k}_{ex}}(\mathbf{r}_1)u_{v\mathbf{k}}(\mathbf{r}_1) \approx \frac{1}{V}i\,\mathbf{k}_{ex} \cdot \mathbf{r}_{cv} , \quad (7c)$$

where

$$\mathbf{r}_{cv} = \int_V u^*_c(\mathbf{k}, \mathbf{r})\,\mathbf{r}\,u_v(\mathbf{k}, \mathbf{r})\,d\mathbf{r} . \quad (7d)$$

We notice explicitly that the wavefunctions $u_c(\mathbf{k}, \mathbf{r})$ and $u_v(\mathbf{k}, \mathbf{r})$ are normalized to one in the volume V of the crystal; thus *the matrix element* $\mathbf{r}_{cv}$, *as well as the dipole* $\mathbf{d}_{cv} = e\mathbf{r}_{cv}$, *is an intrinsic property of the crystal independent of its dimensions.* Because of the periodicity of the wavefunctions $u_c(\mathbf{k}, \mathbf{r})$ and $u_v(\mathbf{k}, \mathbf{r})$, the dipole $\mathbf{d}_{cv}$ can be interpreted as the sum of N identical dipoles of electric moment $\mathbf{d}_{cv}/N$, each of them localized at one of the N unit cells of the crystal. For simplicity, in the following, we neglect the $\mathbf{k}$-dependence of the matrix elements (7d).

We now insert Eqs. (7) and similar ones into Eq. (6), and obtain

$$U(\mathbf{k}, \mathbf{k}') = -\frac{1}{V}\frac{4\pi e^2}{|\mathbf{k} - \mathbf{k}'|^2} + \frac{1}{V}2\delta_{S,0}\frac{4\pi e^2}{k_{\mathrm{ex}}^2}(\mathbf{k}_{\mathrm{ex}} \cdot \mathbf{r}_{cv})(\mathbf{k}_{\mathrm{ex}} \cdot \mathbf{r}_{cv}^*) . \tag{8}$$

The expression (8) of the kernel is suitable for some physical discussion. The first term in the right hand side of Eq. (8) is the Fourier transform of the interaction $-e^2/r$; in the following, we take the liberty to screen it with an appropriate background dielectric constant ε of the medium (for simplicity, ε is assumed to be frequency independent). A full many-body treatment, which decouples the given pair of bands from all the others, justifies the screening of the electron–hole interaction $-e^2/r$ and also suggests more sophisticated screening procedures [see for instance G. D. Mahan "Many-Particle Physics" (Plenum Press, New York 1990)].

For singlet excitons (S = 0), the second term in the right-hand side of Eq. (8) is different from zero. This term is independent of $\mathbf{k}$ (and represents thus a short-range interaction in real space), but does depend on the direction of $\mathbf{k}_{\mathrm{ex}}$ with respect to the dipole $\mathbf{d}_{cv}$ (a fact referred to as *"non-analytic behaviour"*). Let us indicate with $J_{cv}(\mathbf{k}_{\mathrm{ex}})$ the energy

$$J_{cv}(\mathbf{k}_{\mathrm{ex}}) = \frac{1}{V}\frac{4\pi}{k_{\mathrm{ex}}^2}(\mathbf{k}_{\mathrm{ex}} \cdot \mathbf{d}_{cv})(\mathbf{k}_{\mathrm{ex}} \cdot \mathbf{d}_{cv}^*) = \frac{4\pi}{V}(\widehat{\mathbf{k}}_{\mathrm{ex}} \cdot \mathbf{d}_{cv})(\widehat{\mathbf{k}}_{\mathrm{ex}} \cdot \mathbf{d}_{cv}^*) , \tag{9}$$

where $\widehat{\mathbf{k}}_{\mathrm{ex}} = \mathbf{k}_{\mathrm{ex}}/|\mathbf{k}_{\mathrm{ex}}|$. The energy $J_{cv}(\mathbf{k}_{\mathrm{ex}})$ can be interpreted as the classical dipole–dipole interaction energy of a static system of dipoles of electric moment $\mathbf{d}_{cv}/N$, localized on each of the N unit cells of the crystal. Consider in fact the polarization wave $\mathbf{P}(\mathbf{r}) = \mathbf{P}_0 \exp(i\mathbf{k}_{\mathrm{ex}} \cdot \mathbf{r})$ of amplitude $\mathbf{P}_0 = \mathbf{d}_{cv}/V$ (without loss of generality $\mathbf{d}_{cv}$ is assumed real); the microscopic polarization charge is

$$\rho_{\mathrm{micr}}(\mathbf{r}) = -\mathrm{div}\,\mathbf{P}(\mathbf{r}) = -i\mathbf{k}_{\mathrm{ex}} \cdot \mathbf{P}_0 \exp(i\mathbf{k}_{\mathrm{ex}} \cdot \mathbf{r}) .$$

The static electric field $\mathbf{E}$ generated by ρ_{micr} must satisfy the Maxwell equations $\mathrm{div}\,\mathbf{E} = 4\pi\rho_{\mathrm{micr}}$ and $\mathrm{curl}\,\mathbf{E} = 0$; it is immediate to find

$$\mathbf{E} = -4\pi(\widehat{\mathbf{k}}_{\mathrm{ex}} \cdot \mathbf{P}_0)\widehat{\mathbf{k}}_{\mathrm{ex}} \exp(i\mathbf{k}_{\mathrm{ex}} \cdot \mathbf{r}) .$$

The interaction energy $W = -\mathbf{d}_{cv} \cdot \mathbf{E}$ between the dipoles and the electric field (in the long wavelength limit $\mathbf{k}_{\mathrm{ex}} \to 0$) gives Eq. (9).

The above considerations suggest that also the (non-analytic) dipole–dipole interaction should be screened by the background dielectric constant ε of the medium; in fact, when ρ_{micr} is embedded in a continuum of dielectric constant ε, we have div $\mathbf{E} = 4\pi\rho_{\text{micr}}/\varepsilon$ and the field is thus reduced by ε. This heuristic argument is corroborated by rigorous treatments, which decouple the given pair of bands from all the others, and we refer to the literature for further aspects and refinement on dipole-dipole interaction screening [K. Cho, Solid State Commun. **33**, 911 (1980); K. Ehara and K. Cho, Solid State Commun. **44**, 453 (1982); U. Rössler and H. -R. Trebin, Phys. Rev. B**23**, 1961 (1981); L. X. Benedict and E. L. Shirley, Phys. Rev. B**59**, 5441 (1999) and references quoted therein].

We now specify the integral equation (4), with the kernel (8) appropriately screened, in the case the valence and conduction bands are both parabolic in the $\mathbf{k}$-region of interest around $\mathbf{k} = 0$ (see Fig. 3). We have

$$E_{\text{c}}(\mathbf{k}) = E_{\text{G}} + \frac{\hbar^2 k^2}{2m_{\text{c}}} \qquad E_{\text{v}}(\mathbf{k}) = -\frac{\hbar^2 k^2}{2m_{\text{v}}} \qquad E_{\text{c}}(\mathbf{k}) - E_{\text{v}}(\mathbf{k}) = E_{\text{G}} + \frac{\hbar^2 k^2}{2\mu_{\text{ex}}} \,,$$

where $\mu_{\text{ex}}^{-1} = m_{\text{c}}^{-1} + m_{\text{v}}^{-1}$; μ_{ex} is called the reduced effective mass of the exciton. The integral equation (4), for excitons with $\mathbf{k}_{\text{ex}} \approx 0$, takes the form

$$\left[E_{\text{G}} + \frac{\hbar^2 k^2}{2\mu_{\text{ex}}} - E\right] A(\mathbf{k}) - \frac{1}{V}\sum_{\mathbf{k}'} \frac{4\pi e^2}{\varepsilon|\mathbf{k} - \mathbf{k}'|^2} A(\mathbf{k}') + \delta_{\text{S},0} \frac{8\pi}{\varepsilon} |\widehat{\mathbf{k}}_{\text{ex}} \cdot \mathbf{d}_{\text{cv}}|^2 \frac{1}{V}\sum_{\mathbf{k}'} A(\mathbf{k}') = 0 \,.$$

(10)

In order to solve Eq. (10) it is convenient to introduce the so-called *envelope function*, *normalized to unity*, and defined as

$$F(\mathbf{r}) = \frac{1}{\sqrt{V}}\sum_{\mathbf{k}'} A(\mathbf{k}')e^{i\mathbf{k}' \cdot \mathbf{r}} \,.$$

(11)

We can transform the eigenvalue integral equation (10) for the $A(\mathbf{k})$ function into a more familiar differential eigenvalue equation for the envelope function $F(\mathbf{r})$. For this purpose we remark the following relations:

$$A(\mathbf{k}) = \frac{1}{\sqrt{V}}\int F(\mathbf{r})e^{-i\mathbf{k}\cdot\mathbf{r}}\,d\mathbf{r} \qquad k^2 A(\mathbf{k}) = \frac{1}{\sqrt{V}}\int [-\nabla^2 F(\mathbf{r})]e^{-i\mathbf{k}\cdot\mathbf{r}}\,d\mathbf{r}$$

$$-\frac{1}{\sqrt{V}}\sum_{\mathbf{k}'} \frac{4\pi e^2}{|\mathbf{k} - \mathbf{k}'|^2} A(\mathbf{k}') = -\int \frac{e^2}{r}F(\mathbf{r})e^{-i\mathbf{k}\cdot\mathbf{r}}\,d\mathbf{r}$$

$$\frac{1}{\sqrt{V}}\sum_{\mathbf{k}'} A(\mathbf{k}') = \int \delta(\mathbf{r})F(\mathbf{r})e^{-i\mathbf{k}\cdot\mathbf{r}}\,d\mathbf{r}$$

[the above four relations are easily proved putting in their second members the expression (11) for $F(\mathbf{r})$]. We insert the above relations into Eq. (10), and obtain the equation for the envelope function of the form

$$\boxed{\left[-\frac{\hbar^2\nabla^2}{2\mu_{\text{ex}}} - \frac{e^2}{\varepsilon\,r} + \delta_{\text{S},0}\frac{8\pi}{\varepsilon}|\widehat{\mathbf{k}}_{\text{ex}} \cdot \mathbf{d}_{\text{cv}}|^2\delta(\mathbf{r})\right] F(\mathbf{r}) = (E - E_{\text{G}})F(\mathbf{r})}\,.$$

(12)

Equation (12) (omitting momentarily the short-range term) provides a simple description of the exciton states in terms of a hydrogen-like atom in a polarizable medium

(Wannier model); we have

$$\left[-\frac{\hbar^2 \nabla^2}{2\mu_{\text{ex}}} - \frac{e^2}{\varepsilon\, r}\right] F(\mathbf{r}) = (E - E_{\text{G}})F(\mathbf{r}) \ . \tag{13}$$

The effective Rydberg of the exciton problem (i.e. the binding energy $E_{\text{b}}^{(\text{ex})}$ of the lowest exciton state) can be estimated from the relation

$$E_{\text{b}}^{(\text{ex})} = 13.6\, \frac{\mu_{\text{ex}}}{m}\, \frac{1}{\varepsilon^2} \ (\text{in eV}) \ .$$

Typically in semiconductors $\varepsilon \approx 10$ and $\mu_{\text{ex}} \approx 0.1\, m$ and thus the binding energy is of the order of a few millielectronvolt. For strong insulators (such as neon for instance) we have $\varepsilon \approx 2$ and $\mu_{\text{ex}} \approx m$ and the binding energy is of the order of a few electronvolt. In any case, exciton levels below the gap tend to group in well-defined hydrogenic-like series (as seen in Fig. 1 and Fig. 2). The effective radius of the ground exciton state can be estimated from the relation

$$a_{\text{ex}} = a_{\text{B}}\, \frac{m}{\mu_{\text{ex}}}\, \varepsilon \ .$$

With the above estimated values of μ_{ex} and ε, we see that the wavefunction of electrons and holes bound together may extend over a few unit cells in large gap insulators, and over several thousand unit cells in weakly bound excitons. The Schrödinger equation of type (13) was originally introduced in the literature to describe shallow excitons; appropriately implemented by a more accurate description of the electron–hole interaction at small distances, it can be used to describe also strongly bound valence or core excitons [see for instance S. Baroni, G. Grosso, L. Martinelli and G. Pastori Parravicini, Phys. Rev. B**20**, 1713 (1979); Phys. Rev. B**22**, 6440 (1980) and references quoted therein].

We consider now the short-range term in Eq. (12); this term is present only for singlet excitons ($S = 0$), and its expectation value on $F(\mathbf{r})$ is $(8\pi/\varepsilon)|\widehat{\mathbf{k}}_{\text{ex}} \cdot \mathbf{d}_{\text{cv}}|^2 |F(0)|^2$, where $|F(0)|^2$ can be interpreted as the probability that the electron and the hole are in the same position. The energy difference between the transverse exciton ($\mathbf{k}_{\text{ex}} \perp \mathbf{d}_{\text{cv}}$) and the longitudinal exciton ($\mathbf{k}_{\text{ex}} \parallel \mathbf{d}_{\text{cv}}$) is given by

$$\Delta_{\text{LT}} = \frac{1}{\varepsilon}\, \frac{8\pi}{\Omega}\, |\mathbf{d}_{\text{cv}}|^2\, \Omega |F(0)|^2 \ ,$$

where Ω is the volume of the unit cell. The transverse-longitudinal splitting Δ_{LT}, is important mainly for strongly bound exciton states; in the limiting case of excitons localized in the unit cell or so, the value of $\Omega|F(0)|^2$ is about one, and $\Delta_{\text{LT}} \approx (8\pi/\varepsilon)|\mathbf{d}_{\text{cv}}|^2/\Omega$. For higher members of the exciton series, and for shallow excitons, the dimensionless quantity $\Omega|F(0)|^2$ decreases and becomes rapidly negligible.

An exciton can propagate in the crystal and transport energy, without transport of net charge. The treatment for excitons with $\mathbf{k}_{\text{ex}} \neq 0$, requires the appropriate inclusion of the center of mass kinetic energy $\hbar^2 k_{\text{ex}}^2/2(m_{\text{c}} + m_{\text{v}})$ (as can be inferred considering Eq. 4 and keeping $\mathbf{k}_{\text{ex}} \neq 0$). We do not wish to enter here in deeper analysis of screening and local field effects, or consider extensions to multiband models, or discuss the exciton-polariton problem (apart from the few comments on Fig. 12 in Chapter IX); in spite of the several approximations and limits, the two-band model presented here allows a reasonably intuitive picture of the excitonic effects and also a feeling of how to construct more general approaches.

2   Plasmon excitations in crystals

In the above discussion, we have seen that *excitons can be interpreted as collective electron–hole excitations (tendentially) occurring on the low energy side of individual electron–hole excitations.* In this section, we focus on the concept of plasma oscillations, or plasmons, which are longitudinal charge density oscillations, brought about by the Coulomb interaction between the electrons. *Plasmons can be interpreted as collective electron–hole excitations (tendentially) occurring on the high energy side of individual electron–hole excitations.* A common thread exists between these collective phenomena, in spite of the quite different phenomenology. In Fig. 4 we report, as an example, the energy-loss spectrogram of a beam of 20 keV primary electrons, transmitted through an Al thin film; the sharp peaks in inelastic scatterings correspond to plural plasmon excitations.

Plasmon excitations can be intuitively described with the following classical picture. Consider a free-electron gas of average density n, embedded in a uniform and fixed background of neutralizing positive charges. Suppose that the electrons originally at $\mathbf{r}$ undergo a time-dependent longitudinal displacement $u(\mathbf{r}, t)$ of the form

$$\mathbf{u}(\mathbf{r}, t) = \mathbf{u}_0 e^{i(\mathbf{q}\cdot\mathbf{r} - \omega t)} + c.c. \quad \mathbf{u}_0 \parallel \mathbf{q} \tag{14a}$$

(c.c. indicates complex conjugate of the previous term). The polarization of the system due to the electronic displacement is $\mathbf{P}(\mathbf{r}, t) = n(-e)\mathbf{u}(\mathbf{r}, t)$; the associated microscopic charge $\rho_{\mathrm{micr}}(\mathbf{r}, t)$ is

$$\rho_{\mathrm{micr}}(\mathbf{r}, t) = -\mathrm{div}\mathbf{P}(\mathbf{r}, t) = ine\mathbf{q}\cdot\mathbf{u}_0 e^{i(\mathbf{q}\cdot\mathbf{r}-\omega t)} + c.c. \quad \mathbf{u}_0 \parallel \mathbf{q} \;.$$

For longitudinal displacements $\mathbf{u}_0$ and $\mathbf{q}$ are parallel, $\mathbf{u}_0 \cdot \mathbf{q} \neq 0$ and $\rho_{\mathrm{micr}} \neq 0$; the longitudinal electric field satisfying $\mathrm{div}\mathbf{E} = 4\pi\rho_{\mathrm{micr}}$ is given by

$$\mathbf{E}(\mathbf{r}, t) = -4\pi\mathbf{P}(\mathbf{r}, t) = 4\pi ne\mathbf{u}(\mathbf{r}, t) \;. \tag{14b}$$

We can thus write the classical equation of motion for each electron of the gas

$$m\ddot{\mathbf{u}}(\mathbf{r}, t) = -e\mathbf{E}(\mathbf{r}, t) = -4\pi ne^2\mathbf{u}(\mathbf{r}, t) \;. \tag{14c}$$

This equation shows that the longitudinal charge fluctuations occur with the frequency ω_{p} (*plasma frequency*) given by

$$\omega_{\mathrm{p}}^2 = \frac{4\pi ne^2}{m} \;; \tag{14d}$$

typical values of $\hbar\omega_{\mathrm{p}}$ are in the range 10–20 eV. Notice that this classic model fails to predict a wavevector dependence of the frequency of the longitudinal charge fluctuations.

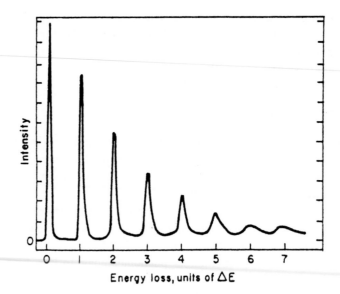

Fig. 4 Energy-loss spectrum (at very small angles) for a beam of 20 keV primary electrons passing through an aluminum film of thickness ≈ 2580 Å. The unit of energy-loss ΔE (≈ 15 eV) is the plasmon excitation energy for Al. Zero-loss, first-loss, second-loss and other plural-loss peaks are clearly detected [from L. Marton, J. A. Simpson, H. A. Fowler and N. Swanson, Phys. Rev. **126**, 182 (1962); copyright 1962 by the American Physical Society].

In the following sections, we consider some essential aspects of plasmons from a quantistic point of view. For this purpose, we establish the microscopic quantum expression of the longitudinal dielectric function of crystals; plasmons are then inferred from the identification of self-sustaining collective motions of electron particles. Plasmons can be considered (ideally) as independent elementary excitations of the interacting electron system, and are a typical many-body longitudinal-wave effect.

3   General considerations on the longitudinal dielectric function

In this section we consider some general features of the longitudinal response function of a quantum system of electrons to driving perturbations, and the connected longitudinal dielectric function. We begin with a few general assumptions on the sample and the field not yet in interaction. The sample (in the absence of perturbations) is assumed to be homogeneous (or nearly so). The "bare" or "external" perturbation (in the absence of the sample) is assumed to be arbitrarily small, and periodic in space and time with wavevector $\mathbf{q}$ and angular frequency ω. The external perturbation energy felt by an electron is thus taken in the form

$$U_{\text{ext}}(\mathbf{r}, t) = A_{\text{ext}}(\mathbf{q}, \omega)e^{i(\mathbf{q}\cdot\mathbf{r} - \omega t)} + c.c. , \tag{15}$$

where $A_{\text{ext}}(\mathbf{q}, \omega)$ is the (infinitesimal) amplitude of the external perturbation, and c.c. indicates the complex conjugate of the previous term. When carrying out calculations, it is always assumed (implicitly or explicitly) that the potential (15) is turned on adiabatically at $t = -\infty$, i.e. it is multiplied by $\exp(\eta t/\hbar)$ with $\eta \to 0^+$ (η is an infinitesimal positive energy).

We notice that the electrostatic potential $\phi_{ext}(\mathbf{r}, t)$ corresponding to the energy potential $U_{ext}(\mathbf{r}, t)$ is given by $\phi_{ext} = U_{ext}/(-e)$, where e is the modulus of the electronic charge. The electric field becomes

$$\mathbf{E}_{ext}(\mathbf{r}, t) = -\nabla \phi_{ext}(\mathbf{r}, t) = \frac{1}{e} A_{ext}(\mathbf{q}, \omega) i\, \mathbf{q}\, e^{i(\mathbf{q} \cdot \mathbf{r} - \omega t)} + c.c. \ . \tag{16}$$

The electric field (16) is of *longitudinal type*, i.e. it is parallel to the propagation wavevector $\mathbf{q}$; it is also irrotational (curl $\mathbf{E}_{ext} \equiv 0$), consistently with the fact that it has been derived by a scalar potential.

When the sample is exposed to the external perturbation (15), a modulation of the electron wavefunctions, and thus a modulation of the electronic charge density, is forced within the sample. In *homogeneous systems* and in the *linear response regime*, the *total self-consistent potential* $U_{tot}(\mathbf{r}, t)$ (born both from *external* and *internal* sources) is expected to depend on space and time with the same wavevector $\mathbf{q}$ and the same frequency ω as the external potential (15). One can thus write for the total driving perturbation

$$U_{tot}(\mathbf{r}, t) = A_{tot}(\mathbf{q}, \omega) e^{i(\mathbf{q} \cdot \mathbf{r} - \omega t)} + c.c. \ . \tag{17}$$

In the linear response regime, the amplitudes $A_{ext}(\mathbf{q}, \omega)$ and $A_{tot}(\mathbf{q}, \omega)$ are proportional to each other, and the macroscopic dielectric function $\varepsilon(\mathbf{q}, \omega)$ can then be defined as

$$A_{tot}(\mathbf{q}, \omega) = \frac{1}{\varepsilon(\mathbf{q}, \omega)} A_{ext}(\mathbf{q}, \omega) \ ; \tag{18}$$

relation (18) is the analogue of the standard relation $\mathbf{E} = (1/\varepsilon)\mathbf{D}$ between the total electric field $\mathbf{E}$ in a medium (due to external and internal sources) and the induction electric field $\mathbf{D}$ (due to external sources only).

The dielectric function $\varepsilon(\mathbf{q}, \omega)$ for a homogeneous system satisfies some general properties. From reality condition of both members of Eq. (15), it follows $A_{ext}(\mathbf{q}, \omega) = A_{ext}^*(-\mathbf{q}, -\omega)$ (for real $\mathbf{q}$ and ω); similarly, from Eq. (17) it holds $A_{tot}(\mathbf{q}, \omega) = A_{tot}^*(-\mathbf{q}, -\omega)$; hence, from Eq. (18), it is inferred $\varepsilon(\mathbf{q}, \omega) = \varepsilon^*(-\mathbf{q}, -\omega)$. Since the system is homogeneous, perturbations of wavevector $\mathbf{q}$ or $-\mathbf{q}$ produce similar effects and we thus expect $\varepsilon(\mathbf{q}, \omega) = \varepsilon(-\mathbf{q}, \omega)$. The two previous properties imply $\varepsilon(\mathbf{q}, \omega) = \varepsilon^*(\mathbf{q}, -\omega)$; this shows that *the real part* $\varepsilon_1(\mathbf{q}, \omega)$ *of the dielectric function is an even function of* ω, *while the imaginary part* $\varepsilon_2(\mathbf{q}, \omega)$ *is an odd function of* ω. The physical requirement that the response is causal, linear and finite, ensures that $\varepsilon(\mathbf{q}, \omega)$ is analytical in the upper half of the complex ω plane and obeys the Kramers–Kronig relations

$$\varepsilon_1(\mathbf{q}, \omega) = 1 + \frac{2}{\pi} P \int_0^\infty \omega' \frac{\varepsilon_2(\mathbf{q}, \omega')}{\omega'^2 - \omega^2} d\omega' \ , \qquad \varepsilon_2(\mathbf{q}, \omega) = -\frac{2\omega}{\pi} P \int_0^\infty \frac{\varepsilon_1(\mathbf{q}, \omega') - 1}{\omega'^2 - \omega^2} d\omega' \ , \tag{19}$$

where P denotes the principal part of the integral. From Eq. (18), in the case $\varepsilon(\mathbf{q},\omega)=0$, we can have $A_{\mathrm{tot}}(\mathbf{q},\omega)\neq 0$ even when $A_{\mathrm{ext}}(\mathbf{q},\omega)=0$ (i.e. there is the possibility of density fluctuation in the system even in the absence of any external perturbation!); thus *the zeroes of the longitudinal dielectric function identify the plasmon-type collective excitations* of the system.

Before considering specific models and applications to crystals for the longitudinal dielectric function, it is important a comment on the assumption of *homogeneity*. Crystals, strictly speaking, are not homogeneous materials; in particular the crystal electron density is invariant under lattice translations, but not under arbitrary displacements. In a periodic medium, the induced electron density, besides a wavevector $\mathbf{q}$ component as the applied field, also may include other wavevector components $\mathbf{q}+\mathbf{g}_m$, with $\mathbf{g}_m$ reciprocal lattice vectors. In general "local field effects" due to inhomogeneity of the crystal within the unit cell are rather complicated to account for; we ignore them in the following and we refer to the literature for their discussion [see for instance S. Adler, Phys. Rev. **126**, 413 (1962); N. Wiser, Phys. Rev. **129**, 62 (1963)].

4   Static dielectric screening in metals with the Thomas–Fermi model

Consider a medium and an extra point charge Ze embedded in it. In a dielectric (semiconductor or insulator) it is well known that the test charge Ze is surrounded by a screening charge equal to $-(1-1/\varepsilon_{\mathrm{s}})Ze$, where ε_{s} is the static dielectric constant of the medium. In a metal instead the screening is complete; the *bare Coulomb potential* (born from the external charge Ze) is modified into a (total) *screened Coulomb potential* (born from the external charge Ze and the screening charge induced in the metal), with the following peculiar properties: (i) The screened Coulomb field is cut off at a characteristic distance of the order of k_{F}^{-1}; the electrons are very effective in screening external charges. (ii) Weakly decaying long-range oscillations of electron density occur (Friedel oscillations); Knight shift in nuclear magnetic resonance of the nuclei near impurities in metals, confirms these oscillations. (iii) The change of electron density must be finite at the origin (where the point charge Ze is located). In fact the lifetime for positron annihilation in metals (which is proportional to the electron density) is finite. Also the Knight shift of the nuclei of impurities is finite.

To understand qualitatively the highly effective shielding in metals, we consider first the Thomas–Fermi model. This model explains quite well the exponential screening at intermediate distances, but fails in predicting finite induced charge at the origin ($r\approx 0$) and long-range oscillations ($r\approx\infty$).

Let us suppose that the electrons of a metal are subjected to a total static potential

$$U(\mathbf{r}) = A(\mathbf{q})\exp(i\mathbf{q}\cdot\mathbf{r}) + c.c. ;$$

the corresponding electric field $\mathbf{E}=-\nabla U(\mathbf{r})/(-e)$ is of longitudinal type, i.e. $\mathbf{E}\parallel\mathbf{q}$,

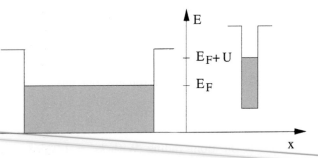

Fig. 5 Schematic representation of the linearized Thomas–Fermi approximation. On the left we have represented a metal with Fermi energy E_F, and density-of-states $n(E_F)$ (per unit volume and both spin directions). On the right we have represented an (isolated) piece of the same metal, with all band structure rigidly shifted by a superimposed energy U. When the contact is established between the reservoir on the left and the (small) sample on the right, the Fermi energy of the sample equalizes E_F; as a consequence, the induced electron density in the sample is $n_{ind} = -n(E_F)U$.

and is given by

$$\mathbf{E}(\mathbf{r}) = \frac{1}{e}\, i\,\mathbf{q}\, A(\mathbf{q}) \exp(i\mathbf{q}\cdot\mathbf{r}) + c.c. \ .$$

In order to determine the static dielectric function $\varepsilon(\mathbf{q}, \omega = 0) = \varepsilon(\mathbf{q})$ of the metal, we have to establish the induced electron density $n_{ind}(\mathbf{r})$ related to the perturbation potential $U(\mathbf{r})$. A very simple answer is provided by the Thomas–Fermi approximation (Fig. 5), that requires that the Fermi level of the material remains constant throughout its volume.

The induced change of electron density is then

$$n_{ind}(\mathbf{r}) = -n(E_F)U(\mathbf{r}) = -n(E_F)A(\mathbf{q})e^{i\mathbf{q}\cdot\mathbf{r}} + c.c. \ ,$$

where $n(E_F)$ is the density-of-states of the metal (per unit volume and both spin directions) at the Fermi energy (we have also assumed that $U(\mathbf{r})$ is small enough that the density-of-states of the metal can be taken as constant). The induced change of electron charge density is $\rho_{ind}(\mathbf{r}) = -e\, n_{ind}(\mathbf{r})$. The macroscopic polarization field $\mathbf{P}(\mathbf{r})$, defined so that $\operatorname{div} \mathbf{P} = -\rho_{ind}$, takes the form

$$\mathbf{P}(\mathbf{r}) = \frac{e\, n(E_F)}{q^2}\, i\,\mathbf{q}\, A(\mathbf{q}) \exp(i\mathbf{q}\cdot\mathbf{r}) + c.c. \ .$$

The static dielectric function of the medium $\varepsilon(\mathbf{q}) = 1 + 4\pi\mathbf{P}/\mathbf{E}$ becomes

$$\varepsilon(q) = 1 + \frac{4\pi e^2 n(E_F)}{q^2} \ , \tag{20}$$

which is known as the Thomas–Fermi dielectric function.

The above result can also be written as

$$\varepsilon(q) = 1 + \frac{k_{TF}^2}{q^2} ,$$

where $k_{TF}^2 = 4\pi e^2 n(E_F)$. For a free-electron gas, one has

$$k_{TF}^2 = 4\pi e^2 \frac{3}{2}\frac{n}{E_F} = 6\pi e^2 \frac{k_F^3}{3\pi^2}\frac{2m}{\hbar^2 k_F^2} = \frac{4}{\pi}\frac{k_F}{a_B} , \tag{21}$$

where $a_B = \hbar^2/me^2$ is the Bohr radius. With $k_F = 1.92/r_s a_B$, and $2 < r_s < 6$ at ordinary metallic densities, the screening wavenumber k_{TF} turns out to be of the order of a_B^{-1}.

To better illustrate the meaning of k_{TF}, consider an extra point charge Ze, at the origin in the medium. The bare Coulomb potential energy is

$$U_{ext}(\mathbf{r}) = -\frac{Ze^2}{r} \equiv -\frac{1}{(2\pi)^3}\int \frac{4\pi Ze^2}{q^2} e^{i\mathbf{q}\cdot\mathbf{r}} \, d\mathbf{q} \tag{22a}$$

(the last passage uses the standard Fourier representation of $1/r$). The total potential energy $U(\mathbf{r})$ becomes

$$U(\mathbf{r}) = -\frac{1}{(2\pi)^3}\int \frac{1}{\varepsilon(q)}\frac{4\pi Ze^2}{q^2} e^{i\mathbf{q}\cdot\mathbf{r}} \, d\mathbf{q} . \tag{22b}$$

Using the Thomas–Fermi dielectric function, one obtains

$$U(\mathbf{r}) = -\frac{1}{(2\pi)^3}\int \frac{4\pi Ze^2}{q^2 + k_{TF}^2} e^{i\mathbf{q}\cdot\mathbf{r}} \, d\mathbf{q} = -\frac{Ze^2}{r} e^{-k_{TF}r} .$$

Thus, the bare long-range Coulomb interaction Ze^2/r is transformed into an exponentially damped interaction with screening length equal to $1/k_{TF}$.

We consider now the induced charge density around the point impurity Ze (inserted at the origin). Using the Poisson equation, we have

$$-\nabla^2 \frac{U(\mathbf{r}) - U_{ext}(\mathbf{r})}{(-e)} = 4\pi \rho_{ind}(\mathbf{r}) .$$

Inserting Eqs. (22) in the above expression, one finds

$$\rho_{ind}(\mathbf{r}) = Ze\frac{1}{(2\pi)^3}\int \left[\frac{1}{\varepsilon(q)} - 1\right] e^{i\mathbf{q}\cdot\mathbf{r}} \, d\mathbf{q} ; \tag{23}$$

the total induced screening charge Q_S is

$$Q_S = \int \rho_{ind}(\mathbf{r}) d\mathbf{r} = Ze\left[\frac{1}{\varepsilon(0)} - 1\right] .$$

This equation shows that in metals, where $\varepsilon(0) \to \infty$, the total displaced charge is $Q_S = -Ze$, and the screening is complete.

In the case of the Thomas–Fermi dielectric function, Eq. (23) gives

$$\rho_{ind}(\mathbf{r}) = -\frac{Ze}{(2\pi)^3}\int \frac{k_{TF}^2}{q^2 + k_{TF}^2} e^{i\mathbf{q}\cdot\mathbf{r}} \, d\mathbf{q} = -e\,n(E_F)\frac{Ze^2}{r}\exp(-k_{TF}r) ;$$

thus $\rho_{\text{ind}}(0)$ is singular at the origin (contrary to experimental evidence, such as finite lifetime of positrons in metals and the finite Knight shift of impurity atoms). In order to have a finite value of $\rho_{\text{ind}}(0)$ in Eq. (23), it is necessary that $\varepsilon(q) - 1$ for large q decreases more rapidly than q^{-2}. The behaviour of $\rho_{\text{ind}}(\mathbf{r})$ at large r furthermore does not contain the Friedel oscillations. Thus the Thomas–Fermi dielectric screening needs improvements to provide the correct behaviour of $\rho_{\text{ind}}(\mathbf{r})$ at small and large distances.

5 Static dielectric screening in metals with the Lindhard model

We consider here the simplest model of a metal, the free-electron gas, described by the Hamiltonian $H_0 = -(\hbar^2/2m)\nabla^2$. The conduction band wavefunctions are plane waves $W_{\mathbf{k}}(\mathbf{r}) = (1/\sqrt{V})\exp(i\mathbf{k}\cdot\mathbf{r})$ (normalized to one in the volume V of the crystal) and the energies are $E(\mathbf{k}) = \hbar^2 k^2/2m$; the occupancy of the states is determined by the Fermi–Dirac distribution function $f(E(\mathbf{k}))$, or for brevity $f(\mathbf{k})$, ordinarily considered at the temperature $T = 0$.

The static dielectric function $\varepsilon(\mathbf{q}, 0)$ can be obtained specifying the general results of the linear response theory, summarized by Eq. (60), to the case of plane waves eigenfunctions and taking the limit $\omega \rightarrow 0$; one finds

$$\varepsilon(\mathbf{q},0) = 1 + \frac{8\pi e^2}{q^2}\frac{1}{V}\sum_{\mathbf{k}} \frac{f(\mathbf{k}) - f(\mathbf{k}+\mathbf{q})}{E(\mathbf{k}+\mathbf{q}) - E(\mathbf{k}) - i\eta} , \tag{24}$$

an expression which is known as the *static Lindhard dielectric function* for the free-electron gas.

It is possible to evaluate Eq. (24) in analytic form, as shown in Appendix A. From Eq. (A6) it is seen that $\varepsilon_1(q, 0)$, denoted as $\varepsilon(q)$, becomes

$$\varepsilon(q) = 1 + \frac{2me^2}{\pi\hbar^2}\frac{k_{\text{F}}}{q^2} + \frac{2me^2}{\pi\hbar^2}\frac{1}{q^3}\left(k_{\text{F}}^2 - \frac{q^2}{4}\right)\ln\left|\frac{2k_{\text{F}}+q}{2k_{\text{F}}-q}\right| . \tag{25a}$$

Introducing the dimensionless quantity $x = q/2k_{\text{F}}$, Eq. (25a) can be recast in the form

$$\varepsilon(q) = 1 + \frac{k_{\text{TF}}^2}{q^2}F(x) , \tag{25b}$$

where the function $F(x)$ is given by

$$F(x) = \frac{1}{2} + \frac{1-x^2}{4x}\ln\left|\frac{1+x}{1-x}\right| . \tag{25c}$$

The function $F(x)$ has been already discussed in Section IV-7, although in a different context. The behaviour of $\varepsilon(q)$ is illustrated in Fig. 6.

In spite of its simplicity, the Lindhard function (25a) already embodies some key features, that are common to more sophisticated many-body treatments. Let us consider in fact the behaviour of $\varepsilon(q)$ in the long wavelength limit, in the short wavelength limit, and in the intermediate region $q \approx 2k_{\text{F}}$.

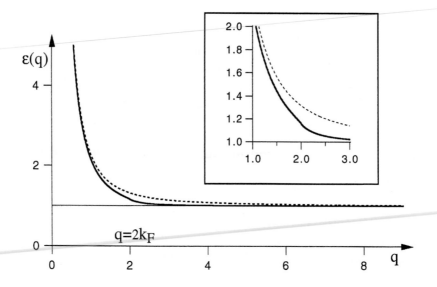

Fig. 6 Behaviour of the static Lindhard (solid line) and Thomas–Fermi (dashed line) dielectric function of a free-electron gas. The inset shows the behaviour of $\varepsilon(q)$ near $q=2k_F$ (we have chosen $k_F=1\,a_B^{-1}$ or equivalently $r_s = 1.92$; q is in units of a_B^{-1}).

For *small q* ($q \ll 2k_F$) the Lindhard function gives the same result as the linearized Thomas–Fermi theory; in fact for small q, we have $x \ll 1$ and $F(x) \approx 1$.

For *large q* ($q \gg 2k_F$), we have $x \gg 1$ and $F(x) \approx 1/3x^2$; the Lindhard function has the asymptotic behaviour

$$\varepsilon(q) \to 1 + \frac{k_{TF}^2}{q^2}\frac{1}{3(q/2k_F)^2} = 1 + \frac{4}{3}\frac{k_{TF}^2 k_F^2}{q^4} \qquad q \gg 2k_F \;.$$

A decrease of $\varepsilon(q) - 1$ as q^{-4} assures a well-behaved screening charge density at the origin (see Eq.23).

For *intermediate q* ($q \approx 2k_F$) the Lindhard dielectric function is continuous for $q = 2k_F$, but with a logarithmic singularity in the derivative. In fact Eq. (25a) for $q \approx 2k_F$ gives

$$\varepsilon(q) \approx \varepsilon(2k_F) + \frac{me^2}{4\pi\hbar^2}\frac{1}{k_F^2}(q - 2k_F)\ln\left|\frac{q - 2k_F}{4k_F}\right| \;.$$

For the derivative we have

$$\frac{d\varepsilon(q)}{dq} \approx \frac{me^2}{4\pi\hbar^2}\frac{1}{k_F^2}\ln\left|\frac{q - 2k_F}{4k_F}\right| \approx -\infty \qquad \text{for} \qquad q \approx 2k_F \;. \tag{26}$$

We can easily understand the origin of the discontinuity in the slope of $\varepsilon(q)$ for $q \approx 2k_F$ considering the energy denominators appearing in the sum over $\mathbf{k}$ in Eq. (24). For $q < 2k_F$ contributions with $E(\mathbf{k} + \mathbf{q}) \approx E(\mathbf{k})$ are possible, but this is no more possible for $q > 2k_F$ (as illustrated in Fig. 7).

The singularity (26) in reciprocal space generates oscillations of the screening charge

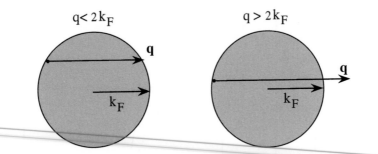

Fig. 7 Fermi sphere and excitations with wavevector transfer $q < 2k_F$ and $q > 2k_F$; $E(\mathbf{k} + \mathbf{q}) \approx E(\mathbf{k})$ is possible in the former case, but not in the latter.

in real space (see Fig. 8). To illustrate this effect, we consider here the screening charge induced by a point charge $+Ze$ inserted in the metal (at the origin). From Eq. (23) and performing the angular integration, we have

$$\rho_{\text{ind}}(\mathbf{r}) = Ze \frac{1}{(2\pi)^3} \int \left[\frac{1}{\varepsilon(q)} - 1 \right] e^{i\mathbf{q}\cdot\mathbf{r}} \, d\mathbf{q} = -Ze \frac{1}{r} \int_0^\infty g(q) \sin qr \, dq \,, \qquad (27a)$$

where the expression of $g(q)$ is

$$g(q) = \frac{q}{2\pi^2} \frac{\varepsilon(q) - 1}{\varepsilon(q)} \,. \qquad (27b)$$

From the asymptotic behaviour of $\varepsilon(q)$ for small and large wavenumbers, we have that $g(q)$ vanishes both for large and small values of q. The derivative $g'(q)$ vanishes for $q = \infty$; for $q \to 2k_F$, from Eq. (26) and Eq. (27b), we have

$$g'(q) \approx C \ln |q - 2k_F| \quad \text{and} \quad g''(q) \approx \frac{C}{q - 2k_F} \qquad \text{for } q \to 2k_F$$

(where C can be considered as constant).

We now perform two successive integrations by parts in Eq. (27a) and obtain

$$\rho_{\text{ind}}(r) = Ze \frac{1}{r^3} \int_0^\infty g''(q) \sin qr \, dq \,. \qquad (28a)$$

We look for the behaviour of $\rho_{\text{ind}}(r)$ at very large r. Because of the rapid oscillations of $\sin qr$ as $r \to \infty$, the most important contribution in the integration (28a) comes from a small region $(2k_F - \Delta < q < 2k_F + \Delta)$ around $2k_F$, where $g''(q)$ is singular. Keeping only the dominant term in Eq. (28a), we have

$$\rho_{\text{ind}}(r) \to Ze \frac{C}{r^3} \int_{2k_F - \Delta}^{2k_F + \Delta} \frac{1}{q - 2k_F} \sin qr \, dq \qquad r \to \infty \,. \qquad (28b)$$

We can write

$$\sin qr = \sin[(q - 2k_F)r] \cos 2k_F r + \cos[(q - 2k_F)r] \sin 2k_F r \,.$$

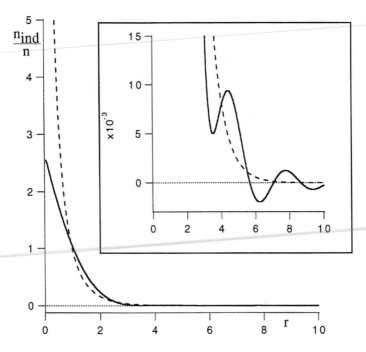

Fig. 8 Behaviour of the induced electron density $n_{\mathrm{ind}}(r)$, relative to the unperturbed density n, when a point charge $+e$ is inserted in a metal at the origin. The Thomas–Fermi approximation is indicated with a dashed line, the Lindhard approximation with a continuous line (r is in units of the Bohr radius a_{B}; $k_{\mathrm{F}} = 1\,a_{\mathrm{B}}^{-1}$; $r_s = 1.92$).

We insert the above expression into (28b) and omit the term that vanishes for parity considerations; we have

$$\rho_{\mathrm{ind}}(r) \to Ze\frac{C\cos 2k_{\mathrm{F}}r}{r^3}\int_{2k_{\mathrm{F}}-\Delta}^{2k_{\mathrm{F}}+\Delta}\frac{\sin(q-2k_{\mathrm{F}})r}{q-2k_{\mathrm{F}}}\,dq = Ze\frac{C\cos 2k_{\mathrm{F}}r}{r^3}\int_{-r\Delta}^{r\Delta}\frac{\sin x}{x}\,dx\ .$$

$$(28c)$$

The integration limits in Eq. (28c) can be extended from $-\infty$ to $+\infty$ (and the integral is π); we thus obtain the asymptotic expression

$$\rho_{\mathrm{ind}}(r) \to \pi Ze\frac{C\cos 2k_{\mathrm{F}}r}{r^3} \qquad r \to \infty\ .\qquad (28d)$$

The charge oscillations of $\rho_{\mathrm{ind}}(r)$, at large r, are known as Friedel oscillations. The singularity in the dielectric function is responsible also for the Kohn anomalies in the vibrational spectrum of metals for $q \approx 2k_{\mathrm{F}}$ [W. Kohn, Phys. Rev. Lett. **2**, 393 (1959)].

Friedel sum rule and Fumi theorem

We can now examine some features of the screening in metals, representing the metal as a *large spherical box of radius R*, in which the charge of positive ions is distributed uniformly and the conduction electrons can move freely. When no impurity is present at the origin of the large box, the free-electron wavefunctions are of the form $\psi_{klm}(\mathbf{r}) = Aj_l(kr)Y_{lm}(\mathbf{r})$, where j_l is the spherical Bessel function of order l ($l = 0, 1, 2, \dots$), $Y_{lm}(\mathbf{r})$ ($m = l, l-1, \dots, -l$) are spherical harmonics, and A is a normalization factor. For large values of the argument, the asymptotic behaviours of the Bessel functions are $j_l(x) \to (1/x)\sin(x - l\pi/2)$. We can thus write for the normalized wavefunctions within the sphere (of large radius R) the following expression

$$\psi_{klm}(\mathbf{r}) = \sqrt{\frac{2}{R}}\frac{1}{r}\sin(kr - \frac{1}{2}l\pi)Y_{lm}(\mathbf{r}) \qquad r \to \infty . \tag{29a}$$

The boundary condition that the unperturbed wavefunctions vanish at $r = R$ gives

$$kR - \frac{1}{2}l\pi = n\pi \qquad (n = 1, 2, \dots) . \tag{29b}$$

The number dn of allowed states between k and $k + dk$ is obtained by differentiating both members of Eq. (29b); we have $R\,dk = \pi\,dn$. Thus the (unperturbed) density-of-states (for any chosen value of l) is

$$D_0(k) = \frac{dn}{dk} = \frac{R}{\pi} . \tag{29c}$$

Consider now the stationary solutions of the Schrödinger equation in the presence of the total spherical potential $U(r)$ (born by the external charge Ze located at the origin, and the accompanying screening charge). The stationary solutions are now

$$\widetilde{\psi}_{klm}(\mathbf{r}) = \sqrt{\frac{2}{R}}\frac{1}{r}\sin\left[kr + \delta_l(k) - \frac{1}{2}l\pi\right]Y_{lm}(\mathbf{r}) \qquad r \to \infty , \tag{30a}$$

where $\delta_l(k)$ is the phase shift of the l wave due to the screened potential. The boundary condition that the perturbed wavefunctions vanish at $r = R$ gives

$$kR + \delta_l(k) - \frac{1}{2}l\pi = n\pi \qquad (n = 1, 2, \dots) . \tag{30b}$$

By differentiating both members of Eq. (30b), we obtain $R\,dk + \delta_l'(k)\,dk = \pi\,dn$. The perturbed density-of-states $D(k) = dn/dk$ is given by

$$D(k) = \frac{R}{\pi} + \frac{1}{\pi}\frac{d\delta_l}{dk} = D_0(k) + \frac{1}{\pi}\frac{d\delta_l}{dk} . \tag{30c}$$

From Eq. (30c), considering that the degeneracy of an l-orbital is $2(2l + 1)$ (spin degeneracy included), we have that the total number of displaced states up to the Fermi wavevector k_F is

$$\Delta n = \sum_l 2(2l + 1)\int_0^{k_F}\frac{1}{\pi}\frac{d\delta_l(k)}{dk}\,dk = \frac{2}{\pi}\sum_l(2l + 1)[\delta_l(k_F) - \delta_l(0)] . \tag{31}$$

If the number of bound states of $U(r)$ is zero, then it is also zero the sum involving the phase shifts $\delta_l(0)$ (Levinson theorem). From the physical requirement of perfect shielding, we have

$$\boxed{\frac{2}{\pi}\sum_l(2l + 1)\delta_l(k_F) \equiv Z} . \tag{32}$$

Relation (32) is the Friedel sum rule [J. Friedel, Phil. Mag. **43**, 153 (1952); Adv. Phys. **3**, 446 (1954)]; it expresses the fact that the total displaced electron charge is equal in magnitude, and opposite in sign, to the external point charge inserted in the metal.

From Eq. (30a) we notice that the electron wavefunctions undergo phase shifts in order to screen the impurity potential; this fact entails also a *change of the kinetic energy of the free electrons* [F. G. Fumi, Phil. Mag. **46**, 1007 (1955)]. From Eq. (30b) and Eq. (29b), we see that the change Δk of the wavenumber of an l electron is $\Delta k \cdot R = -\delta_l(k)$, and the change in energy is $\hbar^2 k \Delta k/m$. Thus the energy change of the free electrons because of the presence of the impurity is

$$\Delta E_{\text{el}} = -\sum_l 2(2l+1) \int_0^{k_{\text{F}}} \frac{\hbar^2 k \, \delta_l(k)}{mR} \frac{R}{\pi} \, dk \;, \tag{33}$$

where R/π is the density-of-states for l electrons. The Friedel and the Fumi theorems have been of major importance for understanding the electronic structure of impurities in metals, and remain useful guidelines also after the advent of the first-principle density functional treatments.

6 Dynamic dielectric screening in metals and plasmon modes

The quantum expression of the dynamical dielectric function $\varepsilon(\mathbf{q}, \omega)$ of the free-electron gas can be obtained inserting plane waves into the general equation (60) for the dielectric function of crystals; one finds

$$\varepsilon(\mathbf{q}, \omega) = 1 + \frac{8\pi e^2}{q^2} \frac{1}{V} \sum_{\mathbf{k}} \frac{f(\mathbf{k}) - f(\mathbf{k}+\mathbf{q})}{E(\mathbf{k}+\mathbf{q}) - E(\mathbf{k}) - \hbar\omega - i\eta} \;. \tag{34}$$

In the previous section, we have discussed the most significant aspects embodied in the static expression $\varepsilon(\mathbf{q}, 0)$. We consider now the most significant dynamical aspects, and especially the behaviour of $\varepsilon(\mathbf{q}, \omega)$ in the long wavelength limit $\mathbf{q} \approx 0$.

It is useful to re-write expression (34) in a slightly different form, by performing the replacement $\mathbf{k}+\mathbf{q} \rightarrow -\mathbf{k}'$ in the term containing $f(\mathbf{k}+\mathbf{q})$ (then $\mathbf{k}'$ is relabelled as $\mathbf{k}$); we obtain

$$\varepsilon(\mathbf{q}, \omega) = 1 + \frac{8\pi e^2}{q^2} \frac{1}{V} \sum_{\mathbf{k}} \left[\frac{f(\mathbf{k})}{E(\mathbf{k}+\mathbf{q}) - E(\mathbf{k}) - \hbar\omega - i\eta} + \frac{f(\mathbf{k})}{E(\mathbf{k}+\mathbf{q}) - E(\mathbf{k}) + \hbar\omega + i\eta} \right] \;. \tag{35}$$

Performing the sum in square brackets, we have

$$\varepsilon(\mathbf{q}, \omega) = 1 + \frac{16\pi e^2}{q^2} \frac{1}{V} \sum_{\mathbf{k}} f(\mathbf{k}) \frac{E(\mathbf{k}+\mathbf{q}) - E(\mathbf{k})}{[E(\mathbf{k}+\mathbf{q}) - E(\mathbf{k})]^2 - (\hbar\omega + i\eta)^2} \;. \tag{36}$$

Expressions (35) and (36) are particularly convenient for analytic investigations; in fact the sum over $\mathbf{k}$ is just confined to the Fermi sphere, centered at the origin of the reciprocal space.

We begin to evaluate the dynamic dielectric function in the long wavelength limit. For $\mathbf{q} \to 0$, the energy difference $E(\mathbf{k}+\mathbf{q}) - E(\mathbf{k}) = (\hbar^2/2m)(2\mathbf{k}\cdot\mathbf{q}+q^2) \to 0$; we can perform a series development in the terms in Eq. (36) and we obtain (omitting contributions that vanish or are of higher order in powers of q):

$$\varepsilon(\mathbf{q} \to 0, \omega)$$

$$= 1 - \frac{16\pi e^2}{q^2}\frac{1}{V}\sum_{\mathbf{k}} f(\mathbf{k})\left[\frac{E(\mathbf{k}+\mathbf{q})-E(\mathbf{k})}{(\hbar\omega+i\eta)^2} + \frac{[E(\mathbf{k}+\mathbf{q})-E(\mathbf{k})]^3}{(\hbar\omega+i\eta)^4} + \cdots\right]$$

$$= 1 - \frac{16\pi e^2}{q^2}\frac{1}{V}\sum_{\mathbf{k}} f(\mathbf{k})\left[\frac{1}{(\hbar\omega+i\eta)^2}\frac{\hbar^2 q^2}{2m} + \frac{1}{(\hbar\omega+i\eta)^4}\frac{\hbar^6}{8m^3}12(\mathbf{k}\cdot\mathbf{q})^2 q^2 + O(q^6)\right]$$

$$= 1 - \frac{8\pi e^2}{m(\omega+i\eta/\hbar)^2}\frac{1}{V}\sum_{\mathbf{k}} f(\mathbf{k}) - \frac{8\pi e^2}{m(\omega+i\eta/\hbar)^4}q^2\frac{1}{V}\sum_{\mathbf{k}} f(\mathbf{k})(\frac{\hbar k}{m})^2 - \cdots \qquad (37a)$$

We notice that

$$\frac{1}{V}\sum_{\mathbf{k}} f(\mathbf{k}) = \frac{n}{2} \qquad \text{and} \qquad \frac{1}{V}\sum_{\mathbf{k}}(\frac{\hbar k}{m})^2 f(\mathbf{k}) = \frac{3}{5}v_{\mathrm{F}}^2\frac{n}{2}$$

(where $v_{\mathrm{F}} = \hbar k_{\mathrm{F}}/m$). Using the above expressions, Eq. (37a) becomes

$$\varepsilon(q \to 0, \omega) = 1 - \frac{\omega_{\mathrm{p}}^2}{(\omega+i\eta/\hbar)^2} - \frac{3}{5}\frac{\omega_{\mathrm{p}}^2}{(\omega+i\eta/\hbar)^4}v_{\mathrm{F}}^2 q^2 - \cdots , \qquad (37b)$$

where ω_{p} is the standard expression of the plasma frequency $\omega_{\mathrm{p}}^2 = 4\pi n e^2/m$.

From Eq. (37b), we see that the dynamic dielectric function for $q \equiv 0$ becomes

$$\boxed{\varepsilon(0, \omega) = 1 - \frac{\omega_{\mathrm{p}}^2}{\omega^2 + i\omega/\tau}}, \qquad (38)$$

where we have done the replacement $2\eta/\hbar \to 1/\tau$, and τ can be interpreted as a relaxation time characteristic of the material.

Equation (38) can be recognized as the Drude model for the dielectric function of a free-electron metal; the Drude model will be discussed in detail in Section XI-2, in connection with the intraband optical properties of metals and the current density response function to transverse electromagnetic fields (in the long wavelength limit, transverse and longitudinal dielectric functions of homogeneous media are numerically equal).

The knowledge of the longitudinal dielectric function $\varepsilon = \varepsilon_1 + i\varepsilon_2$ of a material allows to calculate the energy-loss function $-\mathrm{Im}\,1/\varepsilon = \varepsilon_2/(\varepsilon_1^2 + \varepsilon_2^2)$ and to infer important information on plasmon excitations [the function $-\mathrm{Im}\,1/\varepsilon$ determines the characteristic energy losses suffered by fast charged particles traversing the material, as shown in Section 9]. The reciprocal of the Drude formula (38) gives the so-called "inverted" Drude–Sellmeier formula

$$\frac{1}{\varepsilon(0, \omega)} = 1 - \frac{\omega_{\mathrm{p}}^2}{\omega_{\mathrm{p}}^2 - \omega^2 - i\omega/\tau} ;$$

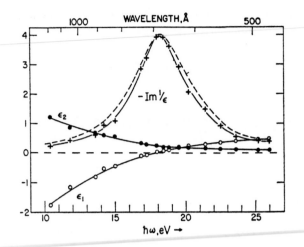

Fig. 9 Dielectric function $\varepsilon_1(0,\omega)$, $\varepsilon_2(0,\omega)$ and $-\mathrm{Im}\,1/\varepsilon(0,\omega)$ for beryllium, obtained from optical measurements. The dashed curve for $-\mathrm{Im}1/\varepsilon(0,\omega)$ is elaborated from electron-scattering experiments [from J. Toots, H. A. Fowler and L. Marton, Phys. Rev. **172**, 670 (1968); copyright 1968 by the American Physical Society].

from the above expression we obtain

$$-\mathrm{Im}\frac{1}{\varepsilon(0,\omega)} = \frac{\pi}{\tau}\frac{\omega_{\mathrm{p}}^2}{(\omega^2 - \omega_{\mathrm{p}}^2)^2 + \omega^2/\tau^2}\,. \tag{39}$$

The energy-loss function (39) has its maximum around $\hbar\omega \approx \hbar\omega_{\mathrm{p}}$ and has approximately a Lorenzian shape of half-width $\Delta = \hbar/\tau$.

In actual metals, the $-\mathrm{Im}\,1/\varepsilon$ plasma resonances have typical widths Δ in the range $[0.05\text{–}0.25]\hbar\omega_{\mathrm{p}}$; narrow-line resonances ($\Delta/\hbar\omega_{\mathrm{p}} \approx 0.05$) occur, for instance, in aluminum, magnesium and tin, while beryllium presents a rather broad-line resonance ($\Delta/\hbar\omega_{\mathrm{p}} \approx 0.26$). As an example, we report in Fig. 9 observed values of $\varepsilon_1(0,\omega)$, $\varepsilon_2(0,\omega)$ and $-\mathrm{Im}1/\varepsilon(0,\omega)$ in beryllium around the plasma resonance. The energy-loss function has its maximum at $\hbar\omega_{\mathrm{p}}{=}18.4\,\mathrm{eV}$, a value corresponding to the plasma frequency of an ideal free-electron gas with 2.0 electrons per atom; the half-width Δ is measured as $\Delta{=}4.7\,\mathrm{eV}$, corresponding to a lifetime $\tau{=}\hbar/\Delta{=}1.4{\times}10^{-16}$ sec. Such a short lifetime of plasmons may entail several decay channels (background of interband single-particle transitions, collisional effects, or other mechanisms) and we refer to the literature for further analysis [see for instance N. Swanson, J. Opt. Soc. Am. **54**, 1130 (1964); H. A. Fowler and J. J. Filliben, J. Appl. Phys. **52**, 6701 (1981) and references quoted therein].

From the dynamical dielectric function of a medium, one can obtain the dispersion curve of plasmons from the locus of points in $(\mathbf{q},\omega)$ space where $\varepsilon(\mathbf{q},\omega) \equiv 0$. In the limit of $\mathbf{q} \to 0$ and $\hbar\omega > E(\mathbf{k}+\mathbf{q})-E(\mathbf{k})$, we have $\varepsilon_2(\mathbf{q},\omega) = 0$; the zeroes of Eq. (37b) occur for

$$1 - \frac{\omega_{\mathrm{p}}^2}{\omega^2} - \frac{3}{5}\frac{\omega_{\mathrm{p}}^2}{\omega^4}v_{\mathrm{F}}^2 q^2 = 0\,. \tag{40a}$$

The solution of equation (40a) provides the dispersion curve of plasmon modes for small q wavevectors

$$\omega_{\text{plasmon}}(q) = \omega_{\text{p}}\left[1 + \frac{3}{10}\frac{v_{\text{F}}^2 q^2}{\omega_{\text{p}}^2} + \cdots\right]. \tag{40b}$$

Notice that the dispersion curve (40b) is rather flat (and this may justify simplified models that neglect dispersion altogether).

It is instructive to consider the occurrence of plasma oscillations looking at the zeroes of expression (36) with a graphic approach. For this purpose, we re-write Eq. (36) in the equivalent form (obtained by adding a part which vanishes identically)

$$\varepsilon(\mathbf{q},\omega) = 1 + \frac{16\pi e^2}{q^2}\frac{1}{V}\sum_{\mathbf{k}} f(\mathbf{k})[1 - f(\mathbf{k}+\mathbf{q})]\frac{E(\mathbf{k}+\mathbf{q}) - E(\mathbf{k})}{[E(\mathbf{k}+\mathbf{q}) - E(\mathbf{k})]^2 - (\hbar\omega + i\eta)^2}. \tag{41a}$$

Expression (41a) makes transparent the general structure of $\varepsilon(\mathbf{q},\omega)$ as a function of ω. We notice that $f(\mathbf{k})[1 - f(\mathbf{k}+\mathbf{q})]$ is different from zero (at $T = 0$) only between occupied states $\mathbf{k}$ below the Fermi level and empty states $\mathbf{k}+\mathbf{q}$ above the Fermi level; then $E(\mathbf{k}+\mathbf{q}) - E(\mathbf{k}) > 0$ and Eq. (41a) takes the general form

$$\varepsilon(\mathbf{q},\omega) = 1 + \sum_n \frac{C_n}{\omega_n^2 - (\omega + i\eta)^2} \tag{41b}$$

where ω_n are the *excitation frequencies of the system in the independent particle approximation* and C_n are *definite-positive quantities, that can be interpreted as the oscillator strengths of the resonant frequencies* ω_n.

For simplicity, and just for qualitative remarks, we consider a system with a finite (although very large) number N of resonant frequencies ω_n in the interval $[\omega_1, \omega_N]$. The real part $\varepsilon_1(\mathbf{q},\omega)$ of Eq. (41b) (in the limit $\eta \to 0^+$) is schematically indicated in Fig. 10. The zeroes of $\varepsilon_1(\mathbf{q},\omega)$ occur in correspondence to the (nearly continuum) of single particle excitations. In addition, there is one discrete root; this collective state represents electron–hole charge fluctuations correlated by the long-range Coulomb interactions.

Before closing, we should notice that the general expression (34) or (35) of the dielectric function for any given frequency ω and wavevector q can be evaluated in closed analytic form (as shown in Appendix A). The availability of closed analytic results is useful to establish guidelines or extrapolations in the study of realistic materials, where laborious numerical calculations become unavoidable. The Lindhard function has been also modified for simple (but still significant) descriptions of the dielectric screening in model semiconductors [see for instance E. Tosatti and G. Pastori Parravicini, J. Phys. Chem. Solids **32**, 623 (1971); Z. H. Levine and S. G. Louie, Phys Rev. B**25**, 6310 (1982) and references quoted therein].

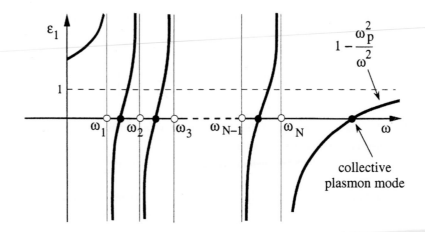

Fig. 10 Qualitative behaviour of the real part of the dielectric function of a system with proper frequencies ω_n (indicated by white circles) in the interval $[\omega_1, \omega_N]$. Since $\varepsilon_1(\mathbf{q}, \omega)$ is an even function of ω, only the part $\omega > 0$ is indicated; the zeroes of $\varepsilon_1(\mathbf{q}, \omega)$ are indicated by black circles; notice the split-off collective state.

7 Quantum expression of the longitudinal dielectric function in materials

In this section we provide the *quantum expression of the longitudinal dielectric function for homogeneous (or nearly homogeneous) materials, within the linear response theory.* We consider an arbitrary (periodic or aperiodic) electron system of volume V, and suppose that it can be described by the one-electron Hamiltonian

$$H_0 = -\frac{\hbar^2 \nabla^2}{2m} + V(\mathbf{r}) . \tag{42}$$

We indicate by $\{\psi_\alpha(\mathbf{r})\}$ and E_α the wavefunctions and the eigenvalues of H_0 (the wavefunctions are normalized to 1 within the volume V of the sample). The occupancy of the states is determined by the Fermi–Dirac function $f(E_\alpha)$ (ordinarily considered at zero temperature). Spin degeneracy is taken into account by appropriate inclusion of a factor two, where required. The general results here obtained are later specified in the case of periodic systems, where band wavefunctions are of Bloch type.

We examine the response of the electron system to a time-dependent perturbation of wavevector $\mathbf{q}$ and frequency ω, of the type

$$U(\mathbf{r}, t) = A_0 e^{i(\mathbf{q} \cdot \mathbf{r} - \omega t)} + c.c. , \tag{43}$$

where $A_0 = A_{\text{tot}}(\mathbf{q}, \omega)$ is the (infinitesimal) amplitude of the driving perturbation, and c.c. indicates the complex conjugate of the previous term. The electrostatic potential corresponding to the perturbation energy term (43) is $\phi(\mathbf{r}, t) = U(\mathbf{r}, t)/(-e)$; the electric field $\mathbf{E} = -\nabla \phi$ is a longitudinal field parallel to $\mathbf{q}$ and given by

$$\mathbf{E}(\mathbf{r}, t) = E_0 \, \mathbf{e} \, e^{i(\mathbf{q} \cdot \mathbf{r} - \omega t)} + c.c. \tag{44}$$

with $E_0 = iqA_0/e$; the versor of the wavevector $\mathbf{q}$, which coincides with the versor of the electric field $\mathbf{E}(\mathbf{r}, t)$, has been denoted by $\mathbf{e}$.

In the presence of the perturbation (43), the Hamiltonian of the system becomes

$$H = H_0 + A_0 e^{i(\mathbf{q}\cdot\mathbf{r}-\omega t)} + A_0^* e^{-i(\mathbf{q}\cdot\mathbf{r}-\omega t)} . \qquad (45)$$

The two time-dependent terms in Eq. (45) induce transitions among the states of H_0, with absorption and emission of energy $\hbar\omega$, respectively. With the standard Fermi golden rule, it is easily seen that the net number of transitions per unit time involving absorption or emission processes of energy $\hbar\omega$ is given by the expression

$$W(\mathbf{q}, \omega) = \frac{2\pi}{\hbar} 2 \sum_{\alpha\beta} \left| \langle \psi_\beta | A_0 e^{i\mathbf{q}\cdot\mathbf{r}} | \psi_\alpha \rangle \right|^2 \delta(E_\beta - E_\alpha - \hbar\omega) \left[f(E_\alpha) - f(E_\beta) \right] , \qquad (46)$$

where the factor 2 in front of the summation takes into account the spin degeneracy. The forthcoming elaboration of Eq. (46) is rather similar to the treatment of Eq. (XII-6), to which we refer for complementary details.

The energy per unit time, i.e. the power, dissipated in the system of volume V is given by

$$P(\mathbf{q}, \omega) = \hbar\omega W(\mathbf{q}, \omega)$$

$$= 4\pi\omega |A_0|^2 \sum_{\alpha\beta} \left| \langle \psi_\beta | e^{i\mathbf{q}\cdot\mathbf{r}} | \psi_\alpha \rangle \right|^2 \delta(E_\beta - E_\alpha - \hbar\omega) \left[f(E_\alpha) - f(E_\beta) \right] ; \qquad (47)$$

since $f(E_\alpha) > f(E_\beta)$ for any $E_\alpha < E_\beta$, it is evident that $P(\mathbf{q}, \omega) \geq 0$ for any value (positive or negative) of the frequency ω.

In an isotropic medium, in the presence of the applied electric field (44) of wavevector $\mathbf{q}$ and frequency ω, it is reasonable to expect an induced current density parallel to the electric field, proportional to it for small field strength, and with the same time and space dependence; we can thus write

$$\mathbf{J}(\mathbf{r}, t) = \sigma(\mathbf{q}, \omega) E_0 \, \mathbf{e} \, e^{i(\mathbf{q}\cdot\mathbf{r}-\omega t)} + c.c. \qquad (\mathbf{e} \parallel \mathbf{q}) ,$$

where $\sigma(\mathbf{q}, \omega)$ defines the *longitudinal conductivity* function. When a current density $\mathbf{J}$ flows in a medium in the presence of an electric field $\mathbf{E}$, the energy per unit time dissipated in the system of volume V is

$$\int_V \mathbf{J} \cdot \mathbf{E} \, d\mathbf{r} = \int_V \left[\sigma(\mathbf{q}, \omega) E_0 e^{i(\mathbf{q}\cdot\mathbf{r}-\omega t)} + c.c. \right] \left[E_0 e^{i(\mathbf{q}\cdot\mathbf{r}-\omega t)} + c.c. \right] d\mathbf{r}$$

$$= \sigma(\mathbf{q}, \omega) |E_0|^2 V + c.c. = 2\sigma_1(\mathbf{q}, \omega) \frac{q^2}{e^2} |A_0|^2 V .$$

This classical expression of the power dissipated in the system, can be identified with the quantum expression $\hbar\omega W(\mathbf{q}, \omega)$; we obtain

$$\sigma_1(\mathbf{q}, \omega) = \frac{1}{2} \frac{e^2}{q^2} \frac{\hbar\omega W(\mathbf{q}, \omega)}{|A_0|^2 V} , \qquad (48)$$

which constitutes the basic relationship linking the real part of the conductivity to the microscopic relation (46).

It is well known that the dielectric function is related to the conductivity function by the expression $\varepsilon = 1 + 4\pi i\sigma/\omega$; this relationship follows from the requirement $\partial\mathbf{D}/\partial t = \partial\mathbf{E}/\partial t + 4\pi\mathbf{J}$, with $\mathbf{D} = \varepsilon\mathbf{E}$ and $\mathbf{J} = \sigma\mathbf{E}$, and the form $\exp(-i\omega t)$ for the time dependence of the fields. We thus have

$$\varepsilon_2(\mathbf{q}, \omega) = \frac{4\pi}{\omega}\sigma_1(\mathbf{q}, \omega) = \frac{2\pi\hbar e^2}{q^2}\frac{1}{V}\frac{W(\mathbf{q}, \omega)}{|A_0|^2} \ .$$

Using Eq. (46), the quantum expression of the imaginary part of the longitudinal dielectric function becomes

$$\varepsilon_2(\mathbf{q}, \omega) = \frac{8\pi^2 e^2}{q^2 V}\sum_{\alpha\beta}\left|\langle\psi_\beta|e^{i\mathbf{q}\cdot\mathbf{r}}|\psi_\alpha\rangle\right|^2 \delta(E_\beta - E_\alpha - \hbar\omega)\left[f(E_\alpha) - f(E_\beta)\right] \ . \qquad (49a)$$

The real part of the dielectric function is obtained inserting Eq. (49a) into the Kramers–Kronig relation

$$\varepsilon_1(\mathbf{q}, \omega) = 1 + \frac{1}{\pi}P\int_{-\infty}^{+\infty}\frac{\varepsilon_2(\mathbf{q}, \omega)}{\omega' - \omega}\,d\omega' \ ;$$

we have

$$\varepsilon_1(\mathbf{q}, \omega) = 1 + \frac{8\pi e^2}{q^2}\frac{1}{V}\sum_{\alpha\beta}\left|\langle\psi_\beta|e^{i\mathbf{q}\cdot\mathbf{r}}|\psi_\alpha\rangle\right|^2\frac{f(E_\alpha) - f(E_\beta)}{E_\beta - E_\alpha - \hbar\omega} \qquad (49b)$$

(if discretized sums are replaced by integrals, the principal part of integrals is to be considered). Equations (49) give the longitudinal dielectric function of a generic system, with eigenfunctions $|\psi_\alpha\rangle$ and eigenvalues E_α, under the assumption that local field effects can be disregarded.

It is immediately seen that Eqs. (49) can be re-written in the compact form

$$\varepsilon(\mathbf{q}, \omega) = 1 + \frac{8\pi e^2}{q^2}\frac{1}{V}\sum_{\alpha\beta}\frac{|\langle\psi_\beta|e^{i\mathbf{q}\cdot\mathbf{r}}|\psi_\alpha\rangle|^2}{E_\beta - E_\alpha - \hbar\omega - i\eta}\left[f(E_\alpha) - f(E_\beta)\right] \ . \qquad (50)$$

The explicit separation of the real and imaginary part of the dielectric function in Eq. (50), and the equivalence with Eq. (49), can be verified either by inspection, or using the formal relation (or its complex conjugate)

$$\lim_{\eta\to 0^+}\frac{1}{x - i\eta} = P\frac{1}{x} + i\pi\delta(x) \ , \qquad (51)$$

where x is real, and P denotes the principal part (i.e. any integration involving the product of $1/x$ by a function of x must be intended in principal part). The relation (51) follows from the algebraic equality

$$\frac{1}{x - i\eta} = \frac{x}{x^2 + \eta^2} + i\pi\frac{1}{\pi}\frac{\eta}{x^2 + \eta^2} \ ,$$

keeping the limit $\eta \to 0^+$.

It is easily seen that Eqs. (49) can be recast in the form

$$\varepsilon_2(\mathbf{q}, \omega) = \frac{8\pi^2 e^2}{q^2} \frac{1}{V} \sum_{\alpha\beta} f(E_\alpha) |\langle \psi_\beta | e^{i\mathbf{q}\cdot\mathbf{r}} | \psi_\alpha \rangle|^2 \left[\delta(E_\beta - E_\alpha - \hbar\omega) - \delta(E_\beta - E_\alpha + \hbar\omega) \right]$$

(52a)

and

$$\varepsilon_1(\mathbf{q}, \omega) = 1 + \frac{16\pi e^2}{q^2} \frac{1}{V} \sum_{\alpha\beta} f(E_\alpha) \frac{|\langle \psi_\beta | e^{i\mathbf{q}\cdot\mathbf{r}} | \psi_\alpha \rangle|^2 (E_\beta - E_\alpha)}{(E_\beta - E_\alpha)^2 - \hbar^2\omega^2} \; ;$$

(52b)

the elaboration from Eqs. (49) to Eqs. (52) can be verified by inspection, choosing the wavefunctions $\psi_\alpha(\mathbf{r})$ to be real valued [this choice is possible because the Hamiltonian operator (42) is invariant under time-reversal symmetry; this entails that both ψ_α and ψ_α^*, whether linear independent or not, correspond to the same eigenvalue E_α].

From Eq. (47) and Eq. (49a), we see that the power dissipated in the medium, due to the driving perturbation (43) or (17), can be written as

$$P(\mathbf{q}, \omega) = V \frac{\omega q^2}{2\pi e^2} \varepsilon_2(\mathbf{q}, \omega) |A_{\text{tot}}(\mathbf{q}, \omega)|^2 \; ,$$

(53a)

where $A_{\text{tot}} = A_0$ is the amplitude of the driving perturbation. Equation (53a) shows that the power dissipated in the medium, due to a driving perturbation of wavevector $\mathbf{q}$, frequency ω, and amplitude $A_{\text{tot}}(\mathbf{q}, \omega)$, is essentially determined by the *imaginary part of the dielectric function* $\varepsilon_2(\mathbf{q}, \omega)$.

Using the relation (18), Eq. (53a) can also be written in the equivalent form

$$P(\mathbf{q}, \omega) = V \frac{\omega q^2}{2\pi e^2} \frac{\varepsilon_2(\mathbf{q}, \omega)}{|\varepsilon(\mathbf{q}, \omega)|^2} |A_{\text{ext}}(\mathbf{q}, \omega)|^2 \; .$$

(53b)

Equation (53b) shows that the power dissipated in the medium, due to an external perturbation of wavevector $\mathbf{q}$, frequency ω, and amplitude $A_{\text{ext}}(\mathbf{q}, \omega)$, is essentially determined by the so-called *energy-loss function*, defined as

$$-\operatorname{Im} \frac{1}{\varepsilon(\mathbf{q}, \omega)} = \frac{\varepsilon_2(\mathbf{q}, \omega)}{\varepsilon_1^2(\mathbf{q}, \omega) + \varepsilon_2^2(\mathbf{q}, \omega)} \; .$$

(54)

Sum rules for the dielectric function

We can now discuss other general properties of the dielectric function with the following argument. The Hamiltonian operator of the form (42) satisfies the general commutation relation

$$[H_0, e^{i\mathbf{q}\cdot\mathbf{r}}] = e^{i\mathbf{q}\cdot\mathbf{r}} \left(\frac{\hbar^2 q^2}{2m} - \frac{\hbar^2}{m} i\mathbf{q} \cdot \nabla \right) .$$

For the double commutator we have

$$\left[e^{-i\mathbf{q}\cdot\mathbf{r}}, \left[H_0, e^{i\mathbf{q}\cdot\mathbf{r}} \right] \right] = \frac{\hbar^2 q^2}{m} \; .$$

Taking the expectation value of the double commutator on any chosen state ψ_α, and inserting appropriately the unit operator $1 = \sum |\psi_\beta\rangle\langle\psi_\beta|$ (and also assuming that the

Hamiltonian is invariant for time reversal symmetry) we obtain

$$\sum_{\beta} |\langle \psi_\beta | e^{i\mathbf{q}\cdot\mathbf{r}} | \psi_\alpha \rangle|^2 (E_\beta - E_\alpha) = \frac{\hbar^2 q^2}{2m} . \tag{55}$$

We sum up the above relation over the occupied orbitals ψ_α, and divide by the volume V of the system; we obtain the useful relation

$$\frac{1}{V} \sum_{\alpha\beta} f(E_\alpha) |\langle \psi_\beta | e^{i\mathbf{q}\cdot\mathbf{r}} | \psi_\alpha \rangle|^2 (E_\beta - E_\alpha) = \frac{\hbar^2 q^2}{2m} \frac{n}{2} , \tag{56}$$

where n is the number of electrons of both spin directions per unit volume, and $n/2$ is the number of occupied orbitals per unit volume.

The exact result (56) allows us to establish some general and significant properties of the real and imaginary part of the dielectric function. From Eq. (52a) and Eq. (56), we obtain the sum rule

$$\int_0^\infty \omega\, \varepsilon_2(\mathbf{q}, \omega)\, d\omega = \frac{\pi}{2}\omega_p^2 , \tag{57a}$$

where ω_p is the plasma frequency $\omega_p^2 = 4\pi n e^2 / m$. From Eq. (52b) and Eq. (56), in the limit of $\hbar\omega$ much larger than any resonant frequency of the medium, we have the exact asymptotic behaviour for the real part of the dielectric function

$$\varepsilon_1(\mathbf{q}, \omega) \to 1 - \frac{\omega_p^2}{\omega^2} \qquad \text{for } \omega \to \infty ; \tag{57b}$$

this relation shows that in the high frequency limit the response of any medium is essentially free-electron like.

Dielectric function in the long wavelength limit

Before closing, we will add a comment on the dielectric function in the long wavelength limit. In the limit $\mathbf{q} \to 0$, the matrix elements appearing in Eq. (50) can be written in the form

$$\langle \psi_\beta | e^{i\mathbf{q}\cdot\mathbf{r}} | \psi_\alpha \rangle = i\mathbf{q} \cdot \langle \psi_\beta | \mathbf{r} | \psi_\alpha \rangle = \frac{\hbar}{m} \frac{i\mathbf{q} \cdot \langle \psi_\beta | \mathbf{p} | \psi_\alpha \rangle}{E_\beta - E_\alpha} \tag{58}$$

(with $E_\alpha \neq E_\beta$). The first passage in Eq. (58) is obtained with the series expansion of the exponential to first order in q; the substitution of the matrix elements of $\mathbf{r}$ with matrix elements of $\mathbf{p}$ is done exploiting the commutation rule $[H_0, \mathbf{r}] = -i(\hbar/m)\mathbf{p}$ of the Hamiltonian operator (42).

Let us indicate by $\mathbf{e}$ the versor in the direction of $\mathbf{q}$ of the applied perturbing longitudinal field. Using Eq. (58), we can recast Eq. (50) into the form

$$\varepsilon(0, \omega) = 1 + \frac{8\pi e^2}{m^2} \frac{1}{V} \sum_{\alpha\beta} \frac{|\langle \psi_\beta | \mathbf{e}\cdot\mathbf{p} | \psi_\alpha \rangle|^2}{[(E_\beta - E_\alpha)/\hbar]^2} \frac{f(E_\alpha) - f(E_\beta)}{E_\beta - E_\alpha - \hbar\omega - i\eta} . \tag{59}$$

In the long wavelength limit, the longitudinal dielectric function given by Eq. (59) obviously coincides with the transverse dielectric function given by Eq. (XII-11); the transverse dielectric function will be thorough discussed in Chapter XII in the framework of the optical properties of materials.

8 Quantum expression of the longitudinal dielectric function in crystals

We specify the general results of the previous section to periodic structures, where the electronic wavefunctions are Bloch functions. We indicate by $\psi_{mk}(\mathbf{r})$ the band wavefunctions (normalized to one in the volume V of the crystal) and by E_{mk} the corresponding band energies; the occupancy of the states is determined by the Fermi–Dirac function $f(E_{mk})$ (ordinarily considered at zero temperature). Spin degeneracy is included with appropriate insertion of a factor two, where required.

The general expression of the dielectric function $\varepsilon(\mathbf{q}, \omega)$ in terms of the wavevector $\mathbf{q}$ and frequency ω can be obtained from Eq. (50), considering the non-vanishing matrix elements of the perturbing wave $\exp(i\mathbf{q} \cdot \mathbf{r})$ between Bloch functions. We have

$$\varepsilon(\mathbf{q}, \omega) = 1 + \frac{8\pi e^2}{q^2} \frac{1}{V} \sum_{mnk} \frac{|\langle \psi_{nk+q}|e^{i\mathbf{q} \cdot \mathbf{r}}|\psi_{mk}\rangle|^2 [f(E_{mk}) - f(E_{nk+q})]}{E_{nk+q} - E_{mk} - \hbar\omega - i\eta} . \tag{60}$$

Expression (60) holds for any type of crystal (metal, semiconductor or insulator); from it, the Lindhard expression (34) is recovered, in the case Bloch functions are simply plane waves and band energies are free-electron like.

In the case of semiconductors and insulators (at $T = 0$) we do not need the explicit presence of the Fermi–Dirac distribution function. For dielectrics, from Eq. (60) or equivalently from Eqs. (52), we obtain for the imaginary part of the dielectric function

$$\varepsilon_2(\mathbf{q}, \omega) = \frac{8\pi^2 e^2}{q^2} \frac{1}{V} \sum_{vck} |\langle \psi_{ck+q}|e^{i\mathbf{q} \cdot \mathbf{r}}|\psi_{vk}\rangle|^2 \delta(E_{ck+q} - E_{vk} - \hbar\omega) \tag{61a}$$

(for positive frequencies), while for the real part we have

$$\varepsilon_1(\mathbf{q}, \omega) = 1 + \frac{16\pi e^2}{q^2} \frac{1}{V} \sum_{vck} \frac{|\langle \psi_{ck+q}|e^{i\mathbf{q} \cdot \mathbf{r}}|\psi_{vk}\rangle|^2 (E_{ck+q} - E_{vk})}{(E_{ck+q} - E_{vk})^2 - \hbar^2\omega^2} , \tag{61b}$$

where v labels the valence bands (fully occupied) and c labels the conduction bands (fully empty).

In the case of a semiconductor or insulator, we can obtain a (crude) model of $\varepsilon(\mathbf{q} \approx 0, \omega)$ by replacing the energy differences $(E_{ck+q} - E_{vk})$ in Eqs. (61) with some finite average excitation energy $E_{av} = \hbar\omega_{av}$; then exploiting the sum rules and the asymptotic behaviour of Eqs. (57), we obtain

$$\varepsilon_1(0, \omega) = 1 + \frac{\omega_p^2}{\omega_{av}^2 - \omega^2} , \qquad \varepsilon_2(0, \omega) = \frac{\pi}{2} \frac{\omega_p^2}{\omega_{av}} \delta(\omega - \omega_{av}) \tag{62}$$

for $\omega > 0$. The dielectric function (62) can be written in the compact form

$$\varepsilon(0, \omega) = 1 + \frac{\omega_p^2}{\omega_{av}^2 - (\omega + i\eta)^2} \qquad \text{with} \quad \eta \to 0^+ , \tag{63}$$

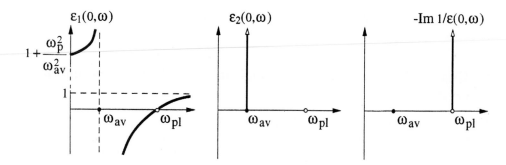

Fig. 11 Schematic behaviour of the Lorentz dielectric function $\varepsilon(0,\omega) = 1 + \omega_p^2/(\omega_{av}^2 - \omega^2 - i\eta)$ (with $\eta \to 0^+$ and $\omega > 0$) and the energy-loss function $-\text{Im}\, 1/\varepsilon(0,\omega)$.

which can be recognized as the Lorentz dielectric function of bound charged carriers of frequency ω_{av}.

The model dielectric function (62) is reported for convenience in Fig. 11. Notice that $\varepsilon_1(0,\omega)$ vanishes for $\omega_{pl} = (\omega_p^2 + \omega_{av}^2)^{1/2}$; there is thus a blue-shift of the plasma frequency ω_{pl} in semiconductors and (especially) in insulators due to the average energy gap, a fact confirmed by more rigorous treatments. The model (62) also predicts a finite static dielectric constant $\varepsilon_s = 1 + \omega_p^2/\omega_{av}^2$ (often used to estimate semi-empirically ω_{av}).

The reciprocal of the Lorentz formula (63) gives

$$\frac{1}{\varepsilon(0,\omega)} = 1 - \frac{\omega_p^2}{\omega_p^2 + \omega_{av}^2 - (\omega + i\eta)^2} = 1 - \frac{\omega_p^2}{\omega_{pl}^2 - (\omega + i\eta)^2}\,.$$

In the limiting case of $\eta \to 0^+$, it is easily seen that

$$-\text{Im}\,\frac{1}{\varepsilon(0,\omega)} = \frac{\pi}{2}\frac{\omega_p^2}{\omega_{pl}}\,\delta(\omega - \omega_{pl})\,;$$

thus the function $-\text{Im}\, 1/\varepsilon(0,\omega)$ exhibits a δ-like spike at the plasmon frequency ω_{pl}, and is zero elsewhere (see Fig. 11). For finite values of the parameter η, the energy-loss function $-\text{Im}\, 1/\varepsilon(0,\omega)$ is peaked at $\omega \approx \omega_{pl}$ with a width of the order of η.

In actual materials, due to the background and damping of single-particle transitions (and other decay mechanisms), the $-1/\varepsilon$ plasma resonances are rather broad, with typical half-widths of the order of several eV, and corresponding lifetimes of the order of 10^{-16} sec. In Fig. 12, as an example, we show the characteristic energy-loss experiments for silicon and germanium; the energy-loss data of Ge are from the work of C. J. Powell, Proc. Phys. Soc. (London) **76**, 593 (1960), who used 1.5-keV electrons; for Si, the data are from the work of H. Dimigen, Z. Physik **165**, 53 (1961), who used 47-keV electrons. Since the longitudinal and transverse dielectric functions of a medium are equal at long wavelengths, the energy-loss function $-\text{Im}\, 1/\varepsilon$ can be also obtained from the optical measurements of ε_1 and ε_2. From Fig. 12, one can notice the satisfactory agreement in position and width of the plasma peaks, measured from

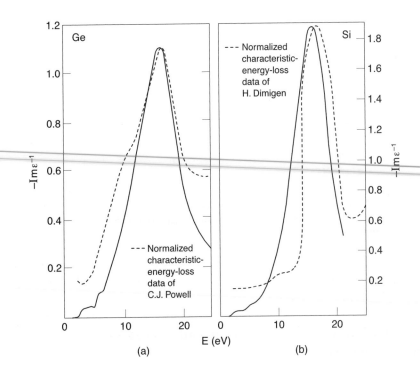

Fig. 12 Comparison of the energy-loss function $-\mathrm{Im}\,1/\varepsilon(0,\omega)$ obtained from the results of optical measurements and characteristic-energy-loss experiments for Ge and Si [from H. R. Philipp and H. Ehrenreich, Phys. Rev. **129**, 1550 (1963); copyright 1963 by the American Physical Society].

energy-loss experiments and optical experiments, in spite of the different sensitivity of these two tools for measuring electronic excitation spectra (the energy-loss function $-\mathrm{Im}\,1/\varepsilon$ is large where the dielectric functions ε_1 and ε_2 are small).

Before closing, we add a few comments on the static dielectric function of semi-conductors and insulators. In the static limit $\omega \to 0$, the imaginary part $\varepsilon_2(\mathbf{q},0)$ of the dielectric function, given by Eq. (61a), vanishes. For the real part $\varepsilon_1(\mathbf{q},0)$, from Eq. (61b) we have

$$\varepsilon_1(\mathbf{q},0) = 1 + \frac{16\pi e^2}{q^2} \sum_{vc} \int_{\text{B.Z.}} \frac{d\mathbf{k}}{(2\pi)^3} \frac{|\langle \psi_{c\mathbf{k}+\mathbf{q}}|e^{i\mathbf{q}\cdot\mathbf{r}}|\psi_{v\mathbf{k}}\rangle|^2}{E_{c\mathbf{k}+\mathbf{q}} - E_{v\mathbf{k}}} , \tag{64}$$

where the discrete sum over $\mathbf{k}$ has been transformed as usual into an integral over the first Brillouin zone, times $V/(2\pi)^3$. The presence of the energy gap makes the denominator in Eq. (64) always finite. In the long wavelength limit, $\varepsilon_1(\mathbf{q},0)$ tends to a finite value $\varepsilon_1(0)$, which represents the electronic contribution to the static dielectric constant of the material. For values of $\mathbf{q}$ larger than the smallest reciprocal lattice vector, $\varepsilon_1(\mathbf{q},0)$ drops rapidly to zero. As an example, in Fig. 13, we report the static dielectric function of silicon and germanium, calculated numerically from Eq. (64); in the same figure different simplified models are also reported.

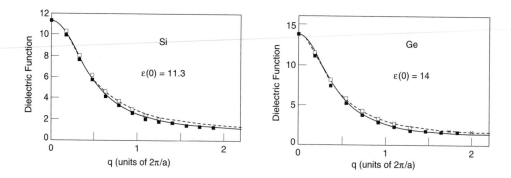

Fig. 13 Static longitudinal dielectric function $\varepsilon_1(\mathbf{q},0)$ for silicon and germanium. Closed boxes [$\mathbf{q}$ along the (1,1,1) direction] and open boxes [$\mathbf{q}$ along the (1,0,0) direction] are from numerical calculations of J. P. Walter and M. L. Cohen, Phys. Rev. B**2**, 1831 (1970). The solid line is the model dielectric function of G. Cappellini, R. Del Sole, L. Reining and F. Bechstedt, Phys. Rev. B**47**, 9892 (1993) (copyright 1993 by the American Physical Society). The dashed curve represents the model of Z. H. Levine and S. G. Louie, Phys. Rev. B**25**, 6310 (1982).

9   Longitudinal dielectric function and energy-loss of a fast charged particle

We consider now the rate by which a charged fast particle (for instance a fast electron of kinetic energy 1–10 keV or so) transfers energy $\hbar\omega$ and momentum $\hbar\mathbf{q}$ to a material. As long as the fractional changes of energy and momentum of the fast electron are small, the electron can be considered as a classical point charge moving with uniform velocity $\mathbf{v}_e$ in the medium. We indicate by $\mathbf{R}_e(t) = \mathbf{v}_e\, t$ its position at time t, and express the perturbation potential in the form

$$U_{\text{ext}}(\mathbf{r} - \mathbf{R}_e(t)) = \frac{e^2}{|\mathbf{r} - \mathbf{R}_e(t)|} = \frac{e^2}{|\mathbf{r} - \mathbf{v}_e\, t|} \, .$$

The bare Coulomb potential U_{ext} of the fast electron can be written in the form

$$U_{\text{ext}}(\mathbf{r} - \mathbf{R}_e(t)) = \frac{1}{V} \sum_{\mathbf{q}} \frac{4\pi e^2}{q^2} e^{i\mathbf{q}\cdot(\mathbf{r} - \mathbf{v}_e\, t)} \, , \tag{65}$$

where the standard expansion of $1/r$ in plane waves has been used; as usual, the discrete sum over $\mathbf{q}$ can be replaced by the integral in $d\mathbf{q}$ times $V/(2\pi)^3$.

The bare Coulomb potential (65) is the sum of longitudinal waves of type (15), of wavevector $\mathbf{q}$, frequency $\omega = \mathbf{q}\cdot\mathbf{v}_e$, and $A_{\text{ext}}(\mathbf{q},\omega) = (1/V)4\pi e^2/q^2$. Using Eq. (53b), the power dissipated in the medium is the sum of contributions of the form

$$P(\mathbf{q},\omega) = \frac{8\pi e^2}{q^2}\, \omega\, \frac{1}{V}\, (-1)\, \text{Im}\frac{1}{\varepsilon(\mathbf{q},\omega)} \, . \tag{66}$$

Thus the rate of energy loss by momentum transfer $\hbar\mathbf{q}$, suffered by a fast particle traversing a material of longitudinal dielectric function $\varepsilon(\mathbf{q}, \omega)$, is primarily determined by the function $-\mathrm{Im}\, 1/\varepsilon(\mathbf{q}, \omega)$, referred to as *the energy-loss function*.

The total energy-loss per unit time of a fast electron, traversing a medium with speed $\mathbf{v}_e$, is obtained summing up the contributions (66) over all possible values of $\mathbf{q}$; the total energy-loss rate is then

$$-\frac{dE}{dt} = \frac{1}{(2\pi)^3} \int \frac{8\pi e^2}{q^2}\, \omega\,(-1)\mathrm{Im}\frac{1}{\varepsilon(\mathbf{q}, \omega)}\, d\mathbf{q}$$

where $\omega = \omega(\mathbf{q}) = \mathbf{q} \cdot \mathbf{v}_e$. Since $-\mathrm{Im}\, 1/\varepsilon(\mathbf{q}, \omega)$ is in general sharply peaked near $\omega = \omega_p$, the mechanism of energy loss of fast electrons essentially occurs via excitations of longitudinal currents, known as plasma oscillations.

Appendix A. Lindhard dielectric function for the free-electron gas

The purpose of this Appendix is to evaluate explicitly the longitudinal dielectric function of the free-electron gas, given by Eq. (35) of the text

$$\varepsilon(\mathbf{q}, \omega) = 1 + \frac{8\pi e^2}{q^2 V} \sum_{\mathbf{k}} \left[\frac{f(\mathbf{k})}{E(\mathbf{k}+\mathbf{q}) - E(\mathbf{k}) - \hbar\omega - i\eta} + \frac{f(\mathbf{k})}{E(\mathbf{k}+\mathbf{q}) - E(\mathbf{k}) + \hbar\omega + i\eta} \right] , \quad (A1)$$

where $E(\mathbf{k}) = \hbar^2 k^2/2m$ is the free-electron energy, $f(\mathbf{k}) = f(E_{\mathbf{k}})$ is the Fermi–Dirac distribution function, spin degeneracy is already included, and the limit $\eta \to 0^+$ is understood. For brevity, the two sums over $\mathbf{k}$ appearing in the right-hand side of Eq. $(A1)$ are denoted as S_1 and S_2, respectively.

We consider the limit of zero temperature, where the Fermi–Dirac distribution function reduces to the step function (with discontinuity at the Fermi energy). As usual, we convert the sum over $\mathbf{k}$ into the integral in $d\mathbf{k}$ times $V/(2\pi)^3$. The expression of S_1 becomes

$$S_1 = \frac{8\pi e^2}{q^2} \frac{1}{(2\pi)^3} \int_{k<k_F} \frac{2m}{\hbar^2} \frac{1}{2\mathbf{q} \cdot \mathbf{k} + q^2 - \widetilde{\omega}}\, d\mathbf{k} , \quad (A2)$$

where $\widetilde{\omega} = (2m/\hbar)\omega + i\widetilde{\eta}$, $\widetilde{\eta} = (2m/\hbar^2)\eta$ and the limit $\widetilde{\eta} \to 0^+$ is understood.

We pass to polar coordinates (without loss of generality we assume $\mathbf{q}$ along the k_z axis) and obtain

$$S_1 = \frac{4me^2}{\pi\hbar^2} \frac{1}{q^2} \int_0^{k_F} k^2\, dk \int_0^{\pi} \frac{1}{2qk\cos\theta + q^2 - \widetilde{\omega}}\, \sin\theta\, d\theta$$

$$= \frac{2me^2}{\pi\hbar^2} \frac{1}{q^3} \int_0^{k_F} k \ln \frac{2qk + q^2 - \widetilde{\omega}}{-2qk + q^2 - \widetilde{\omega}}\, dk \quad (A3)$$

(the integral over the angular variable θ has been performed after the change of variable $x = \cos\theta$).

The integration in Eq. $(A3)$ can again be performed analytically using the following indefinite integral

$$\int x \ln \frac{ax+b}{-ax+b}\, dx = \frac{b}{a}x + \frac{1}{2}\left(x^2 - \frac{b^2}{a^2}\right)\ln\frac{ax+b}{-ax+b} \; ;$$

one obtains

$$S_1 = \frac{2me^2}{\pi\hbar^2}\frac{1}{q^3}\left\{\frac{q^2-\widetilde{\omega}}{2q}k_F + \frac{1}{2}\left[k_F^2 - \frac{(q^2-\widetilde{\omega})^2}{4q^2}\right]\ln\frac{2qk_F+q^2+\widetilde{\omega}}{-2qk_F+q^2-\widetilde{\omega}}\right\} \; .$$

From the above expression of S_1, and a similar expression of S_2, the analytic expression of the dielectric function ($A1$) becomes

$$\varepsilon(q,\omega) = 1 + \frac{2me^2}{\pi\hbar^2}\frac{k_F}{q^2} + \frac{me^2}{\pi\hbar^2}\frac{1}{q^3}\left[k_F^2 - \left(\frac{q}{2} - \frac{\widetilde{\omega}}{2q}\right)^2\right]\ln\frac{2qk_F+q^2-\widetilde{\omega}}{-2qk_F+q^2-\widetilde{\omega}}$$

$$+\frac{me^2}{\pi\hbar^2}\frac{1}{q^3}\left[k_F^2 - \left(\frac{q}{2} + \frac{\widetilde{\omega}}{2q}\right)^2\right]\ln\frac{2qk_F+q^2+\widetilde{\omega}}{-2qk_F+q^2+\widetilde{\omega}} \; . \qquad (A4)$$

In Eq. ($A4$) we separate real and imaginary part taking the limit $\operatorname{Im}\widetilde{\omega}\to 0^+$. The logarithmic function of complex argument is defined with the cut from $-\infty$ to 0; we have $\ln z = \ln|z| + i\arg z$ with $-\pi < \arg z < +\pi$. The real part of the dielectric function becomes

$$\boxed{\begin{aligned}\varepsilon_1(q,\omega) &= 1 + \frac{2me^2}{\pi\hbar^2}\frac{k_F}{q^2} + \frac{me^2}{\pi\hbar^2}\frac{1}{q^3}\left[k_F^2 - \left(\frac{q}{2} - \frac{m}{\hbar}\frac{\omega}{q}\right)^2\right]\ln\left|\frac{2qk_F+q^2-(2m/\hbar)\omega}{-2qk_F+q^2-(2m/\hbar)\omega}\right| \\[2mm] &\quad + \frac{me^2}{\pi\hbar^2}\frac{1}{q^3}\left[k_F^2 - \left(\frac{q}{2} + \frac{m}{\hbar}\frac{\omega}{q}\right)^2\right]\ln\left|\frac{2qk_F+q^2+(2m/\hbar)\omega}{-2qk_F+q^2+(2m/\hbar)\omega}\right|\end{aligned}}$$

$$(A5)$$

In particular, for the static dielectric function we have

$$\varepsilon_1(q,0) = 1 + \frac{2me^2}{\pi\hbar^2}\frac{k_F}{q^2} + \frac{2me^2}{\pi\hbar^2}\frac{1}{q^3}\left(k_F^2 - \frac{q^2}{4}\right)\ln\left|\frac{2k_F+q}{2k_F-q}\right| \; . \qquad (A6)$$

It is also straightforward to obtain the imaginary part of the dielectric function ($A4$); we have

$$\varepsilon_2(q,\omega) = \frac{me^2}{\pi\hbar^2}\frac{1}{q^3}\left[k_F^2 - \left(\frac{q}{2} - \frac{m}{\hbar}\frac{\omega}{q}\right)^2\right](I_1 - I_2) - \frac{me^2}{\pi\hbar^2}\frac{1}{q^3}\left[k_F^2 - \left(\frac{q}{2} - \frac{m}{\hbar}\frac{\omega}{q}\right)^2\right]I_3 \quad (A7)$$

where

$$I_1 = \operatorname{Im}\ln\left[\,2qk_F + q^2 - (2m/\hbar)\omega - i0^+\right]$$

$$I_2 = \operatorname{Im}\ln\left[-2qk_F + q^2 - (2m/\hbar)\omega - i0^+\right]$$

$$I_3 = \operatorname{Im}\ln\left[-2qk_F + q^2 + (2m/\hbar)\omega + i0^+\right] \; .$$

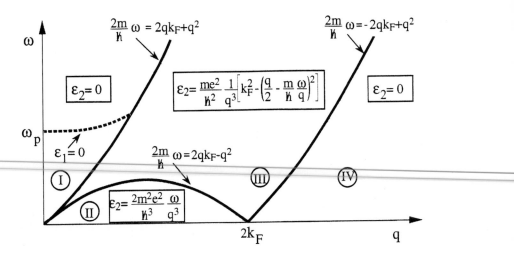

Fig. 14 Imaginary part of the Lindhard dielectric function. The (q, ω) plane is divided into four regions by the three parabolas defined by Eq. ($A9$). The dashed line, where simultaneously $\varepsilon_1(q, \omega)$ and $\varepsilon_2(q, \omega)$ vanish, gives the dispersion curve of plasmon modes.

The explicit expressions of I_1, I_2, I_3 are

$$I_1 = -\pi \quad \text{if} \quad (2m/\hbar)\omega > 2qk_F + q^2; \quad I_1 = 0 \quad \text{otherwise}$$

$$I_2 = -\pi \quad \text{if} \quad (2m/\hbar)\omega > -2qk_F + q^2; \quad I_2 = 0 \quad \text{otherwise} \qquad (A8)$$

$$I_3 = +\pi \quad \text{if} \quad (2m/\hbar)\omega < 2qk_F - q^2; \quad I_3 = 0 \quad \text{otherwise} \ .$$

The expressions ($A8$) are more conveniently visualized plotting in the (q, ω) plane the following three parabolas

$$(2m/\hbar)\omega = 2qk_F + q^2 \ , \quad (2m/\hbar)\omega = -2qk_F + q^2 \ , \quad (2m/\hbar)\omega = 2qk_F - q^2 \ . \ (A9)$$

These parabolas divide the (q, ω) plane (with positive q and ω) into four regions, as indicated in Fig. 14.

Inserting Eqs. ($A8$) into Eq. ($A7$), and with the help of Fig. 14, we can summarize the imaginary part of the dielectric function as follows

$$
\varepsilon_2(q, \omega) = \begin{cases} \dfrac{2m^2 e^2}{\hbar^3} \dfrac{\omega}{q^3} & \text{if } q < 2k_F \text{ and } 0 \le (2m/\hbar)\omega \le 2qk_F - q^2 \\[2ex] \dfrac{me^2}{\hbar^2} \dfrac{1}{q^3}\left[k_F^2 - \left(\dfrac{q}{2} - \dfrac{m}{\hbar}\dfrac{\omega}{q}\right)^2 \right] & \text{if } 2qk_F - q^2 \le (2m/\hbar)\omega \le 2qk_F + q^2 \\[2ex] 0 & \text{in the other cases} \end{cases}
$$

$$(A10)$$

From Eq. ($A10$) and Fig. 14, we see that $\varepsilon_2(q, \omega)$ vanishes in regions I and IV; only in regions II and III is $\varepsilon_2(q, \omega)$ different from zero, and depends linearly on ω in the region II and quadratically in the region III.

We note that in the regions I and IV electron–hole single particle excitations of vector $\mathbf{q}$ are not possible (and thus ε_2 vanishes there). Consider in fact the energy difference

$$\Delta(\mathbf{k}+\mathbf{q},\mathbf{k}) = E(\mathbf{k}+\mathbf{q},\mathbf{k}) - E(\mathbf{k}) = \frac{\hbar^2 \mathbf{q} \cdot \mathbf{k}}{m} + \frac{\hbar^2 q^2}{2m}, \qquad (A11)$$

with $f(\mathbf{k}) \equiv 1$ and $f(\mathbf{k}+\mathbf{q}) \equiv 0$. Expression $(A11)$ represents the energy required to excite an electron within the Fermi sphere to an empty state outside it with wavevector transfer $\mathbf{q}$. It is evident that the maximum value $\Delta_{\max}$, obtained when $\mathbf{k} = k_F$ vers $\mathbf{q}$, is given by

$$\Delta_{\max} = \frac{\hbar^2 q k_F}{m} + \frac{\hbar^2 q^2}{2m}.$$

If $\hbar\omega > \Delta_{\max}$, or equivalently $(2m/\hbar)\omega > 2qk_F + q^2$ (region I), no electron–hole excitation of wavevector transfer q is possible. Similarly for $q > 2k_F$ the minimum value of expression $(A11)$ occurs for $\mathbf{k} = -k_F$ vers $\mathbf{q}$, and it is given by

$$\Delta_{\min} = -\frac{\hbar^2 q k_F}{m} + \frac{\hbar^2 q^2}{2m}.$$

Thus no electron–hole excitation is possible in region IV either.

Further reading

V. M. Agranovich and V. L. Ginzburg "Spatial Dispersion in Crystal Optics and the Theory of Excitons" (Interscience Publishing, New York 1966)

L. C. Andreani "Optical Transitions, Excitons, and Polaritons in Bulk and Low-Dimensional Semiconductor Structures" in "Confined Electrons and Photons: New Physics and Devices" edited by E. Burnstein and C. Weisbuch (Plenum Press, 1995) p.57

N. W. Ashcroft and D. Stroud "Theory of the Thermodynamics of Simple Liquid Metals" in Solid State Physics **33**, 1 (1978), edited by H. Ehrenreich, F. Seitz and D. Turnbull (Academic Press, New York)

M. Balkanski and D. Wallis "Many-Body Aspects of Solid State Spectroscopy" (North-Holland, Amsterdam 1986)

F. Bassani and G. Pastori Parravicini "Electronic States and Optical Transitions in Solids" (Pergamon Press, Oxford 1975)

K. Cho (editor) "Excitons" (Springer, Berlin 1979)

I. Egri "Excitons and Plasmons in Metals, Semiconductors and Insulators: a Unified Approach" Phys. Rep. **119**, 363 (1985)

W. A. Harrison "Solid State Theory" (Dover Publication, New York 1979) chapter 3

C. Kittel "Quantum Theory of Solids" (Wiley, New York 1987) chapters 3 and 6

R. Knox "Theory of Excitons" (Academic Press, New York 1963)

S. Moroni, D. M. Ceperley and G. Senatore "Static Response from Quantum Monte Carlo Calculations" Phys. Rev. Lett. **69**, 1837 (1992); "Static Response and Local Field Factor of the Electron Gas" Phys. Rev. Lett. **75**, 689 (1995)

S. Nakajima, Y. Toyozawa and R. Abe "The Physics of Elementary Excitations" (Springer, Berlin 1980)

D. Pines "Elementary Excitations in Solids" (Benjamin, New York 1963)

H. Raether "Excitation of Plasmons and Interband Transitions by Electrons" (Springer, Berlin 1980)

E. I. Rashba and M. D. Sturge (eds.) "Excitons" (North-Holland, Amsterdam 1982)

S. E. Schnatterly "Inelastic Electron Scattering Spectroscopy" in Solid State Physics **34**, 275 (1979), edited by H. Ehrenreich, F. Seitz and D. Turnbull (Academic Press, New York 1979)

K. S. Singwi and M. P. Tosi "Correlations in Electron Liquids" in Solid State Physics **36**, 177 (1981), edited by H. Ehrenreich, F. Seitz and D. Turnbull (Academic Press, New York)

J. M. Ziman "Principles of the Theory of Solids" (Cambridge University Press, Cambridge 1972) chapter 5

VIII

Interacting electronic–nuclear systems and the adiabatic principle

In the study of the electronic structure of crystals we have so far considered the nuclei fixed in a given spatial configuration, usually the equilibrium one. In this chapter we analyse the consequences related to the fact that the nuclei have indeed a mass much larger than the electron mass, but yet finite. Thus the solution of the electronic problem with the nuclei frozen in a given configuration can only be considered as a preliminary step, from which to start a more realistic investigation.

In this chapter we develop the quantum theory for systems of electrons and nuclei in interaction. We focus on the concepts embodied in the adiabatic approximation and on some aspects beyond it; these ideas are the natural link between the electronic properties of crystals and the lattice dynamics, the two most traditional subjects of solid state physics. The parametric dependence of the crystal Hamiltonian from the nuclear coordinates in the adiabatic approximation offers the occasion to present some general properties of paramater dependent operators, and in particular the Hellmann–Feynman theorem and the concept of geometric Berry phase. The general principles of this chapter are further elaborated and put at work in several other parts of the book; thus, the reader can choose to give initially an overall view to this chapter and later re-examine in depth its topics, as these are encountered in specific contexts.

1 Electronic–nuclear systems and adiabatic potential-energy surfaces

Introductory remarks

A crystal is composed by a large number of interacting particles and consequently its theoretical treatment cannot avoid appropriate approximations. The starting one is suggested by the large difference in the masses of the nuclei and of the electrons; in most situations, this allows one to decouple the dynamics of the fast variables (the electrons) from the dynamics of the slow variables (the nuclei). The decoupling of electron and nuclear dynamics (when it is possible at all) is conceptually achieved by means of the so-called "adiabatic" procedure; before considering its formal treatment, we anticipate here some intuitive aspects.

In essence the adiabatic scheme can be summarized as follows. In a first stage the *nuclei are supposed fixed in selected spatial configurations*, and attention is focused on the electronic eigenvalues as a function of the chosen nuclear coordinates; these curves describe the so-called *adiabatic potential-energy surfaces*, also simply denoted as *potential surfaces* or *adiabatic surfaces* (or *sheets*).

The potential surfaces of electronic–nuclear systems may be non-degenerate or may exhibit degeneracy points (see Fig. 1). In some systems the potential surface under attention is non-degenerate in the whole domain of nuclear coordinates of interest (*ordinary Born–Oppenheimer systems*). In other systems *two or more potential surfaces are degenerate* at some point in the nuclear coordinate space; in general, at and near the degeneracy point the gradients of the potential-energy surfaces are different from zero, in which case the system is called *Jahn–Teller system*.

Once the potential-energy surfaces $E_i(R)$ are known as functions of the multi-dimensional nuclear coordinates (collectively indicated by R), the nuclear dynamics can be studied. From a classical point of view, the forces acting on the nuclei are just given by the negative gradient of the potential-energy surface, to which the system belongs. If the potential surface under consideration (*usually the interest is on the ground adiabatic sheet*), is well separated from all the others, it seems reasonable to assume that the nuclei move (classically or quantistically) on the selected potential surface itself (Fig. 1a). In the case two (or more) adiabatic surfaces are degenerate at some configuration, the nuclear dynamics is determined jointly by the multiple potential-energy surfaces (as schematically indicated in Fig. 1b). In either cases, the limitation to one or a number of potential-energy surfaces is only a useful starting approximation and (whenever necessary) the influence of the other adiabatic surfaces should be taken into account.

It is perhaps appropriate to warn the reader that there is not in the literature a unique recipe for the adiabatic approximation; probably it suffices to say that, just to define the term adiabatic approximation and the various meanings in different contexts, a sort of dictionary has been compiled to avoid misunderstanding among different authors (see for instance Azumi and Matsuzaki 1977); what is defined as "adiabatic" by some authors in some context, may well be referred to as "non-adiabatic"

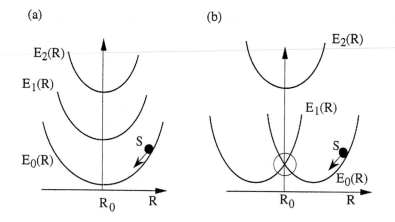

Fig. 1 Schematic representation of adiabatic potential-energy surfaces $E_i(R)$ (the multi-dimensional nuclear variable R is represented as a one-dimensional parameter). In Fig. 1a the adiabatic surfaces are supposed non-degenerate, and the nuclear dynamics of the system S is mostly determined by the single non-degenerate potential surface under consideration. In Fig. 1b two (or more) adiabatic surfaces are degenerate at R_0, and the nuclear dynamics of the system S is determined by the whole set of degenerate adiabatic surfaces.

by others. It is not our intention to follow the historical developments and all the subtleties and technicalities related to the adiabatic approximation; rather, we limit ourselves to an overall account of some basic aspects of the approach.

Adiabatic potential-energy surfaces

At the foundation of the quantum mechanics of any polynuclear system (molecules, clusters, solids) is the large difference of the masses of electrons and nuclei. It has been shown by Born and Oppenheimer how to exploit such a difference, and how to describe the interdependence between the electronic properties and the nuclear dynamics [M. Born and R. Oppenheimer, Ann. Physique **84**, 457 (1927)].

A crystal is composed by mutually interacting electrons and nuclei, and the total Hamiltonian includes the kinetic energy of the nuclei (T_N), the kinetic energy of the electrons (T_e), and all possible electron–electron (V_{ee}), electron–nucleus (V_{eN}), nucleus–nucleus (V_{NN}) interactions. If we consider only Coulomb forces, the expression of the Hamiltonian (in the non-relativistic limit) becomes

$$H_{\text{tot}} = T_N + T_e + V_{ee} + V_{eN} + V_{NN}$$
$$= -\sum_I \frac{\hbar^2 \nabla_I^2}{2M_I} - \sum_i \frac{\hbar^2 \nabla_i^2}{2m} + \frac{1}{2} \sum_{i \neq j} \frac{e^2}{|\mathbf{r}_i - \mathbf{r}_j|} - \sum_{iI} \frac{z_I e^2}{|\mathbf{r}_i - \mathbf{R}_I|} + \frac{1}{2} \sum_{I \neq J} \frac{z_I z_J e^2}{|\mathbf{R}_I - \mathbf{R}_J|} \quad (1)$$

where the indices i, j refer to the electrons, and the indices I, J to the nuclei.

It is convenient to partition the total Hamiltonian (1) in the form

$$H_{\text{tot}} = T_N(R) + T_e(r) + V(r, R) , \quad (2)$$

where T_N is the nuclear kinetic operator, T_e is the electronic kinetic operator, and $V(r, R)$ denotes *all* the electrostatic interactions between particles (i.e. between electrons, between nuclei, and between nuclei and electrons). The multi-dimensional nuclear variable R is a shorthand notation for all the $3N$ coordinates $(R_{1x}, R_{1y}, R_{1z}, R_{2x}, \ldots, R_{Nz})$ of the nuclei of the system; when required by clarity, in place of R we use the notations $\{R_I\}$ $(I = 1, 2, \ldots, 3N)$ or also $\{\mathbf{R}_I\}$ $(I = 1, 2, \ldots, N)$. The nuclear kinetic energy $T_N(R)$ is given by $\sum_I (-\hbar^2/2M_I)(\partial^2 \ldots/\partial R_I^2)$ $(I = 1, 2, \ldots, 3N)$, or in shorthand notation $-(\hbar^2/2M)(\partial^2 \ldots/\partial R^2)$. Similarly r is a shorthand notation for all space and spin electron coordinates of the system. In the following, whenever necessary, we replace shorthand notations with the extended notations specific for the system under investigation. [In actual calculations, it is often convenient to break the crystal into a collection of ions (nuclei plus core electrons) and valence electrons, with appropriate pseudopotentials taking care of the disregarded electronic core wavefunctions; the present discussion also applies to these descriptions].

The Schrödinger equation to be solved for the system of electrons and nuclei in interaction is

$$[T_N(R) + T_e(r) + V(r, R)] \Psi(r, R) = W \Psi(r, R) . \qquad (3)$$

The eigenvalues W and the eigenfunctions $\Psi(r, R)$ of the combined electronic–nuclear system are called "vibronic energies" and "vibronic wavefunctions", respectively.

In order to attack the eigenvalue equation (3), we begin to observe what happens of the total Hamiltonian H_{tot} if the nuclear masses M_I, much larger than the electron mass m, would actually be treated as infinite. *In this case the nuclear kinetic operator T_N could be dropped and the nuclei could be thought as "fixed" in some assigned configuration R.* In the realistic situation of large (and yet finite) nuclear masses M_I, we can estimate qualitatively the order of magnitude of the error in the kinetic energy, introduced in neglecting T_N in Eq. (3), with the following argument. Consider the case of a vibronic wavefunction $\Psi(r, R)$ made up of functions of the type $\phi(\mathbf{r} - \mathbf{R})$, representing an electron $\mathbf{r}$ bound to a nucleus localized at $\mathbf{R}$. The expectation values of the operators $\nabla_{\mathbf{R}}^2$ and $\nabla_{\mathbf{r}}^2$ on the wavefunctions of the type $\phi(\mathbf{r} - \mathbf{R})$ are obviously equal; thus the mean kinetic energy of the nuclei is of the order of $m/M \approx 10^{-3}$ smaller than the mean kinetic energy of the electrons.

In the "fixed" lattice approximation, obtained by ignoring altogether the nuclear kinetic operator in the total Hamiltonian (2), we are left with the so-called electronic adiabatic Hamiltonian $H_e(r; R)$ given by

$$H_e(r; R) = T_e + V(r, R) . \qquad (4)$$

In this Hamiltonian the *variables R appear simply as parameters* (instead of quantum dynamical observables); thus $H_e(r; R)$ belongs to the class of parameter dependent operators; a semicolon is used to denote this parametric dependence. The eigenvalue equation for the electronic Hamiltonian $H_e(r; R)$ is

$$H_e(r; R) \psi_n(r; R) = E_n(R) \psi_n(r; R) . \qquad (5)$$

The electronic wavefunctions $\psi_n(r; R)$, as well as the eigenvalues $E_n(R)$, depend on the parameters R; the suffix n summarizes the electronic quantum numbers.

In Chapter IV we have seen (at least in principle) how to determine the eigenvalues and eigenfunctions $E_n(R)$ and $\psi_n(r; R)$ of a many-electron system with the nuclei fixed in a given spatial configuration R (usually, but non-necessarily, the equilibrium configuration R_0). As we vary the parameters R, the eigenvalues $E_n(R)$ define the so-called *adiabatic potential-energy surfaces*, and $\psi_n(r; R)$ define the set of *adiabatic electronic wavefunctions*. With the help of the adiabatic surfaces and adiabatic wavefunctions, we can now proceed to the study of the *nuclear dynamics* (also called *lattice dynamics* of the atoms or ions composing the crystal); for this purpose, we consider first non-degenerate adiabatic surfaces (Section 2) and then Jahn–Teller systems (Section 3).

2   Non-degenerate adiabatic surface and nuclear dynamics

2.1   Non-degenerate adiabatic surface and classical nuclear dynamics

Consider an electronic–nuclear system initially described by an adiabatic wavefunction $\psi_m(r; R)$ belonging to a non-degenerate adiabatic surface $E_m(R)$ (well separated in energy from all the other adiabatic surfaces). From a classical point of view, the dynamics of the nuclei is that of a set of material points of mass M_I, subject to forces given by the negative gradient of the potential energy $E_m(R)$. In particular when the system explores the ground adiabatic sheet $E_0(R)$, the classical equations of motion for the nuclei are

$$M_I \ddot{R}_I = -\frac{\partial E_0(\{R_J\})}{\partial R_I} . \tag{6}$$

The determination of the trajectories obeying Eq. (6) (with appropriate initial conditions) is the classical subject of *Molecular Dynamics*.

Often one is interested in the behaviour of the ground-adiabatic surface near its absolute minimum at R_0. In crystals we expect that the nuclear displacements $u_I = R_I - R_{I0}$ from the equilibrium positions $\{R_{I0}\}$ are small with respect to the lattice constant. The potential function $E_0(R)$ can then be expanded in Taylor series around the equilibrium positions

$$E_0(R) = E_0(R_0) + \frac{1}{2} \sum_{IJ} \left(\frac{\partial^2 E_0}{\partial R_I \, \partial R_J} \right)_0 u_I u_J + \text{higher order terms} . \tag{7}$$

Notice that terms linear in u_I do not appear, because all the derivatives $\partial E_0 / \partial R_I$ vanish at the equilibrium configuration. If the Taylor series is truncated up to second order and inserted into Eq. (6), we obtain a very simple picture of the lattice dynamics as an ensemble of point masses interacting with harmonic springs, with force constants $(\partial^2 E_0 / \partial R_I \, \partial R_J)_0$. These equations for the "small oscillations" of the nuclei around their equilibrium positions will be extensively studied in the next chapter, in dealing with lattice vibrations and phonons.

Most significant properties of solids (such as the equilibrium configuration, bulk modulus, force constants) are determined from the knowledge of the ground adiabatic potential-energy surface $E_0(\{R_I\})$. The ground-state total energy $E_0(\{R_I\})$ of a given electronic–nuclear system can often be obtained with various degrees of sophistication, ranging from semi-empirical to fully ab initio approaches. Among the semi-empirical approaches, the Lennard–Jones potential or the Born effective inter-ionic potential, discussed in Chapter VI, are perhaps the most elementary examples. With the development of the density functional formalism, the determination of the ground-state total energy $E_0(\{R_I\})$, of the equilibrium configurations and other ground-state properties has been put on a firm theoretical basis. Furthermore, the advent of the Car–Parrinello method has made possible to reach with unprecedent efficiency the equilibrium configurations of polynuclear systems, by simultaneous relaxation of nuclear coordinates and electronic wavefunctions [R. Car and M. Parrinello "Unified Approach for Molecular Dynamics and Density Functional Theory" Phys. Rev. Lett. **55**, 2471 (1985)]. The Car-Parrinello method has opened new perspectives particularly in the physics of complicated systems; major applications concern ab initio studies of liquids, clusters, and solids, also in the presence of impurities and complex surface reconstructions.

2.2   Non-degenerate adiabatic surface and quantum nuclear dynamics

We describe now some relevant aspects of the quantum theory of the lattice dynamics. At first sight one could expect that the quantum mechanics eventually ends up with the quantum transcript of the classical equations of motion of Molecular Dynamics; however, this is not the whole story, and a detailed analysis is needed to determine the circumstances that make this transcription possible.

We focus our considerations on a single non-degenerate adiabatic surface, say the mth adiabatic surface, of potential energy $E_m(R)$ and electronic wavefunctions $\psi_m(r;R)$, and neglect at this stage any other adiabatic surface (the adiabatic surface of actual interest is most often the *ground adiabatic sheet*). We approximate the vibronic wavefunctions of the electronic–nuclear system in the Born–Oppenheimer product form

$$\Psi_{\text{trial}}(r, R) = \chi(R)\, \psi_m(r; R) \;, \tag{8}$$

where $\chi(R)$ depends only on the nuclear coordinates and has to be determined following standard variational principles. The *product wavefunction* $\chi(R)\, \psi_m(r; R)$ *assumes that the electronic system is strictly confined in the given adiabatic surface even when the nuclei are allowed to move*; this approximation is known as *adiabatic approximation*; it pictures the electrons as instantaneously adjusting to the actual nuclear configurations, even though R are thought as dynamical variables and not fixed parameters.

Before using the adiabatic wavefunctions $\psi_m(r; R)$ as basis functions in Eq. (8), it is understood that $\psi_m(r; R)$ (thought as function of R) are *continuous and single-valued in the parameter space R* (as it is routinely required for the dependence on the electronic space coordinates r). The requirements of continuity and single-valuedness do not determine univocally the adiabatic wavefunctions $\psi_m(r; R)$; in fact any change of phases of the type

$$\widetilde{\psi}_m(r; R) = e^{i\alpha(R)} \psi_m(r; R) \tag{9}$$

(where $\alpha(R)$ is any real, continuous and single-valued function) defines a new set of adiabatic wavefunctions $\widetilde{\psi}_m(r; R)$ essentially interchangeable with the set $\psi_m(r; R)$; *this arbitrariness in the choice of phases is called gauge arbitrariness* and any transformation of type (9) is called *gauge transformation* (see Section 5 for further considerations).

In dealing with wavefunctions $\psi(r; R)$ with *two types of arguments r and R*, it is convenient to specify *two types of scalar products*, depending on the fact that both r and R are thought as variables, or r is thought as variable and R as parameter. In order to avoid any possible confusion in handling wavefunctions of *both* arguments r and R, we adopt the following convention. A scalar product $\langle\langle\psi(r; R)|\phi(r; R)\rangle\rangle$ denotes integration on internal and external coordinates r and R; namely

$$\langle\langle\psi(r; R)|\phi(r; R)\rangle\rangle = \int\int \psi^*(r; R)\,\phi(r; R)\,dr\,dR \ . \tag{10a}$$

Instead the notation $\langle\psi(r; R)|\phi(r; R)\rangle$ denotes integration only on the internal coordinates r, while the external coordinates R (or parameters) are kept unchanged; we have

$$\langle\psi(r; R)|\phi(r; R)\rangle = \int \psi^*(r; R)\,\phi(r; R)\,dr \ . \tag{10b}$$

Thus Eq. (10a) defines a number, while Eq. (10b) defines a function of R.

We apply now the variational principle to obtain the equation satisfied by the optimized choice of $\chi(R)$ (supposed to be normalized). The expectation value of H_{tot} on the vibronic wavefunction (8) defines the functional

$$F[\chi] = \langle\langle\,\chi(R)\,\psi_m(r; R)\,|-\frac{\hbar^2}{2M}\frac{\partial^2}{\partial R^2} + H_e(r; R)\,|\,\chi(R)\,\psi_m(r; R)\,\rangle\rangle$$

$$= \langle\langle\,\chi(R)\,\psi_m(r; R)\,|-\frac{\hbar^2}{2M}\frac{\partial^2}{\partial R^2}\,|\,\chi(R)\,\psi_m(r; R)\,\rangle\rangle + \langle\,\chi(R)\,|E_m(R)|\,\chi(R)\,\rangle \ .$$

The explicit expression of this functional is

$$F[\chi] = \langle\chi(R)|-\frac{\hbar^2}{2M}\frac{\partial^2}{\partial R^2}|\chi(R)\rangle + \langle\chi(R)|E_m(R)|\chi(R)\rangle$$

$$-\frac{\hbar^2}{2M}\langle\langle\,\chi(R)\,\psi_m(r; R)\,|\,\chi(R)\frac{\partial^2\psi_m(r; R)}{\partial R^2}\,\rangle\rangle$$

$$-\frac{\hbar^2}{M}\langle\langle\,\chi(R)\,\psi_m(r; R)\,|\,\frac{\partial\chi(R)}{\partial R}\cdot\frac{\partial\psi_m(r; R)}{\partial R}\,\rangle\rangle \ . \tag{11}$$

We now perform the minimization of the functional (11), under the constraint that $\chi(R)$ is normalized; as usual (see for instance Section IV-5.1), the normalization condition can be accounted for by introducing a Lagrange multiplier W, and then minimizing without constraints the functional $F[\chi] - W \langle \chi(R)|\chi(R)\rangle$. The variational procedure produces the eigenvalue equation

$$\left[-\frac{\hbar^2}{2M}\frac{\partial^2}{\partial R^2} + E_m(R) \right] \chi(R) + \Lambda_{mm}(R)\,\chi(R) = W\,\chi(R), \qquad (12)$$

where

$$\Lambda_{mm}(R) = -\frac{\hbar^2}{2M}\langle \psi_m(r;R)| \frac{\partial^2 \psi_m(r;R)}{\partial R^2}\rangle - \frac{\hbar^2}{M}\langle \psi_m(r;R)| \frac{\partial \psi_m(r;R)}{\partial R}\rangle \cdot \frac{\partial}{\partial R}. \qquad (13)$$

The Schrödinger equation (12) shows that the effective total potential that determines the nuclear dynamics is $E_m(R) + \Lambda_{mm}(R)$. Thus *besides the easily predictable "adiabatic potential" $E_m(R)$ we have the additional operator $\Lambda_{mm}(R)$* (sometimes called "non-adiabatic operator"); we notice that the former is a gauge independent potential, while the latter is a gauge dependent operator. [The presence of the operator $\Lambda_{mm}(R)$ is essential to guarantee that the vibronic eigenfunctions $\chi(R)\,\psi_m(r;R)$ and the vibronic eigenvalues W are not affected by gauge transformations of type (9); on the contrary, the adiabatic functions $\psi_m(r;R)$, the functions $\chi(R)$, and the operator $\Lambda_{mm}(R)$ are obviously gauge dependent].

We can better understand the structure and physical meaning of the eigenvalue equation (12) with the following arguments. We notice that at any particular nuclear configuration it is possible to choose the adiabatic electronic wavefunction $\psi_m(r;R)$ to be *real-valued*. In fact, if $\psi_m(r;R)$ is an eigenfunction of the Schrödinger equation (5) with energy $E_m(R)$ also the complex conjugate wavefunction $\psi_m^*(r;R)$ is eigenfunction with the same eigenvalue; since $E_m(R)$ is assumed to be non-degenerate $\psi_m(r;R)$ and $\psi_m^*(r;R)$ must be linearly dependent and can be put in real form. [More generally we notice that the electronic Hamiltonian $H_e(r;R)$ is invariant under time-reversal symmetry. When the electron spin can be disregarded, the time-reversal operator essentially becomes the complex conjugation operation, and the eigenfunctions (whether degenerate or not) can be taken as real (in the following we discuss the implications of time-reversal symmetry for systems of spinless particles). When the electron spin cannot be disregarded, we refer for properties of the time-reversal operator to textbooks on group theory, for instance F. Bassani and G. Pastori Parravicini "Electronic States and Optical Transitions in Solids" (Pergamon Press, Oxford 1975) chapters I and II].

In general, the choice of reality of wavefunctions at all nuclear configurations is of no particular value, unless the choice still preserves continuity with respect to R. In the particular case of a non-degenerate adiabatic surface, the constraints of reality, continuity and single-valuedness for $\psi_m(r;R)$ can all be satisfied simultaneously (the proof of this key issue is postponed to Section 5, where discussing the Berry phase concept). We thus examine Eq. (12) and Eq. (13) in the *"preferential gauge"*, in which the *adiabatic wavefunctions* $\psi_m(r; R)$ *are real-valued, besides being continuous and single-valued.*

The first term in the right-hand side of Eq. (13) is expected to be a small correction (of order m/M) of the electronic kinetic energy $-(\hbar^2/2m)\langle \psi_m(r; R)|\partial^2 \psi_m(r; R)/\partial r^2\rangle$ [the estimate holds under assumption that the wavefunction $\psi_m(r; R)$ is made up of functions of the type $\phi(r - R)$, which depend on the relative coordinates of electronic and nuclear positions]. The electronic kinetic energy enters automatically into the adiabatic potential $E_m(R)$ and it is thus often not essential to include the above-mentioned term.

The second term appearing in the right-hand side of Eq. (13) identically vanishes for real and normalized wavefunctions; in fact we have

$$\langle \psi_m(r; R)| \frac{\partial}{\partial R} \psi_m(r; R)\rangle = \int \psi_m^*(r; R) \frac{\partial}{\partial R} \psi_m(r; R)\, dr$$

$$= \int \psi_m(r; R) \frac{\partial}{\partial R} \psi_m(r; R)\, dr = \frac{1}{2} \frac{\partial}{\partial R} \int \psi_m(r; R)\, \psi_m(r; R)\, dr \equiv 0$$

(the derivation of the normalization constant is trivially zero).

When the preferential gauge of real, continuous and single-valued adiabatic wave-functions, exists and is adopted, Eq. (12) becomes

$$\boxed{\left[-\frac{\hbar^2}{2M} \frac{\partial^2}{\partial R^2} + E_m(R) \right] \chi(R) = W\, \chi(R)} . \qquad (14)$$

According to the Born–Oppenheimer adiabatic approximation, summarized by Eq. (14), the nuclear dynamics is described by a Schrödinger equation with an effective potential $E_m(R)$ given by the adiabatic potential-energy surface. Eq. (14) has the great merit to accomplish a full decoupling between nuclear and electronic degrees of freedom. The electronic coordinates and momenta do not enter directly into Eq. (14), but only indirectly via the adiabatic electronic energy $E_m(R)$, thought of as a function of the nuclear coordinates.

Before concluding, we notice that the Born–Oppenheimer "single-product vibronic wavefunctions" of the form $\chi(R)\,\psi_m(r; R)$, with $\chi(R)$ satisfying Eq. (12) or Eq. (14), are the "best wavefunctions" of H_{tot} in a variational sense, but are not genuine wavefunctions of the combined electronic–nuclear system. The exact vibronic wavefunctions of the total Hamiltonian H_{tot} are in fact appropriate linear combinations of Born–Oppenheimer products (8), associated to different adiabatic surfaces; in other words, *when the nuclei are allowed to move, the electronic–nuclear system cannot be strictly confined to a given adiabatic surface.* As long as the mixing of different adiabatic surfaces can be considered as a small perturbation, the separation of nuclear and electronic wavefunctions is justified; otherwise the mixing of different adiabatic surfaces must be appropriately taken into account. In general, first principle account of the part of the electronic–nuclear interaction that causes the mixing (*electron–phonon interaction*) is rather demanding, and is often estimated on the basis of simplified semi-empirical models.

Case of a single one-dimensional vibrational mode

As an illustration of our discussion, we consider briefly the particular case where the multi-dimensional variables $\{R_I\}$ reduce to a single one-dimensional variable. This happens for instance for a diatomic molecule, where the inter-nuclear distance is the parameter of interest; also in the study of more complicated systems, such as localized impurities in solids, one may have to handle one (or a small number) of (collective) variables.

We consider a dimer species, with the two nuclei at the distance R (in a fixed direction), and we indicate with $\psi_g(r; R)$ and $E_g(R)$ the electronic wavefunction and the energy of the ground-state (assumed non-degenerate). Near the (absolute) minimum R_0 we expand $E_g(R)$ up to second order in the displacements $R - R_0$ in the form

$$E_g(R) = E_g(R_0) + \frac{1}{2} C (R - R_0)^2 \ . \tag{15a}$$

Within the adiabatic approximation for the ground adiabatic surface, the vibronic wavefunctions have the "product form"

$$\Psi_{gm}(r, R) = \chi_m(R)\, \psi_g(r; R) \ ; \tag{15b}$$

the functions $\chi_m(R)$ are the standard Hermite solutions of the harmonic oscillator equation

$$\left[-\frac{\hbar^2}{2M^*} \frac{\partial^2}{\partial R^2} + E_g(R_0) + \frac{1}{2} C (R - R_0)^2 \right] \chi_m(R) = W_m\, \chi_m(R) \ , \tag{15c}$$

where M^* is the reduced mass of the two nuclei. The vibronic energies are given by

$$W_{gm} = E_g(R_0) + (m + \frac{1}{2})\, \hbar\omega \ , \tag{15d}$$

where $m = 0, 1, 2, \ldots$ and $\omega = \sqrt{C/M^*}$. The model is schematically illustrated in Fig. 2.

We consider now the lowest excited energy adiabatic sheet $E_e(R)$ (assumed non-degenerate) and the corresponding electronic eigenfunctions $\psi_e(r; R)$. We suppose that $E_e(R)$ has the minimum at the same value R_0 and with the same curvature C as the ground-energy sheet $E_g(R)$. Within the adiabatic approximation for the excited adiabatic sheet, the vibronic wavefunctions have the product form

$$\Psi_{en}(r, R) = \chi_n(R)\, \psi_e(r; R) \ , \tag{16a}$$

where $\chi_n(R)$ are the standard Hermite functions, solutions of the harmonic oscillator of frequency ω. The quantized energies of the nuclear motion associated with the excited adiabatic sheet are given by

$$W_{en} = E_e(R_0) + (n + \frac{1}{2})\, \hbar\omega \ , \tag{16b}$$

($n = 0, 1, 2, \ldots$) and are schematically shown in Fig. 2a. The separate quantization of the different adiabatic sheets is only a (useful) starting approximation; in fact, it is relaxed by a proper account of the electron–phonon operator.

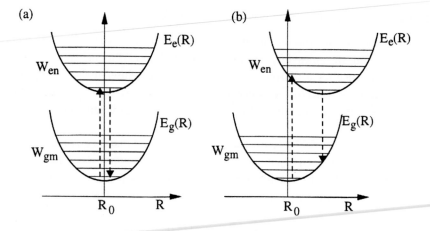

Fig. 2 Schematic behaviour of the adiabatic electronic energies $E_g(R)$ and $E_e(R)$ of the ground-state and lowest excited state of a model dimer system (R is assumed to be a one-dimensional variable). In case (a) the minima of the adiabatic potential sheets coincide. In case (b) the minima are in different configurations (Franck–Condon model). The quantized vibronic energies are also indicated.

In Fig. 2a, the ground adiabatic sheet and the higher one are non-degenerate with minima at the same configuration R_0. The structure of Fig. 2a can be considered as orientative of the electronic–nuclear systems, whose electron wavefunctions are reasonably delocalized (as it happens for valence and conduction states in ordinary crystals). In this case, the electronic charge density of the system is much the same in the ground-state or in the (lowest lying) excited states, and so are the forces $-\partial E_m(R)/\partial R$ [in fact the Hellmann–Feynman theorem of Section 4 shows that the forces equal the negative gradients of the classical electrostatic potential due to the nuclei and electronic charge density].

In Fig. 2b the ground-adiabatic sheet and the higher one are non-degenerate, and their minima occur at different configurations. Fig. 2b is known as *Seitz model*, or *Franck–Condon model*; it is orientative for the adiabatic structure of impurities in crystals with well localized electronic wavefunctions; in fact excitations from localized impurity orbitals may be accompanied by significant local change of electron density and hence of local forces. These situations are of particular interest in optical transitions, and may lead to substantial energy shifts in absorption and luminescence processes, as we shall see in detail in the study of the optical properties of impurities (see Section XII-7).

3   Degenerate adiabatic surfaces and Jahn–Teller systems

3.1   Degenerate adiabatic surfaces and nuclear dynamics

In the previous section, we have assumed that the adiabatic potential-energy surface under attention is non-degenerate. We consider now electronic–nuclear systems in

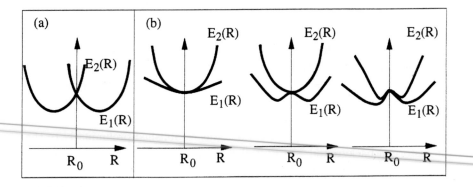

Fig. 3 Schematic representation of possible topologies of two adiabatic potential-energy surfaces near a degeneracy point R_0 in nuclear coordinate space. In Fig. 3a we report the typical Jahn–Teller "conical intersection", with the presence of linear terms in $R - R_0$. In Fig. 3b we report Renner–Teller "glancing intersections", with the absence of linear terms in $R - R_0$, and positive or negative second derivatives. In general, terms linear in certain symmetry-breaking distortions are always possible for any type of degeneracy (except for Kramers degeneracy and some levels in linear molecules).

which two or more electronic wavefunctions $\{\psi_1(r; R_0), \psi_2(r; R_0), \dots, \psi_\nu(r; R_0)\}$ are degenerate (for symmetry reasons) at the high symmetry nuclear configuration R_0, and are thus quasi-degenerate for values of R near to R_0. The behaviour (or the so-called *topology*) of the potential-energy surfaces $E_n(R)$ ($n = 1, 2, \dots, \nu$) near the intersection point R_0 can be distinguished essentially into *two classes, depending whether or not R_0 is a stationary point for all the potential-energy surfaces $E_n(R)$ ($n = 1, 2, \dots, \nu$)*; in other words, the distinction concerns whether or not the gradients $\nabla_R E_n(R) \equiv 0$ for $R \to R_0$. By far the most common situation is that in which the gradients of the potential-energy surfaces are different from zero (at least in some directions in the nuclear coordinate space) and R_0 is not a stationary point: these systems are called *Jahn–Teller systems*.

To help to visualize different topologies, in Fig. 3 we indicate schematically possible behaviour of two Born–Oppenheimer potential-energy surfaces near the degeneracy point R_0 in nuclear coordinate space. In Fig. 3a, we report the typical Jahn–Teller intersection with the presence of linear terms in $R - R_0$; the local topology of the potential-energy surfaces is that of a "double cone" and the name "conical intersection" has been coined for it. In Fig. 3b we show possible kinds of "glancing intersections" where R_0 is a stationary point (Renner–Teller intersections). The basic difference between Jahn–Teller and Renner–Teller intersections, and the consequent implications, can be clarified at the light of the geometric Berry phase, discussed in Section 5.

The topology of the adiabatic potential surfaces near a degeneracy point is regulated by the Jahn–Teller theorem [H. A. Jahn and E. Teller, Proc. Roy. Soc. A**161**, 220 (1937)]. The theorem states that *any electronically degenerate system can lower its energy (and is thus intrinsically unstable) under certain asymmetric distortions of the* *nuclear framework.* The theorem has an obvious corollary: the initially electronically degenerate system undergoes symmetry-breaking distortions and eventually reaches an equilibrium position, in which the symmetry of the nuclear framework is low enough to completely remove the electronic degeneracy. [The only exceptions of the Jahn–Teller theorem concern some levels in linear molecules and the twofold Kramers degeneracy. The former case is of limited interest in solids; the latter case is due to time-reversal symmetry and cannot be lifted reducing the symmetry of the nuclear framework. As a matter of fact, the discovery of exceptions was established preliminarily to the rule; for the historical background, see E. Teller "The Jahn–Teller Effect. Its History and Applicability" Physica **114A**, 14 (1982)].

In order to clarify the physical arguments underlying the Jahn–Teller theorem, we need to describe the behaviour of the adiabatic energy surfaces near the degeneracy point R_0. For this purpose we consider the matrix elements of the electron Hamiltonian $H_e(r; R) = T_e + V(r, R)$ within the degenerate manifold of electronic wavefunctions $\{\psi_n(r; R_0)\}$ $(n = 1, 2, \ldots, \nu)$. For configurations R reasonably near to the high symmetry configuration R_0, the adiabatic energy surfaces $E_n(R)$ and the adiabatic electronic wavefunctions $\{\psi_n(r; R)\}$ $(n = 1, 2, \ldots, \nu)$ are determined from the eigenvalues and eigenvectors of the secular equation

$$||U_{mn}(R) - E\,\delta_{mn}|| = 0 , \tag{17a}$$

where

$$U_{mn}(R) = \langle \psi_m(r; R_0) | H_e(r; R) | \psi_n(r; R_0) \rangle \tag{17b}$$

$(m, n = 1, 2, \ldots, \nu)$. The matrix elements $U_{mn}(R)$ can be recast in the form

$$U_{mn}(R) = \langle \psi_m(r; R_0) | H_e(r; R_0) + H_e(r; R) - H_e(r; R_0) | \psi_n(r; R_0) \rangle$$

$$= E(R_0)\,\delta_{mn} + \langle \psi_m(r; R_0) | H_e(r; R) - H_e(r; R_0) | \psi_n(r; R_0) \rangle$$

$$= E(R_0)\,\delta_{mn} + \langle \psi_m(r; R_0) | V(r, R) - V(r, R_0) | \psi_n(r; R_0) \rangle , \tag{18}$$

where $E(R_0)$ denotes the energy of the adiabatic surfaces at the degeneracy point.

The *topology of the potential sheets* obtained by solving the secular equation (17), and specifically the assessment *whether or not R_0 is a stationary point*, can be established by direct inspection of the matrix elements $U_{mn}(R)$ for $R \approx R_0$; to first order in the displacements R - R_0, the matrix elements (18) become

$$U_{mn}(R) = E(R_0)\delta_{mn} + \langle \psi_m(r; R_0) | \left[\frac{\partial H_e(r; R)}{\partial R} \right]_{R=R_0} | \psi_n(r; R_0) \rangle \cdot (R - R_0) . \tag{19}$$

The topology of the potential sheets is dictated by the presence or absence of terms linear in the displacements in Eq. (19); thus, we have to examine whether the matrix elements

$$M_{mn}(R_0) = \langle \psi_m(r; R_0) | \left[\frac{\partial H_e(r; R)}{\partial R} \right]_{R=R_0} | \psi_n(r; R_0) \rangle$$

$$= \langle \psi_m(r; R_0) | \left[\frac{\partial V(r, R)}{\partial R} \right]_{R=R_0} | \psi_n(r; R_0) \rangle \tag{20}$$

$(m, n = 1, 2, \ldots, \nu)$ can be different from zero or not.

From group theory considerations it has been shown by Jahn and Teller that in general the matrix elements (20) can be different from zero, except the situation of linear molecules and Kramers degeneracy. The proof is done listing one-by-one every irreducible representation, that classifies the basis functions $\psi_n(r; R_0)$ ($n = 1, 2, \ldots, \nu$). Then one considers the representation according to which the product functions $\psi_m^*(r; R_0) \psi_n(r; R_0)$ transform. In general, it is realized by inspection that there are (non-total-symmetric) linear combinations of nuclear coordinates transforming as one (or more) of the irreducible representations contained in the representation of the product functions (Jahn–Teller active modes). Thus, in general, no symmetry reason exists why the matrix elements (20) should vanish (apart the exceptions mentioned above).

Jahn–Teller systems may present different manifestations depending on the strength of the coupling between electronic and nuclear operators. In Jahn–Teller systems, the adiabatic potential surfaces near the confluence point R_0 vary linearly in certain directions of asymmetric nuclear displacements; a number of equivalent minima are thus expected away from the high symmetry point R_0. For strong coupling, the minima of the adiabatic sheets are distant in R space and a large amount of energy is required for the system to migrate from one minimum to another. In this case we have a "permanent" lowering of the local symmetry and the *Jahn–Teller effect is said to be static*. If the coupling is weak, the system may tunnel from a configuration to the others and is thus delocalized on all the equivalent configurations. In this case, the *Jahn–Teller effect is said to be dynamic*; if one considers a reasonable period of time, there is in average no permanent distortion and no lowering of the symmetry. Of course this (sharp) distinction is only qualitative; indeed it is even possible that the same system shows a static or a dynamic Jahn–Teller effect depending for instance from the characteristic time of the specific experiment (such as spin resonance, nuclear magnetic resonance), from the sample temperature, or other conditions.

After this discussion of the topology of the adiabatic potential surfaces at and near degeneracy points, we pass now to the description of the nuclear dynamics in Jahn–Teller systems; we express the electronic–nuclear wavefunctions in the form

$$\Psi(r, R) = \sum_{n=1}^{\nu} \chi_n(R) \psi_n(r; R_0) , \tag{21}$$

where $\chi_n(R)$ are suitable expansion coefficients on the basis of the electronic wavefuntions $\psi_n(r; R_0)$ degenerate at R_0. A vibronic wavefunction of type (21) permits the appropriate mixing of the different potential surfaces. To determine the equations obeyed by the vibrational functions $\chi_n(R)$, we replace Eq. (21) into Eq. (3), multiply on the left by $\psi_m^*(r; R_0)$ and integrate over the electronic coordinates. We obtain for the nuclear motion the following system of coupled differential equations

$$-\frac{\hbar^2}{2M}\frac{\partial^2}{\partial R^2}\chi_m(R) + \sum_n U_{mn}(R)\,\chi_n(R) = W\,\chi_m(R) \ . \tag{22}$$

The above system of equations can also be written in the matrix form

$$H_{JT}\begin{pmatrix} \chi_1(R) \\ \chi_2(R) \\ \cdots \\ \chi_\nu(R) \end{pmatrix} = W \begin{pmatrix} \chi_1(R) \\ \chi_2(R) \\ \cdots \\ \chi_\nu(R) \end{pmatrix} ,$$

where the Jahn–Teller operator H_{JT} is defined as

$$H_{JT} = -\frac{\hbar^2}{2M}\frac{\partial^2}{\partial R^2} + \begin{pmatrix} U_{11}(R) & \cdots & U_{1\nu}(R) \\ \cdots & \cdots & \cdots \\ U_{\nu 1}(R) & \cdots & U_{\nu\nu}(R) \end{pmatrix} \tag{23}$$

[the kinetic energy operator in the right-hand side of Eq. (23) is understood to be multiplied by the unit matrix of dimension ν and then added to the matrix $U_{mn}(R)$].

In the Born–Oppenheimer adiabatic approximation (with a non-degenerate adiabatic surface) the effective potential $E_m(R)$ for the nuclear motion is an ordinary R-dependent function (see Eq. 14). In Eq. (23), the effective potential for the nuclear motion is an R-dependent matrix, of rank equal to the number of degenerate sheets. The non-diagonal terms $U_{mn}(R)$ clearly show that near degeneracy points the nuclear dynamics is determined by all *the adiabatic sheets on the same footing, and not by each of them individually.*

3.2   The Jahn–Teller effect for doubly degenerate electronic states

We now illustrate the key aspects of the Jahn–Teller effect in a few prototype systems, that can be worked out analytically. Among possible examples, we select the regular triangular and the regular octahedral molecules; the interest is not confined to molecules but extends, for instance, to the study of certain defects and complexes in crystals in the "quasi-molecular" approximation. Since degeneracy is a consequence of symmetry, a systematic study of the vibronic systems requires the knowledge of group theory; thus most often a newcomer, not yet familiar with group theory, does not approach the fascinating subject of the Jahn–Teller effect and its manifestations. Our purpose here is not to address the subject in its full generalities; instead, we choose to analyse some typical Jahn–Teller systems, with simple microscopic models and rather intuitive symmetry arguments.

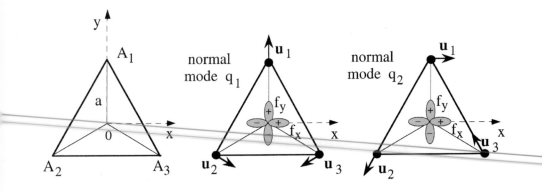

Fig. 4 Geometry of the regular triangular molecule and nuclear displacements of the degenerate vibrational mode ε. The doublet of electronic states (f_x, f_y) centred at the origin is also indicated.

The $E \otimes \varepsilon$ Jahn–Teller system

Probably one of the most studied Jahn–Teller system is the so-called $E \otimes \varepsilon$ system, in which an *electronic doublet* (E) *interacts linearly with a doublet of vibrational degenerate states* (ε) [it is customary to denote the symmetry of the electronic states by upper case Latin letters, and the symmetry of vibrational states by lower case Greek letters]. A possible realization of the $E \otimes \varepsilon$ Jahn–Teller system is given by two degenerate electronic wavefunctions, at the center of a regular triangular molecule, interacting with the doublet of vibrational states of the triagular complex (see Fig. 4).

The geometry of the regular triangular molecule is indicated in Fig. 4. The motion of the nuclei is described in terms of *normal coordinates*; these are combination of Cartesian coordinates determined by the symmetry properties of the system under consideration. For their definition in the case of triangular molecules we refer for instance to the very clear article of A. Nussbaum, Am. J. Phys. **36**, 529 (1968); for other molecules we refer to standard books such as J. Hertzberg "Molecular Spectra and Molecular Structure" Vol.2 , chapter II (Van Nostrand, New York 1966). The normal vibrations of the equilateral molecule consist of a totally symmetric mode, which does not alter the symmetry of the structure, and a doublet of degenerate modes (called ε); the nuclear displacements corresponding to the two partner components q_1 and q_2 of the doubly degenerate mode ε are indicated in Fig. 4. In Table 1 the equilibrium positions of the atomic sites and the normal displacements of the vibrational mode ε are reported.

We determine now the interaction matrix for the $E \otimes \varepsilon$ vibronic system with the following considerations. We consider two degenerate electronic wavefunctions (f_x, f_y), respectively with p_x-like and p_y-like symmetry, localized at the center O of the molecule; we assume that the environment potential felt by the electrons in O can be considered as the sum of atomic-like spherically symmetric potentials $V_a(\mathbf{r})$ centred in the positions $\mathbf{d}_1 + \mathbf{u}_1, \mathbf{d}_2 + \mathbf{u}_2, \mathbf{d}_3 + \mathbf{u}_3$.

Table 1 Equilibrium positions and normal displacements of the vibrational mode ε in the regular triangular molecule. In order to calculate easily the Jahn–Teller matrix of the $E \otimes \varepsilon$ system, we also report the modulus and the direction cosines of the vectors $\mathbf{d}_i + \mathbf{u}_i(q_1)$ and $\mathbf{d}_i + \mathbf{u}_i(q_2)$ (up to terms linear in q_1 and q_2).

	normal mode q_1		normal mode q_2	
$\mathbf{d}_1 = a(0,1)$	$\mathbf{u}_1(q_1) = \frac{q_1}{\sqrt{3}}(0,1)$		$\mathbf{u}_1(q_2) = \frac{q_2}{\sqrt{3}}(1,0)$	
$\mathbf{d}_2 = a(\frac{-\sqrt{3}}{2},\frac{-1}{2})$	$\mathbf{u}_2(q_1) = \frac{q_1}{\sqrt{3}}(\frac{\sqrt{3}}{2},\frac{-1}{2})$		$\mathbf{u}_2(q_2) = \frac{q_2}{\sqrt{3}}(\frac{-1}{2},\frac{-\sqrt{3}}{2})$	
$\mathbf{d}_3 = a(\frac{\sqrt{3}}{2},\frac{-1}{2})$	$\mathbf{u}_3(q_1) = \frac{q_1}{\sqrt{3}}(-\frac{\sqrt{3}}{2},\frac{-1}{2})$		$\mathbf{u}_3(q_2) = \frac{q_2}{\sqrt{3}}(\frac{-1}{2},\frac{\sqrt{3}}{2})$	
$\lvert\mathbf{d}_1+\mathbf{u}_1(q_1)\rvert = a+\frac{q_1}{\sqrt{3}}$	$l_x^2 = 0$	$l_x l_y = 0$	$l_y^2 = 1$	
$\lvert\mathbf{d}_2+\mathbf{u}_2(q_1)\rvert = a-\frac{q_1}{2\sqrt{3}}$	$l_x^2 = \frac{3}{4} - \frac{\sqrt{3}}{4}\frac{q_1}{a}$	$l_x l_y = \frac{\sqrt{3}}{4} + \frac{q_1}{4a}$	$l_y^2 = \frac{1}{4} + \frac{\sqrt{3}}{4}\frac{q_1}{a}$	
$\lvert\mathbf{d}_3+\mathbf{u}_3(q_1)\rvert = a-\frac{q_1}{2\sqrt{3}}$	$l_x^2 = \frac{3}{4} - \frac{\sqrt{3}}{4}\frac{q_1}{a}$	$l_x l_y = \frac{-\sqrt{3}}{4} - \frac{q_1}{4a}$	$l_y^2 = \frac{1}{4} + \frac{\sqrt{3}}{4}\frac{q_1}{a}$	
$\lvert\mathbf{d}_1+\mathbf{u}_1(q_2)\rvert = a$	$l_x^2 = 0$	$l_x l_y = \frac{1}{\sqrt{3}}\frac{q_2}{a}$	$l_y^2 = 1$	
$\lvert\mathbf{d}_2+\mathbf{u}_2(q_2)\rvert = a+\frac{q_2}{2}$	$l_x^2 = \frac{3}{4} - \frac{q_2}{4a}$	$l_x l_y = \frac{\sqrt{3}}{4} + \frac{q_2}{4\sqrt{3}a}$	$l_y^2 = \frac{1}{4} + \frac{q_2}{4a}$	
$\lvert\mathbf{d}_3+\mathbf{u}_3(q_2)\rvert = a-\frac{q_2}{2}$	$l_x^2 = \frac{3}{4} + \frac{q_2}{4a}$	$l_x l_y = \frac{-\sqrt{3}}{4} + \frac{q_2}{4\sqrt{3}a}$	$l_y^2 = \frac{1}{4} - \frac{q_2}{4a}$	

Following Eq. (18) with obvious adjustment of notations to the present problem, we consider the 2×2 matrices of the type

$$U_{\alpha\beta}(q_j) = \int f_\alpha^* \sum_{i=1,3} [V_a(\mathbf{r} - \mathbf{d}_i - \mathbf{u}_i(q_j)) - V_a(\mathbf{r} - \mathbf{d}_i)] \, f_\beta \, d\mathbf{r} \tag{24}$$

$(\alpha, \beta = x, y)$, corresponding to the two normal displacements q_1 and q_2 of the twofold degenerate ε mode.

The matrix elements (24) can be expressed in terms of independent parameters, following the same techniques introduced by Slater and Koster, and explained for the tight-binding method (see Section V-2.1). We summarize the integrals of interest in the following list:

$$\int f_x^* V_a(\mathbf{r} - \mathbf{R}) f_x \, d\mathbf{r} = l_x^2 \, I(pp\sigma, R) + (1 - l_x^2) \, I(pp\pi, R)$$

$$\int f_x^* V_a(\mathbf{r} - \mathbf{R}) f_y \, d\mathbf{r} = l_x l_y [I(pp\sigma, R) - I(pp\pi, R)] \tag{25}$$

$$\int f_y^* V_a(\mathbf{r} - \mathbf{R}) f_y \, d\mathbf{r} = l_y^2 \, I(pp\sigma, R) + (1 - l_y^2) \, I(pp\pi, R) \ .$$

In Eqs.(25), f_x and f_y are p_x-like and p_y-like (real) functions centred at the origin; $V_a(\mathbf{r} - \mathbf{R}) = V_a(\lvert \mathbf{r} - \mathbf{R} \rvert)$ is a localized spherically symmetric potential centered around the vector $\mathbf{R}$, with director cosines l_x, l_y, l_z; $I(pp\sigma, R)$ and $I(pp\pi, R)$ are the only two independent integrals, for a given distance R.

Using Table 1 and Eqs.(25), we can express all matrix elements (24) in terms of the independent integrals. For instance we have

$$\int f_x^* \left[V_a(\mathbf{r} - \mathbf{d}_1 - \mathbf{u}_1(q_1)) - V_a(\mathbf{r} - \mathbf{d}_1) \right] f_x dr = I(pp\pi, a + \frac{q_1}{\sqrt{3}}) - I(pp\pi, a) = \frac{q_1}{\sqrt{3}} I'(pp\pi, a) \,,$$

where $I'(pp\pi, a) = dI(pp\pi, R)/dR$ for $R = a$. Similarly we have

$$\int f_x^* \left[V_a(\mathbf{r} - \mathbf{d}_2 - \mathbf{u}_2(q_1)) - V_a(\mathbf{r} - \mathbf{d}_2) \right] f_x \, dr = (\frac{3}{4} - \frac{\sqrt{3}}{4}\frac{q_1}{a}) I(pp\sigma, a - \frac{1}{2}\frac{q_1}{\sqrt{3}})$$

$$+ (\frac{1}{4} + \frac{\sqrt{3}}{4}\frac{q_1}{a}) I(pp\pi, a - \frac{1}{2}\frac{q_1}{\sqrt{3}}) - \frac{3}{4} I(pp\sigma, a) - \frac{1}{4} I(pp\pi, a)$$

$$= -\frac{\sqrt{3}}{8} q_1 I'(pp\sigma, a) - \frac{\sqrt{3}}{4}\frac{q_1}{a} I(pp\sigma, a) - \frac{\sqrt{3}}{24} q_1 I'(pp\pi, a) + \frac{\sqrt{3}}{4}\frac{q_1}{a} I(pp\pi, a) \,.$$

With similar procedure the matrix elements $U_{\alpha\beta}(q_1)$ and $U_{\alpha\beta}(q_2)$ defined in Eq. (24) can be easily worked out. The explicit expressions for the matrices are

$$U(q_1) = \gamma \begin{pmatrix} -q_1 & 0 \\ 0 & q_1 \end{pmatrix} \quad \text{and} \quad U(q_2) = \gamma \begin{pmatrix} 0 & q_2 \\ q_2 & 0 \end{pmatrix}, \tag{26}$$

where

$$\gamma = \frac{\sqrt{3}}{2a}[I(pp\sigma, a) - I(pp\pi, a)] + \frac{\sqrt{3}}{4}[I'(pp\sigma, a) - I'(pp\pi, a)]$$

is the only independent parameter of the model. From Eqs.(26) we see that the mode q_1 splits the degeneracy of the electronic states (f_x, f_y) without mixing them, while the mode q_2 also acts to mix them.

We now specify the Jahn–Teller operator (23) to the present case; the interaction matrices (26) are used for the coupling linear in q_1 and q_2, and a simple harmonic approximation is assumed for the quadratic terms ("warping" quadratic terms are omitted for simplicity); with these assumptions, the Hamiltonian (23) for the $E \otimes \varepsilon$ vibronic system takes the form

$$H_{JT} = -\frac{\hbar^2}{2M}\frac{\partial^2}{\partial q_1^2} - \frac{\hbar^2}{2M}\frac{\partial^2}{\partial q_2^2} + \gamma \begin{pmatrix} -q_1 & q_2 \\ q_2 & q_1 \end{pmatrix} + \frac{1}{2}C(q_1^2 + q_2^2) \tag{27}$$

(when a scalar operator or function is added to a matrix, it is implicitly understood that the scalar is multiplied by the identity matrix before the algebraic summation is performed). The terms in Eq. (27) that survive when $\gamma \equiv 0$ represent the kinetic energy and the potential energy of two harmonic oscillators with coordinates q_1 and q_2. The matrix in Eq. (27) represents the linear coupling between the pair of degenerate vibrational modes (q_1, q_2) and the pair of degenerate electronic states (f_x, f_y). The 2×2 interacting matrix has been derived by us on the basis of some simplifying assumptions (that preserve the embodied symmetry), but *its structure is prescribed by symmetry and is thus independent of the assumptions made*. Eq. (27) expresses the famous vibronic Hamiltonian of the $E \otimes \varepsilon$ system, which constitutes the simplest example of (non-trivial) Jahn–Teller system.

Adiabatic surfaces of the $E \otimes \varepsilon$ vibronic system

The solution of the $E \otimes \varepsilon$ vibronic system, described by the Hamiltonian (27), has engaged several authors and different mathematical techniques, including the analytic theory of continued fractions (the continued fraction solution is presented in Section XII-7, while studying the manifestations of the Jahn–Teller effect on optical spectra). Here we wish to discuss some aspects of the adiabatic surfaces of this simple and yet surprisingly rich model.

If the nuclei are regarded as fixed and their kinetic energy operator is ignored in Eq. (27), the potential energy matrix becomes

$$U(q_1, q_2) = \gamma \begin{pmatrix} -q_1 & q_2 \\ q_2 & q_1 \end{pmatrix} + \frac{1}{2}C(q_1^2 + q_2^2) . \tag{28}$$

We can diagonalize the above 2×2 matrix and obtain for the two adiabatic potential surfaces, or branches, the expressions

$$\begin{cases} E_1(q_1, q_2) = -\gamma\sqrt{q_1^2 + q_2^2} + \frac{1}{2}C(q_1^2 + q_2^2) \\ E_2(q_1, q_2) = +\gamma\sqrt{q_1^2 + q_2^2} + \frac{1}{2}C(q_1^2 + q_2^2) \end{cases} \tag{29}$$

(from now on we assume $\gamma > 0$; a change of sign of γ simply would interchange the upper and lower branches). A section of the two branches along the q_1 axis is plotted in Fig. 5, together with the rotation figure generated by it and known as the "Mexican Hat".

It is convenient to introduce the polar coordinates (q, θ) in the $\mathbf{q} \equiv (q_1, q_2)$ plane, and express q_1 and q_2 in polar form

$$q_1 = q\cos\theta \qquad \text{and} \qquad q_2 = q\sin\theta .$$

We remember that the operators $\nabla_{\mathbf{q}}$ and $\nabla_{\mathbf{q}}^2$ in polar coordinates are given by

$$\nabla_{\mathbf{q}} = (\frac{\partial}{\partial q}, \frac{1}{q}\frac{\partial}{\partial\theta}) \quad \text{and} \quad \nabla_{\mathbf{q}}^2 = \frac{\partial^2}{\partial q^2} + \frac{1}{q}\frac{\partial}{\partial q} + \frac{1}{q^2}\frac{\partial^2}{\partial\theta^2} . \tag{30}$$

Using polar coordinates, the adiabatic potential sheets (29) become

$$E_1(q) = -\gamma q + \frac{1}{2}Cq^2 , \qquad E_2(q) = \gamma q + \frac{1}{2}Cq^2 . \tag{31}$$

The minimum of $E_1(q)$ occurs for $q_0 = \gamma/C$ and its value is $E_0 = -\gamma^2/2C$; the depth $|E_0|$ of the minimum is called *Jahn–Teller energy* E_{JT} of the system.

The adiabatic electronic eigenfunctions corresponding to the adiabatic eigenvalues

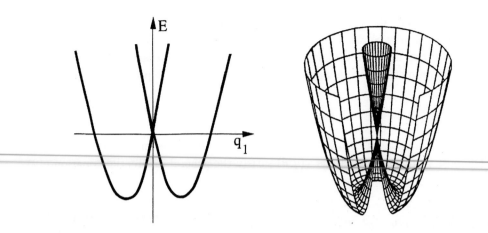

Fig. 5 Adiabatic potential-energy surfaces of the $E \otimes \varepsilon$ Jahn–Teller system. Fig. 5a shows the section of the two branches with the $q_2 = 0$ plane. When rotation by 2π through the energy axis is performed, the lower branch takes the shape of a "Mexican Hat" and the upper branch takes the shape of a conical "Wizard Hat" [from Lai-Shengh Wang, B. Niu, Y. T. Lee, D. A. Shirley, E. Ghelichkhani and E. R. Grant, J. Chem. Phys. **93**, 6318 (1990); copyright 1990 by the American Physical Society].

(29) are easily obtained analytically (apart an arbitrary choice of phase factors); a possible option is

$$\begin{cases} |\overline{\psi}_1(r;q,\theta)\rangle = \cos\frac{\theta}{2}|f_x\rangle + \sin\frac{\theta}{2}|f_y\rangle \\ |\overline{\psi}_2(r;q,\theta)\rangle = -\sin\frac{\theta}{2}|f_x\rangle + \cos\frac{\theta}{2}|f_y\rangle \end{cases} \tag{32}$$

(r stands for all the electronic coordinates). The wavefunctions (32) are obviously determined within an arbitrary phase factor (in principle, any arbitrary regular function of q, θ). The choice done in Eq. (32) guarantees reality and continuity of the adiabatic wavefunctions ψ_1 and ψ_2. With this two requirements, however, the wavefunctions (32) change sign under rotation by 2π, and are thus double-valued.

It is possible to restore single-valuedness in Eqs.(32), for instance multiplying them by the phase factor $\exp(i\theta/2)$; the adiabatic wavefunctions

$$\begin{cases} |\psi_1(r;q,\theta)\rangle = e^{i\theta/2}\cos\frac{\theta}{2}|f_x\rangle + e^{i\theta/2}\sin\frac{\theta}{2}|f_y\rangle \\ |\psi_2(r;q,\theta)\rangle = -e^{i\theta/2}\sin\frac{\theta}{2}|f_x\rangle + e^{i\theta/2}\cos\frac{\theta}{2}|f_y\rangle \end{cases} \tag{33}$$

are continuous and single-valued, but this has been achieved at expense of reality. Thus a *"preferential gauge"* in which the adiabatic wavefunctions are real and at the same time continuous and single-valued does not exist for the $E \otimes \varepsilon$ vibronic system.

The dynamic problem described by the Hamiltonian (27) can be dealt exactly from a mathematical point of view for any value of the coupling constant γ (see Section XII-7). Here we focus specifically on large values of γ, and discuss the $E \otimes \varepsilon$ model with an instructive "ad hoc" procedure, which holds only in the strong coupling limit.

In this case the circular trough of the "Mexican Hat" is deep (Fig. 5) and the nuclear motion is essentially confined there. For low-lying vibronic levels, far from the conical intersection, the nuclei never move on the upper potential-energy surface; thus the vibronic wavefunction can be approximated with the Born–Oppenheimer product of the type

$$\Psi_{\text{trial}} = \chi(q, \theta)\, \psi_1(r; q, \theta) \ ,$$

where both ψ_1 and χ are *single-valued functions* of the vibrational coordinates (q, θ).

The best vibrational functions $\chi(q, \theta)$ satisfy the general differential equation (12), which in the present case takes the form

$$-\frac{\hbar^2}{2M}\left[\frac{\partial^2}{\partial q^2} + \frac{1}{q}\frac{\partial}{\partial q} + \frac{1}{q^2}\frac{\partial^2}{\partial \theta^2}\right]\chi(q,\theta) + E_1(q)\,\chi(q,\theta) + \Lambda_{11}(q,\theta)\chi(q,\theta) = W\,\chi(q,\theta) \ .$$
(34a)

The first operator in Eq. (34a) is the nuclear kinetic energy in polar coordinates; $E_1(q)$ is the adiabatic potential energy surface of the lowest branch (the Mexican Hat); the remaining operator is the non-adiabatic operator $\Lambda_{11}(q,\theta)$, defined by Eq. (13) and given by

$$\Lambda_{11}(q,\theta) = -\frac{\hbar^2}{2M}\langle\psi_1|\nabla_{\mathbf{q}}^2\psi_1\rangle - \frac{\hbar^2}{M}\langle\psi_1|\nabla_{\mathbf{q}}\psi_1\rangle \cdot \nabla_{\mathbf{q}} \ .$$

Using Eqs. (30) for the operators $\nabla_{\mathbf{q}}$ and $\nabla_{\mathbf{q}}^2$, and Eq. (33) for the wavefunction ψ_1, it is easily seen that

$$\Lambda_{11}(q,\theta) = -\frac{\hbar^2}{2M}\langle\psi_1|\nabla_{\mathbf{q}}^2\psi_1\rangle - \frac{\hbar^2}{M}\langle\psi_1|\frac{1}{q}\frac{\partial}{\partial\theta}\psi_1\rangle\frac{1}{q}\frac{\partial}{\partial\theta} = \frac{\hbar^2}{4Mq^2} - \frac{\hbar^2}{2Mq^2}i\frac{\partial}{\partial\theta} \ . \quad (34b)$$

The solutions of Eq. (34a) can be factorized into the product of a θ-dependent function and a q-dependent function in the form

$$\chi(q,\theta) = e^{im\theta}\frac{g(q)}{\sqrt{q}} \quad (m = 0, \pm 1, \pm 2, \ldots) \ , \quad (35a)$$

where the function $1/\sqrt{q}$ has been introduced in view of further elaborations. Inserting Eq. (35a) into Eq. (34a), and straightforward elaboration, we obtain the one-dimensional Schrödinger equation

$$\left[-\frac{\hbar^2}{2M}\frac{\partial^2}{\partial q^2} + E_1(q)\right]g(q) + \frac{\hbar^2}{2Mq^2}\left(m + \frac{1}{2}\right)^2 g(q) = W g(q) \ . \quad (35b)$$

We notice that

$$E_1(q) = -\gamma q + \frac{1}{2}Cq^2 \equiv -\frac{\gamma^2}{2C} + \frac{1}{2}C\left(q - \frac{\gamma}{C}\right)^2 \ .$$

In the strong-coupling limit the trough minimum $E_0 = -\gamma^2/2C$ is very deep; the wavefunctions of Eq. (35b) for low-lying energy levels are concentrated at $q \approx q_0 = \gamma/C$ and the term $\hbar^2(m + 1/2)^2/2Mq^2$ can be replaced by $\hbar^2(m + 1/2)^2/2Mq_0^2$. The eigenvalues of Eq. (35b) then become

$$W_{n,j} = E_0 + (n + \frac{1}{2})\hbar\omega + \frac{\hbar^2}{2Mq_0^2}j^2 \quad (n = 0, 1, 2, \ldots ; j = \pm\frac{1}{2}, \pm\frac{3}{2}, \ldots), \qquad (35c)$$

where $\omega = \sqrt{C/M}$, and j stands for the half odd integer $m+1/2$. The above eigenvalues and the corresponding wavefunctions intuitively describe the combination of a radial harmonic oscillator centred around q_0 and a two-dimensional rigid rotor, moving in the lowest adiabatic sheet.

It is interesting to notice that the ground-state of the $E \otimes \varepsilon$ vibronic system is characterized by the quantum numbers $n=0$ and $j=\pm 1/2$, and is thus doubly degenerate. The responsibility of this twofold degeneracy can be traced back to the conical intersection of the upper and lower adiabatic surfaces at the origin of the (q_1, q_2) plane. It is the conical intersection that prevents the existence of a "preferential gauge" of real-valued, continuous and single-valued adiabatic functions; this is so regardless of the fact that the nuclei never move on the upper potential-energy surface. Notice that in the hypothetical case of existence of the "preferential gauge", the quantum number j would take the integer values $0, \pm 1, \pm 2, \ldots$ and the ground state would be non-degenerate. More detailed theoretical solutions of the dynamic vibronic problem, as well as experimental evidence, confirm that the ground state of the $E \otimes \varepsilon$ model with linear coupling is a doublet [for the more general case of the presence of quadratic coupling see H. Koizumi and I. B. Bersuker, Phys. Rev. Letters **83**, 3009 (1999)].

Before concluding this section, we notice that the Hamiltonian (27), obtained with reference to the triangular molecule geometry of Fig. 4, also describes the interaction of a doublet of electronic states with a doublet of vibrational modes in other high symmetry environments. Here we wish to mention the classical case of Cu^{2+} ions in various materials. The Cu^{2+} ion has the electronic configuration d^9, and can be pictured as a single hole in a closed d shell. When the ion is embedded in an octahedral cage of negative charges, the d-states split into a triplet (at lower energy) and a doublet (at higher energy); the hole occupies the doublet and suffers a Jahn–Teller interaction with the doublet of vibrational modes of the surrounding octahedron (see Fig. 6). It is believed that the Jahn–Teller Cu^{2+} ion plays an important role in the physical and structural properties in the high-temperature copper oxide superconductors, which typically contain CuO_6 octahedra or CuO_5 pyramids; a wide debate in the literature concerns a possible contribution of the Jahn–Teller effect in the pairing mechanism of carriers in high-temperature superconductivity [see for instance the review by M. C. M. O'Brien and C. C. Chancey, Am. J. Phys. **61**, 688 (1993) and references quoted therein].

3.3   The Jahn–Teller effect for triply degenerate electronic states

The $T \otimes \varepsilon$ Jahn–Teller system

The $T \otimes \varepsilon$ Jahn–Teller system concerns an *electronic triplet interacting with a doublet of vibrational modes* in full cubic symmetry. The geometry of an octahedral complex (the

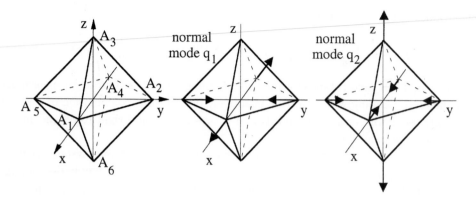

Fig. 6 Geometry of the octahedral complex and degenerate vibrational modes q_1 and q_2, partner of the ε representation.

Table 2 Equilibrium positions and normal displacements of the ε vibrational modes q_1 and q_2 in the cubic symmetry environment.

$\mathbf{d}_1 = a(+1,0,0)$	$\mathbf{u}_1(q_1) = q_1(\frac{1}{2},0,0)$	$\mathbf{u}_1(q_2) = q_2(\frac{-1}{2\sqrt{3}},0,0)$
$\mathbf{d}_2 = a(0,+1,0)$	$\mathbf{u}_2(q_1) = q_1(0,-\frac{1}{2},0)$	$\mathbf{u}_2(q_2) = q_2(0,\frac{-1}{2\sqrt{3}},0)$
$\mathbf{d}_3 = a(0,0,+1)$	$\mathbf{u}_3(q_1) = 0$	$\mathbf{u}_3(q_2) = q_2(0,0,\frac{1}{\sqrt{3}})$
$\mathbf{d}_4 = a(-1,0,0)$	$\mathbf{u}_4(q_1) = q_1(-\frac{1}{2},0,0)$	$\mathbf{u}_4(q_2) = q_2(\frac{1}{2\sqrt{3}},0,0)$
$\mathbf{d}_5 = a(0,-1,0)$	$\mathbf{u}_5(q_1) = q_1(0,\frac{1}{2},0)$	$\mathbf{u}_5(q_2) = q_2(0,\frac{1}{2\sqrt{3}},0)$
$\mathbf{d}_6 = a(0,0,-1)$	$\mathbf{u}_6(q_1) = 0$	$\mathbf{u}_6(q_2) = q_2(0,0,\frac{-1}{\sqrt{3}})$

molecule of SF_6 for instance) is indicated in Fig. 6. In the same figure we also indicate the normal mode displacements q_1 and q_2 of the nuclei, corresponding to the double degenerate ε mode. In Table 2 the equilibrium positions and normal displacements of the vibrational mode ε in the cubic symmetry are reported.

We can easily establish the matrix for the linear interaction of the triplet electronic states with the doublet of normal modes; we have to evaluate the 3×3 matrices of the type

$$U_{\alpha\beta}(q_j) = \int f_\alpha^* \sum_{i=1,6} [V_a(\mathbf{r} - \mathbf{d}_i - \mathbf{u}_i(q_j)) - V_a(\mathbf{r} - \mathbf{d}_i)] \, f_\beta \, d\mathbf{r} \qquad (36)$$

$(\alpha, \beta = x, y, z)$ corresponding to the normal displacements q_1 and q_2. Using Table 2 and Eqs.(25), we express all matrix elements (36) in terms of independent integrals, and consider contributions linear in q_1 and q_2. The matrix potential interaction (18)

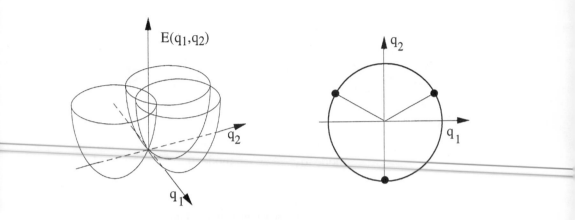

Fig. 7 Schematic representation of the adiabatic-potential energy surfaces in the case of a triply degenerate state T interacting with the ε modes (q_1, q_2) in the cubic symmetry environment. The coordinates of the minima of the potential are also shown in a top view.

for the $T \otimes \varepsilon$ model becomes

$$U(q_1, q_2) = \gamma \begin{pmatrix} q_1 - \dfrac{1}{\sqrt{3}} q_2 & 0 & 0 \\ 0 & -q_1 - \dfrac{1}{\sqrt{3}} q_2 & 0 \\ 0 & 0 & +\dfrac{2}{\sqrt{3}} q_2 \end{pmatrix} + \frac{1}{2} C(q_1^2 + q_2^2) \qquad (37)$$

The interaction constant is $\gamma = [I'(pp\sigma, a) - I'(pp\pi, a)]$ and the spring constant is C; the form of the interaction matrix (37) depends only on symmetry and not on the simplified model, adopted here.

The potential matrix (37) is diagonal and the ε distortion fails to mix electronic states. The energies of the adiabatic sheets and the coordinates of the minima are

$$E_1(q_1, q_2) = \gamma(q_1 - \frac{1}{\sqrt{3}} q_2) + \frac{1}{2} C(q_1^2 + q_2^2) \qquad q_1^0 = -\frac{\gamma}{C} \; ; \quad q_2^0 = \frac{\gamma}{\sqrt{3}C}$$

$$E_2(q_1, q_2) = -\gamma(q_1 + \frac{1}{\sqrt{3}} q_2) + \frac{1}{2} C(q_1^2 + q_2^2) \qquad q_1^0 = \frac{\gamma}{C} \; ; \quad q_2^0 = \frac{\gamma}{\sqrt{3}C}$$

$$E_3(q_1, q_2) = \gamma \frac{2}{\sqrt{3}} q_2 + \frac{1}{2} C(q_1^2 + q_2^2) \qquad q_1^0 = 0 \; ; \quad q_2^0 = -\frac{2\gamma}{\sqrt{3}C} \; .$$

From the values q_1^0 and q_2^0 and Table 2, the Cartesian components of the displacements can be obtained; it is seen by inspection that the three minima describe a distorted octahedron, extended along one axis and compressed along the other two.

The minimum energy is $E_{\min} = -2\gamma^2/3C$ for any of the three valleys. The adiabatic surfaces in the (q_1, q_2) space are the three "disjoint" paraboloids indicated in Fig. 7. Thus a triplet electronic state should always suffer a static Jahn–Teller distortion (but inclusion of spin–orbit coupling may act to mix the different valleys).

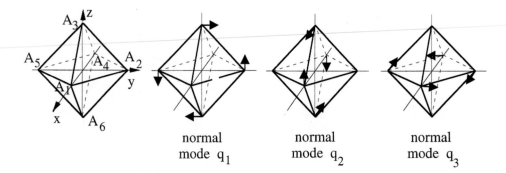

normal
mode q_1

normal
mode q_2

normal
mode q_3

Fig. 8 Geometry of the octahedral complex and normal coordinates q_1, q_2, q_3 of the triply degenerate vibrational mode τ (even under space inversion symmetry).

Table 3 Normal displacements of the vibrational mode τ in the cubic symmetry environment.

normal mode q_1	normal mode q_2	normal mode q_3
$\mathbf{u}_1(q_1) = 0$	$\mathbf{u}_1(q_2) = q_2(0, 0, \frac{1}{2})$	$\mathbf{u}_1(q_3) = q_3(0, \frac{1}{2}, 0)$
$\mathbf{u}_2(q_1) = q_1(0, 0, \frac{1}{2})$	$\mathbf{u}_2(q_2) = 0$	$\mathbf{u}_2(q_3) = q_3(\frac{1}{2}, 0, 0)$
$\mathbf{u}_3(q_1) = q_1(0, \frac{1}{2}, 0)$	$\mathbf{u}_3(q_2) = q_2(\frac{1}{2}, 0, 0)$	$\mathbf{u}_3(q_3) = 0$
$\mathbf{u}_4(q_1) = 0$	$\mathbf{u}_4(q_2) = q_2(0, 0, -\frac{1}{2})$	$\mathbf{u}_4(q_3) = q_3(0, -\frac{1}{2}, 0)$
$\mathbf{u}_5(q_1) = q_1(0, 0, -\frac{1}{2})$	$\mathbf{u}_5(q_2) = 0$	$\mathbf{u}_5(q_3) = q_3(-\frac{1}{2}, 0, 0)$
$\mathbf{u}_6(q_1) = q_1(0, -\frac{1}{2}, 0)$	$\mathbf{u}_6(q_2) = q_2(-\frac{1}{2}, 0, 0)$	$\mathbf{u}_6(q_3) = 0$

The $T \otimes \tau$ Jahn–Teller system

In the $T \otimes \tau$ Jahn–Teller system, an *electronic triplet T interacts linearly with a triplet of vibrational modes τ in a full cubic symmetry*. The triplet of vibrational modes τ (even under space inversion symmetry) is indicated in Fig. 8, while in Table 3 we give the displacements corresponding to each degenerate vibrational mode.

The matrix potential describing the $T \otimes \tau$ model can be worked out using the same procedure adopted before; we easily obtain

$$U(q_1, q_2, q_3) = \gamma \begin{pmatrix} 0 & q_3 & q_2 \\ q_3 & 0 & q_1 \\ q_2 & q_1 & 0 \end{pmatrix} + \frac{1}{2}C(q_1^2 + q_2^2 + q_3^2), \tag{38}$$

where $\gamma = \frac{2}{a}[I(pp\sigma, a) - I(pp\pi, a)]$ (in the following we suppose $\gamma > 0$).

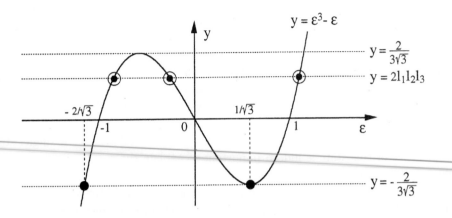

Fig. 9 Graphical solution of the equation $\varepsilon^3 - \varepsilon = 2l_1l_2l_3$; the roots are obtained from the intersection of the cubic curve $y = \varepsilon^3 - \varepsilon$ with the straight line $y = 2l_1l_2l_3$. In particular at the point $P = (1/\sqrt{3})(-1,-1,-1)$ we have $l_1l_2l_3 = -1/(3\sqrt{3})$; the equation $\varepsilon^3 - \varepsilon = -2/(3\sqrt{3})$ has the single root $\varepsilon_1 = -2/\sqrt{3}$ and the doubly degenerate roots $\varepsilon_2 = \varepsilon_3 = 1/\sqrt{3}$.

The eigenvalues of the 3×3 matrix appearing in the right-hand side of Eq. (38) are determined by the equation

$$E^3 - \gamma^2(q_1^2 + q_2^2 + q_3^2)E - 2\gamma^3 q_1 q_2 q_3 = 0 . \tag{39a}$$

If $E(q_1, q_2, q_3)$ indicate the solutions of Eq. (39a), the adiabatic potential-energy surfaces are given by

$$E^{(\mathrm{ad})}(q_1, q_2, q_3) = E(q_1, q_2, q_3) + \frac{1}{2}C(q_1^2 + q_2^2 + q_3^2) . \tag{39b}$$

In order to obtain in a more explicit way the adiabatic potential-energy surfaces, we consider in the q_1, q_2, q_3 space the sphere of radius q, given by $q_1^2 + q_2^2 + q_3^2 = q^2$. We indicate with l_1, l_2, l_3 the director cosines of a point P in q-space, and introduce the dimensionless variable $\varepsilon = E/\gamma q$. Equation (39a) takes now the form

$$\varepsilon^3 - \varepsilon - 2l_1l_2l_3 = 0 , \tag{40a}$$

and its graphical solution is shown in Fig. 9.

It can then immediately be seen that the lowest root of Eq. (40a) is $\varepsilon = -2/\sqrt{3}$; it occurs in any of the four directions

$$\frac{1}{\sqrt{3}}(-1,-1,-1) , \quad \frac{1}{\sqrt{3}}(-1,1,1) , \quad \frac{1}{\sqrt{3}}(1,-1,1) , \quad \frac{1}{\sqrt{3}}(1,1,-1) . \tag{40b}$$

The energy of the ground adiabatic sheet for q along any of the directions (40b) is

$$E_0^{(\mathrm{ad})}(q_1, q_2, q_3) = -\frac{2\gamma}{\sqrt{3}}q + \frac{1}{2}Cq^2 . \tag{40c}$$

Eq. (40c) takes its minimum value $E_0 = -2\gamma^2/3C$ for $q_0 = 2\gamma/C\sqrt{3}$. The eigenvalue equation (40a) can be solved for any other point in q-space of interest and the adiabatic potential-energy branches can be worked out.

Among the numerous problems investigated with the $T \otimes \tau$ vibronic model, we briefly mention here the so-called EL2 impurity center in GaAs. An accredited model in the literature is that this impurity is an isolated antisite defect, with As anion replacing Ga in the cation lattice [see M. Kaminska, Phys. Scripta **T19**, 551 (1987)]. From the four valley structure of the $T \otimes \tau$ ground adiabatic sheet, it is possible to infer that the sequence of vibronic states is a triplet state followed by a total symmetric singlet state; the presence of this state is indeed important to explain the non-linear behaviour of splitting of the ground triplet under uniaxial stress.

We have introduced so far a few typical examples of Jahn–Teller systems. These simple models have been considered because of their own interest and also because they may provide qualitative guidelines in much more complicated and demanding situations. In general the coupling between electronic and nuclear vibrational degrees of freedom are much more complicated than those here introduced, because they may involve several Jahn–Teller active modes of different frequency or symmetry. Another interesting extension is the pseudo Jahn–Teller effect that concerns almost degenerate adiabatic sheets, that nevertheless have to be considered on the same footing. For instance it occurs between the s and p levels of the relaxed F center and other color centers; the vibronic model is essential to explain the luminescence of these centers and the decay times. Further phenomenology and properties of vibronic systems can be found in the wide literature; see for instance G. Bevilacqua, L. Martinelli and N. Terzi (1999) and references quoted therein.

4   The Hellmann–Feynman theorem and electronic–nuclear systems

4.1   General considerations on the Hellmann–Feynman theorem

Parametric dependent operators present some general aspects, that we are going to examine; in this section we consider the Hellmann–Feynman theorem and, in the next section, we consider the geometric Berry phase.

Consider an operator $H(\lambda)$ which depends parametrically on one parameter, or a set of parameters, collectively indicated as λ. The electron Hamiltonian operator $H_e(r; R)$ defined by Eq. (4) is just an example, in which the set of parameters λ are the coordinates R of the nuclei fixed in a given configuration. Consider the eigenvalue equation

$$H(\lambda)\, \psi_n(r; \lambda) = E_n(\lambda)\, \psi_n(r; \lambda) \; , \tag{41}$$

where the variable r denotes all the internal coordinates of the system. We imagine to vary continuously λ in a given region of the parameter space and follow the eigenvalues $E_n(\lambda)$ and the stationary states $\psi_n(r; \lambda)$, obtaining *parameter dependent energy sheets;*

for simplicity we suppose that the sheet under attention has no degeneracy in the domain of parameters λ of interest.

The representation of the operator $H(\lambda)$ in terms of its eigenstates is

$$\langle \psi_m(r; \lambda)|H(\lambda)|\psi_n(r; \lambda) \rangle = E_n(\lambda)\,\delta_{mn} \ . \tag{42a}$$

We have also the orthonormality relations

$$\langle \psi_m(r; \lambda)|\psi_n(r; \lambda) \rangle = \delta_{mn} \ . \tag{42b}$$

We now derive both members of Eqs.(42) with respect to some component, say λ_i, of the set of parameters λ; in doing this we assume that it is possible to interchange integration on the internal coordinates r and partial differentiation with respect to the parameters λ.

The derivation of both members of Eq. (42b) with respect to λ_i gives

$$\langle \frac{\partial}{\partial \lambda_i}\psi_m(r; \lambda)|\psi_n(r; \lambda) \rangle = -\langle \psi_m(r; \lambda)|\frac{\partial}{\partial \lambda_i}\psi_n(r; \lambda) \rangle \ ; \tag{43a}$$

the above equation is a trivial consequence of orthonormality of wavefunctions. In the particular case $m = n$, Eq. (43a) implies

$$\boxed{\langle \psi_m(r; \lambda)|\frac{\partial}{\partial \lambda_i}\psi_m(r; \lambda) \rangle = \text{pure imaginary quantity}} \ . \tag{43b}$$

The derivation of both members of Eq. (42a) with respect to λ_i gives

$$\langle \frac{\partial \psi_m}{\partial \lambda_i}|H|\psi_n \rangle + \langle \psi_m|\frac{\partial H}{\partial \lambda_i}|\psi_n \rangle + \langle \psi_m|H|\frac{\partial \psi_n}{\partial \lambda_i} \rangle = \frac{\partial E_n}{\partial \lambda_i}\delta_{mn} \ .$$

Since $\psi_m(r; \lambda)$ and $\psi_n(r; \lambda)$ are eigenfunctions of $H(\lambda)$ with energy $E_m(\lambda)$ and $E_n(\lambda)$, respectively, it follows

$$\langle \frac{\partial \psi_m}{\partial \lambda_i}|\psi_n \rangle E_n + \langle \psi_m|\frac{\partial H}{\partial \lambda_i}|\psi_n \rangle + E_m\langle \psi_m|\frac{\partial \psi_n}{\partial \lambda_i} \rangle = \frac{\partial E_n}{\partial \lambda_i}\delta_{mn} \ .$$

With the help of Eq. (43a), the above relation becomes

$$\langle \psi_m|\frac{\partial H}{\partial \lambda_i}|\psi_n \rangle + [E_m - E_n]\langle \psi_m|\frac{\partial \psi_n}{\partial \lambda_i} \rangle = \frac{\partial E_n}{\partial \lambda_i}\delta_{mn} \ . \tag{44}$$

The specification of Eq. (44) to the case $m \equiv n$, gives the Hellmann–Feynman theorem

$$\boxed{\langle \psi_m(r; \lambda)|\frac{\partial H(\lambda)}{\partial \lambda_i}|\psi_m(r; \lambda) \rangle = \frac{\partial E_m(\lambda)}{\partial \lambda_i}} \ . \tag{45}$$

This theorem is very important because it provides the gradient along which to move in the case we wish to minimize the energy on a given energy sheet; the gradient on a given energy sheet along λ_i can conveniently be obtained from the expectation value of the quantum mechanical operator $\partial H(\lambda)/\partial \lambda_i$ on the appropriate wavefunction.

The specification of Eq. (44) to the case $m \neq n$, gives the Epstein generalization of the Hellmann–Feynman theorem

$$\langle \psi_m(r;\lambda)| \frac{\partial}{\partial \lambda_i} \psi_n(r;\lambda)\rangle = \frac{1}{E_n(\lambda) - E_m(\lambda)} \langle \psi_m(r;\lambda)| \frac{\partial H(\lambda)}{\partial \lambda_i}|\psi_n(r;\lambda)\rangle \qquad (46)$$

with $E_n(\lambda) \neq E_m(\lambda)$. The theorems (45) and (46) are completely general, and apply whatever is the physical meaning of the external parameters λ (and internal coordinates r). Notice that in Section II-6.2 we discussed an elementary application of the Hellmann–Feynman theorem; in that case the parametric Hamiltonian was the so-called $\mathbf{k} \cdot \mathbf{p}$ Hamiltonian and the parameters λ were just the components of the wavevector $\mathbf{k}$.

4.2   Charge density and atomic forces

We apply now the Hellmann–Feynman theorem to interacting electronic–nuclear systems, with the nuclei fixed in a given configuration; in this case the parameters λ are just the coordinates R of the nuclei in the electron Hamiltonian $H_e(r; R)$ of Eq. (4). The "atomic force" $\mathbf{F}_K = -\partial E/\partial \mathbf{R}_K$ on the nucleus at $\mathbf{R}_K$ is given by differentiation of the adiabatic potential-energy function $E(R)$ (supposed non-degenerate) with respect to the nuclear coordinate $\mathbf{R}_K$. With straightforward elaborations, we can now prove that the "atomic forces" are just the "electrostatic forces", born from all the other nuclear charges and from the total one-body electronic charge density.

Consider the electronic problem with the nuclei fixed in a given configuration R; the electronic Hamiltonian $H_e(r; R)$ is here rewritten for convenience in the form

$$H_e(r;R) = -\sum_i \frac{\hbar^2 \nabla_i^2}{2m} + \frac{1}{2}\sum_{i \neq j} \frac{e^2}{|\mathbf{r}_i - \mathbf{r}_j|} - \sum_{iI} \frac{z_I e^2}{|\mathbf{r}_i - \mathbf{R}_I|} + \frac{1}{2}\sum_{I \neq J} \frac{z_I z_J e^2}{|\mathbf{R}_I - \mathbf{R}_J|} . \qquad (47)$$

Let $\psi(r; R)$ denote the wavefunction of a given adiabatic surface (supposed to be non-degenerate) and $E(R)$ the corresponding energy. It holds

$$\langle \psi(r;R)|H_e(r;R)|\psi(r;R)\rangle = E(R) .$$

Derivation of both members of the above equation with respect to the nuclear coordinate $\mathbf{R}_K$ of the Kth nucleus, and application of the Hellmann–Feynman theorem, gives

$$\langle \psi(r;R)| \frac{\partial}{\partial \mathbf{R}_K} H_e(r;R)|\psi(r;R)\rangle = \frac{\partial E(R)}{\partial \mathbf{R}_K} . \qquad (48)$$

The gradient of the operator $H_e(r; R)$ is given by

$$\frac{\partial}{\partial \mathbf{R}_K} H_e(r;R) = -\frac{\partial}{\partial \mathbf{R}_K}\sum_i \frac{z_K e^2}{|\mathbf{r}_i - \mathbf{R}_K|} + \frac{\partial}{\partial \mathbf{R}_K}\sum_{I(\neq K)} \frac{z_I z_K e^2}{|\mathbf{R}_I - \mathbf{R}_K|} . \qquad (49)$$

From Eq. (48) and Eq. (49),the force $\mathbf{F}_K$ on the nucleus K becomes

$$\mathbf{F}_K = -\frac{\partial E}{\partial \mathbf{R}_K} = \langle \psi(r;R)| \frac{\partial}{\partial \mathbf{R}_K} \sum_i \frac{z_K e^2}{|\mathbf{r}_i - \mathbf{R}_K|} |\psi(r;R)\rangle - \frac{\partial}{\partial \mathbf{R}_K} \sum_{I(\neq K)} \frac{z_I z_K e^2}{|\mathbf{R}_I - \mathbf{R}_K|} \; . \tag{50}$$

Let $n(\mathbf{r};R)$ denote the one-body electron density corresponding to the many-body wavefunction $\psi(r;R)$ (for the definition see for instance Section IV-8); expression (50) reads

$$\boxed{\mathbf{F}_K = -\frac{\partial E}{\partial \mathbf{R}_K} = \int n(\mathbf{r};R) \frac{\partial}{\partial \mathbf{R}_K} \frac{z_K e^2}{|\mathbf{r} - \mathbf{R}_K|} \, d\mathbf{r} - \frac{\partial}{\partial \mathbf{R}_K} \sum_{I(\neq K)} \frac{z_I z_K e^2}{|\mathbf{R}_I - \mathbf{R}_K|} \; . } \tag{51}$$

Thus the force acting on a given nucleus equals the negative gradient of the classical electrostatic potential energy, originated from all the other point charged nuclei and from the quantum mechanical electronic charge distribution.

Equation (51) is rich of physical consequences; once $n(\mathbf{r};R)$ is known, one can determine the forces on the nuclei and then the nuclear motion itself; this is indeed the microscopic foundation of the (classical or quantum) lattice dynamics. Equation (51) contains in germ the working principle of the atomic force microscope [G. Binning, C. F. Quate and Ch. Gerber, Phys. Rev. Lett. **12**, 930 (1986)]. This type of atomic resolution microscope maps the force between the sample surface and a sharp tip probe, and is presently one of the most useful tools for visualization and investigation of surface profiles; among its advantages, we notice that the sample under investigation is not required to be an electric conductor (differently from the scanning tunneling microscope), as the force is directly related to the charge density of the material.

5   Parametric Hamiltonians and Berry phase

Consider the most generic quantum Hamiltonian $H(\mathbf{R})$, that depends on a set of parameters $(R_1, R_2, \dots)$ collectively indicated as $\mathbf{R}$. In this section, for sake of simplicity, the parameter space $\mathbf{R}$ is assumed to span an ordinary three-dimensional space (we can thus use familiar vector relations, without need of generalization of vector algebra to a multi-dimensional space). We consider the eigenfunctions $\psi_n(r;\mathbf{R})$ and the eigenvalues $E_n(\mathbf{R})$ of the Schrödinger equation

$$H(\mathbf{R})\, \psi_n(r;\mathbf{R}) = E_n(\mathbf{R})\, \psi_n(r;\mathbf{R}) \; , \tag{52}$$

where all the internal variables of the system are collectively indicated as r. We discuss here some *general properties of parameter dependent eigenvalue equations*, which are independent on the specific meaning of the "internal variables r" and the "external variables (or parameters) $\mathbf{R}$".

It is evident that the eigenvalue equation (52) at different $\mathbf{R}$ does not determine univocally the wavefunctions $\psi_n(r;\mathbf{R})$, because an $\mathbf{R}$-dependent arbitrariness in the phases of $\psi_n(r;\mathbf{R})$ remains (gauge arbitrariness). Suppose that $\psi_n(r;\mathbf{R})$ constitute a

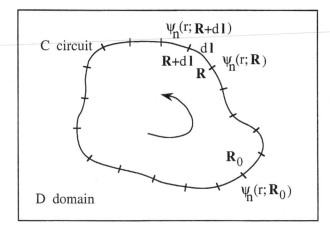

Fig. 10 Schematic representation of the sequence of states $\psi_n(r; \mathbf{R})$ as $\mathbf{R}$ is changed along a circuit C drawn in the parameter space D; it is assumed that no degeneracy point is encountered along the circuit.

set of continuous and single-valued functions in the parameter domain D of interest; then any gauge transformation of the type

$$\widetilde{\psi}_n(r; \mathbf{R}) = e^{i\alpha_n(\mathbf{R})} \psi_n(r; \mathbf{R}) \tag{53}$$

(where $\alpha_n(\mathbf{R})$ are real, continuous and single-valued functions) defines an equivalent set of continuous and single-valued wavefunctions in the domain D. However, in spite of this gauge arbitrariness, it is possible to identify gauge invariant geometric phases, following a procedure initiated by Berry.

In the parameter space $\mathbf{R}$, let us draw a circuit C and let us focus on the nth adiabatic sheet of energy $E_n(\mathbf{R})$ and eigenfunctions $\psi_n(r; \mathbf{R})$; we assume that no degeneracy point is encountered along the circuit C in the chosen adiabatic sheet (see Fig. 10). Along the circuit C, we consider two contiguous points $\mathbf{R}$ and $\mathbf{R} + d\mathbf{l}$, and the corresponding wavefunctions $\psi_n(r; \mathbf{R})$ and $\psi_n(r; \mathbf{R} + d\mathbf{l})$. *The infinitesimal phase difference $d\phi$ between the two functions $\psi_n(r; \mathbf{R})$ and $\psi_n(r; \mathbf{R} + d\mathbf{l})$ is most naturally defined as the argument of their scalar product* (omitting of course any integer multiple of 2π); namely the phase change $d\phi$ is defined by the relation

$$\boxed{d\phi \equiv \arg\langle \psi_n(r; \mathbf{R}) | \psi_n(r; \mathbf{R} + d\mathbf{l}) \rangle} . \tag{54}$$

The real quantity $d\phi$ can be referred as the infinitesimal phase acquired by the quantum system moving from $\mathbf{R}$ to $\mathbf{R} + d\mathbf{l}$ in the given adiabatic sheet along the circuit C. However, the quantity $d\phi$ cannot have a precise physical meaning; in fact $d\phi$ *is clearly a gauge dependent quantity*, i.e. it depends on the actual phases embodied in the wavefunctions $\psi_n(r; \mathbf{R})$ and $\psi_n(r; \mathbf{R} + d\mathbf{l})$.

It is convenient to express $d\phi$ in a slightly different form, by expanding $\psi_n(r; \mathbf{R} + d\mathbf{l})$

in Taylor series up to first order in $d\mathbf{l}$; we obtain from Eq. (54)

$$d\phi = \arg\left[1 + \langle \psi_n(r;\mathbf{R})|\frac{\partial}{\partial\mathbf{R}}\psi_n(r;\mathbf{R})\rangle \cdot d\mathbf{l}\right] .$$

As shown by Eqs.(43), the quantity multiplying $d\mathbf{l}$ in the above equation is *purely imaginary* (this is nothing more than a trivial consequence of normalisation of wave-functions). We can thus write

$$d\phi = (-i)\,\langle \psi_n(r;\mathbf{R})|\frac{\partial}{\partial\mathbf{R}}\psi_n(r;\mathbf{R})\rangle \cdot d\mathbf{l} , \tag{55a}$$

or equivalently

$$d\phi = \mathrm{Im}\,\langle \psi_n(r;\mathbf{R})|\frac{\partial}{\partial\mathbf{R}}\psi_n(r;\mathbf{R})\rangle \cdot d\mathbf{l} , \tag{55b}$$

where Im stands for "imaginary part of"; notice that Im(...) denotes a *real* quantity. Expression (55b) for the phase change $d\phi$ has the advantage to exhibit explicitly that $d\phi$ is a real and infinitesimal quantity.

The *total geometric phase* $\gamma_n(C)$, or *Berry phase*, introduced in completing a circuit C in the parameter space is defined as

$$\boxed{\gamma_n(C) = \mathrm{Im}\,\oint_C \langle \psi_n(r;\mathbf{R})|\frac{\partial}{\partial\mathbf{R}}\psi_n(r;\mathbf{R})\rangle \cdot d\mathbf{l}} . \tag{56}$$

The "Berry phase" is thus the phase acquired by a quantum system moving along a circuit C on a given adiabatic surface (it is assumed that the circuit C does not touch any degeneracy point). It is convenient to re-write the expression of the Berry phase in the equivalent form

$$\gamma_n(C) = \oint_C \mathbf{A}_n(\mathbf{R}) \cdot d\mathbf{l} , \tag{57a}$$

where $\mathbf{A}_n(\mathbf{R})$ is defined as

$$\mathbf{A}_n(\mathbf{R}) = \mathrm{Im}\,\langle \psi_n(r;\mathbf{R})|\nabla_\mathbf{R}\psi_n(r;\mathbf{R})\rangle , \tag{57b}$$

and the gradient $\partial \ldots /\partial\mathbf{R}$ is now denoted by $\nabla_\mathbf{R}$ for convenience.

It is now easy to show that the geometric phase $\gamma_n(C)$ is *gauge invariant*, in spite of the fact that every infinitesimal contribution is *gauge dependent*. Consider in fact the gauge transformation defined by Eq. (53); we have

$$\widetilde{\mathbf{A}}_n(\mathbf{R}) = \mathrm{Im}\,\langle \widetilde{\psi}_n(r;\mathbf{R})|\nabla_\mathbf{R}\widetilde{\psi}_n(r;\mathbf{R})\rangle = \mathrm{Im}\,\langle e^{i\alpha_n(\mathbf{R})}\psi_n(r;\mathbf{R})|\nabla_\mathbf{R} e^{i\alpha_n(\mathbf{R})}\psi_n(r;\mathbf{R})\rangle$$

$$= \mathrm{Im}\,\langle \psi_n(r;\mathbf{R})|\nabla_\mathbf{R}\psi_n(r;\mathbf{R})\rangle + \nabla_\mathbf{R}\alpha_n(\mathbf{R}) .$$

The relation so obtained can be written as

$$\widetilde{\mathbf{A}}_n(\mathbf{R}) = \mathbf{A}_n(\mathbf{R}) + \nabla_\mathbf{R}\alpha_n(\mathbf{R}) . \tag{58}$$

Hence

$$\widetilde{\gamma}_n(C) = \oint_C \widetilde{\mathbf{A}}_n(\mathbf{R}) \cdot d\mathbf{l} = \oint_C [\mathbf{A}_n(\mathbf{R}) + \nabla_\mathbf{R}\alpha_n(\mathbf{R})] \cdot d\mathbf{l} \equiv \gamma_n(C) ; \tag{59}$$

the last passage holds because the circuitation on a closed loop of the gradient of the regular single-valued function $\alpha_n(\mathbf{R})$ vanishes.

The expression (56) of the Berry phase defines a gauge independent quantity as a sum of contributions, which taken one-by-one are gauge dependent. It would be desirable to express the Berry phase directly in terms of gauge invariant contributions; this is possible by using the Stokes theorem to transform the line integral along C into an integral on a surface enclosed by C.

Consider a circuit C and a surface S of contour C in the parameter domain D ; we fix our attention to the nth adiabatic functions, and *suppose that no degeneracy point exists on the circuit C and on the surface S of contour C* (in other words we assume that the circuit C *does not touch* and *does not encircle any degeneracy point*). Exploiting Stokes theorem, Eq. (57a) can be written in the form

$$\gamma_n(C) = \oint_C \mathbf{A}_n(\mathbf{R}) \cdot d\mathbf{l} = \int_S [\mathrm{curl}\, \mathbf{A}_n(\mathbf{R})] \cdot d\mathbf{S} \ . \tag{60}$$

From Eq. (58), it is seen that $\mathrm{curl}\, \mathbf{A}_n(\mathbf{R}) = \mathrm{curl}\, \widetilde{\mathbf{A}}_n(\mathbf{R})$, and the invariant character of the Berry phase to gauge transformations is thus confirmed.

It is of interest to derive an explicit expression of the quantity

$$\mathbf{B}_n(\mathbf{R}) = \mathrm{curl}\, \mathbf{A}_n(\mathbf{R}) \ . \tag{61}$$

Using elementary vector algebra, we have

$$\mathbf{B}_n(\mathbf{R}) = \mathrm{Im}\, \mathrm{curl}\, \langle \psi_n(r; \mathbf{R}) | \nabla_{\mathbf{R}} \psi_n(r; \mathbf{R}) \rangle = \mathrm{Im} \langle \nabla_{\mathbf{R}} \psi_n(r; \mathbf{R}) | \times | \nabla_{\mathbf{R}} \psi_n(r; \mathbf{R}) \rangle$$

$$= \mathrm{Im} \sum_m{}' \langle \nabla_{\mathbf{R}} \psi_n(r; \mathbf{R}) | \psi_m(r; \mathbf{R}) \rangle \times \langle \psi_m(r; \mathbf{R}) | \nabla_{\mathbf{R}} \psi_n(r; \mathbf{R}) \rangle \ .$$

The exclusion of the term $m = n$ in the summation is justified by the fact that it vanishes identically (it involves the vectorial product of two vectors, which differ only for the sign). Since for $m \neq n$ we can use the Epstein generalization of the Hellmann–Feynman theorem (see Eq. 46), we obtain the expression

$$\mathbf{B}_n(\mathbf{R}) = \mathrm{Im} \sum_m{}' \frac{\langle \psi_n(r; \mathbf{R}) | \nabla_{\mathbf{R}} H | \psi_m(r; \mathbf{R}) \rangle \times \langle \psi_m(r; \mathbf{R}) | \nabla_{\mathbf{R}} H | \psi_n(r; \mathbf{R}) \rangle}{[E_n(\mathbf{R}) - E_m(\mathbf{R})]^2} \tag{62}$$

and

$$\gamma_n(C) = \int_S \mathbf{B}_n(\mathbf{R}) \cdot d\mathbf{S} \ . \tag{63}$$

Thus the Berry phase $\gamma_n(C)$ can be interpreted as the flux of $\mathbf{B}_n(\mathbf{R})$ across a surface of contour C.

The microscopic expression (62) for $\mathbf{B}_n(\mathbf{R})$ is clearly independent on the phases embodied in the wavefunctions, and has been of value to calculate the Berry phase in a number of systems. Furthermore, Eq. (61) and Eq. (63) make apparent some analogy with the electromagnetic theory, where $\mathbf{A}_n(\mathbf{R})$ plays the role of a "fictitious vector potential" and $\mathbf{B}_n(\mathbf{R}) = \mathrm{curl}\, \mathbf{A}_n(\mathbf{R})$ the role of a "fictitious magnetic field".

Considerations on the Berry phase for adiabatic potential-energy surfaces

We consider now the particularly important case in which the parameter dependent Hamiltonian in consideration is the electron Hamiltonian $H_e(r; R)$. This Hamiltonian is invariant under time-reversal symmetry, as discussed in Section 2.2. In the case of spinless particles, the adiabatic wavefunctions $\psi_m(r; \mathbf{R})$ can be taken in real form; it follows that expression (62) for $\mathbf{B}_n(\mathbf{R})$ vanishes identically. Thus the Berry phase is zero for any circuit fully embedded in a domain with no degeneracy point; this result is compatible with the existence in the domain of interest of a "preferential gauge" of real-valued, continuous and single-valued adiabatic wavefunctions.

We wish to stress that the above discussion, and in particular Eq. (62), is valid provided the nth adiabatic sheet under attention is *never degenerate* at any point belonging to the circuit C or to the surface S (of contour C). If the circuit C does encircle a point of degeneracy $\mathbf{R}_0$, and we attempt to apply the Stokes theorem for transforming the line integral (56) into a surface integral, we face the problem of how to handle the degeneracy point $\mathbf{R}_0$ present in the adiabatic surface. It is evident that Eq. (62) for $\mathbf{B}_n(\mathbf{R})$ presents vanishingly small denominators for $\mathbf{R} \approx \mathbf{R}_0$; thus $\mathbf{B}_n(\mathbf{R})$ is expected to become singular, unless all the matrix elements $\langle \psi_m(r; \mathbf{R}_0) | \nabla_{\mathbf{R}} H_e(r; \mathbf{R}) | \psi_n(r; \mathbf{R}_0) \rangle \equiv M_{mn}(\mathbf{R})$ vanish for $\mathbf{R} = \mathbf{R}_0$. The Jahn–Teller systems are just characterized by non-vanishing of some (or all) of the matrix elements $M_{mn}(\mathbf{R}_0)$ (see Eq. 20). In the case of Jahn–Teller systems, we can infer that the "magnetic field" $\mathbf{B}_n(\mathbf{R})$ is zero everywhere except at the conical intersection, where it has a delta-function singularity.

The $E \otimes \varepsilon$ system, as well as other Jahn–Teller systems, constitute a nice working demonstration of the Berry phase procedure, and are illuminating on several deductions. The treatment of the $E \otimes \varepsilon$ Jahn–Teller system has been discussed in Section 3.2. It is straightforward to see that the Berry phase (56) evaluated by means of the eigenfunctions (33) along a circuit that surrounds the degeneracy point $q_1 = q_2 = 0$ is π; the whole discussion of Section 3.2 can be revisited at the light of the geometric Berry phase and appreciated in its model transcendent aspects (see Ham 1987). A quite similar analysis can be carried out for other systems; in particular for the $T \otimes \tau$ system we refer to the elegant works of O'Brien and Ham [M. C. M. O'Brien, J. Phys. A**22**, 1779 (1989); F. S. Ham, J. Phys. Condens. Matter **2**, 1163 (1990)].

6 Macroscopic electric polarization in crystals and Berry phase

The macroscopic electric polarization of materials plays a fundamental role in the phenomenological description of dielectrics. In the literature, most often, simplified and intuitive models are adopted for the electric polarization of systems of charged particles. More recently, the progress of the methods of electronic state calculations, and in particular of the density functional method, have made possible accurate first-principle investigations of the ground-state properties of interacting electronic–nuclear systems. In this section, we discuss some aspects of the quantum theory of polarization of crystalline solids and the role assumed in the theory by the geometric Berry phase.

Consider a crystal of volume $V = N\Omega$, formed by an arbitrary (large) number N of identical unit cells of volume Ω, and let $\rho_{\text{tot}}(\mathbf{r})$ denote the total electronic and nuclear (or ionic) charge density. The charge distribution $\rho_{\text{tot}}(\mathbf{r})$ is neutral and averages to zero within any unit cell, as well as within the whole crystal (composed by an integer number of unit cells). The *average electric polarization* of the crystal (i.e. the electric dipole per unit volume) is related to the crystal charge density by the expression

$$\mathbf{P} = \frac{1}{N\Omega} \int_{\text{crystal}} \mathbf{r}\, \rho_{\text{tot}}(\mathbf{r})\, d\mathbf{r} = \frac{1}{N\Omega} \left[\sum_j z_j e \mathbf{R}_j - e \int_{\text{crystal}} n_{\text{el}}(\mathbf{r})\, \mathbf{r}\, d\mathbf{r} \right] , \qquad (64)$$

where e is the absolute value of the electronic charge, $\mathbf{R}_j$ are the positions within the crystal of the pointwise nuclei (or ions) of charge $z_j e$, and $n_{\text{el}}(\mathbf{r})$ is the electronic density. The average polarization (64) is also called macroscopic polarization, or simply polarization, of the crystal.

It is almost superfluous to remark that the polarization vector $\mathbf{P}$, defined by Eq. (64), *depends on the details* (shape and location) *of the unit cell, chosen to build up the crystal* (see Fig. 11 for an illustration); on the other hand, it is shown below that infinitesimal (and hence also finite) changes of polarization are independent of how the crystal has been assembled and are thus genuine bulk properties. Thus all the physical effects related to *changes of polarization* can be evaluated unambiguously and compared with experimental measurements. Experimental means to produce changes of the polarization of crystals include application of a stress (piezoelectric crystals), or changes of temperature (pyroelectric crystals), or spontaneous polarization reversal by electric fields (ferroelectric crystals).

To keep things at the essential, consider a dielectric composed by two sublattices (GaAs for instance), and let λ indicate a parameter that regulates with continuity the relative position of one sublattice with respect to the other. The parameter λ is arranged to take values 0 and 1 at the initial and final configuration. For simplicity, it is also assumed that the fundamental translation vectors remain unchanged, so that the Brillouin zone remains unchanged too, and that the material remains a semiconductor or an insulator for all the values of λ in the interval $[0, 1]$. For any assigned value of λ, we indicate by $\psi_n(\mathbf{k}, \mathbf{r}; \lambda)$ the Kohn–Sham orbitals, and we focus now on the change of electronic polarization as λ is varied (the change of ionic polarization with λ is trivial, because of the pointwise nature of ionic charges).

From the knowledge of the parameter-dependent orbitals $\psi_n(\mathbf{k}, \mathbf{r}; \lambda)$ (normalized to unity in the volume $V = N\Omega$ of the crystal), we can express the electronic contribution to the average crystal polarization in the form

$$\mathbf{P}_{\text{el}}(\lambda) = \frac{1}{V} \int_{\text{crystal}} \mathbf{r}\, \rho_{\text{el}}(\mathbf{r}; \lambda)\, d\mathbf{r} = \frac{2(-e)}{V} \sum_{n\mathbf{k}} \langle \psi_n(\mathbf{k}, \mathbf{r}; \lambda)|\, \mathbf{r}\, |\psi_n(\mathbf{k}, \mathbf{r}; \lambda)\rangle ,$$

where the factor 2 takes into account spin degeneracy and the sum is over all occupied bands of the semiconductor or insulator, under attention. We indicate, as usual, the

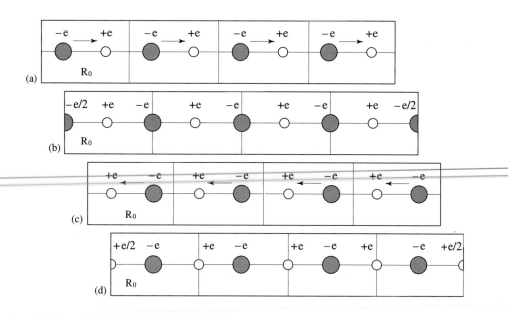

Fig. 11 Schematic illustration of the electric polarization of a one-dimensional crystal, composed by a number of unit cells of volume Ω. The crystal is constituted by uniformly charged spheres of charge $+e$ and $-e$ (white and black circles) at distance R_0. Four different ways to choose the unit cell and to assemble the crystal are indicated; arrows indicate the electric dipole vector in the unit cell. In case (a) the polarization is $P = +eR_0/\Omega$; in case (b) and (d) it is zero; in case (c) $P = -eR_0/\Omega$. If all the negative charges of the crystal are displaced by δ, while the positive charges are kept fixed, the *change of polarization* is in any case given by $\Delta P = -e\delta/\Omega$, regardless of the different ways the crystal has been assembled.

periodic part of the Bloch functions by $u_n(\mathbf{k}, \mathbf{r}; \lambda)$ and obtain

$$\mathbf{P}_{\text{el}}(\lambda) = \frac{2(-e)}{V} \sum_{n\mathbf{k}} \langle u_n(\mathbf{k}, \mathbf{r}; \lambda) | \mathbf{r} | u_n(\mathbf{k}, \mathbf{r}; \lambda) \rangle . \tag{65}$$

The above expression for $\mathbf{P}_{\text{el}}(\lambda)$ defies a clear physical interpretation, as expected from the fact that (obviously) the electronic charge density $\rho_{\text{el}}(\mathbf{r}; \lambda)$, considered by itself, is not neutral. Actually, Eq. (65) is even dependent on a shift of origin in $\mathbf{r}$-space. We thus focus on the derivative $\partial \mathbf{P}_{\text{el}}(\lambda)/\partial \lambda$, which represents the change of polarization and is directly related to a macroscopic polarization current; we obtain

$$\boxed{\frac{\partial \mathbf{P}_{\text{el}}(\lambda)}{\partial \lambda} = \frac{2(-e)}{V} \sum_{n\mathbf{k}} 2 \operatorname{Re} \langle u_n(\mathbf{k}, \mathbf{r}; \lambda) | \mathbf{r} | \frac{\partial}{\partial \lambda} u_n(\mathbf{k}, \mathbf{r}; \lambda) \rangle ,} \tag{66}$$

where Re stands for "real part of".

It is possible to express Eq. (66) in terms of the geometric Berry phase acquired by cell-periodic wavefunctions $u_n(\mathbf{k}, \mathbf{r}; \lambda)$ moving along an appropriate closed loop in the $(\lambda, \mathbf{k})$ parameter space. For this purpose we consider the quantity

$$\mathbf{B}_{n\mathbf{k}}(\lambda) = 2 \operatorname{Re} \langle u_n(\mathbf{k}, \mathbf{r}; \lambda) | \mathbf{r} | \frac{\partial}{\partial \lambda} u_n(\mathbf{k}, \mathbf{r}; \lambda) \rangle . \tag{67a}$$

We insert appropriately the partition operator $\sum_\alpha |u_\alpha(\mathbf{k}, \mathbf{r}; \lambda)\rangle\langle u_\alpha(\mathbf{k}, \mathbf{r}; \lambda)|$ and also exploit Eq. (II-36) (with some obvious change in notation); Eq. (67a) can be recast in the form

$$\mathbf{B}_{n\mathbf{k}}(\lambda) = 2\,\mathrm{Re}\sum_\alpha \langle u_n(\mathbf{k}, \mathbf{r}; \lambda)|\,\mathbf{r}\,|u_\alpha(\mathbf{k}, \mathbf{r}; \lambda)\rangle\langle u_\alpha(\mathbf{k}, \mathbf{r}; \lambda)|\frac{\partial}{\partial\lambda}u_n(\mathbf{k}, \mathbf{r}; \lambda)\rangle$$

$$= 2\,\mathrm{Re}(-i)\sum_\alpha\langle\frac{\partial}{\partial\mathbf{k}}u_n(\mathbf{k}, \mathbf{r}; \lambda)|u_\alpha(\mathbf{k}, \mathbf{r}; \lambda)\rangle\langle u_\alpha(\mathbf{k}, \mathbf{r}; \lambda)|\frac{\partial}{\partial\lambda}u_n(\mathbf{k}, \mathbf{r}; \lambda)\rangle$$

$$= 2\,\mathrm{Im}\,\langle\frac{\partial}{\partial\mathbf{k}}u_n(\mathbf{k}, \mathbf{r}; \lambda)|\frac{\partial}{\partial\lambda}u_n(\mathbf{k}, \mathbf{r}; \lambda)\rangle \tag{67b}$$

(it is irrelevant whether the term $\alpha = n$ is included or not, as its contribution to the real part is zero in any case). Eq. (66) for the polarization derivative becomes

$$\boxed{\frac{\partial\mathbf{P}_{\mathrm{el}}(\lambda)}{\partial\lambda} = \frac{2(-e)}{V}\sum_{n\mathbf{k}}2\,\mathrm{Im}\,\langle\frac{\partial}{\partial\mathbf{k}}u_n(\mathbf{k}, \mathbf{r}; \lambda)|\frac{\partial}{\partial\lambda}u_n(\mathbf{k}, \mathbf{r}; \lambda)\rangle}\,. \tag{67c}$$

The total change $\Delta\mathbf{P}_{\mathrm{el}}$ in polarization is obtained by integrating in $d\lambda$ within the range $0 \le \lambda \le 1$ and in $d\mathbf{k}$ in the Brillouin zone; the Brillouin zone is taken to be a rectangular parallelepiped with $-\pi/a \le k_x < \pi/a$, $-\pi/b \le k_y < \pi/b$, and $-\pi/c \le k_z < \pi/c$; for simplicity we assume that the k_x, k_y dependence of the cell-periodic functions can be neglected [these and other assumptions are not strictly necessary and are relaxed in realistic calculations. In practice, the integration over the three-dimensional Brillouin zone is carried out performing integrations over one variable (say k_z), once a number of discretized "special points" are chosen for the other two variables k_x, k_y]. We obtain for the contribution of the nth band to the change of electronic polarization

$$\Delta P_{\mathrm{el}} = \frac{2(-e)}{V}\frac{V}{(2\pi)^3}\frac{2\pi}{a}\frac{2\pi}{b}\int_0^1 d\lambda\int_{-\pi/c}^{+\pi/c}2\,\mathrm{Im}\,\langle\frac{\partial}{\partial k}u_n(k, \mathbf{r}; \lambda)|\frac{\partial}{\partial\lambda}u_n(k, \mathbf{r}; \lambda)\rangle dk \tag{68}$$

where the component k_z is indicated by k.

The integrand in Eq. (68) can be elaborated using the straightforward identity $2\,\mathrm{Im}\,z = \mathrm{Im}\,z - \mathrm{Im}\,z^*$; it holds

$$2\,\mathrm{Im}\,\langle\frac{\partial}{\partial k}u_n(k, \mathbf{r}; \lambda)|\frac{\partial}{\partial\lambda}u_n(k, \mathbf{r}; \lambda)\rangle$$

$$= \mathrm{Im}\,\left[\langle\frac{\partial}{\partial k}u_n(k, \mathbf{r}; \lambda)|\frac{\partial}{\partial\lambda}u_n(k, \mathbf{r}; \lambda)\rangle - \langle\frac{\partial}{\partial\lambda}u_n(k, \mathbf{r}; \lambda)|\frac{\partial}{\partial k}u_n(k, \mathbf{r}; \lambda)\rangle\right]$$

$$= \mathrm{Im}\,\left[\frac{\partial}{\partial k}\langle u_n(k, \mathbf{r}; \lambda)|\frac{\partial}{\partial\lambda}u_n(k, \mathbf{r}; \lambda)\rangle - \frac{\partial}{\partial\lambda}\langle u_n(k, \mathbf{r}; \lambda)|\frac{\partial}{\partial k}u_n(k, \mathbf{r}; \lambda)\rangle\right]$$

$$= \mathrm{Im}\,\mathrm{curl}\,\langle u_n(k, \mathbf{r}; \lambda)|\nabla u_n(k, \mathbf{r}; \lambda)\rangle\,,$$

where the gradient is respect to the variables (k, λ), and curl $\mathbf{A}$ denotes the component $\partial A_\lambda/\partial k - \partial A_k/\partial\lambda$.

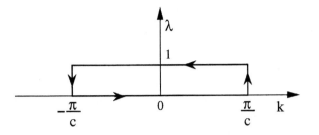

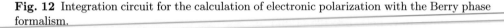

Fig. 12 Integration circuit for the calculation of electronic polarization with the Berry phase formalism.

The integration (68) then becomes

$$\Delta P_{\mathrm{el}} = \frac{(-e)}{\pi\, ab}\mathrm{Im}\int_0^1 d\lambda \int_{-\pi/c}^{+\pi/c} \mathrm{curl}\,\langle u_n(k,\mathbf{r};\lambda)|\nabla u_n(k,\mathbf{r};\lambda)\rangle\, dk\, d\lambda$$

$$= \frac{(-e)}{\pi\, ab}\mathrm{Im}\oint_C \langle u_n(k,r;\lambda)|\nabla u_n(k,r;\lambda)\rangle \cdot d\mathbf{l}\;, \tag{69}$$

where the integration in the (k,λ) space has been reduced, by Stokes circuitation theorem, to the evaluation of a numerical integral along the circuit indicated in Fig. 12.

Comparison with the general equation (56) shows that the polarization change can be expressed in the elegant and compact form

$$\Delta P_{\mathrm{el}} = \frac{(-e)}{\pi\, ab}\gamma_n(C)\;, \tag{70}$$

where $\gamma_n(C)$ is the Berry phase of the cell-periodic wavefunctions moving along the circuit of Fig. 12. For more details on the formalism based on the Berry phase, and for workable microscopic calculations of polarization changes in ferroelectric and piezo-electric crystals we refer to the works of King-Smith and Vanderbilt (1993) and of Resta (1994).

Further reading

M. P. Allen and D. J. Tildesley "Computer Simulation of Liquids" (Clarendon, Oxford 1989)

T. Azumi and K. Matsuzaki "What Does the Term 'Vibronic Coupling' Mean?" Photochemistry and Photobiology **25**, 315 (1977)

M. V. Berry "Quantal Phase Factors Accompanying Adiabatic Changes" Proc. Roy. Soc. London **392**, 451 (1984)

I. Bersuker and V. Z. Polinger "Vibronic Interactions in Molecules and Crystals" (Springer, New York 1989)

G. Bevilacqua, L. Martinelli and N. Terzi (eds.) "Electron–Phonon Dynamics and Jahn–Teller Effect" (World Scientific, Singapore 1999)

M. Born and K. Huang "Dynamical Theory of Crystal Lattices" (Oxford 1954)

R. Englman "The Jahn–Teller Effect in Molecules and Crystals" (Wiley-Interscience, New York 1972)

F. S. Ham "Berry's Geometrical Phase and the Sequence of States in the Jahn–Teller Effect" Phys. Rev. Lett. **58**, 725 (1987)

R. D. King-Smith and D. Vanderbilt "Theory of Polarization of Crystalline Solids" Phys. Rev. B**47**, 1651 (1993)

H. C. Longuet-Higgins "Some Recent Developments in the Theory of Molecular Energy Levels" in Advances in Spectroscopy vol. 2 (Wiley-Interscience, New York 1961)

A. A. Maradudin, E. W. Montroll, G. H. Weiss and I. P. Ipatova "Theory of Lattice Dynamics in the Harmonic Approximation" (Academic Press, New York 1971); A. A. Maradudin in "Dynamical Properties of Solids" edited by G. K. Horton and A. A. Maradudin (North-Holland, Amsterdam 1974)

Mary C. M. O'Brien and C. C. Chancey "The Jahn–Teller Effect: an Introduction and Current Review" Am. J. Phys. **61**, 688 (1993)

M. C. Payne, M. P. Teter, D. C. Allan, T. A. Arias and J. D. Joannopoulos "Iterative Minimization Techniques for ab-Initio Total-Energy Calculations: Molecular Dynamics and Conjugate Gradients" Rev. Mod. Phys. **64**, 1045 (1992)

Yu. E. Perlin and M. Wagner eds. "The Dynamical Jahn–Teller Effect in Localized Systems" (North-Holland, Amsterdam 1984)

R. Resta "Macroscopic Polarization in Crystalline Dielectrics: the Geometric Phase Approach" Rev. Mod. Phys. **66**, 899 (1994)

A. Shapere and F. Wilczek eds. "Geometric Phases in Physics" (World Scientific, Singapore 1989)

M. D. Sturge "The Jahn–Teller Effect in Solids" Solid State Physics **20**, 92 (1967), edited by F. Seitz, D. Turnbull and H. Ehrenreich (Academic, New York)

D. Y. Yarkony "Diabolical Conical Intersections" Rev. Mod. Phys. **68**, 985 (1996)

IX

Lattice dynamics of crystals

In the previous chapter we have laid down the basic concepts of the quantum mechanical theory of electrons and nuclei in mutual interaction. According to the adiabatic principle, the system of light particles (the electrons) is preliminarly studied with the heavy particles (the nuclei) fixed in a given spatial configuration; the total ground-state energy of the electronic system, thought of as a function of the nuclear coordinates, becomes then the potential energy for the nuclear motion. We are interested here in small displacements around the equilibrium configuration of the ground adiabatic energy surface (supposed to be non-degenerate). Within the harmonic approximation, we describe the dispersion curves for normal mode propagation in crystals, and introduce the concept of *phonons*, as travelling quanta of vibrational energy. In the study of the lattice vibrations of ionic or partially ionic crystals (polar crystals), essentially new features appear in the long-wavelength limit; in polar crystals, the coupling of optical vibrational branches with the electromagnetic field leads to the concept of mixed phonon–photon quasiparticles, known as *polaritons*.

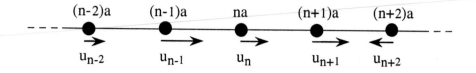

Fig. 1 Longitudinal displacements in a one-dimensional monatomic lattice. The equilibrium positions $t_n = na$ are indicated by circles; the displacements u_n at a given instant are indicated by arrows.

1　Dynamics of monatomic one-dimensional lattices

To describe the lattice vibrations of crystals, we consider first linear chains of equal atoms (present section), then linear chains with a basis of different atoms (Section 2), and finally general three-dimensional structures (Section 3). This sequence of increasing sophistication is adopted because some physical concepts are better illustrated in one-dimensional situations, where notations and technicalities can be kept at the essential.

So, we begin by considering a one-dimensional chain, of lattice constant a, formed by a (large) number N of atoms of mass M. We indicate by u_n the (longitudinal) displacement of the n-th atom from the equilibrium position $t_n = na$, at a particular time (see Fig. 1). We denote by $E_0(\{u_n\})$ the total ground-state energy of the crystal Hamiltonian, with the *nuclei fixed in the positions* $R_n = na + u_n$; the energy $E_0(\{u_n\})$ is also called *static lattice energy*. The ground state of the crystal is supposed to be non-degenerate for all configurations $\{u_n\}$ of interest.

In agreement with the general adiabatic principles of Section VIII-2, the total ground-state energy $E_0(\{u_n\})$ of the interacting electronic-nuclear system, with the nuclei fixed in the configuration u_n, becomes the "potential energy" for the nuclear motion. It is also assumed that, the forces $F_i = -\partial E_0(\{u_n\})/\partial u_i$ acting on the nuclei depend on the instantaneous nuclear positions $\{u_n\}$; retardation effects due to the finite propagation velocity of light are neglected at this stage and must be properly included, whenever necessary (see Section 7).

In the study of small oscillations, it is convenient to expand the total ground-state energy $E_0(\{u_n\})$ in increasing powers of the displacements u_n; we have the Taylor expansion

$$E_0(\{u_n\}) = E_0(0) + \frac{1}{2} \sum_{nn'} \left(\frac{\partial^2 E_0}{\partial u_n \partial u_{n'}} \right)_0 u_n u_{n'}$$

$$+ \frac{1}{3!} \sum_{nn'n''} \left(\frac{\partial^3 E_0}{\partial u_n \partial u_{n'} \partial u_{n''}} \right)_0 u_n u_{n'} u_{n''} + \dots \quad (1)$$

(derivatives evaluated at the equilibrium configuration carry the subscript 0). In the expansion (1), the linear terms in the displacements are not present since $\partial E_0/\partial u_n = 0$ at the equilibrium configuration. The total ground-state energy $E_0(0)$ at the equilibrium configuration is important in the discussion of the cohesive energy, but irrelevant in the discussion of lattice vibrations. The truncation to quadratic terms in the expression (1) is called "harmonic approximation"; in general the anharmonic terms (cubic, quartic terms and so on) are taken into account only after the harmonic approximation has been carried out.

In the harmonic approximation, which is appropriate for sufficiently small displacements, the total crystal energy (1) becomes

$$E_0^{(\text{harm})}(\{u_n\}) = E_0(0) + \frac{1}{2} \sum_{n\,n'} D_{nn'} u_n u_{n'} , \tag{2}$$

where

$$D_{nn'} = \left(\frac{\partial^2 E_0}{\partial u_n \partial u_{n'}} \right)_0 \tag{3}$$

denote the second derivatives of the static lattice energy $E_0(\{u_n\})$ evaluated at the equilibrium configuration. The quantities $D_{nn'}$ are called *force constants*, and the matrix D formed with the force constants $D_{nn'}$ is called *force-constant matrix*. The force constants $D_{nn'}$ represent the proportionality coefficients connecting the forces acting on the nuclei with the displacements suffered by the nuclei; in the harmonic approximation, in fact we have

$$F_n = -\frac{\partial E_0^{(\text{harm})}}{\partial u_n} = -\sum_{n'} D_{nn'} u_{n'} . \tag{4}$$

There are some general symmetries and constraints that must be obeyed by the force-constant matrix D. From the definition (3), it follows that the force-constant matrix D is real and symmetric

$$D_{nn'} = D_{n'n} . \tag{5a}$$

The translational symmetry of the lattice requires

$$D_{nn'} = D_{mm'} \quad \text{if} \quad t_n - t_{n'} = t_m - t_{m'} . \tag{5b}$$

Furthermore we have the general and important "sum rule"

$$\sum_{n'} D_{nn'} \equiv 0 \quad \text{for any } n \; ; \tag{5c}$$

this is a trivial consequence of the fact that the forces, given by Eq. (4), vanish not only when all nuclear displacements are zero, but also when all nuclear displacements are equal. Eq. (5a) and Eq. (5c) show that the sum of the matrix elements of any row or any column of the force-constant matrix D vanishes.

We consider now the classical equation of motion for the nth nucleus of mass M in the position $R_n = na + u_n$ under the force F_n; we have

$$M \ddot{u}_n = -\sum_{n'} D_{nn'} u_{n'} \tag{6}$$

with $n = 1, 2, \ldots, N$. The set of coupled differential equations (6) can be solved, in general, looking for solutions periodic in time of the form $u_n(t) = A_n \exp(-i\omega t)$. In the present case, we can take advantage of the translational symmetry of the force-constant matrix D in real space; this suggests to solve Eqs. (6) looking for solutions in the form of travelling waves, periodic in space and time, of the type

$$u_n(t) = A \, e^{i(qna - \omega t)} \, . \tag{7}$$

Replacing Eq. (7) into Eqs. (6), we have that the phonon frequencies ω are given by the relation

$$-M\omega^2 A = -\sum_{n'} D_{nn'} \, e^{-iq(na - n'a)} A \, .$$

The above expression can be recast in the form

$$\boxed{M\omega^2(q) = D(q)} \, , \tag{8a}$$

where

$$\boxed{D(q) = \sum_{n'} D_{nn'} \, e^{-iq(na - n'a)}} \; ; \tag{8b}$$

notice that the Fourier transform $D(q)$ of the force-constant matrix elements $D_{nn'}$ does not depend on the specific value of n because of the property (5b).

Equation (8a) provides the dispersion relation $\omega = \omega(q)$ connecting the phonon frequency to the phonon wavenumber of the travelling plane wave of type (7). Notice that the displacements (7), are unaffected by changes of q by integer multiples of $2\pi/a$; the independent values of q are confined within the Brillouin zone $-\pi/a < q \leq \pi/a$. When standard Born–von Karman boundary conditions are applied, i.e. one requires $u_n(t) \equiv u_{n+N}(t)$, the allowed values of q within the Brillouin zone become discretized with values $m(2\pi/Na)$ and m integer; the number of allowed wavenumbers within the Brillouin zone equals the number of unit cells of the crystal (see Section I-1). We now apply our analysis to the specific case of a linear chain of atoms with nearest neighbour interactions only.

Linear monatomic chain with nearest neighbour interactions

To show the essential aspects of the lattice vibrations in the linear chain, we suppose that the only relevant inter-atomic interactions occur between nearest neighbour atoms; in other words, we assume that the only force constants different from zero are D_{nn}, $D_{n\,n+1}$ and $D_{n-1\,n}$. From the general properties of the force constants summarized by Eqs. (5), it is seen that there is a unique independent parameter (denoted below by C), and it holds

$$D_{nn} = 2C \, , \qquad D_{n\,n+1} = D_{n-1\,n} = -C \, . \tag{9}$$

The energy (2) of the linear chain, in the harmonic approximation and nearest

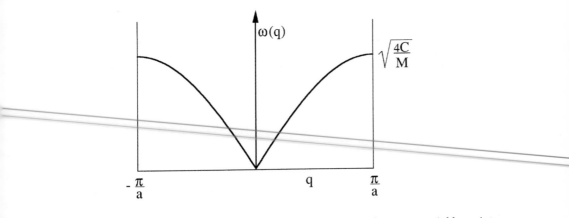

Fig. 2 Phonon dispersion curve for a monatomic linear lattice with nearest neighbour interactions only; the Brillouin zone is the segment between $-\pi/a$ and $+\pi/a$.

neighbour interaction becomes (taking $E_0(0)$ as the reference energy)

$$E_0^{(\text{harm})} = \frac{1}{2}C\sum_n(2u_n^2 - u_nu_{n+1} - u_nu_{n-1}) \equiv \frac{1}{2}C\sum_n(u_n - u_{n+1})^2 . \qquad (10)$$

Equation (10) is quite intuitive and represents the elastic energy of a chain of atoms, connected to nearest neighbours with springs of constant C.

The classical equations of motion (6) for the nuclear vibrations are thus

$$\boxed{M\ddot{u}_n = -C(2u_n - u_{n+1} - u_{n-1})} \qquad (11)$$

for any n integer number. The set of discrete coupled differential equations (11) can be solved looking for travelling waves, periodic in space and time, of the form $u_n(t) = A\exp(iqna - i\omega t)$. By direct substitution, or equivalently from Eqs. (8), one obtains

$$-M\omega^2 = -C(2 - e^{iqa} - e^{-iqa}) = -4C\sin^2\frac{1}{2}qa . \qquad (12a)$$

The dipersion relation for normal modes is thus

$$\omega = \sqrt{\frac{4C}{M}}\,|\sin\frac{1}{2}qa| , \qquad (12b)$$

and is illustrated in Fig. 2; we see that the spectrum of vibrational frequencies extends from zero to a cutoff frequency $\omega_{\max} = \sqrt{4C/M}$.

It is interesting to consider the normal modes in the *long wavelength limit* $qa \ll 1$. The dispersion relation (12b) takes the form

$$\omega = \sqrt{\frac{C}{M}}\,aq \equiv v_sq \qquad (qa \ll 1) ;$$

the proportionality coefficient v_s between phonon frequency ω and phonon wavenum-

Table 1 Energy units of most frequent use for phonons.

	$h\nu \; (\nu = 10^{12} \, \text{hertz}) = 4.1357 \, \text{meV}$	$\hbar\omega \; (\omega = 10^{13} \, \text{rad/sec}) = 6.5822 \, \text{meV}$
	$1 \, \text{eV}/hc = 8065 \, \text{cm}^{-1}$	$1 \, \text{eV}/k_B = 11605 \, \text{K}$
for brevity:	$1 \, \text{eV} \longleftrightarrow 8065 \, \text{cm}^{-1}$	$1 \, \text{eV} \longleftrightarrow 11605 \, \text{K}$
	$1 \, \text{meV} \longleftrightarrow 8.065 \, \text{cm}^{-1}$	$1 \, \text{cm}^{-1} \longleftrightarrow 0.124 \, \text{meV}$

ber q represents the velocity of the sound in the medium, and is given by

$$v_s \equiv \sqrt{\frac{C}{M}} \, a \; . \tag{12c}$$

From Eq. (12c), we can estimate the value of the cutoff angular frequency

$$\omega_{\text{max}} = 2\sqrt{\frac{C}{M}} = \frac{2v_s}{a} \approx \frac{10^5 \, \text{cm/sec}}{10^{-8} \, \text{cm}} = 10^{13} \, \text{rad/sec} \; .$$

A typical vibration spectrum extends in general to the infrared region, up to energies $\hbar\omega$ of several tens of meV (see Table 1).

In the long wavelength limit, we can perform a continuous approximation to the set of discretized coupled differential equations (11); in fact $(-2u_n + u_{n+1} + u_{n-1})/a^2$ can be considered as the finite difference expression of the second order derivative $\partial^2 u/\partial x^2$. Equation (11) is thus equivalent to $M\ddot{u} = Ca^2 \partial^2 u/\partial x^2$ and the propagation velocity of the elastic wave is again given by Eq. (12c).

2   Dynamics of diatomic one-dimensional lattices

We consider now the dynamics of a diatomic linear chain, of lattice constant a_0, with two atoms of mass M_1 and M_2 in the unit cell; this model can be considered as the prototype of a crystal with basis. In the equilibrium configuration, we assume that the atoms of mass M_1 occupy the sublattice positions $R_n^{(1)} = na_0$, while the atoms of mass M_2 occupy the sublattice positions $R_n^{(2)} = (n + 1/2)a_0$ (see Fig. 3).

We denote by u_n the displacements of the atoms of mass M_1 and with v_n the displacements of the atoms of mass M_2. For simplicity we assume that only nearest neighbour atoms interact with elastic forces of spring constant C. The classical equations of motion for the two types of particles are

$$\boxed{\begin{aligned} M_1 \ddot{u}_n &= -C(2u_n - v_{n-1} - v_n) \\ M_2 \ddot{v}_n &= -C(2v_n - u_n - u_{n+1}) \end{aligned}} \tag{13}$$

for any integer number n.

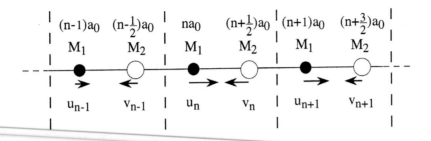

Fig. 3 Longitudinal displacements in a one-dimensional diatomic lattice. The equilibrium positions of the two sublattices of atoms, of mass M_1 and M_2, are indicated by black and white circles, respectively; the displacements u_n and v_n at a given instant are indicated by arrows.

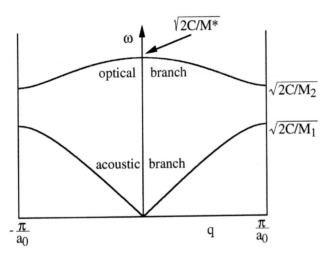

Fig. 4 Phonon dispersion curves of a diatomic linear chain, with nearest neighbour atoms interacting with spring constant C. The masses of the atoms are M_1 and M_2 (with $M_1 > M_2$); M^* is the reduced mass.

To solve the set of discrete coupled differential equations (13), we look for travelling waves, periodic in space and time, of the form

$$u_n(t) = A_1\, e^{i(qna_0 - \omega t)} \quad \text{and} \quad v_n(t) = A_2\, e^{i(qna_0 + qa_0/2 - \omega t)} \,. \tag{14}$$

We notice that the vibrations of atoms of the same sublattice in different cells have the same amplitudes and phase relations of Bloch type.

Replacing (14) into (13), we obtain

$$-M_1 \omega^2 A_1 = -C(2A_1 - A_2\, e^{-iqa_0/2} - A_2\, e^{iqa_0/2})$$

$$-M_2 \omega^2 A_2 = -C(2A_2 - A_1\, e^{-iqa_0/2} - A_1\, e^{iqa_0/2}) \,.$$

The two linear homogeneous equations in the two unknown amplitudes A_1 and A_2 have a non-trivial solution if the determinant of the coefficients of A_1 and A_2 is zero, namely

$$\begin{vmatrix} 2C - M_1\omega^2 & -2C\cos\frac{qa_0}{2} \\ -2C\cos\frac{qa_0}{2} & 2C - M_2\omega^2 \end{vmatrix} = 0 .$$

The eigenvalues are given by

$$\omega^2 = C\left(\frac{1}{M_1} + \frac{1}{M_2}\right) \pm C\sqrt{\left(\frac{1}{M_1} + \frac{1}{M_2}\right)^2 - \frac{4\sin^2(qa_0/2)}{M_1 M_2}} , \tag{15a}$$

and the corresponding amplitudes satisfy the relation

$$\frac{A_1}{A_2} = \frac{2C\cos(qa_0/2)}{2C - M_1\omega^2} . \tag{15b}$$

The dispersion relations (15a) are illustrated in Fig. 4. We have now two branches, the lower one called "acoustic" and the upper one called "optical", with a frequency gap between them. In the particular case that $M_1 \equiv M_2$, the gap between the acoustic branch and the optical branch disappears and we recover the result of the previous section [in the case $M_1 \equiv M_2$, Fig. 4 and Fig. 2 coincide exactly, once the trivial folding due to the fact that $a_0 = 2a$ is performed].

Long wavelength limit of a diatomic linear chain: acoustic and optical modes

We consider now the normal modes of a diatomic chain in the long wavelength limit $qa_0 \ll 1$. For small qa_0, the two roots ω^2 and the corresponding amplitudes (given by Eqs. 15) are

$$\omega^2 = \frac{2C}{M_1 + M_2}\left(\frac{a_0}{2}\right)^2 q^2 + O(q^4) \qquad A_1 = A_2 \quad \text{acoustic branch}$$

and

$$\omega^2 = \frac{2C}{M^*} + O(q^2) \qquad M_1 A_1 = -M_2 A_2 \quad \text{optical branch} ,$$

where M^* is the reduced mass given by $1/M^* = (1/M_1) + (1/M_2)$.

In the acoustic branch, in the long wavelength limit, the atoms vibrate in phase and with the same amplitude; the frequency ω is proportional to the wavenumber q, and the proportionality coefficient, the sound velocity v_s, is given by

$$v_s = \sqrt{\frac{C}{(M_1 + M_2)/2}}\,\frac{a_0}{2} .$$

The above relation is the obvious counterpart of expression (12c), if we notice that the average mass $(M_1 + M_2)/2$ replaces the mass M; a and $a_0/2$ denote nearest neighbour distance in the monatomic and diatomic models, respectively; C is the spring constant.

In the upper branch, A_1 and A_2 have opposite signs and absolute values inversely proportional to atomic masses; this means that the two atoms in the unit cell move in opposite directions, while the "center of mass" of the unit cell remains fixed. As $q \to 0$, the frequency $\omega(q)$ of the optical branch tends to the finite value $\omega_0 = \sqrt{2C/M^*}$.

It is instructive to consider the optical modes of the diatomic chain in the continuous approximation; this is useful also to justify the origin of the name attached to such a branch. For optical modes in the long wavelength limit, we can assume that nearest neighbour atoms on the same sublattice have the same displacements. Eqs. (13) can be recast in the form

$$M_1 \ddot{u} = -2C(u - v) \tag{16a}$$

$$M_2 \ddot{v} = -2C(v - u) , \tag{16b}$$

where the discretized index n is now replaced by the continuous variable x in the argument of the functions $u(x, t)$ and $v(x, t)$. In optical modes, the atoms in the two sublattices move against each other; so it is convenient to discuss optical vibrations via the *relative displacement variable* $w = u - v$. If we divide both members of Eq. (16a) by M_1, both members of Eq. (16b) by M_2, and subtract, we obtain

$$\ddot{w} = -\omega_0^2 w \tag{17}$$

where $\omega_0 = \sqrt{2C/M^*}$; the relative motion of the atoms M_1 and M_2 around the center of mass of the unit cell is harmonic with frequency ω_0.

The optical modes owe their name to the fact that they are expected to couple strongly with electromagnetic fields (of appropriate frequency). Suppose in fact that the diatomic crystal can be pictured as composed by ionic (or partially ionic) units, with effective net charge $\pm e^*$ (*polar crystals*). In the presence of electric fields, the equations of motion (16) have to be modified in the form

$$M_1 \ddot{u} = -2C(u - v) + e^* E_{\text{loc}} \tag{18a}$$

$$M_2 \ddot{v} = -2C(v - u) - e^* E_{\text{loc}} . \tag{18b}$$

We have indicated with E_{loc} the local electric field acting at the lattice sites, as this field may be different from the average macroscopic electromagnetic field; the distinction between local and average fields (if any) is discussed in Section 7-3. In terms of the relative displacement w we have

$$\ddot{w} = -\omega_0^2 w + \frac{e^*}{M^*} E_{\text{loc}} ; \tag{19}$$

the above equation describes the forced oscillations of a mechanical system of proper frequency ω_0, coupled to a driving field of some frequency ω; coupling effects are expected to be particularly significant when the frequency ω is resonant or almost resonant with the optical mode frequency ω_0. A detailed analysis of the coupling of photon and phonon modes in polar crystals, and the resulting polariton effects are discussed in Section 7.

3 Dynamics of general three-dimensional crystals

Crystal dynamical matrix and phonon frequencies

In the previous two sections, we have discussed the lattice dynamics of one-dimensional crystals, with or without a basis. We complete now the subject, addressing the general problem of lattice dynamics of three-dimensional crystals, with or without a basis; for this purpose, we follow step-by-step the script of the previous two sections.

Consider a general three-dimensional crystal, with N unit cells, translational vectors $\mathbf{t}_n$, and a basis of atoms in the positions $\mathbf{d}_1, \mathbf{d}_2, \ldots, \mathbf{d}_{n_b}$. We label atoms with two indices $(n\nu)$, where the (Latin) index n denotes the unit cells of the crystal and the (Greek) index ν the atoms inside the unit cell. According to the general principles of the adiabatic approximation, we consider first *the nuclei fixed in the positions* $\mathbf{t}_n + \mathbf{d}_\nu + \mathbf{u}_{n\nu}$, and we denote by $E_0(\{\mathbf{u}_{n\nu}\})$ the total ground-state energy (or *static lattice energy*) of the electronic-nuclear system. The expansion of E_0 up to second order in the displacements from the equilibrium positions (*harmonic approximation*) gives

$$E_0^{(\text{harm})}(\{\mathbf{u}_{n\nu}\}) = E_0(0) + \frac{1}{2} \sum_{n\nu\alpha, n'\nu'\alpha'} D_{n\nu\alpha, n'\nu'\alpha'} u_{n\nu\alpha} u_{n'\nu'\alpha'} \tag{20a}$$

where $\alpha, \alpha' = x, y, z$; $\nu, \nu' = 1, 2, \ldots, n_b$; $n = 1, 2, \ldots, N$. The "force constants" are defined as the second derivative of $E_0(\{\mathbf{u}_{n\nu}\})$ evaluated at the equilibrium configuration

$$D_{n\nu\alpha, n'\nu'\alpha'} = \left(\frac{\partial^2 E_0}{\partial u_{n\nu\alpha} \, \partial u_{n'\nu'\alpha'}} \right)_0 . \tag{20b}$$

In the expansion (20a), the linear terms in the displacements are not present since $\partial E_0 / \partial u_{n\nu\alpha} \equiv 0$ at the equilibrium configuration.

The matrix D, formed with the "force constants" elements $D_{n\nu\alpha, n'\nu'\alpha'}$, obeys some general properties and constraints. From the definition (20b), we have immediately that the matrix D is real and symmetric, with

$$D_{n\nu\alpha, n'\nu'\alpha'} = D_{n'\nu'\alpha', n\nu\alpha} . \tag{21a}$$

The translation symmetry of the lattice implies

$$D_{n\nu\alpha, n'\nu'\alpha'} = D_{m\nu\alpha, m'\nu'\alpha'} \quad \text{if} \quad \mathbf{t}_n - \mathbf{t}_{n'} = \mathbf{t}_m - \mathbf{t}_{m'} ; \tag{21b}$$

(the presence of point symmetry operations may imply further constraints, that can be analysed with group theory considerations). Furthermore we have the general and important "sum rule"

$$\sum_{n'\nu'} D_{n\nu\alpha, n'\nu'\alpha'} \equiv 0 \tag{21c}$$

[the proof is a straightforward generalization of what already done for the demonstration of Eq. (5c)].

The classical equations of motion for the nuclei in the instantaneous positions $\mathbf{t}_n + \mathbf{d}_\nu + \mathbf{u}_{n\nu}$ under the forces $\mathbf{F}_{n\nu} = -\partial E_0^{(\text{harm})} / \partial \mathbf{u}_{n\nu}$ are

$$M_\nu \ddot{u}_{n\nu\alpha} = - \sum_{n'\nu'\alpha'} D_{n\nu\alpha, n'\nu'\alpha'} \, u_{n'\nu'\alpha'} \tag{22}$$

where $n, n' = 1, 2, \ldots, N$; $\nu, \nu' = 1, 2, \ldots, n_b$; $\alpha, \alpha' = x, y, z$. The presence of translational symmetry suggests to solve the set of coupled differential equations (22) looking for solutions in the form of travelling waves of the type

$$\mathbf{u}_{n\nu}(t) = \mathbf{A}_\nu(\mathbf{q}, \omega) \, e^{i(\mathbf{q} \cdot \mathbf{t}_n - \omega t)} \; . \tag{23a}$$

The *polarization vectors* $A_{\nu\alpha}(\mathbf{q}, \omega)$ $(\nu = 1, 2, \ldots, n_b; \alpha = x, y, z)$ of the vibrations of the nuclei within the primitive cell are left unspecified and are determined below by solution of an appropriate secular equation. Replacing (23a) into (22), we have

$$-M_\nu \omega^2 A_{\nu\alpha} = - \sum_{n'\nu'\alpha'} D_{n\nu\alpha, n'\nu'\alpha'} \, e^{-i\mathbf{q} \cdot (\mathbf{t}_n - \mathbf{t}_{n'})} A_{\nu'\alpha'} \; . \tag{23b}$$

Non trivial solutions of Eq. (23b) are obtained by solving the determinantal equation

$$\boxed{\left\| D_{\nu\alpha, \nu'\alpha'}(\mathbf{q}) - M_\nu \omega^2 \delta_{\alpha\alpha'} \delta_{\nu,\nu'} \right\| = 0} \; , \tag{23c}$$

where

$$\boxed{D_{\nu\alpha, \nu'\alpha'}(\mathbf{q}) = \sum_{n'} D_{n\nu\alpha, n'\nu'\alpha'} \, e^{-i\mathbf{q} \cdot (\mathbf{t}_n - \mathbf{t}_{n'})}} \; . \tag{23d}$$

The matrix $D(\mathbf{q})$, with elements $D_{\nu\alpha, \nu'\alpha'}(\mathbf{q})$ is called *"the dynamical matrix of the crystal"* in reciprocal space.

Equation (23c) is the fundamental eigenvalue equation for the normal modes of a crystal. The dynamical matrix $D(\mathbf{q})$ has dimension $3n_b$, where n_b is the number of atoms forming the basis of the unit cell. The secular equation (23c) produces $3n_b$ eigenvalues (called phonons or normal modes); at every vector $\mathbf{q}$ we have thus $3n_b$ normal modes, giving rise to $3n_b$ phonon branches as $\mathbf{q}$ is varied within the first Brillouin zone. For a crystal with N unit cells, the total number of normal modes equals $3n_b N$, i.e. 3 times the total number of atoms. Let $\omega(\mathbf{q}, p)$ $(p = 1, 2, \ldots 3n_b)$ denote the frequency of the p-th normal mode of wavevector $\mathbf{q}$, and $\mathbf{A}_\nu(\mathbf{q}, p)$ $(\nu = 1, 2, \ldots, n_b)$ the corresponding polarization vectors. A mode $\omega(\mathbf{q}, p)$ is called "longitudinal" (or "transverse") in the case the polarization vectors $\mathbf{A}_\nu(\mathbf{q}, p)$ are parallel to $\mathbf{q}$ (or perpendicular to $\mathbf{q}$). Modes which involve oscillating electric dipoles are called "optically active" since they can couple directly with electromagnetic fields.

From the standard algebraic properties of matrix eigenvalue equations, the eigenvectors of the secular equation (23c) satisfy the orthogonality relations

$$\sum_{\nu\alpha} M_\nu A_{\nu\alpha}^*(\mathbf{q}, p) A_{\nu\alpha}(\mathbf{q}, p') = \delta_{p,p'} \; .$$

The above relation can be written in extended form

$$M_1 \mathbf{A}_1^*(\mathbf{q}, p) \cdot \mathbf{A}_1(\mathbf{q}, p') + \ldots + M_{n_b} \mathbf{A}_{n_b}^*(\mathbf{q}, p) \cdot \mathbf{A}_{n_b}(\mathbf{q}, p') = \delta_{p,p'} \; . \tag{24}$$

In the case of a monatomic three-dimensional Bravais lattice, $n_b = 1$ in Eq. (24), and for the three branches it holds $\mathbf{A}^*(\mathbf{q}, p) \cdot \mathbf{A}(\mathbf{q}, p') = \delta_{p,p'}$.

It is interesting to consider the eigenvalue equation (23c) in the long wavelength limit $\mathbf{q} \to 0$. From the sum rule (21c) and Eq. (23d) it follows

$$\sum_{\nu'} D_{\nu\alpha,\nu'\alpha'}(\mathbf{q}=0) \equiv 0 \; ;$$

in particular we have

$$\sum_{\nu'\alpha'} D_{\nu\alpha,\nu'\alpha'}(\mathbf{q}=0)\, A_{\alpha'} \equiv 0 \; . \qquad (25a)$$

The above equation shows that vibrations, with amplitudes $A_{\nu'\alpha'} \equiv A_{\alpha'}$ equal for all the atoms of the basis, satisfy the secular equation (23c) with frequency $\omega = 0$. Thus *any three-dimensional crystal presents three acoustic branches* with $\omega(\mathbf{q}) \to 0$ as $\mathbf{q} \to 0$, with all atoms in the unit cell vibrating in phase and with the same amplitude. The remaining $3n_b - 3$ "optical" modes vibrate in such a way that the motion of the center of mass of the cell is unaltered; in fact the orthogonality relation (24) to the acoustic modes, with polarization vectors $\mathbf{A}^*$ independent from the atomic index, gives

$$M_1 \mathbf{A}_1 + M_2 \mathbf{A}_2 + \ldots + M_{n_b} \mathbf{A}_{n_b} = 0 \qquad (25b)$$

for any of the optical modes.

Before considering a few examples of phonon branches in crystals, it is convenient to summarize the basic approximations contained in the dynamical matrix approach. According to the general adiabatic principle (see Section VIII-2), one preliminarily considers the nuclei fixed in a given configuration $\{R_I\}$, and determines (ab initio or semiempirically) the total ground-state energy $E_0(\{R_I\})$ versus $\{R_I\}$ (the ground state is supposed to be non-degenerate). Then, the crucial assumption is done that the static lattice energy $E_0(\{R_I\})$, which is just a static property, actually controls the nuclear dynamics. To justify this crucial point, it is required that transitions from the ground adiabatic surface to the excited adiabatic surfaces, induced by the nuclear motion, can be neglected. It is also required that the forces acting on the nuclei depend on the instantaneous nuclear positions, so that retardation effects can be neglected too. Finally, it is assumed that the displacements from equilibrium positions are sufficiently small to justify the expansion of $E_0(\{R_I\})$ to second order in the displacements. In summary, the basic approximations of the crystal dynamical matrix approach include (*i*) *the adiabatic approximation*, (*ii*) *the harmonic approximation*, and (*iii*) *instantly interparticle interactions*.

Phonon dispersion curves with the crystal dynamical matrix and short-range or long-range nature of force constants

We wish now to provide a few illustrative examples of phonon dispersion curves in crystals, obtained with the force constant approach. First principle methods based on the density functional theory have permitted the accurate calculation of the total ground-state energy of several materials (with the nuclei fixed in chosen configurations); this has made it possible the more ambitious project of calculating the inter-atomic force constants in a number of cases. First principle calculations of force constants are in

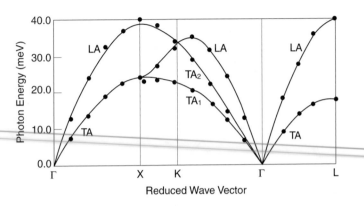

Fig. 5 Phonon dispersion curves of aluminum along symmetry directions. The solid lines represents the calculations of A. A. Quong and B. M. Klein, Phys. Rev. B**46**, 10734 (1992) (copyright 1992 by the American Physical Society). Longitudinal and transverse acoustic branches are indicated by LA and TA (or TA$_1$ and TA$_2$), respectively. The experimental points are from the papers of G. Gilat and R. M. Nicklow, Phys. Rev. **143**, 487 (1966) and R. Stedman, S. Almqvist and G. Nilsson, Phys. Rev. **162**, 549 (1967).

general very demanding; thus most often, empirical assumptions and special models of interaction are made. The assumption of "two-body" central forces, acting between pairs of atoms composing the lattice is the most common; however in many cases this approximation is too drastic, and angular forces and torsional forces are to be included appropriately. Often the force constants of the dynamical matrix are considered as disposable or semi-empirical parameters.

Once a specific model of interatomic forces has been chosen, the dynamical matrix in reciprocal space can be set up and diagonalization at several points of the Brillouin zone provides the phonon dispersion curves of the crystal. Some crystals can be intuitively described as consisting of neutral atoms interacting with *short-range forces*; the force constants extend to a reasonably small number of shells (say nearest neighbours, second nearest neighbours and possibly a few more up to ten shells or so) and become safely negligible afterwards. Examples of crystals with short-range nature of inter-atomic forces include simple Bravais lattices, metals, homopolar elemental semiconductors (such as silicon and germanium).

In polar crystals, such as ionic crystals and heteropolar semiconductors, the crystal lattice can be intuitively described as constituted by charged ions interacting both with *short-range forces and long-range Coulomb forces*. In the ionic picture of polar crystals, appropriate site dependent "Born effective charges" are attributed to the ions of the different sublattices, to mimic long-range interactions.

The crystal dynamical matrix for polar crystals is obtained by direct summation of short-range terms and Ewald method for long-range Coulomb terms. It should be noticed that the dynamical matrix of polar crystals at small q is particularly vulnerable to the long-range nature of the force constants in real space; actually, the dynamical

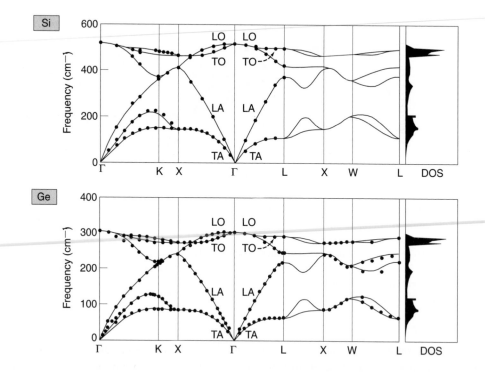

Fig. 6 Phonon dispersion curves and density-of-states of Si and Ge calculated by P. Gian-nozzi, S. de Gironcoli, P. Pavone and S. Baroni, Phys. Rev. B**43**, 7231 (1991) (copyright 1991 by the American Physical Society). Longitudinal and transverse acoustic (or optical) modes are indicated by LA and TA (LO and TO), respectively. The experimental points are from G. Dolling, in "Inelastic Scattering of Neutrons in Solids and Liquids" edited by S. Ekland (IAEA, Vienna 1963) Vol.II, p.37; G. Nilsson and G. Nelin, Phys. Rev. B**3**, 364 (1971) and Phys. Rev. B**6**, 3777 (1972). Conversion to meV units can be done noting that 1 cm^{-1}= 0.124 meV.

matrix has a pathological behaviour (called "non-analyticity"), which is responsible of the transverse-longitudinal splitting of optical phonons (see Section 7 for further aspects). The effects due to the long-range nature of inter-atomic forces are evident in the optical phonon branches of polar semiconductors (see Fig. 7), and are even more important in typical ionic crystals (see Fig. 8).

In Fig. 5 we show the phonon dispersion curves of aluminum, together with the experimental measurements obtained with coherent inelastic scattering of slow neutrons. Since aluminum crystallizes in a simple fcc Bravais lattice, the phonon dipersion curves consist of three acoustic branches. The acoustic modes are degenerate at $\mathbf{q} = 0$; along the high symmetry directions ΓL and ΓX the two transverse modes are degenerate; in directions of low or no symmetry, the three branches are non-degenerate.

In Fig. 6 we show the phonon dispersion curves of silicon and germanium. Since these elemental semiconductors crystallize in a fcc lattice with two atoms per unit cell, we have now three acoustic branches and three optical branches. The acoustic

Table 2 Frequencies ω_{LO} and ω_{TO} (in cm^{-1}) of longitudinal optical and transverse optical phonons for six semiconductors. The calculations are taken from P. Giannozzi, S. de Gironcoli, P. Pavone and S. Baroni, Phys. Rev. **B43**, 7231 (1991), to which we refer for further details; experimental data are in parentheses. The static and the high frequency dielectric constants are also given; notice that the ratio $\omega_{LO}^2/\omega_{TO}^2$ equals (within experimental error) the ratio $\varepsilon_s/\varepsilon_\infty$.

	Si	Ge	GaAs	AlAs	GaSb	AlSb
ω_{LO}	517 (517)	306 (304)	291 (291)	400 (402)	237 (233)	334 (344)
ω_{TO}	517 (517)	306 (304)	271 (271)	363 (361)	230 (224)	316 (323)
$\dfrac{\omega_{LO}^2}{\omega_{TO}^2}$	1	1	1.15 (1.17)	1.22 (1.24)	1.06 (1.08)	1.12 (1.14)
ε_s	12.1	16.5	12.40	10.06	15.69	12.04
ε_∞	12.1	16.5	10.60	8.16	14.44	10.24
$\dfrac{\varepsilon_s}{\varepsilon_\infty}$	1	1	1.17	1.23	1.09	1.17

branches as well as the optical branches are degenerate at $\mathbf{q} = 0$. Along the high symmetry directions ΓX and ΓL the two transverse acoustic modes and the two transverse optical modes are degenerate; in direction of low or no symmetry, all modes are non-degenerate.

In Fig. 7 we show the phonon dispersion curves for heteropolar semiconductors GaAs, AlAs, GaSb and AlSb. In these crystals the inter-atomic forces include long-range Coulomb interaction, because of the partial ionic nature of the chemical bond. Since these heteropolar semiconductors crystallize in a fcc lattice with two atoms per unit cell, the phonon curves present three acoustic and three optical branches; as expected, the acoustic and optical branches are well separated in crystals where the mass difference of the two atoms in the unit cell is large. A most important feature of Fig. 7 is the longitudinal-transverse splitting of optical modes at $\mathbf{q} \approx 0$; this splitting is the fingerprint of the long-range nature of inter-atomic forces, and is connected with the break of cubic symmetry, due to the induced dipoles accompanying the vibrational modes. A simplified modellistic study of the long-wavelength optical phonons is given in Section 7. In Table 2 we report for convenience the frequencies of the optical phonons at the center of the Brillouin zone for the elemental semiconductors Si and Ge, and for the polar semiconductors GaAs, AlAs, GaSb, AlSb.

As a final example, we report in Fig. 8 the phonon branches of LiF. Lithium fluoride is a typical ionic material with the NaCl structure; there are two ions in the unit cell, and there are thus three acoustic and three optical branches. The long-range nature of interionic forces produces a strong longitudinal-transverse splitting of optical modes at $\mathbf{q} \approx 0$. In LiF the ratio of the low-frequency dielectric constant ($\varepsilon_s = 8.9$) and high-frequency dielectric constant ($\varepsilon_\infty = 1.9$) is relatively large, and so is the squared ratio of the measured LO and TO mode frequencies.

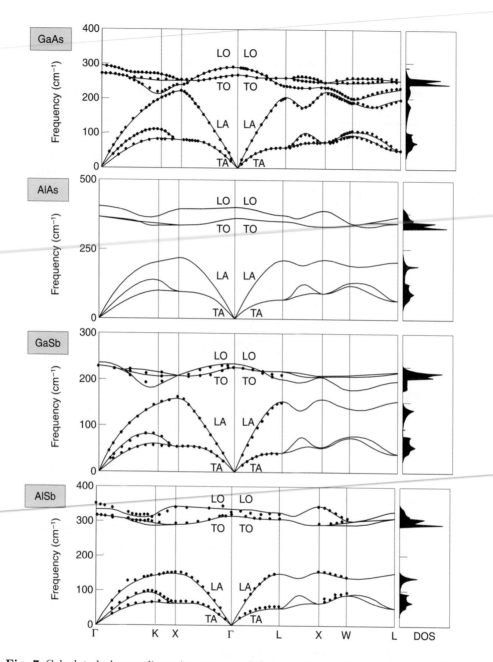

Fig. 7 Calculated phonon dispersion curves and density-of-states for binary semiconductors GaAs, AlAs, GaSb and AlSb [from P. Giannozzi, S. de Gironcoli, P. Pavone and S. Baroni, Phys. Rev. B**43**, 7231 (1991); copyright 1991 by the American Physical Society]. Longitudinal and transverse acoustic (or optical) modes are indicated by LA and TA (LO and TO), respectively.

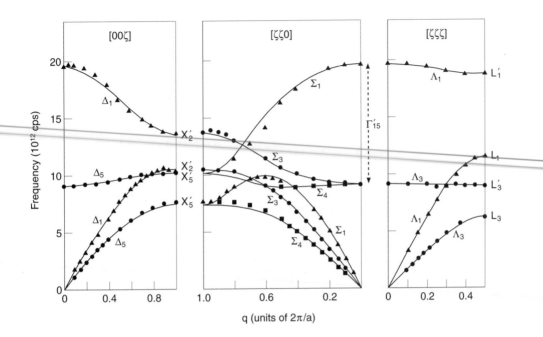

Fig. 8 Measured phonon dispersion curves along three directions of high symmetry in LiF; the solid curves are a best least-squares fit of a parameter model [from G. Dolling, H. G. Smith, R. M. Nicklow, P. R. Vijayaraghavan and M. K. Wilkinson, Phys. Rev. **168**, 970 (1968); copyright 1968 by the American Physical Society].

4 Quantum theory of the harmonic crystal

In the previous sections, we have discussed the lattice vibrations by means of the classical equations of motion. We can reconsider the problem from a quantum mechanical point of view, and show that the classical and quantum treatments are completely equivalent as far as dispersion curves are concerned; on the other hand, the quantum treatment of the elastic field shows that energies are discretized into quanta, called phonons.

In Section 1, we have considered the classical dynamics of a monatomic linear chain; we consider now the quantum mechanical counterpart of the same problem. In the harmonic approximation, and nearest neighbour interactions (see Eq. 10), the Hamiltonian of the linear chain becomes

$$H = \sum_n \frac{1}{2M}p_n^2 + \frac{1}{2}C\sum_n (2u_n^2 - u_n u_{n+1} - u_n u_{n-1}) , \qquad (26)$$

where u_n and p_n are the coordinate and conjugate moment of the nucleus at the nth site; these observables obey the commutation rules

$$[u_n, p_{n'}] = i\hbar\delta_{n,n'} , \qquad [u_n, u_{n'}] = [p_n, p_{n'}] = 0 . \qquad (27)$$

Instead of the dynamical variables u_n and p_n, it is convenient to perform a canonical transformation, with the final aim to put in diagonal form the Hamiltonian (26). We define the "phonon annihilation operator" a_q and the "phonon creation operator" $a_q^\dagger$ as the following linear combination of displacements and momenta of the nuclei

$$a_q = \frac{1}{\sqrt{N}} \sum_{t_n} e^{-i q t_n} \left[\sqrt{\frac{M\omega(q)}{2\hbar}} \, u_n + i \sqrt{\frac{1}{2\hbar M\omega(q)}} \, p_n \right] \qquad (28a)$$

$$a_q^\dagger = \frac{1}{\sqrt{N}} \sum_{t_n} e^{i q t_n} \left[\sqrt{\frac{M\omega(q)}{2\hbar}} \, u_n - i \sqrt{\frac{1}{2\hbar M\omega(q)}} \, p_n \right] . \qquad (28b)$$

The angular frequency $\omega(q)$ is at the moment left unspecified, and determined later so that to diagonalize the operator H. According to Eqs. (28), the phonon annihilation and creation operators are defined as linear combinations of the dynamical variables of all the nuclei, with appropriate phase factors $\exp(iqt_n)$ of the Bloch form.

It is easily seen that the transformations (28) from the set of operators u_n, p_n, to the set of operators a_q, $a_q^\dagger$ are canonical, i.e. the commutation rules are preserved. For this purpose we remember the standard properties

$$\frac{1}{N} \sum_{t_n} e^{-i(q-q')t_n} = \delta_{q,q'} \qquad \text{and} \qquad \frac{1}{N} \sum_{q} e^{-iq(t_n - t_{n'})} = \delta_{n,n'} . \qquad (29)$$

From the first of relations (29) and commutation rules (27), it is immediate to verify that

$$\left[a_q, a_{q'}^\dagger \right] = \delta_{q,q'} \qquad \text{and} \qquad \left[a_q, a_{q'} \right] = \left[a_q^\dagger, a_{q'}^\dagger \right] = 0 ;$$

thus the transformation (28) is canonical.

With the use of the second of the relations (29), we can invert Eqs. (28) and obtain

$$u_n = \frac{1}{\sqrt{N}} \sum_{q} \sqrt{\frac{\hbar}{2M\omega(q)}} \, e^{i q t_n} \left[a_q + a_{-q}^\dagger \right] \qquad (30a)$$

$$p_n = \frac{-i}{\sqrt{N}} \sum_{q} \sqrt{\frac{\hbar M\omega(q)}{2}} \, e^{i q t_n} \left[a_q - a_{-q}^\dagger \right] . \qquad (30b)$$

Insertion of Eqs. (30) into Eq. (26) gives

$$H = -\frac{1}{4} \sum_{q} \hbar\omega(q) \left[a_q - a_{-q}^\dagger \right] \left[a_{-q} - a_q^\dagger \right]$$

$$+ \frac{1}{4} \sum_{q} \frac{\hbar}{\omega(q)} \frac{C}{M} \left[a_q + a_{-q}^\dagger \right] \left[a_{-q} + a_q^\dagger \right] \left[2 - e^{iqa} - e^{-iqa} \right] .$$

We now exploit the arbitrariness in the frequency $\omega(q)$ by choosing

$$\boxed{\omega(q) = \frac{1}{\omega(q)} \frac{C}{M} (2 - e^{iqa} - e^{-iqa})} ; \qquad (31a)$$

notice the equivalence of Eq. (31a) and Eq. (12). The Hamiltonian (26) of the linear chain then becomes

$$H = \sum_q \hbar\omega(q)(a_q^\dagger a_q + \frac{1}{2}) , \tag{31b}$$

which is the sum of the Hamiltonians of N independent linear oscillators of frequency $\omega(q)$. The quanta of energy $\hbar\omega(q)$ are called phonons.

Completely similar analysis and conclusions could be performed for the diatomic linear chain of Section 2, and the general three-dimensional crystal of Section 3; the dispersion relations provided by the quantum mechanical treatment and by the classical treatment are the same, since the unitary transformation from localized variables to itinerant (or collective) variables are the same in the classical and quantum treatment. The quantum theory thus recovers the same $\omega = \omega(\mathbf{q}, p)$ dispersion curves of the classical theory; however the quantum theory leads to the quantization of the elastic field in terms of phonons, which can be considered as travelling quanta of energy $\hbar\omega = \hbar\omega(\mathbf{q}, p)$, wavevector propagation $\mathbf{q}$ and branch index p ($p = 1, 2, \ldots 3n_b$).

5 Lattice heat capacity. Einstein and Debye models

Consider a crystal composed by N unit cells and a basis of n_b atoms in the unit cell; the crystal volume is $V = N\Omega$ and the total number of atoms is $N_a = N n_b$. In the harmonic approximation, the system of N_a vibrating atoms is equivalent to a system of $3N_a$ independent (one-dimensional) oscillators of frequency $\omega = \omega(\mathbf{q}, p)$, where $\mathbf{q}$ assumes N allowed values in the first Brillouin zone, and p runs over the $3n_b$ branches of the phonon dispersion curves. The average vibrational energy of the harmonic crystal is the sum of independent phonon contributions; according to the Bose–Einstein statistics, we have

$$U_{\text{vibr}}(T) = \sum_{\mathbf{q}p} \left[\frac{\hbar\omega(\mathbf{q}, p)}{e^{\hbar\omega(\mathbf{q},p)/k_B T} - 1} + \frac{1}{2}\hbar\omega(\mathbf{q}, p) \right] .$$

The *lattice heat capacity at constant volume*, using Eq. (III-23a), is given by

$$C_V(T) = \frac{\partial U_{\text{vibr}}}{\partial T} = \frac{\partial}{\partial T} \sum_{\mathbf{q}p} \frac{\hbar\omega(\mathbf{q}, p)}{e^{\hbar\omega(\mathbf{q},p)/k_B T} - 1} . \tag{32}$$

The above expression allows the numerical calculation of the lattice heat capacity of crystals, once the phonon dispersion curves $\omega(\mathbf{q}, p)$ are known. In the following we consider the application of Eq. (32) to simple models of dispersion curves, that can be worked out analytically.

Einstein model

In the Einstein model, the actual frequencies of the normal modes are replaced by a unique (average) frequency ω_e (Einstein frequency). If N_a is the total number of atoms, Eq. (32) for the heat capacity at constant volume becomes

$$C_V(T) = 3N_a \frac{\partial}{\partial T} \frac{\hbar\omega_e}{e^{\hbar\omega_e/k_BT} - 1} = 3N_a k_B \left(\frac{\hbar\omega_e}{k_BT}\right)^2 \frac{e^{\hbar\omega_e/k_BT}}{(e^{\hbar\omega_e/k_BT} - 1)^2} .$$

The behaviour of $C_V(T)$ in the low and high temperature limits is

$$C_V \to e^{-\hbar\omega_e/k_BT} \quad \text{for} \quad k_BT \ll \hbar\omega_e ,$$

and

$$C_V \to 3N_a k_B \quad \text{for} \quad k_BT \gg \hbar\omega_e .$$

In the high temperature limit the Einstein model recovers the Dulong and Petit value $3N_a k_B$. In the low temperature limit, the Einstein model predicts for $C_V(T)$ an exponentially vanishing behaviour, contrary to the T^3 experimental law. The origin of this discrepancy is the presence in crystals of the phonon acoustic branches, which cannot be mimicked by a unique Einstein frequency, and actually need a more realistic description.

Debye model

Any three-dimensional crystal, with or without a basis, presents three acoustic branches with linear dispersion $\omega = v_s q$ for small q. For simplicity we assume the same sound velocity v_s for each of the three acoustic branches and extend the linear dispersion relation to the whole Brillouin zone. To avoid inessential details, we approximate the Brillouin zone with a sphere (Debye sphere) of equal volume (in order to preserve the total number of allowed wavevectors); we indicate with q_D the radius of the Debye sphere and define $\omega_D = v_s q_D$ as the *cutoff Debye frequency*. We notice that $(4/3)\pi q_D^3 = (2\pi)^3/\Omega$, where Ω is the volume of the unit cell in the direct space.

The density of phonon states corresponding to a branch with linear dispersion relation $\omega = v_s q$ is easily obtained. In fact the number of states $D(\omega)\,d\omega$ with frequency in the interval $[\omega, \omega + d\omega]$ equals the number of states in the reciprocal space with wavevector between $[q, q + dq]$; namely:

$$D(\omega)\,d\omega = \frac{V}{(2\pi)^3} 4\pi q^2 dq = \frac{V}{(2\pi)^3} 4\pi \left(\frac{\omega}{v_s}\right)^2 d\frac{\omega}{v_s} .$$

It follows

$$D(\omega) = \frac{V}{(2\pi)^3} 4\pi \frac{\omega^2}{v_s^3} = \frac{N\Omega}{(2\pi)^3} \frac{4\pi q_D^3}{3} \frac{3\omega^2}{\omega_D^3} = N \frac{3\omega^2}{\omega_D^3} \qquad 0 \le \omega \le \omega_D \qquad (33)$$

(N is the number of unit cells of the crystal).

The contribution of the three acoustic branches to the average vibrational energy (apart the constant zero point energy) is

$$U_{\text{vibr}}^{(\text{acoustic})}(T) = 3 \int_0^{\omega_D} N \frac{3\omega^2}{\omega_D^3} \frac{\hbar\omega}{e^{\hbar\omega/k_BT} - 1}\,d\omega . \qquad (34)$$

It is convenient to perform the change of variables $x = \hbar\omega/k_BT$ and define $x_D =$

$\hbar\omega_D/k_BT = T_D/T$, where $T_D = \hbar\omega_D/k_B$ is called Debye temperature. The expression (34) becomes

$$U_{\text{vibr}}^{(\text{acoustic})}(T) = 9Nk_BT\left(\frac{T}{T_D}\right)^3 \int_0^{x_D} \frac{x^3}{e^x - 1}\, dx \ . \tag{35}$$

In the high temperature limit $T_D \ll T$, $x_D \ll 1$ and $e^x - 1 \approx x$. The integral in Eq. (35) then gives $x_D^3/3$ and hence $U_{\text{vibr}}^{(\text{acoustic})}(T) = 3Nk_BT$; thus for $T \gg T_D$, the heat capacity of the three acoustic branches approaches the value $3Nk_B$ of the Dulong and Petit law.

In the low temperature limit $T \ll T_D$ we can replace $x_D = \infty$, and the integral in Eq. (35) equals $\pi^4/15$ [for a simple and instructive demonstration of the value of this integral see for instance B. D. Sukheeja, Am. J. Phys. **38**, 923 (1970)]. Equation (35) thus gives

$$U_{\text{vibr}}^{(\text{acoustic})}(T) = \frac{3}{5}\pi^4 Nk_B \frac{T^4}{T_D^3} \qquad T \ll T_D \ .$$

Correspondingly, in the low temperature region the heat capacity becomes

$$C_V(T) = \frac{12}{5}\pi^4 Nk_B \frac{T^3}{T_D^3} \qquad T \ll T_D$$

and the correct experimental T^3 behaviour is reproduced.

The Debye model can be refined in several ways. For instance the three acoustic branches could be treated with different sound velocities. In the case of crystals with a basis, one could use the Debye model for the acoustic modes and the Einstein model for the optical modes. We notice that, in the high temperature limit, anharmonic effects are of increasing importance, and corrections to the Dulong and Petit value are likely to be of significance. In metals, besides the vibrational contribution to the internal energy, we have to consider the electronic contribution; the electronic contribution to the heat capacity is proportional to T at any temperature and may become the dominant term at very low temperatures (see Section III-3). We notice finally that the T^3 law depends on the crystal dimensionality. In a two-dimensional crystal, instead of Eq. (33), the density-of-states $D(\omega)$ is proportional to ω and the low temperature lattice heat capacity is characterized by a T^2 power law. Similarly, in an ideal one-dimensional crystal, one would obtain a lattice heat capacity linear in the temperature.

6 Considerations on anharmonic effects and melting of solids

So far we have confined our attention to the harmonic approximation for the lattice vibrations; in this approximation, phonons are elementary excitations of the elastic field, which do not decay and cannot interact. The anharmonic terms, which correspond to cubic, quartic and successive terms in the series expansion of the crystal potential energy, have quite important consequences; for instance, cubic terms make possible three-phonon processes in which one-phonon decays into two phonons or two phonons merge into one. Among the physical effects of anharmonicity, we mention the thermal expansion of solids, the change of normal mode frequencies with temperature (or other parameters), the thermal resistivity, the broadening of one-phonon peaks in neutron scattering experiments, the solid–liquid transition. It is not our intention to discuss the wealth of problems related to anharmonicity; here we simply provide some intuitive remarks concerning the amplitude of localized motions of the atoms and the Lindemann criterion of melting.

The mean quadratic displacement of a given atom about its equilibrium position is an important quantity, which influences X-ray scattering, cold neutron scattering, Mössbauer effect, and also determines the solid–liquid transition. We wish thus to give an estimate of the mean quadratic displacement of an atom around its equilibrium position as a function of temperature.

For simplicity we consider a three-dimensional crystal with N unit cells and one atom per unit cell. In this case the normal modes consist of three acoustic branches, of frequency $\omega(\mathbf{q}, p)$ and polarization vectors $\mathbf{A}(\mathbf{q}, p)$ ($p = 1, 2, 3$). We can expand the displacement $\mathbf{u}_n$ of the atom at $\mathbf{t}_n$ in normal modes; as a straight generalization of Eq. (30a), we have

$$\mathbf{u}_n = \sum_{\mathbf{q}p} \sqrt{\frac{\hbar}{2NM\omega(\mathbf{q}, p)}}\, \mathbf{A}(\mathbf{q}, p)\, e^{i\mathbf{q}\cdot\mathbf{t}_n} \left[a_{\mathbf{q}p} + a^{\dagger}_{-\mathbf{q}p}\right]. \tag{36}$$

To avoid unessential details, we assume that the three acoustic branches are degenerate, so that $\omega(\mathbf{q}, p)$ does not depend on the polarization index p. Furthermore, for any wavevector, the three polarization vectors $\mathbf{A}(\mathbf{q}, p)$ form an orthonormal triad, that we intend to orient parallel to some fixed reference frame. Then the component of $\mathbf{u}_n$ along a direction, say z, dropping the now unnecessary polarization index p in Eq. (36) becomes

$$u_{nz} = \sum_{\mathbf{q}} \sqrt{\frac{\hbar}{2NM\omega(\mathbf{q})}}\, e^{i\mathbf{q}\cdot\mathbf{t}_n} \left[a_{\mathbf{q}} + a^{\dagger}_{-\mathbf{q}}\right].$$

We can now calculate the ensemble average $\langle u_{nz}^2 \rangle$ of the quadratic displacement u_{nz}^2. Without loss of generality we take $\mathbf{t}_n = 0$, and use the standard results

$$\langle a^{\dagger}_{\mathbf{q}} a_{\mathbf{q}} \rangle = \frac{1}{\exp(\hbar\omega_{\mathbf{q}}/k_B T) - 1}, \quad \langle a_{\mathbf{q}} a^{\dagger}_{\mathbf{q}} \rangle = \langle a^{\dagger}_{\mathbf{q}} a_{\mathbf{q}} \rangle + 1,$$

where $\langle a^{\dagger}_{\mathbf{q}} a_{\mathbf{q}} \rangle$ is the well known Bose population factor (see Appendix A). The average square displacement $\langle u_z^2 \rangle$ of each atom thus becomes

$$\langle u_z^2 \rangle = \sum_{\mathbf{q}} \frac{\hbar}{2NM\omega(\mathbf{q})} \left[\frac{2}{\exp(\hbar\omega(\mathbf{q})/k_B T) - 1} + 1\right]. \tag{37}$$

It is instructive to calculate the average square displacement for the Debye model of the phonon spectrum. We have already seen that the density-of-states for any of the three branches is given by Eq. (33). Equation (37) thus becomes

$$\langle u_z^2 \rangle = \int_0^{\omega_D} \frac{\hbar}{2NM\omega} \left(\frac{2}{e^{\hbar\omega/k_BT} - 1} + 1 \right) N \frac{3\omega^2}{\omega_D^3} \, d\omega \ .$$

If we introduce the dimensionless variable $x = \hbar\omega/k_BT$ and define $\hbar\omega_D = k_BT_D$, we have

$$\langle u_z^2 \rangle = 3 \frac{\hbar^2 T^2}{Mk_BT_D^3} \int_0^{T_D/T} \left(\frac{1}{e^x - 1} + \frac{1}{2} \right) x \, dx \ . \tag{38}$$

The integral can be easily performed in the limits of very high temperatures $(T \gg T_D)$ using a series development of the exponential, and of very low temperatures $(T \ll T_D)$. We have respectively

$$\langle u_z^2 \rangle = \begin{cases} \dfrac{3}{4} \dfrac{\hbar^2}{Mk_BT_D} & \text{for} \quad T \ll T_D \\[3mm] 3 \dfrac{\hbar^2 T}{Mk_BT_D^2} & \text{for} \quad T \gg T_D \ . \end{cases} \tag{39}$$

From Eq. (39), it is seen that the value of $\langle u_z^2 \rangle$ at zero temperature is only somewhat smaller than the value of $\langle u_z^2 \rangle$ at the Debye temperature; in fact $\langle u_z^2 \rangle_{T=T_D} \approx 4\langle u_z^2 \rangle_{T=0}$.

We can now establish a simple qualitative criterion for melting. Let r_0 be the mean radius of the unit cell, and consider the ratio

$$f = \frac{\sqrt{\langle u_x^2 \rangle + \langle u_y^2 \rangle + \langle u_z^2 \rangle}}{r_0} = \sqrt{\frac{9\hbar^2 T}{Mk_BT_D^2 r_0^2}} \qquad T \gg T_D \ . \tag{40a}$$

When the ratio f reaches a critical value f_c (almost independent from the specific solid in consideration), melting is expected to occur. The melting temperature is thus given by the Lindemann formula

$$T_m = \frac{f_c^2}{9\hbar^2} Mk_BT_D^2 r_0^2 \ ; \tag{40b}$$

the critical value f_c turns out to be of the order of 0.2–0.3 in many solids.

Another interesting conclusion can be done on the stability of one-dimensional and two-dimensional crystals. In these cases, the calculation of the mean quadratic displacement in the plane or in the chain (using the appropriate density of phonon states), leads to a divergent value at any temperature. Thus one-dimensional and two-dimensional crystals are unstable in the harmonic approximation; some three-dimensional interaction (whatever small with respect to intralayer or intrachain interaction) is necessary to stabilize low-dimensional structures.

7 Optical phonons and polaritons in polar crystals

7.1 General considerations

In Section 3, we have studied the crystal lattice vibrations by means of the dynamical matrix formalism. The dynamical matrix treatment implicitly assumes that the inter-

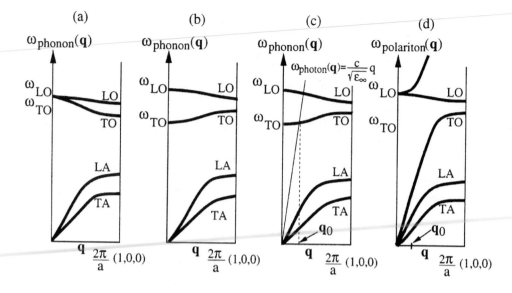

Fig. 9 (a) Schematic behaviour of phonon dispersion curves of a cubic *homopolar* semiconductor (or insulator) with two equal atoms per unit cell, along a high symmetry direction. (b) Schematic behaviour of phonon dispersion curves of a cubic *heteropolar* semiconductor (or insulator) with two different atoms per unit cell, when retardation effects are ignored. (c) Superimposed to the previous phonon dispersion curves, there is now the photon dispersion curve $w(\mathbf{q}) = c\, q/\sqrt{\varepsilon_\infty}$, assuming momentarily no coupling between electromagnetic waves and lattice vibrations. We have indicated by $q_0 \approx \sqrt{\varepsilon_\infty}\,\omega_{TO}/c \approx (\sqrt{\varepsilon_\infty}\,v_s/c)(2\pi/a)$ the point at which the photon-like dispersion curve crosses the transverse phonon dispersion curve. In the figure, the slope of the photon dispersion curve (of the order of the velocity of light) and the slope of the acoustic modes (of the order of sound velocity) could not be drawn in scale. (d) Schematic picture of *polariton effects*. Phonons and photons with nearly equal wavevectors and energies interact and determine the polariton dispersion curve. *Polariton effects extend from $q = 0$ to approximately q_0, which is a fraction of the order of $v_s/c \approx 10^{-5}$ of the Brillouin zone dimension* (see Fig. 10 and Fig. 11 for an expanded scale and further details). Notice that polariton effects restore the threefold degeneracy of the state of frequency ω_{LO} at $q = 0$.

atomic interactions are instantaneous. For polar crystals (such as ionic crystals and heteropolar semiconductors), the long-range nature of inter-atomic Coulomb interactions requires a proper account of retardation effects due to the finite velocity of light. The coupling of transverse mechanical waves and electromagnetic waves is particularly important for wavevectors q in the range from $q=0$ to approximately the value q_0, which denotes the crossing point of the dispersion curves of (uncoupled) photons and phonons (see Fig. 9). Before considering a continuous model to describe the optical phonons in polar crystals, we briefly summarize some relevant phenomenological aspects.

As a preliminary to further considerations, let us compare the vibrational curves of homopolar and heteropolar cubic crystals with two atoms per unit cell (see Fig. 6, Fig. 7 and Fig. 8 for specific examples; see also Fig. 9 for a schematic summary of the important features). From the general discussion of Section 3, we have seen that the dispersion curves of a diatomic crystal consist of three acoustic and three optical branches. In cubic crystals, for vectors $\mathbf{q}$ along high symmetry directions, there are two degenerate transverse acoustic modes (TA) and one longitudinal acoustic mode (LA). Similarly, there are two degenerate transverse optical modes (TO) and one longitudinal optical mode (LO), whose frequencies go to finite values ω_{TO} and ω_{LO} in the long wavelength limit. In nonpolar diatomic cubic crystals, the optical modes for small wavelengths are degenerate, i.e. $\omega_{LO} \equiv \omega_{TO}$ (Fig. 9a). On the contrary, the frequencies ω_{LO} and ω_{TO} of longitudinal and transverse optical phonons are different in polar crystals, in spite of the cubic symmetry (Fig. 9b); this symmetry-breaking effect is due to the long-range nature of the electrostatic forces.

The longitudinal-transverse splitting has implications also on the infrared dielectric properties of polar semiconductors and insulators. These materials are strongly reflecting in the frequency region $\omega_{TO} < \omega < \omega_{LO}$; repeated reflexions can be used to select a band of wavelengths of infrared radiation, which is known as Reststrahlen (residual rays) radiation. Furthermore, the frequencies ω_{LO} and ω_{TO} satisfy the Lyddane–Sachs–Teller relation

$$\boxed{\frac{\omega_{LO}^2}{\omega_{TO}^2} = \frac{\varepsilon_s}{\varepsilon_\infty}} , \tag{41}$$

where ε_s is the static dielectric constant and ε_∞ is the high-frequency dielectric constant. By ε_∞ we mean the infrared dielectric constant at frequencies much higher than a typical phonon frequency (so that ionic displacement contribution can be neglected) and much smaller than any electronic transition frequency; ε_∞ is determined by the electronic contribution to the static dielectric constant. If ω_{LO} and ω_{TO} are significantly different, the same occurs for ε_s and ε_∞ (and vice versa); in some materials (such as ferroelectric ionic crystals), ω_{TO} is anomalously small and ε_s anomalously large.

The electromagnetic coupling between a radiation field and transverse optical phonons leads to the concept of new quasiparticles, known as *polaritons*; photons and transverse phonons strongly interact near the crossing of the corresponding dispersion curves, which are modified into polariton dispersion curves (see Fig. 9c and Fig. 9d). We pass now to interpret the phenomenological properties of polar crystals mentioned above, with a suitable continuous model.

7.2   Lattice vibrations in polar crystals and polaritons

The continuous approximation for optical vibrational modes in isotropic materials

We can establish a reasonable simple model for polaritons, combining a continuous approximation for the description of the mechanical waves of the optical modes and the Maxwell equations for the description of the electromagnetic waves [J. J. Hopfield and D. G. Thomas, Phys. Rev. **132**, 563 (1963)]. We confine our attention here to polar cubic crystals with two atoms (cation and anion) in the unit cell, of effective charge $\pm e^*$, mass M_1 and M_2 (and reduced mass M^*). In optical vibrational modes, cations and anions move against each other, so we can discuss the motion of the ions in the unit cell by the relative displacement variable $\mathbf{w}$. In analogy with Eq. (19), we expect for isotropic cubic crystals that the relative displacement variable $\mathbf{w}$ obeys the equation of motion

$$\ddot{\mathbf{w}} = -\omega_0^2 \mathbf{w} + \frac{e^*}{M^*}\mathbf{E} \ . \tag{42}$$

Notice that in the case the effective charge e^* vanishes, the three optical modes become degenerate, with frequency $\omega_{TO} = \omega_{LO} = \omega_0$, and are dispersionless. The continuous model, summarized by Eq. (42), is rather simplified, but it has the merit to allow analytic elaborations and to provide guidelines for more complicated situations. Among the limitations of the model, we notice that the local electric field acting on the site effective charge and the average electric field (due to any internal or external sources) are supposed to coincide; local field effects are introduced in Section 7.3.

In order to estimate semi-empirically the quantity e^*/M^*, consider Eq. (42) in the presence of a static electric field $\mathbf{E}_s$; the static ionic displacement is given by $\mathbf{w}_s = (e^*/\omega_0^2 M^*)\mathbf{E}_s$. The average ionic polarization of a crystal, of volume V and N unit cells, is $\mathbf{P}_{\text{ion,s}} = (N/V)e^*\mathbf{w}_s = (N/V)(e^{*2}/\omega_0^2 M^*)\mathbf{E}_s$. The static dielectric constant ε_s and the high-frequency dielectric constant ε_∞ are related by the expression

$$\varepsilon_s = \varepsilon_\infty + 4\pi\frac{P_{\text{ion,s}}}{E_s} = \varepsilon_\infty + \frac{4\pi(N/V)e^{*2}}{\omega_0^2 M^*} \ .$$

Using the above equation, we can re-write Eq. (42) in the form

$$\ddot{\mathbf{w}} = -\omega_0^2 \mathbf{w} + \omega_0^2\frac{\varepsilon_s - \varepsilon_\infty}{4\pi(N/V)e^*}\mathbf{E} \ . \tag{43}$$

The ionic contribution to the polarization (dipole per unit volume) of the specimen is given by

$$\mathbf{P}_{\text{ion}} = \frac{N}{V}e^*\mathbf{w} \ .$$

Multiplying both members of Eq. (43) by $(N/V)e^*$ we obtain

$$\boxed{\ddot{\mathbf{P}}_{\text{ion}} = -\omega_0^2\mathbf{P}_{\text{ion}} + \omega_0^2\frac{\varepsilon_s - \varepsilon_\infty}{4\pi}\mathbf{E}} \ . \tag{44}$$

Equation (44) is the very useful "constitutive" equation of polar crystals; it couples the electric polarization, produced by the vibrating lattice of ions, to the electric field in the crystal; the phenomenological coupling constant is given by $\omega_0^2(\varepsilon_s - \varepsilon_\infty)/4\pi$.

We consider first the propagation in the medium of *longitudinal optical vibrations*, in which case the polarization field and the electric field are also expected to be of longitudinal type with the form

$$\mathbf{P}_{\text{ion}}(\mathbf{r}, t) = \mathbf{P}_0\, e^{i(\mathbf{q}\cdot\mathbf{r} - \omega t)} \quad \text{with} \quad \mathbf{P}_0 \| \mathbf{q} \tag{45a}$$

$$\mathbf{E}(\mathbf{r}, t) = \mathbf{E}_0\, e^{i(\mathbf{q}\cdot\mathbf{r} - \omega t)} \quad \text{with} \quad \mathbf{E}_0 \| \mathbf{q} \ . \tag{45b}$$

We can easily show that for longitudinal fields we have

$$\boxed{\mathbf{E}_0 = -\frac{4\pi}{\varepsilon_\infty}\mathbf{P}_0} \qquad \mathbf{E}_0 \| \mathbf{P}_0 \| \mathbf{q} \; . \qquad (45c)$$

In fact the microscopic charge density accompanying the polarization field (45a) is given by $\rho_{\mathrm{micr}} = -\mathrm{div}\,\mathbf{P}_{\mathrm{ion}} = -iqP_0\exp(i\mathbf{q}\cdot\mathbf{r} - i\omega t)$; the value $\mathbf{E}_0$ is determined so that $\mathrm{div}\,\mathbf{E} = iqE_0\exp(i\mathbf{q}\cdot\mathbf{r} - i\omega t) \equiv 4\pi\rho_{\mathrm{micr}}/\varepsilon_\infty$. Also notice that the curl of the longitudinal fields (45) vanish identically.

Inserting Eqs. (45) into Eq. (44), it is found that the frequency ω for the longitudinal waves satisfies

$$-\omega^2 = -\omega_0^2 - \omega_0^2\frac{\varepsilon_s - \varepsilon_\infty}{\varepsilon_\infty} \equiv -\omega_0^2\frac{\varepsilon_s}{\varepsilon_\infty} \; .$$

Thus the longitudinal waves are characterized by the dispersionless relation

$$\omega^2 = \omega_0^2\frac{\varepsilon_s}{\varepsilon_\infty} \equiv \omega_{LO}^2 \; . \qquad (46)$$

The effect of the long-range Coulomb field on the longitudinal optical vibrations does not introduce any dispersion; it however does increase the "restoring forces" and does increase the oscillation frequency from ω_0 (the value neglecting long-range contributions) to ω_{LO}.

We consider now the propagation in the medium of *transverse optical vibrations*, in which case the polarization field and the electric field are of transverse type, with the form

$$\mathbf{P}_{\mathrm{ion}}(\mathbf{r}, t) = \mathbf{P}_0\, e^{i(\mathbf{q}\cdot\mathbf{r} - \omega t)} \qquad \text{with} \quad \mathbf{P}_0 \perp \mathbf{q} \qquad (47a)$$

$$\mathbf{E}(\mathbf{r}, t) = \mathbf{E}_0\, e^{i(\mathbf{q}\cdot\mathbf{r} - \omega t)} \qquad \text{with} \quad \mathbf{E}_0 \perp \mathbf{q} \; . \qquad (47b)$$

It is shown below that for transverse fields we have

$$\boxed{\mathbf{E}_0 = \frac{4\pi\omega^2}{c^2q^2 - \varepsilon_\infty\omega^2}\mathbf{P}_0} \qquad \mathbf{E}_0 \| \mathbf{P}_0 \perp \mathbf{q} \; . \qquad (47c)$$

From Eq. (47c) it is seen that the ratio E_0/P_0 for transverse fields depends both on q and ω, contrary to the situation for longitudinal fields expressed by Eq. (45c); also notice that for $q \to 0$ and finite ω, Eq. (47c) coincide with Eq. (45c), and the transverse and longitudinal polaritons become thus degenerate.

It is worthwhile to remark that the divergence of the transverse fields vanishes identically (in particular $\rho_{\mathrm{micr}} = -\mathrm{div}\,\mathbf{P}_{\mathrm{ion}} \equiv 0$ means that no microscopic charge is accompanying the polarization wave, and thus also $\mathrm{div}\,\mathbf{E} \equiv 0$). For transverse fields $\mathrm{curl}\,\mathbf{E} \neq 0$; before considering the correct treatment of $\mathrm{curl}\,\mathbf{E}$ with the Maxwell equations, we examine the so called *electrostatic limit* (also called the $c \to \infty$ limit, or instantaneous interaction limit, or omission of retardation effects), which just consists in taking $\mathrm{curl}\,\mathbf{E} = 0$. Since also $\mathrm{div}\,\mathbf{E} = 0$, we conclude that the electric field accompanying a transverse optical vibration vanishes; the frequency of the optical transverse modes would be unchanged with respect to ω_0 in the electrostatic approximation.

For the correct determination of curl $\mathbf{E}$, we have to resort to the appropriate Maxwell equations

$$\text{curl}\,\mathbf{E} = -\frac{1}{c}\frac{\partial \mathbf{B}}{\partial t} \tag{48a}$$

$$\text{curl}\,\mathbf{H} = \frac{1}{c}\frac{\partial \mathbf{D}}{\partial t} = \frac{\varepsilon_\infty}{c}\frac{\partial \mathbf{E}}{\partial t} + \frac{4\pi}{c}\frac{\partial \mathbf{P}_{\text{ion}}}{\partial t} \ . \tag{48b}$$

From the curl of both members of Eq. (48a), and using Eq. (48b) (in non-magnetic materials $\mathbf{B} = \mathbf{H}$), we have

$$\text{curl}\,\text{curl}\,\mathbf{E} = -\frac{\varepsilon_\infty}{c^2}\ddot{\mathbf{E}} - \frac{4\pi}{c^2}\ddot{\mathbf{P}}_{\text{ion}} \ .$$

For transverse fields $\text{div}\,\mathbf{E} = 0$; using the vectorial identity $\text{curl}\,\text{curl} = \text{grad}\,\text{div} - \nabla^2$, it follows

$$-\nabla^2\mathbf{E} = -\frac{\varepsilon_\infty}{c^2}\ddot{\mathbf{E}} - \frac{4\pi}{c^2}\ddot{\mathbf{P}}_{\text{ion}} \ ; \tag{49}$$

it is then seen by inspection that Eq. (47a) and Eq. (47b) imply Eq. (47c).

We insert now Eqs. (47) into the constitutive Eq. (44) and obtain the compatibility condition

$$\varepsilon_\infty \omega^4 - (\omega_0^2 \varepsilon_s + c^2 q^2)\omega^2 + \omega_0^2 c^2 q^2 = 0 \ . \tag{50}$$

We can solve for ω^2 and obtain the dispersion relation for polaritons

$$\omega^2 = \frac{1}{2\varepsilon_\infty}\left[\omega_0^2 \varepsilon_s + c^2 q^2 \pm \sqrt{(\omega_0^2 \varepsilon_s + c^2 q^2)^2 - 4\omega_0^2 c^2 q^2 \varepsilon_\infty}\right] \ . \tag{51}$$

The dispersion curves for polaritons, given by Eq. (51), are schematically shown in Fig. 10.

It is interesting to examine the lower and upper polariton branches in the limit of $q < q_0$ and $q > q_0$, where q_0 is the point for which $(c/\sqrt{\varepsilon_\infty})\,q_0 \equiv \omega_{TO}$. For $q \gg q_0$, the two solutions of Eq. (51), and the corresponding amplitudes $\mathbf{E}_0$ and $\mathbf{P}_0$ given by Eq. (47c) are

$$\omega^2 = \omega_0^2 = \omega_{TO}^2 \qquad \mathbf{E}_0 = 0 \qquad \mathbf{P}_0 \neq 0$$

and

$$\omega^2 = \frac{c^2}{\varepsilon_\infty}q^2 \qquad \mathbf{P}_0 = 0 \qquad \mathbf{E}_0 \neq 0 \ ;$$

thus for $q \gg q_0$ the lower branch is a pure mechanical wave (with $\mathbf{E}_0 = 0$) and the upper branch is a pure electromagnetic wave (with $\mathbf{P}_0 = 0$).

For $q \ll q_0$, the two solutions of Eq. (51), and the corresponding amplitudes $\mathbf{E}_0$ and $\mathbf{P}_0$ given by Eq. (47c) are

$$\omega^2 = \omega_0^2\frac{\varepsilon_s}{\varepsilon_\infty} = \omega_{LO}^2 \qquad \mathbf{E}_0 = -\frac{4\pi}{\varepsilon_\infty}\mathbf{P}_0$$

and

$$\omega^2 = \frac{c^2}{\varepsilon_s}q^2 \qquad \mathbf{P}_0 = \frac{\varepsilon_s - \varepsilon_\infty}{4\pi}\mathbf{E}_0 \ .$$

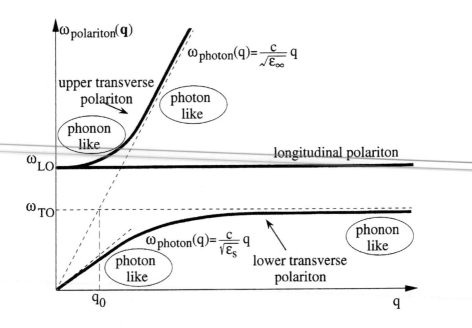

Fig. 10 Schematic description of polaritons. The dispersion curves of uncoupled transverse phonons and photons are shown by dashed lines; q_0 is the crossing point $q_0 = \omega_{TO}/(c/\sqrt{\varepsilon_\infty})$; dispersion curves of the longitudinal phonon and transverse polaritons are shown by solid lines. In the lower polariton branch, the character of the dispersion curve changes from photon-like for $q < q_0$ to phonon-like for $q > q_0$; in the upper branch, the character changes from phonon-like to photon-like as q increases.

Thus for q smaller or near q_0 the polariton modes have the character of coupled mechanical–electromagnetic waves, with $\mathbf{E}_0$ and $\mathbf{P}_0$ *simultaneously* different from zero. Notice in particular that for $\omega = \omega_{LO}$ we have $\mathbf{E}_0 = -(4\pi/\varepsilon_\infty)\mathbf{P}_0$ both for transverse and longitudinal waves (see Eq. 45c); thus the degeneracy of transverse and longitudinal modes is restored at $\mathbf{q} = 0$.

As an illustrative example of the concepts developed so far, we report in Fig. 11 the polariton dispersion curves in GaP. The dispersion curves for optical phonons of long wavelength, in the absence of coupling to photons, are horizontal straight lines; transverse optical phonons and photons with nearly the same energy and wavevector are strongly coupled by the phonon–photon interaction and lead to the polariton dispersion curves.

We have seen that transverse optical phonons and photons with nearly the same energy and wavevector are strongly coupled; we notice that quite similar coupling effects occur also for photons and transverse excitons (exciton states have been studied in Section VII-1). The mixed exciton–photon states are called exciton–polaritons and their dispersion curves have a behaviour qualitatively similar to the polariton curves so far discussed [see for instance L. C. Andreani in "Confined Electrons and Photons: New Physics and Devices" edited by E. Burnstein and C. Weisbuch, Plenum Press

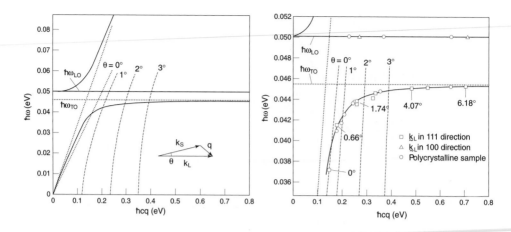

Fig. 11 Polariton dispersion curves in GaP. In Fig. 11a the vector diagram of Raman spectroscopy measurements is also indicated; $\mathbf{k}_L$, $\mathbf{k}_S$ and $\mathbf{q}$ are the wavevectors of the incident laser photon, scattered Stokes photon, and polariton; θ is the scattering angle. Values of energies and wavevectors which are kinematically possible at angle θ are shown by long-dashed lines. In Fig. 11b the plot of the observed energies and wavectors of polaritons and LO phonons are given. The figures are taken from C. H. Henry and J. J. Hopfield, Phys. Rev. Lett. **15**, 964 (1965); copyright 1965 by the American Physical Society.

1995, p.57, and references quoted therein]. An example of dispersion curves of exciton–polaritons in CuCl is reported in Fig. 12.

Infrared dielectric properties of polar crystals

We can now discuss the infrared dielectric properties of polar crystals exploiting the "constitutive" equation (44), that couples the ionic polarization to the electric field in the medium. We consider the response of the system to a time-dependent driving electric field, periodic in space and time, of the form

$$\mathbf{E}(\mathbf{r}, t) = \mathbf{E}_0 \, e^{i(\mathbf{q}\cdot\mathbf{r} - \omega t)} e^{\eta t} \; ; \tag{52a}$$

the electric field is turned on adiabatically from $t = -\infty$, and this is achieved through the exponential factor $\exp(\eta t)$ with $\eta \to 0^+$. By analogy to Eq. (52a), we assume for $\mathbf{P}_{\text{ion}}$ the expression

$$\mathbf{P}_{\text{ion}}(\mathbf{r}, t) = \mathbf{P}_0 \, e^{i(\mathbf{q}\cdot\mathbf{r} - \omega t)} e^{\eta t} \; . \tag{52b}$$

Replacing Eqs. (52) into Eq. (44), one obtains

$$\mathbf{P}_0 = \frac{\omega_0^2}{\omega_0^2 - (\omega + i\eta)^2} \frac{\varepsilon_s - \varepsilon_\infty}{4\pi} \mathbf{E}_0 \tag{53}$$

(for the present model under consideration, it is irrelevant whether $\mathbf{E}_0$ and $\mathbf{P}_0$ are parallel or orthogonal to the vector $\mathbf{q}$).

The dielectric function is given by $\varepsilon(\omega) = \varepsilon_\infty + 4\pi P_0/E_0$, where as before ε_∞ denotes the dielectric constant due to the electronic polarizability (at frequencies well

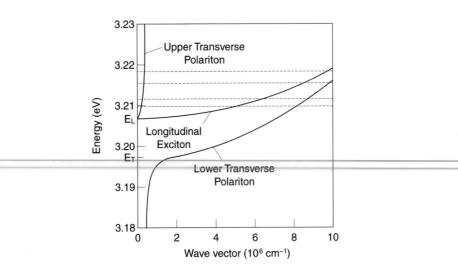

Fig. 12 Dispersion curves of the transverse exciton–polaritons of CuCl in the anomalous dispersive region. The dashed lines refer to the laser energies used to study luminescence line shape due the decay from bound pairs of excitons [from J. K. Pribram, G. L. Koos, F. Bassani and J. P. Wolfe, Phys. Rev. B**28**, 1048 (1983); copyright 1983 by the American Physical Society].

below any electronic transition resonance). Using Eq. (53), we obtain for the dielectric function of the polar crystal

$$\varepsilon(\omega) = \varepsilon_\infty + \frac{\omega_0^2}{\omega_0^2 - \omega^2 - i\,\eta\,\omega}(\varepsilon_s - \varepsilon_\infty) \qquad (54)$$

(where 2η has been relabelled as η). The real and imaginary parts of the dielectric function are schematically indicated in Fig. 13.

In the limit of $\eta \to 0^+$ the real and imaginary parts of Eq. (54) become

$$\varepsilon_1(\omega) = \varepsilon_\infty + \frac{\omega_0^2}{\omega_0^2 - \omega^2}(\varepsilon_s - \varepsilon_\infty) = \frac{\varepsilon_s\,\omega_0^2 - \varepsilon_\infty\,\omega^2}{\omega_0^2 - \omega^2} \qquad (55a)$$

and

$$\varepsilon_2(\omega) = \frac{\pi\omega_0(\varepsilon_s - \varepsilon_\infty)}{2}\left[\delta(\omega - \omega_0) - \delta(\omega + \omega_0)\right] . \qquad (55b)$$

At positive frequencies, $\varepsilon_1(\omega)$ exhibits a pole for $\omega = \omega_0 = \omega_{TO}$ (transverse phonon frequency) and has a zero for $\omega = \omega_0\sqrt{\varepsilon_s/\varepsilon_\infty} = \omega_{LO}$ (longitudinal phonon frequency); the values ω_{TO} and ω_{LO} satisfy the Lyddane–Sachs–Teller relation (41).

The dielectric function $\varepsilon_1(\omega)$ is negative for $\omega_{TO} < \omega < \omega_{LO}$. In this region the reflectivity equals one, and the electromagnetic propagation in the crystal is forbidden. Outside the interval $[\omega_{TO}, \omega_{LO}]$ the dielectric function $\varepsilon_1(\omega)$ is positive and $\varepsilon_2(\omega)$ vanishes (when $\eta \to 0^+$); in this region, the dispersion relations for electromagnetic

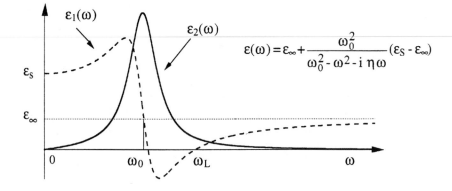

Fig. 13 Schematic behaviour of the real and imaginary part of the dielectric function $\varepsilon(\omega)$ of a polar crystal in the infrared region. Since $\varepsilon_1(\omega)$ and $\varepsilon_2(\omega)$ are even and odd functions of ω, respectively, only the part $\omega > 0$ is indicated. For $\eta \to 0^+$, the real part $\varepsilon_1(\omega)$ presents a pole at $\omega = \omega_0$, while the imaginary part $\varepsilon_2(\omega)$ presents a δ-like singularity at $\omega = \omega_0$.

waves propagating in a medium are determined by the requirement

$$\omega = \frac{c}{\sqrt{\varepsilon_1(\omega)}} q \tag{56a}$$

(see also the discussion at the end of Section XI-1). We have $\omega^2 \varepsilon_1(\omega) = c^2 q^2$; using for $\varepsilon_1(\omega)$ the expression (55a), we obtain

$$\omega^2 \frac{\varepsilon_s \omega_0^2 - \varepsilon_\infty \omega^2}{\omega_0^2 - \omega^2} = c^2 q^2 \ . \tag{56b}$$

It can be immediately seen that Eq. (56b) coincides exactly with Eq. (50), and defines thus the dispersion curves of the polaritons in the crystal.

7.3 Local field effects on polaritons

The internal field according to Lorentz

In the discussion of Section 7.2 we have assumed that the macroscopically averaged electric field $\mathbf{E}$ and the local field $\mathbf{E}_{\mathrm{loc}}$ are the same. In solids, however, there can be significant differences betweeen the two fields, and a central (and not easy) problem in the theory of dielectrics is the calculation of the electric field at the position of a given atom or molecule. Without entering in all the subtleties of this problem, we here briefly discuss the internal field according to Lorentz.

Consider an isotropic dielectric crystal with the shape of a bar, very long in the z-direction (see Fig. 14). Imagine that microscopic electric dipoles are set up at the lattice points so to give rise to a uniform polarization $\mathbf{P}$ in the z direction. We notice that since $\mathbf{P}$ is uniform, we have $\rho_{\mathrm{micr}} = -\operatorname{div}\mathbf{P} = 0$ and no volume microscopic charge density is accompanying the polarization. We also notice that the geometry of the (thin and very long) bar is chosen so that $\mathbf{P}$ is parallel to the surface of the sample;

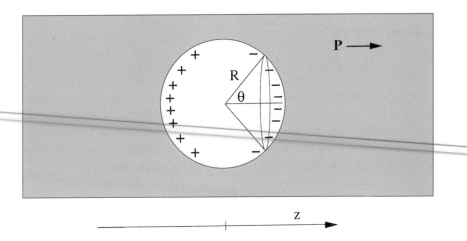

Fig. 14 Schematic representation of the Lorentz cavity for the calculation of the local electric field. The sample (with the ideal shape of a thin and infinitely long bar) is supposed to be uniformly polarized with **P** parallel to the surface of the sample.

thus no discontinuity of the normal component of **P** occurs at the surfaces, and no microscopic surface charge density occurs either. Because of the absence of any internal and external charges, we have div $\mathbf{E} = 0$; in stationary situations also curl $\mathbf{E} = 0$ and the electric field is thus zero.

Although the uniformly polarized specimen in the geometry of Fig. 14 is in a *null electric field*, it is easily realized that the *local electric field acting on the microscopic dipoles*, which are the origin of the macroscopic polarization field **P**, is in general different from zero. For simplicity, we confine our attention to the extreme tight-binding limit, in which the crystal can be viewed as a collection of microscopic electric dipoles, well localized around the lattice sites; in this case, the local field can be obtained with the following arguments.

Imagine we carve a small spherical region around the site at which the local field is to be evaluated. The medium contained in this region is considered as a discretized collection of dipoles and we determine the electric field at the center of the cavity by summing up the electric fields generated by every dipole (except the one at the origin); when certain conditions of symmetry are fulfilled, the sum may vanish. For simplicity, we focus our attention on structures with sufficiently high local symmetry (some cubic structures, for instance), so that the electric field generated by the point-like dipoles within the cavity vanishes at its center.

The uniformly polarized medium outside the Lorentz cavity is dealt with in the continuum approximation (Fig. 14). The contribution due to the dipoles outside the ideal cavity, of radius R, can be obtained noticing that the discontinuity of the component of **P** normal to the surface implies a microscopic density of surface polarization charge given by

$$\sigma_p = P_n = -P \cos\theta .$$

The electric field in the z direction at the center of the cavity due to the polarization charges at the cavity surface is

$$E_0 = \int_0^\pi (-P\cos\theta)\cdot R\,d\theta \cdot 2\pi R\sin\theta \cdot \frac{-1}{R^2}\cos\theta = \frac{4\pi}{3}P \ . \tag{57}$$

From Eq. (57), we see that in (above specified) isotropic materials in null electric field, the local field is $(4\pi/3)\mathbf{P}$. If the electric field $\mathbf{E}$ applied to the material is different from zero, we have that the local field acting on the dipoles is the sum of $\mathbf{E}$ and $(4\pi/3)\mathbf{P}$. In summary, in the Lorentz model, the relationship between local field, average macroscopic field and electric polarization is given by

$$\mathbf{E}_{\text{loc}} = \mathbf{E} + \frac{4\pi}{3}\mathbf{P} \ . \tag{58a}$$

The above expression has several limitations. These may be due to the overlapping of electronic clouds, or dipolar fields far from homogeneity. At times the Lorentz field is better approximated by the generalized form

$$\mathbf{E}_{\text{loc}} = \mathbf{E} + \gamma\frac{4\pi}{3}\mathbf{P} \ , \tag{58b}$$

where γ is a semi-empirical parameter. The case $\gamma = 0$ indicates no distinction between local and average field (this is the case of free electrons or essentially spread out wavefunctions), while $\gamma = 1$ is the case of strong localized dipoles in highly symmetric crystals. In even more refined models, γ may be different for different sublattices.

Internal field, polarizability and dielectric constant of materials

Consider a system that can be visualized as constituted by N atoms (or molecules) in the volume V, and suppose for simplicity that the interaction between different atoms (or molecules) can be neglected. We wish to express the dielectric constant ε of the material in terms of the polarizability α of the composing units.

In the presence of an applied field $\mathbf{E}$, the average polarization due to induced dipoles of polarizability α is

$$\mathbf{P} = \frac{N}{V}\alpha\,\mathbf{E}_{\text{loc}} = \frac{N}{V}\alpha(\mathbf{E} + \gamma\frac{4\pi}{3}\mathbf{P}) \ .$$

Hence

$$\mathbf{P} = \frac{(N/V)\alpha}{1 - \gamma(4\pi/3)(N/V)\alpha}\mathbf{E} \ . \tag{59}$$

It is interesting a brief discussion of Eq. (59). In the case the local field and the macroscopic field are the same ($\gamma = 0$), we have $\mathbf{P} = (N/V)\alpha\mathbf{E}$; the polarization $\mathbf{P}$ is thus finite for any finite polarizability. In the case the local field and the macroscopic field are different ($\gamma \neq 0$), the polarization tends to diverge if $\gamma(4\pi/3)(N/V)\alpha \to 1$; this condition is known as "polarization catastrophe". For ordinary dielectrics, the denominator in Eq. (59) is safely far from vanishing condition. For very special crystals, candidate to become ferroelectric, the polarization catastrophe considerations are basic for understanding physical and structural properties near phase transition. Notice that the polarization catastrophe concept is inherent to the local field theory ($\gamma \neq 0$) and is essentially a cooperative effect.

From Eq. (59), we obtain for the dielectric constant ($\varepsilon = 1 + 4\pi P/E$) the expression

$$\boxed{\varepsilon = 1 + \frac{4\pi(N/V)\alpha}{1 - \gamma(4\pi/3)(N/V)\alpha}} \; . \tag{60}$$

Equation (60) expresses the dielectric constant in terms of the polarizability α of the composing units and of the parameter γ, which characterizes the local field. In the specific case $\gamma = 1$ (Lorentz field), Eq. (60) becomes

$$\varepsilon = 1 + \frac{4\pi(N/V)\alpha}{1 - (4\pi/3)(N/V)\alpha} \; . \tag{61a}$$

The Lorentz formula (61a) can also be written in the form

$$\frac{\varepsilon - 1}{\varepsilon + 2} = \frac{4\pi}{3} \frac{N}{V} \alpha \; , \tag{61b}$$

which is named Lorentz–Lorenz (or Clausius–Mossotti) relation.

Infrared dielectric function and polaritons in polar crystals in the presence of local field effects

In Section 7.2 the study of polaritons and optical properties of polar crystals in the infrared region has been done starting from the equation of motion (42). In the presence of local field effects we have rather to consider the equation of motion of the type

$$\ddot{\mathbf{w}} = -\omega_0^2 \mathbf{w} + \frac{e^*}{M^*} \mathbf{E}_{\text{loc}} \; . \tag{62}$$

We now study the consequences brought about by the fact that the local electric field and the macroscopic electric field may be different.

Consider Eq. (62) when the electric field and the relative displacement are periodic in space and time with the form

$$\mathbf{E}_{\text{loc}} = \mathbf{E}_0 \, e^{i(\mathbf{q}\cdot\mathbf{r} - \omega t)} e^{\eta t} \quad \text{and} \quad \mathbf{w} = \mathbf{w}_0 \, e^{i(\mathbf{q}\cdot\mathbf{r} - \omega t)} e^{\eta t} \; ;$$

the exponential $\exp(\eta t)$ with $\eta \to 0^+$ has been included, so that the electric field is turned on adiabatically at $t = -\infty$. We obtain

$$-(\omega + i\eta)^2 \, \mathbf{w}_0 = -\omega_0^2 \, \mathbf{w}_0 + \frac{e^*}{M^*} \mathbf{E}_0 \; .$$

Thus the ionic polarizability becomes

$$\alpha_{\text{ion}}(\omega) \equiv \frac{e^* w_0}{E_0} = \frac{e^{*2}}{M^*} \frac{1}{\omega_0^2 - (\omega + i\eta)^2} = \frac{e^{*2}}{M^* \omega_0^2} \frac{\omega_0^2}{\omega_0^2 - \omega^2 - i\eta\omega}$$

(where 2η has been relabeled as η). The ionic polarizability has a significant frequency dependence in the infrared region.

Let us indicate with α_+ and α_- the electronic polarizabilities of the cation and the

anion of the polar crystal; in the infrared region we can neglect any frequency dependence of electronic polarizabilities. Assuming that electronic and ionic polarizabilities
add up, the total polarizability (per unit cell) becomes

$$\alpha_{\text{tot}}(\omega) = \alpha_+ + \alpha_- + \frac{e^{*2}}{M^*\omega_0^2} \frac{\omega_0^2}{\omega_0^2 - \omega^2 - i\eta\omega} \cdot \tag{63}$$

Inserting Eq. (63) into Eq. (60), we obtain for the dielectric function

$$\varepsilon(\omega) = 1 + \frac{4\pi \frac{N}{V}\left[\alpha_+ + \alpha_- + \dfrac{e^{*2}}{M^*\omega_0^2} \dfrac{\omega_0^2}{\omega_0^2 - \omega^2 - i\eta\omega}\right]}{1 - \gamma\frac{4\pi}{3}\frac{N}{V}\left[\alpha_+ + \alpha_- + \dfrac{e^{*2}}{M^*\omega_0^2} \dfrac{\omega_0^2}{\omega_0^2 - \omega^2 - i\eta\omega}\right]} \cdot \tag{64}$$

It is convenient to introduce the quantities A_{el} and A_{ion} defined as

$$A_{\text{el}} = \frac{4\pi}{3}\frac{N}{V}(\alpha_+ + \alpha_-) \quad \text{and} \quad A_{\text{ion}} = \frac{4\pi}{3}\frac{N}{V}\frac{e^{*2}}{M^*\omega_0^2} \cdot$$

We can re-write Eq. (64) in the form

$$\varepsilon(\omega) = 1 + \frac{3(A_{\text{el}} + A_{\text{ion}})\omega_0^2 - 3A_{\text{el}}(\omega^2 + i\eta\omega)}{(1 - \gamma A_{\text{el}} - \gamma A_{\text{ion}})\omega_0^2 - (1 - \gamma A_{\text{el}})(\omega^2 + i\eta\omega)} \cdot \tag{65}$$

With an eye to the denominator of Eq. (65) we define the "renormalized transverse
frequency" ω_{TO} as

$$\omega_{TO}^2 = \omega_0^2\frac{1 - \gamma A_{\text{el}} - \gamma A_{\text{ion}}}{1 - \gamma A_{\text{el}}} \tag{66}$$

(notice that in the case local field effects are negligible $\gamma = 0$ and $\omega_{TO} = \omega_0$). From
Eq. (65) we obtain

$$\varepsilon(\omega) = 1 + \frac{3(A_{\text{el}} + A_{\text{ion}})\omega_0^2 - 3A_{\text{el}}(\omega^2 + i\eta\omega)}{1 - \gamma A_{\text{el}}} \frac{1}{\omega_{TO}^2 - (\omega^2 + i\eta\omega)} \cdot \tag{67}$$

The above expression, in the limiting case of static and high-frequency regions, takes
the values

$$\varepsilon_{\text{s}} = 1 + \frac{3(A_{\text{el}} + A_{\text{ion}})\omega_0^2}{(1 - \gamma A_{\text{el}})\omega_{TO}^2} \quad \text{and} \quad \varepsilon_\infty = 1 + \frac{3A_{\text{el}}}{1 - \gamma A_{\text{el}}} \cdot$$

Eq. (67) thus becomes

$$\varepsilon(\omega) = 1 + \frac{(\varepsilon_{\text{s}} - 1)\omega_{TO}^2 - (\varepsilon_\infty - 1)(\omega^2 + i\eta\omega)}{\omega_{TO}^2 - (\omega^2 + i\eta\omega)} \equiv \varepsilon_\infty + \frac{\omega_{TO}^2}{\omega_{TO}^2 - \omega^2 - i\eta\omega}(\varepsilon_{\text{s}} - \varepsilon_\infty) \cdot \tag{68}$$

Comparison of Eq. (68) with Eq. (54) is self-explanatory; we see that local field effects do not change the form of $\varepsilon(\omega)$, except for the "renormalization" of the transverse
and longitudinal frequencies ω_{TO} and ω_{LO}. In particular the transverse frequency (66)
decreases with respect to the short-range value ω_0 as an effect of long-range Coulomb
interaction and tends to become *soft*. In any case the renormalized transverse and
longitudinal frequencies are still related by the Lyddane–Sachs–Teller relation, as this
depends on the analytic structure of the response function, rather than on the details

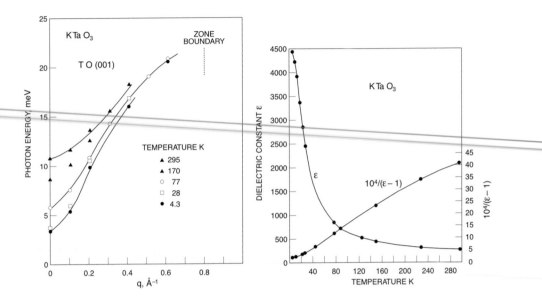

Fig. 15 (a) Temperature dependence of the soft transverse-optical branch in KTaO$_3$ [from G. Shirane, R. Nathans and V. J. Minkiewicz, Phys. Rev. **157**, 396 (1967); copyright 1967 by the American Physical Society]. (b) Dielectric constant and reciprocal susceptibility of KTaO$_3$ as a function of temperature [from S. H. Wemple, Phys. Rev. **137**, A1575 (1965); copyright 1965 by the American Physical Society].

of the orienting fields. We do not have to discuss again the polariton dispersion curves, as the treatment can be performed following step-by-step the previous section, once transverse and longitudinal frequencies are renormalized (as specified by Eq. (66)).

An interesting implication of the local field theory is the possible occurrence of *soft phonon modes*. From Eq. (66) we see that ω_{TO} is reduced with respect to ω_0, since the long-range Coulomb interaction tends to counteract the short-range restoring forces. In the case, due to some mechanism, the frequency ω_{TO} tends to zero, from the Lyddane–Sachs–Teller relation we expect that ε_s tends to infinity. Thus a polar crystal, which exhibits a transverse optical branch with a low frequency mode ω_{TO}, is candidate to develop an extraordinary large polarization. Eventually the crystal might undergo a phase transition and acquire a spontaneous polarization, even in the absence of external fields. Neutron measurements and far infrared optical measurements well support the role of a soft transverse branch in some perovskite ionic crystals.

As an example, we consider the case of perovskite potassium tantalite, and we report in Fig. 15 the temperature dependence of the soft transverse-optical branch (studied by inelastic neutron scattering techniques), as well as the dielectric constant measurements. From Fig. 15 it can be seen that the phonon energy of the soft mode at $q = 0$ is 10.7 meV at 295 K, and decreases to 3.1 meV at 4 K; correspondingly the dielectric constant passes from the value $\varepsilon = 243$ at 295 K to very large values (exceeding several thousands) at low temperatures.

Appendix A. Quantum theory of the linear harmonic oscillator

Creation and annihilation operators

We summarize here some results of the quantum theory of the linear harmonic oscillator, that are preliminary and useful for the discussion of lattice vibrations of crystals.

Consider a one-dimensional harmonic oscillator, of angular frequency ω, described by the Hamiltonian

$$H = \frac{1}{2M}p_x^2 + \frac{1}{2}M\omega^2 x^2 . \tag{A1}$$

The lowering (or annihilation) operator and the raising (or creation) operator are defined by the following linear transformations of the observables x and p_x

$$a = \sqrt{\frac{M\omega}{2\hbar}}\, x + i\sqrt{\frac{1}{2\hbar M\omega}}\, p_x$$

$$a^\dagger = \sqrt{\frac{M\omega}{2\hbar}}\, x - i\sqrt{\frac{1}{2\hbar M\omega}}\, p_x . \tag{A2}$$

The operators a and $a^\dagger$ satisfy the commutation rule

$$[a, a^\dagger] = 1 . \tag{A3}$$

In terms of a and $a^\dagger$, the Hamiltonian $(A1)$ takes the form

$$H = \hbar\omega(a^\dagger a + \frac{1}{2}) , \tag{A4}$$

as can be easily verified inserting expression $(A2)$ into Eq. $(A4)$.

In order to work out eigenvalues and eigenfunctions of the Hamiltonian $(A4)$, we note a few relationships from the commutation relation $(A3)$. We have

$$aa^\dagger = a^\dagger a + 1 , \qquad aa^{\dagger 2} = (a^\dagger a + 1)a^\dagger = a^{\dagger 2}a + 2a^\dagger ,$$

and in general

$$aa^{\dagger n} = a^{\dagger n}a + na^{\dagger n-1} . \tag{A5}$$

Let $|0\rangle$ denote the normalized state that satisfies the equation $a|0\rangle = 0$; and let $|n\rangle$ indicate the normalized state

$$|n\rangle = \frac{1}{\sqrt{n!}}a^{\dagger n}|0\rangle . \tag{A6}$$

The correctness of the normalization follows from the observation that

$$\langle 0|a^n a^{\dagger n}|0\rangle = \langle 0|a^{n-1}aa^{\dagger n}|0\rangle = n\,\langle 0|a^{n-1}a^{\dagger n-1}|0\rangle = n! ,$$

where use has been done of Eq. $(A5)$. With similar procedures, we have that

$$a^\dagger a|n\rangle = n|n\rangle .$$

The number operator $a^\dagger a$ indicates the number of quanta (phonons) in the state $|n\rangle$. The eigenvalues of the Hamiltonian $(A4)$ are thus $E_n = (n+\frac{1}{2})\hbar\omega$ with $n = 0, 1, 2 \ldots$.

From the expression $(A6)$ of the normalized eigenstates of the harmonic oscillator, we see that the operators a and $a^\dagger$ satisfy the relations

$$a\,|n\rangle = \sqrt{n}\,|n-1\rangle \qquad a^\dagger\,|n\rangle = \sqrt{n+1}\,|n+1\rangle \ .$$

We also notice that

$$\langle n|a^{\dagger p}a^p|n\rangle = \langle n|a^{\dagger p}|n-p\rangle\langle n-p|a^p|n\rangle = (\sqrt{n-p+1}\cdot\ldots\cdot\sqrt{n})^2 \qquad n\ge p\ .$$

Thus

$$\langle n|a^{\dagger p}a^p|n\rangle = \begin{cases} n!/(n-p)! & \text{if } n\ge p \\ 0 & \text{if } n<p \end{cases} \ . \tag{A7}$$

Statistical average of operators

At thermodynamic equilibrium, the statistical average of an operator A is defined as

$$\langle A\rangle = \sum_{n=0}^{\infty} P_n\langle n|A|n\rangle \ , \tag{A8}$$

where

$$P_n = \frac{e^{-(n+\frac{1}{2})\hbar\omega/k_B T}}{\sum_m e^{-(m+\frac{1}{2})\hbar\omega/k_B T}} = \frac{e^{-n\hbar\omega/k_B T}}{\sum_m \left(e^{-\hbar\omega/k_B T}\right)^m} \ .$$

Summing up the geometric series in the denominator, and replacing into Eq. $(A8)$, we have

$$\boxed{\langle A\rangle = (1-z)\sum_{n=0}^{\infty} z^n\langle n|A|n\rangle \quad \text{with} \quad z = \exp(-\hbar\omega/k_B T)} \ . \tag{A9}$$

Using Eq. $(A9)$ we can obtain the thermal average of operators of interest. For instance, for the thermal average of the number operator we have

$$\langle a^\dagger a\rangle = (1-z)\sum_{n=0}^{\infty} z^n\langle n|a^\dagger a|n\rangle = (1-z)\sum_{n=0}^{\infty} n\,z^n$$

$$= (1-z)z\frac{\partial}{\partial z}\sum_{n=0}^{\infty} z^n = \frac{z}{1-z} = \frac{1}{e^{\hbar\omega/k_B T}-1} \ , \tag{A10}$$

which expresses the standard Bose–Einstein statistics. We also have

$$\langle a\,a^\dagger\rangle = \langle a^\dagger a\rangle + 1$$

$$\langle a\,a\rangle = \langle a^\dagger a^\dagger\rangle = 0\ .$$

With a little of algebra, we can prove the following relation

$$\boxed{\langle a^{\dagger p}a^p\rangle = p!\,\langle a^\dagger a\rangle^p} \tag{A11}$$

for any $p = 0, 1, 2, \ldots$. In fact, from Eq. $(A9)$ and Eq. $(A7)$ we have

$$\langle a^{\dagger p} a^p \rangle = (1-z) \sum_{n(\geq p)} \frac{n!}{(n-p)!} z^n = (1-z) \sum_{n=0}^{\infty} \frac{(n+p)!}{n!} z^{n+p} . \qquad (A12)$$

From Eq. $(A10)$ it follows

$$\langle a^{\dagger} a \rangle^p = \frac{z^p}{(1-z)^p} = \frac{(1-z) z^p}{(1-z)^{p+1}}$$

$$= (1-z) z^p \left[1 + (p+1)z + \frac{(p+1)(p+2)}{2!} z^2 + \ldots \right] = (1-z) z^p \sum_{n=0}^{\infty} \frac{(n+p)!}{n! \, p!} z^n . \qquad (A13)$$

From comparison of Eq. $(A12)$ and Eq. $(A13)$, we obtain Eq. $(A11)$.

Weyl identity

We establish now two identities (the Weyl identity and the Bloch identity), which are very useful in the study of the correlation functions and Debye–Waller factor in the scattering theory of the harmonic crystal (see Chapter X). We remember the Weyl identity, discussed in standard quantum mechanics textbooks [see for instance A. S. Davydov "Quantum Mechanics" (Pergamon Press, Oxford 1965) p.132].

Consider any two operators A and B, that commute with their commutator $[A, B]$; then we have

$$\boxed{e^A e^B = e^{A+B} e^{[A,B]/2}} \quad \text{if} \quad [A, [A, B]] \equiv [B, [A, B]] \equiv 0 . \qquad (A14)$$

The proof of the above identity can be performed, for instance, following a procedure due to Glauber. We replace momentarily the operators A and B by xA and xB, respectively, where the parameter x will be set equal to 1 at the end of the reasoning. We consider then the following two operators depending from the parameter x:

$$F_1(x) = e^{xA} e^{xB} \quad \text{and} \quad F_2(x) = e^{x(A+B)} e^{x^2[A,B]/2} . \qquad (A15)$$

We show below that both functions $F_1(x)$ and $F_2(x)$ satisfy the differential equation

$$\frac{dF}{dx} = (A + B + x[A, B]) F(x) . \qquad (A16)$$

Eq. $(A16)$, together with the boundary condition $F_1(0) = F_2(0)$, implies $F_1(x) = F_2(x)$ for any x; in particular for $x = 1$ we have Eq. $(A14)$.

It is immediately seen that the function $F_2(x)$ satisfies Eq. $(A16)$; thus, we have only to prove that also $F_1(x)$ satisfies it. In fact

$$\frac{dF_1}{dx} = A e^{xA} e^{xB} + e^{xA} B e^{xB} = \left[A + e^{xA} B e^{-xA} \right] F_1(x) . \qquad (A17)$$

We now use the operatorial identity

$$e^{-S} O e^{S} = \left[1 - S + \frac{1}{2!} S^2 - \frac{1}{3!} S^3 + \ldots \right] O \left[1 + S + \frac{1}{2!} S^2 + \frac{1}{3!} S^3 + \ldots \right]$$

$$= O + [O, S] + \frac{1}{2!} [[O, S], S] + \frac{1}{3!} [[[O, S], S], S] + \cdots , \tag{A18}$$

which holds for any operator O and S. The particular case $S = -x A$ and $O = B$ gives $\exp(x A) B \exp(-x A) = B + x[A, B]$; this result, together with Eq. $(A17)$, proves that $F_1(x)$ satisfies the differential equation $(A16)$.

Bloch identity

We now prove that any operator C, arbitrary linear combination of phonon operators a and $a^\dagger$, satisfies the Bloch identity

$$\boxed{\langle e^C \rangle = e^{\langle C^2 \rangle / 2}} ; \tag{A19}$$

this theorem states that the thermal average of the exponential of an operator, linear in a and $a^\dagger$, is just the exponential of half of the thermal average of the squared operator itself.

Consider in fact the linear combination of the phonon operators a and $a^\dagger$ of the form

$$C = c_1 a^\dagger + c_2 a$$

with c_1 and c_2 arbitrary complex numbers. We remark that

$$\langle C^2 \rangle = c_1 c_2 \left[2 \langle a^\dagger a \rangle + 1 \right] .$$

Using the Weyl identity it follows

$$e^C = e^{c_1 a^\dagger + c_2 a} = e^{c_1 a^\dagger} e^{c_2 a} e^{c_1 c_2 / 2} .$$

Performing the thermal average one gets

$$\langle e^C \rangle = e^{c_1 c_2 / 2} \langle e^{c_1 a^\dagger} e^{c_2 a} \rangle = e^{c_1 c_2 / 2} \sum_{mn} \frac{c_1^m c_2^n}{m! \, n!} \langle a^{\dagger m} a^n \rangle .$$

In the double sum only the terms with $m = n$ survive, and using Eq. $(A11)$ it follows

$$\langle e^C \rangle = e^{c_1 c_2 / 2} \sum_m \frac{(c_1 c_2)^m}{m!} \langle a^\dagger a \rangle^m = e^{c_1 c_2 [2 \langle a^\dagger a \rangle + 1] / 2} ,$$

and the Bloch identity is thus proved.

From the Bloch identity and the Weyl identity, we can obtain the following important result. Let A and B indicate two operators linear in creation and annihilation operators; it holds

$$\boxed{\langle e^A e^B \rangle = \langle e^{A+B} \rangle e^{[A,B]/2} = e^{\langle A^2 + 2AB + B^2 \rangle / 2}} . \tag{A20}$$

This relation will be used in Chapter X, in the study of the dynamical structure factor for the scattering of particles from harmonic crystals.

Further reading

H. Bilz and W. Kress "Phonon Dispersion Relations in Insulators" (Springer Verlag, Berlin 1979)

M. Born and K. Huang "Dynamical Theory of Crystal Lattices" (Oxford University Press 1954)

L. Brillouin "Wave Propagation in Periodic Structures" (Dover, New York 1953)

P. F. Choquard "The Anharmonic Crystal" (Benjamin, New York 1967)

B. Di Bartolo and R. C. Powell "Phonons and Resonances in Solids" (Wiley, New York 1976)

M. Dominoni and N. Terzi "How to Play with Springs and Pulses in a Classical Harmonic Crystal" in "Advances in Non-Radiative Processes in Solids" edited by B. Di Bartolo (Plenum Press, New York 1991).

H. Frölich "Theory of Dielectrics" (Clarendon Press, Oxford 1958)

G. Grimvall "The Electron – Phonon Interaction in Metals" (North-Holland, Amsterdam 1981)

M. E. Lines and A. M. Glass "Principles and Applications of Ferroelectric and Related Materials" (Clarendon Press, Oxford 1977)

H. A. Lorentz "The Theory of Electrons" (Dover, New York 1952)

A. A. Maradudin, E. W. Montroll, G. H. Weiss and I. P. Ipatova "Theory of Lattice Dynamics in the Harmonic Approximation" (Academic Press, New York 1971)

G. P. Srivastava "The Physics of Phonons" (Adam Hilger, Bristol 1990)

J. M. Ziman "Electrons and Phonons" (Clarendon Press, Oxford 1960)

X
Scattering of particles by crystals

1 General considerations

Elastic and inelastic scattering of particles (such as photons, neutrons and electrons) by a crystal may provide important information on the crystal structure, electron density, energy–wavevector dispersion laws of elementary excitations (phonons, polaritons, excitons, plasmons, magnons etc.). Generally X-rays and neutrons are well suited for analysis of bulk properties because they easily penetrate in the crystals, while electrons are used for analysis of surfaces and thin films, due to their poor penetration in the sample. The scattering of X-rays is essentially produced by the interaction of the electric field of the radiation with the electronic charge density of the crystal. In contrast with X-rays, electrons are efficiently scattered by the nuclei as well by the electrons of the sample, because of the strong Coulomb interactions with charged particles. Neutrons interact via nuclear forces with the nuclei, and thus are sensitive to their spatial distribution and vibrations; neutrons also interact with electronic magnetic moments and thus give useful information on magnetic materials.

A schematic representation of the experimental set-up for scattering measurements is indicated in Fig. 1. A beam of incident particles (photons, or neutrons, or electrons)

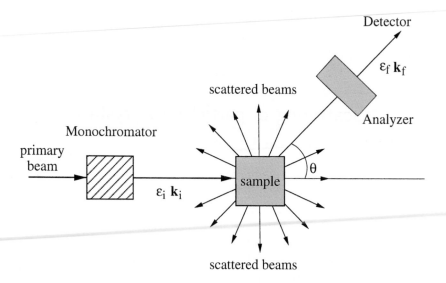

Fig. 1 Schematic experimental set-up for scattering measurements.

is collimated and monochromatized, so as to select particles with an initial momentum $\hbar\mathbf{k}_i$ and initial energy ε_i (often the impinging beam is also polarized in one of the possible polarization states). The monochromatic beam impinges on the sample and is diffused (partially elastically and partially inelastically) in the space. An analyser selects particles of final energy ε_f and final momentum $\hbar\mathbf{k}_f$, which are received by a detector. From the measurements of the momentum transfer $\hbar\Delta\mathbf{k} = \hbar\mathbf{k}_f - \hbar\mathbf{k}_i$, of the energy transfer $\Delta\varepsilon = \varepsilon_f - \varepsilon_i$, and of the intensity of the scattered beam, information on the structural and dynamical properties of the sample can be inferred. For this purpose it is convenient to consider first some general aspects on photonic, neutronic and electronic beams, their energy–wavelength relations, and their use as probes in scattering experiments.

Photons

The elastic scattering by X-rays is the most traditional tool to obtain information on the crystal structure and electronic charge distribution. An electromagnetic wave is scattered by electrons, which act as oscillating dipoles under the influence of the wave electric field. For photons, the energy–wavevector relation is $\hbar\omega = \hbar c k = \hbar c\, 2\pi/\lambda$; with $\hbar\omega$ expressed in electron-volt and λ in Angstrom, we obtain

$$photons: \quad \lambda = \frac{12398.5}{\hbar\omega} \qquad (\lambda \text{ in Å and } \hbar\omega \text{ in eV}) . \tag{1a}$$

Thus, for the determination of the crystal structures, one usually works in the $\approx 10-50$ keV energy range, for which λ is of the order of inter-atomic spacing.

Inelastic scattering of photons is also a very important tool for the investigation of crystals. Very accurate information on phonons (or polaritons) is obtained by inelastic scattering of visible light (Brillouin and Raman scattering), often from a laser beam; however, the wavevector of visible light is very small on the scale of the Brillouin zone dimension, and thus only the $q \approx 0$ region can be explored. Information on phonons throughout the Brillouin zone can be inferred from inelastic scattering of X-rays. Typically, the energy of the incident photons is in the range ≈ 10 keV, while the energy shift due to emission or absorption of phonons is as small as ≈ 10 meV; thus high resolution experiments techniques are required for detecting so small energy shifts. Finally, we notice that X-rays can be scattered inelastically by electrons, through the Compton mechanism (the energy shift of photons in typical experiments is of the order of a few keV); the analysis of Compton profiles provides important information on the ground-state electron momentum density.

Neutrons

The interaction of neutrons with solids is an important area of research for the determination of crystal structures, lattice dynamics and magnetic properties of materials. For neutrons, the energy–wavevector relationship is $E = \hbar^2 k^2 / 2 M_n$, where $k = 2\pi/\lambda$ and M_n is the neutron mass; with E expressed in electron-volt and λ in Angstrom, we obtain

$$neutrons : \quad \lambda = \frac{0.2862}{\sqrt{E}} \quad (\lambda \text{ in Å and } E \text{ in eV}) . \tag{1b}$$

Neutrons with wavelength of the order of one Å have energy of about 80 meV, a value of the same order of magnitude as $k_B T$ at room temperature ($k_B T = 23.538$ meV for $T = 273.15$ K); thus, thermal neutrons, obtained after moderation from high flux nuclear reactors or pulsed sources, have momenta and energies comparable with those of phonons; for this reason, inelastic scattering of neutrons is the most natural and accurate method for investigation of phonon dispersion curves.

Neutrons interact with nucleons through nuclear forces. The interaction between a neutron at position $\mathbf{r}$ and a nucleus located at $\mathbf{R}_0$ is usually described by the Fermi pseudo-potential

$$V(\mathbf{r}, \mathbf{R}_0) = \frac{2\pi\hbar^2}{M_n} b \, \delta(\mathbf{r} - \mathbf{R}_0) ,$$

where b is a phenomenological parameter, called Fermi length (or scattering length) of the given nucleus. The scattering length b varies irregularly across the periodic table, can assume positive or negative values, and its order of magnitude is 10^{-13} cm $= 10^{-5}$ Å, the range of nuclear forces.

The scattering amplitude for X-rays is related to the number Z of electrons of the atoms; the scattering amplitude for neutrons is related to the scattering length b, which varies in an erratic way with the mass number of the nuclei; thus neutron diffraction is often used as a precious complementary tool to the X-ray diffraction experiments. For instance, light elements can be investigated in the presence of heavy ones (a difficult or impossible job with X-rays). Isotopes of the same element are equally efficient as X-ray scatterers, but may behave differently as neutron scatterers;

in particular the large difference in the scattering length for hydrogen and deuterium can be exploited for the study of complicated molecular systems by appropriately replacing one isotope with the other one. We also notice that a nucleus acts as a point scatterer for thermal neutrons (the wavelength of thermal neutrons is of the order of $1\,\text{Å}$, while the scattering length of the nuclei is of the order of $10^{-5}\,\text{Å}$); thus the scattering amplitude for neutrons does not depend on the scattering angle; this isotropy leads improved resolution capability in the study of diffraction beams with high values of momentum transfer, and is particularly important in the study of complex systems.

Neutrons are particles with magnetic moment and interact with matter also through electromagnetic forces. Thus neutrons also take notice of unpaired spin electrons and are an invaluable tool for the characterization of magnetic structures.

Electrons

For electrons the relationship between energy and wavevector is $E = \hbar^2 k^2/2m$, where $k = 2\pi/\lambda$ and m is the electron mass; we have

$$electrons:\ \ \lambda = \frac{12.264}{\sqrt{E}} \quad (\lambda\ \text{in Å and } E\ \text{in eV}) . \tag{1c}$$

Electrons with wavelength of the order of one Å have energy of about 150 eV; the energy region $10(\text{eV}) < E < 10^3(\text{eV})$ is particularly suitable for diffraction experiments and is called low energy electron diffraction (LEED) region.

Electrons are charged particles and thus they interact strongly with the nuclei and the electrons of the crystal via Coulomb forces. Electrons penetrate into the crystal for small distances and elastic electron diffraction is particularly convenient for investigation of surfaces and thin layers. Inelastic scattering of electrons with appropriate incident kinetic energy gives information on the energy spectrum of the excitations localized at the surface; energy-loss experiments with fast electrons also provide information on excitons and plasmons in the crystal.

2   Elastic scattering of X-rays from crystals

In this section we discuss the *elastic scattering or diffraction* of X-rays from crystals, assuming the atoms fixed at the equilibrium positions (the effect of the thermal motion of atoms and inelastic scattering processes are studied in the following sections). We shall see that the geometry of the diffracted beams gives direct information on the reciprocal lattice vectors of the crystal, while the intensity of diffracted beams provides information on the contents of the unit cell. Although we discuss here the specific case of elastic scattering of X-rays, the results that depend essentially on the wave nature of the incident particle apply as well to other incident beams.

2.1   Elastic scattering of X-rays and Bragg diffraction condition

Consider an incident radiation beam of frequency ω, propagation vector $\mathbf{k}_i$ ($\omega = ck_i$), polarization versor $\mathbf{e}_i$ ($\mathbf{e}_i \perp \mathbf{k}_i$), and amplitude E_0; the electric field of the

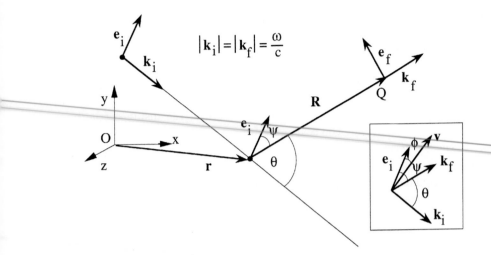

Fig. 2 Schematic representation of an incident electromagnetic wave (of frequency ω, propagation vector $\mathbf{k}_i$, polarization versor $\mathbf{e}_i$) elastically scattered from a free electron at the point $\mathbf{r}$ and detected at the point $Q \equiv \mathbf{r} + \mathbf{R}$. In the inset, the angles of geometrical interest are indicated. The scattering angle is θ ($\cos\theta = \widehat{\mathbf{k}}_i \cdot \widehat{\mathbf{k}}_f$, where $\widehat{\mathbf{k}}_i$ and $\widehat{\mathbf{k}}_f$ denote versors); ψ denotes the angle between $\mathbf{e}_i$ and $\mathbf{R}$ or $\mathbf{k}_f$ ($\cos\psi = \mathbf{e}_i \cdot \widehat{\mathbf{k}}_f$); $\mathbf{e}_i = \mathbf{e}_f \sin\psi + \widehat{\mathbf{k}}_f \cos\psi$; $\mathbf{v} = (\widehat{\mathbf{k}}_f - \widehat{\mathbf{k}}_i \cos\theta)/\sin\theta$ is the versor that lies in the $(\mathbf{k}_i, \mathbf{k}_f)$ plane and is perpendicular to $\mathbf{k}_i$; ϕ is the polarization angle ($\cos\phi = \mathbf{e}_i \cdot \mathbf{v} = \cos\psi/\sin\theta$).

monochromatic electromagnetic wave can be expressed as

$$\mathbf{E}(\mathbf{r}, t) = \mathbf{e}_i E_0 e^{i(\mathbf{k}_i \cdot \mathbf{r} - \omega t)} \ . \tag{2a}$$

A free electron, located at the point $\mathbf{r}$, under the influence of the impinging (high frequency) electric field (2a) is accelerated according to the classical equation of motion $m\ddot{\mathbf{u}} = (-e)\mathbf{E}(\mathbf{r}, t)$; the electron acceleration due to the electric field is thus

$$\ddot{\mathbf{u}}(t) = \frac{(-e)}{m} \mathbf{e}_i E_0 e^{i(\mathbf{k}_i \cdot \mathbf{r} - \omega t)} \ . \tag{2b}$$

The electron follows the periodic variations of $\mathbf{E}$, acts as an oscillating dipole and radiates electromagnetic waves at the same frequency as the incident one.

According to elementary electrodynamics, the scattered electric field at the point Q at distance $\mathbf{R}$ from the diffusion center at $\mathbf{r}$ (when R is much larger than the wavelength of the incident radiation) is given by the expression

$$\mathbf{E}_s(\mathbf{R}, t) = \mathbf{e}_f \frac{(-e)}{c^2 R} \ddot{\mathbf{u}}\left(t - \frac{R}{c}\right) \cdot \mathbf{e}_f = \mathbf{e}_f \frac{1}{R} \frac{e^2}{mc^2} E_0\, e^{i\mathbf{k}_i \cdot \mathbf{r}}\, e^{-i\omega(t - R/c)} \sin\psi \ ,$$

where ψ is the angle between $\mathbf{R}$ and $\mathbf{e}_i$, and the versor $\mathbf{e}_f$ is perpendicular to $\mathbf{R}$ and lies in the plane formed by $\mathbf{R}$ and $\mathbf{e}_i$ (see Fig. 2). For elastic scattering $\omega/c = k_f$ and we can also write

$$\mathbf{E}_s(\mathbf{R}, t) = \mathbf{e}_f \frac{1}{R} \frac{e^2}{mc^2} E_0\, e^{i\mathbf{k}_i \cdot \mathbf{r}}\, e^{i(\mathbf{k}_f \cdot \mathbf{R} - \omega t)} \sin\psi \ . \tag{3}$$

From the modulus squared of the above expression, we obtain for the intensity of the scattered field

$$I_s(\mathbf{R}) = I_0 \frac{1}{R^2} \left(\frac{e^2}{mc^2}\right)^2 \sin^2 \psi \,, \tag{4a}$$

where I_0 is the intensity of the incident field. From Eq. (4a) we see that the intensity of the scattered wave decreases as the inverse square of the distance (the mass at the denominator ensures that the scattering from nuclei can be safely ignored, with respect to the scattering from electrons).

The expression (4a) can be given in terms of the scattering angle θ and the polarization angle ϕ. The geometrical relation between ψ, θ and ϕ is shown in Fig. 2, and we have $\cos \psi = \cos \phi \sin \theta$; Eq. (4a) then becomes

$$I_s(\mathbf{R}) = I_0 \frac{1}{R^2} \left(\frac{e^2}{mc^2}\right)^2 (1 - \cos^2 \phi \sin^2 \theta) \,.$$

In the case the incident field is not polarized, the above expression must be averaged over ϕ (i.e. $\cos^2 \phi$ must be replaced by $1/2$); one obtains the well-known Thomson formula

$$I_s(R,\theta) = I_0 \frac{1}{R^2} \left(\frac{e^2}{mc^2}\right)^2 \frac{1 + \cos^2 \theta}{2} \equiv I_0 \frac{1}{R^2} \left(\frac{d\sigma}{d\Omega}\right)_T \,, \tag{4b}$$

where $e^2/mc^2 = r_0 = 2.82 \cdot 10^{-13}$ cm is the classical radius of the electron; the quantity $(d\sigma/d\Omega)_T = (1/2)r_0^2(1+\cos^2 \theta)$ is the Thomson differential cross-section, per unit solid angle, for scattering of unpolarized photons from free electrons (the above classical treatment only considers the coupling of the electronic charge with the electromagnetic field, while the interaction between the electron spin and the electromagnetic field has been disregarded). The Thomson cross-section is as small as ≈ 0.1 barn (1 barn=10^{-24} cm^2) and the scattering can be safely termed as weak.

Let us now consider the Thomson scattering by two electrons, one at the origin O and the other located at the point P (see Fig. 3). The two electrons are both accelerated by the electric field of the impinging electromagnetic wave, and both radiate. Using Eq. (3), the scattered field from the electron located at the origin and from the electron located at P are given by:

$$\text{(electron at } \mathbf{r} = 0) \quad \mathbf{E}_s(\mathbf{R}_1, t) = \mathbf{e}_f \frac{1}{R_1} \frac{e^2}{mc^2} E_0 \, e^{i(\mathbf{k}_f \cdot \mathbf{R}_1 - \omega t)} \sin \psi \tag{5a}$$

$$\text{(electron at } \mathbf{r} \neq 0) \quad \mathbf{E}_s(\mathbf{R}_2, t) = \mathbf{e}_f \frac{1}{R_2} \frac{e^2}{mc^2} E_0 \, e^{i\mathbf{k}_i \cdot \mathbf{r}} e^{i(\mathbf{k}_f \cdot \mathbf{R}_2 - \omega t)} \sin \psi \,. \tag{5b}$$

From Fig. 3, we see that $\mathbf{k}_f \cdot \mathbf{R}_2 = \mathbf{k}_f \cdot \mathbf{R}_1 - \mathbf{k}_f \cdot \mathbf{r}$. At large distance $1/R_1 \approx 1/R_2 \approx 1/R$ (where R is the average distance between the scatterers and the detector); thus, expression (5b) coincides with expression (5a) times the phase factor $\exp(-i\Delta\mathbf{k} \cdot \mathbf{r})$, where $\Delta\mathbf{k} = \mathbf{k}_f - \mathbf{k}_i$ is the wavevector change in the scattering process; the phase factor can be traced back to the difference in optical length from the wave scattered from P and the wave scattered from O (see Fig. 3).

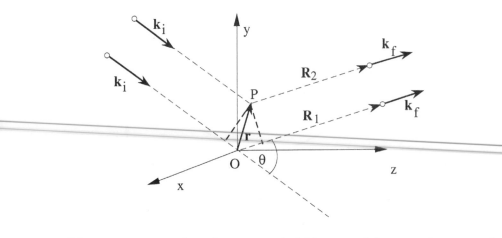

Fig. 3 Schematic geometry of an electromagnetic field scattered from one electron at the origin O and one electron at the point P; the difference in optical length between the wave scattered from O and the wave scattered from P is $\mathbf{k}_f \cdot \mathbf{r} - \mathbf{k}_i \cdot \mathbf{r} = \Delta\mathbf{k} \cdot \mathbf{r}$.

We sum up the electric field contributions of Eq. (5a) and Eq. (5b), and perform the modulus squared; the intensity of the scattered radiation (when the incident radiation is not polarized) becomes

$$I_s(R, \theta) = I_0 \frac{1}{R^2} \left| 1 + e^{-i\Delta\mathbf{k}\cdot\mathbf{r}} \right|^2 \left(\frac{d\sigma}{d\Omega} \right)_T , \tag{6}$$

where the quantity within modulus squared takes proper account of the effect of geometrical phases on the scattered waves.

We can generalize Eq. (6) to a quantum system with electron density $n_{el}(\mathbf{r})$. The scattering of sufficiently hard X-rays is then given by

$$I_s(R, \theta) = I_0 \frac{1}{R^2} \left| \int n_{el}(\mathbf{r}) e^{-i\Delta\mathbf{k}\cdot\mathbf{r}} \, d\mathbf{r} \right|^2 \left(\frac{d\sigma}{d\Omega} \right)_T . \tag{7}$$

The quantity within modulus squared in Eq. (7) is the Fourier transform of the electronic density

$$\boxed{F(\Delta\mathbf{k}) = \int n_{el}(\mathbf{r}) e^{-i\Delta\mathbf{k}\cdot\mathbf{r}} \, d\mathbf{r}} . \tag{8}$$

We can summarize the results so far obtained by observing that the *intensity of X-ray scattering is related to the modulus squared of the Fourier transform of the electron density of the system.*

Condition for elastic scattering in periodic systems

In crystals $n_{el}(\mathbf{r})$ is a periodic function; thus its Fourier coefficients (also called *X-ray crystal structure factors* or *crystal form factors*), defined by Eq. (8), can be different from zero only in correspondence to reciprocal lattice vectors; this means that

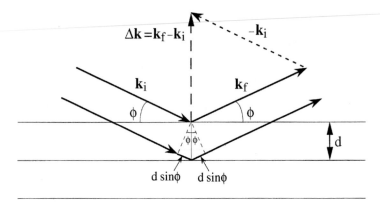

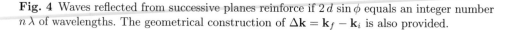

Fig. 4 Waves reflected from successive planes reinforce if $2\,d\,\sin\phi$ equals an integer number $n\,\lambda$ of wavelengths. The geometrical construction of $\Delta\mathbf{k} = \mathbf{k}_f - \mathbf{k}_i$ is also provided.

occurrence of diffraction peaks requires

$$\boxed{\Delta\mathbf{k} = \mathbf{G}} \;, \tag{9}$$

where $\mathbf{G}$ is a reciprocal lattice vector. Thus a necessary condition for X-ray diffraction is that the difference $\Delta\mathbf{k} = \mathbf{k}_f - \mathbf{k}_i$ between the scattered and the incident wavevectors equals a reciprocal lattice vector. Equation (9) can also be written in the form

$$\hbar\mathbf{k}_f = \hbar\mathbf{k}_i + \hbar\mathbf{G} \; ;$$

thus in the scattering process the momentum is preserved within reciprocal lattice vectors times $\hbar$.

It is instructive to show that the diffraction condition (9), deduced by von Laue, is equivalent to the intuitive description by Bragg, who considers specular reflection of by a family of lattice planes; as illustrated in Fig. 4, the geometrical condition for the coherent scattering from two successive planes (and hence from the whole sequence of parallel planes) requires

$$\boxed{2\,d\,\sin\phi = n\,\lambda} \;, \tag{10}$$

where n is an integer, λ is the wavelength of the scattered (and incident) wave, 2ϕ is the scattering angle and d is the distance between adjacent planes of the family.

For convenience in Fig. 4 we also report the geometrical construction of the vector $\Delta\mathbf{k} = \mathbf{k}_f - \mathbf{k}_i$. Since in elastic scattering $|\mathbf{k}_i| = |\mathbf{k}_f| = 2\pi/\lambda$, we have

$$|\Delta\mathbf{k}| = 2|\mathbf{k}_i| \sin\phi = 2\frac{2\pi}{\lambda}\sin\phi = n\frac{2\pi}{d} \;, \tag{11}$$

where the last equality follows from the Bragg law (10). From Fig. 4 we see that $\Delta\mathbf{k}$ is perpendicular to the family of lattice planes with distance d; from Eq. (11) we see that the magnitude $|\Delta\mathbf{k}|$ is an integer multiple of the quantity $2\pi/d$; these two observations (together with the general properties of real and reciprocal spaces studied in Section II-4.2) allow us to conclude that $\Delta\mathbf{k}$ must be a reciprocal lattice vector.

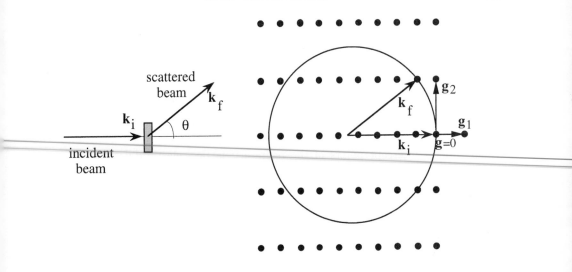

Fig. 5 Scattering of X-rays and Ewald construction.

The Laue condition (9) for occurrence of "diffracted beams" is thus fully equivalent to the Bragg condition (10) for occurrence of "reflected beams", and the expressions diffracted beam and reflected beam become synonymous.

From the Bragg condition (10), we see that since $\sin \phi < 1$ the possibility of elastic scattering occurs only if $\lambda < 2d$. Thus λ must be of the order of the Å or less. Furthermore λ cannot be much smaller than the interatomic distance, otherwise experimental arrangements at glancing angles are necessary to detect diffraction peaks with small momentum transfer. This restricts the ordinary frequency range of interest to the X-ray region.

Ewald construction

We can consider the following geometrical construction due to Ewald for an easy determination of the occurrence of (elastic) diffraction peaks. We suppose to have an incident monochromatic beam of particles of wavevector $\mathbf{k}_i$ and a diffracted beam with propagation wavevector $\mathbf{k}_f$. For elastic scattering, the conservation of energy and the conservation of momentum (within reciprocal lattice vectors times $\hbar$) implies

$$|\mathbf{k}_i| = |\mathbf{k}_f| \quad \text{and} \quad \mathbf{k}_f = \mathbf{k}_i + \mathbf{G} \ . \tag{12}$$

The Ewald construction (see Fig. 5) permits a simple geometrical interpretation of Eqs. (12). In the reciprocal space, the vector $\mathbf{k}_i$ is drawn in such a way that its tip terminates at the reciprocal lattice vector $\mathbf{G} = 0$. With the center at the origin of the vector $\mathbf{k}_i$, the sphere of radius k_i is also drawn; if this sphere intersects, besides $\mathbf{G} = 0$, one (or more) points of the reciprocal lattice, conditions (12) are satisfied and Bragg diffractions are possible. It is evident that diffractions may occur only if the magnitude k_i exceeds one half of the magnitude of the smallest $\mathbf{G}$ vector.

Methods of X-ray scattering

In general, if we consider a *monochromatic X-ray beam and a fixed crystal orientation*, we have no possibility of elastic diffraction. Some experimental methods are thus adopted to perform X-ray diffraction measurements.

(a) *The Laue method.* In the Laue method the incident radiation covers a wide band of wavelengths. Correspondingly, all Ewald spheres with radii within the appropriate range of incident wavevectors are to be drawn in the reciprocal lattice; whenever the Ewald spheres intercept points of the reciprocal lattice, elastic diffractions become possible.

(b) *The Bragg method* (rotating crystal method). In the Bragg method, the incident radiation is monochomatic, but the crystal is capable of rotational motion around a given axis. The radius of the Ewald sphere is thus fixed, but the reciprocal space rotates following the crystal rotation in direct space; when points of the reciprocal lattice intercept the Ewald sphere, elastic diffractions become possible.

(c) *The Debye–Scherrer method* (powder method). The incident radiation is monochromatic, but now the single (rotating) crystal is replaced by a polycrystalline specimen (powder) with random orientation of the composing crystallites.

We shall not enter into the details of the various experimental methods. We simply notice that *from each elastically diffracted beam a specific reciprocal lattice vector* $\mathbf{G} = \mathbf{k}_f - \mathbf{k}_i$ *is individuated*. A sufficiently detailed map of the reciprocal lattice vectors permits to infer the fundamental vectors $\mathbf{g}_1, \mathbf{g}_2, \mathbf{g}_3$ (and hence the fundamental translation vectors of the direct lattice, via the standard relation $\mathbf{g}_i \cdot \mathbf{t}_j = 2\pi \delta_{ij}$).

2.2   Elastic scattering of X-rays and intensity of diffracted beams

Until now we have focused our attention on the geometrical aspects of diffraction (i.e. on the change $\Delta \mathbf{k} = \mathbf{k}_f - \mathbf{k}_i$ between the scattered and incident wavevectors). Very important information on the electronic charge distribution can be obtained by the measurements of the intensity of the diffracted beams, and comparison with appropriate theoretical models. In the recent literature, ab initio calculations of crystal ground-state electronic density $n_{el}(\mathbf{r})$ and structure factors $F(\Delta \mathbf{k})$ have been performed for an increasing number of crystals. Here, however, we focus on simplified theoretical frameworks, which allow us to interpret relevant features of Bragg reflections intensities, without need of demanding computational labour.

Consider first an atom with electron density $n_a(\mathbf{r})$; the *atomic form factor* $f_a(\Delta \mathbf{k})$ is defined as the Fourier transform

$$f_a(\Delta \mathbf{k}) = \int e^{-i\Delta \mathbf{k} \cdot \mathbf{r}} n_a(\mathbf{r}) \, d\mathbf{r} \ . \tag{13a}$$

In isolated atoms, the electron cloud $n_a(\mathbf{r})$ is spherically symmetric, and Eq. (13a) can be simplified to an integral over the radial variable

$$f_a(\Delta \mathbf{k}) = \int_0^\infty n_a(r) \frac{\sin |\Delta \mathbf{k}| r}{|\Delta \mathbf{k}| r} 4\pi r^2 \, dr \ . \tag{13b}$$

The atomic form factor depends only on the magnitude (but not on the orientation) of the scattering vector $\Delta\mathbf{k}$; for an atom with Z electrons, the atomic form factor changes monotonically from Z to zero, as $|\Delta\mathbf{k}|$ increases from zero to large values.

When atoms are assembled to build a solid, their electronic clouds are no more spherically symmetric. However, for a preliminary analysis of the crystal structure factors, it is useful to approximate the crystal electron density $n_{\mathrm{el}}(\mathbf{r})$ as *sum of spherically symmetric contributions, centered at the various atomic positions* (for brevity we call this the *local spherical model* for electron distribution). We write

$$n_{\mathrm{el}}(\mathbf{r}) = \sum_{\mathbf{t}_n}\sum_{\mathbf{d}_\nu} n_{a\nu}(\mathbf{r} - \mathbf{t}_n - \mathbf{d}_\nu) \, , \tag{14}$$

where $\mathbf{t}_n$ are lattice translational vectors, $\mathbf{d}_\nu$ are the positions of the atoms in the unit cell, and $n_{a\nu}(\mathbf{r} - \mathbf{t}_n - \mathbf{d}_\nu) \equiv n_{a\nu}(|\mathbf{r} - \mathbf{t}_n - \mathbf{d}_\nu|)$ indicates the contribution of the atom at the site $\mathbf{t}_n + \mathbf{d}_\nu$ to the total electron density. Notice that Eq. (14) is a fairly accurate description of the "core electronic contribution", while for the "valence electronic contribution" some degree of non-sphericity is expected to occur (the environment of a lattice point, in fact, has the local point crystal symmetry, and not the full rotational symmetry). In particular, bonding charge tends to accumulate midway between interacting atoms and cannot be accurately described by the decomposition (14). Charge asymmetry is at the origin of subtle but interesting effects determining a (weak) intensity for some "forbidden reflections" of the local spherical model, as we shall discuss below.

We now replace the approximate form (14) of the crystal electronic density into Eq. (8) and we have

$$F(\Delta\mathbf{k}) = \sum_{\mathbf{t}_n}\sum_{\mathbf{d}_\nu} \int e^{-i\Delta\mathbf{k}\cdot\mathbf{r}} n_{a\nu}(\mathbf{r} - \mathbf{t}_n - \mathbf{d}_\nu)\, d\mathbf{r} \, . \tag{15}$$

We insert in Eq. (15) the unit quantity $\exp[-i\Delta\mathbf{k}\cdot(\mathbf{t}_n + \mathbf{d}_\nu)]\,\exp[+i\Delta\mathbf{k}\cdot(\mathbf{t}_n + \mathbf{d}_\nu)]$, and with straightforward manipulations we obtain

$$F(\Delta\mathbf{k}) = \sum_{\mathbf{t}_n} e^{-i\Delta\mathbf{k}\cdot\mathbf{t}_n} \sum_{\mathbf{d}_\nu} e^{-i\Delta\mathbf{k}\cdot\mathbf{d}_\nu} f_{a\nu}(\Delta\mathbf{k}) \, , \tag{16a}$$

where

$$f_{a\nu}(\Delta\mathbf{k}) = \int e^{-i\Delta\mathbf{k}\cdot\mathbf{r}} n_{a\nu}(\mathbf{r})\, d\mathbf{r} \, . \tag{16b}$$

In Eq. (16a), the sum over the lattice translations is different from zero if $\Delta\mathbf{k} = \mathbf{G}$, in agreement with the Laue condition (9). Once the Laue condition is satisfied, the crystal structure factors $F(\mathbf{G})$ become

$$\boxed{F(\mathbf{G}) = N \sum_{\mathbf{d}_\nu} e^{-i\mathbf{G}\cdot\mathbf{d}_\nu} f_{a\nu}(\mathbf{G})} \, . \tag{17}$$

We see that the crystal structure factors $F(\mathbf{G})$ are expressed in terms of the atomic form factors $f_{a\nu}(\mathbf{G})$ of the atoms in the position $\mathbf{d}_\nu$ of the unit cell and of the phase factors $\exp(-i\mathbf{G} \cdot \mathbf{d}_\nu)$.

In the case of crystals with a single atom per unit cell, the structure factors $F(\mathbf{G})$ are proportional to the atomic form factors of the only atom composing the crystal; thus $|F(\mathbf{G})|^2$ are expected to decrease monotonically as $|\mathbf{G}|$ increases, this being the behaviour of the atomic form factors.

A more interesting situation occurs in crystals with a basis of two or more atoms in the unit cell; in this case, systematic enhancement or weakening (or absence) of diffraction beams may occur. If all the atoms in the unit cell are equal and contribute the same spherically-symmetric electron cloud around the appropriate centers, the form factors $f_{a\nu}(\mathbf{G}) \equiv f_a(\mathbf{G})$ can be factorized out from the sum in Eq. (17) and we obtain

$$F(\mathbf{G}) = N\, f_a(\mathbf{G})\, S(\mathbf{G}) \; ; \tag{18a}$$

the *geometrical structure factor*

$$S(\mathbf{G}) = \sum_{\mathbf{d}_\nu} e^{-i\mathbf{G}\cdot\mathbf{d}_\nu} \tag{18b}$$

may lead to systematic absence of diffraction beams ("forbidden reflexions").

As an application of Eqs. (18), we consider the elemental semiconductors with the diamond structure. As described in Fig. II-9, the diamond structure consists of a fcc Bravais lattice with a basis of two equal atoms in the unit cell in the positions $\mathbf{d}_1 = 0$ and $\mathbf{d}_2 = (a/4)(1,1,1)$. The fundamental vectors of the reciprocal lattice are $\mathbf{g}_1 = (2\pi/a)(-1,1,1)$, $\mathbf{g}_2 = (2\pi/a)(1,-1,1)$ and $\mathbf{g}_3 = (2\pi/a)(1,1,-1)$. It is easily seen by inspection that the most general vector $\mathbf{G}$ of the reciprocal lattice has the form $\mathbf{G} = (2\pi/a)(h_1, h_2, h_3)$ with h_1, h_2, h_3 *all odd or all even* integers.

For a preliminary analysis of the structure factors, we assume that the crystal electron density is made by the superposition of identical spherically symmetric contributions, centered at the various atomic positions (local spherical approximation). The geometrical factor (18b) in the case of the diamond structure becomes

$$S(\mathbf{G}) = e^{-i\mathbf{G}\cdot\mathbf{d}_1} + e^{-i\mathbf{G}\cdot\mathbf{d}_2} = 1 + e^{-i\pi(h_1+h_2+h_3)/2} \; . \tag{19a}$$

The possible values of $S(\mathbf{G})$ are the following:

$$S(\mathbf{G}) = \begin{cases} 1-i & \text{if} \quad h_1, h_2, h_3 \ \text{are all odd and} \quad h_1+h_2+h_3 = 4n+1 \\ 1+i & \text{if} \quad h_1, h_2, h_3 \ \text{are all odd and} \quad h_1+h_2+h_3 = 4n+3 \\ 2 & \text{if} \quad h_1, h_2, h_3 \ \text{are all even and} \quad h_1+h_2+h_3 = 4n \\ 0 & \text{if} \quad h_1, h_2, h_3 \ \text{are all even and} \quad h_1+h_2+h_3 = 4n+2 \end{cases} \tag{19b}$$

From Eqs. (19b), we see that the reflections with h_1, h_2, h_3 odd numbers have the same value of $|S(\mathbf{G})|^2$. On the contrary, the reflections with h_1, h_2, h_3 even numbers and with $h_1+h_2+h_3 = 4n+2$ (n integer) *are forbidden in the local spherical approximation*. The first few "forbidden" reflections (ordered in increasing values of $|\mathbf{G}|$) correspond to

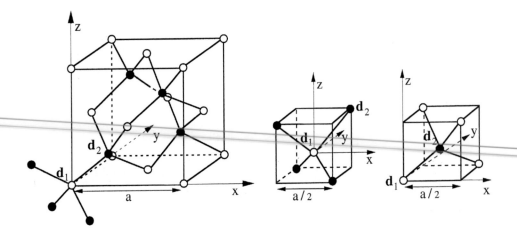

Fig. 6 Diamond structure with the two fcc composing sublattices represented with white and shaded circles, respectively. The inversion of bond orientations in the two fcc sublattices is illustrated in the two smaller cubes, which provide the four atoms surrounding $\mathbf{d}_1 = 0$ and $\mathbf{d}_2 = (a/4)(1,1,1)$.

reciprocal lattice vectors $(2\pi/a)(2,0,0)$, $(2\pi/a)(2,2,2)$, $(2\pi/a)(4,2,0)$, $(2\pi/a)(4,4,2)$ etc. (see Table 1).

In reality the charge cloud around each site has tetrahedral symmetry (and not spherical symmetry); because of the particular orientation of the tetrahedra with respect to each other, tetrahedral deformations invert at each successive atomic plane in the (1 1 1) direction, as schematically indicated in Fig. 6. This charge asymmetry transforms some of the "forbidden reflections" into "weakly allowed reflections" [notice that also anharmonic effects could have similar consequences, but we shall not concern ourselves with this mechanism].

A simple qualitative model that allows to mimic the tetrahedral symmetry of the covalent charge, and thus to distinguish "forbidden reflections" from "weakly allowed reflections" consists in simulating the bond charge in the unit cell as the sum of eight spherically symmetric contributions centred in the positions

$$
\begin{aligned}
\mathbf{d}_{1b} &= \tfrac{a}{4}(\gamma, \ \ \gamma, \ \ \gamma) & \mathbf{d}_{5b} &= \tfrac{a}{4}(1-\gamma, 1-\gamma, 1-\gamma) \\
\mathbf{d}_{2b} &= \tfrac{a}{4}(\gamma, -\gamma, -\gamma) & \mathbf{d}_{6b} &= \tfrac{a}{4}(1-\gamma, 1+\gamma, 1+\gamma) \\
\mathbf{d}_{3b} &= \tfrac{a}{4}(-\gamma, \gamma, -\gamma) & \mathbf{d}_{7b} &= \tfrac{a}{4}(1+\gamma, 1-\gamma, 1+\gamma) \\
\mathbf{d}_{4b} &= \tfrac{a}{4}(-\gamma, -\gamma, \gamma) & \mathbf{d}_{8b} &= \tfrac{a}{4}(1+\gamma, 1+\gamma, 1-\gamma) \, .
\end{aligned}
\tag{20a}
$$

The four centers $\mathbf{d}_{\nu b}$ $(\nu = 1, \ldots, 4)$ surround the site $\mathbf{R} = 0$ along the four nearest neighbour directions; similarly, the four centers $\mathbf{d}_{\nu b}$ $(\nu = 5, \ldots, 8)$ surround the site $\mathbf{R} = (a/4)(1,1,1)$ along the four directions of formation of the covalent bond; the fraction of electrons taking part in it, or the specific value of the dimensionless parameter

γ $(0 < \gamma < 1/2)$ are irrelevant, as far as we limit ourselves to work out the "selection rules".

We now calculate the geometrical structure factor $S_b(\mathbf{G}) = \sum \exp(-i\mathbf{G} \cdot \mathbf{d}_{\nu b})$, where the sum runs over the eight vectors given by Eq. (20a), and the reciprocal lattice vectors $\mathbf{G} = (2\pi/a)(h_1, h_2, h_3)$ satisfy $h_1 + h_2 + h_3 = 4n + 2$; we obtain

$$
\begin{aligned}
S_b(\mathbf{G}) = -2i \Big[&\sin \frac{\pi\gamma}{2}(h_1 + h_2 + h_3) + \sin \frac{\pi\gamma}{2}(h_1 - h_2 - h_3) \\
&+ \sin \frac{\pi\gamma}{2}(-h_1 + h_2 - h_3) + \sin \frac{\pi\gamma}{2}(-h_1 - h_2 + h_3) \Big] \\
= 8i \sin &\frac{\pi\gamma}{2} h_1 \, \sin \frac{\pi\gamma}{2} h_2 \, \sin \frac{\pi\gamma}{2} h_3 \, .
\end{aligned}
\tag{20b}
$$

From Eq. (20b) we see that $S_b(\mathbf{G}) \equiv 0$ if one (or more) of the integers h_1 or h_2 or h_3 is zero, while $S_b(\mathbf{G}) \neq 0$ if none of the integers h_1 or h_2 or h_3 is zero. We thus see, for instance, that the forbidden reflections $(2\pi/a)(2,2,2)$, $(2\pi/a)(4,4,2)$ etc. are (weakly) allowed in tetrahedral symmetry, while the reflections $(2\pi/a)(2,0,0)$, $(2\pi/a)(4,2,0)$ etc. remain forbidden also in the adopted tetrahedral symmetry model. The study of weakly allowed reflexions appears thus an invaluable tool for investigating wavefunctions, responsible of the chemical bond.

In Table 1, we give theoretical and experimental X-ray structure factors for silicon and germanium. We notice in particular that the structure factor for the $(2\pi/a)(2,2,2)$ reflection (forbidden in the local spherical approximation) is indeed different from zero (although very weak) in tetrahedral symmetry. Notice also that the lack of local spherical symmetry explains why the reflections $(2\pi/a)(3,3,3)$ and $(2\pi/a)(5,1,1)$ may have (slightly) different intensities, at parity of modulus $(2\pi/a)\sqrt{27}$.

The study of the structure factors of elemental semiconductors is just an example of the wealth of information provided by X-ray scattering experiments. In Table 1, we report for a useful comparison the X-ray structure factors of GaP, a binary III-V compound with zincblende structure; the two atoms in the unit cell have rather different atomic numbers ($Z = 15$ for P and $Z = 31$ for Ga) and atomic form factors; in this case, it can be seen that the crystal form factors for $(2\pi/a)(2,0,0)$, $(2\pi/a)(2,2,2)$, $(2\pi/a)(4,2,0)$ reflections are somewhat smaller, but of the same order of magnitude than the other ones.

A similarly fruitful analysis could be carried out for other classes of composite crystals. In particular for lithium hydride (with two core electrons and two valence electrons per unit cell), the ground-state density matrix and theoretical X-ray structure factors can be examined with relatively elementary techniques; comparison with experimental data can be used to test the quantum mechanical models for the valence wavefunctions associated to the s-like orbitals of the hydride ions [see for instance G. Grosso and G. Pastori Parravicini, Phys. Rev. **B17**, 3421 (1978) and references quoted therein].

Table 1 Theoretical and experimental X-ray structure factors (absolute values) for Si, Ge and GaP (in units of electrons per unit cell). For elemental semiconductors, reflections forbidden in the local spherical approximation are denoted by (*); reflections forbidden in the local tetrahedral symmetry are denoted by (**). The theoretical results are taken from C. S. Wang and B. M. Klein, Phys. Rev. B **24**, 3393 (1981) (copyright 1981 by the American Physical Society) and we refer to this paper for further details and comments on experiments.

Reciprocal lattice vectors (in units $2\pi/a$)	Number of vectors in the shell	Si Theory	Si Exp.	Ge Theory	Ge Exp.	GaP Theory	GaP Exp.
(0,0,0)	1	28.00	28.00	64.00	64.00	46.00	46.00
(1,1,1)	8	15.11	15.19	38.83	39.42	28.84	28.83
(2,0,0)(*)(**)	6	-	-	-	-	14.63	14.40
(2,2,0)	12	17.26	17.30	47.23	47.44	31.77	32.19
(3,1,1)	24	11.37	11.35	31.29	31.37	22.89	22.92
(2,2,2) (*)	8	0.25	0.38	0.22	0.26	12.45	12.79
(4,0,0)	6	14.92	14.89	40.56	40.50	26.95	26.19
(3,3,1)	24	10.17	10.25	27.39	27.72	19.69	19.43
(4,2,0)(*)(**)	24	-	-	-	-	10.44	10.48
(4,2,2)	24	13.37	13.42	35.91	36.10	23.79	23.86
(3,3,3)	8	9.07	9.08	24.35	24.50	17.34	17.24
(5,1,1)	24	9.08	9.11	24.35	-	17.35	17.24
(4,4,0)	12	12.04	12.08	32.23	32.34	21.32	21.01

3 Inelastic scattering of particles and phonon spectra of crystals

Up to now we have considered elastic scattering of photons by the electron charge density of crystals, assuming the atoms fixed at the equilibrium configuration. In reality, even at zero absolute temperature, the nuclei move around their equilibrium positions. The vibrational motion modifies the intensity of the elastically scattered waves and also makes it possible inelastic processes with phonon emission or absorption.

In this section we focus on conservation laws of energy and crystal momentum for one-phonon processes, which are at the basis of the experimental determination of phonon spectra of solids. We do not need to be very specific about the coupling mechanism between the probe beam and the crystal, except for the assumption that the coupling is invariant under uniform translations of the whole probe–crystal system, so that momentum is preserved to within reciprocal lattice vectors times $\hbar$; in general, processes involving $\mathbf{G} = 0$ and $\mathbf{G} \neq 0$ are referred as *normal* and *umklapp* processes, respectively. We also do not need a detailed analysis of the relative role of elastic and inelastic processes; this is regulated by the Debye–Waller factor, which is discussed in Section 5 and following.

Inelastic scattering of X-rays by phonons

In addition to elastic scattering, X-rays can be scattered inelastically with the absorption or emission of one (or more) phonons. Away from directions satisfying the Bragg condition, one can measure a diffuse background of radiation with a range of photon energies, which reflect multiphonon processes. Since phonon energies (≈ 10 meV) are much smaller than X-ray photon energies (≈ 10 keV), in most experiments scattered beams are not analysed in energy; information on the phonon dispersion curves can be disentangled from experimental data only indirectly and with limited accuracy.

With the development of synchrotron radiation sources with high photon flux, inelastic X-ray scattering experiments with meV energy resolution have become possible [see for instance F. Sette, G. Ruocco, M. Krisch, U. Bergmann, C. Masciovecchio, V. Mazzacurati, G. Signorelli and R. Verbeni, Phys. Rev. Lett. **75**, 850 (1995)]. High resolution is achieved by varying with millikelvin precision the temperature of the analyzer (or polarizer) crystal, so controlling with high accuracy the distance between the family of reflecting planes and thus the Bragg selected wavelength. Also notice that the momentum transfer is determined only by the scattering angle θ, via the relation $q = 2k_i \sin(\theta/2)$, where $\mathbf{k}_i$ is the wavevector of the incident photon; this is so because the fractional change of energy of the X-ray photon is very small ($\approx 10^{-6}$), and $|\mathbf{k}_i|$ and $|\mathbf{k}_f|$ are practically equal.

As an example, we report in Fig. 7 inelastic X-ray scattering measurements of a monochromatic X-ray beam of energy ≈ 18 keV, focused on ice samples, taken at the indicated wavevector transfer values. The spectrum of the scattered light clearly shows peaks, at frequencies below and above the frequency of the incident radiation, corresponding to one-phonon emission processes (*Stokes processes*) and one-phonon absorption processes (*anti-Stokes processes*). The observed asymmetry in the spectra is related to the fact that phonon annihilation is proportional to the Bose population factor $\langle n \rangle$, while the creation of phonons is proportional to $\langle n \rangle + 1$. The non-dispersive peak at ≈ 7 meV is the transverse optical (TO) phonon mode. The peak, whose energy increases with the wavevector, is the longitudinal acoustic (LA) phonon mode; for $q = 12$ nm$^{-1} = 1.2$ Å^{-1}, the observed energy of the LA mode is $\hbar\omega \approx 0.025$ eV, and the sound velocity in ice can be estimated as $v_s = \hbar\omega/\hbar q \approx 3200$ m/sec ($\hbar = 6.58 \cdot 10^{-16}$ eV·sec). Inelastic X-ray scattering measurements have become a precious tool for obtaining information on the phonon dynamics in solids, and also on the collective excitations in liquids.

Brillouin and Raman scattering

The inelastic scattering of optical beams is a very sensitive tool for the experimental determination of phonon energies in the small wavevector limit. Consider a monochromatic optical beam (usually a laser beam in the visible region) that propagates in a crystal and undergoes inelastic scattering with emission of a phonon (Stokes process) or absorption of a phonon (anti-Stokes process). In the photon frequency region of interest, the crystal is assumed to be transparent and characterized by refractive index

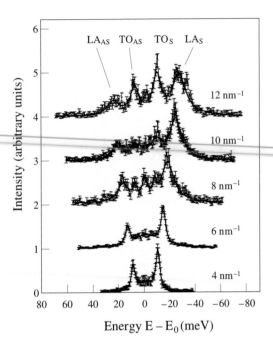

Fig. 7 Inelastic X-ray scattering spectra of polycrystalline H_2O ice at $-20\,^\circ$C. The reference energy E_0 is that of the incident beam; the intensity of the scattered beam versus E is reported at the indicated values of the wavevector transfer. The labels TO, LA, S and AS stand for transverse optical, longitudinal acoustic, Stokes and anti-Stokes, respectively [from G. Ruocco, F. Sette, U. Bergmann, M. Krisch, C. Masciovecchio, V. Mazzacurati, G. Signorelli and R. Verbeni, Nature **379**, 521 (1996); reprinted with permission from Nature, copyright 1996 Macmillan Magazines Limited].

n, so that the dispersion relation for photons $\omega = ck/n$ holds for both incident and scattered radiation.

Consider a scattering process (see Fig. 8), in which a photon propagating in the medium with wavevector $\mathbf{k}_i$ and frequency $\omega_i = ck_i/n$ is scattered into the state $\mathbf{k}_f$ and $\omega_f = ck_f/n$, with emission of a phonon of wavevector $\mathbf{q}$ and frequency $\Omega(\mathbf{q}, p)$, where p labels a phonon branch (scattering processes with absorption of a phonon could be treated similarly). The laws of conservation of momentum (divided by $\hbar$) and energy are

$$\mathbf{k}_i = \mathbf{k}_f + \mathbf{q} \tag{21a}$$

and

$$\hbar\omega_i = \hbar\omega_f + \hbar\Omega(\mathbf{q}, p) \quad . \tag{21b}$$

In the visible region, both $\mathbf{k}_i$ and $\mathbf{k}_f$ are much smaller than the Brillouin zone dimensions; thus the vector $\mathbf{q}$ defined by Eq. (21a) is already within the Brillouin zone (very near to the center) without need of adding a reciprocal lattice vector $\mathbf{G}$ in the second member of Eq. (21a). The conservation relations (21) are very stringent indeed: from

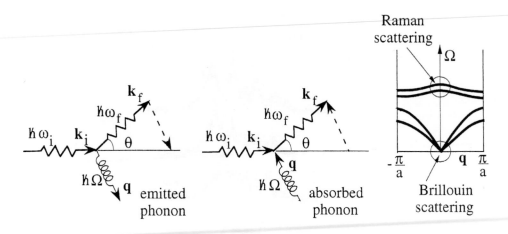

Fig. 8 Scattering of a photon through an angle θ with creation of a phonon (Stokes process) or annihilation of a phonon (anti-Stokes process); $\mathbf{k}_i$ and $\mathbf{k}_f$ denote the propagation wavevectors of the photon in the medium, before and after the scattering process. For photons in the visible region, a very small part of the phonon dispersion curves can be explored.

the determination of $\Delta\mathbf{k} = \mathbf{k}_f - \mathbf{k}_i$ and the measure of $\hbar\Delta\omega = \hbar\omega_f - \hbar\omega_i$ one can obtain the wavevector $\mathbf{q}$ and the energy $\hbar\Omega(\mathbf{q}, p)$ of the phonon taking part to the scattering process.

The conservation relations (21) can be elaborated taking into account that the phonon energies (≈ 0.01 eV) are much smaller than photon energies (≈ 1 eV) used in typical experimental situations; when the fractional change of the photon energy is very small, the photon scattering is termed "quasi-elastic", and it holds $|\mathbf{k}_i| \approx |\mathbf{k}_f|$. In the quasi-elastic scattering process, the wavevector transfer q is controlled only by the scattering angle θ and is given with good approximation by

$$\boxed{q = 2k_i \sin \frac{\theta}{2}} \,, \tag{22a}$$

as shown in Fig. 8. From the relation $\omega_i = ck_i/n$ it follows

$$q \approx 2\omega_i \frac{n}{c} \sin \frac{\theta}{2} \,. \tag{22b}$$

The light scattering with the emission or absorption of one (or more) acoustic phonons is called *Brillouin scattering*. For acoustic phonons of small wavevector $\Omega(q) = v_s q$ (where v_s is the sound velocity), and Eq. (22b) gives

$$\boxed{\Omega = 2\omega_i \, n \frac{v_s}{c} \sin \frac{\theta}{2}} \,. \tag{22c}$$

If the velocity v_s depends on the different acoustic branches, the Brillouin spectrum exhibits a characteristic structure, reflecting the different acoustic phonons. The measurements of the frequency shifts $\pm\Omega$ of the Brillouin components of the scattered

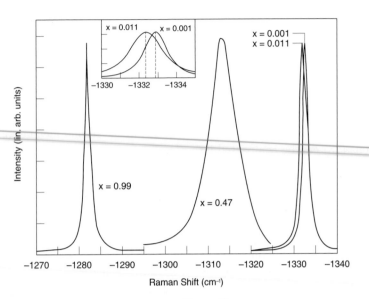

Fig. 9 Raman spectra of diamond specimens $^{12}C_{1-x}{}^{13}C_x$ for $x = 0.001$, $x = 0.011$ (natural composition), $x = 0.47$ and $x = 0.99$; the change of the optical mode frequency ω_0 with the isotopic composition x follows approximately the $M^{-1/2}$ dependence, where M is the average atomic mass [from R. Vogelgesang, A. K. Ramdas, S. Rodriguez, M. Grimsditch and T. R. Anthony, Phys. Rev. B**54**, 3989 (1996); copyright 1996 by the American Physical Society].

light yield accurate information on the acoustic phonons with wavevector q close to the zone center.

The light scattering with the emission or the absorption of optical phonons is called *Raman scattering*. In this case, the dispersion relation of optical phonons can be neglected and the Raman spectra reveal peaks corresponding to optical phonons with wavevector q essentially at the center of the Brillouin zone. In Fig. 9 we report as an example the Raman spectra of diamond for several isotopic compositions; diamond is a non-polar crystal with the zone center optical modes triply degenerate and infrared inactive; the large value of the optical mode frequency ω_0 is produced by the strong covalent bond and the light mass of ^{12}C or ^{13}C constituent atoms. In polar crystals, the Raman scattering of laser light has become a precious tool for investigating polariton dispersion curves (see for instance Fig. IX-11).

Inelastic scattering of neutrons by phonons

The inelastic scattering of neutrons by phonons is a powerful tool for investigating the phonon dispersion curves over the whole Brillouin zone. Consider a neutron of wavevector $\mathbf{k}_i$ and energy E_i, which is scattered into the state $\mathbf{k}_f$ and energy E_f, with emission of a phonon of wavevector $\mathbf{q}$ and frequency $\Omega(\mathbf{q}, p)$, where p labels a phonon branch (scattering processes with absorption of a phonon could be handled similarly).

The conservation laws for momentum (divided by $\hbar$) and energy give

$$\mathbf{k}_i = \mathbf{k}_f + \mathbf{q} + \mathbf{G} \tag{23a}$$

and

$$\frac{\hbar^2 k_i^2}{2M_n} = \frac{\hbar^2 k_f^2}{2M_n} + \hbar\Omega(\mathbf{q}, p) \; ; \tag{23b}$$

a reciprocal lattice vector $\mathbf{G}$ is in general required in the second member of Eq. (23a).

Equations (23) can be compacted into a unique equation

$$\frac{\hbar^2 k_i^2}{2M_n} = \frac{\hbar^2 k_f^2}{2M_n} + \hbar\Omega(\mathbf{k}_i - \mathbf{k}_f, p) \; . \tag{24}$$

The highly restrictive aspect of the relation (24) can be understood imagining to fix $\mathbf{k}_i$ and also the direction of observation $\widehat{\mathbf{k}}_f = \mathbf{k}_f/k_f$; thus only k_f remains as a disposable parameter and presumably only one (or a few) discretized values of k_f can satisfy the relation (24). In particular, in the case $\hbar^2 k_i^2/2M_n > \hbar\Omega(\mathbf{k}_i, p)$, it is easily seen by inspection that there is (at least) a solution of Eq. (24) in any pth branch and for any given direction of observation $\widehat{\mathbf{k}}_f$.

Neutron scattering measurements are typically performed by means of a "triple-axis spectrometer" (or some appropriate implementation). A beam of collimated and polyenergetic neutrons strikes a (orientable) single-crystal; by means of Bragg reflection, neutrons of a given wavelength are selected. The resulting beam of monoenergetic neutrons strikes a single-crystal sample (capable of different orientations). The scattered neutrons are then analysed by means of a second (orientable) Bragg scatterer. By use of conservation laws, one can determine the phonon wavevector $\mathbf{q}$, besides the measured energy change $\hbar\Omega(\mathbf{q}, p)$; thus the dispersion relations $\hbar\Omega(\mathbf{q}, p)$ can be worked out over the Brillouin zone.

4   Compton scattering and electron momentum density

The Compton effect is the inelastic scattering of a photon (usually X-ray or γ-ray) by an electron; when the target electron is moving, the Compton-scattered radiation is also Doppler-broadened, and its energy distribution at a given scattering angle is called Compton profile. Measurements of Compton profiles of materials give information on the electron momentum density, projected along the scattering direction. The technique is particularly sensitive to the most external electronic wavefunctions, which describe slowly moving electrons and are responsible of the chemical bond. A closely related method which provides similar information is the study of the angular correlation in positron annihilation.

When a photon is scattered by an electron, its wavelength shift can be obtained from energy and momentum conservation laws. The scattering process is schematically represented in Fig. 10; $\mathbf{k}_i$ and $\mathbf{k}_f$ denote the wavevector of the incident and scattered photon, respectively, and $\hbar\omega_i$ and $\hbar\omega_f$ their energy; θ is the scattering angle and $\mathbf{q}_f$ the

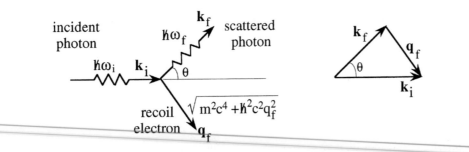

Fig. 10 Schematic representation of the Compton scattering, through an angle θ, of a photon impinging on an electron initially at rest.

final wavevector of the electron (the initial wavevector of the electron is momentarily supposed to be $\mathbf{q}_i = 0$). The conservation laws of momentum and energy require

$$\mathbf{k}_i = \mathbf{k}_f + \mathbf{q}_f \tag{25a}$$

and

$$\hbar\omega_i + mc^2 = \hbar\omega_f + \sqrt{m^2c^4 + \hbar^2c^2q_f^2} \,, \tag{25b}$$

where the relativistic expression of the electron energy is used.

We now eliminate from Eqs. (25) the final electron wavevector $\mathbf{q}_f$; for this purpose we compare the value of q_f^2 obtained from Eq. (25a) and the one obtained from Eq. (25b). From Eq. (25a) we have

$$q_f^2 = (\mathbf{k}_i - \mathbf{k}_f)^2 = k_i^2 - 2k_ik_f\cos\theta + k_f^2 = (k_i - k_f)^2 + 4k_ik_f\sin^2\frac{\theta}{2} \,. \tag{26a}$$

From Eq. (25b) we have

$$\hbar c(k_i - k_f) + mc^2 = \sqrt{m^2c^4 + \hbar^2c^2q_f^2} \;;$$

squaring both members of the above relation, one obtains

$$q_f^2 = (k_i - k_f)^2 + \frac{2mc}{\hbar}(k_i - k_f) \,. \tag{26b}$$

Direct comparison of Eqs. (26) gives

$$k_i - k_f = \frac{2\hbar}{mc}k_ik_f\sin^2\frac{\theta}{2} \,.$$

With the replacement $k_i = 2\pi/\lambda_i$ and $k_f = 2\pi/\lambda_f$, we obtain for the increase of the wavelength of the scattered radiation the well known Compton relation

$$\boxed{\Delta\lambda = \lambda_f - \lambda_i = \frac{2h}{mc}\sin^2\frac{\theta}{2}} \,, \tag{27}$$

where $2h/mc = 0.0485$ Å.

Using the relations $\lambda_i = 2\pi c/\omega_i$ and $\lambda_f = 2\pi c/\omega_f$, Eq. (27) can be recast into the equivalent form

$$\omega_f = \omega_i \frac{1}{1 + \frac{\hbar \omega_i}{mc^2} 2\sin^2 \frac{\theta}{2}} . \tag{28a}$$

At low energies such that $\hbar\omega_i \ll mc^2$ ($=0.511$ MeV), the above expression becomes

$$\frac{\omega_i - \omega_f}{\omega_i} = \frac{\hbar \omega_i}{mc^2} 2\sin^2 \frac{\theta}{2} . \tag{28b}$$

Experiments are usually performed at high scattering angles (150 – 170°), and the photon energy shifts are quite measurable. For instance for the Kα radiation of molybdenum ($\hbar\omega_i \approx 17.6$ keV), Eq. (28b) gives for the energy shift of the backscattered photon the value ≈ 1.2 keV; this is also the energy transferred to the recoil electron. The numerical example considered shows that as long as the energy of X-ray is small with respect to mc^2, we have $k_i \approx k_f$ and the scattering of photons can be considered "quasi-elastic". Moreover, if the energy of the photon is in the hard X-ray region (10 – 50 keV) the electron recoil energy is of the order of a few keV; this energy greatly exceeds the binding energy of outer electrons and justifies the impulse approximation in the quantitative treatment of the collision between photon and electron.

We consider now the Compton scattering between an incoming photon and a moving electron. In this case, the scattered photon beam is also Doppler-broadened, and the wavelength shift can be obtained from the standard conservation laws of energy and momentum. In view of the relatively small initial and final kinetic energy of electrons (and also for sake of simplicity) we use the non-relativistic expression of the electron kinetic energy and we have

$$\mathbf{k}_i + \mathbf{q}_i = \mathbf{k}_f + \mathbf{q}_f \tag{29a}$$

and

$$\hbar\omega_i + \frac{\hbar^2 q_i^2}{2m} = \hbar\omega_f + \frac{\hbar^2 q_f^2}{2m} . \tag{29b}$$

As before, we now proceed to eliminate the final wavevector $\mathbf{q}_f$ from the above equations. From Eq. (29a) we have:

$$q_f^2 = [(\mathbf{k}_i - \mathbf{k}_f) + \mathbf{q}_i]^2 = (\mathbf{k}_i - \mathbf{k}_f)^2 + q_i^2 - 2\mathbf{q}_i \cdot (\mathbf{k}_f - \mathbf{k}_i)$$

$$= (k_i - k_f)^2 + 4k_i k_f \sin^2 \frac{\theta}{2} + q_i^2 - 2\mathbf{q}_i \cdot (\mathbf{k}_f - \mathbf{k}_i) . \tag{30a}$$

This relation is exact; in the "quasi-elastic scattering approximation" (when it holds) we neglect the difference $k_i - k_f$ with respect to the modulus k_i or k_f; for what concerns the scattering vector $\mathbf{S} = \mathbf{k}_f - \mathbf{k}_i$ we can approximate its modulus in the form $S \approx 2k_f \sin(\theta/2)$. Indicating with z the direction of the scattering vector, Eq. (30a) simplifies in the form

$$q_f^2 = 4k_i k_f \sin^2 \frac{\theta}{2} + q_i^2 - 4q_{iz} k_f \sin \frac{\theta}{2} . \tag{30b}$$

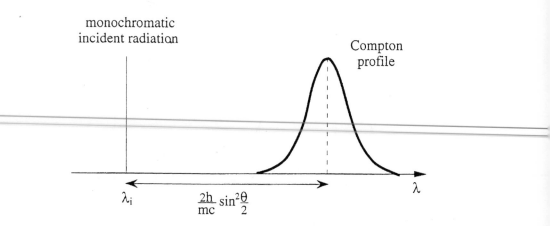

Fig. 11 Schematic representation of the profile of the Compton-scattered radiation. The incident radiation is supposed to be monochromatic, with wavelength λ_i; the scattered radiation is shifted by $(2h/mc)\sin^2(\theta/2)$ and also Doppler-broadened.

Comparing Eq. (29b) and Eq. (30b) we obtain for the change of the photon wavelength

$$\lambda_f - \lambda_i = \frac{2h}{mc}\sin^2\frac{\theta}{2} - \frac{2\lambda_i}{mc}p_z\sin\frac{\theta}{2}, \tag{31}$$

where $p_z = \hbar q_{iz}$ is the component of the momentum of the electron *before* the scattering along the X-ray scattering vector $\mathbf{k}_f - \mathbf{k}_i$; the term linear in p_z in Eq. (31) expresses the Doppler broadening of the scattered radiation (see Fig. 11).

Using the relations $\lambda_i = 2\pi c/\omega_i$ and $\lambda_f = 2\pi c/\omega_f$, we can recast Eq. (31) into the equivalent form

$$\omega_f = \omega_i \frac{1}{1 + \frac{\hbar\omega_i}{mc^2}2\sin^2\frac{\theta}{2} - \frac{p_z}{mc}2\sin\frac{\theta}{2}} \, .$$

At low energies $\hbar\omega_i \ll mc^2$, and low electronic velocities $v_z = p_z/m \ll c$, the above expression becomes

$$\frac{\omega_i - \omega_f}{\omega_i} = \frac{\hbar\omega_i}{mc^2}2\sin^2\frac{\theta}{2} - \frac{p_z}{mc}2\sin\frac{\theta}{2} \, . \tag{32}$$

As an estimation of the Doppler shift, consider for instance electrons at the Fermi surface in simple metals; we have $p_F/mc = v_F/c \approx 1/100$; thus also the Doppler shift $\Delta\omega/\omega_i$ is of the order of $1/100$ and can be easily detected.

The above semi-classical analysis shows that *the Compton profile is proportional to the projection of the electron momentum along the scattering direction* (say the z direction). Thus if $n(\mathbf{p})$ denotes the electron momentum density of the ground-state

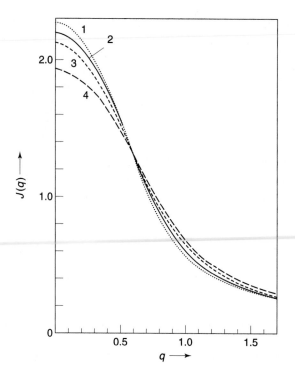

Fig. 12 Compton profile of lithium hydride (q is in units a_B^{-1}, inverse Bohr radius). Curves 1 and 2 correspond to the free ion model with overlap and to the cluster model with overlap of Li$^+$ and H$^-$ wavefunctions; the calculations are taken from G. Grosso, G. Pastori Parravicini and R. Resta, Phys. Stat. Sol. (b) **73**, 371 (1976). Curves 3 and 4 give the experimental results of W. C. Phillips and R. W. Weiss, Phys. Rev. **182**, 923 (1969) and those of J. Felsteiner, R. Fox and S. Kahane, Phys. Rev. B**6**, 4689 (1972), respectively.

of the sample, the theoretical Compton profile $J(p_z)$ is calculated as

$$J(p_z) = \int\int n(\mathbf{p})\, dp_x\, dp_y \; . \tag{33}$$

A more rigorous theory of Compton lineshapes, done in terms of second-order time-dependent perturbation theory, gives formal support and establishes validity limits of the semiclassical analysis, summarized by expression (33).

As an example, we report in Fig. 12 experimental and computed isotropic Compton profile of LiH crystals; the Compton profile at small wavevectors, and in particular $J(0)$, is dominated by the behaviour of the hydride ion wavefunctions; a proper account of the overlap of orbitals on different centers is necessary to correctly describe the ground-state electron momentum density of lithium hydride crystals.

The complete relativistic treatment of the Compton amplitude does depend also on the spin of the electron (in the semiclassical treatment, justified in the extreme non-relativistic limit, X-rays couple exclusively with the charge of the electrons by means of the electric field of the electromagnetic wave). In particular, in the case of circularly polarized photons, aligned spins induce an asymmetry in the Compton scattered radiation. In principle, from the change of the spectral distribution of scattered circularly polarized light when the polarization (or the magnetic field) is reversed, it is possible to infer the "magnetic Compton profiles" given by

$$I_{\text{magnetic}}(p_z) = \int\int [n_\uparrow(\mathbf{p}) - n_\downarrow(\mathbf{p})]\, dp_x\, dp_y\ ,$$

where $n_\uparrow(\mathbf{p})$ and $n_\downarrow(\mathbf{p})$ are the electron momentum densities for spin-up and spin-down electrons, respectively. With the development of synchrotron radiation facilities, intense beams of circularly polarized X-rays have become available for the investigation of the Compton profiles of unpaired electrons in magnetic materials (magnetic Compton scattering). Notice that also the elastic scattering of X-rays may provide information on the electronic magnetic structure of a crystal, since the scattering amplitude in a relativistic treatment depends on the spins of the electrons (magnetic X-ray scattering).

5 Diffusion of particles by a single elastically-bound scatterer

5.1 Dynamical structure factor of a single scattering center

In the previous sections, we have considered the elastic scattering of particles from a *static crystal*, with the nuclei fixed in their equilibrium positions. We have then taken into account the lattice vibrations, and we have discussed inelastic processes with one-phonon absorption or emission, essentially from the point of view of energy and momentum conservation laws. We wish now to study the general effects of lattice vibrations on the (elastic or inelastic) diffusion of particles by crystals. For this aim, we start considering the *idealized model of scattering from a single atom, or a single nucleus, of mass M*. We shall consider first the *static situation* (corresponding to $M \to \infty$), and then the *dynamical situation*, with M finite and the scatterer performing harmonic oscillations with frequency ω_0. The results of this section are presented so to make it possible a straightforward generalization to actual crystals, where all the N nuclei of the lattice are undergoing normal modes oscillations; the present results thus pave the way to the discussion of the Debye–Waller factor in crystals (Section 6) and the γ-emission or absorption from nuclei in solids (Section 7).

Consider a beam of particles impinging on a "scatterer" (or "target") and suppose that the probe–target interaction is invariant under uniform translations of the probe–target system and has thus the form

$$\boxed{V_{\text{probe–target}}(\mathbf{r}; \mathbf{R}_0) = V_a(\mathbf{r} - \mathbf{R}_0)}\ , \tag{34}$$

where $\mathbf{r}$ is the coordinate of the impinging particle (a neutron or an electron, for instance), $\mathbf{R}_0$ is the coordinate of the scatterer, and $V_a(\mathbf{r} - \mathbf{R}_0)$ represents a localized potential of the scatterer in its actual position $\mathbf{R}_0$.

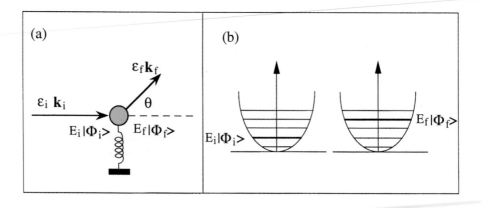

Fig. 13 (a) Diffusion of particles by a scatterer of finite mass M (an atom or a nucleus connected by a spring to a fixed center). (b) Schematic representation of initial and final energy levels and states of the harmonic oscillator scatterer, of given frequency ω_0.

We do not need to be specific on the details of the probe–target interaction, *provided the form* (34) *is preserved*. For instance, in the case the impinging particle is an electron, $V_a(\mathbf{r} - \mathbf{R}_0)$ can be taken as the atomic potential of the atom at $\mathbf{R}_0$ (assuming that the atomic electronic cloud follows adiabatically the nuclear position $\mathbf{R}_0$, without readjustements). Similarly, in the scattering of X-ray photons, the probe–target interaction is proportional to the atomic electron density $n_a(\mathbf{r} - \mathbf{R}_0)$. In the case of thermal neutrons impinging on a nucleus located at $\mathbf{R}_0$ the probe–target interaction is the contact Fermi interaction $(2\pi\hbar^2/M_n)\, b\, \delta(\mathbf{r} - \mathbf{R}_0)$, where b is the scattering length of the nucleus.

Let the impinging particle be described by a monochromatic plane wave $|W_{\mathbf{k}_i}\rangle$ of vector $\mathbf{k}_i$ and energy ε_i; we are interested in the transition probability rate of the impinging particle into a final state $|W_{\mathbf{k}_f}\rangle$ of vector $\mathbf{k}_f$ and energy ε_f. In the scattering process, the energy change of the probe is

$$\hbar\Delta\omega = \varepsilon_f - \varepsilon_i \,, \tag{35a}$$

and the momentum change of the probe is

$$\hbar\Delta\mathbf{k} = \hbar\mathbf{k}_f - \hbar\mathbf{k}_i \,. \tag{35b}$$

The energies and wavefunctions of the target (a harmonic oscillator) before and after collision are indicate by E_i, $|\Phi_i\rangle$, E_f and $|\mathfrak{I}_f\rangle$, respectively. The essential kinematic aspects of the scattering process are indicated in Fig. 13.

We consider first the *static situation*, with the variable $\mathbf{R}_0$ considered simply as a fixed classical parameter, instead of a quantum observable. In the static approximation, the scattering is elastic; at the lowest order of perturbation theory (Born approximation), the probability per unit time that the perturbation potential (34) induces a transition from the initial state $|W_{\mathbf{k}_i}\rangle$ to a final state $|W_{\mathbf{k}_f}\rangle$ is given by the

Fermi golden rule

$$P_{\mathbf{k}_f \leftarrow \mathbf{k}_i}^{\text{(fixed scatt.)}} = \frac{2\pi}{\hbar} \left| \langle W_{\mathbf{k}_f} | V_a(\mathbf{r} - \mathbf{R}_0) | W_{\mathbf{k}_i} \rangle \right|^2 \delta(\varepsilon_f - \varepsilon_i)$$

$$= \frac{2\pi}{\hbar^2} \left| V_a(\Delta\mathbf{k}) \, e^{-i\,\Delta\mathbf{k}\cdot\mathbf{R}_0} \right|^2 \delta(\Delta\omega) = \frac{2\pi}{\hbar^2} |V_a(\Delta\mathbf{k})|^2 \delta(\Delta\omega) , \qquad (36)$$

where $V_a(\Delta\mathbf{k})$ is the Fourier transform of the interaction potential

$$V_a(\Delta\mathbf{k}) = \frac{1}{V} \int e^{-i(\mathbf{k}_f - \mathbf{k}_i)\cdot\mathbf{r}} V_a(\mathbf{r}) \, d\mathbf{r} \qquad (37)$$

(V is the volume over which the initial and final plane waves are normalized). Eq. (36) shows that the transition rate is determined by the modulus squared of the Fourier transform of the interaction potential.

We analyse now the *dynamic situation*, with the variable $\mathbf{R}_0$ correctly considered as a quantum observable and the scatterer as a quantum system. The probability per unit time of a transition of the probe–target system from an initial state $|W_{\mathbf{k}_i}\Phi_i\rangle$ to a final state $|W_{\mathbf{k}_f}\Phi_f\rangle$ is given by the Fermi golden rule

$$P_{\mathbf{k}_f \Phi_f \leftarrow \mathbf{k}_i \Phi_i} = \frac{2\pi}{\hbar} \left| \langle W_{\mathbf{k}_f}\Phi_f | V_a(\mathbf{r} - \mathbf{R}_0) | W_{\mathbf{k}_i}\Phi_i \rangle \right|^2 \delta(\varepsilon_f + E_f - \varepsilon_i - E_i)$$

$$= \frac{2\pi}{\hbar} \left| \langle \Phi_f | V_a(\Delta\mathbf{k}) e^{-i\Delta\mathbf{k}\cdot\mathbf{R}_0} | \Phi_i \rangle \right|^2 \delta(E_f - E_i + \hbar\Delta\omega) . \qquad (38a)$$

We perform the sum over all the states $|\Phi_f\rangle$ of the scatterer, and we obtain the probability rate for a scattering to a final state of the probe (regardless of the final state of the target)

$$P_{\mathbf{k}_f \leftarrow \mathbf{k}_i \Phi_i} = \frac{2\pi}{\hbar} |V_a(\Delta\mathbf{k})|^2 \sum_{\Phi_f} \langle \Phi_i | e^{i\Delta\mathbf{k}\cdot\mathbf{R}_0} | \Phi_f \rangle \langle \Phi_f | e^{-i\Delta\mathbf{k}\cdot\mathbf{R}_0} | \Phi_i \rangle \delta(E_f - E_i + \hbar\Delta\omega) ,$$

$$(38b)$$

where the modulus squared of the matrix elements has been written in extended form for convenience.

With the help of the integral representation for the delta function

$$\delta(E_f - E_i + \hbar\Delta\omega) = \frac{1}{\hbar} \int_{-\infty}^{+\infty} \frac{dt}{2\pi} \exp\left[\frac{i(E_f - E_i + \hbar\Delta\omega)t}{\hbar} \right] ,$$

we can write Eq. (38b) in the form

$$P_{\mathbf{k}_f \leftarrow \mathbf{k}_i \Phi_i} = \frac{2\pi}{\hbar^2} |V_a(\Delta\mathbf{k})|^2 \int_{-\infty}^{+\infty} \frac{dt}{2\pi} e^{i\Delta\omega t}$$

$$\cdot \sum_{\Phi_f} \langle \Phi_i | e^{i\Delta\mathbf{k}\cdot\mathbf{R}_0} | \Phi_f \rangle \langle \Phi_f | e^{iE_f t/\hbar} e^{-i\Delta\mathbf{k}\cdot\mathbf{R}_0} e^{-iE_i t/\hbar} | \Phi_i \rangle .$$

It is convenient to consider in the Heisenberg representation the following operator

$$e^{-i\Delta\mathbf{k}\cdot\mathbf{R}_0(t)} = e^{iHt/\hbar} e^{-i\Delta\mathbf{k}\cdot\mathbf{R}_0} e^{-iHt/\hbar}$$

where H is the Hamiltonian of the target. An operator whose time dependence is not explicitly indicated, is intended at $t = 0$, the instant at which the Heisenberg and the Schrödinger representations coincide. From the matrix elements of the operator $\exp\left[-i\Delta\mathbf{k}\cdot\mathbf{R}_0(t)]\right]$ on the eigenfunctions of H, and from the closure relation $\sum_f |\Phi_f\rangle\langle\Phi_f| = 1$, it follows

$$P_{\mathbf{k}_f \leftarrow \mathbf{k}_i, \Phi_i} = \frac{2\pi}{\hbar^2}\,|V_a(\Delta\mathbf{k})|^2 \int_{-\infty}^{+\infty} \frac{dt}{2\pi}\, e^{i\Delta\omega\, t}\langle\Phi_i|e^{i\Delta\mathbf{k}\cdot\mathbf{R}_0}\, e^{-i\Delta\mathbf{k}\cdot\mathbf{R}_0(t)}|\Phi_i\rangle \,. \qquad (38c)$$

Generally the target is not in a pure state $|\Phi_i\rangle$ but rather it is spread over all its accessible states in thermal equilibrium. Averaging Eq. (38c) over a distribution of equilibrium states gives

$$P_{\mathbf{k}_f \leftarrow \mathbf{k}_i} = \frac{2\pi}{\hbar^2}|V_a(\Delta\mathbf{k})|^2 \int_{-\infty}^{+\infty} \frac{dt}{2\pi} e^{i\Delta\omega\, t}\langle e^{i\Delta\mathbf{k}\cdot\mathbf{R}_0(0)} e^{-i\Delta\mathbf{k}\cdot\mathbf{R}_0(t)}\rangle \,; \qquad (39)$$

the quantity $\langle A^\dagger(0)\,A(t)\rangle$ denotes the thermal average of $A^\dagger(0)\,A(t)$, i.e. the *autocorrelation function of the operator* A, defined as

$$\langle A^\dagger(0)\,A(t)\rangle = \frac{\sum_i e^{-E_i/k_B T}\langle\Phi_i|A^\dagger(0)\,A(t)|\Phi_i\rangle}{\sum_i e^{-E_i/k_B T}} \,.$$

In the case the scatterer is "fixed" and $\mathbf{R}_0(t) \approx \mathbf{R}_0(0)$, we see that Eq. (39) coincides with the expression (36) for the static scatterer.

Scattering cross-section and dynamical structure factor

Expression (39) gives the probability per unit time that an incident particle is scattered from $\mathbf{k}_i$ to $\mathbf{k}_f$ because of the interaction with the target at the canonical thermal equilibrium. The probability per unit time that an incident particle is scattered from $\mathbf{k}_i$ to a volume element $d\mathbf{k}_f$ around $\mathbf{k}_f$, is

$$P_{\mathbf{k}_f \leftarrow \mathbf{k}_i}\frac{V}{(2\pi)^3}\,d\mathbf{k}_f = P_{\mathbf{k}_f \leftarrow \mathbf{k}_i}\frac{V}{(2\pi)^3}\,k_f^2\,dk_f\,d\Omega = P_{\mathbf{k}_f \leftarrow \mathbf{k}_i}\frac{V}{(2\pi)^3}\,k_f\frac{M}{\hbar^2}\,d\varepsilon\,d\Omega \,,$$

where $V/(2\pi)^3$ represents as usual the density-of-states in $\mathbf{k}$ space; for the last passage, use has been made of the energy–wavevector relation $\varepsilon = (\hbar^2/2M)k_f^2$ for a particle of mass M, and hence of the relation $d\varepsilon = (\hbar^2/M)k_f\,dk_f$.

The differential cross-section, which is the quantity measured in scattering experiments, can now be obtained as follows. The flux of incident particles of mass M, initial momentum $\hbar\mathbf{k}_i$, and wavefunction $\psi_i = (1/\sqrt{V})\exp(i\mathbf{k}_i \cdot \mathbf{r})$ is

$$\mathbf{J}_i = \frac{\hbar\mathbf{k}_i}{M}\,|\psi_i|^2 = \frac{\hbar\mathbf{k}_i}{M}\frac{1}{V} \,.$$

The number of particles per unit time intercepted by a surface $d\sigma$ orthogonal to $\mathbf{J}_i$ is $J_i\,d\sigma = (\hbar k_i/MV)\,d\sigma$. We can make the identification

$$\frac{\hbar k_i}{M}\frac{1}{V}\,d\sigma = P_{\mathbf{k}_f \leftarrow \mathbf{k}_i}\frac{V}{(2\pi)^3}\,k_f\frac{M}{\hbar^2}\,d\varepsilon\,d\Omega \,;$$

then $d\sigma/d\varepsilon d\Omega$, which represents the differential cross-section for scattering of particles in the energy interval $d\varepsilon$ and in the solid angle $d\Omega$, becomes

$$\frac{d\sigma}{d\varepsilon d\Omega} = \frac{V^2}{(2\pi)^3}\frac{M^2}{\hbar^3}\frac{k_f}{k_i}P_{\mathbf{k}_f \leftarrow \mathbf{k}_i}.$$

With the help of Eq. (39), we can express the differential cross-section in the form

$$\frac{d\sigma}{d\varepsilon d\Omega} = \frac{V^2}{(2\pi)^2}\frac{M^2}{\hbar^5}|V_a(\Delta\mathbf{k})|^2\frac{k_f}{k_i}S(\Delta\mathbf{k}, \Delta\omega), \tag{40}$$

where the so-called *dynamical structure factor* $S(\Delta\mathbf{k}, \Delta\omega)$ is defined as

$$S(\Delta\mathbf{k}, \Delta\omega) = \int_{-\infty}^{+\infty}\frac{dt}{2\pi}e^{i\Delta\omega t}\langle\exp(i\Delta\mathbf{k}\cdot\mathbf{R}_0)\cdot\exp(-i\Delta\mathbf{k}\cdot\mathbf{R}_0(t))\rangle. \tag{41}$$

Apart for kinematic factors and for the Fourier transform of the interaction potential (in general a smooth quantity or even a constant in the case of contact interaction in real space), the essential features of the scattering are thus determined by the *dynamical structure factor* $S(\Delta\mathbf{k}, \Delta\omega)$; this is the basic tool for investigating the scattering of an incident particle of wavevector $\mathbf{k}_i$ and energy ε_i into a final state of wavevector $\mathbf{k}_f = \mathbf{k}_i + \Delta\mathbf{k}$ and energy $\varepsilon_f = \varepsilon_i + \hbar\Delta\omega$. *Notice that the dynamical structure factor depends only on the scatterer and not on the impinging beam or the specific details of the probe–scattering coupling.* The dynamical structure factor (41) satisfies the sum-rule

$$\int_{-\infty}^{+\infty} S(\Delta\mathbf{k}, \Delta\omega)\,d(\Delta\omega) = 1.$$

Furthermore, in the limiting case of a "fixed scatterer" we have $\mathbf{R}_0(t) \approx \mathbf{R}_0(0)$ and thus $S(\Delta\mathbf{k}, \Delta\omega) = \delta(\Delta\omega)$.

In the specific case of neutron scattering, $V_a(\mathbf{r})$ is the contact interaction potential $(2\pi\hbar^2/M_n)\,b\,\delta(\mathbf{r})$, where b is the scattering length of the nucleus, and Eq. (37) gives

$$|V_a(\Delta\mathbf{k})|^2 = \frac{1}{V^2}\frac{4\pi^2\hbar^4}{M_n^2}b^2 ;$$

the expression (40) for the differential cross-section of a single scatterer thus becomes

$$\frac{d\sigma}{d\varepsilon d\Omega} = \frac{k_f}{k_i}\frac{b^2}{\hbar}S(\Delta\mathbf{k}, \Delta\omega). \tag{42}$$

In the case kinematic factors and kinematic restrictions are disregarded (i.e. it is assumed $k_f/k_i \approx 1$ and $\Delta\omega$ varies in the whole frequency region), the total scattering cross-section becomes $\sigma_{\text{tot}} = 4\pi b^2$.

5.2 Dynamical structure factor of a three-dimensional harmonic oscillator

It is instructive to calculate explicitly the dynamical structure factor (41) in the case the target is described by an isotropic three-dimensional harmonic oscillator of mass

M and angular frequency ω_0. We indicate with $\mathbf{R}_0 = (u_{0x}, u_{0y}, u_{0z})$ the displacement from the equilibrium position, and we use creation and annihilation operators, because they are very convenient in the calculation of thermal averages. We have

$$u_{0x} = \sqrt{\frac{\hbar}{2M\omega_0}}\,[a_x + a_x^\dagger]\,, \tag{43a}$$

and similar expressions for u_{0y} and u_{0z}. From elementary properties of harmonic oscillators (see Appendix IX-A) we obtain

$$\langle u_{0x}^2 \rangle = \frac{\hbar}{2M\omega_0}\langle (a_x + a_x^\dagger)(a_x + a_x^\dagger)\rangle = \frac{\hbar}{2M\omega_0}(2\langle n \rangle + 1)\,,$$

where the average number of phonons $\langle n \rangle = \langle a_x^\dagger a_x \rangle$ is given by the Bose–Einstein distribution function $1/(\exp(\hbar\omega_0/k_B T) - 1)$. Then

$$\langle u_{0x}^2 \rangle = \frac{\hbar}{2M\omega_0}\left(\frac{2}{e^{\hbar\omega_0/k_B T} - 1} + 1\right). \tag{43b}$$

In the Heisenberg representation we have

$$u_{0x}(t) = \sqrt{\frac{\hbar}{2M\omega_0}}\,[a_x e^{-i\omega_0 t} + a_x^\dagger e^{i\omega_0 t}]\,, \tag{43c}$$

and also

$$\langle u_{0x}^2(t) \rangle = \langle u_{0x}^2 \rangle\,. \tag{43d}$$

According to Eq. (IX-A20), it is convenient to remind the following identity

$$\langle e^A e^B \rangle = e^{\langle A^2 + 2AB + B^2\rangle/2}\,, \tag{44}$$

where A and B are any two operators linear in creation and annihilation operators. We use the above expression with $A = i\,\Delta\mathbf{k}\cdot\mathbf{u}_0$ and $B = -i\,\Delta\mathbf{k}\cdot\mathbf{u}_0(t)$ (without loss of generality, whenever useful, we can suppose that $\Delta\mathbf{k}$ is in the x direction). Eq. (41) thus takes the form

$$\boxed{S(\Delta\mathbf{k}, \Delta\omega) = e^{-2W}\int_{-\infty}^{+\infty}\frac{dt}{2\pi}e^{i\Delta\omega t}\exp[\langle \Delta\mathbf{k}\cdot\mathbf{u}_0\,\Delta\mathbf{k}\cdot\mathbf{u}_0(t)\rangle]\,,} \tag{45a}$$

where

$$\boxed{2W = \langle (\Delta\mathbf{k}\cdot\mathbf{u}_0)^2\rangle}\,. \tag{45b}$$

The quantity $\exp(-2W)$ is called the *Debye–Waller factor*; its physical consequences will be apparent soon.

Using Eq. (43b), the explicit expression of $2W$ reads

$$2W = \frac{\hbar^2(\Delta k)^2}{2M}\frac{1}{\hbar\omega_0}\left(\frac{2}{e^{\hbar\omega_0/k_B T} - 1} + 1\right).$$

In the limiting case of temperatures much smaller or much higher than the reference temperature $T_0 = \hbar\omega_0/k_B$ it holds

$$2W \equiv \langle (\Delta\mathbf{k} \cdot \mathbf{u}_0)^2 \rangle = \begin{cases} \dfrac{\hbar^2(\Delta\mathbf{k})^2}{2M}\dfrac{1}{\hbar\omega_0} & \text{for} \quad T \ll T_0 \\[2ex] \dfrac{2T}{T_0}\dfrac{\hbar^2(\Delta\mathbf{k})^2}{2M}\dfrac{1}{\hbar\omega_0} & \text{for} \quad T \gg T_0 \end{cases} . \tag{46}$$

Thus, for all the temperatures of usual interest (ranging from $T = 0$ to temperatures smaller or near T_0), we see that the important physical effect determining the quantity $2W$ is just *the ratio $E_R/\hbar\omega_0$ between the target recoil energy $E_R = \hbar^2(\Delta\mathbf{k})^2/2M$ and the harmonic oscillator quantum of energy $\hbar\omega_0$*.

It is convenient to break the dynamical structure factor (45a) into a series of displacement-displacement correlation functions representing zero-phonon, one-phonon and multi-phonon scattering processes

$$S(\Delta\mathbf{k}, \Delta\omega) = e^{-2W}\int_{-\infty}^{+\infty}\frac{dt}{2\pi}e^{i\Delta\omega t}\sum_{m=0}^{\infty}\frac{1}{m!}[\langle\Delta\mathbf{k}\cdot\mathbf{u}_0\,\Delta\mathbf{k}\cdot\mathbf{u}_0(t)\rangle]^m . \tag{47}$$

We consider now in particular the zero-phonon and the one-phonon contributions to the expansion (47), given by the $m = 0$ and $m = 1$ terms, respectively.

Zero-phonon contribution (elastic scattering)

The zero-phonon contribution to the dynamical structure factor describes elastic scattering processes and is given by

$$S_0(\Delta\mathbf{k}, \Delta\omega) = e^{-2W}\int_{-\infty}^{+\infty}\frac{dt}{2\pi}e^{i\Delta\omega t} = e^{-2W}\delta(\Delta\omega) . \tag{48}$$

The above expression shows that *the intensity for elastic scattering from an oscillating target with respect to the intensity of elastic scattering from a fixed target is reduced by the Debye–Waller factor* $\exp(-2W)$. Thus, the fraction f of elastic (or "recoil-free") scattering processes, using Eq. (45b), can be written as

$$f = \exp(-2W) = \exp(-|\Delta\mathbf{k}|^2\langle u_{0x}^2\rangle) ,$$

where $\langle u_{0x}^2\rangle$ is the mean square vibrational amplitude of the scatterer in the direction of $\Delta\mathbf{k}$. Notice that the recoilless fraction f equals one for a "rigid" scatterer, i.e. for a scatterer where $(\langle u_{0x}^2\rangle)^{1/2}$ is much smaller than the wavelength $2\pi/|\Delta\mathbf{k}|$.

In the temperature range $0 \le T \approx T_0$, using Eq. (46), the recoil-free fraction can be expressed as

$$f = \exp(-E_R/\hbar\omega_0)$$

where $E_R = \hbar^2(\Delta\mathbf{k})^2/2M$ is the recoil energy of the scatterer. In the case $E_R \ll \hbar\omega_0$, we see that most of the scattering takes place as if *the target is held fixed*; in the case $E_R > \hbar\omega_0$ the elastic scattering is strongly reduced with respect to inelastic processes.

One-phonon contribution

The one-phonon contribution to the dynamical structure factor is

$$S_1(\Delta\mathbf{k}, \Delta\omega) = e^{-2W} \int_{-\infty}^{+\infty} \frac{dt}{2\pi} e^{i\Delta\omega t} \langle \Delta\mathbf{k} \cdot \mathbf{u}_0 \, \Delta\mathbf{k} \cdot \mathbf{u}_0(t) \rangle \ .$$

Using Eqs. (43) we have

$$\langle \Delta\mathbf{k} \cdot \mathbf{u}_0 \, \Delta\mathbf{k} \cdot \mathbf{u}_0(t) \rangle = \frac{\hbar(\Delta\mathbf{k})^2}{2M\omega_0} \langle (a_x + a_x^\dagger)(a_x e^{-i\omega_0 t} + a_x^\dagger e^{i\omega_0 t}) \rangle$$

$$= \frac{\hbar^2(\Delta\mathbf{k})^2}{2M} \frac{1}{\hbar\omega_0} \left[\langle n \rangle e^{-i\omega_0 t} + \langle n + 1 \rangle e^{i\omega_0 t} \right] \ .$$

With the help of the above expression, we obtain

$$\boxed{S_1(\Delta\mathbf{k}, \Delta\omega) = e^{-2W} \frac{\hbar^2(\Delta\mathbf{k})^2}{2M} \frac{1}{\hbar\omega_0} \left[\langle n \rangle \delta(\Delta\omega - \omega_0) + \langle n + 1 \rangle \delta(\Delta\omega + \omega_0) \right]} \ . \qquad (49)$$

The dynamical structure factor $S_1(\Delta\mathbf{k}, \Delta\omega)$ consists of sharp delta-function peaks describing one-phonon absorption and emission processes; the Bose thermal factors $\langle n \rangle$ and $\langle n \rangle + 1$ are appropriate for processes in which phonons are absorbed or emitted, respectively (notice that at $T = 0$ absorption processes are suppressed). The relative importance of zero-phonon versus one-phonon contributions is in essence dictated by the ratio between the recoil energy and the quantum of energy $\hbar\omega_0$. This same ratio controls the relative importance of multiphonon scattering processes in which two or more phonons are absorbed or emitted.

6   Diffusion of particles by a crystal and effects of lattice vibrations

In the previous section we have analysed the diffusion of particles of a single dynamic scatterer, schematized as an atom or a nucleus performing harmonic oscillations, and we have seen that the relative importance of *elastic* and *inelastic processes* is essentially determined by the Debye–Waller factor. We wish now to study the scattering of particles (electrons, neutrons, X-rays) by an actual crystal with N nuclei in thermal vibration, and to assess the relative importance of elastic and inelastic processes.

The essential kinematic aspects of the scattering process are indicated in Fig. 14. The crystal can be pictured as a collection of scatterers in vibration around their equilibrium lattice positions. We make the basic assumption that the probe–sample interaction is invariant under uniform translations of the probe–sample system, and is given by

$$\boxed{V_{\text{probe–sample}}(\mathbf{r}; \{\mathbf{R}_m\}) = \sum_m V_a(\mathbf{r} - \mathbf{R}_m)} \ . \qquad (50)$$

This form, that generalizes Eq. (34), represents the sum of interaction potentials $V_a(\mathbf{r})$ centred at the atomic positions $\mathbf{R}_m$. To avoid inessential details, we assume that all atoms are equal and their equilibrium positions describe a Bravais lattice (in the case of neutron scattering, we also assume that all the nuclei are of the same isotope).

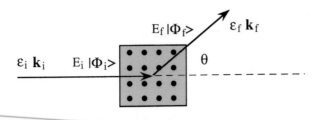

Fig. 14 Diffusion of particles from a crystal and effect of the thermal motion of the atoms; E_i, $|\Phi_i\rangle$, E_f and $|\Phi_f\rangle$ denote energies and vibrational states of the whole crystal lattice before and after the scattering; ε_i, $\mathbf{k}_i$, ε_f and $\mathbf{k}_f$ are the energies and the wavevectors of the probe particle before and after the scattering.

We now follow step-by-step the treatment of Section 5, inserting the appropriate adjustments whenever needed. Similarly to what done in Section 5.1 the "kinematic approximation" (or first-order Born approximation) is supposed to hold; this implies that scattering cross-sections are sufficiently small and multiple scattering negligible. We indicate with E_i, $|\Phi_i\rangle$, E_f and $|\Phi_f\rangle$ the vibrational energies and states of the whole crystal lattice, and with ε_i, $|W_{\mathbf{k}_i}\rangle$, ε_f and $|W_{\mathbf{k}_f}\rangle$, the energies and the wavefunctions of the probe particles before and after the collision, respectively. The transition probability per unit time from an initial state $|W_{\mathbf{k}_i}\Phi_i\rangle$ to a final state $|W_{\mathbf{k}_f}\Phi_f\rangle$ is given by the Fermi golden rule

$$P_{\mathbf{k}_f\Phi_f\leftarrow\mathbf{k}_i\Phi_i} = \frac{2\pi}{\hbar}\left|\langle W_{\mathbf{k}_f}\Phi_f|\sum_m V_a(\mathbf{r}-\mathbf{R}_m)|W_{\mathbf{k}_i}\Phi_i\rangle\right|^2 \delta(\varepsilon_f+E_f-\varepsilon_i-E_i)$$

$$= \frac{2\pi}{\hbar}\left|\langle\Phi_f|V_a(\Delta\mathbf{k})\sum_m e^{-i\,\Delta\mathbf{k}\cdot\mathbf{R}_m}|\Phi_i\rangle\right|^2 \delta(\varepsilon_f+E_f-\varepsilon_i-E_i)\,, \quad (51)$$

where

$$V_a(\Delta\mathbf{k}) = \frac{1}{V}\int e^{-i(\mathbf{k}_f-\mathbf{k}_i)\cdot\mathbf{r}}V_a(\mathbf{r})\,d\mathbf{r}$$

is the Fourier transform of the interaction potential $V_a(\mathbf{r})$.

Apart for the replacement of Eq. (38a) with Eq. (51), we can now follow, mutatis mutandis, all the reasoning of Section 5.1 concerning the scattering cross-section and the dynamical structure factor. In particular, with the help of Eq. (42), it easily seen that the differential cross-section for neutron scattering from a crystal is expressed as

$$\boxed{\frac{d\sigma}{d\varepsilon d\Omega} = \frac{k_f}{k_i}\frac{b^2}{\hbar}S(\Delta\mathbf{k},\Delta\omega)}\,, \quad (52a)$$

where the so-called *dynamical structure factor of the crystal* $S(\Delta\mathbf{k},\Delta\omega)$ is given by

$$\boxed{S(\Delta\mathbf{k},\Delta\omega) = \int_{-\infty}^{+\infty}\frac{dt}{2\pi}e^{i\Delta\omega t}\sum_{mn}\langle\exp(i\Delta\mathbf{k}\cdot\mathbf{R}_m)\cdot\exp(-i\Delta\mathbf{k}\cdot\mathbf{R}_n(t))\rangle}\,; \quad (52b)$$

as previously, the symbol $\langle\ldots\rangle$ indicates the thermal average. The basic expression (52b) is related to the probability that an incident particle of wavevector $\mathbf{k}_i$ and energy ε_i is scattered into the state with wavevector $\mathbf{k}_f$ and energy ε_f, with a consequent change $\hbar\,\Delta\mathbf{k}$ and $\hbar\,\Delta\omega$ in momentum and energy. We stress again that the dynamical structure factor depends only on the scatterer and not on the impinging beam or the specific details of the probe–scattering coupling.

Let us now exploit in the general expression (52b) the *translational symmetry* of the Bravais lattice. We indicate the atomic positions as $\mathbf{R}_n = \mathbf{t}_n + \mathbf{u}_n$ where $\mathbf{t}_n$ is a lattice translation vector and $\mathbf{u}_n$ is the displacement of the atom from its equilibrium position. We can simplify the double sum in (52) with N times a unique sum with fixed $\mathbf{R}_m$ (for instance $\mathbf{R}_m = 0$). We have

$$S(\Delta\mathbf{k}, \Delta\omega) = N \sum_{\mathbf{t}_n} e^{-i\Delta\mathbf{k}\cdot\mathbf{t}_n} \int_{-\infty}^{+\infty} \frac{dt}{2\pi} e^{i\Delta\omega t} \langle \exp(i\Delta\mathbf{k}\cdot\mathbf{u}_0)\exp(-i\Delta\mathbf{k}\cdot\mathbf{u}_n(t))\rangle \ . \quad (53)$$

We now use for *harmonic crystals* the identity (44), with $A = i\Delta\mathbf{k}\cdot\mathbf{u}_0$ and $B = -i\Delta\mathbf{k}\cdot\mathbf{u}_n(t)$. The dynamical structure factor takes the form

$$S(\Delta\mathbf{k}, \Delta\omega) = N e^{-2W} \sum_{\mathbf{t}_n} e^{-i\Delta\mathbf{k}\cdot\mathbf{t}_n} \int_{-\infty}^{+\infty} \frac{dt}{2\pi} e^{i\Delta\omega t} \exp(\langle \Delta\mathbf{k}\cdot\mathbf{u}_0\,\Delta\mathbf{k}\cdot\mathbf{u}_n(t)\rangle) \ , \quad (54a)$$

where

$$2W = \langle(\Delta\mathbf{k}\cdot\mathbf{u}_0)^2\rangle \qquad (54b)$$

and $\exp(-2W)$ is the Debye–Waller factor. The quantity $\langle(\Delta\mathbf{k}\cdot\mathbf{u}_0)^2\rangle$ for a Bravais crystal lattice has been calculated in Chapter IX; using Eq. (IX-39), we have

$$2W \equiv \langle(\Delta\mathbf{k}\cdot\mathbf{u}_0)^2\rangle = \begin{cases} \dfrac{3}{2}\dfrac{\hbar^2(\Delta\mathbf{k})^2}{2M}\dfrac{1}{k_BT_D} & \text{for} \quad T \ll T_D \\[3mm] \dfrac{6T}{T_D}\dfrac{\hbar^2(\Delta\mathbf{k})^2}{2M}\dfrac{1}{k_BT_D} & \text{for} \quad T \gg T_D \end{cases} \qquad . \quad (55)$$

Apart for some numerical factors, it is apparent the perfect analogy between Eq. (55) and Eq. (46) of the simplified treatment of the previous section.

The meaning of Eq. (54a) becomes evident if the second exponential occurring under integral sign is expanded in series; we have

$$S(\Delta\mathbf{k}, \Delta\omega) = \sum_{m=0}^{\infty} S_m(\Delta\mathbf{k}, \Delta\omega) \ , \qquad (56a)$$

where

$$S_m(\Delta\mathbf{k}, \Delta\omega) = N e^{-2W} \sum_{\mathbf{t}_n} e^{-i\Delta\mathbf{k}\cdot\mathbf{t}_n} \int_{-\infty}^{+\infty} \frac{dt}{2\pi} e^{i\Delta\omega t} \frac{1}{m!} \left[\langle \Delta\mathbf{k}\cdot\mathbf{u}_0\,\Delta\mathbf{k}\cdot\mathbf{u}_n(t)\rangle\right]^m \ . \quad (56b)$$

We see that the mth term in the expansion gives the contribution of the processes involving m phonons (emitted or absorbed); we consider now in particular the zero-phonon and one-phonon contributions.

Zero-phonon contribution (elastic scattering)

The zero-phonon contribution to the dynamical structure factor is

$$S_0(\Delta\mathbf{k}, \Delta\omega) = N\,e^{-2W} \sum_{\mathbf{t}_n} e^{-i\Delta\mathbf{k}\cdot\mathbf{t}_n}\,\delta(\Delta\omega)\ . \tag{57}$$

The sum over $\mathbf{t}_n$ is different from zero only if $\Delta\mathbf{k}$ equals a reciprocal lattice vector $\mathbf{G}$, in this case it gives N. We thus recover the standard result that the Bragg diffraction pattern for elastic scattering maps the reciprocal lattice vectors. Notice that $S_0(\mathbf{G}, \Delta\omega)$ is proportional to N^2 (and not to N), as expected from the fact that the Bragg scattering is a coherent process. We also see that the intensity of the diffracted beams is reduced by the Debye–Waller factor $\exp(-2W(\mathbf{G}))$, which decreases as the magnitude of momentum transfer increases.

One-phonon contribution

The one-phonon contribution to the dynamical structure factor is

$$S_1(\Delta\mathbf{k}, \Delta\omega) = N\,e^{-2W} \sum_{\mathbf{t}_n} e^{-i\Delta\mathbf{k}\cdot\mathbf{t}_n} \int_{-\infty}^{+\infty} \frac{dt}{2\pi} e^{i\Delta\omega\,t} \langle \Delta\mathbf{k}\cdot\mathbf{u}_0\,\Delta\mathbf{k}\cdot\mathbf{u}_n(t)\rangle\ . \tag{58}$$

To calculate explicitly the thermal average in the above equation, we express the operators $\mathbf{u}_n(t)$ in terms of the normal modes operators. For a simple lattice, the normal modes consist of three acoustic branches; for simplicity we assume that the three acoustic branches are degenerate. The component $u_{nx}(t)$ along a given direction (say x) of the displacement of the atom in the n-th cell is given by

$$u_{nx}(t) = \frac{1}{\sqrt{N}} \sum_{\mathbf{q}} \sqrt{\frac{\hbar}{2M\omega_{\mathbf{q}}}} \left[a_{\mathbf{q}} e^{i(\mathbf{q}\cdot\mathbf{t}_n - \omega_{\mathbf{q}}t)} + a_{\mathbf{q}}^{\dagger} e^{-i(\mathbf{q}\cdot\mathbf{t}_n - \omega_{\mathbf{q}}t)} \right]\ .$$

In particular for $t = 0$ and $\mathbf{t}_n = 0$, we have

$$u_{0x} = \frac{1}{\sqrt{N}} \sum_{\mathbf{q}} \sqrt{\frac{\hbar}{2M\omega_{\mathbf{q}}}} \left[a_{\mathbf{q}} + a_{\mathbf{q}}^{\dagger} \right]\ .$$

Hence

$$\langle \Delta\mathbf{k}\cdot\mathbf{u}_0\,\Delta\mathbf{k}\cdot\mathbf{u}_n(t)\rangle = \frac{\hbar^2(\Delta\mathbf{k})^2}{2M} \frac{1}{N} \sum_{\mathbf{q}} \frac{1}{\hbar\omega_{\mathbf{q}}} \left[n_{\mathbf{q}} e^{i(\mathbf{q}\cdot\mathbf{t}_n - \omega_{\mathbf{q}}t)} + (n_{\mathbf{q}}+1) e^{-i(\mathbf{q}\cdot\mathbf{t}_n - \omega_{\mathbf{q}}t)} \right]$$

where $n_{\mathbf{q}} = 1/\left[\exp\left(\hbar\omega_{\mathbf{q}}/k_B T\right) - 1\right]$ is the Bose population factor.

Inserting the above expression into Eq. (58), we obtain

$$S_1(\Delta\mathbf{k}, \Delta\omega) = e^{-2W} \sum_{\mathbf{q}\,\mathbf{t}_n} \frac{\hbar^2(\Delta\mathbf{k})^2}{2M} \frac{1}{\hbar\omega_{\mathbf{q}}} \exp[-i(\Delta\mathbf{k} - \mathbf{q})\cdot\mathbf{t}_n]\,n_{\mathbf{q}}\,\delta(\Delta\omega - \omega_{\mathbf{q}})$$

$$+ e^{-2W} \sum_{\mathbf{q}\,\mathbf{t}_n} \frac{\hbar^2(\Delta\mathbf{k})^2}{2M} \frac{1}{\hbar\omega_{\mathbf{q}}} \exp[-i(\Delta\mathbf{k} + \mathbf{q})\cdot\mathbf{t}_n]\,(n_{\mathbf{q}}+1)\,\delta(\Delta\omega + \omega_{\mathbf{q}})\ . \tag{59}$$

The dynamical structure factor $S_1(\Delta\mathbf{k}, \Delta\omega)$ consists of sharp delta-function peaks describing one-phonon absorption and emission processes, compatible with conservation laws of energy and momentum. The thermal factors $n_{\mathbf{q}}$ and $n_{\mathbf{q}}+1$ are appropriate for processes in which phonons are absorbed or emitted, respectively. Notice that phonon annihilation processes are proportional to the Bose population factor $n_{\mathbf{q}}$ and vanishes as the temperature is decreased; on the contrary, phonon creation processes are proportional to $n_{\mathbf{q}}+1$, and can even occur in the lattice at zero temperature.

Multiphonon contributions

We can proceed with higher order terms in the exponential and describe scattering processes in which two or more phonons are absorbed or emitted. The conservation laws for these higher order processes are less selective, and multiphonon effects in many practical situations provide a rather unstructured background that does not smear out the important information provided by zero-phonon and one-phonon contribution. The effect of masking of course increases for high $|\Delta\mathbf{k}|$ and at high temperature, due to the Debye–Waller factor.

7   Mössbauer effect

Consider the emission of a low energy γ-ray ($E_\gamma = 10 \div 100\,\text{keV}$) from a nucleus, in the hypothetical situation that the emitting nucleus is *held fixed*; a typical emission lineshape could be (approximately) expected to have a Lorentzian form

$$I(E) = \frac{1}{\pi} \frac{\Gamma}{(E - E_0)^2 + \Gamma^2} , \tag{60}$$

where E is the emitted energy, E_0 is the energy separation between the two nuclear states of interest, and Γ is the linewidth broadening (essentially determined by the natural radiative width). The uncertainty Γ in the energy transition corresponds to a lifetime $\tau = \hbar/\Gamma$ of the excited state of the nucleus. For instance, in the case of the isotope ^{57}Fe, the parameters of the emission spectrum are taken to be $E_0 = 14.4\,\text{keV}$, $\tau = 141\,\text{ns}$, $\Gamma = \hbar/\tau \approx 4.66 \cdot 10^{-9}\,\text{eV}$. In the case of ^{191}Ir, we have $E_0 = 129\,\text{keV}$, $\tau = 0.13\,\text{ns}$, $\Gamma \approx 5 \cdot 10^{-6}\,\text{eV}$.

If the nucleus is free and is initially at rest, after the emission of a γ-ray of momentum $\hbar\mathbf{k}_\gamma$ ($k_\gamma = \omega_\gamma/c$) it recoils with momentum $-\hbar\mathbf{k}_\gamma$. The recoil kinetic energy of a nucleus of mass M is

$$E_R = \frac{\hbar^2 k_\gamma^2}{2M} = \frac{\hbar^2 \omega_\gamma^2}{2Mc^2} \approx \frac{E_0^2}{2Mc^2} \tag{61}$$

(notice that $M_{\rm p}c^2 = 938$ MeV for the proton mass $M_{\rm p}$); thus, the emission spectrum of gammas from a *free nucleus at rest* would be centered at the energy $E_0 - E_R$. The energy shift E_R is in general orders of magnitude larger than the natural linewidth Γ of the nucleus. For instance, in the case of ^{57}Fe we have $MC^2 = 57\,M_{\rm p}c^2 = 57 \cdot 938$ MeV, and the recoil energy becomes $E_R = 1.9$ meV, while $\Gamma = 4.66 \cdot 10^{-9}$ eV; in the case of ^{191}Ir, the recoil energy is $E_R = 46$ meV, and $\Gamma = 5 \cdot 10^{-6}$ eV.

If the emitting nuclei are supposed to be free and in thermal equilibrium at the temperature T, we expect for their mean quadratic velocity $(1/2)M\langle v^2\rangle = (3/2)k_BT$; the γ-ray emission line is broadened by longitudinal Doppler effect, which gives for the percentual variation of the energy of the emitted line (to first order in v/c)

$$\frac{\Delta E}{E_0} \approx \frac{\sqrt{<v^2>}}{c} = \sqrt{\frac{3k_BT}{Mc^2}} \ . \tag{62}$$

From Eq. (61) and Eq. (62), the energy broadening due to the Doppler effect is estimated $\Delta E \approx \sqrt{E_R\,k_BT}$, a value in general orders of magnitude higher than the natural width.

The above considerations show that the γ-ray emission spectrum from *free nuclei* exhibits a shift and a broadening much larger than the natural width. On the contrary, and in striking contrast with the free nuclei situation, we can now see that the γ-ray *emission spectrum from nuclei bound in a crystal lattice contains a sharp emission line, which is neither shifted nor broadened with respect to the natural width (Mössbauer effect)*.

The formal theory for γ-light emission intensity from a nucleus, bound elastically to a fixed center or (more realistically) bound elastically in a crystal lattice, is completely similar to what has been done in the previous two sections, except for some minor adjustments. As in the general theory of particle scattering by crystals, the fraction of elastic (i.e. zero-phonon) emission processes, which produce the recoilless Mössbauer line, is controlled by the Debye–Waller factor $\exp(-2W)$. From Eq. (54b), with the replacement $|\Delta \mathbf{k}| = k_\gamma$, the recoil-free fraction of emitted gammas can be written as

$$f = \exp(-2W) = \exp(-k_\gamma^2\langle u_{0x}^2\rangle) \ , \tag{63}$$

where $\langle u_{0x}^2\rangle$ is the mean square vibrational amplitude of the emitting nucleus (in the direction of the emission of the gamma ray). Notice that the recoiless fraction f drops to zero in the case $\langle u_{0x}^2\rangle$ is not bound or is large with respect to $1/k_\gamma$. The recoil fraction of emitted gammas, involving creation or annihilation of one or more phonons, is regulated by the supplemental fraction $1 - f$.

For temperatures lower than the Debye temperature, from Eq. (55) and Eq. (63) we have

$$f = \exp\left[-\frac{3}{2}\frac{\hbar^2 k_\gamma^2}{2M}\frac{1}{k_BT_D}\right] = \exp\left[-\frac{3}{2}\frac{E_R}{k_BT_D}\right] \ .$$

Thus if the recoil energy E_R is less (or not much higher) than k_BT_D, we have that a substantial part of the emission process occurs with *no broadening and no shift* of the emission line, described by Eq. (60). Typical values of f are 0.92 for the 14.4 keV gamma ray of ^{57}Fe ($E_R = 1.9$ meV, $T_D = 400$ K $= 34.5$ meV), and $f = 0.06$ for the 129 keV gamma ray of ^{191}Ir ($E_R = 46$ meV, $T_D = 285$ K $= 24.6$ meV). The condition $E_R \leq k_BT_D$ limits the Mössbauer effect to γ-rays up to ≈ 150 keV or so.

In addition to the sharp peak at energy E_0 and width Γ, the emission spectrum presents a broad background (of the order of tens of meV) due to one-, two- and multiphonon processes. The presence of the so sharp central line in low energy γ-ray

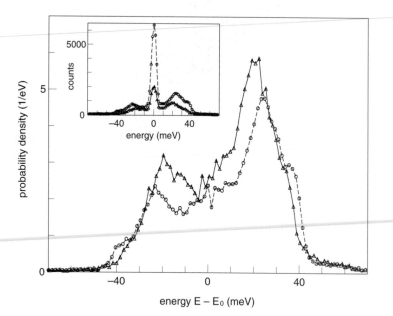

Fig. 15 Absorption probability density for α-iron (circles, dashed line) and stainless steel (triangles, solid line), after removing the elastic peak. The inset shows the raw data [from W. Sturhahn, T. S. Toellner, E. E. Alp, X. Zhang, M. Ando, Y. Yoda, S. Kikuta, M. Seto, C. W. Kimball and B. Dabrowski, Phys. Rev. Lett. **74**, 3832 (1995); copyright 1995 by the American Physical Society].

emission is very appealing for spectroscopists and has found numerous applications in fundamental problems of solid state physics, chemical physics and relativity tests.

The above considerations on the photon emission can be extended to the phonon absorption from the nuclear resonances of nuclei bound in crystals. The absorption spectrum consists of a narrow recoil-free peak at the energy of the nuclear transition (controlled by the recoilless fraction f), plus a broad background due to phonon assisted transitions (controlled by the recoil fraction $1 - f$). Important information on the lattice dynamics can be obtained by the analysis of the recoil fraction of resonant nuclear absorptions.

In Fig. 15, we report a typical example of measured absorption spectrum of X-rays from the 14.413 keV nuclear resonance of ^{57}Fe [X-rays are produced by synchrotron radiation; the intensity of nuclear absorption is measured by counting the fluorescence events, resulting from nuclear internal conversion processes]. The observed absorption spectrum consists of an elastic peak and sidebands at lower and higher energiy (insert of Fig. 15). The central peak corresponds to the nuclear absorption without recoil. The sidebands above (below) the central peak corresponds to inelastic absorption accompanied by creation (annihilation) of phonons. The observed asymmetry in the spectra, as usual, is related to the fact that phonon annihilation probability is proportional to the number $\langle n \rangle$ of existing phonons, while phonon creation probability is proportional to $\langle n \rangle + 1$. The inelastic nuclear absorption by synchrotron radiation has found several successful applications in the study of lattice dynamics and in providing direct measurements of the phonon density-of-states [for further aspects see A. I. Chumakov, R. Rüffer, A. Q. R. Baron, H. Grünsteudel, H. F. Grünsteudel and V. G. Kohn, Phys. Rev. B**56**, 10758 (1997) and references quoted therein].

Further reading

N. W. Ashcroft and N. D. Mermin "Solid State Physics" (Holt, Rinehart and Winston, New York 1976)

G. E. Bacon "Neutron Diffraction" (Clarendon Press, Oxford 1975)

B. N. Brockhouse "Slow Neutron Spectroscopy and the Grand Atlas of the Physical World" Rev. Mod. Phys. **67**, 735 (1995)

C. R. Brundle and A. D. Baker eds. "Electron Spectroscopy, Theory, Techniques and Applications" (Academic Press, London 1977, 1978, 1981 Vols 1, 2, 3)

M. Cardona and G. Guntherodt eds. "Light Scattering in Solids" Vols. I-VI (Springer, Berlin 1991)

M. J. Cooper "Compton Scattering and Electron Momentum Determination" Rep. Progr. Phys. **48**, 415 (1985)

J. M. Cowley "Diffraction Physics" (North-Holland, Amsterdam 1975, third edition)

B. Di Bartolo and R. C. Powell "Phonons and Resonances in Solids" (Wiley, New York 1976)

W. Hayes and R. Loudon "Scattering of Light by Crystals" (Wiley, New York 1987)

S. Hüfner "Photoelectron Spectroscopy" (Springer, Berlin 1996)

H. Ibach and H. Lüth "Solid-State Physics" (Springer, Berlin 1995)

H. Ibach and D. L. Mills "Electron Energy Loss Spectroscopy and Surface Vibrations" (Academic Press, New York 1982)

J. D. Jackson "Classical Electrodynamics" (Wiley, New York 1975)

D. A. Long "Raman Spectroscopy" (McGraw Hill, New York 1977)

S. W. Lovesey "Theory of Neutron Scattering from Condensed Matter" (Clarendon Press, Oxford 1984)

W. K. H. Panofsky and M. Phillips "Classical Electricity and Magnetism" (Addison-Wesley, Reading, Mass. 1962)

D. Schwarzenbach "Crystallography" (Wiley, Chichester 1996)

G. K. Wertheim "Mössbauer Effect: Principles and Applications" (Academic Press, New York 1964)

P. Y. Yu and M. Cardona "Fundamentals of Semiconductors" (Springer, Berlin 1996)

XI

Optical and transport properties in metals

This is the first of four interrelated chapters, in which we study the optical and transport properties of crystals. Similarly to its historical role in atomic and molecular physics, optical spectroscopy has always accompanied and stimulated major progresses also in the physics of crystals. The importance of transport phenomena for the development of solid state electronics and devices is also well known. In this chapter we consider optical and transport properties of metals. The optical constants of semiconductors and insulators are discussed in Chapter XII; the transport properties of homogeneous and inhomogeneous semiconductors are presented in Chapters XIII and XIV, respectively.

This chapter begins with a section, rather general in nature, that describes the phenomenological optical constants of homogeneous materials. We then consider the optical properties due to *intraband transitions* of carriers; this contribution is typical of the conduction electrons in metals, but also semiconductors (with small bandgap or strongly doped) may present a significant intraband contribution at finite temperature. The optical properties due to *interband transitions* are discussed in the next chapter; interband transitions are basic in insulators and semiconductors, where bands are either fully occupied or fully empty at zero temperature, and are of relevance also in metals at sufficiently high excitation energies.

In the study of the optical properties of free carriers, classical and semiclassical approaches maintain their usefulness, and are preliminary to full quantum mechanical treatments. We thus consider the optical properties of free carriers within the classical theory of Drude, because of its simplicity and pedagogical value. Successively, we consider static and dynamic conductivity of free carriers by means of the semiclassical transport equation of Boltzmann; this treatment is then compared with the linear response formalism of the quantum mechanical theory. The transport equations in the presence of electric fields and temperature gradients are also examined, and the thermoelectric effects of metals are discussed.

1   Macroscopic theory of optical constants in homogeneous materials

The propagation of electromagnetic waves in materials is described by the Maxwell's equations, supplemented by appropriate constitutive equations of the specific media. In this section we summarize the phenomenological constants, more commonly investigated in experiments. The considerations of the present section are rather general and may apply to homogeneous materials of different nature (metals, semiconductors and insulators), while the following sections refer almost exclusively to simple metals.

We consider for simplicity non-magnetic media ($\mathbf{B} = \mathbf{H}$; $\mu = 1$) in the absence of external charges and currents ($\rho_{\text{ext}} = 0$; $\mathbf{J}_{\text{ext}} = 0$). We focus our attention on (macroscopically) *homogeneous materials, in the linear response regime*; this means that changes of the internal charge density ρ_{ind} and of the internal current density $\mathbf{J}_{\text{ind}}$ induced by a driving electromagnetic field (of small intensity) are proportional to the field; furthermore, if the driving field is periodic in space and time, the same space and time dependence holds for the response of the system. [We notice that even an ideally perfect crystal is not rigorously homogeneous on a microscopic scale; for instance, the internal charge density has the crystal periodicity, and varies sharply in the unit cell. In general, the response of a crystal to a driving field of wavevector $\mathbf{q}$ includes not only a plane wave of vector $\mathbf{q}$, but also plane waves of vectors $\mathbf{q} + \mathbf{g}_m$, where $\mathbf{g}_m$ are reciprocal lattice vectors. In the following we adopt the standard simplification of neglecting "local field effects" due to inhomogeneity of the sample; in other words, we assume that if the driving field is characterized by a wavevector $\mathbf{q}$, the same vector characterizes also the response of the system. See a similar discussion in the study of the longitudinal dielectric function in Chapter VII].

The Maxwell's equations in a non-magnetic medium ($\mathbf{B} = \mathbf{H}$), in the absence of external charges and currents, in the presence of internal charge density ρ_{ind} and internal current density $\mathbf{J}_{\text{ind}}$ are (in Gauss units)

$$
\boxed{
\begin{array}{ll}
\text{div } \mathbf{E} = 4\pi \rho_{\text{ind}} & \text{curl } \mathbf{E} = -\dfrac{1}{c}\dfrac{\partial \mathbf{B}}{\partial t} \\[2mm]
\text{div } \mathbf{B} = 0 & \text{curl } \mathbf{B} = \dfrac{1}{c}\dfrac{\partial \mathbf{E}}{\partial t} + \dfrac{4\pi}{c}\mathbf{J}_{\text{ind}}
\end{array}
}
\tag{1}
$$

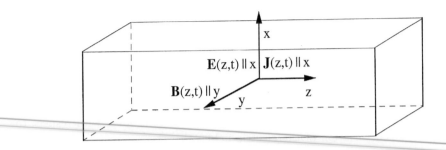

Fig. 1 Geometry chosen for the description of transverse electromagnetic fields in isotropic materials. The electric field and the internal density current are in the x-direction, the magnetic field is in the y-direction, the wave propagation is along the z-direction.

We consider the propagation of *transverse electromagnetic waves of given angular frequency ω in homogeneous materials*, and specify the Maxwell's equations in the geometry of Fig. 1. We take the electric field $\mathbf{E}$ in the x direction, the magnetic field $\mathbf{B}$ in the y-direction, and the current density $\mathbf{J}_{\text{ind}}$ parallel to the electric field in the isotropic medium; we can write

$$\mathbf{E}(\mathbf{r}, t) = E(z)\, e^{-i\omega t}(1, 0, 0) \qquad \mathbf{E} \parallel x \tag{2a}$$

$$\mathbf{B}(\mathbf{r}, t) = B(z)\, e^{-i\omega t}(0, 1, 0) \qquad \mathbf{B} \parallel y \tag{2b}$$

$$\mathbf{J}_{\text{ind}}(\mathbf{r}, t) = J(z)\, e^{-i\omega t}(1, 0, 0) \qquad \mathbf{J}_{\text{ind}} \parallel x \ . \tag{2c}$$

From Eq. (2a) we have div $\mathbf{E} = 0$; combined with the first Maxwell's equation, it entails $\rho_{\text{ind}} = 0$. From Eq. (2b) we have div $\mathbf{B} = 0$, consistently with the second Maxwell's equation. The remaining two Maxwell's equations give

$$\frac{dE(z)}{dz} = \frac{i\omega}{c} B(z) \tag{3a}$$

$$-\frac{dB(z)}{dz} = -\frac{i\omega}{c} E(z) + \frac{4\pi}{c} J(z) \ . \tag{3b}$$

By eliminating $B(z)$ we obtain

$$\boxed{\frac{d^2 E(z)}{dz^2} = -\frac{\omega^2}{c^2} E(z) - \frac{4\pi i\omega}{c^2} J(z)} \ . \tag{4}$$

This is the basic information, provided by the Maxwell's equations, on the electric field and current density; to proceed further we must establish between them a phenomenological or microscopic constitutive relationship.

Local and non-local constitutive relationship for the conductivity

The simplest and often adopted constitutive equation, linking the current density to the electric field, is a local relationship of the type

$$J(z) = \sigma(\omega)\, E(z) \ , \tag{5}$$

where the frequency-dependent proportionality constant $\sigma(\omega)$ is called the complex *conductivity constant*, or complex *conductivity function*, of the medium (the time dependence of the fields $\mathbf{J}$ and $\mathbf{E}$ is assumed to be characterized by a given angular frequency ω). According to Eq. (5), the current density at a given point in the material is proportional to the value of the electric field at the same point: this is the so-called *local-response regime*. A local relationship between current and field can be justified when the average distance travelled by the carriers is small with respect to the length of spatial variation of the electric field; otherwise, one should follow the trajectory of the carriers to find the effect of the spatially varying electric field: this is the so-called *non-local-response regime*.

In the non-local-response regime, the relationship between current density and electric field must be generalized in the form

$$J(z) = \int \sigma(z - z', \omega)\, E(z')\, dz' , \tag{6a}$$

where $\sigma(z, z', \omega) = \sigma(z - z', \omega)$ represents the generalized complex conductivity function of the *homogeneous* medium, at a given frequency ω. If we multiply both members of Eq. (6a) by $\exp(iqz)$ and integrate in dz, we obtain

$$J(q) = \sigma(q, \omega)\, E(q) , \tag{6b}$$

where

$$\sigma(q, \omega) = \int \sigma(z - z', \omega)\, e^{iq(z-z')}\, dz = \int \sigma(z, \omega)\, e^{iqz}\, dz \tag{6c}$$

and, similarly, $J(q)$ and $E(q)$ are the Fourier transforms of $J(z)$ and $E(z)$. Notice that the local-response regime is recovered when the generalized conductivity becomes a δ-like function of the form $\sigma(z - z', \omega) = \sigma(\omega)\, \delta(z - z')$; in this case the conductivity $\sigma(q, \omega)$ given by Eq. (6c) is independent of q, and one has $\sigma(q, \omega) = \sigma(q=0, \omega) = \sigma(\omega)$. Thus, the *local-response regime corresponds to neglecting the wavevector dependence of the conductivity.*

Local relationship for the conductivity and expression of the optical constants

In the rest of this section, we assume that a local constitutive equation holds between the current density and the electric field. Once the local approximation (5) between J and E is assumed, we can easily obtain the other frequency-dependent optical constants (or optical functions), which are commonly used for the description of the optical properties of matter. The basic equation (4), together with the assumption of Eq. (5), gives

$$\frac{d^2 E(z)}{dz^2} = -\frac{\omega^2}{c^2}\left[1 + \frac{4\pi i\omega\, \sigma(\omega)}{\omega} \right] E(z) . \tag{7}$$

The solution of this equation inside the material is a damped (or undamped) wave, which can be written in the form

$$E(z) = E_0\, e^{i(\omega/c)\, N z} , \tag{8a}$$

where the *complex refractive index* $N(\omega)$ is given by

$$N^2 = 1 + \frac{4\pi i \sigma(\omega)}{\omega} .$$
(8b)

The magnetic field associated to the electric field (8a) can be obtained from Eq. (3a), and is given by

$$B(z) = N E_0 \, e^{i(\omega/c) N z} = N E(z) .$$
(8c)

It is customary to write the complex refractive index in the form $N = n + ik$, where n is the ordinary *refractive index* and k the *extinction coefficient*; Eq. (8a) takes the form

$$E(z) = E_0 \, e^{i(\omega/c) n z} e^{-(\omega/c) k z} .$$
(9)

From Eq. (9), we see that the velocity of the electromagnetic wave in the medium is c/n; the penetration depth (*classical skin depth*, defined as the distance at which the field amplitude drops of $1/e$) is

$$\delta(\omega) = \frac{c}{\omega \, k(\omega)} .$$
(10a)

The damping of the electromagnetic wave is related to the absorption of electromagnetic energy. Since the intensity of an electromagnetic field is proportional to $|E(z)|^2$, from Eq. (9) we have $I(z) = I_0 \exp(-2\omega k z/c)$. The absorption coefficient is thus

$$\alpha(\omega) = \frac{2\omega \, k(\omega)}{c} \equiv \frac{2}{\delta(\omega)} .$$
(10b)

Equation (8b) can also be written in the more effective form $N^2 = \varepsilon$, where ε is the *complex dielectric function*. Indicating $\varepsilon = \varepsilon_1 + i\varepsilon_2$, we have

$$\varepsilon_1 = n^2 - k^2 \quad \text{and} \quad \varepsilon_2 = 2nk ;$$

and inversely

$$n^2 = \frac{1}{2}\left(\varepsilon_1 + \sqrt{\varepsilon_1^2 + \varepsilon_2^2}\right) \quad \text{and} \quad k^2 = \frac{1}{2}\left(-\varepsilon_1 + \sqrt{\varepsilon_1^2 + \varepsilon_2^2}\right) .$$

From Eq. (8b), and $N^2 = \varepsilon$, it is seen that the conductivity and the dielectric function satisfy the relation

$$\varepsilon(\omega) = 1 + \frac{4\pi i \sigma(\omega)}{\omega} .$$
(11)

Separation of the real and imaginary part gives

$$\varepsilon_1(\omega) = 1 - \frac{4\pi \sigma_2(\omega)}{\omega} \quad \text{and} \quad \varepsilon_2(\omega) = \frac{4\pi \sigma_1(\omega)}{\omega} ;$$

these equations link the real and imaginary part of the dielectric function to the imaginary and real part of the conductivity.

We consider now an electromagnetic wave that impinges at normal incidence on

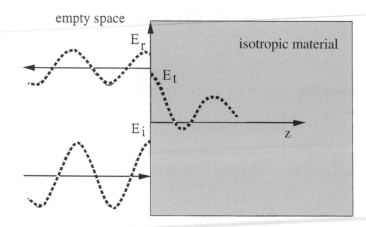

Fig. 2 Schematic representation of incident, reflected and transmitted electromagnetic waves at the surface (the $z = 0$ plane) of an isotropic material; the normal incident situation is considered.

the surface of an isotropic material, where it is partially transmitted and partially reflected. In the geometry of Fig. 2 we have for the spatial part

$$E(z) = E_t\, e^{i(\omega/c)N z} \qquad z > 0 \tag{12a}$$

$$E(z) = E_i\, e^{i(\omega/c)z} + E_r\, e^{-i(\omega/c)z} \qquad z < 0 . \tag{12b}$$

From Eqs. (12), the continuity condition at $z = 0$ of the electric field $E(z)$ parallel to the surface gives $E_t = E_i + E_r$. From Eq. (3a) and Eqs. (12), the continuity condition of the y-components of the magnetic field gives $N E_t = E_i - E_r$. The ratio E_r/E_i equals $(1 - N)/(1 + N)$; thus *the reflectivity R at normal incidence is*

$$R = \left|\frac{E_r}{E_i}\right|^2 = \left|\frac{1 - N}{1 + N}\right|^2 = \frac{(n - 1)^2 + k^2}{(n + 1)^2 + k^2} . \tag{13}$$

Another convenient measurable quantity for describing a surface is the *surface impedance* Z defined as

$$Z = \frac{4\pi}{c}\frac{E(0)}{B(0)} . \tag{14a}$$

From Eq. (3a) and Eq. (8a) we have also

$$Z = \frac{4\pi i \omega}{c^2}\left(\frac{E(z)}{dE(z)/dz}\right)_{z=0} = \frac{4\pi}{c N} . \tag{14b}$$

The relationships between optical constants are summarized in Table 1.

Before closing, it is very instructive to consider the implications of the occurrence $\varepsilon_1(\omega) > 0$ and $\varepsilon_2(\omega) = 0$ for some given ω region; then $n(\omega) = \sqrt{\varepsilon_1(\omega)}$ and $k(\omega) = 0$. From Eqs. (2) and Eq. (9), it is seen that the medium can sustain *undamped transverse*

Table 1 Relationships between optical constants

Conductivity	$\sigma = \sigma_1 + i\,\sigma_2$
Dielectric constant	$\varepsilon = \varepsilon_1 + i\,\varepsilon_2 \quad \varepsilon = 1 + \frac{4\pi i \sigma}{\omega} \quad \varepsilon_1 = 1 - \frac{4\pi\sigma_2}{\omega} \quad \varepsilon_2 = \frac{4\pi\sigma_1}{\omega}$
Refractive index	$N = n + i\,k \quad \varepsilon = N^2 \quad \varepsilon_1 = n^2 - k^2 \quad \varepsilon_2 = 2\,n\,k$
	$n^2 = (\varepsilon_1 + \sqrt{\varepsilon_1^2 + \varepsilon_2^2})/2 \quad k^2 = (-\varepsilon_1 + \sqrt{\varepsilon_1^2 + \varepsilon_2^2})/2$
Absorption coefficient	$\alpha(\omega) = 2\omega\,k/c \equiv \omega\varepsilon_2/nc$
Classical skin depth	$\delta = c/\omega\,k$
Surface impedance	$Z = 4\pi/cN$
Reflectivity	$R = \dfrac{(n-1)^2 + k^2}{(n+1)^2 + k^2} \quad$ (at normal incidence)

waves (constituted by an electromagnetic field and an induced current density) *of frequency ω and propagation vector* $\mathbf{q}$, with dispersion relation

$$q = \frac{\omega}{c}\,n\,, \quad \text{or equivalently} \quad \omega = \frac{c}{n}\,q = \frac{c}{\sqrt{\varepsilon_1(\omega)}}\,q\,;$$

notice that in the local response regime we are considering, it holds $\varepsilon_1(q,\omega) = \varepsilon_1(0,\omega) = \varepsilon_1(\omega)$. The undamped transverse waves of the system can be interpreted as elementary excitations of energy $\hbar\omega$ and momentum $\hbar\mathbf{q}$. [An example of these concepts has been considered in Section IX-7.2, in the study of the infrared dielectric properties of polar crystals and *polaritons*. We should also remember that in the case of longitudinal waves and longitudinal dielectric function, *plasmon* excitations are inferred from the vanishing of both ε_1 and ε_2, as discussed in particular in Section VII-6].

2 The Drude theory of the optical properties of free carriers

Consider a free-electron gas with n carriers per unit volume, each with effective mass m and charge $(-e)$; the carriers are embedded in a uniform background of neutralizing positive charge. The classical equation for the motion of an electron in a dissipative medium in the presence of an electric field $\mathbf{E}(\mathbf{r}, t) = \mathbf{E}_0 \exp(i\mathbf{q}\cdot\mathbf{r} - i\omega t)$ of frequency ω and wavevector $\mathbf{q}$ is

$$m\,\ddot{\mathbf{r}} = -\frac{m}{\tau}\dot{\mathbf{r}} + (-e)\,\mathbf{E}_0 e^{i(\mathbf{q}\cdot\mathbf{r}-\omega t)}\,, \tag{15}$$

where $\mathbf{r}(t)$ is the coordinate of the particle, and τ is a phenomenological relaxation time. The viscous damping term $-(m/\tau)\dot{\mathbf{r}}$ is introduced without specifying the dissipative mechanisms; in a wide sense it is due to the random collisions between the electron and whatever kind of impurities, phonons and imperfections in the crystal.

[Strictly speaking, the second term in the right-hand side of Eq. (15) should be accompanied by its complex conjugate, so that the electric field acting on the electron is real; for simplicity, we omit the complex conjugate term, since nothing would change either in the formal treatment or in the final results].

The term *spatial dispersion* has been coined to refer to the wavevector dependence of the electric field and optical constants. Suppose the spatial excursions of $\mathbf{r}(t)$, in the whereabouts of some point $\mathbf{r}_0$, are much smaller than the wavelength of the driving field; in this case, in the exponential in the right-hand side of Eq. (15) we can replace $\mathbf{r}(t)$ with $\mathbf{r}_0$ (without loss of generality we can take $\mathbf{r}_0 \equiv 0$), and we obtain

$$m\ddot{\mathbf{r}} = -\frac{m}{\tau}\dot{\mathbf{r}} - e\,\mathbf{E}_0 e^{-i\omega t} \ . \tag{16}$$

We give now a criterion to establish when the electron dynamics can be described by the simple Eq. (16), which neglects the spatial dispersion of the oscillating electric field. Consider first the case $\omega\tau \ll 1$, corresponding to a relaxation time τ much smaller than the period $T = 2\pi/\omega$ of the electromagnetic field. The displacement of an electron with typical Fermi velocity v_F between two collisions is about the mean free path $\Lambda_F = v_F\tau$; the spatial dispersion of the electromagnetic field of wavelength $\lambda = 2\pi/q$ is irrelevant if $\Lambda_F \ll \lambda$, or equivalently

$$v_F\tau \ll \frac{1}{q} \quad (\omega\tau \ll 1) \ . \tag{17a}$$

In the case $\omega\tau \gg 1$, the displacement of an electron with velocity v_F during the period $T = 2\pi/\omega$ of the electromagnetic field is $\approx v_F/\omega$; when $\omega\tau \gg 1$ the spatial dispersion of the electromagnetic field becomes irrelevant if

$$v_F\frac{1}{\omega} \ll \frac{1}{q} \quad (\omega\tau \gg 1) \ . \tag{17b}$$

We can thus conclude that Eq. (16) is justified when condition (17a), or condition (17b), is verified.

The conditions (17), here obtained on the basis of simple qualitative considerations, are confirmed by the study of the optical properties of metals with the Boltzmann equation (see Section 4.2). Such a study shows that the spatial dispersion of the electromagnetic field can be neglected provided the dimensionless parameter

$$s(q,\omega) = \frac{iq\,v_F\tau}{1 - i\omega\tau} \tag{18}$$

satisfies the condition

$$|s(q,\omega)| \ll 1 \ ; \tag{19}$$

the equivalence of (19) with (17a) and (17b) can be seen by inspection.

Having established the criterion (19) for the validity of Eq. (16), we insert $\mathbf{r}(t)$ in the form $\mathbf{r}(t) = \mathbf{A}_0 \exp(-i\omega t)$ into Eq. (16) and obtain

$$\mathbf{A}_0 = \frac{e\tau}{m}\frac{1}{\omega\,(i + \omega\tau)}\mathbf{E}_0 \ .$$

The free-carrier contribution to the current density is

$$\mathbf{J} = n\,(-e)\,\dot{\mathbf{r}} = n\,(-e)(-i\omega)\,\mathbf{A}_0\,e^{-i\omega t} = \frac{n\,e^2\tau}{m}\frac{1}{1-i\omega\tau}\mathbf{E}_0 e^{-i\omega t}\ .$$

The frequency dependent complex conductivity is thus

$$\sigma(\omega) = \frac{n\,e^2\tau}{m}\frac{1}{1-i\omega\tau} = \sigma_0\frac{1}{1-i\omega\tau}\ , \qquad (20)$$

where $\sigma_0 = n\,e^2\tau/m$ is the static conductivity. Notice that in general the conductivity $\sigma(q,\omega)$ should depend both on the frequency ω and on the wavevector q of the driving electric field; the Drude theory neglects spatial dispersion and thus provides only $\sigma(q \to 0, \omega)$, also denoted $\sigma(0,\omega)$ or simply $\sigma(\omega)$.

From Eq. (20) and Eq. (11), the complex dielectric constant becomes

$$\boxed{\varepsilon(\omega) = 1 - \frac{\omega_p^2}{\omega(\omega + i/\tau)}}\ , \qquad (21)$$

where ω_p denotes the free-electron plasma frequency

$$\omega_p^2 = \frac{4\pi\,n\,e^2}{m}\ , \qquad (22)$$

and n is the electron density. From Eq. (21) the real and imaginary parts of the dielectric function are

$$\boxed{\varepsilon_1(\omega) = 1 - \frac{\omega_p^2\tau^2}{1+\omega^2\tau^2}} \quad \text{and} \quad \boxed{\varepsilon_2(\omega) = \frac{\omega_p^2\tau}{\omega\,(1+\omega^2\tau^2)}}\ . \qquad (23)$$

Inserting into Eq. (22) the dimensionless parameter r_s, related to the electron density by the expression $(4/3)\pi r_s^3 a_B^3 = 1/n$, we obtain

$$\hbar^2\omega_p^2 = \frac{12}{r_s^3}\frac{\hbar^2}{2m\,a_B^2}\frac{e^2}{2\,a_B} \quad \text{and} \quad \hbar\omega_p = \sqrt{\frac{12}{r_s^3}}\ \text{Rydberg}\ .$$

For ordinary metals with $2 < r_s < 6$, typical values of $\hbar\omega_p$ are in the range 3 - 17 eV. For relaxation times we have typically $\tau \approx 10^{-14}$ sec, and the quantity $\gamma = \hbar/\tau$ is of the order of 0.1 eV.

In the study of the optical properties of free carriers in metals, we can roughly distinguish three frequency regions, called non-relaxation region ($\omega \ll 1/\tau$), relaxation region ($1/\tau \ll \omega \ll \omega_p$) and ultraviolet region ($\omega \approx \omega_p$ and $\omega > \omega_p$).

Non-relaxation region ($\omega\tau \ll 1 \ll \omega_p\tau$ or equivalently $\hbar\omega \ll \gamma \ll \hbar\omega_p$)

In the non-relaxation region $\omega\tau$ is negligible with respect to 1, and the dielectric function (23) becomes

$$\varepsilon_1(\omega) \approx -\omega_p^2\tau^2 \quad \text{and} \quad \varepsilon_2(\omega) \approx \frac{\omega_p^2\tau}{\omega}\ .$$

We see that $\varepsilon_1(\omega)$ is negative and constant, while $\varepsilon_2(\omega)$ is singular for $\omega \to 0$. Notice that $|\varepsilon_2| \gg |\varepsilon_1|$.

From the relation $N^2 = \varepsilon \approx i\varepsilon_2$, it follows $N \approx \varepsilon_2(1+i)/\sqrt{2}$. The refractive index n and the extinction coefficient k are almost equal and given by

$$n(\omega) \approx k(\omega) \approx \sqrt{\frac{\varepsilon_2(\omega)}{2}} = \sqrt{\frac{\omega_p^2 \tau}{2\omega}} .$$

Thus also $n(\omega)$ and $k(\omega)$ are singular for $\omega \to 0$. Since $n(\omega) \gg 1$ the reflectivity is almost 1 and the metal is strongly reflecting.

The penetration depth is

$$\delta(\omega) = \frac{c}{\omega\, k(\omega)} \approx \frac{c}{\omega_p}\sqrt{\frac{2}{\omega\tau}} \gg \frac{c}{\omega_p} .$$

To estimate the penetration depth in ordinary metals, assume as typical values $\tau \approx 10^{-14}$ sec, $\hbar\omega_p \approx 10$ eV and then $\omega_p \approx 10^{16}$ rad/sec ($\hbar = 6.852 \cdot 10^{16}$ eV·sec). The Debye length, defined as c/ω_p, is of the order of 10^{-6} cm; for $\omega \approx 100$ rad/sec the penetration depth is of the order of one cm; the penetration depth is of the order of the Debye length at the high frequency limit of the non-relaxation region.

Relaxation region ($1 \ll \omega\tau \ll \omega_p\tau$ or equivalently $\gamma \ll \hbar\omega \ll \hbar\omega_p$)

When $\omega\tau \gg 1$, we can perform a series expansion of $(1 + i/\omega\tau)^{-1}$ in Eq. (21) and the dielectric function becomes

$$\varepsilon(\omega) \approx 1 - \frac{\omega_p^2}{\omega^2} + i\frac{\omega_p^2}{\omega^3\tau} . \tag{24}$$

Thus $\varepsilon_1(\omega) \approx -\omega_p^2/\omega^2$ is still negative, and the imaginary part of the dielectric function is much smaller than the real part. From $N^2 = (n + ik)^2 = \varepsilon_1 + i\varepsilon_2$, and taking into account that $\varepsilon_1 < 0$ and $|\varepsilon_2| \ll |\varepsilon_1|$, we obtain

$$k(\omega) \approx \sqrt{-\varepsilon_1} = \frac{\omega_p}{\omega} \quad \text{and} \quad n(\omega) \approx \frac{1}{2}\frac{\varepsilon_2}{\sqrt{-\varepsilon_1}} = \frac{1}{2}\frac{\omega_p}{\omega^2\tau} .$$

The above relations show that $k(\omega) \gg n(\omega) \gg 1$; since the refractive index $n(\omega)$ is still much larger than 1, the reflectivity is almost 1 and the metal is strongly reflecting also in the relaxation region. The penetration depth is now approximately a constant, given by the Debye length.

Ultraviolet region ($\omega \approx \omega_p$ and $\omega > \omega_p$)

In this region, Eqs. (23) for the dielectric function read

$$\varepsilon_1(\omega) \approx 1 - \frac{\omega_p^2}{\omega^2} \quad \text{and} \quad \varepsilon_2(\omega) \approx \frac{\omega_p^2}{\omega^3\tau} \approx 0 .$$

We see that $\varepsilon_1(\omega)$ is positive for $\omega \geq \omega_p$; the reflectivity changes from (almost) one to (almost) zero, when ω increases above the plasma frequency, and the metal becomes transparent for $\omega > \omega_p$. The schematic behaviour of the dielectric function of a metal in the Drude model is given in Fig. 3.

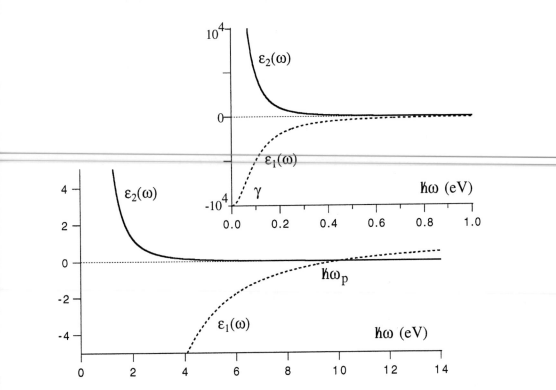

Fig. 3 Behaviour of real and imaginary part of the dielectric function of a free-electron gas with the Drude model; we have taken $\hbar\omega_p = 10\,\text{eV}$ and $\gamma = \hbar/\tau = 0.1\,\text{eV}$.

Illustrative examples of application of the Drude theory

The Drude model of intraband transitions describes an ideal system of free electrons, with spherical Fermi surface, a single relaxation time, in the local-response regime. In spite of these restrictions, the Drude theory applies reasonably well for the description of the optical properties of a number of metals in the red and infrared region, sufficiently below the threshold of interband electronic transitions. As an illustration, we consider here the optical constants of copper (Fig. 4 and Fig. 5) and alkali metals (Fig. 6); in these materials the Fermi surfaces are almost spheroid single sheets, and at sufficiently long wavelengths the optical properties are determined almost entirely by the free electrons of the sample.

In Fig. 4, the measured imaginary part of the dielectric function of copper is reported in the wavelength range from 0.365 to 2.5 micron ($\hbar\omega$ is in the energy range from 3.4 to 0.5 eV). At the low frequencies, $\varepsilon_2(\omega)$ decreases rapidly with increasing ω; then there is a sudden increase for $\hbar\omega \approx 2.1\,\text{eV}$, which marks the onset of interband electronic transitions; below this threshold, the Drude theory is very useful for an overall understanding of the optical constants.

The complex dielectric function $\varepsilon(\omega)$ of a metal, can be expressed as the sum of a Drude free-carrier contribution, given by Eq. (24) for $\omega\tau \gg 1$, and an interband contri-

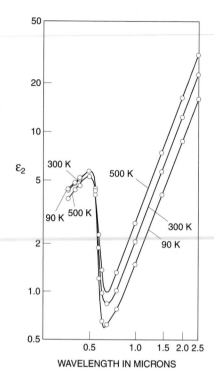

Fig. 4 Imaginary part of the dielectric function of copper at different temperatures [from S. Roberts, Phys. Rev. **118**, 1509 (1960); copyright 1960 by the American Physical Society].

bution; this latter, for frequencies sufficiently below the onset of interband transitions can be considered as a constant, here denoted as ε_{inter}; we have thus

$$\varepsilon(\omega) = 1 - \frac{\omega_p^2}{\omega^2} + i\,\frac{\omega_p^2}{\omega^3\tau} + \varepsilon_{inter}\ .$$

With the replacement $\omega = ck = 2\pi c/\lambda$, it follows

$$\varepsilon_1(\lambda) = 1 + \varepsilon_{inter} - \frac{\omega_p^2}{4\pi^2 c^2}\,\lambda^2 \quad \text{and} \quad \varepsilon_2(\lambda) = \frac{\omega_p^2}{8\pi^3 c^3\tau}\,\lambda^3\ ,$$

where λ is the free-space wavelength of light. From the above expressions, we expect that $\varepsilon_1(\lambda)$ varies linearly with λ^2, and the same occurs for $\varepsilon_2(\lambda)/\lambda$; we also expect that ε_2 is rather sensitive to the temperature and increases with temperature, since the relaxation time is expected to decrease with temperature. In Fig. 5 we report the experimental data of ε_1 and ε_2/λ versus λ^2 for copper in the infrared and red region: it can be noticed that the data fall almost exactly on straight lines, and also the temperature dependence of ε_2/λ is in agreement with what expected. The fact that the straight lines representing ε_2/λ do not pass through the origin, however, needs a proper account of the anomalous skin effect, and we refer to the paper of Roberts for specific considerations on this point.

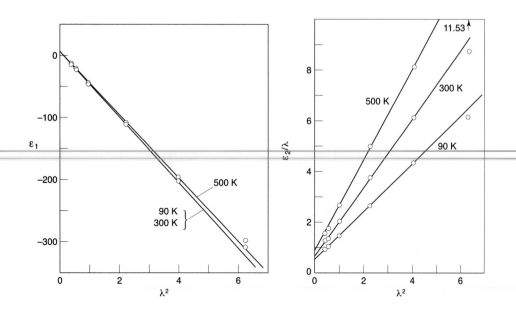

Fig. 5 Dielectric function of copper in the infrared and red region at different temperatures. (a) ε_1 versus square of wavelength (with λ in micron); (b) ε_2/λ versus square of wavelength (with λ in micron) [from S. Roberts, Phys. Rev. **118**, 1509 (1960); copyright 1960 by the American Physical Society].

As another example, we consider the optical properties of alkali metals in the region from 0.5 eV to 4 eV. In the relaxation region $\omega\tau \gg 1$, the optical conductivity $\sigma_1(\omega)$ for intraband transitions becomes

$$\sigma_1(\omega) = \frac{\omega\varepsilon_2(\omega)}{4\pi} \approx \frac{\omega_{\mathrm{p}}^2}{4\pi\tau\omega^2} \ .$$

The experimental values of the optical conductivity for four alkali metals are reported in Fig. 6; it can be seen that in the infrared region the simple dependence $\sigma_1(\omega) \sim \omega^{-2}$ is closely followed. Defining a phenomenological lifetime

$$\tau = \frac{\omega_{\mathrm{p}}^2}{4\pi\sigma_1(\omega)\omega^2} \ ,$$

then, if the Drude theory model is applicable, one expects that τ should be sensibly constant. Typical values of τ are of the order of 10^{-14} sec; however, from a close analysis of experimental data, it is seen that τ is not exactly a constant quantity, and depends somewhat on frequency. The optical absorption of alkali metals has been interpreted considering intraband and interband transitions within the nearly free-electron model, and we refer to the paper of Smith for further considerations.

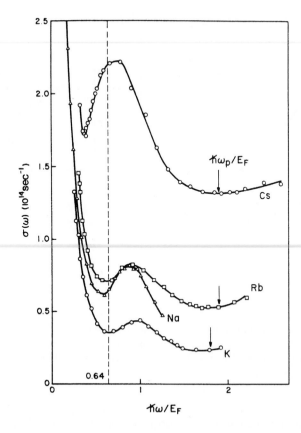

Fig. 6 Optical conductivity of the four alkali metals Na, K, Rb, Cs plotted against the normalized frequency $\hbar\omega/E_F$, where E_F is the free-electron Fermi energy [from N. V. Smith, Phys. Rev. B**2**, 2840 (1970); copyright 1970 by the American Physical Society].

Range of validity of the Drude theory

The Drude theory gives an overall comprehension of the intraband contribution to the optical constants of metals; however seldom if ever it is sufficient for a quantitative account of experimental observations (even in the frequency regions where interband contributions can be safely ignored or worked out separately). There are several reasons for this. (i) The conduction band structure of the electrons in actual metals may be significantly different from the spherical energy band assumption (with effective mass m), implicit in the Drude model. (ii) The relaxation time appears as an energy independent phenomenological parameter, while an analysis of scattering mechanisms imply an appropriate energy dependence. (iii) A more basic reason is the assumption of local-response regime. In fact the Drude theory is based on a local relationship between the density current and the electric field; this is justified only if the electric field is slowly varying in space, so that spatial dispersion can be neglected.

To put to a test the local-response assumption, let us consider the penetration depth of an electromagnetic wave in a metal, when the frequency ω is at the end of the non-relaxation region or at the beginning of the relaxation region. We have seen that the penetration depth (when the local-response assumption holds) is of the order of c/ω_p, a quantity that depends only on the electron density (via the plasma frequency); in ordinary metals we have $\delta \approx 10^{-6}$ cm. In a very pure specimen at helium temperature we can have $\tau \approx 10^{-11}$ sec or so (instead of $\tau \approx 10^{-14}$ sec in ordinary situations). The electron mean free path in very pure samples at very low temperature may become of the order $\Lambda_F = v_F \tau \approx 10^8 \times 10^{-11} \approx 10^{-3}$ cm. When $\Lambda_F \gg \delta$, the current density at the point $\mathbf{r}$ can no longer be determined by the electric field at the point $\mathbf{r}$ and the local approximation thus far considered is no more applicable.

We have seen that the Drude theory is justified in the regions (q, ω) for which $|s(q, \omega)| \ll 1$ (see Eq. 19); when this condition is violated and we have $|s(q, \omega)| \gg 1$, we are in the so-called *extreme anomalous region*. In very pure and good conductors at very low temperatures (copper for instance), this extreme anomalous region may extend from frequencies ω in the higher part of the non-relaxation region to include the lower part of the relaxation region. The determination of the optical constants when spatial dispersion has to be taken into account requires more sophisticated semiclassical approaches based on the Boltzmann transport equation, or the quantum mechanical treatment; these are the subjects of the next sections.

3 Transport properties and Boltzmann equation

In the study of intraband transport processes, we focus for simplicity on metals with a unique partially filled conduction band of energy $E=E(\mathbf{k})$ (multi-band situations, where two or more bands are partially filled with electrons, would require appropriate elaboration of the one-band model we are going to describe). At thermodynamic equilibrium, the probability of occupation of an energy level $E(\mathbf{k})$ is given by the Fermi–Dirac distribution function

$$f_0(\mathbf{k}) = \frac{1}{e^{(E(\mathbf{k})-\mu)/k_B T} + 1} , \qquad (25)$$

where T is the temperature of the sample and μ is the chemical potential (the chemical potential will be denoted by μ or by E_F, indifferently); $f_0(\mathbf{k})$ is a shorthand notation for $f_0(E(\mathbf{k}))$.

When external perturbations (electric fields, magnetic fields, temperature gradients) are applied to the sample, the electron distribution is disturbed from the equilibrium Fermi–Dirac function. In general the disturbed distribution function $f(\mathbf{r}, \mathbf{k}, t)$, in addition to $\mathbf{k}$, depends also on the real space coordinate $\mathbf{r}$, and on time t; the quantity $f(\mathbf{r}, \mathbf{k}, t)\, d\mathbf{r}\, d\mathbf{k}/4\pi^3$ gives the number of electrons at time t in the element of volume $d\mathbf{r}\, d\mathbf{k}$ around the point $(\mathbf{r}, \mathbf{k})$ of the "phase space".

According to the semiclassical dynamics of carriers in a given energy band, an electron in the point $(\mathbf{r}, \mathbf{k})$ at time t evolves toward the point $(\mathbf{r} + \mathbf{v}_\mathbf{k} dt, \mathbf{k} + (\mathbf{F}/\hbar)dt)$ at time $t + dt$, where

$$\mathbf{v}_\mathbf{k} = \frac{1}{\hbar} \frac{\partial E(\mathbf{k})}{\partial \mathbf{k}} , \qquad \frac{d(\hbar \mathbf{k})}{dt} = \mathbf{F} , \qquad (26)$$

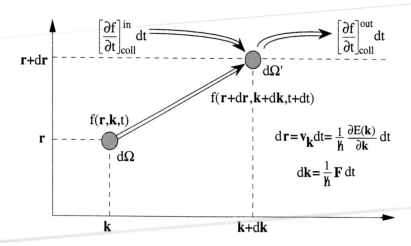

Fig. 7 Schematic representation of the conservation of the number of electrons moving in the space phase $\mathbf{r}, \mathbf{k}$. The region $d\Omega$ around $\mathbf{r}, \mathbf{k}$ at time t evolves into a new region $d\Omega'$, whose volume is the same as $d\Omega$ (Liouville theorem). The distribution function $f(\mathbf{r}+d\mathbf{r}, \mathbf{k}+d\mathbf{k}, t+dt)$ must thus equal $f(\mathbf{r}, \mathbf{k}, t)$ supplemented by the net change $[\partial f/\partial t]_{\text{coll}} dt = [\partial f/\partial t]^{\text{in}}_{\text{coll}} dt - [\partial f/\partial t]^{\text{out}}_{\text{coll}} dt$ of the number of electrons forced in and ejected out because of collision processes.

and $\mathbf{F}$ denotes the external force acting on the carriers. During the motion, collision processes (due for instance to lattice vibrations, impurities, boundaries) may cause a net rate of change $[\partial f/\partial t]_{\text{coll}}$ of the number of electrons in the phase space volume $d\mathbf{r}\, d\mathbf{k}$. Using Liouville theorem (volumes in phase space are preserved by the semiclassical equations of motion) we must have for the distribution function

$$f\left(\mathbf{r} + \mathbf{v}\, dt, \mathbf{k} + \frac{\mathbf{F}}{\hbar}\, dt, t + dt\right) \equiv f(\mathbf{r}, \mathbf{k}, t) + \left[\frac{\partial f}{\partial t}\right]_{\text{coll}} dt \; . \tag{27a}$$

Eq. (27a) expresses the detailed balance in each volume $d\mathbf{r}\, d\mathbf{k}$ of the number of carriers, when moving in the phase space under the action of external fields and in the presence of collision processes (as indicated in pictorial form in Fig. 7). The left member of Eq. (27a) can be expanded in Taylor series up to first order, and we obtain the Boltzmann equation

$$\frac{\partial f}{\partial \mathbf{r}} \cdot \mathbf{v} + \frac{\partial f}{\partial \mathbf{k}} \cdot \frac{\mathbf{F}}{\hbar} + \frac{\partial f}{\partial t} = \left[\frac{\partial f}{\partial t}\right]_{\text{coll}} , \tag{27b}$$

where $\partial f/\partial \mathbf{r}$ and $\partial f/\partial \mathbf{k}$ stand for $\nabla_{\mathbf{r}} f$ and $\nabla_{\mathbf{k}} f$, respectively.

A crucial aspect in the transport theory is just the collision term, which makes the Boltzmann equation (27b) a formidable integro-differential equation. When the deviation of f from the thermal equilibrium distribution f_0 is small, it is customary to assume that the rate of change of f due to collisions is proportional to the deviation itself, i.e.

$$\left[\frac{\partial f}{\partial t}\right]_{\text{coll}} = -\frac{f - f_0}{\tau} \; ,$$

where τ denotes the appropriate proportionality coefficient and is called *relaxation time*; in general $\tau = \tau(\mathbf{k}, \mathbf{r})$ depends on the energy $E(\mathbf{k})$ and on position, and is often considered a semi-empirical parameter. In the *relaxation time approximation*, the Boltzmann equation (27b) becomes the rather manageable partial differential equation

$$\boxed{\frac{\partial f}{\partial \mathbf{r}} \cdot \mathbf{v} + \frac{1}{\hbar}\frac{\partial f}{\partial \mathbf{k}} \cdot \mathbf{F} + \frac{\partial f}{\partial t} = -\frac{f - f_0}{\tau}} \; . \tag{28}$$

The role of the relaxation time can be further clarified by considering the case of sudden removal of external forces ($\mathbf{F} = 0$) at $t = 0$ in a uniform system ($\partial f / \partial \mathbf{r}) = 0$; from Eq. (28) one obtains

$$f(\mathbf{k}, t) = f_0 + [f(\mathbf{k}, t{=}0) - f_0]e^{-t/\tau} \; ,$$

and the distribution function moves to its equilibrium value f_0 exponentially with time constant τ.

The non-equilibrium distribution function f, evaluated with the Boltzmann equation, permits the investigation of a number of transport phenomena due to intraband electronic processes. Transport coefficients can be inferred from the general expression of electron current density and the energy flux density

$$\mathbf{J} = \frac{1}{4\pi^3} \int (-e)\, \mathbf{v_k}\, f\, d\mathbf{k} \tag{29a}$$

$$\mathbf{U} = \frac{1}{4\pi^3} \int E_\mathbf{k}\, \mathbf{v_k}\, f\, d\mathbf{k} \; , \tag{29b}$$

where the factor $2/(2\pi)^3$ takes into account spin degeneracy and density of allowed points in $\mathbf{k}$ space per unit volume. In the next section we apply the transport equations to establish the electrical conductivity in constant and oscillating electric fields; in particular we take into account the spatial dispersion of electric field and examine the extreme anomalous region. Successively, we describe the thermoelectric effects, due to the simultaneous presence of electric fields and temperature gradients.

As we shall see in the next sections, the Boltzmann transport equation describes a variety of transport phenomena; however, the description of transport phenomena by means of a distribution function is doomed to fail for dimension scales of the order of the de Broglie wavelength of the particles of the distribution. We do not dwell on this and other shortcomings of the Boltzmann transport equation, and refer to the literature for alternatives and techniques of quantum transport [see for instance M. A. Stroscio in "Introduction to Semiconductor Technology" edited by Cheng T. Wang (Wiley, New York 1990) and references quoted therein; see also Datta (1995)].

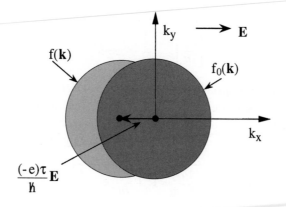

Fig. 8 Schematic representation of the equilibrium distribution function $f_0(\mathbf{k})$ and of the non-equilibrium distribution function $f(\mathbf{k}) \approx f_0(\mathbf{k} + e\tau\hbar^{-1}\mathbf{E})$ in the presence of a static electric field $\mathbf{E}$. The effect of the field is to shift the whole Fermi sphere by $-e\tau\hbar^{-1}\mathbf{E}$ in the $\mathbf{k}$ space. For sake of clarity of the figure, the value of $-e\tau\hbar^{-1}\mathbf{E}$ has been magnified and taken of the same order as k_F.

4   Static and dynamic conductivity in metals

4.1   Static conductivity with the Boltzmann equation

We consider the static conductivity of a metal by using the Boltzmann approach. For a homogeneous material in a uniform and steady electric field $\mathbf{E}$, the distribution function f depends only on $\mathbf{k}$ and the Boltzmann equation (28) becomes

$$\frac{1}{\hbar}\frac{\partial f}{\partial \mathbf{k}} \cdot (-e)\,\mathbf{E} = -\frac{f - f_0}{\tau} \ . \tag{30a}$$

For low electric fields, we can assume that $f - f_0$ is linear in the field strength, and we can thus put $f \approx f_0$ in the first member of Eq. (30a); we obtain

$$f = f_0 + \frac{e\tau}{\hbar}\frac{\partial f_0}{\partial \mathbf{k}} \cdot \mathbf{E} \ . \tag{30b}$$

In ordinary situations $e\tau E \ll \hbar k_\mathrm{F}$ (take for instance $E \approx 10^4\,\mathrm{volt/cm} = 10^{-4}\,\mathrm{volt/\mathring{A}}$, $\hbar/\tau \approx 0.1\,\mathrm{eV}$ and $k_\mathrm{F} \approx 1\,\mathring{A}^{-1}$). Thus Eq. (30b), at the lowest order in $e\tau E/\hbar k_\mathrm{F}$, can be written as $f(\mathbf{k}) \approx f_0(\mathbf{k} + e\tau\mathbf{E}/\hbar)$. The non-equilibrium distribution function $f(\mathbf{k})$ and the equilibrium distribution function $f_0(\mathbf{k})$ are schematically indicated in Fig. 8.

It is convenient to write $f = f_0 + f_1$, and recast Eq. (30b) in the form

$$f_1 = \frac{e\tau}{\hbar}\frac{\partial f_0}{\partial \mathbf{k}} \cdot \mathbf{E} = \frac{e\tau}{\hbar}\frac{\partial f_0}{\partial E}\frac{\partial E(\mathbf{k})}{\partial \mathbf{k}} \cdot \mathbf{E} = e\tau\frac{\partial f_0}{\partial E}\mathbf{v} \cdot \mathbf{E} \ ,$$

where $\mathbf{v} = (1/\hbar)\,\partial E(\mathbf{k})/\partial \mathbf{k}$ is the semiclassical expression of the velocity. Inserting f_1 in the expression (29a) of the current density (f_0 gives zero contribution), one obtains

$$\mathbf{J} = \frac{1}{4\pi^3}\int(-e)\,\mathbf{v}\,f_1\,d\mathbf{k} = \frac{e^2}{4\pi^3}\int\tau\left(-\frac{\partial f_0}{\partial E}\right)\mathbf{v}\,(\mathbf{v} \cdot \mathbf{E})\,d\mathbf{k} \ .$$

For simplicity we assume that the material is isotropic so that $\mathbf{J}$ and $\mathbf{E}$ are parallel, and the proportionality constant is the static conductivity σ_0; indicating by $\mathbf{e} = \mathbf{E}/|\mathbf{E}|$ the versor in the direction of the electric field, and projecting $\mathbf{J}$ along $\mathbf{e}$, we obtain

$$\sigma_0 = \frac{e^2}{4\pi^3} \int \tau \, (\mathbf{e} \cdot \mathbf{v})^2 (-\frac{\partial f_0}{\partial E}) \, d\mathbf{k} \; . \tag{31}$$

It is well known that the Fermi distribution function $f_0(E)$ changes sharply from unity to zero within a small interval ($\approx k_B T$) around the Fermi level, and that $(-\partial f_0/\partial E)$ is significantly different from zero in this same interval. Thus the conductivity σ_0 is basically determined from the features of the conduction bands in the thermal shell $k_B T$ around the Fermi energy E_F (this conclusion holds not only for the conductivity but also for any other transport coefficient in metals).

To estimate the conductivity given by expression (31), one can replace $(\mathbf{e} \cdot \mathbf{v})^2$ with $v^2/3$, and approximate $(-\partial f_0/\partial E)$ with $\delta(E(\mathbf{k}) - E_F)$; it follows

$$\sigma_0 = \frac{e^2}{12\pi^3} \int \tau \, v^2 \delta(E(\mathbf{k}) - E_F) \, d\mathbf{k} \; .$$

The presence of the δ-function in the integrand simplifies the three-dimensional integration in $d\mathbf{k}$ into the following integration over the Fermi surface

$$\sigma_0 = \frac{e^2}{12\pi^3} \int_{\substack{\text{Fermi} \\ \text{surface}}} \tau \, v^2 \frac{dS}{|\nabla_{\mathbf{k}} E(\mathbf{k})|} = \frac{e^2}{12\pi^3\hbar} \int_{\substack{\text{Fermi} \\ \text{surface}}} \tau \, v \, dS \; . \tag{32}$$

In the particular case of a parabolic conduction band with effective mass m^*, Eq. (32) gives

$$\sigma_0 = \frac{e^2}{12\pi^3\hbar} \tau_F \, v_F \, 4\pi k_F^2 = \frac{ne^2}{m^*} \tau_F \tag{33}$$

($v_F = \hbar k_F/m^*$ and $k_F^3 = 3\pi^2 n$). Notice that the rather naive result (33), in which *all the electrons seem to take part to the transport*, is valid only for parabolic bands; in reality, the generally valid Eq. (32) shows that *in any case* (parabolic or non-parabolic bands) only *the electrons at (or near) the Fermi surface can change their state under perturbations and are relevant in the transport phenomena.*

4.2   Frequency and wavevector dependence of the conductivity

In the previous section we have considered the conductivity in the presence of a steady and uniform electric field. We consider now the conductivity function in the presence of an electric field, periodic in space and time, given by

$$\mathbf{E}(\mathbf{r}, t) = \mathbf{E}_0 \, e^{i(\mathbf{q} \cdot \mathbf{r} - \omega t)} \; ;$$

at the moment we need not specify the relative orientations of $\mathbf{E}_0$ and $\mathbf{q}$ (for transverse fields $\mathbf{E}_0 \perp \mathbf{q}$, while for longitudinal fields $\mathbf{E}_0 \parallel \mathbf{q}$).

We start again from the Boltzmann equation (28) that now becomes

$$\frac{\partial f}{\partial \mathbf{r}} \cdot \mathbf{v} + \frac{1}{\hbar}\frac{\partial f}{\partial \mathbf{k}} \cdot (-e)\,\mathbf{E} + \frac{\partial f}{\partial t} = -\frac{f - f_0}{\tau} \ . \tag{34a}$$

As before, we write $f = f_0 + f_1$, and assume that f_1 is linear in the applied field; for consistency, we neglect the term $(1/\hbar)\,(\partial f_1/\partial \mathbf{k}) \cdot (-e)\,\mathbf{E}$ in Eq. (34a) and we have

$$\frac{\partial f_1}{\partial \mathbf{r}} \cdot \mathbf{v} + \frac{1}{\hbar}\frac{\partial f_0}{\partial \mathbf{k}} \cdot (-e)\,\mathbf{E} + \frac{\partial f_1}{\partial t} = -\frac{f_1}{\tau} \ . \tag{34b}$$

In an isotropic material, in the linear response regime, f_1 is assumed to have the same space and time dependence as the electric field; inserting

$$f_1(\mathbf{r}, \mathbf{k}, t) = \Phi(\mathbf{k})\,e^{i(\mathbf{q}\cdot\mathbf{r} - \omega t)}$$

into Eq. (34b), one obtains

$$i\,\mathbf{q}\cdot\mathbf{v}\,\Phi(\mathbf{k}) + \frac{1}{\hbar}\frac{\partial f_0}{\partial \mathbf{k}}(-e)\cdot\mathbf{E}_0 - i\,\omega\,\Phi(\mathbf{k}) = -\frac{\Phi(\mathbf{k})}{\tau} \ .$$

Thus, the function $\Phi(\mathbf{k})$ is given by

$$\Phi(\mathbf{k}) = \frac{e\,\tau\,\mathbf{v}\cdot\mathbf{E}_0}{1 - i\tau\,(\omega - \mathbf{q}\cdot\mathbf{v})}\,\frac{\partial f_0}{\partial E} \ .$$

The current density, evaluated with Eq. (29a), becomes

$$\mathbf{J} = \frac{1}{4\pi^3}\int (-e)\,\mathbf{v}\,f_1\,d\mathbf{k} = \frac{e^2}{4\pi^3}\int \tau\,(-\frac{\partial f_0}{\partial E})\,\mathbf{v}\,\frac{\mathbf{v}\cdot\mathbf{E}}{1 - i\tau\,(\omega - \mathbf{q}\cdot\mathbf{v})}\,d\mathbf{k} \ . \tag{35}$$

In the case of isotropy $\mathbf{J}$ and $\mathbf{E}$ are parallel, and the proportionality constant is the conductivity $\sigma(\mathbf{q}, \omega)$; indicating by $\mathbf{e} = \mathbf{E}/|\mathbf{E}|$ the polarization versor in the direction of the electric field, from Eq. (35) we obtain for the conductivity

$$\boxed{\sigma(\mathbf{q}, \omega) = \frac{e^2}{4\pi^3}\int \frac{\tau\,(\mathbf{e}\cdot\mathbf{v})^2}{1 - i\tau\,(\omega - \mathbf{q}\cdot\mathbf{v})}\,(-\frac{\partial f_0}{\partial E})\,d\mathbf{k}} \ . \tag{36}$$

Expression (36) for the frequency and wavevector dependent conductivity $\sigma(\mathbf{q}, \omega)$ is the natural generalization of Eq. (31), to which it reduces in the long wavelength limit $\mathbf{q} \to 0$ and static limit $\omega \to 0$.

Transverse conductivity for a spherical energy band

We now evaluate the conductivity in the case of a transverse electric field (with polarization vector $\mathbf{e}$ in x-direction and propagation vector $\mathbf{q}$ in z-direction) and spherical energy band; Eq. (36) becomes

$$\sigma(q, \omega) = \frac{e^2}{4\pi^3}\int \frac{\tau\,v_x^2}{1 - i\tau\,(\omega - q\,v_z)}\,(-\frac{\partial f_0}{\partial E})\,d\mathbf{k} \ . \tag{37a}$$

We approximate $(-\partial f_0/\partial E) \approx \delta(E(\mathbf{k}) - E_F)$, similarly to what done in Eq. (32), and obtain

$$\sigma(q,\omega) = \frac{e^2}{4\pi^3} \int_{\substack{\text{Fermi} \\ \text{surface}}} \frac{\tau\, v_x^2}{1 - i\omega\tau + iq\,v_z\,\tau} \frac{dS}{\hbar v_F} \ . \tag{37b}$$

The integral over the Fermi surface in Eq. (37b) can be performed analytically using polar coordinates and the standard relations

$$dS = k_F^2 \sin\theta\, d\theta\, d\phi \ , \qquad v_x = v_F \sin\theta \cos\phi \ , \qquad v_z = v_F \cos\theta \ .$$

Performing the integral over ϕ (which gives π) we have

$$\sigma(q,\omega) = \frac{e^2\tau}{4\pi^3\hbar} v_F\, k_F^2 \pi \int_0^\pi \frac{\sin^3\theta}{1 - i\omega\tau + iq\,v_F\,\tau \cos\theta}\, d\theta \ .$$

With the change of variable $x = \cos\theta$ the above expression becomes

$$\sigma(q,\omega) = \frac{3}{4}\sigma_0 \int_{-1}^{+1} \frac{1 - x^2}{1 - i\omega\tau + iq\,v_F\,\tau\,x}\, dx \ ,$$

where use has been made of the relationship $k_F^3 = 3\pi^2 n$ and $\sigma_0 = ne^2\tau/m$.

It is convenient to define

$$\boxed{ s = \frac{i\,q\,v_F\,\tau}{1 - i\omega\tau} } \ ,$$

a quantity whose physical meaning has been already discussed in Eq. (18) and Eq. (19). Using the indefinite integral

$$\int \frac{1 - x^2}{1 + s\,x}\, dx = -\frac{x^2}{2s} + \frac{x}{s^2} + \frac{s^2 - 1}{s^3}\ln(1 + s\,x) \ ,$$

the conductivity becomes

$$\boxed{ \sigma(q,\omega) = \frac{3}{4}\frac{\sigma_0}{1 - i\omega\tau}\left[\frac{2}{s^2} + \frac{s^2 - 1}{s^3}\ln\frac{1 + s}{1 - s}\right] } \ . \tag{38}$$

Expression (38) for the transverse conductivity holds on the whole (q,ω) plane.

It is very instructive to consider Eq. (38) in the limiting cases in which the discriminating parameter s takes the values $|s| \ll 1$ (Drude or normal q, ω region) and $|s| \gg 1$ (extreme anomalous q, ω region). In the normal region we have $|s| \ll 1$, $\ln[(1 + s)/(1 - s)] = 2s + (2/3)s^3 + \cdots$ and the expression in square brackets in Eq. (38) equals $4/3$. Thus we obtain for the conductivity the Drude result

$$\sigma(q,\omega) = \frac{\sigma_0}{1 - i\omega\tau} \qquad |s| \ll 1 \ . \tag{39a}$$

In the extreme anomalous region $|s| \gg 1$, the expression in square brackets in Eq. (38) equals $i\pi/s$, and the conductivity becomes

$$\sigma(q,\omega) = \frac{3}{4}\sigma_0 \frac{\pi}{v_F\,\tau\,q} \qquad |s| \gg 1 \ , \tag{39b}$$

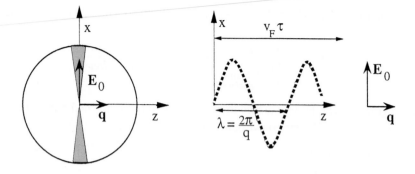

Fig. 9 Illustration of Pippard ineffectiveness concept. Out of all the electrons within the Fermi sphere, only those travelling (almost) parallel to $\mathbf{E}_0$ can absorb energy and contribute to the conductivity, in the extreme anomalous region.

a result which is *real, and independent both from frequency and from relaxation time* (we recall that $\sigma_0 = n e^2 \tau / m$). The conductivity in the extreme anomalous region is quite different from the conductivity of the Drude theory; the independence from ω and τ can be exploited to put in evidence features related to the shape of the Fermi surface.

The result (39b) can be qualitatively understood using Pippard ineffectiveness concept (see Fig. 9). Consider first a static and uniform electric field $\mathbf{E}_0$ (in the x-direction); in this case the conductivity is just $\sigma_0 = n e^2 \tau / m$, and all the electrons seem to participate to the transport. Consider now an oscillatory electric field with amplitude $\mathbf{E}_0$ and wavevector $\mathbf{q}$ (in the z-direction). We notice that $|s| \gg 1$ implies $q v_F \tau \gg 1$. It is evident that an electron with velocity $\mathbf{v}_F$ parallel to $\mathbf{q}$ sees an electric field that has changed its sign $\approx q v_F \tau$ times, before a collision takes place. Thus, if $q v_F \tau \gg 1$ we expect that the *effective electrons capable to absorb energy* are only the very electrons with velocity almost parallel to $\mathbf{E}_0$ such that $0 \leq \mathbf{q} \cdot \mathbf{v}_F \tau \leq 1$. These electrons are a fraction $f \approx 1/q v_F \tau$ of all the electrons and correspondingly σ_0 is reduced to $\approx \sigma_0 f$, in qualitative agreement with Eq. (39b).

4.3   Anomalous skin effect

We can now analyse quantitatively the propagation of an electromagnetic wave in a metal in general situations, including the *extreme anomalous region*. The basic relation between current density and electric field, derived from Maxwell's equations, is given by Eq. (4), here rewritten for convenience

$$\frac{d^2 E(z)}{dz^2} = -\frac{\omega^2}{c^2} E(z) - \frac{4\pi i \omega}{c^2} J(z) \ . \tag{40a}$$

We have already noticed that the local relationship $J(z) = \sigma(0, \omega) E(z)$ between current and field, expressed by Eq. (5), does not account for the spatial dispersion; for a proper account of spatial dispersion, one must consider the more general Eq. (6a) in real space, or Eq. (6b) in q space; the latter is here rewritten for convenience

$$J(q) = \sigma(q, \omega)\, E(q) \ , \tag{40b}$$

where $\sigma(q, \omega)$ is the frequency and wavevector dependent conductivity of the system, and $E(q)$ and $J(q)$ are the Fourier components of the electric field and current density, respectively.

The relation (40a) refers to the geometry of Fig. 2, where a metal fills half of the space $(z > 0)$. For simplicity we assume specular reflection of the electrons approaching the surface of the metal. The assumption of completely specular reflection is equivalent to consider the remaining half of space $(z < 0)$ filled with another piece of the same metal. The two pieces of metal are mirror images one of the other. The electric field is damped in both $+z$ and $-z$ directions and the boundary condition on the electric field is

$$\left(\frac{dE(z)}{dz}\right)_{0+} = -\left(\frac{dE(z)}{dz}\right)_{0-} \ . \tag{41}$$

A simple trick to account for the boundary condition (41) consists in adding to Eq. (40a) a delta-like term in the form

$$\frac{d^2 E(z)}{dz^2} = -\frac{\omega^2}{c^2} E(z) - \frac{4\pi\, i\, \omega}{c^2} J(z) + 2\left(\frac{dE(z)}{dz}\right)_{0+} \delta(z) \ . \tag{42}$$

In fact, if we multiply both members of Eq. (42) by dz, integrate from $-\varepsilon$ to $+\varepsilon$, and let $\varepsilon \to 0$, we obtain a result in agreement with Eq. (41).

It is convenient to expand $E(z)$, $J(z)$ and $\delta(z)$ in plane waves:

$$E(z) = \int_{-\infty}^{+\infty} E(q)\, e^{i q z}\, dq \tag{43a}$$

$$J(z) = \int_{-\infty}^{+\infty} J(q)\, e^{i q z}\, dq \tag{43b}$$

$$\delta(z) = \frac{1}{2\pi} \int_{-\infty}^{+\infty} e^{i q z}\, dq \ . \tag{43c}$$

We replace Eqs. (43) into Eq. (42), also use Eq. (40b) and obtain

$$E(q)\left[-q^2 + \frac{\omega^2}{c^2} + \frac{4\pi\, i\, \omega\, \sigma(q, \omega)}{c^2}\right] = \frac{1}{\pi}\left(\frac{dE(z)}{dz}\right)_{0+} \ .$$

Inserting the above result into Eq. (43a), it follows

$$E(z) = \frac{1}{\pi}\left(\frac{dE(z)}{dz}\right)_{0+} \int_{-\infty}^{+\infty} \frac{e^{i q z}}{-q^2 + \frac{\omega^2}{c^2} + \frac{4\pi\, i\, \omega\, \sigma(q, \omega)}{c^2}}\, dq \ . \tag{44}$$

Consider first Eq. (44) in the case σ does not depend on q. In this case the denominator in Eq. (44) is conveniently written as $-q^2 + (\omega^2/c^2)\, N^2$, where N^2 denotes the quantity $1 + 4\pi\, i\, \sigma/\omega$; the integral can be performed with the method of residues and the standard result of Eq. (8a) is recovered.

In the extreme anomalous region, the integral in Eq. (44) is more complicated; to give an idea of the results, we evaluate Eq. (44) in the case $z = 0$ and the displacement current term $(\omega/c)^2$ is negligible. Using expression (39b) for $\sigma(q,\omega)$, and the general property $\sigma(q,\omega) = \sigma(-q,\omega)$, one obtains

$$E(0) = \frac{2}{\pi}\left(\frac{dE}{dz}\right)_{0^+}\int_0^{+\infty}\frac{1}{-q^2 + i\frac{A}{q}}\,dq = \frac{2}{\pi}\left(\frac{dE}{dz}\right)_{0^+}\frac{1}{A^{1/3}}\int_0^{+\infty}\frac{x}{-x^3 + i}\,dx \quad (45a)$$

with

$$A = \frac{4\pi\omega}{c^2}\frac{3\pi\sigma_0}{4v_F\tau} = \frac{3\pi^2\omega\sigma_0}{c^2 v_F\tau}\,.$$

The integral in Eq. (45a) can be carried out with the help of the definite integral

$$\int_0^{+\infty}\frac{x^{\mu-1}}{x^\nu + 1}\,dx = \frac{\pi}{\nu}\frac{1}{\sin(\mu\pi/\nu)}\qquad \mathrm{Re}\,\nu \geq \mathrm{Re}\,\mu > 0$$

[see I. S. Gradshtein and I. M. Ryzhik "Table of Integrals, Series and Products" (Academic Press, New York 1980) page 292]. We have

$$\int_0^\infty\frac{x}{-x^3 + i}\,dx = -\int_0^\infty\frac{x^4}{x^6 + 1}\,dx - i\int_0^\infty\frac{x}{x^6 + 1}\,dx = -\frac{\pi\sqrt{3}}{9}(\sqrt{3} + i)\,. \quad (45b)$$

From Eqs. (45) and Eq. (14b), one obtains for the surface impedance in the anomalous region the expression

$$Z_{an} = \frac{4\pi i\omega}{c^2}\left(\frac{E}{dE(z)/dz}\right)_{0^+} = \frac{8}{9}\left(\frac{\sqrt{3}\pi\,\omega^2\,v_F\,\tau}{c^4\,\sigma_0}\right)^{1/3}(1 - i\sqrt{3})\,. \quad (46)$$

The most interesting aspect of expression (46) is that it does not depend on the relaxation time; in fact

$$\frac{\sigma_0}{v_F\tau} = \frac{n\,e^2}{m^*\,v_F} = \frac{k_F^3}{3\pi^2\,\hbar\,k_F}\frac{e^2}{} = \frac{e^2}{3\pi^2\,\hbar}k_F^2\,.$$

In the extreme anomalous region, measurements of the surface resistance (real part of the surface impedance) provide an accurate tool for the investigation of the Fermi wavevector.

Thus far, only spherical Fermi surfaces have been discussed. The general problem of an arbitrary Fermi surface has been treated by Pippard using the ineffectiveness concept. As one might expect, a great deal can be learned on the shapes of the Fermi surfaces from measurements of surface resistance for different crystallographic orientations. As an example, we report in Fig. 10 the surface resistance of copper for different orientations, in the extreme anomalous region. In spite of the fact that copper is an fcc cubic crystal, the surface resistances with orientation (100) and (110) differ by about a factor two. As pointed out by Pippard, the shape of the Fermi surface, that best fits the measurements of Fig. 10, is an almost spheroidal single sheet, which makes contact with the Brillouin zone hexagonal faces (see Fig. VI-14); the study of the anomalous skin effect has provided a direct experimental evidence of the open nature of the Fermi surface of copper.

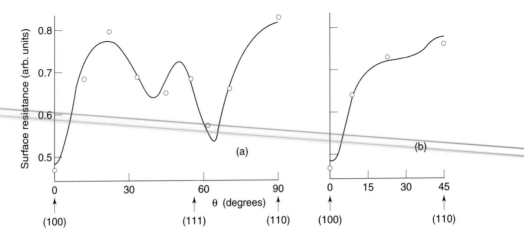

Fig. 10 Surface resistance (arbitrary units) versus orientation for copper, in the extreme anomalous region. The angle θ subtends from (100) to the (110) direction, following two different arrangements (a) and (b) [from A. B. Pippard, Phil. Trans. Roy. Soc. (London) A**250**, 325 (1957)].

5 Boltzmann treatment and quantum treatment of intraband transitions

The Boltzmann treatment of intraband transport phenomena is based on the semiclassical equations of motion, given by Eqs. (26), and proper account of collision effects; the Boltzmann transport equation has thus a semiclassical nature. We wish now to compare the Boltzmann treatment of intraband transport phenomena with a full quantum mechanical treatment. This comparison is very instructive because it clarifies the (explicit or tacit) simplifications of the semiclassical theory and thus puts in evidence both its merits and its limitations. In particular it is instructive to analyse under which circumstances a semiclassical treatment, which *takes into account the band energy dispersion but ignores wavefunctions*, may become equivalent to a correct quantum mechanical treatment, whose ingredients are *both energies and wavefunctions*.

The linear response of a system to an electromagnetic field of frequency ω, wavevector $\mathbf{q}$ and polarization vector $\mathbf{e}$, can be obtained quantum mechanically by first-order perturbation theory, and is reported in Chapter XII in its general form. The quantum theory describes on the same footing intraband and interband transitions; here we wish to show that the Boltzmann semiclassical approach constitutes a significant elaboration and simplification of the full quantum mechanical treatment of intraband transitions.

To avoid inessential details, consider the simplest model for a metal, with a unique conduction band of energy dispersion $E_\mathbf{k}$ and Bloch wavefunctions $\psi_\mathbf{k}$. In this specific one-band model, the transverse conductivity $\sigma(\mathbf{q}, \omega)$ provided by the Kubo linear response theory is given by Eq. (XII-15), and reads

$$\sigma(\mathbf{q}, \omega) = \frac{2e^2}{m^2} \int \frac{d\mathbf{k}}{(2\pi)^3} \frac{|\langle \psi_{\mathbf{k}+\mathbf{q}} | e^{i\,\mathbf{q}\cdot\mathbf{r}}\,\mathbf{e}\cdot\mathbf{p} | \psi_{\mathbf{k}} \rangle|^2}{(E_{\mathbf{k}+\mathbf{q}} - E_{\mathbf{k}})/\hbar} \frac{(-i)\,[f_0(E_{\mathbf{k}}) - f_0(E_{\mathbf{k}+\mathbf{q}})]}{E_{\mathbf{k}+\mathbf{q}} - E_{\mathbf{k}} - \hbar\omega - i\eta} \;, \quad (47)$$

where f_0 is the Fermi–Dirac equilibrium distribution function, and the limit $\eta \to 0^+$ is understood.

We now make on Eq. (47) some approximations, which are justified for sufficiently small wavevectors. The change of the distribution function is approximated with the series development

$$f_0(E_{\mathbf{k}+\mathbf{q}}) - f_0(E_{\mathbf{k}}) \approx \frac{\partial f_0}{\partial E}(E_{\mathbf{k}+\mathbf{q}} - E_{\mathbf{k}}) \;; \quad (48a)$$

the matrix elements in Eq. (47) can be simplified assuming $\psi_{\mathbf{k}+\mathbf{q}} \approx \exp(i\,\mathbf{q}\cdot\mathbf{r})\,\psi_{\mathbf{k}}$ so to obtain

$$\frac{1}{m}\langle \psi_{\mathbf{k}+\mathbf{q}} | e^{i\,\mathbf{q}\cdot\mathbf{r}}\,\mathbf{p} | \psi_{\mathbf{k}} \rangle \approx \frac{1}{m}\langle \psi_{\mathbf{k}} | \mathbf{p} | \psi_{\mathbf{k}} \rangle = \mathbf{v} \;. \quad (48b)$$

With the approximations (48), the conductivity becomes

$$\sigma(\mathbf{q}, \omega) = \frac{e^2\,\hbar}{4\,\pi^3} \int d\mathbf{k}\,(\mathbf{e}\cdot\mathbf{v})^2 \frac{-i}{E_{\mathbf{k}+\mathbf{q}} - E_{\mathbf{k}} - \hbar\omega - i\eta}\left(-\frac{\partial f_0}{\partial E}\right) \;.$$

We now expand $E_{\mathbf{k}+\mathbf{q}} - E_{\mathbf{k}} \approx \mathbf{q}\cdot(\partial E/\partial \mathbf{k}) = \mathbf{q}\cdot\mathbf{v}\hbar$, and make the identification $\eta = \hbar/\tau$; the expression for the conductivity becomes

$$\sigma(\mathbf{q}, \omega) = \frac{e^2}{4\,\pi^3} \int \frac{\tau\,(\mathbf{e}\cdot\mathbf{v})^2}{1 - i\,\tau\,(\omega - \mathbf{q}\cdot\mathbf{v})}\left(-\frac{\partial f_0}{\partial E}\right) d\mathbf{k} \;. \quad (49)$$

The frequency and wavevector dependent conductivity (49), obtained with the above elaboration of the quantum mechanical expression, coincides with Eq. (36) obtained on the basis of semiclassical Boltzmann equation; the present discussion has the merit to show the limits of applicability and validity of the semiclassical picture. With similar considerations, it is possible (and instructive) to compare the longitudinal conductivity produced by the Boltzmann equation with the longitudinal conductivity (or longitudinal dielectric function) calculated quantum mechanically in Chapter VII.

6   The Boltzmann equation in electric fields and temperature gradients

6.1   The transport equations in general form

In the previous sections we have studied transport effects due to the presence of electric fields in samples at uniform temperature (i.e. in *isothermal* conditions); we consider now transport equations in the presence of electric fields and temperature gradients. As usual, we consider the simplest possible electronic structure of the metal with a unique conduction band of interest of energy $E(\mathbf{k})$; the influence (if any) of the temperature on the energy band structure $E(\mathbf{k})$ is assumed to be negligible.

In a crystal kept at non-uniform temperature, it is convenient to define the local equilibrium distribution function $f_0(\mathbf{k}, \mathbf{r})$ as

$$f_0(\mathbf{k}, \mathbf{r}) = \frac{1}{\exp[(E(\mathbf{k}) - \mu(\mathbf{r}))/k_B T(\mathbf{r})] + 1} \; ; \tag{50a}$$

the local equilibrium distribution function $f_0(\mathbf{k}, \mathbf{r})$, in addition to $\mathbf{k}$, depends implicitly on $\mathbf{r}$ since the local temperature $T = T(\mathbf{r})$ is a function of $\mathbf{r}$, and the chemical potential $\mu = \mu(T(\mathbf{r}), n(\mathbf{r})) = \mu(\mathbf{r})$ depends on $\mathbf{r}$ via the local temperature $T(\mathbf{r})$ and the local electron density $n(\mathbf{r})$. Notice that the local chemical potential $\mu(\mathbf{r})$ at the point $\mathbf{r}$, entering in Eq. (50), is just the *chemical potential of an ideal infinite sample at thermodynamic equilibrium, characterized by band structure $E(\mathbf{k})$, uniform temperature T equal to $T(\mathbf{r})$, and uniform electron density n equal to $n(\mathbf{r})$.*

In the following we need the gradients of f_0 with respect to $\mathbf{k}$ and with respect to $\mathbf{r}$; these are given by

$$\frac{\partial f_0}{\partial \mathbf{r}} = \frac{\partial f_0}{\partial E} k_B T \frac{\partial}{\partial \mathbf{r}} \frac{E - \mu}{k_B T} = \frac{\partial f_0}{\partial E} \left[-\frac{E}{T} \frac{\partial T}{\partial \mathbf{r}} - T \frac{\partial}{\partial \mathbf{r}} \frac{\mu}{T} \right] , \tag{50b}$$

and

$$\frac{\partial f_0}{\partial \mathbf{k}} = \frac{\partial f_0}{\partial E} \frac{\partial E(\mathbf{k})}{\partial \mathbf{k}} = \frac{\partial f_0}{\partial E} \hbar \mathbf{v} . \tag{50c}$$

The Boltzmann equation (28) for the stationary distribution $f(\mathbf{r}, \mathbf{k})$ in the presence of an electric field $\mathbf{E}$ and temperature gradient is

$$\frac{\partial f}{\partial \mathbf{r}} \cdot \mathbf{v} + \frac{1}{\hbar} \frac{\partial f}{\partial \mathbf{k}} \cdot (-e) \mathbf{E} = -\frac{f - f_0}{\tau} = -\frac{f_1}{\tau} . \tag{51}$$

Since the electric field and temperature gradient are usually small, we can assume that f_1 is linear in these variables; then we can put $f = f_0$ on the left hand side of Eq. (51) and obtain

$$\frac{\partial f_0}{\partial \mathbf{r}} \cdot \mathbf{v} + \frac{1}{\hbar} \frac{\partial f_0}{\partial \mathbf{k}} \cdot (-e) \mathbf{E} = -\frac{f_1}{\tau} .$$

Using Eqs. (50), the stationary non-equilibrium distribution function becomes

$$f_1 = \left(-\frac{\partial f_0}{\partial E} \right) \tau \left[-e \mathbf{E} - T \nabla \frac{\mu}{T} \right] \cdot \mathbf{v} + \left(-\frac{\partial f_0}{\partial E} \right) \tau(-E) \frac{\nabla T}{T} \cdot \mathbf{v} ,$$

where $\nabla = \partial/\partial \mathbf{r}$ indicates the gradient with respect to space variable $\mathbf{r}$.

We can now use the general expressions (29) for the current density and the energy flux. Furthermore (though not strictly necessary) we suppose that the system is isotropic, so that transport kinetic parameters become scalar quantities, rather than tensors. We obtain

$$\mathbf{J} = e K_0 \left[e \mathbf{E} + T \nabla \frac{\mu}{T} \right] + e K_1 \frac{\nabla T}{T} \tag{52a}$$

$$\mathbf{U} = -K_1 \left[e \mathbf{E} + T \nabla \frac{\mu}{T} \right] - K_2 \frac{\nabla T}{T} . \tag{52b}$$

The expressions of the *kinetic coefficients* K_0, K_1, K_2 are

$$K_n = \frac{1}{4\pi^3} \int \tau\,(\mathbf{e}\cdot\mathbf{v})^2\, E^n \left(-\frac{\partial f_0}{\partial E}\right) d\mathbf{k} \quad (n = 0, 1, 2) ,$$

where $\mathbf{e}$ is the versor in the direction of the electric field.

Evaluation of the transport coefficients

The transport coefficients K_n can be evaluated replacing $(\mathbf{e}\cdot\mathbf{v})^2$ by $v^2/3$, and then breaking the three-dimensional integral in $d\mathbf{k}$ into a two-dimensional integral on constant energy surfaces and an integration on the energy variable; we have

$$K_n = \frac{1}{12\pi^3} \int \left(-\frac{\partial f_0}{\partial E}\right) E^n\, dE \int_{E=\text{const}} \frac{\tau\, v^2}{|\nabla_{\mathbf{k}} E(\mathbf{k})|}\, dS , \tag{53}$$

where $|\nabla_{\mathbf{k}} E(\mathbf{k})| = \hbar v$. With an eye to Eq. (32), it is convenient to define the generalized conductivity $\sigma(E)$ as

$$\sigma(E) = \frac{e^2}{12\pi^3\hbar} \int_{E=\text{const}} \tau\, v\, dS ; \tag{54}$$

notice that $\sigma(E_{\text{F}}) \equiv \sigma_0$ is the standard conductivity of a metal. Eq. (53) takes the form

$$e^2 K_n = \int \left(-\frac{\partial f_0}{\partial E}\right) E^n\, \sigma(E)\, dE .$$

For the calculation of the integrals appearing in the second member of the above equation, we use the Sommerfeld expansion of Section III.2, here rewritten as

$$\int_0^\infty \left(-\frac{\partial f_0}{\partial E}\right) g(E)\, dE = g(\mu) + \frac{\pi^2}{6} k_B^2 T^2 \left(\frac{d^2 g}{dE^2}\right)_{E=\mu} + O(T^4) ,$$

with $g(E) = E^n\sigma(E)$. The transport coefficients to $O(T^4)$ then become

$$e^2 K_0 = \sigma(\mu) + \frac{\pi^2}{6} k_B^2 T^2 \sigma''(\mu) \tag{55a}$$

$$e^2 K_1 = \mu\,\sigma(\mu) + \frac{\pi^2}{6} k_B^2 T^2 \left[2\sigma'(\mu) + \mu\sigma''(\mu)\right] \tag{55b}$$

$$e^2 K_2 = \mu^2\,\sigma(\mu) + \frac{\pi^2}{6} k_B^2 T^2 \left[2\,\sigma(\mu) + 4\mu\sigma'(\mu) + \mu^2\sigma''(\mu)\right] , \tag{55c}$$

where first and second derivatives of $\sigma(E)$ are calculated at the Fermi energy $E = \mu$.

Now that we have established the expressions (55) for the transport coefficients, it is convenient to rewrite the two basic transport equations (52) in a slightly different form, which is more convenient for the interpretation of the thermoelectric phenomena. Eq. (52a) can be written in the more significant form

$$\boxed{\mathbf{J} = e^2 K_0 \left[\mathbf{E} + \frac{1}{e}\nabla\mu - S(T)\,\nabla T\right]} , \tag{56a}$$

where the transport coefficient $S(T)$ is given by

$$S(T) = \frac{1}{(-e)T} \left(\frac{K_1}{K_0} - \mu \right) = \frac{\pi^2}{3} \frac{k_B^2 T}{(-e)} \frac{\sigma'(\mu)}{\sigma(\mu)} \;. \tag{56b}$$

The last passage in Eq. (56b) is obtained using Eqs. (55), and considering only the leading term in temperature. The coefficient $S(T)$ is called *absolute thermoelectric power or Seebeck coefficient.*

From Eq. (56a) we see that the current density $\mathbf{J}$ consists of three contributions. The first term $e^2 K_0 \mathbf{E}$ is the standard drift term $\sigma_0 \mathbf{E}$, where σ_0 is the conductivity of the metal ($e^2 K_0 = \sigma_0$ apart small corrections of order T^2/T_F^2). The second one is due to the inhomogeneity (i.e. to the $\mathbf{r}$-dependence) of the chemical potential entering into Eq. (50a). The third is due to the presence of a temperature gradient. It is interesting to notice that the energy dissipated per unit time and unit volume $\mathbf{E} \cdot \mathbf{J}$, besides the essentially positive joule term J^2/σ_0 (irreversible heat), contains two additional terms linear in J, which can be either positive or negative (reversible heat).

For what concerns the transport equation for the energy flux, it is convenient to obtain $[e \, \mathbf{E} + T \, \nabla \, (\mu/T)]$ from Eq. (52a) and replace it in Eq. (52b). We have

$$\boxed{\mathbf{U} = \frac{K_1}{(-e) \, K_0} \mathbf{J} - k_e \, \nabla T} \;, \tag{57a}$$

where the transport parameter k_e (which is called *electron thermal conductivity*) is given by

$$k_e = \frac{1}{T} \left(K_2 - \frac{K_1^2}{K_0} \right) \equiv \frac{\pi^2}{3} \frac{k_B^2}{e^2} T \sigma_0 \;. \tag{57b}$$

We discuss now some applications of Eqs. (56) and Eqs. (57), which are the basic equations of transport in isotropic materials.

Drift and diffusion currents in isothermal conditions

As a first application, consider the electron current density in a metal in isothermal conditions, but with a non-uniform carrier concentration $\nabla n \neq 0$; this implies $\nabla \mu \neq 0$. Putting $\nabla T = 0$ into Eq. (56a) we have

$$\mathbf{J} = \sigma_0 \left[\mathbf{E} + \frac{1}{e} \nabla \mu \right] \;. \tag{58}$$

We can thus distinguish a *drift current density* $\mathbf{J}_{\text{drift}} = \sigma_0 \, \mathbf{E}$ and a *diffusion current density* $\mathbf{J}_{\text{diff}} = \sigma_0 \nabla \mu / e$.

Let us consider for sake of simplicity a free-electron-like conduction band for a metal. We have in this case $\sigma_0 = n \, e^2 \tau / m^*$, $\mu = (\hbar^2/2m^*)(3\pi^2 n)^{2/3}$ and $\nabla \mu / \mu = (2/3) \, \nabla n / n$. The current density (58) in the metal can thus be written as

$$\boxed{\mathbf{J} = n \, e \, \mu_e \, \mathbf{E} + e \, D \, \nabla n} \;, \tag{59}$$

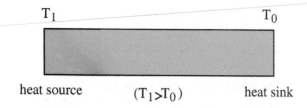

Fig. 11 Schematic representation of a bar of homogeneous material, whose ends are kept at different temperatures.

where $\mu_e = e\tau/m^*$ is the *electron mobility*, and D is the *diffusion coefficient*

$$D = \frac{2}{3}\frac{E_F}{e}\mu_e \; .$$
(60a)

In the case the free-electron gas is non degenerate and follows the Boltzmann distribution, we have $\nabla n/n = \nabla\mu/k_B T$. The current density is again given by Eq. (59), but now the diffusion coefficient becomes

$$D = \frac{k_B T}{e}\mu_e \; .$$
(60b)

Relations (60a) and (60b) are the *Einstein relations* between mobility and diffusion coefficient for the degenerate and non-degenerate electron gas, respectively.

6.2   Thermoelectric phenomena

Thermal conductivity of electrons in metals

Consider a metal in the presence of a uniform temperature gradient ∇T and in open circuit situation, so that $\mathbf{J} = 0$ (see Fig. 11); in this case, Eqs. (57) take the form

$$\mathbf{U} = -k_e\,\nabla T \quad \text{with} \quad k_e = \frac{\pi^2}{3}\frac{k_B^2}{e^2}T\,\sigma_0 \; .$$
(61)

Thus the energy flows in the direction opposite to ∇T. From the expression of the electron thermal conductivity k_e, we see that the ratio of thermal to electrical conductivity is proportional to T (Wiedemann–Franz law).

Consider now the ratio

$$\frac{k_e}{T\,\sigma_0} = L \equiv \frac{\pi^2}{3}\frac{k_B^2}{e^2} \; ,$$

which is known as *Lorentz number*. The Lorentz number would actually be a universal constant (independent from the specific metal, temperature and relaxation time), provided the approximations done in the transport equations are justified. If one goes over the whole treatment, one realizes that the most vulnerable point is the relaxation time approximation of the collision term. This approximation is justified above the Debye temperature, where the electron-phonon scattering is the dominant process,

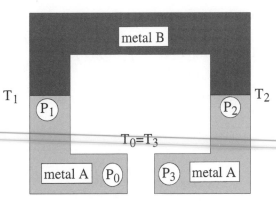

Fig. 12 Standard bimetallic circuit to measure the thermoelectric effect. The two junctions between the metals are kept at different temperatures $(T_1 \neq T_2)$; a voltage appears between points P_0 and P_3.

and also at very low temperature, where the impurity scattering is dominant. In both temperature regimes, the ratio $k_e/T\,\sigma_0$ is approximately the same for all metals. At intermediate temperatures, however significant deviations may occur.

Seebeck effect and thermoelectric power

When a temperature gradient is established in a long bar (in open circuit situation) an electric field has to set in, so to prevent any net carrier flux. Consider in fact a specimen with a cool end at temperature T_0 and a hot end at temperature T_1 (see Fig. 11). In open circuit situation $\mathbf{J} = 0$ and the electric field can be obtained from Eq. (56a) in the form

$$\mathbf{E} = -\frac{1}{e}\nabla\mu + S(T)\,\nabla T \ , \tag{62}$$

where $S(T)$ is given by Eq. (56b).

The thermoelectric power $S(T)$ of a material can be measured using the standard bimetallic circuit of Fig. 12, in which the two juntions are kept at different temperatures. Using Eq. (62), it is easy to evaluate the potential difference at the extremal points P_0 and P_3 (kept at the same temperature). We have

$$V_3 - V_0 = -\int_{P_0}^{P_3} \mathbf{E}\cdot d\mathbf{l} = \frac{1}{e}\int_{P_0}^{P_3}\nabla\mu\cdot d\mathbf{l} - \int_{P_0}^{P_3} S(T)\,\nabla T\cdot d\mathbf{l} \ , \tag{63}$$

where the integral can be performed along any line going from P_0 to P_3 within the circuit. In Eq. (63), the line integral involving the gradient of the chemical potential from the point P_0 to the point P_3 vanishes, since $T_3 \equiv T_0$. For the line integral involving the thermoelectric power we have:

$$\int_{P_0}^{P_3} S(T)\,\nabla T \cdot d\mathbf{l} = \int_{P_0}^{P_1} S_A(T)\,dT + \int_{P_1}^{P_2} S_B(T)\,dT + \int_{P_2}^{P_3} S_A(T)\,dT$$

$$= \int_{T_0}^{T_1} S_A(T)\,dT + \int_{T_1}^{T_2} S_B(T)\,dT + \int_{T_2}^{T_3 \equiv T_0} S_A(T)\,dT = \int_{T_1}^{T_2} [S_B(T) - S_A(T)]\,dT \ .$$

We obtain

$$\boxed{V_3 - V_0 = \int_{T_1}^{T_2} S_A(T)\,dT - \int_{T_1}^{T_2} S_B(T)\,dT} \ . \tag{64}$$

Thus if we choose a material A with $S_A(T)$ known (often lead is taken because its thermoelectric power is negligible) and vary T_2 with respect to T_1 we can obtain an experimental determination of $S_B(T)$, by measuring the potential difference $V_3 - V_0$.

We give now an order of magnitude estimate of $S(T)$ in metals, noticing however that the thermoelectric powers of actual materials are rather sensitive to the energy-dependence of the relaxation time and to the peculiarities of the Fermi surface. From Eq. (56b), we have that the thermoelectric power can be either negative or positive depending on the sign of $d\sigma/dE$ at the Fermi energy. To evaluate the order of magnitude of the thermoelectric coefficient, given by Eq. (56b), let us suppose that the generalized conductivity has a power law form of the type $\sigma(E) \approx C\,E^p$ ($p = 3/2$ for the free-electron gas). Then we can estimate

$$\frac{\sigma'(\mu)}{\sigma(\mu)} \approx \frac{1}{\mu} = \frac{1}{k_B T_F} \quad \text{and} \quad S(T) \approx \frac{k_B}{(-e)}\frac{T}{T_F} \ .$$

With $T \approx 300\,\mathrm{K}$, $T_F \approx 100\,T$, $k_B/(-e) = -0.48 \times 10^{-4}$ volt/K we expect a thermoelectric power of the order of -10^{-6} volt/K in normal metals at room temperature.

Thomson effect

When an electric current flows in a given homogeneous material in the presence of a temperature gradient, heat is released or absorbed reversibly at a rate depending on the current density and on the nature of the material; if the direction of current is reversed, the Thomson effect also changes sign (contrary to the Joule heating effect).

To study the Thomson effect, we imagine that temperature gradient, electric field and density current depend on a single direction (say x), and we consider a small cylinder of section $d\Sigma$ and length dl with its axis parallel to $\mathbf{J}$, and two sections at the temperatures T_A and T_B, respectively; for simplicity we also suppose that the temperature is kept constant and equal to T_A on the left side of the cylinder, while it is kept constant and equal to T_B ($= T_A + dT$) on the right side of the cylinder. The geometry is schematically indicated in Fig. 13.

When a current $\mathbf{J}$ flows from a point at temperature T_A to a point at temperature

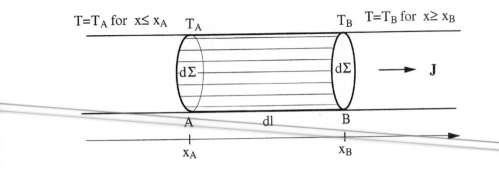

Fig. 13 Schematic figure for the calculation of the Thomson coefficient.

T_B reversible heat is generated in the cylinder at the rate

$$\frac{\delta Q}{dt\,d\Sigma} = -J \int_{T_A}^{T_B} K_{\text{Thomson}}(T)\,dT \ , \tag{65}$$

where $K_{\text{Thomson}}(T)$ is known as Thomson coefficient. We now prove that the *Thomson coefficient* is related to the absolute thermoelectric power through the relationship

$$\boxed{K_{\text{Thomson}}(T) = T\frac{dS(T)}{dT}} \ . \tag{66}$$

The internal energy fluxes across the basis of the cylinder at sections A and B are respectively

$$\mathbf{U}_A = -\frac{K_1(T_A)}{e\,K_0(T_A)}\mathbf{J} \quad \text{and} \quad \mathbf{U}_B = -\frac{K_1(T_B)}{e\,K_0(T_B)}\mathbf{J} \ ,$$

as can be seen from Eq. (57a) (taking into account that temperature gradients at the left and right sides of the cylinder are assumed to be zero).

The heat δQ generated in the time dt in the cylinder $dl\,d\Sigma$ is given by

$$\delta Q = dU + \delta L \ ,$$

where dU is the energy which accumulates in the time dt because of the unbalance between energy flowing in and out of the considered cylinder, and δL is the work performed by the electric field in the time dt. We have

$$\frac{dU}{dt\,d\Sigma} = U_A - U_B = \left[-\frac{K_1(T_A)}{e\,K_0(T_A)} + \frac{K_1(T_B)}{e\,K_0(T_B)} \right] J \ . \tag{67a}$$

Similarly, using Eq. (56a) for the electric field, we have

$$\frac{\delta L}{dt\,d\Sigma} = \int_A^B \mathbf{J}\cdot\mathbf{E}\,dl = \int_A^B J\left[\frac{1}{\sigma_0}J - \frac{1}{e}\nabla\mu + S(T)\,\nabla T \right]\cdot d\mathbf{l}$$

$$= \frac{1}{\sigma_0}J^2\,dl - \frac{J}{e}(\mu_B - \mu_A) + J\int_{T_A}^{T_B} S(T)\,dT \ . \tag{67b}$$

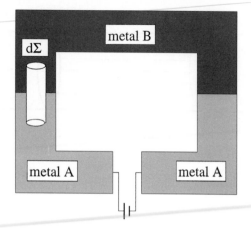

Fig. 14 Standard bimetallic circuit for illustration of the Peltier effect; the temperature is uniform throughout the whole circuit.

From Eqs. (67), and disregarding the Joule heating $\frac{1}{\sigma_0} J^2 \, dl$, we obtain for the reversible heat generation rate

$$\frac{\delta Q}{dt \, d\Sigma} = \left[-\frac{K_1(T_A)}{e \, K_0(T_A)} + \frac{\mu_A}{e} \right] J - \left[-\frac{K_1(T_B)}{e \, K_0(T_B)} + \frac{\mu_B}{e} \right] J + J \int_{T_A}^{T_B} S(T) \, dT$$

$$= [T_A \, S_A(T) - T_B \, S_B(T)] \, J + J \int_{T_A}^{T_B} S(T) \, dT \; .$$

Performing an integration by parts, it follows

$$\frac{\delta Q}{dt \, d\Sigma} = -J \int_{T_A}^{T_B} T \, dS(T) = -J \int_{T_A}^{T_B} T \, \frac{dS}{dT} \, dT \; . \tag{68}$$

Comparison of Eq. (68) and Eq. (65) proves expression (66) for the Thomson coefficient.

Peltier effect

Heat is generated reversibly not only when current flows in a given homogeneous material in the presence of temperature gradient, but also when current flows across a junction between two contacting materials (Peltier effect). If the direction of current changes, the Peltier effect changes sign (contrary to the Joule heating effect).

For a quantitative analysis consider the standard bimetallic circuit of Fig. 14 in isothermal conditions ($\nabla T = 0$) and with a current density J flowing throughout the circuit. Across the contact between metal A and metal B the rate (per unit time and unit section) of reversible heat released or absorbed is

$$\frac{\delta Q}{dt \, d\Sigma} = \Pi_{AB} \, J \; , \tag{69}$$

where J is supposed to flow from metal A to metal B. The Peltier coefficient of a given metal is connected to the Seebeck coefficient by the simple relationship

$$\boxed{\Pi(T) = T\, S(T)}\,. \tag{70}$$

To show this, we can use Eq. (68) (with a trivial extension of its meaning) keeping T constant, and S changing not because of temperature but because of inhomogeneity in the material. We have

$$\frac{\delta Q}{dt\, d\Sigma} = -J\,T\,[S_B(T) - S_A(T)]\,.$$

The above relation, together with Eq. (69), proves Eq. (70).

Considerations on other transport effects

We have seen that the Boltzmann equation is very useful for the description of transport effects in metals. The transport phenomena we have investigated have been confined to the simplest situations in which the driving perturbation is a static or oscillating electric field (electrical conductivity effects), or a static electric field and a temperature gradient (thermal conductivity, Seebeck, Peltier and Thomson effects).

The Boltzmann equation is of major help for several other transport phenomena. These include transport effects in the presence of electric and magnetic fields (Hall and magnetoresistivity effects), galvanomagnetic effects (Righi–Leduc effect, Nernst effect, Etthingshausen effect, etc.), "anomalies" or "giant effects" (in particular situations, for instance in the presence of magnetic impurities). A variety of challenging situations occur when the "phonon thermal bath", that usually ensures relaxation towards equilibrium of the electron distribution function, is itself dragged out from thermal equilibrium.

The Boltzmann equation has been widely applied to describe transport properties in semiconductors, following essentially the same semiclassical concepts given for metals, but keeping in mind some obvious differences. Among these, the fact that the distribution of conduction electrons in a semiconductor is in general non-degenerate, because of the low density of carriers. Furthermore we have to consider both electrons in the conduction band and holes in the valence band; and in general their contribution to a given transport phenomenon is not simply additive. These differences may lead to profound effects with respect to metals. For instance the electronic conductivity of a semiconductor is in general several orders of magnitude lower than that of a metal; nevertheless the thermoelectric power of a semiconductor, due to the presence of the energy gap, is in general much higher than typical thermopowers of metals. We do not dwell on other semiclassical aspects of transport properties, except for the discussion in Section XV-5 of magnetoresistivity and the Hall effect.

Further reading

F. Abelès (editor) "Optical Properties of Solids" (North-Holland, Amsterdam 1972)

A. A. Abrikosov "Fundamentals of the Theory of Normal Metals" (North-Holland, Amsterdam 1988)

M. Born and E. Wolf "Principles of Optics" (Pergamon Press, Oxford 1980)

S. Datta "Electron Transport in Mesoscopic Systems" (Cambridge University Press, 1995)

W. A. Harrison and M. B. Webb (editors) "The Fermi Surface" (Wiley, New York 1960)

E. D. Palik (editor) "Handbook of the Optical Constants of Solids" vols. I and II (Academic Press, New York 1985 and 1991)

A. B. Pippard "Dynamics of Conduction Electrons" (Gordon and Breach, New York 1965)

A. C. Smith, J. F. Janak and R. B. Adler "Electronic Conduction in Solids" (McGraw Hill, New York 1967)

R. A. Smith "Semiconductors" (Cambridge University Press, 2nd edition 1978)

F. Stern "Elementary Optical Properties of Solids" in Solid State Physics **15**, 299 (1963) (edited by H. Ehrenreich, F. Seitz and D. Turnbull, Academic Press, New York)

A. H. Wilson "The Theory of Metals" (Cambridge University Press, 2nd edition 1953)

F. Wooten "Optical Properties of Solids" (Academic Press, New York 1972)

XII

Optical properties of semiconductors and insulators

In the previous chapter we have considered the role of *intraband electronic transitions* in the optical and transport properties of materials with partially occupied energy bands (metals, or also semiconductors with a sufficiently high number of free carriers); here, we consider *interband electronic transitions* in materials with fully occupied or fully empty energy bands (insulators, and also semiconductors apart a possible free carrier contribution). Interband transitions are also of relevance in metals, usually at energies above the intraband contribution.

This chapter begins with a section, rather general in nature, that provides the quantum mechanical expression of the transverse optical constants in homogeneous materials within the linear response formalism (Kubo–Greenwood theory). The quantum theory treats on the same footing intraband and interband transitions in crystals. For the intraband contribution, the quantum theory provides a justification of the semiclassical treatment based on the Boltzmann equation; for interband optical transitions, the theory is developed within the dipole approximation and the role of the critical points in the joint density-of-states is analysed.

Aspects beyond the linear response theory are also discussed, in order to interpret phonon-assisted indirect interband transitions and two-photon absorption; transition rates for absorption or emission processes at first and at higher orders of perturbation theory are summarized for convenience in Appendix A.

The remaining sections of this chapter contain a selection of significant applications of the linear response theory. We consider first excitonic effects on the optical transitions; excitons are particularly important in large gap insulators, where the dielectric constant is small and the effective masses are large. We describe then resonant states coupled to a continuum and the Fano absorption profiles. Finally we consider the optical properties of (strongly) coupled electron-phonon systems; we focus on the Franck–Condon vibronic model and on typical Jahn–Teller systems.

1   Quantum expression of the transverse dielectric function in materials

The general properties of the longitudinal dielectric function have been considered in Section VII-3 for homogeneous media, in the linear response regime, and neglecting local field effects. In this section, within the same approximations of homogeneity, linearity and negligible local field effects, we consider the transverse dielectric function of an electron system. The general analytic properties of the dielectric function (either longitudinal or transverse) related to the physical requirement that the response is causal, linear and finite, have been discussed in Section VII-3 and need not be repeated here; we simply remind the reader that the real part of the dielectric function is an even function of the frequency, while the imaginary part is an odd function; real and imaginary parts of the dielectric function are linked by the Kramers–Kronig relations.

1.1   Optical constants of homogeneous media in the linear response theory

Consider a general (periodic or aperiodic) electronic system, described by the one-electron Hamiltonian

$$H_0 = \frac{\mathbf{p}^2}{2m} + V(\mathbf{r}) , \tag{1}$$

where $V(\mathbf{r})$ is the energy potential for an electron in the sample. We indicate by $\psi_n(\mathbf{r})$ and E_n the eigenfunctions and eigenvalues of H_0 (the eigenfunctions are normalized to unity in the volume V of the crystal); spin degeneracy is taken into account by including a factor 2, whenever needed. The occupancy of the states is determined by the Fermi–Dirac function $f(E_n)$ (in most situations, the temperature $T = 0$ is considered). The theory of the optical properties of the system is quite general in nature, and its explicit specification to periodic systems will be done only in a later stage.

In the presence of an electromagnetic field, the Hamiltonian H of the system is obtained from the unperturbed Hamiltonian H_0 replacing the momentum operator $\mathbf{p}$ with the generalized momentum $\mathbf{p} + (e/c)\,\mathbf{A}(\mathbf{r}, t)$, where $\mathbf{A}$ is the vector potential of the electromagnetic field and e is the modulus of the electronic charge. Without loss of generality, we take the scalar potential as zero and we adopt the Coulomb gauge div $\mathbf{A} = 0$, so that the terms $\mathbf{A} \cdot \mathbf{p}$ and $\mathbf{p} \cdot \mathbf{A}$ coincide. The Hamiltonian of the system in the presence of a radiation field thus becomes

$$H = \frac{1}{2m}[\mathbf{p} + \frac{e}{c}\,\mathbf{A}(\mathbf{r},t)]^2 + V(\mathbf{r}) = H_0 + \frac{e}{mc}\,\mathbf{A}(\mathbf{r},t)\cdot\mathbf{p} + \frac{e^2}{2mc^2}\,\mathbf{A}^2(\mathbf{r},t)\,. \qquad (2)$$

The last two terms in the right-hand side of Eq. (2) represent the quantum mechanical interaction between the radiation and the charge of the electron; for simplicity, any other interaction term, as for instance the interaction of the electromagnetic field with the electron spin, is ignored in the following.

We consider a *transverse electromagnetic plane wave* of frequency ω, wavevector $\mathbf{q}$ and polarization vector $\mathbf{e}$, described by the vector potential

$$\mathbf{A}(\mathbf{r},t) = A_0\,\mathbf{e}\,e^{i(\mathbf{q}\cdot\mathbf{r}-\omega t)} + c.c. \qquad \mathbf{e} \perp \mathbf{q}\,, \qquad (3)$$

where c.c. indicates the complex conjugate of the previous term, A_0 is chosen to be real, and the frequency ω is assumed positive. In the following, we neglect non-linear effects by disregarding the term $\mathbf{A}^2$ in Eq. (2); this is rigorously justified in the dipole approximation. Under the stated approximations the Hamiltonian H becomes

$$H = H_0 + \frac{e\,A_0}{m\,c}\,e^{i(\mathbf{q}\cdot\mathbf{r}-\omega t)}\,\mathbf{e}\cdot\mathbf{p} + \frac{e\,A_0}{m\,c}\,e^{-i(\mathbf{q}\cdot\mathbf{r}-\omega t)}\,\mathbf{e}\cdot\mathbf{p}\,; \qquad (4)$$

at this point the treatment of the transverse dielectric function can be done following (*mutatis mutandis*) a procedure similar to the one adopted for the longitudinal dielectric function of Section VII-7.

The time dependent terms in the right-hand side of Eq. (4) induce transitions among the states of H_0, and can be treated according to standard time-dependent perturbation theory (see Appendix A). The first time-dependent term in Eq. (4) gives rise to absorption of radiation, while the second term gives rise to emission. The probability per unit time that an electron initially in the state $|\psi_i\rangle$ is transferred to the final state $|\psi_j\rangle$ with *absorption of one photon of energy* $\hbar\omega$ is given by the Fermi golden rule

$$P_{j\leftarrow i} = \frac{2\pi}{\hbar}\left(\frac{e\,A_0}{m\,c}\right)^2 |\langle\psi_j|e^{i\mathbf{q}\cdot\mathbf{r}}\,\mathbf{e}\cdot\mathbf{p}|\psi_i\rangle|^2\,\delta(E_j - E_i - \hbar\omega)\,. \qquad (5a)$$

Quite similarly, the transition probability that an electron initially in the state $|\psi_j\rangle$ is transferred to the final state $|\psi_i\rangle$ with *emission of one photon of energy* $\hbar\omega$ is

$$P_{i\leftarrow j} = \frac{2\pi}{\hbar}\left(\frac{e\,A_0}{m\,c}\right)^2 |\langle\psi_i|e^{-i\mathbf{q}\cdot\mathbf{r}}\,\mathbf{e}\cdot\mathbf{p}|\psi_j\rangle|^2\,\delta(E_i - E_j + \hbar\omega)\,. \qquad (5b)$$

The two processes are schematically represented in Fig. 1.

We can now obtain the net number of transitions per unit time, each involving a single photon of energy $\hbar\omega$. We suppose that the electronic states are occupied in agreement to the standard Fermi–Dirac distribution function $f(E)$. We multiply Eq. (5a) by $f(E_i)\,(1 - f(E_j))$ and Eq. (5b) by $f(E_j)\,(1 - f(E_i))$, subtract the two expressions and sum up over all initial and final states; we obtain that the net number

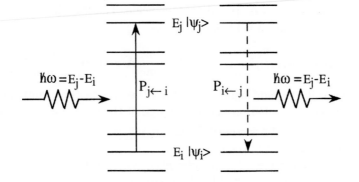

Fig. 1 Schematic representation of the radiatiative transition probability between a pair of electronic states $|\psi_i\rangle$ and $|\psi_j\rangle$, of energy E_i and E_j; any transition between these two states involves the absorption or the emission of a photon with energy $\hbar\omega = E_j - E_i$.

of transitions per unit time involving energy $\hbar\omega$ is given by the expression

$$W(\mathbf{q},\omega) = \frac{2\pi}{\hbar}\left(\frac{eA_0}{mc}\right)^2 2\sum_{ij}|\langle\psi_j|e^{i\mathbf{q}\cdot\mathbf{r}}\,\mathbf{e}\cdot\mathbf{p}|\psi_i\rangle|^2\,\delta(E_j-E_i-\hbar\omega)\left[f(E_i)-f(E_j)\right],\quad(6)$$

where the factor 2 in front of the summation takes into account the spin degeneracy. The energy per unit time dissipated in the system is $P = \hbar\omega W(\mathbf{q},\omega)$, a positive quantity for any value (positive or negative) of the frequency ω.

The microscopic expression of the optical constants can be obtained with the following considerations. The electric field in the medium associate to the vector potential $\mathbf{A}(\mathbf{r},t)$, given by Eq. (3), is

$$\mathbf{E}(\mathbf{r},t) = -\frac{1}{c}\frac{\partial\mathbf{A}}{\partial t} = E_0\,\mathbf{e}\,e^{i(\mathbf{q}\cdot\mathbf{r}-\omega t)} + c.c. \quad\text{with}\quad E_0 = i\omega\frac{A_0}{c}\,.\quad(7a)$$

In an isotropic medium (or also in an anisotropic medium when the electric field is along a principal axis), the induced current density is parallel to the electric field and proportional to it for small field strength; we assume for the current density the same time and space dependence as the electric field (7a) and write

$$\mathbf{J}(\mathbf{r},t) = \sigma(\mathbf{q},\omega)\,E_0\,\mathbf{e}\,e^{i(\mathbf{q}\cdot\mathbf{r}-\omega t)} + c.c. \quad(\mathbf{e}\perp\mathbf{q})\,,\quad(7b)$$

where $\sigma(\mathbf{q},\omega)$ defines the *transverse conductivity* function. When a current density $\mathbf{J}$ flows in a medium in the presence of an electric field $\mathbf{E}$, the energy per unit time dissipated in the system is $\int_V \mathbf{J}\cdot\mathbf{E}\,d\mathbf{r}$; using Eq. (7a) for $\mathbf{E}(\mathbf{r},t)$ and Eq. (7b) for $\mathbf{J}(\mathbf{r},t)$, we obtain

$$\int_V \mathbf{J}\cdot\mathbf{E}\,d\mathbf{r} = 2\,\sigma_1(\mathbf{q},\omega)\,|E_0|^2\,V = 2\,\sigma_1(\mathbf{q},\omega)\,\frac{1}{c^2}\omega^2 A_0^2\,V\,.$$

This expression of the power dissipated in the system can be identified with the quantum expression $\hbar\omega\, W(\mathbf{q},\omega)$; the real part of the conductivity function thus becomes

$$\sigma_1(\mathbf{q},\omega) = \frac{c^2}{2V}\frac{\hbar\omega\, W(\mathbf{q},\omega)}{\omega^2 A_0^2}\ .$$

It is convenient to continue our elaborations considering the dielectric function $\varepsilon(\mathbf{q},\omega)$ linked to the conductivity by the expression

$$\varepsilon(\mathbf{q},\omega) = 1 + \frac{4\pi i}{\omega}\sigma(\mathbf{q},\omega) \tag{8}$$

[this relationship follows from the requirement $\partial\mathbf{D}/\partial t = \partial\mathbf{E}/\partial t + 4\pi\mathbf{J}$, with $\mathbf{D} = \varepsilon\,\mathbf{E}$, $\mathbf{J} = \sigma\,\mathbf{E}$, and the form $\exp(-i\omega t)$ for the time-dependence of the fields]. From Eq. (8), it follows that $\varepsilon_2(\mathbf{q},\omega) = (4\pi/\omega)\,\sigma_1(\mathbf{q},\omega)$, and then

$$\boxed{\varepsilon_2(\mathbf{q},\omega) = \frac{2\pi\hbar c^2}{\omega^2}\frac{1}{V}\frac{W(\mathbf{q},\omega)}{A_0^2}}\ . \tag{9}$$

Using Eq. (6), the quantum expression of the imaginary part of the transverse dielectric function becomes

$$\varepsilon_2(\mathbf{q},\omega) = \frac{8\pi^2 e^2}{m^2\omega^2}\frac{1}{V}\sum_{ij}\left|\langle\psi_j|e^{i\mathbf{q}\cdot\mathbf{r}}\,\mathbf{e}\cdot\mathbf{p}|\psi_i\rangle\right|^2\delta(E_j - E_i - \hbar\omega)[f(E_i) - f(E_j)]\ . \tag{10a}$$

Notice that $\varepsilon_2(\mathbf{q},\omega)$ obeys the general relation $\varepsilon_2(\mathbf{q},-\omega) = -\varepsilon_2(\mathbf{q},\omega)$. The real part of the dielectric function is obtained inserting Eq. (10a) into the Kramers–Kronig relation

$$\varepsilon_1(\mathbf{q},\omega) = 1 + \frac{1}{\pi}P\int_{-\infty}^{+\infty}\frac{\varepsilon_2(\mathbf{q},\omega')}{\omega' - \omega}\,d\omega' = 1 + \frac{2}{\pi}P\int_0^{+\infty}\frac{\omega'\,\varepsilon_2(\mathbf{q},\omega')}{\omega'^2 - \omega^2}\,d\omega'\ . \tag{10b}$$

Eq. (10a) and Eq. (10b) give separately the imaginary part and the real part of the dielectric function.

It is easily seen that the complete expression of the complex dielectric function can be written in the form

$$\boxed{\varepsilon(\mathbf{q},\omega) = 1 + \frac{8\pi e^2}{m^2}\frac{1}{V}\sum_{ij}\frac{\left|\langle\psi_j|e^{i\mathbf{q}\cdot\mathbf{r}}\,\mathbf{e}\cdot\mathbf{p}|\psi_i\rangle\right|^2}{(E_j - E_i)^2/\hbar^2}\frac{[f(E_i) - f(E_j)]}{E_j - E_i - \hbar\omega - i\eta}} \tag{11}$$

where $\eta \to 0^+$. To separate the real and imaginary part of Eq. (11) one can use the well-known identity

$$\lim_{\eta\to 0^+}\frac{1}{x - i\eta} = P\frac{1}{x} + i\pi\,\delta(x)\ , \tag{12}$$

where x is real and P denotes the principal part (see Eq. VII-51).

Equation (11) is the basic expression of the *transverse dielectric function* of an electronic system of volume V, with one-electron wavefunctions $\psi_i(\mathbf{r})$ and eigenvalues E_i (the factor two taking account of the spin degeneracy is already included).

It is worthwhile a comparison with the *longitudinal dielectric function*, whose expression is given by Eq. (VII-50) in general, and by Eq. (VII-59) in the particular case $\mathbf{q} = 0$. It is seen that transverse and longitudinal dielectric functions of isotropic media are equal (as expected) in the long wavelength limit $\mathbf{q} \to 0$, while they are in general different for $\mathbf{q} \neq 0$.

It is of interest to specify the general results so far obtained to the case of insulators or semiconductors, characterized (at $T=0$) by *fully occupied or fully empty* one-electron states; in this situation, we can avoid the explicit presence of the Fermi–Dirac distribution function, and Eq. (10a) gives

$$\varepsilon_2(\mathbf{q}, \omega) = \frac{8\pi^2 e^2}{m^2 \omega^2} \frac{1}{V} \sum_i^{(occ)} \sum_j^{(unocc)} \left| \langle \psi_j | e^{i\mathbf{q}\cdot\mathbf{r}} \, \mathbf{e}\cdot\mathbf{p} | \psi_i \rangle \right|^2 \delta(E_j - E_i - \hbar\omega) . \tag{13}$$

The above relation holds for positive frequencies, while for negative frequencies we resort to the general property $\varepsilon_2(\mathbf{q}, -\omega) = -\varepsilon_2(\mathbf{q}, \omega)$.

Transverse conductivity function and intraband transitions

We consider now the expression of the transverse conductivity $\sigma(\mathbf{q}, \omega)$; from Eq. (8) and Eq. (11) we have

$$\sigma(\mathbf{q}, \omega) = \frac{2e^2}{m^2} \frac{1}{V} \sum_{ij} \frac{\left| \langle \psi_j | e^{i\mathbf{q}\cdot\mathbf{r}} \, \mathbf{e}\cdot\mathbf{p} | \psi_i \rangle \right|^2}{(E_j - E_i)^2/\hbar^2} \frac{(-i)\,\omega \, [f(E_i) - f(E_j)]}{E_j - E_i - \hbar\omega - i\eta} .$$

We exploit the identity

$$\frac{\hbar\omega}{E_j - E_i - \hbar\omega - i\eta} = \frac{E_j - E_i}{E_j - E_i - \hbar\omega - i\eta} - 1 \qquad \eta \to 0^+ ,$$

which can be easily proved using Eq. (12). We obtain for the conductivity

$$\sigma(\mathbf{q}, \omega) = \frac{2e^2}{m^2} \frac{1}{V} \sum_{ij} \frac{\left| \langle \psi_j | e^{i\mathbf{q}\cdot\mathbf{r}} \, \mathbf{e}\cdot\mathbf{p} | \psi_i \rangle \right|^2}{(E_j - E_i)/\hbar} \frac{(-i)\,[f(E_i) - f(E_j)]}{E_j - E_i - \hbar\omega - i\eta} . \tag{14}$$

Equation (14) is the *basic quantum mechanical expression of the transverse conductivity function of homogeneous materials*; in crystals, it treats on the same footing intraband and interband contributions.

It is of interest to compare the quantum mechanical treatment of the intraband contribution to the conductivity with the semiclassical approach based on the transport Boltzmann equation. For instance, in the specific case of a simple metal with a unique conduction band of energy dispersion $E_\mathbf{k}$ and Bloch wavefunctions $|\psi_\mathbf{k}\rangle$, the transverse conductivity (14) becomes

$$\sigma(\mathbf{q}, \omega) = \frac{2e^2}{m^2} \int \frac{d\mathbf{k}}{(2\pi)^3} \frac{\left| \langle \psi_{\mathbf{k}+\mathbf{q}} | e^{i\mathbf{q}\cdot\mathbf{r}} \, \mathbf{e}\cdot\mathbf{p} | \psi_\mathbf{k} \rangle \right|^2}{(E_{\mathbf{k}+\mathbf{q}} - E_\mathbf{k})/\hbar} \frac{(-i)[f(E_\mathbf{k}) - f(E_{\mathbf{k}+\mathbf{q}})]}{E_{\mathbf{k}+\mathbf{q}} - E_\mathbf{k} - \hbar\omega - i\eta} . \tag{15}$$

As discussed in Section XI-5, the quantum expression (15), for sufficiently small wavevectors, well justifies the Boltzmann transport treatment of intraband optical transitions.

The purpose of this chapter is to apply the quantum theory of interband transitions to investigate the optical properties of semiconductors and insulators; however, before doing this, it is of major interest for some applications to express the dielectric function, or the conductivity, in terms of the Green's function of the Hamiltonian H_0 of the system.

1.2 Optical constants and Green's function of the electronic system

According to Eq. (11) or Eq. (14), *the calculation of the optical constants of a system requires the knowledge of the eigenfunctions and eigenvalues of its Hamiltonian H_0.* In problems where the explicit diagonalization of H_0 poses difficult computational problems, it may be convenient to express the optical properties using the Green's function operator $G(E) = (E + i\eta - H_0)^{-1}$ with $\eta \to 0^+$. The reason is that the evaluation of Green's function matrix elements does not necessarily require the diagonalization of H_0 as an intermediate step (see for instance the discussion of the recursion method in Section V-8); this advantage may become particularly rewarding in the treatment of disordered systems, since the Green's function formalism allows often to perform averaging procedures in a very effective way.

The (retarded) Green's function $G(E)$ of the Hamiltonian H_0 is defined as

$$G(E) = \frac{1}{E + i\eta - H_0} \qquad \eta \to 0^+ \ ;$$

as usual, it is understood that the real energy E is accompanied by a small imaginary part $i\eta$ and the limit $\eta \to 0^+$ is taken. Using the complex conjugate of Eq. (12), we can formally define the operator $\text{Im}\, G(E)$ as

$$\text{Im}\, G(E) = \text{Im}\, \frac{1}{E + i\eta - H_0} = -\pi\, \delta(E - H_0) \ .$$

Consider Eq. (14) for the conductivity in the long wavelength limit $\mathbf{q} \to 0$ (for simplicity); the real part $\sigma_1(0, \omega)$ is given by

$$\sigma_1(0, \omega) = \frac{2\pi e^2}{m^2 \omega} \frac{1}{V} \sum_{ij} |\langle \psi_j | \mathbf{e} \cdot \mathbf{p} | \psi_i \rangle|^2\, \delta(E_j - E_i - \hbar\omega)\, [f(E_i) - f(E_j)] \ . \qquad (16a)$$

We now show that $\sigma_1(0, \omega)$, when expressed by means of the Green's function, takes the Kubo–Greenwood form

$$\sigma_1(0, \omega) = \frac{2e^2 \hbar}{\pi m^2} \frac{1}{V} \int_{-\infty}^{+\infty} dE\, \frac{f(E) - f(E + \hbar\omega)}{\hbar\omega}\, \text{Tr}\, \{\mathbf{e} \cdot \mathbf{p}\, \text{Im}\, G(E + \hbar\omega)\, \mathbf{e} \cdot \mathbf{p}\, \text{Im}\, G(E)\} \ ,$$
$$(16b)$$

where Tr stands for the trace operation, to be performed on any chosen complete set. Notice that the ingredients of Eq. (16a) are the eigenfunctions and eigenvalues of H_0; on the other hand, the ingredients of Eq. (16b) are the matrix elements of the Green's function of H_0.

To prove Eq. (16b), it is convenient to carry out the trace on the complete set of the eigenfunctions ψ_i of H_0, and also exploit the closure property $\sum |\psi_j\rangle\langle\psi_j| = 1$; we have

$$\text{Tr}\left\{\mathbf{e}\cdot\mathbf{p}\,\text{Im}\,G(E + \hbar\omega)\,\mathbf{e}\cdot\mathbf{p}\,\text{Im}\,G(E)\right\}$$

$$= \sum_{ij}\langle\psi_i|\mathbf{e}\cdot\mathbf{p}\,\text{Im}\,G(E + \hbar\omega)|\psi_j\rangle\langle\psi_j|\mathbf{e}\cdot\mathbf{p}\,\text{Im}\,G(E)|\psi_i\rangle$$

$$= \pi^2 \sum_{ij}|\langle\psi_j|\mathbf{e}\cdot\mathbf{p}|\psi_i\rangle|^2\,\delta(E + \hbar\omega - E_j)\,\delta(E - E_i) \; ;$$

insertion of the above expression into Eq. (16b) shows its equivalence to Eq. (16a).

It is instructive to obtain from Eq. (16b) the static conductivity $\sigma_1(\mathbf{q}=0, \omega=0)=\sigma_0$ in the long wavelength limit. For $\omega \to 0$, the first term within the integral in Eq. (16b) becomes $(-\partial f/\partial E)$; in a metal $(-\partial f/\partial E) \approx \delta(E-E_F)$. The integration on the energy can then be performed and the static electron conductivity becomes

$$\sigma_0 = \frac{2e^2\hbar}{\pi m^2}\frac{1}{V}\,\text{Tr}\left\{\mathbf{e}\cdot\mathbf{p}\,\text{Im}\,G(E_F)\,\mathbf{e}\cdot\mathbf{p}\,\text{Im}\,G(E_F)\right\} \; .$$

The above expression has been widely applied for the evaluation of the static electronic conductivity in disordered metals and alloys.

We provide another useful elaboration of the dielectric function, which applies in the particular case that *all optical transitions originate from the ground state* $|\Phi_g\rangle$ (supposed to be non degenerate) of an electronic system. From Eq. (13), we see that the imaginary part $\varepsilon_2(0, \omega)$ of the dielectric function at positive frequencies can be written as

$$\varepsilon_2(0, \omega) = \frac{8\pi^2 e^2}{m^2\omega^2}\frac{1}{V}\,I(\hbar\omega) \tag{17a}$$

with

$$I(\hbar\omega) = \sum_f |\langle\psi_f|T|\Phi_g\rangle|^2\,\delta(E_f - E_g - \hbar\omega) \; , \tag{17b}$$

where T is the dipole operator $\mathbf{e}\cdot\mathbf{p}$, and $|\psi_f\rangle$ are the eigenstates of the Hamiltonian H_0 of the system. From Eq. (17a), we see that the structure of $\varepsilon_2(0, \omega)$ is essentially determined by the *lineshape function* $I(\hbar\omega)$, defined by Eq. (17b).

A convenient expression of Eqs. (17) in terms of the Green's function can be obtained as follows. Let us introduce the *dipole carrying state* defined as

$$|\chi\rangle = T|\Phi_g\rangle \; . \tag{18a}$$

The diagonal matrix element of the Green's function $G(E_g+\hbar\omega)$ on the dipole carrying state is

$$G_{\chi\chi}(E_g + \hbar\omega) = \langle\chi|\frac{1}{E_g + \hbar\omega + i\eta - H_0}|\chi\rangle \; . \tag{18b}$$

Inserting the unit operator $1 \equiv \sum |\psi_f\rangle\langle\psi_f|$ in the above expression, and using Eq. (18a) and Eq. (12), we have that the $I(\hbar\omega)$ can be written in the convenient and compact form (the limit $\eta \to 0^+$ is implicitly understood)

$$I(\hbar\omega) = -\frac{1}{\pi} \operatorname{Im} G_{\chi\chi}(E_g + \hbar\omega) .$$ (19)

When the focus is on the *lineshape* $I(E)$ (where E stands for $\hbar\omega$) and not on its actual location on the energy axis, we can disregard E_g in Eq. (19), and write

$$I(E) = -\frac{1}{\pi} \operatorname{Im} G_{\chi\chi}(E) .$$ (20)

Equation (20) describes optical transitions originating from a unique initial ground state, and links the shape of the imaginary part of the dielectric function with the imaginary part of the diagonal matrix element of the Green's function on the dipole carrying state.

2 Quantum theory of band-to-band optical transitions and critical points

We apply now the quantum theory of optical transitions to semiconductors and insulators with fully occupied valence bands and fully empty conduction bands. Consider a radiation field, of wavevector $\mathbf{q}$ and frequency ω, and let us examine the matrix elements for the electronic transitions from an initial valence state $\psi_{v\mathbf{k}_i}$ of wavevector $\mathbf{k}_i$ (and given spin) into a final conduction state $\psi_{c\mathbf{k}_j}$ of wavevector $\mathbf{k}_j$ (and same spin):

$$\langle \psi_{c\mathbf{k}_j} | e^{i\mathbf{q}\cdot\mathbf{r}} \mathbf{e}\cdot\mathbf{p} | \psi_{v\mathbf{k}_i} \rangle = \int \psi_{c\mathbf{k}_j}^*(\mathbf{r}) \, \mathbf{e}\cdot\mathbf{p} \, e^{i\mathbf{q}\cdot\mathbf{r}} \, \psi_{v\mathbf{k}_i}(\mathbf{r}) \, d\mathbf{r} .$$ (21a)

We notice that $\exp(i\mathbf{q}\cdot\mathbf{r})\,\psi_{v\mathbf{k}_i}$ is a Bloch function of wavevector $\mathbf{q}+\mathbf{k}_i$; furthermore the operator $\mathbf{p}$ applied to a Bloch function of given wavevector gives a Bloch function with the same wavevector. Thus the matrix element (21a) can be different from zero only if

$$\mathbf{k}_j = \mathbf{q} + \mathbf{k}_i + \mathbf{g} ,$$ (21b)

where $\mathbf{g}$ is a vector of the reciprocal lattice (including the null vector).

In ordinary experimental situations (infrared region, visible, up to near and far ultraviolet region) the *wavelength of the incident radiation is much larger than the lattice parameter*; in these situations, the photon wavevector $\mathbf{q}$ of the incident radiation is small compared to the range of values of $\mathbf{k}$ within the first Brillouin zone; thus we may neglect $\mathbf{q}$ in Eq. (21b) (*dipole approximation*), and we may put $\mathbf{k}_j = \mathbf{k}_i$ (*direct or vertical transitions*). The *optical transitions are thus vertical* in an energy band diagramme, as schematized in Fig. 2.

We now apply Eq. (13) to the case of semiconductors and insulators with fully occupied valence bands and fully empty conduction bands; we denote $\varepsilon_2(\mathbf{q}=0,\omega)$ by

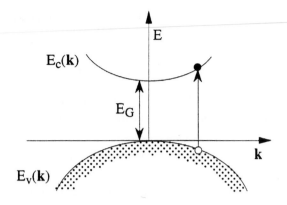

Fig. 2 Schematic representation of vertical transitions.

$\varepsilon_2(\omega)$, and take into account that only vertical transitions are possible; we obtain

$$\varepsilon_2(\omega) = \frac{8\pi^2 e^2}{m^2\omega^2}\frac{1}{V}\sum_{cv}\sum_{\mathbf{k}}|\langle\psi_{c\mathbf{k}}|\,\mathbf{e}{\cdot}\mathbf{p}\,|\psi_{v\mathbf{k}}\rangle|^2\,\delta(E_{c\mathbf{k}} - E_{v\mathbf{k}} - \hbar\omega)$$

$$= \frac{8\pi^2 e^2}{m^2\omega^2}\sum_{cv}\int_{B.Z}\frac{d\mathbf{k}}{(2\pi)^3}|\mathbf{e}\cdot\mathbf{M}_{cv}(\mathbf{k})|^2\,\delta(E_{c\mathbf{k}} - E_{v\mathbf{k}} - \hbar\omega)\ ; \qquad (22)$$

in the above expression ω is positive, the sum runs over every couple of valence and conduction bands, and $\mathbf{M}_{cv}(\mathbf{k})$ denotes the dipole matrix element $\langle\psi_{c\mathbf{k}}|\mathbf{p}|\psi_{v\mathbf{k}}\rangle$.

The knowledge of $\varepsilon_2(\omega)$ (for a sufficiently wide frequency range) allows to obtain the real part $\varepsilon_1(\omega)$ of the dielectric function via the Kramers–Kronig relations (10b); then any other optical constant of interest can be obtained (see Table XI-1). Among the phenomenological constants of frequent use in the description of the optical properties of matter, we are reminded here of the absorption coefficient $\alpha(\omega)$, which is related to $\varepsilon_2(\omega)$ and to the refractive index $n(\omega)$ via the expression

$$\alpha(\omega) = \frac{\omega}{c\,n(\omega)}\varepsilon_2(\omega)\ . \qquad (23a)$$

Using Eq. (9), we can give for the absorption coefficient the following quantum mechanical expression

$$\alpha(\omega) = \frac{2\pi\hbar c}{n(\omega)\,\omega}\frac{1}{V}\frac{W(\omega)}{A_0^2}\ . \qquad (23b)$$

In most practical situations, the structure and the peaks of $\varepsilon_2(\omega)$ and $\alpha(\omega)$ are rather similar.

It is well known that in atoms, molecules and (small) clusters, the absorption of radiation (at least below the photoionization edge) exhibits sharp lines, which correspond to transitions to discrete excited states, allowed in the dipole approximation. Also in the expression of band-to-band transitions the absorption spectrum can have

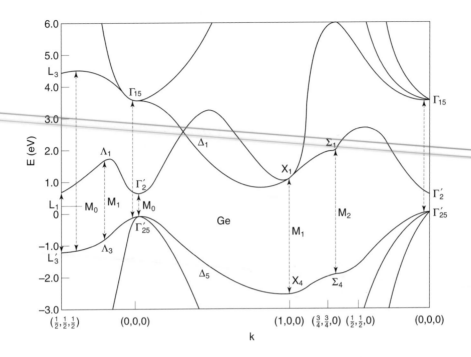

Fig. 3 Energy bands of germanium along symmetry directions, with the empirical pseudopotential method. The **k** wavevector is in units of $2\pi/a$. Relevant direct interband edges are indicated by arrows [from D. Brust, J. C. Phillips and F. Bassani, Phys. Rev. Lett. **9**, 94 (1962); copyright 1962 by the American Physical Society].

(reasonably) sharp structures, which however correspond in this case to critical points in the joint density-of-states.

Consider in fact a given couple of valence and conduction bands; in general the dipole matrix element $\mathbf{e} \cdot \mathbf{M}_{cv}(\mathbf{k})$ is a smooth function of $\mathbf{k}$ over the Brillouin zone (unless zero at some symmetry point for group theory considerations), and its average value can be factorized out of Eq. (22). Then, the contribution to the optical constants from a couple of valence and conduction bands is determined by the so called *joint density-of-states*

$$J_{cv}(\omega) = \int_{B.Z.} \frac{d\mathbf{k}}{(2\pi)^3} \, \delta(E_{c\mathbf{k}} - E_{v\mathbf{k}} - \hbar\omega) . \tag{24a}$$

Critical points in the joint density-of-states are by definition those for which

$$\nabla_{\mathbf{k}}(E_{c\mathbf{k}} - E_{v\mathbf{k}}) \equiv 0 . \tag{24b}$$

For these points the joint density-of-states exhibits rapid variation versus energy. This can be easily seen integrating analytically Eq. (24a) in the neighbourhood of a critical point (24b).

The type of critical points, the singular behaviour in the joint density-of-states $J_{cv}(\omega)$ and hence of the imaginary part $\varepsilon_2(\omega)$ of the dielectric function, the depen-

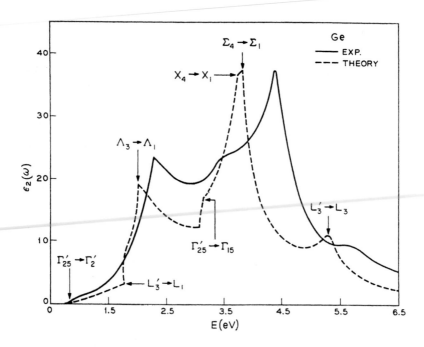

Fig. 4 Spectral structure of $\varepsilon_2(\omega)$ (solid line) compared with the theoretical results of interband transitions for Ge (dashed line with edges emphasized to account for critical points) [from D. Brust, J. C. Phillips and F. Bassani, Phys. Rev. Lett. **9**, 94 (1962) and D. Brust, Phys. Rev. **134**, A1337 (1964); copyright 1964 by the American Physical Society].

dence on the dimensionality, are the same as those worked out in Section II-7 (in the discussion of the band density-of-states), and are not repeated here. We remember that three-dimensional crystals present four types of singularities (minimum, two types of saddle points, and maximum) denoted M_0, M_1 and M_2, M_3, respectively. The singularities become sharper as the dimensionality decreases; in particular the joint density-of-states at the saddle point of two-dimensional crystal presents a logarithmic-like singularity (for details see Section II-7). The singular part of $\varepsilon_1(\omega)$ at the critical points can be obtained using the Kramers–Kronig transformation (10b) of the singular part of $\varepsilon_2(\omega)$.

In order to give an idea of the role of the critical points on the optical constants in semiconductors, we discuss the classic example of the optical properties of germanium. In Fig. 3, we report the band structure of germanium, together with important critical points for the interband transitions. The first direct transition in Ge is $\Gamma'_{25} \rightarrow \Gamma'_2$, which contributes an M_0 singularity around $\approx 1\,\text{eV}$. In the energy region around $\approx 2\,\text{eV}$, the transition $L'_3 \rightarrow L_1$ contributes an M_0 critical point, and the $\Lambda_3 \rightarrow \Lambda_1$ transitions along the Λ direction contribute an M_1 critical point; other relevent singularities at higher energies are indicated in Fig. 3. In Fig. 4, we report the imaginary part of the dielectric constant of Ge, and the overall satisfactory comparison with experimental

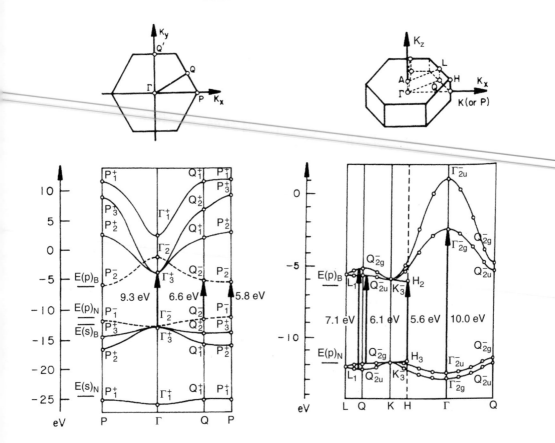

Fig. 5 Two-dimensional (left) and three-dimensional (right) band structures and corresponding Brillouin zones of hexagonal boron nitride. Relevant optical transitions are also indicated [the calculations are due to E. Doni and G. Pastori Parravicini, Nuovo Cimento **64**B, 117 (1969); interpretation of experimental data is due to D. M. Hoffman, G. L. Doll and P. C. Eklund, Phys. Rev. B**30**, 6051 (1984)].

results. It is apparent that the interband transition picture is of major importance for the interpretation of the optical properties of semiconductors and especially of the wealth of experimental features obtained with modulation spectroscopy. In fact, modulated techniques allow to pick out with high accuracy the discontinuities in the optical structures, and these critical points provide invaluable information on specific electronic transitions between energy bands of the crystal.

We have already remarked that the behaviour of the singularities at the critical points depends on the dimensionality; thus for two-dimensional crystals and layered materials we expect sharper structures in the optical properties. Nice examples are offered by graphite and hexagonal boron nitride, and we consider here some theoretical and experimental data for the latter. In Fig. 5, we report the two- and three-dimensional band structure of the hexagonal BN, with indication also of the most important transitions. In Fig. 6 the experimental imaginary dielectric constant $\varepsilon_2(\omega)$

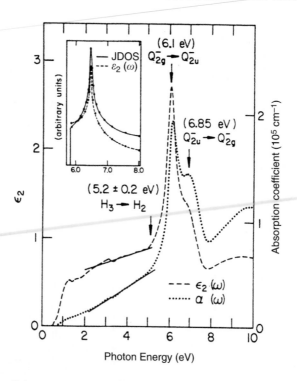

Fig. 6 Imaginary dielectric constant $\varepsilon_2(\omega)$ and absorption coefficient $\alpha(\omega)$ due to interband electronic transitions in hexagonal boron nitride from experimental measurements [from to D. M. Hoffman, G. L. Doll and P. C. Eklund, Phys. Rev. **B30**, 6051 (1984); copyright 1984 by the American Physical Society]. The inset is the joint density-of-states and $\varepsilon_2(\omega)$ calculated in the two-dimensional approximation near the saddle point singularity.

and the absorption coefficient $\alpha(\omega)$ are reported; it can be noticed the reasonable overall agreement between experiments and theoretical assignement of direct interband transitions.

3   Indirect phonon-assisted transitions

We have seen that the absorption of photons in crystals entails vertical transitions in the energy band diagram, because of the small momentum carried by photons (in the ordinary visible and near ultraviolet region). In several semiconductors and insulators, the conduction band and valence band extrema occur at different points of the Brillouin zone (*indirect gap materials*). Non-vertical transitions near the energy gap may occur provided some source supplies the momentum needed for total crystal momentum conservation; phonon gas or impurity centers can accommodate any appropriate momentum and thus indirect optical transitions assisted by phonons or impurities become possible. In the following we treat explicitly only the phonon case.

We illustrate the essential aspects of indirect transitions, considering the simplest model of indirect gap semiconductor (or insulator) with the top of the valence band at

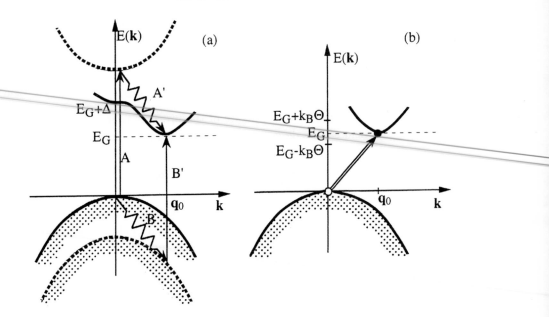

Fig. 7 Model band structure for estimating the absorption coefficient due to indirect transitions. In Fig. 7a we indicate schematically possible second-order processes contributing to the scattering of an electron from the top of the valence band $\psi_{v\mathbf{k}_1}$ ($\mathbf{k}_1 \approx 0$) to the bottom of the conduction band $\psi_{c\mathbf{k}_2}$ ($\mathbf{k}_2 \approx \mathbf{q}_0$). The mechanism $A\,A'$ consists of a direct transition from $\psi_{v\mathbf{k}_1}$ to a virtual state $\psi_{\alpha\mathbf{k}_1}$ (occupied or empty), followed by a phonon-assisted scattering to $\psi_{c\mathbf{k}_2}$. The mechanism $B\,B'$ consists of a phonon-assisted scattering from $\psi_{v\mathbf{k}_1}$ to a virtual state $\psi_{\beta\mathbf{k}_2}$ (occupied or empty), followed by a direct transition to $\psi_{c\mathbf{k}_2}$ (energy conservation and Pauli exclusion principle can be ignored in the virtual states, as discussed in the text). In Fig. 7b we indicate only the highest valence band and the lowest conduction band; near their respective edges, the second-order scattering amplitude (indicated with a double line) is taken as approximately constant. The energy thresholds for photon absorption, with absorption or emission of a phonon of energy $k_B\Theta$, are also indicated.

the center of the Brillouin zone, and the minimum of the conduction band at the point $\mathbf{q}_0 \neq 0$ of the Brillouin zone; often $\mathbf{q}_0$ is at or near the zone boundary. [In general, a number of equivalent minima occur in the conduction band structure $E_c(\mathbf{k})$ depending on the point group symmetry of the crystal, and the various contributions must be added together]. An indirect band structure is schematically indicated in Fig. 7.

The highest valence band and the lowest conduction band are supposed to be parabolic near their extrema, with isotropic effective masses. As shown in Fig. 7, the band gap E_G of the semiconductor is indirect, and is smaller than the first allowed direct energy gap $E_G + \Delta$. Thus when the photon energy $\hbar\omega$ of the incident beam is approximately in the energy range $[E_G, E_G + \Delta]$, optical transitions can occur only through indirect processes.

In order to evaluate some relevant aspects of indirect phonon assisted transitions, we use second-order perturbation theory considering both the optical perturbation $H_{em}(\mathbf{r}, t)$ due to the electromagnetic field and the phonon perturbation $H_{ep}(\mathbf{r}, t)$, due to the lattice vibrations. We have already seen that an electromagnetic field of frequency ω (in the dipole approximation) implies a time-dependent perturbation Hamiltonian of the form

$$H_{\text{em}}(\mathbf{r}, t) = \frac{e\, A_0}{mc} \mathbf{e} \cdot \mathbf{p}\, e^{-i\omega t} + h.c. \,, \tag{25a}$$

where h.c. indicates the hermitian conjugate of the previous term. The perturbation due to the phonon field can be described (in its essential lines) in the form

$$H_{\text{ep}}(\mathbf{r}, t) = \frac{1}{\sqrt{N}} \sum_{\mathbf{q}}^{B.Z.} V_p(\mathbf{q}, \mathbf{r})\, e^{-i\omega_{\mathbf{q}} t}\, a_{\mathbf{q}} + h.c. \,, \tag{25b}$$

where N is the number of unit cells of the crystal, $V_p(\mathbf{q}, \mathbf{r})$ denotes a perturbation function of wavevector $\mathbf{q}$, $\omega_{\mathbf{q}}$ is the frequency of a normal mode of wavevector $\mathbf{q}$ (to avoid inessential details we consider a unique phonon branch) and $a_{\mathbf{q}}$ is the annihilation operator for a phonon. For what concerns the operator $V_p(\mathbf{q}, \mathbf{r})$ we do not need to be very specific; in fact we simply exploit the property that it may scatter electrons from an initial wavevector to a final wavevector differing by $\mathbf{q}$. Whenever needed, the explicit expression of $V_p(\mathbf{q}, \mathbf{r})$ can be inferred (at least in principle) considering the change in the crystal potential when the nuclei are fixed in their equilibrium positions and when the nuclei are displaced according to the phonon mode.

With the use of optical and phonon perturbations (25), it is now straightforward to apply *second-order perturbation theory* (along the lines described in Appendix A). In the presence of both photon and phonon perturbation fields, we can consider second-order processes in which a photon and a phonon are both absorbed, or both emitted, or one is absorbed and the other emitted. We consider first the processes in which a photon and a phonon are both absorbed (and thus we neglect the hermitian conjugate partners in Eqs. (25), because they would give rise to photon and phonon emission).

The second-order transition amplitude that an electron, initially in the valence state $|\psi_{v\mathbf{k}_1}\rangle$ is scattered into the conduction state $|\psi_{c\mathbf{k}_2}\rangle$ *with absorption of a photon and a phonon* is given by the expression

$$t_{c\mathbf{k}_2 \leftarrow v\mathbf{k}_1} = \frac{e\, A_0}{m\, c} \sqrt{n_{\mathbf{q}}}\, \frac{1}{\sqrt{N}} \left[\sum_{\alpha} \frac{\langle \psi_{c\mathbf{k}_2} | V_p(\mathbf{q}, \mathbf{r}) | \psi_{\alpha\mathbf{k}_1} \rangle \langle \psi_{\alpha\mathbf{k}_1} | \mathbf{e} \cdot \mathbf{p} | \psi_{v\mathbf{k}_1} \rangle}{E_v(\mathbf{k}_1) - E_\alpha(\mathbf{k}_1) + \hbar\omega} \right.$$

$$\left. + \sum_{\beta} \frac{\langle \psi_{c\mathbf{k}_2} | \mathbf{e} \cdot \mathbf{p} | \psi_{\beta\mathbf{k}_2} \rangle \langle \psi_{\beta\mathbf{k}_2} | V_p(\mathbf{q}, \mathbf{r}) | \psi_{v\mathbf{k}_1} \rangle}{E_v(\mathbf{k}_1) - E_\beta(\mathbf{k}_2) + \hbar\omega_{\mathbf{q}}} \right] \,, \tag{26}$$

where $n_{\mathbf{q}}$ is the Bose population factor, $n_{\mathbf{q}} = 1/[\exp(\hbar\omega_{\mathbf{q}}/k_B T) - 1]$, and matrix elements involving phonon annihilation processes are proportional to $\sqrt{n_{\mathbf{q}}}$. The first sum in square brackets in Eq. (26) includes all the channels in which H_{em} induces a transition from the initial band state $\psi_{v\mathbf{k}_1}$ to a virtual band state $\psi_{\alpha\mathbf{k}_1}$; the phonon perturbation completes the transition by taking the electron from $\psi_{\alpha\mathbf{k}_1}$ to the final state $\psi_{c\mathbf{k}_2}$. Alternatively, the second sum in square brackets considers all the channels, in which the first step is performed by the phonon field and is then completed by the optical perturbation. A schematic representation of the second-order perturbation channels is indicated in Fig. 7a. [Notice that the Pauli exclusion principle in the intermediate states indicated in Fig. 7a is not operative; this is so because the operators H_{em} and H_{ep} commute; a quite different situation will be encountered in the Kondo effect, illustrated in Section XVI-6].

It is convenient to denote the expression within square brackets in Eq. (26) by C, a quantity which can be taken as approximately constant, as far as $\mathbf{k}_1$ and $\mathbf{k}_2$ are near their respective edges. We have

$$t_{c\mathbf{k}_2 \leftarrow v\mathbf{k}_1} = \frac{e\,A_0}{m\,c} \sqrt{n_{\mathbf{q}}} \frac{C\sqrt{\Omega}}{\sqrt{V}}$$

where $V = N\Omega$ is the volume of the sample. The second-order transition probability per unit time from an initial state $\psi_{v\mathbf{k}_1}$ to a final state $\psi_{c\mathbf{k}_2}$ is

$$P_{c\mathbf{k}_2 \leftarrow v\mathbf{k}_1} = \frac{2\pi}{\hbar} \left| t_{c\mathbf{k}_2 \leftarrow v\mathbf{k}_1} \right|^2 \delta(E_{c\mathbf{k}_2} - E_{v\mathbf{k}_1} - \hbar\omega - \hbar\omega_{\mathbf{q}})$$

$$= \frac{2\pi}{\hbar} \left(\frac{e\,A_0}{m\,c}\right)^2 n_{\mathbf{q}} |C|^2 \frac{\Omega}{V} \delta(E_{c\mathbf{k}_2} - E_{v\mathbf{k}_1} - \hbar\omega - \hbar\omega_{\mathbf{q}}) . \quad (27a)$$

The total number $W(\omega)$ of transitions per unit time involving a photon of energy $\hbar\omega$ is obtained summing Eq. (27a) over all initial states in the valence band and all final states in the conduction band; we have

$$W(\omega) = \frac{2\pi}{\hbar} \left(\frac{e\,A_0}{m\,c}\right)^2 n_{\mathbf{q}} |C|^2 \frac{\Omega}{V} 2 \sum_{\mathbf{k}_1} \sum_{\mathbf{k}_2} \delta(E_{c\mathbf{k}_2} - E_{v\mathbf{k}_1} - \hbar\omega - \hbar\omega_{\mathbf{q}}) ,$$

where a factor 2 due to spin degeneracy has been included. The absorption coefficient, related to $W(\omega)$ through Eq. (23b), becomes

$$\alpha_{(\text{phonon abs})}(\omega) = \frac{8\pi^2 e^2}{m^2 c\,n(\omega)\,\omega} n_{\mathbf{q}} |C|^2 \frac{\Omega}{V^2} \sum_{\mathbf{k}_1} \sum_{\mathbf{k}_2} \delta(E_{c\mathbf{k}_2} - E_{v\mathbf{k}_1} - \hbar\omega - \hbar\omega_{\mathbf{q}}) . \quad (27b)$$

In the above equation, the most important energy dependence is determined by the density-of-states for indirect transitions defined as

$$J^{\text{indirect}}_{(\text{phonon abs})}(\omega) = \int_{\text{B.Z.}} \int_{\text{B.Z.}} \frac{1}{(2\pi)^3 (2\pi)^3} \, d\mathbf{k}_1 \, d\mathbf{k}_2 \, \delta(E_{c\mathbf{k}_2} - E_{v\mathbf{k}_1} - \hbar\omega - k_B\Theta) ; \quad (28a)$$

we have assumed $\hbar\omega_{\mathbf{q}} \approx \hbar\omega_{\mathbf{q}_0}$ for indirect transitions at or near the threshold, and the phonon energy $\hbar\omega_{\mathbf{q}_0}$ is indicated by $k_B\Theta$.

We replace the values of $E_{c\mathbf{k}}$ and $E_{v\mathbf{k}}$ in agreement with the spherical behaviour (assumed in Fig. 7) and obtain

$$J^{\text{indirect}}_{(\text{phonon abs})}(\omega) = \frac{(4\pi)^2}{(2\pi)^6} \int_0^\infty \int_0^\infty k_1^2 \, k_2^2 \, \delta\left(\frac{\hbar^2 k_2^2}{2m_c^*} + \frac{\hbar^2 k_1^2}{2m_v^*} + E_G - \hbar\omega - k_B\Theta\right) dk_1 \, dk_2 .$$

To perform the integral, we make the substitutions

$$\frac{\hbar^2 k_2^2}{2m_c^*} = x, \qquad \frac{\hbar^2 k_1^2}{2m_v^*} = y, \qquad k_2 \, dk_2 = \frac{m_c^*}{\hbar^2} \, dx, \qquad k_1 \, dk_1 = \frac{m_v^*}{\hbar^2} \, dy ,$$

and obtain

$$J^{\text{indirect}}_{(\text{phonon abs})}(\omega) = \frac{(m_c^* m_v^*)^{3/2}}{2\pi^4 \hbar^6} \int_0^\infty \int_0^\infty \sqrt{x}\,\sqrt{y}\,\delta(x + y - b)\,dx\,dy \; ,$$

where $b = \hbar\omega - (E_G - k_B\Theta)$. The δ-function under the sign of integral can be different from zero only for positive values of b. Performing the integral over dy one gets

$$J^{\text{indirect}}_{(\text{phonon abs})}(\omega) = \frac{(m_c^* m_v^*)^{3/2}}{2\pi^4 \hbar^6} \int_0^b \sqrt{x}\,\sqrt{b - x}\,dx \; . \tag{28b}$$

We now use the following indefinite integral

$$\int \sqrt{bx - x^2}\,dx = \frac{2x - b}{4}\sqrt{bx - x^2} + \frac{b^2}{8}\arcsin\frac{2x - b}{b} \qquad (b > 0) \; .$$

The definite integral in Eq. (28b) equals $\pi b^2/8$. The joint density-of-states for indirect transitions thus becomes

$$\boxed{J^{\text{indirect}}_{(\text{phonon abs})}(\omega) = \frac{(m_v^* m_c^*)^{3/2}}{16\pi^3 \hbar^6}(\hbar\omega - E_G + k_B\Theta)^2 \qquad \hbar\omega > E_G - k_B\Theta} \; . \tag{28c}$$

A similar treatment holds for the phonon emission, with the replacement of $+k_B\Theta$ (phonon absorption) by $-k_B\Theta$ (phonon emission). We also take into account that the transition probabilities involving one-phonon annihilation or creation are proportional to

$$n_{\mathbf{q}} = \frac{1}{\exp(\Theta/T) - 1} \qquad \text{and} \qquad n_{\mathbf{q}} + 1 = \frac{1}{1 - \exp(-\Theta/T)} \; ,$$

respectively. We thus obtain for the absorption coefficient for indirect transitions the following law

$$\boxed{\alpha(\omega) = A\left[\frac{(\hbar\omega - E_G + k_B\Theta)^2}{e^{\Theta/T} - 1} + \frac{(\hbar\omega - E_G - k_B\Theta)^2}{1 - e^{-\Theta/T}}\right]} \; , \tag{29}$$

where A is approximately a constant independent from energy and temperature; it is also understood that the the first term in square bracket is present only if $\hbar\omega > E_G - k_B\Theta$, and the second term is present only if $\hbar\omega > E_G + k_B\Theta$.

The first term in brackets in Eq. (29) refers to photon absorption *with annihilation of one phonon* of energy $k_B\Theta$, and the threshold is just $\hbar\omega = E_G - k_B\Theta$. The second term in brackets refers to photon absorption with *creation of one phonon* of energy $k_B\Theta$; the energy threshold for this process is $\hbar\omega = E_G + k_B\Theta$. Thus when $\alpha^{1/2}(\omega)$ is plotted against $\hbar\omega$, a straight line is expected in the energy range $E_G - k_B\Theta < \hbar\omega < E_G + k_B\Theta$; and (approximately) a steeper straight line is expected for $\hbar\omega > E_G + k_B\Theta$.

In Fig. 8 we indicate the behaviour of the square root of the absorption coefficients with the photon energy at various temperatures for silicon and germanium. With $\Theta = 600\,\text{K}$ for silicon and $\Theta = 260\,\text{K}$ for germanium, a reasonable agreement with the experimental results is obtained using the law (29) for indirect transitions [for further aspects see also G. G. Macfarlane, T. P. McLean, V. Roberts and J. E. Quarrington, Phys. Rev. **108**, 1377 (1957); Phys. Rev. **111**, 1245 (1958)].

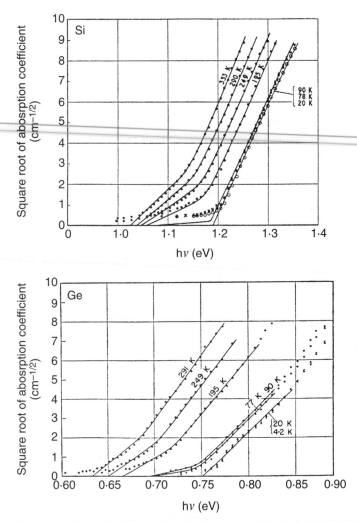

Fig. 8 (a) Indirect transitions in silicon; dots indicate experimental measurements; the full lines are calculated using Eq. (29) of the text [from G. G. Macfarlane and V. Roberts, Phys. Rev. **98**, 1865 (1955); copyright 1955 by the American Physical Society]. (b) Indirect transitions in germanium [from G. G. Macfarlane and V. Roberts, Phys. Rev. **97**, 1714 (1955); copyright 1955 by the American Physical Society].

Before concluding, we mention the effect of dimensionality on the indirect transitions; the joint density-of-states (28a) presents a linear dependence on the photon energy for two-dimensional bands and steps for one-dimensional bands; this behaviour should be observable in indirect transitions in quantum well structures and quantum wires, respectively.

4   Two-photon absorption

The advent of very intense radiation sources, and the development of sophisticated spectroscopic techniques, have made it possible to study experimentally processes in which two (or more) quanta are simultaneously absorbed to promote an electron transition. From a theoretical point of view the possibility of two photon absorption was investigated by Göppert-Mayer in 1931, about thirty years before the development of laser light made experiments possible.

We consider the optical constants of a solid in the presence of two radiation beams, of frequency ω_1 and ω_2, with N_1 and N_2 photons per unit volume. In order to avoid one-photon absorption of either beam, the frequencies ω_1 and ω_2 are chosen in such a way that $\hbar\omega_1 < E_G$, $\hbar\omega_2 < E_G$, but $\hbar\omega_1 + \hbar\omega_2 > E_G$. Thus two-photon spectroscopy is usually adopted to investigate excited states in the approximate energy range $E_G < E < 2E_G$ (with one-photon spectroscopy one can explore in principle the whole energy range). Despite this restriction, two-photon spectroscopy is an invaluable complementary tool to the ordinary one-photon spectroscopy. The symmetry of the accessible final states, and a variety of selection rules, can provide detailed information on the electronic band structure around the fundamental energy gap. Furthermore, the relatively low value of the absorption coefficient makes easier to detect bulk effects, with a decreased importance of surface contamination.

More recently, with the advent of synchrotron radiation, it has also been possible to extend two-photon spectroscopy to explore edges, originated by transitions from core states to the conduction band; for this purpose one can use synchrotron radiation with frequency somewhat lower than the one corresponding to core-band conduction-band threshold (where the material is reasonably transparent) and another beam (with frequency lower than the energy gap). In this way detailed information on transitions from core states in semiconductors and insulators have been obtained.

The theoretical treatment of two-photon absorption requires the application of second-order perturbation theory (along the lines presented in Appendix A). We write the vector potentials of the two incident beams in the form

$$\mathbf{A}_1(\mathbf{r}, t) = A_{01}\, \mathbf{e}_1\, e^{i(\mathbf{q}_1 \cdot \mathbf{r} - \omega_1 t)} + h.c.$$
$$\mathbf{A}_2(\mathbf{r}, t) = A_{02}\, \mathbf{e}_2\, e^{i(\mathbf{q}_2 \cdot \mathbf{r} - \omega_2 t)} + h.c. \tag{30}$$

The probability per unit time of a transition from the state $\psi_{v\mathbf{k}}$ (and given spin direction) to the state $\psi_{c\mathbf{k}}$ (and same spin orientation) considering any intermediate state $\psi_{\beta\mathbf{k}}$ (with the same spin orientation) is

$$P_{c\mathbf{k} \leftarrow v\mathbf{k}} = \frac{2\pi}{\hbar} \left(\frac{e\, A_{01}}{mc}\right)^2 \left(\frac{e\, A_{02}}{mc}\right)^2$$
$$\cdot \left| \sum_{\beta} (1 + P_{12}) \frac{\mathbf{e}_2 \cdot \mathbf{M}_{c\beta}(\mathbf{k})\, \mathbf{e}_1 \cdot \mathbf{M}_{\beta v}(\mathbf{k})}{E_v(\mathbf{k}) - E_\beta(\mathbf{k}) + \hbar\omega_1} \right|^2 \delta(E_{c\mathbf{k}} - E_{v\mathbf{k}} - \hbar\omega_1 - \hbar\omega_2) \tag{31}$$

where P_{12} is the operator which exchanges $\mathbf{e}_1$ and $\hbar\omega_1$ with $\mathbf{e}_2$ and $\hbar\omega_2$. A schematic representation of two-photon absorption processes is provided in Fig. 9.

The total number $W(\omega_1; \omega_2)$ of transitions per unit time involving the simultaneous absorption of a photon ω_1 and a photon ω_2 is obtained summing Eq. (31) over all $\mathbf{k}$ vectors of the first Brillouin zone, and inserting a factor 2 to account for spin

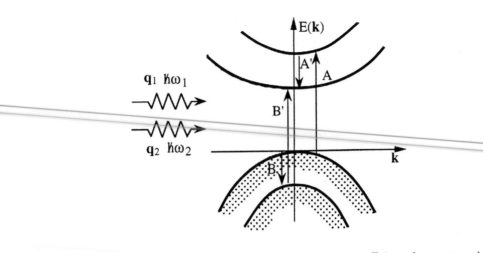

Fig. 9 Model band structure for estimating the absorption coefficient due to two-photon absorption. We indicate schematically by AA' and BB' possible second order processes contributing to the transition of an electron from the top of the valence band to the bottom of the conduction band (energy conservation and Pauli exclusion principle in the intermediate states are not operative, similarly to the discussion of Fig. 7).

degeneracy. The absorption coefficient is related to $W(\omega_1; \omega_2)$ by Eq. (23b); we thus have that the absorption coefficient for photons of energy $\hbar\omega_1$, in the presence of N_2 photons per unit volume of energy $\hbar\omega_2$, is

$$\alpha(\omega_1; \omega_2) =$$

$$A \int_{\text{B.Z.}} \frac{d\mathbf{k}}{(2\pi)^3} \left| \sum_\beta (1 + P_{12}) \frac{\mathbf{e}_2 \cdot \mathbf{M}_{c\beta}(\mathbf{k}) \mathbf{e}_1 \cdot \mathbf{M}_{\beta v}(\mathbf{k})}{E_v(\mathbf{k}) - E_\beta(\mathbf{k}) + \hbar\omega_1} \right|^2 \delta(E_{c\mathbf{k}} - E_{v\mathbf{k}} - \hbar\omega_1 - \hbar\omega_2) \quad (32)$$

where

$$A = \frac{8\pi^2 e^2}{m^2 c \, n(\omega_1) \omega_1} \left(\frac{e \, A_{02}}{m \, c} \right)^2 = \frac{16\pi^3 \hbar \, e^4 \, N_2}{m^4 c \, n(\omega_1) \, n^2(\omega_2) \, \omega_1 \, \omega_2} \; ;$$

in the last passage for obtaining A, use has been made of the relation

$$n^2(\omega_2) \, A_{02}^2 \, \omega_2^2 / 2\pi c^2 = N_2 \, \hbar\omega_2$$

that connects the classical energy density of the electromagnetic field with the number of photons per unit volume.

An analysis of the matrix elements appearing in Eq. (32) is particularly rich and instructive from the point of view of group theory. We notice that *dipole selection rules* for two-photon absorption and for one-photon absorption are in general different; for instance in crystals with inversion symmetry, accessible final states have the same parity as the initial state in two-photon spectroscopy and opposite parity in one-photon spectroscopy. Once a final state of a given symmetry is dipole allowed in two-photon transitions, the actual matrix elements depend on the direction of $\mathbf{e}_1$ and

$\mathbf{e}_2$ relative to each other or relative to the crystallographic axes (*geometrical selection rules*). Furthermore, the actual matrix elements also depend on the frequencies ω_1 and ω_2, and may vanish when ω_1 is equal (or near) to ω_2 (*dynamical selection rules*); this is a kind of quantum effect related to the operator P_{12} which exchanges photon 1 with photon 2. [For further aspects see E. Doni, R. Girlanda and G. Pastori Parravicini, Phys. Stat. Sol. (b) **65**, 203 (1974); (b) **88**, 773 (1978) and references quoted therein].

5   Exciton effects on the optical properties

In Section 2, we have studied the optical properties of crystals within the one-electron approximation for the electronic states. Exciton states are beyond the independent particle approximation, and we wish now to study their effects on the optical properties of crystals. The theory of excitons in semiconductors and insulators has been considered in Section VII-1; in the simplest picture, excitons arise as a consequence of the interaction between the electron promoted in the conduction band and the hole left behind in the valence band. For sake of simplicity, we consider here a two-band model semiconductor, as schematized in Fig. 2. As a preliminary step, we outline the optical properties of this model semiconductor in the independent particle approximation; then we consider the many-body exciton corrections following the classic treatment of R. J. Elliott, Phys. Rev. **108**, 1384 (1957).

Isotropic two-band model semiconductor and band-to-band transitions

Consider an *isotropic two-band model semiconductor* (as schematized in Fig. 2), with valence and conduction band extrema at the same point (for instance $\mathbf{k} = 0$) of the Brillouin zone, and spherical dispersion energy curves. We have

$$E_c(\mathbf{k}) = E_G + \frac{\hbar^2 k^2}{2\,m_c^*}\,, \qquad E_v(\mathbf{k}) = -\frac{\hbar^2 k^2}{2\,m_v^*}\,, \qquad E_c(\mathbf{k}) - E_v(\mathbf{k}) = E_G + \frac{\hbar^2 k^2}{2\,\mu}\,,$$

where $m_c^* > 0, m_v^* > 0$ and $1/\mu = (1/m_c^*) + (1/m_v^*)$. The dielectric function (22), in the specific case of the isotropic two-band model, becomes

$$\varepsilon_2(\omega) = \frac{8\pi^2 e^2}{m^2\omega^2} \int_{B.Z.} \frac{d\mathbf{k}}{(2\pi)^3} |\mathbf{e} \cdot \mathbf{M}_{cv}(\mathbf{k})|^2\, \delta(E_G + \frac{\hbar^2 k^2}{2\,\mu} - \hbar\omega)\,. \qquad (33)$$

We consider first the case of *dipole-allowed* interband transitions between the valence and the conduction band extrema (*first-class transitions*). When the dipole matrix element $\mathbf{e} \cdot \mathbf{M}_{cv}(\mathbf{k})$ is different from zero at the point $\mathbf{k} = 0$, it can be taken as approximately constant in the whereabouts, i.e. we can assume $\mathbf{e} \cdot \mathbf{M}_{cv}(\mathbf{k}) \approx C_1$; Eq. (33) becomes

$$\varepsilon_2(\omega) = \frac{8\pi^2 e^2}{m^2\omega^2} |C_1|^2 \int_0^\infty \frac{1}{(2\pi)^3} 4\pi\, k^2\, \delta(E_G + \frac{\hbar^2 k^2}{2\,\mu} - \hbar\omega)\, dk\,.$$

The integral can be performed trivially with the substitution $x = \hbar^2 k^2/2\mu$ and one obtains

$$\boxed{\textbf{first} - \textbf{class transitions}: \quad \varepsilon_2(\omega) = \frac{A_1}{\omega^2}\,(\hbar\omega - E_G)^{1/2} \qquad \hbar\omega > E_G}\,, \qquad (34)$$

where the constant is $A_1 = (2e^2/m^2)\,|C_1|^2\,(2\mu/\hbar^2)^{3/2}$. Near the threshold of allowed direct transitions, the singular part of ε_2 is of the type $(\hbar\omega - E_G)^{1/2}$.

We consider now the case of *dipole-forbidden* transitions between the valence and the conduction band extrema (*second-class transitions*). When the dipole matrix element $\mathbf{M}_{cv}(\mathbf{k})$ vanishes at $\mathbf{k} = 0$, it can be taken approximately linear in $\mathbf{k}$ in the whereabouts, i.e. we can assume $\mathbf{M}_{cv}(\mathbf{k}) \approx C_2\,\mathbf{k}$. With straightforward elaboration, we obtain from Eq. (33) the expression

$$\boxed{\textbf{second} - \textbf{class transitions}: \quad \varepsilon_2(\omega) = \frac{A_2}{\omega^2}\,(\hbar\omega - E_G)^{3/2} \qquad \hbar\omega > E_G}\,, \qquad (35)$$

where the constant is $A_2 = (2e^2/3m^2)\,|C_2|^2\,(2\mu/\hbar^2)^{5/2}$. Near the threshold of forbidden direct transitions, the singular part of ε_2 is of the type $(\hbar\omega - E_G)^{3/2}$.

Exciton effects in isotropic two-band model semiconductor

The theory of exciton states in semiconductors and insulators has been considered in Section VII-1; the theory requires (often demanding) solutions of appropriate integral equations. In several crystals, and especially in semiconductors with small effective masses and high dielectric constant, the description of excitons can be done in terms of the hydrogen atom model in a polarizable medium; the envelope function $F(\mathbf{r})$, that describes the relative motion of the electron–hole pair, satisfies the Schrödinger equation

$$\left[-\frac{\hbar^2 \nabla^2}{2\mu} - \frac{e^2}{\varepsilon\,r} \right] F(\mathbf{r}) = (E - E_G)\,F(\mathbf{r})\,. \qquad (36a)$$

The effective Rydberg energy R_{ex} and the effective radius a_{ex} of the hydrogenic atom (36a) are given by the scaling laws

$$R_{\mathrm{ex}} = R\,\frac{\mu}{m}\,\frac{1}{\varepsilon^2} \qquad \text{and} \qquad a_{\mathrm{ex}} = a_B\,\frac{\mu}{m}\,\varepsilon\,, \qquad (36b)$$

where $R{=}13.606\,\mathrm{eV}$ is the Rydberg energy and $a_B{=}0.529\,\mathrm{\AA}$ is the Bohr radius.

As discussed in Section VII-1, the singlet exciton states with vanishing wavevector (the very ones of interest in optical transitions) can be expressed as a linear combination of trial excited states $\Psi_{c\mathbf{k},v\mathbf{k}}^{(S=0)}$ (of total spin $S = 0$ and total wavevector equal to zero) in the form

$$\Psi_{\mathrm{ex}} = \sum_{\mathbf{k}} A(\mathbf{k})\,\Psi_{c\mathbf{k},v\mathbf{k}}^{(S=0)}\,; \qquad (37a)$$

the coefficients $A(\mathbf{k})$ are related to the normalized envelope function $F(\mathbf{r})$ by the Fourier transform

$$F(\mathbf{r}) = \frac{1}{\sqrt{V}} \sum_{\mathbf{k}} A(\mathbf{k})\,e^{i\,\mathbf{k}\cdot\mathbf{r}}\,, \qquad (37b)$$

where V is the volume of the system, and $F(\mathbf{r})$ is normalized to one. The probability per unit time of a transition from the fundamental ground-state Ψ_0 of the semiconductor to the exciton state Ψ_{ex}, induced by an electromagnetic wave of angular frequency ω and polarization versor $\mathbf{e}$, is given in the dipole approximation by the expression

$$P_{\Psi_{\text{ex}} \leftarrow \Psi_0} = \frac{2\pi}{\hbar} \left(\frac{e\,A_0}{m\,c} \right)^2 2 \left| \sum_{\mathbf{k}} A(\mathbf{k}) \langle \psi_{c\mathbf{k}} | \mathbf{e} \cdot \mathbf{p} | \psi_{v\mathbf{k}} \rangle \right|^2 \delta(E_{\text{ex}} - E_0 - \hbar\omega) \qquad (37c)$$

(where a factor 2 takes into account spin degeneracy).

We specify Eq. (37c) for *first-class transitions*, and take $\langle \psi_{c\mathbf{k}} | \mathbf{e} \cdot \mathbf{p} | \psi_{v\mathbf{k}} \rangle = C_1$ as a constant. From Eq. (37b) and Eq. (37c), we obtain

$$P_{\Psi_{\text{ex}} \leftarrow \Psi_0} = \frac{2\pi}{\hbar} \left(\frac{e\,A_0}{m\,c} \right)^2 2 |C_1|^2 V |F(0)|^2 \delta(E_{\text{ex}} - E_0 - \hbar\omega) . \qquad (38)$$

It is well known that the envelope function $F(\mathbf{r})$, satisfying the hydrogenic Schrödinger equation (36a), is different from zero at the origin only for spherically symmetric wavefunctions (i.e. for s-states); thus only s-like excitons can be excited in the case of first-class transitions. For bound s-states of principal quantum number $n = 1, 2, 3, \ldots$ we have $F(0) = 1/\sqrt{\pi a_{\text{ex}}^3 n^3}$. The absorption spectrum for energies below the energy gap consists of a series of lines with energies

$$E_n = E_G - \frac{R_{\text{ex}}}{n^2} \quad (n = 1, 2, 3, \ldots) \quad \text{and intensities} \quad I_n \div \frac{1}{\pi a_{\text{ex}}^3 n^3} . \qquad (39a)$$

For photon energies $\hbar\omega > E_G$, we consider the envelope function $F(E, \mathbf{r})$, with $E = \hbar\omega - E_G$, describing the s-like ionized states of the hydrogenic system (36); as an effect of the electron–hole Coulomb interaction, the dielectric function (34) of interband transitions is modified by the factor $|F(E, \mathbf{r} = 0)|^2 = \pi x \exp(\pi x)/\sinh(\pi x)$, where $x = \sqrt{R_{\text{ex}}/E}$ [for the properties of the hydrogen wavefunctions for bound and unbound states see for instance N. F. Mott and H. S. W. Massey "Theory of Atomic Collisions" (Clarendon Press, Oxford 1949)]. The dielectric function including exciton effects thus becomes

$$\varepsilon_2^{(\text{ex})}(\omega) = \varepsilon_2(\omega) \frac{\pi x\, e^{\pi x}}{\sinh \pi x} \qquad x = \sqrt{\frac{R_{\text{ex}}}{\hbar\omega - E_G}} , \qquad (39b)$$

where $\hbar\omega > E_G$, and $\varepsilon_2(\omega)$ is given by Eq. (34). Thus, exciton effects not only introduce discrete levels at energies below the energy gap, but also modify the optical properties in the continuum in an energy region extending for several R_{ex} above the energy gap.

In Fig. 10a, we give a schematic diagram of the imaginary part of the dielectric function for first-class transitions. Experimental data of the absorption spectrum at the fundamental edge of rare gas solids (dipole-allowed insulators) and gallium arsenide (dipole-allowed semiconductor) are reported in Fig. VII-1 and Fig. VII-2; it can be seen that the relevant experimental features of the excitonic absorption spectrum are well understood qualitatively within the simplified model so far discussed; furthermore, whenever necessary, the model can be appropriately implemented to improve quantitative agreement with experiments.

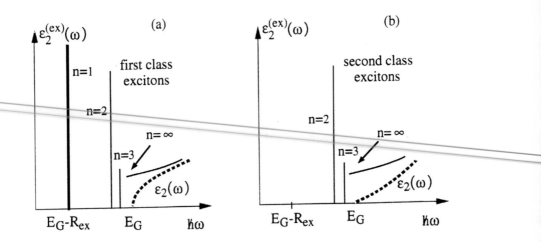

Fig. 10 Schematic diagram of the imaginary part of the dielectric function for first-class and second-class optical transitions neglecting exciton effects (dashed lines) and including exciton effects (solid lines).

A similar analysis can be carried out for *second-class transitions*. In this case the dipole matrix element can be taken in the form $\langle \psi_{c\mathbf{k}} | \mathbf{p} | \psi_{v\mathbf{k}} \rangle = C_2 \, \mathbf{k}$; from Eq. (37b) and Eq. (37c), we obtain

$$P_{\Psi_{\mathrm{ex}} \leftarrow \Psi_0} = \frac{2\pi}{\hbar} \left(\frac{e\,A_0}{m\,c} \right)^2 2\,|C_2|^2\,V\,|\mathbf{e} \cdot \nabla F(\mathbf{r})_{\mathbf{r}=0}|^2\,\delta(E_{\mathrm{ex}} - E_0 - \hbar\omega)\;.$$

Notice that $\nabla F(\mathbf{r})$ can be different from zero at $\mathbf{r}=0$ only for envelope functions with p character. The absorption below the energy gap consists of a set of lines with energies

$$E_n = E_G - \frac{R_{\mathrm{ex}}}{n^2} \quad (n = 2, 3, 4, \ldots) \quad \text{and intensities} \quad I_n \div \frac{n^2 - 1}{\pi\,a_{\mathrm{ex}}^5\,n^5}\;. \qquad (40a)$$

Notice that the $n = 1$ line is missing in this theory and indeed, when experimentally detected, the $n = 1$ line is much weaker than the following members of the series. For energies higher than the energy gap we have

$$\varepsilon_2^{(\mathrm{ex})}(\omega) = \varepsilon_2(\omega) \frac{\pi\,x\,(1 + x^2)\,e^{\pi x}}{\sinh \pi x} \qquad x = \sqrt{\frac{R}{\hbar\omega - E_G}}\,, \qquad (40b)$$

where $\hbar\omega > E_G$, and $\varepsilon_2(\omega)$ is given by Eq. (35). In Fig. 10b we give a schematic diagram of the dielectric function for second-class transitions.

As an example of second-class exciton transitions we consider here the case of excitons at the fundamental edge of Cu_2O crystals (see Fig. 11). Cuprous oxide is a direct gap material with the minimum of the conduction band (Γ_6^+) and the maxima of the valence bands (Γ_7^+ and Γ_8^+) at the point $\mathbf{k} = 0$ of the Brillouin zone; the energy gap is about $E_G = 2.172\,\mathrm{eV}$; the splitting $\Delta_{SO} = 0.130\,\mathrm{eV}$ between the valence states Γ_7^+ and Γ_8^+ is due to the spin–orbit interacton. Two exciton series have been reported, originating from the Γ_7^+ and Γ_8^+ valence bands, and called the yellow and green exciton

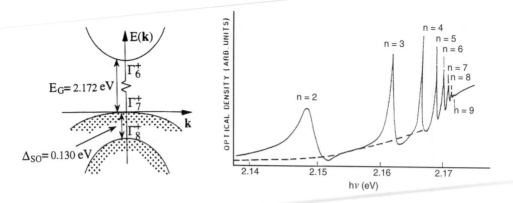

Fig. 11 (a) Schematic band structure of Cu_2O showing the conduction band Γ_6^+, and the valence bands Γ_7^+ and Γ_8^+, split by spin–orbit interaction. (b) Absorption spectrum of the yellow exciton series in Cu_2O at 1.8 K [from K. Shindo, T. Goto and T. Anzai, J. Phys. Soc. Japan **36**, 753 (1974)]. Since the band-to-band transition is dipole forbidden at the symmetry point $\mathbf{k} = 0$, the exciton series begins with the line $n = 2$.

series, respectively. Since the conduction and the valence bands have the same parity (the even parity with respect to the spatial inversion is denoted by the superscript $+$), the dipole matrix element between them vanishes, and the results of second-class transitions apply. Experimental data of the yellow exciton series in cuprous oxide are reported in Fig. 11b, where transitions to exciton levels from $n = 2$ to $n = 9$ are detected. The asymmetry of the absorption lines is due to their interaction with some background continuum (see the Fano effect in next section).

We conclude these considerations on the optical properties of Cu_2O by comparing the one-photon and two-photon absorption spectra to exciton states. In one-photon spectroscopy of Cu_2O, we have seen that the accessible final states are p-like excitons; hence, following the considerations on the dipole selection rules of Section 4, we infer that in two-photon spectroscopy the accessible final states are s-like and d-like excitons, as nicely shown by the experimental data of Fig. 12.

Considerations on excitonic-polaritons and other remarks

It is of interest to comment briefly on the behaviour of the optical constants of a material near an extremely narrow exciton line, say the $n = 1$ exciton line of a first-class exciton series. The exciton line of interest, at energy $\hbar\omega_{ex} = E_{ex} - E_0$, is supposed to be well separated from all the other partners of the series and from the continuum. From Eq. (38), we notice explicitly that *the transition rate to exciton lines is proportional to the volume V of the system* because so is the modulus squared of the dipole matrix element [on the contrary, the transition rate to one-electron excited states is independent of the volume of the crystal because so is the dipole matrix element between two Bloch functions; for band-to-band transitions, it is the density-of-states that is proportional to the volume]. Inserting Eq. (38) into Eq. (9), we see that $\varepsilon_2(\omega)$

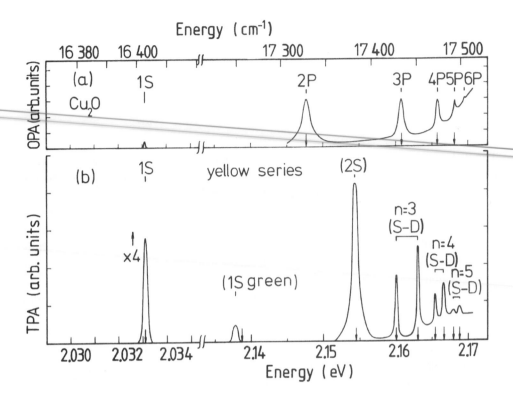

Fig. 12 One-photon absorption (OPA) and two-photon absorption (TPA) spectra of Cu₂O. One-photon data at 4.2 K are from J. B. Grun and S. Nikitine, J. Phys. (Paris) **23**, 159 (1962). Two-photon data at 4.5 K are from Ch. Uihlein, D. Fröhlich and R. Kenklies, Phys. Rev. **B23**, 2731 (1981) (copyright 1981 by the American Physical Society). The arrows indicate the calculated exciton energies in an appropriate multiband model.

for $\omega = \omega_{\rm ex}$ takes the form of a delta-like function with *finite strength* and becomes

$$\varepsilon_2(\omega) = A\,\delta(\omega_{\rm ex} - \omega)\ , \tag{41a}$$

where A is given by $(8\pi^2 e^2/m^2\omega_{\rm ex}^2\,\hbar)\,|C_1|^2\,|F(0)|^2$.

Using the Kramers–Kronig dispersion relation (10b), we obtain for the real part of the dielectric function

$$\varepsilon_1(\omega) = 1 + \frac{2}{\pi}\,P\int_0^{\omega_{\rm ex}+\delta} \frac{\omega'\,\varepsilon_2(\omega')}{\omega'^2 - \omega^2}\,d\omega' + \frac{2}{\pi}\,P\int_{\omega_{\rm ex}+\delta}^{+\infty} \frac{\omega'\,\varepsilon_2(\omega')}{\omega'^2 - \omega^2}\,d\omega'$$

$$= 1 + \frac{2}{\pi}\,A\,\frac{\omega_{\rm ex}}{\omega_{\rm ex}^2 - \omega^2} + \frac{2}{\pi}\,P\int_{\omega_{\rm ex}+\delta}^{+\infty} \frac{\omega'\,\varepsilon_2(\omega')}{\omega'^2 - \omega^2}\,d\omega'\ ,$$

where δ is a (small) positive quantity. We suppose that the contribution of the last integral is approximately constant in the frequency region below $\omega \approx \omega_{\rm ex}$; such a contribution, added to 1, is referred as the background dielectric constant ε_b; in the frequency range extending from zero to about ω_{ex} we have

$$\varepsilon_1(\omega) = \varepsilon_b + \frac{2}{\pi} A \frac{\omega_{\mathrm{ex}}}{\omega_{\mathrm{ex}}^2 - \omega^2} . \tag{41b}$$

We indicate by $\varepsilon_s \equiv \varepsilon_1(0)$ the static dielectric function, and re-write Eq. (41b) in the form

$$\varepsilon_1(\omega) = \varepsilon_b + (\varepsilon_s - \varepsilon_b) \frac{\omega_{\mathrm{ex}}^2}{\omega_{\mathrm{ex}}^2 - \omega^2} = \frac{\varepsilon_s \omega_{\mathrm{ex}}^2 - \varepsilon_b \omega^2}{\omega_{\mathrm{ex}}^2 - \omega^2} . \tag{41c}$$

The dielectric function of Eqs. (41) (apart for some obvious change in notations) has the same form as the dielectric function of Eqs. (IX-55), which directly led to the concept of polariton states in polar crystals. The analogy, of course, is not only formal but also substantial, and the mixed exciton-photon modes propagating in the crystal are called "exciton-polaritons". The experimental dispersion curves of polaritons in GaP in the infrared region have been reported in Fig. IX-11; the figure can be usefully compared with the experimental dispersion curves of exciton-polaritons in CuCl in the visible region reported in Fig. IX-12.

The study of the many-body effects in the optical spectra of solids gives a wealth of experimental and theoretical information. The interpretation of interband spectra is often a challenging problem, which is complicated and enriched by the presence of degenerate bands, spin–orbit interaction, anisotropy effects, autoionization effects with some background continuum. The exciton binding energies vary from a few electronvolts in large gap insulators (where the dielectric constant is relatively small and effective masses relatively large) to a few milli-electronvolts in small gap semiconductors. Core excitons in semiconductors and insulators (and in particular comparison of binding energies and oscillator strengths with valence excitons) are also of major interest. We wish also to mention that in metals the many-body exciton effects are responsible of the sharpness of the absorption coefficient from certain core states to the conduction band, at or near the Fermi level [for the soft X-ray edge problem and the Mahan–Nozières–DeDominicis theory, see for instance K. Ohtaka and Y. Tanabe, Rev. Mod. Phys. **62**, 929 (1990) and references quoted therein].

6   Fano resonances and absorption lineshapes

In this overview on the optical properties of solids, we consider now the absorption lineshapes occurring when one or more discrete levels interact with a continuum of states. The problem was originally considered by Fano in 1935 and later rivisited by the same author within the Green's function formalism [U. Fano, Nuovo Cimento **12**, 156 (1935); Phys. Rev. **124**, 1866 (1961)]. The most remarkable feature of the absorption spectrum is the occurrence of peculiar "asymmetric" and "window" lineshapes, generated by the configuration interaction (mixing) between the discrete and continuum states.

Let us consider a discrete state $|\Phi_e\rangle$ interacting ("resonant") with a continuum set of states $|\Phi_n\rangle$ (also called "ionization channel of scattering states"). This system can be

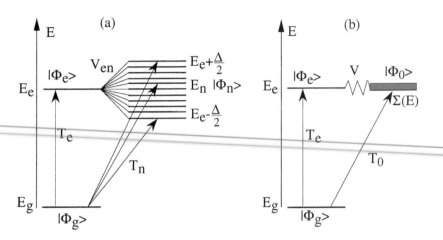

Fig. 13 (a) Schematic representation of a discrete state $|\Phi_e\rangle$ (of energy E_e) interacting with a continuum of states $|\Phi_n\rangle$ ($n = 1, 2, \dots, N$) with energies in the interval $[E_e - \Delta/2, E_e + \Delta/2]$; $|\Phi_g\rangle$ is the ground state from which transitions take place; T_e and T_n denote the dipole matrix elements connecting the ground state to the excited states. (b) Schematic representation of the reduced two-level Hamiltonian describing the electronic system; E_e and $\Sigma(E)$ are the site energies of $|\Phi_e\rangle$ and $|\Phi_0\rangle$, respectively; V is the interaction matrix element.

schematized as in Fig. 13a, where also the ground state $|\Phi_g\rangle$ from which the transitions take place is indicated. Physical systems that can be reduced to such a schematization are numerous: for instance, $|\Phi_g\rangle$ may represent the ground electronic state of a semi-conductor (or insulator); $|\Phi_e\rangle$ may represent a discrete member of an exciton Rydberg series involving a core band and the conduction band of lowest energy; the states $|\Phi_n\rangle$ ($n = 1, 2, \dots, N$ and N very large) may mimic a continuum of excitations from valence states to conduction states at energies far above the fundamental threshold.

The Hamiltonian of a system constituted by a discrete state $|\Phi_e\rangle$ of energy E_e, the continuum of states $|\Phi_n\rangle$ and their interactions, can be represented as follows

$$H = E_e |\Phi_e\rangle\langle\Phi_e| + \sum_{n=1}^{N} E_n |\Phi_n\rangle\langle\Phi_n| + \frac{1}{\sqrt{N}} \sum_{n=1}^{N} [V_{en} |\Phi_e\rangle\langle\Phi_n| + V_{en}^* |\Phi_n\rangle\langle\Phi_e|] \quad (42)$$

(the states Φ_e and Φ_n are normalized to unity). In order to avoid inessential details, we assume for simplicity that the interaction matrix elements $V_{en} = V$ are independent on the index n labelling the states of the continuum. The Hamiltonian (42) can then be written as

$$H = E_e |\Phi_e\rangle\langle\Phi_e| + H_{cont} + V |\Phi_e\rangle\langle\Phi_0| + V^* |\Phi_0\rangle\langle\Phi_e| , \quad (43a)$$

where

$$H_{cont} = \sum_{n=1}^{N} E_n |\Phi_n\rangle\langle\Phi_n| \quad \text{and} \quad |\Phi_0\rangle = \frac{1}{\sqrt{N}} \sum_{n=1}^{N} |\Phi_n\rangle ; \quad (43b)$$

in the above expressions, H_{cont} denotes the Hamiltonian corresponding to the continuum of states $|\Phi_n\rangle$ $(n = 1, 2, 3, \dots)$, and $|\Phi_0\rangle$ is a linear combination (normalized to unity) of them.

The basic obstacle in dealing with the Hamiltonian (43) is the arbitrary large number of states of the continuum to be handled. A possible way for a proper account of the continuum exploits the Green's function $G(E) = (E - H_{\text{cont}})^{-1}$ and specifically its diagonal matrix element

$$G_{00}(E) = \langle \Phi_0 | \frac{1}{E - H_{\text{cont}}} | \Phi_0 \rangle = \frac{1}{E - \Sigma(E)} \; ; \tag{44}$$

as usual, it is understood that the energy E is accompanied by a small imaginary part $i\eta$ and the limit $\eta \to 0^+$ is taken. The explicit calculation of the self-energy $\Sigma(E)$ can be done (for instance) with the recursion method, constructing the semi-infinite chain of states beginning with the seed state $|\Phi_0\rangle$ and applying the operator H_{cont}; also notice that in the case the dipole matrix elements T_n from the ground state to the continuum are independent from the index n, all the hierarchical states of the semi-infinite chain (but the first one) are dipole forbidden (as a consequence of their orthogonalization to any other state of the chain and to $|\Phi_0\rangle$ in particular). The Hamiltonian (43) can thus be replaced by the effective equivalent Hamiltonian with two states

$$H_{\text{eff}} = E_e |\Phi_e\rangle\langle\Phi_e| + \Sigma(E) |\Phi_0\rangle\langle\Phi_0| + V |\Phi_e\rangle\langle\Phi_0| + V^* |\Phi_0\rangle\langle\Phi_e| \; ; \tag{45}$$

the reduced two-level Hamitonian H_{eff}, which mimics the excited states of the electron system, is indicated schematically in Fig. 13b.

In the following, we do not need to be very specific about $\Sigma(E)$ and $G_{00}(E)$. We simply observe that in the case of a constant density-of-states of the continuum in the interval $[-\Delta/2, +\Delta/2]$ we have

$$\begin{cases} -\frac{1}{\pi} \operatorname{Im} G_{00}(E) = \frac{1}{\Delta} & -\Delta/2 \le E \le +\Delta/2 \\ \operatorname{Re} G_{00}(E) = 0 & \text{for } E \approx 0 \end{cases}. \tag{46}$$

The first of the above equations follows from the assumption that the density-of-states of the continuum $n(E) = -(1/\pi) \operatorname{Im} G_{00}(E)$ is constant in the energy interval Δ around the resonant state energy E_e (we have assumed without loss of generality that $E_e = 0$); the second is due to the fact that the ionization channel is taken symmetric with respect to the energy E_e of the resonant state.

On the basis $(|\Phi_e\rangle, |\Phi_0\rangle)$, the Hamiltonian H_{eff} and the operator $E - H_{\text{eff}}$ can be written in the matrix form

$$H_{\text{eff}} = \begin{pmatrix} 0 & V \\ V^* & \Sigma(E) \end{pmatrix}, \qquad E - H_{\text{eff}} = \begin{pmatrix} E & -V \\ -V^* & 1/G_{00}(E) \end{pmatrix}.$$

From a straightforward inversion of the above two-by-two matrix, the Green's function $G_{\text{eff}}(E) = (E - H_{\text{eff}})^{-1}$ is obtained; we have

$$G_{\text{eff}}(E) = \frac{G_{00}(E)}{E - V V^* G_{00}(E)} \begin{pmatrix} 1/G_{00}(E) & V \\ V^* & E \end{pmatrix}. \tag{47a}$$

We consider the dipole carrying state

$$|\chi\rangle = T_e\,|\Phi_e\rangle + T_0\,|\Phi_0\rangle \;,$$

and with straightforward algebra obtain

$$\langle\chi|G_{\text{eff}}(E)|\chi\rangle = \frac{|T_e|^2 + \left[\,T_e^*\,T_0\,V + T_e\,T_0^*\,V^* + |T_0|^2\,E\,\right]\,G_{00}(E)}{E - V\,V^*\,G_{00}(E)} \;; \tag{47b}$$

the imaginary part of the above expression is

$$\text{Im}\,\langle\chi|G_{\text{eff}}(E)|\chi\rangle = \frac{|T_e|^2\,|V|^2 + (T_e^*\,T_0\,V + T_e\,T_0^*\,V^*)\,E + |T_0|^2\,E^2}{|E - V\,V^*\,G_{00}(E)|^2}\,\text{Im}\,G_{00}(E)\;.$$

We now use expression (20) and obtain for the lineshape

$$\boxed{\;I(E) = \frac{|T_0\,E + T_e\,V^*|^2}{|E - V\,V^*\,G_{00}(E)|^2}\left(-\frac{1}{\pi}\right)\text{Im}\,G_{00}(E)\;.\;} \tag{48}$$

We can give a simple appropriate elaboration of Eq. (48), using Eq. (46) for $G_{00}(E)$; for E around the resonance at $E \approx 0$, we have

$$I(E) = \frac{1}{\Delta}\,\frac{|T_0\,E + T_e\,V^*|^2}{E^2 + \pi^2\,|V|^4/\Delta^2}\;. \tag{49}$$

It is convenient to define the *broadening parameter* Γ (and express the photon energy in units of $\Gamma/2$) and *the profile index q* as follows

$$\Gamma = 2\pi\frac{|V|^2}{\Delta}\;, \qquad \varepsilon = \frac{E}{\Gamma/2}\;, \qquad q = \frac{1}{\pi}\frac{T_e}{T_0}\frac{\Delta}{V}\;.$$

Eq. (49) becomes

$$\boxed{\;I(E) = |T_0|^2\,\frac{1}{\Delta}\,\frac{|\varepsilon + q|^2}{\varepsilon^2 + 1}\;.\;} \tag{50}$$

In most ordinary situations the Hamiltonian of the system is invariant under time-reversal symmetry, wavefunctions can be taken as real and so also the profile index q is real. The family of curves giving the natural lineshape $I(E)$ for different values of the ratio q is given in Fig. 14.

In order to clarify the meaning of Eq. (50), we consider the limiting case in which only the discrete state $|\Phi_e\rangle$ is dipole connected to the ground state (i.e. $T_e \neq 0$ and $T_0 \equiv 0$). In this case the profile index $q \to \infty$ and, in this limit, Eq. (50) gives

$$I(E) = \frac{1}{\Delta}\,\frac{|q\,T_0|^2}{\varepsilon^2 + 1} = |T_e|^2\,\frac{1}{\pi}\,\frac{\Gamma/2}{E^2 + (\Gamma/2)^2}\;. \tag{51a}$$

Eq. (51a) is a symmetric Lorentzian curve, with broadening parameter Γ. We consider now the opposite limiting case in which only the seed state $|\Phi_0\rangle$ is dipole connected to the ground state (i.e. $T_e \equiv 0$ and $T_0 \neq 0$). In this case the index profile q vanishes and Eq. (50) becomes

$$I(E) = |T_0|^2\,\frac{1}{\Delta}\,\frac{\varepsilon^2}{\varepsilon^2 + 1} = |T_0|^2\,\frac{1}{\Delta}\left(1 - \frac{1}{\varepsilon^2 + 1}\right)\;. \tag{51b}$$

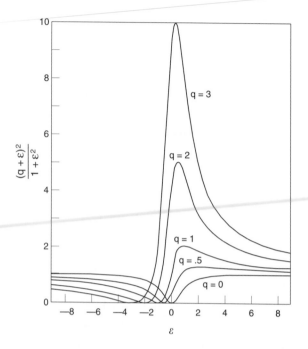

Fig. 14 Natural lineshapes for different values of q (reverse the scale of abscissas for negative q); asymmetry and antiresonances caused by interference between the discrete state and the background are shown [from U. Fano, Phys. Rev. **124**, 1866 (1961); copyright 1961 by the American Physical Society].

The above equation shows that $I(E)$ vanishes for $\varepsilon = 0$, i.e. at the resonance energy of the discrete state; the dipole-forbidden discrete state makes its presence felt because it carves a "window" of width Γ in the background absorption $|T_0|^2/\Delta$ (*spectral repulsion principle*).

In general the dipole transitions are allowed both to the discrete and to the continuum, and the profile index q is finite. Eq. (50) for real q, can be written as

$$I(E) = |T_0|^2 \frac{1}{\Delta}\left[1 + \frac{q^2}{\varepsilon^2 + 1} + \frac{2q\,\varepsilon - 1}{\varepsilon^2 + 1}\right] . \tag{52}$$

The three terms in the right-hand side of the above equation are the background absorption, a symmetric Lorentzian peak, and a term that represents the effect of interference; note in particular that for $\varepsilon = -q$ no transition takes place (*antiresonance*).

As an illustration of Fano profiles, we report in Fig. 15 and Fig. 16 the absorption spectra of neon and argon rare gas atoms and solids in the soft X-ray region; in Table 1 we report the line shape parameters, that best fit the absorption cross-section of the Fano theory. Notice the asymmetric lines occurring in the absorption spectrum of neon, and the window lines occurring in the absorption spectrum of argon.

The model considered so far for a single resonant state interacting with one ionization channel of scattering states can be generalized; in the case of several resonances

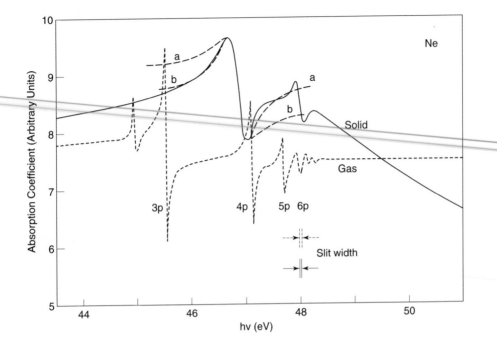

Fig. 15 Absorption spectrum of solid Ne (solid curve) and gaseous Ne (dashed curve). The theoretical curves a and b are fitted on the absorption cross-section of Fano, with two different sets of parameters (see Table 1) [from R. Haensel, G. Keitel, C. Kunz and P. Schreiber, Phys. Rev. Lett. **25**, 208 (1970); copyright 1970 by the American Physical Society].

Table 1 Line shape parameters of the X-ray absorption resonance in neon and argon (E_0 is the resonance energy, Γ the broadening parameter, q the line shape parameter), as reported from R. Haensel, G. Keitel, C. Kunz and P. Schreiber, Phys. Rev. Lett. **25**, 208 (1970) (copyright 1970 by the American Physical Society).

	E_0 (eV)	Γ (eV)	q
Ne gas	45.55	0.013	-1.6
Ne solid (fit a)	46.90	0.34	-0.76
Ne solid (fit b)	46.81	0.36	-1.31
Ar gas	26.614	0.08	-0.22
Ar solid (fit a)	27.515	0.12	-0.39
Ar solid (fit b)	27.525	0.13	-0.20

much care is required to discuss the interference effects. For appropriate procedures that adopt the recursion method to map the ionization channel into an appropriate hierarchy of chain-like states we refer to the literature [see for instance V. Dolcher, G. Grosso and G. Pastori Parravicini, Phys. Rev. B**46**, 9312 (1992) and references quoted therein].

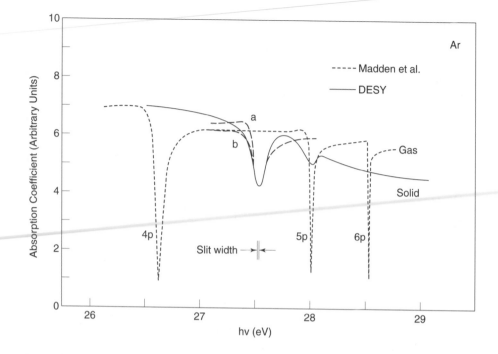

Fig. 16 Absorption spectrum of solid Ar (solid curve) and gaseous Ar (dashed curve). The theoretical curves a and b are fitted on the absorption cross-section of Fano, with two different sets of parameters (see Table 1) [from R. Haensel, G. Keitel, C. Kunz and P. Schreiber, Phys. Rev. Lett. **25**, 208 (1970) (copyright 1970 by the American Physical Society); the absorption spectrum of gaseous Ar is given by R. P. Madden, D. L. Ederer and K. Codling, Phys. Rev. **177**, 136 (1969)].

7   Optical properties of vibronic systems

Thus far, in the study of the optical transitions, we have tacitly assumed that the *nuclear equilibrium positions are the same, before and after the electronic excitations take place* (this is equivalent to assume that the adiabatic potential surfaces of interest have their minima at the same nuclear configuration with the same curvature). Although this is the ordinary situation, we have seen in Chapter VIII examples of vibronic systems of more general nature. When electronic transitions occur between adiabatic potential sheets with minima in different nuclear configurations (or between degenerate adiabatic sheets), the lattice cannot be disregarded and important effects on the optical properties arise. We thus complete this chapter with a survey of the basic principles of the optical constants of vibronic systems.

7.1   Optical properties of the Franck–Condon vibronic model

We begin with the simplest vibronic model, i.e. the Franck–Condon model, which is constituted by an *electronic system with only two levels for each lattice configuration*

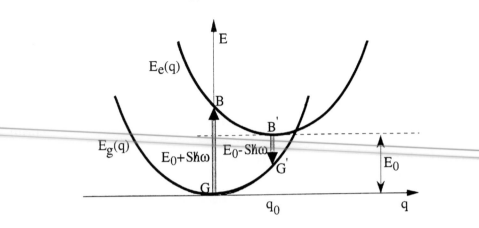

Fig. 17 Configuration-coordinate diagram for allowed transitions between two non-degenerate electronic states (Franck–Condon model). The adiabatic potential surfaces for the ground state and the lowest excited state are indicated as a function of a single normal coordinate q.

and a one-dimensional vibrational mode for the lattice (as schematized in Fig. VIII-2b, and reconsidered in further detail in Fig. 17). Several physical systems can be schematized, at least as a first approach, with such a model. The simplest case is that of a diatomic molecule with two (non-degenerate) electronic states, between which optical transitions can occur. Also in the study of localized impurities in solids, the Franck–Condon model has found wide applications; it is the basis for a simple semi-quantitative explanation for the lineshapes for photon absorption and for luminescence (i.e. photon emission after excitation) in the optical properties of localized centers.

Consider an electronic system with two (non-degenerate) levels for any given lattice configuration, the ground state ψ_g and the excited state ψ_e. Let us indicate with $E_g(q)$ and $E_e(q)$ the energies of the corresponding adiabatic potential surfaces, respectively; the normal coordinate q may represent, for instance, the deviation of the nearest neighbour nuclear distance from its equilibrium value. In the ground state, the equilibrium value of the (collective) coordinate q is assumed to be zero; in the excited state, the equilibrium value is assumed to be q_0; upper and lower adiabatic potential surfaces are assumed to have the same curvature C. The schematic representation of the model is indicated in Fig. 17. With the choice of axes and origin as in Fig. 17, we have

$$\begin{cases} E_g(q) = \frac{1}{2}C\,q^2 \\ E_e(q) = E_0 + \frac{1}{2}C\,(q - q_0)^2 \end{cases} \tag{53}$$

Notice that E_0 represents the energy difference between the minima of the ground and the excited adiabatic potentials, and is called *"zero-phonon transition energy"*; the elastic constant C is related to the frequency ω of the nuclear motion, and the mass M by the usual equation $C = M\,\omega^2$.

It is useful to characterize the model by means of the dimensionless Huang–Rhys parameter S defined as

$$S\,\hbar\omega \equiv \frac{1}{2}C\,q_0^2 = \frac{1}{2}M\,\omega^2\,q_0^2\ ,\tag{54a}$$

or equivalently

$$q_0 = \sqrt{\frac{2\hbar S}{M\,\omega}}\ .\tag{54b}$$

We observe that the energy difference between lower and upper levels at $q = 0$ and $q = q_0$ is given by

$$\begin{aligned}
E_e(0) - E_g(0) &= E_0 + S\,\hbar\omega \\
E_e(q_0) - E_g(q_0) &= E_0 - S\,\hbar\omega\ .
\end{aligned}\tag{54c}$$

We now examine the features of the photon absorption spectrum and luminescence, preliminarily from the semiclassical point of view, and then with the quantum mechanical treatment.

In accordance with an intuitive picture, known as "Franck–Condon principle", the nuclear configuration of a system (molecules, clusters, impurities, crystals) cannot change during the short time of an electron transition; in other terms, optical transitions are *vertical* in the configuration diagram, because the nuclei do not move of any significant amount during the electronic transitions. When the system is initially at the minimum of the ground adiabatic sheet of Fig. 17, we expect that the *absorption spectrum* is peaked at the energy

$$E_{\mathrm{abs}} = E_0 + S\,\hbar\omega\ .\tag{55a}$$

Similarly, when the system is initially at the minimum of the excited adiabatic sheet, we expect that the *emission spectrum* is peaked at the energy

$$E_{\mathrm{emiss}} = E_0 - S\,\hbar\omega\ .\tag{55b}$$

The difference between the peaks of the absorption and emission bands is thus expected to be $2\,S\,\hbar\omega$.

We can examine more closely the photon absorption and photon emission processes. Consider the system initially at the minimum of the ground adiabatic sheet (point G in Fig. 17). During the optical transition from G to B the nuclei remain at rest, leading to an absorption energy $E_0 + S\,\hbar\omega$. After the absorption, in the excited adiabatic curve, the nuclei are no more in the equilibrium position, and the system moves to the minimum B' at energy E_0 releasing an energy equal to $S\,\hbar\omega$; this relaxation process occurs in a time of the order of $\approx 10^{-13}$ sec, the typical time of lattice vibrations. We thus see that the dimensionless parameter S takes the meaning of the average number of phonons accompanying the optical transition; $S\,\hbar\omega$ is the energy transferred into vibrational energy (and then into heat). The system in the excited state B' survives for a lifetime of the order of $\approx 10^{-8}$ sec (in typical ordinary situations), which is about 10^5 times the period of the lattice vibrations. The emission from B' to G' takes place vertically leading to emission energy $E_0 - S\,\hbar\omega$; the system then relaxes from

G' to G releasing an energy $S\hbar\omega$ (transformed into heat). Thus luminescence band and absorption band are shifted by $2S\hbar\omega$, which is the energy transformed into heat in the cycle $G\rightarrow B\rightarrow B'\rightarrow G'\rightarrow G$.

Finally, we note that in the particular case that the Huang–Rhys parameter S vanishes (or equivalently $q_0 \equiv 0$), the optical absorption and the optical emission consist of a unique sharp line at the resonance energy E_0, and the nuclear lattice can be disregarded. We pass now to study quantitatively the lineshapes predicted within the Franck–Condon model when S is different from zero (or equivalently $q_0 \neq 0$).

Quantum treatment of the optical properties of the Franck–Condon vibronic model

We consider the Franck–Condon model from a quantum mechanical point of view, taking into account the motion of the nuclei; the energy due to this motion is quantized in phonon quanta, which are in general much smaller than the transition electronic energies. We introduce the standard phonon creation and annihilation operators, $a^\dagger$ and a, corresponding to the harmonic oscillator of potential energy $(1/2)M\omega^2 q^2$, centred at the origin $q = 0$ of the configurational coordinate q (the *oscillator at the origin* is also referred as *undisplaced oscillator*). From Appendix IX-A, and minor changes in notations, we have

$$a = \sqrt{\frac{M\omega}{2\hbar}}\, q + i\sqrt{\frac{1}{2M\hbar\omega}}\, p \,, \qquad a^\dagger = \sqrt{\frac{M\omega}{2\hbar}}\, q - i\sqrt{\frac{1}{2M\hbar\omega}}\, p \,. \qquad (56a)$$

The states of the undisplaced harmonic oscillator are

$$|\phi_n\rangle = \frac{1}{\sqrt{n!}}\, (a^\dagger)^n |\phi_0\rangle \,; \qquad (56b)$$

the eigenstates and eigenvalues of the vibronic system in the ground adiabatic surface are

$$|\psi_g, \phi_n\rangle \quad \text{and} \quad E_{gn} = E_g(0) + \left(n + \frac{1}{2}\right)\hbar\omega \quad n = 0, 1, 2, \dots \qquad (56c)$$

For the excited adiabatic surface we can follow a similar treatment, except for a proper account of the displacement at $q = q_0$ of the minimum of the potential energy surface. We introduce the phonon creation and annihilation operators, $\tilde{a}^\dagger$ and $\tilde{a}$, corresponding to the harmonic oscillator of potential energy $(1/2)M\omega^2(q - q_0)^2$, centred at the point q_0 of the configurational coordinate q (*displaced oscillator*). For the displaced oscillator we have

$$\begin{aligned}
\tilde{a} &= \sqrt{\frac{M\omega}{2\hbar}}\, (q - q_0) + i\sqrt{\frac{1}{2M\hbar\omega}}\, p \equiv a - \sqrt{S} \\
\tilde{a}^\dagger &= \sqrt{\frac{M\omega}{2\hbar}}\, (q - q_0) - i\sqrt{\frac{1}{2M\hbar\omega}}\, p \equiv a^\dagger - \sqrt{S} \,.
\end{aligned} \qquad (57a)$$

The states of the displaced harmonic oscillator are

$$|\tilde{\phi}_n\rangle = \frac{1}{\sqrt{n!}}\, (\tilde{a}^\dagger)^n |\tilde{\phi}_0\rangle \,; \qquad (57b)$$

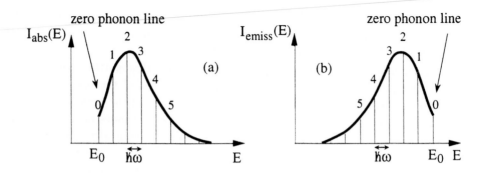

Fig. 18 (a) Palisade spectrum of optical absorption for the two-level single-mode model (the system is initially in the ground electronic state and ground vibrational state); the zero-phonon line is indicated by an arrow. (b) Palisade spectrum of optical emission for the two-level single-mode model (the system is initially in the excited electronic state and ground vibrational state); the zero-phonon line is indicated by an arrow. In the figures we have chosen $S = 2.5$.

the eigenstates and eigenvalues of the vibronic system in the excited adiabatic surface are

$$|\psi_e, \widetilde{\phi}_n\rangle \quad \text{and} \quad E_{en} = E_0 + \left(n + \frac{1}{2}\right)\hbar\omega \quad n = 0, 1, 2, \ldots \tag{57c}$$

To connect the states of the displaced and undisplaced oscillators, we notice the general property of translation operators

$$f(q + q_0) = f(q) + q_0 f'(q) + \frac{1}{2!}q_0^2 f''(q) + \ldots = e^{q_0 \partial/\partial q} f(q) .$$

From Eq. (56a) and Eq. (54b) we have $\sqrt{S}(a - a^\dagger) = i\sqrt{2S/M\hbar\omega}\, p = q_0 \partial/\partial q$; it follows

$$|\widetilde{\phi}_n\rangle = e^{-q_0 \partial/\partial q}|\phi_n\rangle = e^{\sqrt{S}(a^\dagger - a)}|\phi_n\rangle .$$

We also notice the following matrix elements between displaced and undisplaced harmonic oscillator wavefunctions

$$\langle\widetilde{\phi}_n|\phi_0\rangle = \langle\phi_n|e^{-\sqrt{S}(a^\dagger - a)}|\phi_0\rangle = \langle\phi_n|e^{-\sqrt{S}a^\dagger}e^{\sqrt{S}a}|\phi_0\rangle e^{-[-\sqrt{S}a^\dagger, \sqrt{S}a]/2}$$

$$= \langle\phi_n|e^{-\sqrt{S}a^\dagger}|\phi_0\rangle e^{-S/2} = (-1)^n \sqrt{\frac{S^n}{n!}} e^{-S/2} , \tag{58}$$

where use has been made (among others) of the Weil identity of Eq. (IX-A14).

The absorption lineshape at zero temperature (i.e. when the system is initially in the ground electronic state and ground vibational state) is obtained via the golden rule

$$I_{abs}(E) = \frac{2\pi}{\hbar} \left|\langle\psi_e, \widetilde{\phi}_n|T|\psi_g, \phi_0\rangle\right|^2 \delta(E_{en} - E_{g0} - E)$$

$$= \frac{2\pi}{\hbar}|T_{eg}|^2 \left|\langle\widetilde{\phi}_n|\phi_0\rangle\right|^2 \delta(E_0 + n\hbar\omega - E) , \tag{59}$$

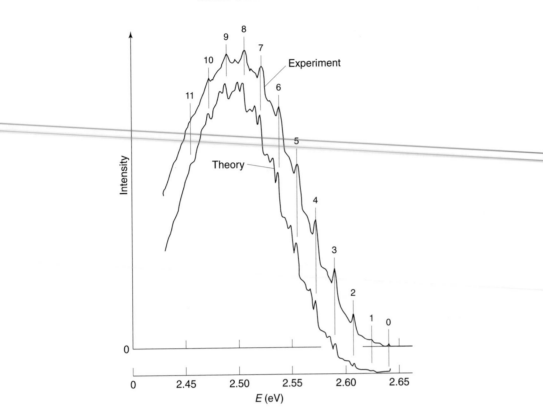

Fig. 19 Comparison of the theoretical and experimental luminescence intensity of AgBr at 2 K containing residual iodine atoms. [Experimental results are taken from W. Czaja, J. Phys. C**16**, 3197 (1983); theoretical results and interpretation are taken from A. Testa, C. Czaja, A. Quattropani and P. Schwendimann, J. Phys. C**20**, 1253 (1987)].

where T_{eg} is the dipole matrix element between the ground electronic state and the excited electronic state. Using Eq. (58), the lineshape (59) (normalized to one) takes the form

$$I_{\mathrm{abs}}(E) = \frac{1}{n!} S^n e^{-S} \delta(E_0 + n\hbar\omega - E) \qquad (n = 0, 1, 2, \dots) \,. \qquad (60)$$

The absorption spectrum $I_{\mathrm{abs}}(E)$ is a palisade of equally spaced lines (at energies E_0, $E_0 + \hbar\omega$, $E_0 + 2\hbar\omega \dots$) with intensities varying according to the Poisson distribution, as schematically shown in Fig. 18a. The factor $\exp(-S)$ gives the fractional intensity of the *zero phonon line*, and is the optical analog of the Debye–Waller factor. The maximum of the Poisson distribution occurs for $n \approx S$, and the absorption spectrum is then peaked at energy $E_0 + S\hbar\omega$, in agreement with Eq. (55a).

A quite similar analysis can be carried out for the luminescence spectrum occuring when the system is initially in the excited electronic state and ground vibrational

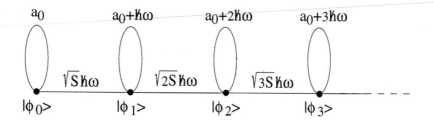

Fig. 20 Graphic representation of the Hamiltonian of the displaced oscillator on the basis of the states $\{\phi_n\}$ of the oscillator at the origin; $a_0 = (S + 1/2)\,\hbar\omega$.

state. The emission spectrum is given by

$$I_{\text{emiss}}(E) = \frac{1}{n!} S^n\, e^{-S}\, \delta(E_0 - n\hbar\omega - E) \qquad (n = 0, 1, 2, \dots)\,. \tag{61}$$

The emission spectrum $I_{\text{emiss}}(E)$ is a palisade of equally spaced lines at energies E_0, $E_0 - \hbar\omega$, $E_0 - 2\hbar\omega$ (see Fig. 18b), and it is peaked at the energy $E_0 - S\hbar\omega$, in agreement with Eq. (55b).

As an illustration of the above considerations, we report in Fig. 19 the phonon-assisted luminescence of iodine impurities in AgBr. The experiments are interpreted in the framework of a conventional configuration coordinate model, in which both optical and acoustic phonons are coupled to the excited electronic states.

Solution of the Franck–Condon model by continued fractions

We can describe the optical absorption spectrum (or also the emission spectrum) of the Franck–Condon model with the tool of continued fractions. Consider the Hamiltonian $\widetilde{H}$ for the displaced oscillator in the excited adiabatic sheet

$$\widetilde{H} = \hbar\omega\,\left(\widetilde{a}^\dagger\,\widetilde{a} + \frac{1}{2}\right)\,. \tag{62a}$$

Such a Hamiltonian can be expressed in terms of annihilation and creation operators of the undisplaced harmonic oscillator. Since $\widetilde{a}^\dagger = a^\dagger - \sqrt{S}$ we have

$$\widetilde{H} = \hbar\omega\,a^\dagger a - \sqrt{S}\,\hbar\omega\,(a^\dagger + a) + \left(S + \frac{1}{2}\right)\hbar\omega\,. \tag{62b}$$

We now represent $\widetilde{H}$ on the basis of the states $\{\phi_n\}$ of the oscillator at the origin; we have

$$\widetilde{H} = \sum_{m=0}^{\infty}\sum_{n=0}^{\infty} |\phi_m\rangle\langle\phi_m|\widetilde{H}|\phi_n\rangle\langle\phi_n| = \left(S + \frac{1}{2}\right)\hbar\omega + \sum_{n=0}^{\infty} n\,\hbar\omega|\phi_n\rangle\langle\phi_n|$$

$$+ \sqrt{S}\,\hbar\omega \sum_{n=0}^{\infty} \sqrt{n+1}\,\left[\,|\phi_n\rangle\langle\phi_{n+1}| + |\phi_{n+1}\rangle\langle\phi_n|\,\right]\,. \tag{62c}$$

The above Hamiltonian has the graphical representation given in the Fig. 20.

The Hamiltonian $\widetilde{H}$ is already in tridiagonal form (see Section I-4.2); thus the diagonal matrix element of the Green's function $G(E) = (E - \widetilde{H})^{-1}$ on the state ϕ_0 is straightforwardly given by

$$G_{00}(E) = \langle \phi_0 | \frac{1}{E - \widetilde{H}} | \phi_0 \rangle = \cfrac{1}{E - a_0 - \cfrac{b_1^2}{E - a_1 - \cfrac{b_2^2}{E - a_2 - \dots}}} \qquad (63a)$$

with

$$a_n = \left(S + \frac{1}{2}\right) \hbar\omega + n\,\hbar\omega \quad (n = 0, 1, 2 \dots) \quad \text{and} \quad b_n^2 = n\,S\,(\hbar\omega)^2 \quad (n = 1, 2, 3 \dots).$$
$$(63b)$$

The ground state $|\psi_g, \phi_0\rangle$ of the Franck–Condon model is dipole connected only to the state $|\psi_e, \phi_0\rangle$, which is thus the dipole carrying state; according to Eq. (20) the absorption spectrum is then given by $-(1/\pi) \operatorname{Im} G_{00}(E + i\eta)$ with $\eta \to 0^+$.

The Green's function $G_{00}(E)$ can be determined with any desired accuracy considering a sufficient number of steps of the continued fraction (63); the absorption spectrum can then be worked out and the results of Fig. 18 recovered. The method of continued fractions (an optional in the case of the Franck–Condon model) becomes particularly convenient in the case of generalized vibronic models, because of its flexibility in the elaboration of the Green's function on the dipole carrying states of interest.

7.2 Optical properties of typical Jahn–Teller systems

We consider now optical transitions involving degenerate vibronic systems and examine the consequences of the Jahn–Teller effect on the absorption spectrum. Several typical Jahn–Teller systems have been discussed in Section VIII-3; here, we focus on the simplest Jahn–Teller system, the $E \otimes \varepsilon$ vibronic system, because of its own interest and also because it gives a feeling of how to treat more complicated situations. We consider thus a system, consisting of a non-degenerate ground-state adiabatic sheet and an orbital doublet E of excited states, interacting with a two-dimensional vibrational mode of symmetry ε. The adiabatic potential surfaces are indicated schematically in Fig. 21.

We have already seen in Eq. (VIII-27) that the full Hamiltonian of the vibronic $E \otimes \varepsilon$ system is

$$H = -\frac{\hbar^2}{2M} \left(\frac{\partial^2}{\partial q_1^2} + \frac{\partial^2}{\partial q_1^2} \right) + \gamma \begin{pmatrix} -q_1 & q_2 \\ q_2 & q_1 \end{pmatrix} + \frac{1}{2} M\omega^2 (q_1^2 + q_2^2), \qquad (64)$$

where γ is the linear coupling constant, ω the frequency of the vibrational mode, q_1 and q_2 are the vibrational normal coordinates.

Let us indicate the vibrational states by $|lm\rangle$ (the integer numbers l and m denote phonon occupation numbers), and the degenerate electronic states by $|\psi_1\rangle$ and $|\psi_2\rangle$. In the basis $|\psi_i, l\,m\rangle$ ($i = 1, 2; l, m = 0, 1, 2 \dots$) the Hamiltonian (64) can be conveniently

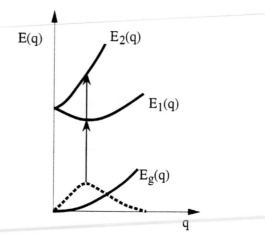

Fig. 21 Adiabatic surfaces $E_1(q)$ and $E_2(q)$ of the vibronic $E \otimes \varepsilon$ system as a function of the radial coordinate $q = (q_1^2 + q_2^2)^{1/2}$; the non-degenerate ground adiabatic surface $E_g(q)$, from which transition begins, is also indicated. The radial density distribution for a two-dimensional oscillator is shown by a dotted line. The origin of the two principal maxima in the semiclassical band shape is indicated by two solid arrows.

written as

$$H = H_e + H_L + H_{eL} , \tag{65a}$$

where

$$H_e + H_L = \sum_{l\,m} \left[E_e + (l + m + 1)\,\hbar\omega \right] \left[|\psi_1, l\,m\rangle\langle\psi_1, l\,m| + |\psi_2, l\,m\rangle\langle\psi_2, l\,m| \right] , \tag{65b}$$

and

$$H_{eL} = k_E\,\hbar\omega \sum_{l\,m} \left[-|\psi_1, l\,m\rangle\langle\psi_1, l\,m| + |\psi_2, l\,m\rangle\langle\psi_2, l\,m| \right] (a_1 + a_1^\dagger)$$

$$+ k_E\,\hbar\omega \sum_{l\,m} \left[|\psi_1, l\,m\rangle\langle\psi_1, l\,m| + |\psi_2, l\,m\rangle\langle\psi_2, l\,m| \right] (a_2 + a_2^\dagger) , \tag{65c}$$

with $k_E = \gamma/[(2M\hbar)^{1/2}\,\omega^{3/2}]$.

The dynamical problem can be solved for instance by applying the recursion method (see Section V-8.2). If we start with the seed state $|f_0\rangle = |\psi_1, 00\rangle$ (or with the partner state $|\psi_2, 00\rangle$), we can easily verify by inspection that the parameters of the recursion method can be obtained *exactly and analytically*; they are

$$a_n = (n + 1)\,\hbar\omega \ \ (n = 0, 1, 2\ldots), \qquad b_n^2 = 2\,\text{Int}\left(\frac{n+1}{2}\right) (k_E\,\hbar\omega)^2 \ \ (n = 1, 2, 3\ldots)$$

where $\text{Int}(x)$ indicates the integer part of x. We define as usual the Huang–Rhys factor S in the form $S = k_E^2$ (or equivalently $S\,\hbar\omega = E_{JT}$ where E_{JT} is the Jahn–

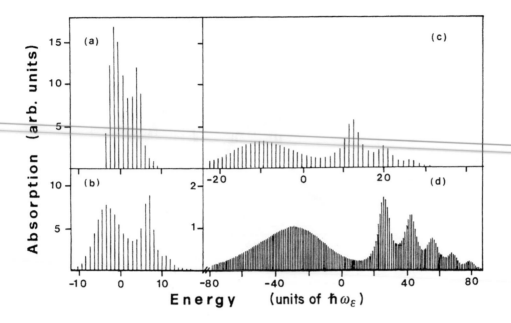

Fig. 22 Absorption spectrum for a transition $A \to E$ for a vibronic system $E \otimes \varepsilon$ (A denotes the non-degenerate electronic ground-state and E denotes the doubly-degenerate excited electronic state). The values of the dimensionless Huang–Rhys parameter S are 4, 16, 100, 900 in Figures (a), (b), (c) and (d). [from L. Martinelli, M. Passaro and G. Pastori Parravicini, Phys. Rev. B**43**, 8395 (1991); copyright 1991 by the American Physical Society].

Teller energy of the vibronic system); we obtain the Green's function

$$G_{00}(E) = \cfrac{1}{E - \hbar\omega - \cfrac{2S\,(\hbar\omega)^2}{E - 2\hbar\omega - \cfrac{2S\,(\hbar\omega)^2}{E - 3\hbar\omega - \cfrac{4S\,(\hbar\omega)^2}{E - 4\hbar\omega - \ldots}}}} . \qquad (66)$$

In agreement with Eq. (20), the absorption spectrum is given by $(-1/\pi)\,\mathrm{Im}\,G_{00}(E+i\eta)$ with $\eta \to 0^+$.

The lineshapes can be easily obtained numerically from the Green's function (66) and are shown in Fig. 22 for several values of the dimensionless parameter S. It can be seen that the absorption spectrum consists of two main bands, in qualitative agreement with the semi-classical considerations of Fig. 21; the oscillations in energy appearing for large S are due to the resonances between the vibrational levels of the two adiabatic surfaces, and are known as Slonzewski resonances. Many other Jahn–Teller systems have been studied in terms of appropriate continued fractions. Also the modified Lanczos method for excited states (see Section V-8.3) has been very fruitful to properly evaluate absorption lineshapes [see for instance G. Bevilacqua, L. Martinelli and G. Pastori Parravicini, Phys. Rev. B**54**, 7626 (1996) and references quoted therein].

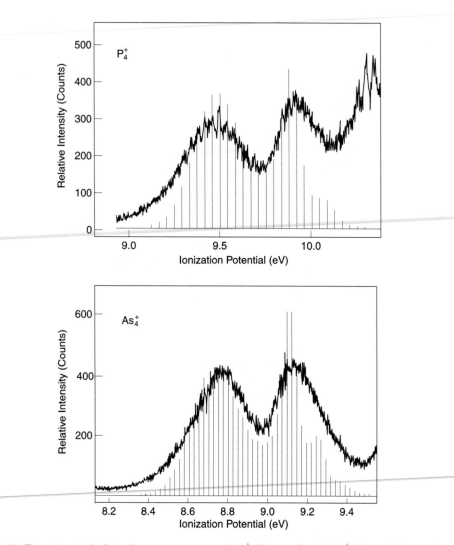

Fig. 23 Experimental photoelectron spectrum of P_4^+ (figure a) and As_4^+ (figure b), and best fit with a linear $E \otimes \varepsilon$ Jahn–Teller model [from Lai-Sheng Wang, B. Niu, Y. T. Lee, D. A. Shirley, E. Ghelichkhani and E. R. Grant, J. Chem. Phys. **93**, 6318 (1990); copyright 1990 by the American Physical Society].

As an illustration of optical transitions involving $E \otimes \varepsilon$ Jahn–Teller systems, we report in Fig. 23 high resolution photoelectron spectra of tetramers of group V elements. The tetramers of P and As have tetrahedral symmetry with a (non-degenerate) total symmetric ground-state; the lowest excited state of the tetrameric cations is an electronic orbital doublet, subject to Jahn–Teller distortion. In Fig. 23, the experimental photoelectron spectra are reasonably well understood on the basis of the $E \otimes \varepsilon$ Jahn–Teller model, at least for what concerns the features of the two main bands.

Appendix A. Transitions rates at first and higher orders of perturbation theory

In the study of optical transitions in solids we have used the time-dependent perturbation theory. It may be useful to recall here in a self-contained way some relevant aspects of the procedure, because we need transition rates at first and also at higher orders of perturbation.

Consider a system described by the Hamiltonian H_0, with eigenfunctions ψ_n and eigenvalues E_n. Consider the time evolution equation

$$i\hbar \frac{\partial \psi}{\partial t} = [H_0 + V(t)]\,\psi(t) \tag{A1}$$

where $V(t)$ is a time-dependent perturbation operator. We assume that the operator $V(t)$ has the form

$$V(t) = A\,e^{-i\omega t}\,e^{(\eta/\hbar)\,t} = A\,e^{-(i/\hbar)\,(\hbar\omega + i\eta)\,t}\,, \tag{A2}$$

where A is an operator independent from time, and the energy η ($\eta > 0$, $\eta \to 0^+$) is introduced so that the perturbation is switched on gently starting from $t = -\infty$.

The general solution of Eq. $(A1)$ can be written as

$$\psi(t) = \sum_m a_m(t)\,e^{-(i/\hbar)\,E_m t}\,|\psi_m\rangle\,. \tag{A3}$$

Replacing Eq. $(A3)$ into Eq. $(A1)$, we obtain

$$i\hbar \sum_m \frac{da_m(t)}{dt}\,e^{-(i/\hbar)\,E_m t}\,|\psi_m\rangle = V(t) \sum_\alpha a_\alpha(t)\,e^{-(i/\hbar)\,E_\alpha t}\,|\psi_\alpha\rangle\,;$$

by taking the scalar product of both sides of the above equation with a given eigenstate $\langle\psi_f|$ we obtain

$$\boxed{\frac{da_f(t)}{dt} = \frac{(-i)}{\hbar} \sum_\alpha a_\alpha(t)\,e^{-(i/\hbar)\,(E_\alpha - E_f + \hbar\omega + i\eta)\,t}\,\langle\psi_f|A|\psi_\alpha\rangle\,.} \tag{A4}$$

The set of coupled equations $(A4)$ together with the initial conditions at some instant, constitute an exact description of the time evolution of the system.

We now assume that at time $t = t_0$ (with $t_0 \to -\infty$) the system is prepared in a given initial state, that we denote as ψ_i. At $t = t_0$ we have the initial conditions

$$\begin{cases} a_i(t_0) = 1 \\ a_\alpha(t_0) = 0 \quad \text{for } \alpha \neq i \end{cases}. \tag{A5}$$

We attempt an iterative solution of Eqs. $(A4)$. We replace Eqs. $(A5)$ (that can be considered as the "zero-order approximation" for the expansion coefficients) into the right-hand side of Eq. $(A4)$, and we obtain in the left-hand side the first-order approximation $a^{(1)}(t)$ for the expansion coefficients; later we replace the first-order approximation for the coefficients in the right-hand side of Eq. $(A4)$ and obtain in the left-hand side the second-order approximation $a^{(II)}(t)$ for the expansion coefficients, and so on.

The first-order approximation, using Eq. $(A4)$ and Eq. $(A5)$ becomes

$$\frac{da_f^{(I)}(t)}{dt} = \frac{(-i)}{\hbar} e^{-(i/\hbar)\,(E_i - E_f + \hbar\omega + i\eta)\,t} \, \langle \psi_f | A | \psi_i \rangle \;.$$

Integration on the time variable gives (for $\psi_f \neq \psi_i$)

$$a_f^{(I)}(t) = \frac{(-i)}{\hbar} \int_{-\infty}^t e^{-(i/\hbar)\,(E_i - E_f + \hbar\omega + i\eta)\,t} \, dt \; \langle \psi_f | A | \psi_i \rangle$$

$$= \frac{e^{-(i/\hbar)\,(E_i - E_f + \hbar\omega + i\eta)\,t}}{E_i - E_f + \hbar\omega + i\eta} \langle \psi_f | A | \psi_i \rangle \;. \tag{A6}$$

We can now replace Eq. $(A6)$ (after appropriate relabelling of indices) into the right-hand side of Eq. $(A4)$ and obtain the expression for the second-order coefficients

$$\frac{da_f^{(II)}(t)}{dt} = \frac{(-i)}{\hbar} \sum_\alpha a_\alpha^{(I)}(t) \, e^{-(i/\hbar)\,(E_\alpha - E_f + \hbar\omega + i\eta)\,t} \, \langle \psi_f | A | \psi_\alpha \rangle$$

$$= \frac{(-i)}{\hbar} e^{-(i/\hbar)\,(E_i - E_f + 2\hbar\omega + 2i\eta)\,t} \sum_\alpha \frac{\langle \psi_f | A | \psi_\alpha \rangle \langle \psi_\alpha | A | \psi_i \rangle}{E_i - E_\alpha + \hbar\omega + i\eta} \;.$$

Integration on the time variable from $-\infty$ to t gives

$$a_f^{(II)}(t) = \frac{e^{-(i/\hbar)\,(E_i - E_f + 2\hbar\omega + 2i\eta)\,t}}{E_i - E_f + 2\hbar\omega + 2i\eta} \sum_\alpha \frac{\langle \psi_f | A | \psi_\alpha \rangle \langle \psi_\alpha | A | \psi_i \rangle}{E_i - E_\alpha + \hbar\omega + i\eta} \;. \tag{A7}$$

Similarly, one could go on to obtain the amplitudes to any desired higher order.

The results $(A6)$ and $(A7)$ can be interpreted in a more effective way. We can define the *first-order transition probability* per unit time of a transition from an initial state ψ_i to a final state ψ_f as

$$P_{f \leftarrow i}^{(I)} = \frac{d}{dt} \, |a_f^{(I)}(t)|^2 \;.$$

Using the relation

$$|a_f^{(I)}(t)|^2 = |\langle \psi_f | A | \psi_i \rangle|^2 \frac{e^{(2\eta/\hbar)\,t}}{(E_f - E_i - \hbar\omega)^2 + \eta^2} \;,$$

the transition probability per unit time becomes

$$P_{f \leftarrow i}^{(I)} = \frac{d}{dt} \, |a_f^{(I)}(t)|^2 = \frac{2}{\hbar} |\langle \psi_f | A | \psi_i \rangle|^2 \frac{\eta \, e^{(2\eta/\hbar)\,t}}{(E_f - E_i - \hbar\omega)^2 + \eta^2} \;. \tag{A8}$$

It is convenient to use the representation of the δ-function in the form

$$\delta(E - E_0) = \frac{1}{\pi} \lim_{\eta \to 0^+} \frac{\eta}{(E - E_0)^2 + \eta^2} \;.$$

In the limit of $\eta \to 0^+$ the expression $(A8)$ can be recast in the form

$$P_{f \leftarrow i}^{(I)} = \frac{2\pi}{\hbar} |\langle \psi_f | A | \psi_i \rangle|^2 \, \delta(E_f - E_i - \hbar\omega) \; ; \qquad (A9)$$

this result represents the well known *Fermi golden rule for first order transition probability rate* in steady state situation.

We now proceed to second order; from Eq. $(A7)$ and similar elaborations, it follows

$$P_{f \leftarrow i}^{(II)} = \frac{2\pi}{\hbar} \left| \sum_\alpha \frac{\langle \psi_f | A | \psi_\alpha \rangle \langle \psi_\alpha | A | \psi_i \rangle}{E_i + \hbar\omega - E_\alpha} \right|^2 \delta(E_f - E_i - \hbar\omega - \hbar\omega) . \qquad (A10)$$

Similarly, the expression for the third-order transition probability is given by

$$P_{f \leftarrow i}^{(III)} = \frac{2\pi}{\hbar} \left| \sum_{\alpha\beta} \frac{\langle \psi_f | A | \psi_\beta \rangle \langle \psi_\beta | A \psi_\alpha \rangle \langle \psi_\alpha | A | \psi_i \rangle}{(E_i + 2\hbar\omega - E_\beta)(E_i + \hbar\omega - E_\alpha)} \right|^2 \delta(E_f - E_i - \hbar\omega - \hbar\omega - \hbar\omega) .$$
$$(A11)$$

A note of warning must be considered when a denominator is zero or near to zero. From a qualitative point of view these situations give rise to resonant effects; from a quantitative point of view the series development should be replaced by a more accurate analysis.

The above considerations can be extended in the case the perturbation has the form

$$V(t) = A \, e^{-i\omega t} \, e^{(\eta/\hbar)t} + A^\dagger \, e^{+i\omega t} \, e^{(\eta/\hbar)t} \; .$$

Besides the general expressions of Eqs. $(A9)$, $(A10)$ and $(A11)$ for the absorption of one or more quanta of energy, one also obtains the general expressions for the emission of one or more quanta, or for mixed absorption–emission processes; the expressions are similar to Eqs. $(A9)$, $(A10)$ and $(A11)$ with appropriate replacements of $+\hbar\omega$ by $-\hbar\omega$ for emission processes.

Further reading

V. M. Agranovich and V. L. Ginzburg "Crystal Optics with Spatial Dispersion and Excitons" (Springer, Berlin 1984)

M. Balkanski "Optical Properties of Semiconductors" (North-Holland, Amsterdam 1994)

F. Bassani and M. Altarelli "Interaction of Radiation with Matter" in Handbook on Synchrotron Radiation, Vol.I, edited by E. E. Koch (North-Holland, Amsterdam 1983)

F. Bassani and G. Pastori Parravicini "Electronic States and Optical Transitions in Soilds" (Pergamon Press, Oxford 1975)

I. B. Bersuker and V. Z. Polinger "Vibronic Interactions in Molecules and Crystals" (Springer, Berlin 1989)

M. Born and E. Wolf "Principles of Optics" (Pergamon Press, Oxford 1980)

M. Cardona "Modulation Spectroscopy" Solid State Physics, Suppl. **11** (1969) (edited by F. Seitz, D. Turnbull and H. Ehrenreich, Academic Press, New York)

U. Fano and J. W. Cooper "Spectral Distribution of Atomic Oscillator Strength" Rev. Mod. Phys. **40**, 441 (1968)

D. L. Greenaway and Harbeke "Optical Properties and Band Structure of Semiconductors" (Pergamon Press, Oxford 1968)

J. N. Hodson "Optical Absorption and Dispersion in Solids" (Chapman and Hall, New York 1970)

H. Huang and S. Koch "Quantum Theory of the Optical and Electronic Properties of Semiconductors" (World Scientific, Singapore 1990)

E. D. Palik (editor) "Handbook of the Optical Constants of Solids" Vols.I and II (Academic Press, New York 1985 and 1991)

J. I. Pankove "Optical Processes in Semiconductors" (Dover, New York 1972)

J. C. Phillips "The Fundamental Optical Spectra of Solids" in Solid State Physics **18**, 55 (1966) (edited by F. Seitz and D. Turnbull, Academic Press, New York).

B. K. Ridley "Quantum Processes in Semiconductors" (Clarendon Press, Oxford 1993)

A. Stahl and I. Balslev "Electrodynamics of the Semiconductor Band Edge" (Springer, Berlin 1987)

F. Stern "Elementary Theory of the Optical Properties of Solids" in Solid State Physics **15**, 331 (1968) (edited by F. Seitz and D. Turnbull, Academic, New York).

F. Wooten "Optical Properties of Solids" (Academic Press, New York 1972)

P. Y. Yu and M. Cardona "Fundamentals of Semiconductors" (Springer, Berlin 1996)

XIII

Transport in intrinsic and homogeneously doped semiconductors

This is the first of two partner chapters concerning the electronic transport phenomena in semiconductors. In this chapter we consider transport processes in homogeneous semiconductors, which are free from impurities or uniformly doped with (donor or acceptor) impurities. In the next chapter we consider semiconductor structures that are inhomogeneous, either in the doping or in the chemical composition. The presence of a finite energy gap between valence and conduction states has a profound effect on carrier density and transport properties. In particular, the Fermi level within the energy gap can be easily engineered by the presence of even a small number of appropriate impurities; this fact ultimately opens the wide field of solid state electronics and device applications.

1 Fermi level and carrier density in intrinsic semiconductors

A semiconductor (or insulator) at zero temperature is constituted by fully occupied valence bands and fully empty conduction bands. An energy gap $E_G = E_c - E_v$ separates the lowest energy level E_c of the conduction bands from the topmost energy level E_v of the valence bands. The fundamental energy gap E_G and the lattice constant of some semiconductors are reported in Fig. 1.

In ideal defect-free semiconductors (*intrinsic semiconductors*), no level exists in the forbidden energy gap. On the contrary in semiconductors containing defects (*extrinsic semiconductors*), impurity levels may be introduced within the energy gap, with

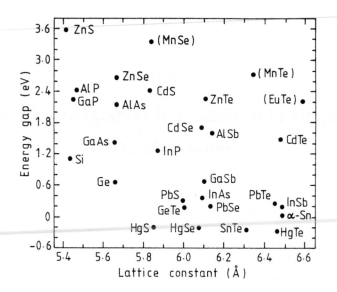

Fig. 1 Energy gap and lattice constant of some semiconductor crystals at room temperature [from M. Jaros "Physics and Applications of Semiconductor Microstructures" (Clarendon Press, Oxford 1989); by permission of Oxford University Press].

significant modifications of the carrier concentration and transport properties. Our purpose is to study the carrier distribution at thermal equilibrium, first in intrinsic semiconductors and then in extrinsic ones.

Consider an intrinsic semiconductor, and let $n_c(E) = D_c(E)/V$ and $n_v(E) = D_v(E)/V$ indicate the *density-of-states per unit volume in the conduction and in the valence bands*, respectively. At zero temperature, all the valence states are occupied and all the conduction states are empty, as schematically indicated in Fig. 2. At finite temperature T, a number of electrons from the valence bands are thermally excited to the conduction bands; the occupancy probability of the allowed band structure levels of energy E is given by the Fermi–Dirac distribution function

$$f(E) = \frac{1}{e^{(E-\mu)/k_BT} + 1} , \tag{1}$$

where μ is the chemical potential (the terms "chemical potential μ" and "Fermi level E_F" are used by us as synonymous).

The density of electrons at temperature T in the conduction bands is given by the expression

$$n_0(T) = \int_{E_c}^{\infty} n_c(E) \, f(E) \, dE = \int_{E_c}^{\infty} n_c(E) \, \frac{1}{e^{(E-\mu)/k_BT} + 1} \, dE , \tag{2a}$$

where the subscript 0 to the electron concentration is just to remind us that the quantity refers to thermal equilibrium. Similarly, the density of missing electrons (or equivalently the density of holes) at temperature T in the valence bands is determined

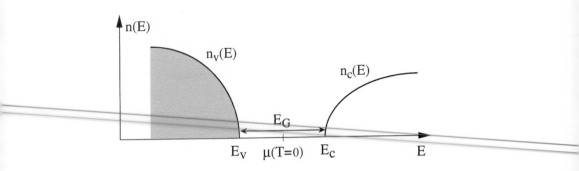

Fig. 2 Schematic representation of the density-of-states of an intrinsic semiconductor. The top of the valence band is E_v and the bottom of the conduction band is E_c. The energy gap is $E_G = E_c - E_v$; the Fermi level at zero temperature lies at the middle of the band gap.

by the density-of-states in the valence bands and the complementary to 1 of the Fermi–Dirac distribution function. We have the expression

$$p_0(T) = \int_{-\infty}^{E_v} n_v(E)\left(1 - f(E)\right) dE = \int_{-\infty}^{E_v} n_v(E)\, \frac{1}{e^{(\mu - E)/k_B T} + 1}\, dE \ . \tag{2b}$$

The integrals in Eqs. (2) extend (in principle) to the whole energy regions where $n_c(E)$ and $n_v(E)$ are different from zero; in practice, as seen below, only the energy regions close to the band edges E_c and E_v are of relevance.

For an intrinsic semiconductor, the chemical potential $\mu(T)$ is determined by the requirement that the number of electrons in the conduction bands equals the number of holes left in the valence bands

$$n_0(T) = p_0(T) \ . \tag{3}$$

Before solving explicitly the balance equation (3), we make the following considerations. Since the distribution function $f(E)$ is a step function around $E = \mu$ at zero temperature, and at the same time $n_0(T{=}0) \equiv p_0(T{=}0) \equiv 0$ in an intrinsic semiconductor, we have from Eqs. (2) that the chemical potential must *lie within the energy gap* at zero temperature, i.e. $E_v < \mu\,(T = 0) < E_c$. Furthermore, in the particular case that the density-of-states of the semiconductor is symmetric with respect to the middle of the energy gap, the balance of electrons and holes obviously requires that the chemical potential, at any temperature, coincides with the middle of the band gap. Even if the density-of-states in the valence and conduction bands (in the energy range of order $k_B T$ around the band edges) are not specular, it is evident that a small shift of the chemical potential of order $k_B T$ (towards the edge with lower density-of-states) is sufficient to equalize the number of electrons and the number of holes.

A semiconductor (either intrinsic or extrinsic) is said to be *non-degenerate* if the chemical potential $\mu(T)$ lies within the energy gap and is separated from the band edges by several $k_B T$ (say $\approx 5\,k_B T$ or more); the *non-degeneracy conditions* are

$$\boxed{E_v < \mu(T) < E_c \ , \quad \text{with} \quad E_c - \mu(T) \gg k_B T \quad \text{and} \quad \mu(T) - E_v \gg k_B T}\ . \tag{4}$$

When conditions (4) are satisfied, the Fermi–Dirac distribution function $f(E)$ in Eq. (2a), as well as the complementary distribution function $1 - f(E)$ in Eq. (2b), can be simplified with their corresponding Maxwell–Boltzmann exponential distributions.

We pass now to a quantitative analysis of Eq. (3). From the previous discussion, we expect that in general an intrinsic semiconductor is non-degenerate at any temperature of interest. For a non-degenerate semiconductor, the expression (2a) for the electrons in the conduction bands simplifies in the form

$$n_0(T) = N_c(T) \, e^{-(E_c - \mu)/k_B T} \ , \tag{5a}$$

where

$$N_c(T) \equiv \int_{E_c}^{\infty} n_c(E) \, e^{-(E - E_c)/k_B T} \, dE \ . \tag{5b}$$

Similarly, for a non-degenerate semiconductor, the expression (2b) for the holes left in the valence bands takes the simplified form

$$p_0(T) = N_v(T) \, e^{-(\mu - E_v)/k_B T} \ , \tag{5c}$$

where

$$N_v(T) \equiv \int_{-\infty}^{E_v} n_v(E) \, e^{-(E_v - E)/k_B T} \, dE \ . \tag{5d}$$

The quantities $N_c(T)$ and $N_v(T)$ are referred as the *effective conduction band and valence band density-of-states*, respectively. Thus a non-degenerate semiconductor can be schematized as a two-level system, where the whole conduction bands can be replaced by a single level of energy E_c and degeneracy $N_c(T)$, and the whole valence bands can be replaced by a single level of energy E_v and degeneracy $N_v(T)$.

The chemical potential in an intrinsic semiconductor is obtained by the requirement that expressions (5a) and (5c) coincide:

$$N_c(T) \, e^{-(E_c - \mu)/k_B T} = N_v(T) \, e^{-(\mu - E_v)/k_B T} \ ;$$

taking the logarithms of both members we have

$$\boxed{\mu(T) = \frac{1}{2}(E_v + E_c) + \frac{1}{2} k_B T \ln \frac{N_v(T)}{N_c(T)}} \ . \tag{6}$$

From this equation we see that the chemical potential of the intrinsic semiconductor at zero temperature is located at the middle of the energy gap; at finite temperature T, the change of μ is of the order of $k_B T$. Inserting Eq. (6) into Eq. (5a) and Eq. (5c), we obtain for *the intrinsic concentrations* $n_i(T)$ and $p_i(T)$ of electrons and holes

$$n_i(T) = p_i(T) = \sqrt{N_c(T) \, N_v(T)} \, e^{-E_G/2 k_B T} \ .$$

The temperature dependence of the intrinsic carrier concentration has (approximately) the exponential form $\exp(-\Delta/k_B T)$, where Δ is *half the energy gap*.

From equations (5a) and (5c), we obtain for the product $n_0(T) \, p_0(T)$ the expression

$$\boxed{n_0(T) \, p_0(T) = N_c(T) \, N_v(T) \, e^{-E_G/k_B T} \equiv n_i^2(T) = p_i^2(T)} \ , \tag{7}$$

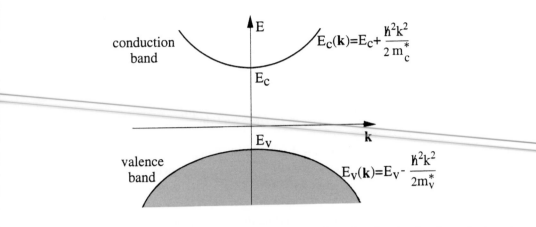

Fig. 3 Isotropic two-band model semiconductor; the valence band region has been shadowed to remind its full occupancy by electrons at $T = 0$.

which is known as *mass-action law*. It is important to notice that the product $n_0(T)\,p_0(T)$ does not depend on the value of the chemical potential; thus, at fixed temperature, the product $n_0(T)\,p_0(T)$ keeps the same value regardless the impurity concentration in the semiconductor; while the actual values of $n_0(T)$ and $p_0(T)$ depend on the chemical potential (which is different for intrinsic or extrinsic semiconductors) their product does not.

It is instructive to specify the above results in the case of the isotropic and parabolic two-band model semiconductor, schematized in Fig. 3. The density-of-states per unit volume in the conduction band for $E > E_c$ is

$$n_c(E) = \int \frac{2}{(2\pi)^3}\, \delta(E_c(\mathbf{k}) - E)\, d\mathbf{k} = \int_0^\infty \frac{2}{(2\pi)^3}\, \delta\Big(\frac{\hbar^2 k^2}{2m_c^*} + E_c - E\Big)\, 4\pi\, k^2\, dk \ ,$$

where the factor 2 takes into account spin degeneracy; the integral is easily carried out after performing the change of variable $x = \hbar^2 k^2 / 2m_c^*$ and gives

$$n_c(E) = \frac{1}{2\pi^2} \Big(\frac{2m_c^*}{\hbar^2}\Big)^{3/2} (E - E_c)^{1/2} \ . \tag{8a}$$

We insert Eq. (8a) into Eq. (5b), and use the relation $\int_0^\infty \sqrt{t}\, \exp(-t)\, dt = \sqrt{\pi}/2$; we obtain for the effective density-of-states in the conduction band

$$
\begin{aligned}
N_c(T) &= \frac{\sqrt{\pi}}{2} \frac{1}{2\pi^2} \Big(\frac{2m_c^*}{\hbar^2}\Big)^{3/2} (k_B T)^{3/2} \\
&= \frac{1}{4\pi^{3/2}} \Big(\frac{m_c^*}{m}\frac{T}{T_0}\Big)^{3/2} \Big(\frac{k_B T_0}{\hbar^2/2ma_B^2}\Big)^{3/2} \frac{1}{a_B^3} = 2.534 \Big(\frac{m_c^*}{m}\frac{T}{T_0}\Big)^{3/2} 10^{19}\ \mathrm{cm}^{-3} \tag{8b}
\end{aligned}
$$

where $T_0 = 300\,\mathrm{K}$ has been chosen as reference temperature, $k_B T_0 \approx 0.026\,\mathrm{eV}$, $\hbar^2/2ma_B^2 = 1\,\mathrm{Ryd} = 13.606\,\mathrm{eV}$ and $a_B = 0.529 \times 10^{-8}\,\mathrm{cm}$.

Similar expressions can be worked out for the holes in the valence band; the effective density-of states for the valence bands becomes

$$N_v(T) = \frac{\sqrt{\pi}}{2} \frac{1}{2\pi^2} \left(\frac{2m_v^*}{\hbar^2} \right)^{3/2} (k_B T)^{3/2} \ . \tag{8c}$$

In the case of the isotropic two-band model semiconductor, Eq. (6) takes the form

$$\mu(T) = \frac{1}{2}(E_v + E_c) + \frac{3}{4} k_B T \ln \frac{m_v^*}{m_c^*} \ . \tag{8d}$$

From Eq. (8d) we see that the chemical potential is independent of temperature if the hole and electron effective masses are equal; otherwise it changes by an amount of the order of $k_B T$ approaching the band edge with lower effective mass (i.e. lower density of states).

The above considerations for a two-band model semiconductor are useful to estimate the order of magnitude of relevant quantities, and can be implemented to describe realistic multiband semiconductors. Some properties of germanium, silicon, and gallium arsenide at 300 K are summarized for convenience in Table 1; for a more complete list of data, we refer for instance to the book of Smith and to the book of Bhattacharya, quoted in the references of this chapter.

2   Impurity levels in semiconductors

Band states and impurity levels

In the band theory of crystals (Chapter V), we have considered in detail the Schrödinger equation for periodic materials

$$\left[\frac{\mathbf{p}^2}{2m} + V(\mathbf{r}) \right] \psi_n(\mathbf{k}, \mathbf{r}) = E_n(\mathbf{k}) \psi_n(\mathbf{k}, \mathbf{r}) \ ; \tag{9}$$

we have also described the general methods to obtain the band energies $E_n(\mathbf{k})$ and the band wavefunctions $\psi_n(\mathbf{k}, \mathbf{r})$, exploiting the translational invariance of the crystal Hamiltonian.

A perfect periodic crystal, however, is an ideal entity. Real crystals contain a number of defects (substitutional or interstitial impurities, vacancies, dislocations, physical surfaces etc.). The presence of defects implies, in general, a (total or partial) breaking of the translational symmetry of the crystal; in the allowed energy regions of the host crystals, the density-of-states function may be modified and resonances may appear; more important, in the forbidden energy gap of the host crystals, new energy levels can occur with wavefunctions localized around the impurity region.

The electronic structure of a crystal, in the presence of an impurity, can be studied by means of the Schrödinger equation

$$\left[\frac{\mathbf{p}^2}{2m} + V(\mathbf{r}) + V_I(\mathbf{r}) \right] \phi(\mathbf{r}) = E \, \phi(\mathbf{r}) \ , \tag{10}$$

where $V(\mathbf{r})$ is the periodic crystal potential, $V_I(\mathbf{r})$ is the modification due to the

Table 1 Some properties of Si, Ge and GaAs at 300 K.

	Si	Ge	GaAs
Crystal structure and lattice constant (Å)	diamond 5.431	diamond 5.646	zinc blende 5.653
Energy gap (eV) (direct or indirect)	1.12 (indirect)	0.66 (indirect)	1.43 (direct)
Electron affinity χ(eV)	4.01	4	4.07
Electron effective mass m^* in units of the free-electron mass m (l=longitudinal, t=transversal)	$m_l^*=0.98$ $m_t^*=0.19$	$m_l^*=1.64$ $m_t^*=0.082$	$m^*=0.063$
Hole effective mass m^*/m (lh=light hole, hh=heavy hole)	$m_{lh}^*=0.16$ $m_{hh}^*=0.49$	$m_{lh}^*=0.044$ $m_{hh}^*=0.28$	$m_{lh}^*=0.09$ $m_{hh}^*=0.48$
Number of equivalent conduction edge valleys	6	4	1
Effective density-of-states in the conduction band N_c (cm^{-3})	2.9×10^{19}	1.04×10^{19}	4.4×10^{17}
Effective density-of-states in the valence band N_v (cm^{-3})	1.1×10^{19}	6.1×10^{18}	8.2×10^{18}
Intrinsic carrier concentration n_i or p_i (cm^{-3})	1.02×10^{10}	2.4×10^{13}	5×10^6
Static dielectric constant	11.7	16	13.2
Carrier mobility (cm^2/volt sec) μ_n (electrons) μ_p (holes)	1450 500	3800 1800	8500 400
Diffusion constant (cm^2/sec) D_n (electrons) D_p (holes)	37.5 13	98 47	220 10
Linear thermal expansion coefficient (in units 10^{-6}/°C)	2.6	5.7	5

impurity, and the one-electron approximation is assumed to hold. The direct solution of Eq. (10) appears a quite demanding (if not impossible) job, since the break of periodicity due to $V_I(\mathbf{r})$ does not allow to use the basic Bloch theorem. However, in general, it is not necessary to solve "ex novo" Eq. (10); rather it is convenient to use the band wavefunctions as starting basis set on which to represent the wavefunctions of the impurity problem.

Within this line of approach, we expand the wavefunctions of the system on the complete set of band wavefunctions of the perfect crystal in the form

$$\phi(\mathbf{r}) = \sum_{n'\,\mathbf{k}'} A_{n'}(\mathbf{k}')\,\psi_{n'}(\mathbf{k}',\mathbf{r}) \ . \tag{11}$$

We insert Eq. (11) into Eq. (10), project on $\langle\psi_n(\mathbf{k},\mathbf{r})|$ and use Eq. (9). We have that the coefficients $A_n(\mathbf{k})$ of the linear expansion (11) satisfy the set of coupled integral equations

$$[E_n(\mathbf{k}) - E]\,A_n(\mathbf{k}) + \sum_{n'\,\mathbf{k}'} U_{nn'}(\mathbf{k},\mathbf{k}')\,A_{n'}(\mathbf{k}') = 0 \ , \tag{12a}$$

where $\mathbf{k}$ and $\mathbf{k}'$ can be thought as continuous variables within the Brillouin zone, and

$$U_{nn'}(\mathbf{k},\mathbf{k}') = \int \psi_n^*(\mathbf{k},\mathbf{r})\,V_I(\mathbf{r})\,\psi_{n'}(\mathbf{k}',\mathbf{r})\,d\mathbf{r} \ . \tag{12b}$$

The general equations (12) become of practical value when a small number of bulk bands are actually involved in the description of the impurity levels. Thus we consider appropriate simplifications of Eqs. (12) with the purpose to obtain (whenever possible) intuitive descriptions of impurity levels.

Single conduction band model and shallow donor levels

We consider here the particular case in which only *one band of the perfect crystal*, suppose the lowest conduction band, is of relevance for the description of the impurity electronic structure. We further assume that the conduction band of interest is parabolic, with a unique valley of effective mass $m_c^* > 0$ at $\mathbf{k} = 0$ and energy dispersion $E_c(\mathbf{k}) = E_c + \hbar^2 k^2/2m_c^*$. With the above assumptions, Eqs. (12) become

$$\left[\frac{\hbar^2 k^2}{2m_c^*} + E_c - E\right] A(\mathbf{k}) + \sum_{\mathbf{k}'} U(\mathbf{k},\mathbf{k}')\,A(\mathbf{k}') = 0 \ , \tag{13a}$$

with the kernel of the integral equation given by

$$U(\mathbf{k},\mathbf{k}') = \int \psi_c^*(\mathbf{k},\mathbf{r})\,V_I(\mathbf{r})\,\psi_c(\mathbf{k}',\mathbf{r})\,d\mathbf{r} \ . \tag{13b}$$

In order to put in evidence relevant physical effects avoiding inessential details, we simplify the kernel $U(\mathbf{k},\mathbf{k}')$ with considerations similar to what done in Section VII-1 for the integral equation of excitons. In Eq. (13b) we write the Bloch wavefunctions as the product of plane waves and periodic parts and obtain

$$U(\mathbf{k},\mathbf{k}') = \int u_c^*(\mathbf{k},\mathbf{r})\,u_c(\mathbf{k}',\mathbf{r})\,e^{-i(\mathbf{k}-\mathbf{k}')\cdot\mathbf{r}}\,V_I(\mathbf{r})\,d\mathbf{r} \ .$$

In the case the impurity potential is reasonably smooth on the crystal unit cell, we can replace the product $u_c^*(\mathbf{k},\mathbf{r})\,u_c(\mathbf{k}',\mathbf{r})$ by its average value on the volume V of the sample (the average value is just $1/V$ for $\mathbf{k} \approx \mathbf{k}'$), and we have

$$U(\mathbf{k},\mathbf{k}') = \frac{1}{V} \int e^{-i(\mathbf{k}-\mathbf{k}')\cdot\mathbf{r}}\,V_I(\mathbf{r})\,d\mathbf{r} \ ; \tag{13c}$$

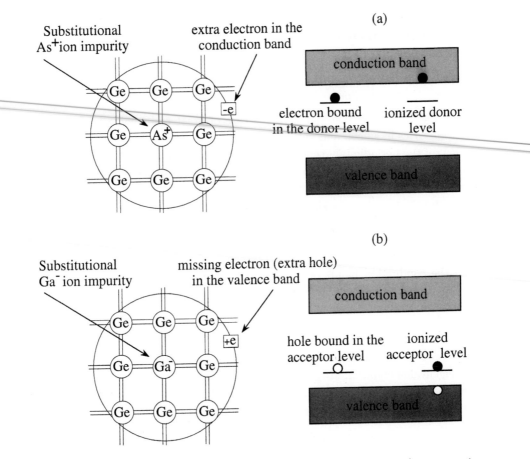

Fig. 4 (a) Schematic picture of a substitutional donor impurity (i.e. arsenic) in germanium; the pentavalent impurity is embodied in the lattice as positive As$^+$ ion, with its four external electrons forming four covalent bonds with the nearest neighbour Ge atoms; each As$^+$ ion introduces a shallow donor level, below the bottom of the conduction band; the fifth electron, not incorporated in the bonding, is bound to the As$^+$ ion by the Coulomb attraction (at low temperature) and is donated to the conduction band (at sufficiently high temperature). (b) Schematic picture of a substitutional acceptor impurity (i.e gallium) in germanium; the trivalent impurity is embodied in the lattice as negative Ga$^-$ ion, with its four external electrons forming four covalent bonds with the nearest neighbour Ge atoms; each Ga$^-$ impurity introduces a shallow acceptor level, above the top of the valence band. The "missing electron" or "hole", created for the formation of the tetrahedral bonding, is bound to the Ga$^-$ ion by Coulomb attraction (at low temperature) and is given to the valence band (at sufficiently high temperature).

the above expression can be recognized as the Fourier transform of the impurity potential.

We now introduce the "envelope function" $F(\mathbf{r})$ defined as

$$F(\mathbf{r}) = \frac{1}{\sqrt{V}} \sum_{\mathbf{k}} A(\mathbf{k}) \, e^{i \mathbf{k} \cdot \mathbf{r}} \, ,$$

and normalized to unity. The integral eigenvalue equation (13a), with the kernel (13c), can be conveniently transformed into an ordinary differential eigenvalue equation for the envelope function

$$\left[-\frac{\hbar^2}{2m_c^*} \nabla^2 + V_I(\mathbf{r}) \right] F(\mathbf{r}) = (E - E_c)\, F(\mathbf{r}) \ . \tag{14}$$

As an illustration, we consider here the case of great technological importance of *donor impurities* in silicon and germanium; donor impurities are formed by atoms of column V of the periodic table entering substitutionally into the silicon or germanium crystal. We can intuitively schematize the long-range part of the impurity potential in the form $V_I(\mathbf{r}) \approx -e^2/\varepsilon_s\, r$, where ε_s is the static dielectric constant of the host crystal (see Fig. 4a). Notice that the short-range part (the so called "*central-cell-correction*") of the impurity potential would require a detailed (and demanding) analysis of the electronic structure before and after the substitution has taken place. Neglecting for simplicity central-cell-corrections (and also disregarding the actual multi-valley structure of the conduction bands of Si and Ge), the Schrödinger equation (14) for the donor impurity levels takes the form

$$\left[-\frac{\hbar^2}{2m_c^*} \nabla^2 - \frac{e^2}{\varepsilon_s\, r} \right] F(\mathbf{r}) = (E - E_c)\, F(\mathbf{r}) \ , \tag{15}$$

which represents a hydrogenic atom characterized by ε_s and m_c^*.

The eigenvalue equation (15) supports the picture of a donor impurity as an extra electron of effective mass m_c^* bound to the defect by a screened Coulomb interaction. The binding energy $\varepsilon_d = E_c - E_d$ of the donor ground-state is

$$\varepsilon_d = E_c - E_d = 13.606\, \frac{m_c^*}{m} \frac{1}{\varepsilon_s^2}\ (\text{eV}) \ , \tag{16a}$$

and the effective radius of the donor ground-state wavefunction is

$$a_d = a_B\, \frac{m}{m_c^*} \varepsilon_s \ . \tag{16b}$$

For instance in the case of Si, with $\varepsilon_s = 11.7$ and average effective mass $m_c^*/m \approx 0.3$, we expect $\varepsilon_d \approx 30\,\text{meV}$ and $a_d \approx 40\, a_B$. For Ge with $\varepsilon_s = 16$ and average effective mass $m_c^*/m \approx 0.2$, we expect $\varepsilon_d \approx 10\,\text{meV}$ and $a_d \approx 80\, a_B$. We thus see that the binding energy of donors in Si and Ge is rather small with respect to the energy gap of the host material, while the effective radius is rather large with respect to the lattice parameter (*shallow impurity level*). Further refinements are necessary to describe the short-range effects of the impurity, or the multi-valley conduction band structure of the host semiconductors. In Fig. 5, we report the experimental ionization energies for various donor impurities in silicon and germanium.

Single valence band model and shallow acceptor levels

We consider now the particular case in which only *one valence band of the perfect crystal* is of relevance for the description of the electronic structure of the impurity. We

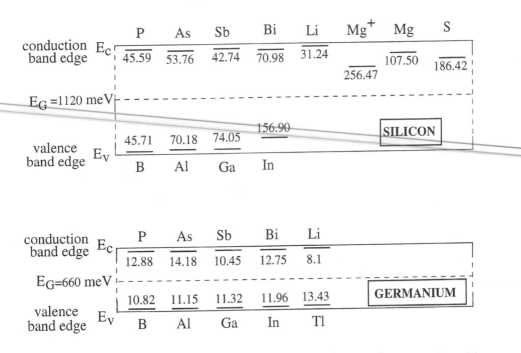

Fig. 5 Experimental ionization energies (meV) for various donor and acceptor impurities (upper and lower part of the panels, respectively) in silicon and germanium. For donor impurities, the electron binding energy $\varepsilon_d = E_c - E_d$ is reported; for acceptor impurities the hole binding energy $\varepsilon_a = E_a - E_v$ is reported. The data are taken from A. K. Ramdas and S. Rodriguez, Rep. Progr. Phys. **44**, 1297 (1981).

also assume that the topmost valence band of interest is isotropic, with the maximum at $\mathbf{k} = 0$ and energy dispersion $E_v(\mathbf{k}) = E_v - \hbar^2 k^2 / 2 m_v^*$ ($m_v^* > 0$). The effect of a (reasonably smooth) impurity potential $V_I(\mathbf{r})$ on the valence band can be described within the envelope function formalism, in a way essentially similar to the above discussion of shallow donor levels. With appropriate relabelling of quantities entering in Eq. (14), we obtain that the envelope function for the impurity states associated to the valence band satisfies the Schrödinger equation

$$\left[\frac{\hbar^2}{2m_v^*}\nabla^2 + V_I(\mathbf{r})\right] F(\mathbf{r}) = (E - E_v) F(\mathbf{r}) . \tag{17a}$$

As an illustration, we consider the case of *acceptor impurities* in silicon and germanium; these impurities are formed when atoms of column III of the periodic table enter substitutionally into the silicon or germanium crystal. We can intuitively schematize the impurity potential felt by an electron because of the negatively charged impurity in the form $V_I(\mathbf{r}) = +e^2/\varepsilon_s r$, where ε_s is the static dielectric constant of the host crystal (see Fig. 4b); from Eq. (17a), with both members multiplied by -1, we obtain

$$\left[-\frac{\hbar^2}{2m_v^*}\nabla^2 - \frac{e^2}{\varepsilon_s r}\right] F(\mathbf{r}) = -(E - E_v) F(\mathbf{r}) . \tag{17b}$$

The binding energy of the ground acceptor level is

$$\varepsilon_a = E_a - E_v = 13.606 \, \frac{m_v^*}{m} \frac{1}{\varepsilon_s^2} \; (\text{eV}) \; ,$$

and the effective radius of the acceptor ground-state wavefunction is $a_B \, (m/m_v^*) \, \varepsilon_s$.

For more refined treatments of acceptor levels in Si and Ge, one has to describe accurately the short-range part of the impurity potential and the fact that the group of valence bands with the same energy at $\mathbf{k} = 0$ contribute on the same footing to the formation of the acceptor levels. In Fig. 5, we report the experimental ionization energies for various acceptor impurities in silicon and germanium.

The concept of donor and acceptor impurities can be extended to III-V compounds (such as GaAs). A group VI impurity substituting a group V atom, or a group IV impurity substituting a group III atom, act as a donor; similarly, a group II impurity replacing a group III atom, or a group IV impurity replacing a group V atom, act as an acceptor.

Considerations on deep impurity levels in semiconductors

In most situations, the set of integral equations (12) cannot be decoupled in the form discussed above, and the analysis of impurity levels becomes more demanding, both from the point of view of the physical picture and from the technical point of view of an adequate solution. Consider for instance group II or group VI substitutional impurities in group IV elemental semiconductors silicon and germanium; in the case of doubly ionized impurities, the radius of the ground impurity orbit decreases, the dielectric screening is less effective at small distances and some of the assumptions inherent the envelope function formalism are at least doubtful. Levels tend to become deep in the forbidden gap, and their description in terms of linear combination of band wavefunctions may include a number of conduction and valence bands. Deep levels are relatively difficult to be ionized thermally, but are important as recombination–generation centers of electron and hole carriers.

Just to give a feeling and extract an orientative model of deep impurity levels in semiconductors, let us consider the general integral equations (12) in the case the kernel $U_{nn'}(\mathbf{k}, \mathbf{k}')$ can be considered independent of $\mathbf{k}$ and $\mathbf{k}'$, and also independent of the band index for the group of bands under consideration; in this case we can write $U_{nn'}(\mathbf{k}, \mathbf{k}') = U_0/N$ (where N is the number of unit cells of the crystal), and Eq. (12a) reads

$$[E_n(\mathbf{k}) - E] \, A_n(\mathbf{k}) + \frac{U_0}{N} \sum_{n' \, \mathbf{k}'} A_{n'}(\mathbf{k}') = 0 \; .$$

Dividing by $[E_n(\mathbf{k}) - E]$ (which is different from zero for E in the forbidden energy gap) and summing up over n and $\mathbf{k}$ we obtain the eigenvalue compatibility equation

$$\frac{1}{N} \sum_{n \, \mathbf{k}} \frac{1}{E - E_n(\mathbf{k})} = \frac{1}{U_0} \; . \tag{18}$$

The first member of Eq. (18) represents a Green's function matrix element, say $G_{00}(E)$;

the structure of the equation $G_{00}(E) = 1/U_0$ has been discussed in Section V-8.4 and already encountered in a number of problems; its graphical solution is schematically shown in Fig. 6. The compatibility equation (18) summarizes the tight-binding Koster–Slater model of impurities in crystals [G. F. Koster and J. C. Slater, Phys. Rev. **95**, 1167 (1954)].

The Green's function approach, with appropriate implementations, has been widely applied to investigate a number of defects, and in particular the substitutional deep traps in covalent semiconductors [H. P. Hjalmarson, P. Vogl, D. J. Wolford and J. D. Dow, Phys. Rev. Lett. **44**, 810 (1980)]. The theory of substitutional deep traps in covalent semiconductors has been extended to treat on the same footing short-range and long-range part of the defect potential [see for instance D. J. Lohrmann, L. Resca, G. Pastori Parravicini and R. D. Graft, Phys. Rev. B**40**, 8404, 8410 (1989) and references quoted therein]. For further information on the vast theoretical and experimental investigations regarding the impurity levels in solids we refer to the literature.

3   Fermi level and carrier density in doped semiconductors

Carrier concentration in n-type semiconductors

We consider now extrinsic semiconductors, containing donor impurities, or acceptor impurities, or both, and we wish to study their influence on the Fermi level and the free carrier concentrations. We consider first the case of semiconductors in which only donor impurities are present (n-*type semiconductors*). The density N_d of donor impurities is supposed to be uniform in the sample, and the binding energy of the donor levels is ε_d. The schematic representation of the energy levels and occupancy (at $T = 0$) is given in Fig. 7a.

In intrinsic semiconductors we have seen that the Fermi level lies (basically) at the middle of the energy gap (see Eq. 6). Doping with donors (or acceptor) levels is the most common method to change in a controlled way the position of the Fermi level within the energy gap. The presence of donor levels shifts the Fermi level from the middle of the energy gap toward the edge of the conduction band. Let us in fact define the temperature

$$k_B T_d \equiv \varepsilon_d \ ,$$

where T_d can be considered as the "ionization temperature" of the donor levels. If $T \ll T_d$ we expect that practically all donor levels are occupied and thus the chemical potential must be located in the energy range $E_d < \mu(T) < E_c$. If T is comparable with T_d we expect that most donor levels are ionized and $\mu(T)$ lies somewhat below the donor energy E_d, but still very near to the conduction band edge. At temperatures so high that the intrinsic carriers are much larger than the concentration of donor impurities, doping becomes uninfluential and we expect that the chemical potential approaches the middle of the bandgap. The chemical potential and the carrier concentration can be determined quantitatively from the knowledge of donor concentration,

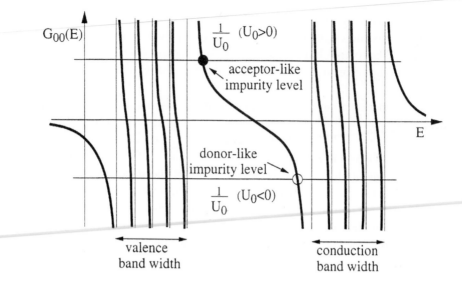

Fig. 6 Schematic graphical solution of the compatibility equation $G_{00}(E) = 1/U_0$ for impurity levels in semiconductors. The vertical lines correspond to (arbitrarily discretized) energies within the allowed energy bands. The intersection of the Green's function $y = G_{00}(E)$ with the horizontal line $y = 1/U_0$ yields the energies of the impurity states (we focus here on the states within the energy gap). Small attractive values of U_0, produce a donor-like impurity level, split-off from the conduction band; small repulsive values of U_0 produce an acceptor-like impurity level, split-off from the valence band. Impurity states, deep in the forbidden energy gap, are controlled jointly by the conduction and valence band states.

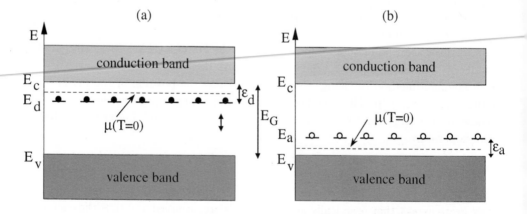

Fig. 7 (a) Schematic representation of the energy levels of a homogeneously doped n-type semiconductor at $T = 0$ (in abscissa any arbitrary direction in the homogeneous material can be considered). Typical energy values are $E_G = E_c - E_v \approx 1\,\text{eV}$ and $\varepsilon_d = E_c - E_d \approx 10\,\text{meV}$. The Fermi level at zero temperature lies at $(1/2)(E_d + E_c)$, which is the middle point between E_d and E_c. (b) Schematic representation of the energy levels of a homogeneously doped p-type semiconductor at $T = 0$; typical values of $\varepsilon_a = E_a - E_v$ are of the order of 10 meV. The Fermi level at zero temperature lies at $(1/2)(E_v + E_a)$, which is the middle point between E_v and E_a.

density-of-states of the bulk crystal, and appropriate Fermi–Dirac statistics for band levels and donor levels.

The impurity states within the energy gap are described by localized wavefunctions; a donor level can thus be empty, or occupied by one electron of either spin, but not by two electrons (of opposite spin) because of the penalty in the electrostatic repulsion energy. Due to this, the probability $P(E_d)$ that the level E_d is occupied by an electron of either spin is given by

$$P(E_d) = \frac{1}{(1/2)\, e^{(E_d - \mu)/k_B T} + 1} \; ; \tag{19}$$

the above expression has been derived in Appendix III-C in the same way as the fundamental Fermi–Dirac statistics (1).

The chemical potential of the doped semiconductor is determined by enforcing the conservation of the total number of electrons as the temperature changes. In a semiconductor with N_d donor impurities per unit volume, the density $n_0(T)$ of electrons in the conduction band must satisfy the relation

$$\boxed{n_0(T) = N_d \left[1 - P(E_d)\right] + p_0(T)} \tag{20}$$

where n_0 and p_0 are given by expressions (2). Eq. (20) is the straightforward generalization of Eq. (3); it states that the free electrons in the conduction bands are supplied by the thermal ionization of donor levels and by the thermal excitation of valence electrons. Eq. (20) can also be interpreted as an overall *charge neutrality condition* in the sample: the concentration n_0 of negative charges equals the concentration of ionized donor impurities plus the concentration of holes.

Equation (20) can be solved (numerically) to obtain the Fermi level and hence the free carrier concentration. In the case the n-type semiconductor is non-degenerate (which is the ordinary situation, except for extremely high concentration of dopants), Eq. (20) can be simplified using Eqs. (5). We have:

$$N_c(T)\, e^{-(E_c - \mu)/k_B T} = N_d \frac{(1/2)\, e^{(E_d - \mu)/k_B T}}{(1/2)\, e^{(E_d - \mu)/k_B T} + 1} + N_v(T)\, e^{-(\mu - E_v)/k_B T} \; . \tag{21}$$

This is a third order algebraic expression in $x = \exp(\mu/k_B T)$ that could be easily solved. We prefer to consider Eq. (21) in different regions of physical interest and handle it analytically.

(i) *Very low temperatures (or "freezing out region")*. Consider the semiconductor at very low temperatures $T \ll T_d$. In this temperature region we certainly have

$$E_d < \mu(T) < E_c \; .$$

Thus the second term in the right hand side of Eq. (21) can safely be neglected; furthermore the denominator in the first term in the right-hand side of Eq. (21) can be taken as unity. We have thus

$$N_c(T)\, e^{-(E_c - \mu)/k_B T} = \frac{1}{2} N_d\, e^{(E_d - \mu)/k_B T} \; ; \tag{22a}$$

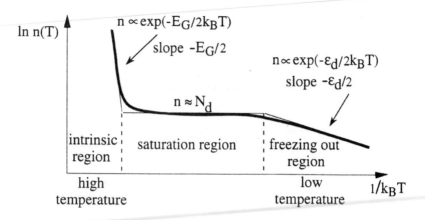

Fig. 8 Schematic variation of the electron concentration as a function of $1/k_B T$ in an n-type semiconductor with N_d donor impurities per unit volume.

taking the logarithm of both members we obtain for the Fermi level

$$\mu(T) = \frac{1}{2}(E_d + E_c) + \frac{1}{2}k_B T \ln \frac{N_d}{2\,N_c(T)} \; . \tag{22b}$$

We can replace expression (22b) into equation (22a), and we obtain that the carrier density in the conduction band is

$$n_0(T) = N_c(T)\, e^{-(E_c - \mu)/k_B T} = \sqrt{N_c(T)\, \frac{N_d}{2}}\, e^{-\varepsilon_d/2\,k_B T} \; . \tag{23}$$

Thus, the temperature dependence of the free electron carriers in n-type semiconductors at temperatures $T \ll T_d$ has (approximately) the exponential form $\exp(-\Delta/k_B T)$, where Δ is half the binding energy of the donor levels. Notice that for high doping, Eq. (22b) shows a tendency of $\mu(T)$ to increase and possibly to invade the conduction band; in this situation we must consider directly the implicit equation (20) for the determination of the chemical potential.

(ii) *Saturation region.* Consider the semiconductor in the temperature region $T_d < T \ll E_G/k_B$; we expect that (almost) all donor levels are ionized, while the thermal excitation of valence electrons is still negligible. We have

$$n_0(T) = N_c(T)\, e^{-(E_c - \mu)/k_B T} \cong N_d \; ; \tag{24a}$$

from the logarithm of both members, we have for the chemical potential

$$\mu(T) = E_c + k_B T \ln \frac{N_d}{N_c(T)} \; . \tag{24b}$$

While the number $n_0(T)$ of majority carriers is essentially constant and equal N_d, the number of minority carriers is obtained by considering the mass-action law (7). In the *saturation region*, characterized by all donor levels ionized, and at temperatures where

$n_i(T) \ll N_d$, we have

$$n_0(T) \cong N_d \quad \text{and} \quad p_0(T) \cong \frac{n_i^2(T)}{N_d} . \tag{24c}$$

For instance, the intrinsic carrier concentration of silicon at room temperature is $n_i(T) \approx 10^{10} \, \text{cm}^{-3}$. In n-type silicon with donor concentration $N_d \approx 10^{14} \, \text{cm}^{-3}$, we have $n_0 \approx 10^{14} \, \text{cm}^{-3}$ and $p_0 \approx 10^6 \, \text{cm}^{-3}$; in the above situation there are eight orders of magnitude in the difference between the concentration of majority carriers and of minority carriers. Notice also that in silicon $N_c(T) \approx 10^{19} \, \text{cm}^{-3}$; the chemical potential (24b) remains near the conduction band edge, but safely below it, so that the non-degeneracy conditions (4) are justified. As another example, consider an n-type GaAs crystal at room temperature with $n_i(T) \approx 10^7 \, \text{cm}^{-3}$ and $n_0 \approx N_d \approx 10^{14} \, \text{cm}^{-3}$; in this case we have $p_0 \approx 1 \, \text{cm}^{-3}$, a value fourteen orders of magnitude less than the majority carrier concentration.

(iii) *Intrinsic region.* If we increase further the temperature, the thermal excitation of valence electrons into the conduction band increases, and eventually the intrinsic situation is recovered. The temperature dependence of the density of free electron carriers in an n-type semiconductor is schematically summarized in Fig. 8.

Up to this point, impurities have been (tacitly) considered as isolated and independent; furthermore the doped semiconductor is assumed to remain non-degenerate, i.e. the Fermi level is several $k_B T$ away from the band edges. As the concentration of dopants is increased new phenomena occur; for instance, the Fermi level may approach and invade the energy bands; the density-of-states of the semiconductor may be perturbed near the edges and a bandgap narrowing may result; the impurity levels may interact forming an impurity band, with effects on the conductivity of the sample; here, we do not enter in these and other interesting consequences of heavy doping in semiconductors.

Carrier concentration in p-type semiconductors

We consider now the case of semiconductors in which only acceptor impurities are present (p-*type semiconductors*). The density N_a of acceptor impurities is supposed to be uniform on the sample, and the binding energy of the acceptor levels is ε_a. The schematic representation of the energy levels and occupancy (at $T=0$) is given in Fig. 7b.

The chemical potential and the carrier concentration of p-type semiconductors can be determined from the knowledge of the acceptor concentration, density-of-states of the bulk crystal, and appropriate Fermi–Dirac statistics for band levels and acceptor levels. Concerning the statistics of acceptor levels, we notice that an acceptor level can be occupied by two paired electrons, or one of either spin, but cannot be empty because of the penalty in electrostatic repulsion energy between the two holes. Due to this, the probability $P(E_a)$ that an acceptor level of energy E_a is occupied by a hole is

$$P(E_a) = \frac{1}{(1/2) \, e^{(\mu - E_a)/k_B T} + 1} \, ;$$

the above expression has been derived in Appendix III-C in the same way as expression (19). It is seen by inspection that the carrier concentration in p-type semiconductors may be stated in terms of holes in the same form as the carrier concentration in n-type semiconductors is stated in terms of electrons; thus there is no need to repeat and report here the explicit results.

Up to now we have considered the cases of semiconductors doped either with donors or with acceptors. In actual semiconductors, beside the intentionally introduced donor (or acceptor) impurities, some concentration of uncontrolled acceptor (or donor) impurities may be present. A semiconductor, containing both donors with concentration N_d and acceptors with concentration N_a, is said to be *partially compensated* if $N_a \neq N_d$; the compensation is said to be ideal if $N_a = N_d$. The equilibrium carrier statistics of compensated semiconductors can be carried out with appropriate extension of the treatment developed so far. We do not enter in details, as the general equations of carriers balance can be easily established and solved in the actual situations of interest; here we only remark that it is essentially the heavier doping that dictates the p or n character of compensated semiconductors.

4   Thermionic emission in semiconductors

The thermionic emission from metals has been considered in Section III-4, and it has been shown that the current density of escaping electrons is given by the Richardson expression

$$J_s = \frac{(-e)\, m\, k_B^2}{2\, \pi^2\, \hbar^3}\, T^2\, e^{-W/k_B T} \,, \tag{25}$$

where the work function W is the energy difference between the vacuum level and the Fermi level. We now show that a completely similar expression holds also for semiconductors.

We consider for simplicity a model semiconductor (schematized in Fig. 9), with an ideally sharp energy potential barrier at the semiconductor–vacuum surface. For our semiquantitative analysis, we assume that all conduction electrons that arrive at the surface with an energy sufficient to overcome the surface barrier χ are actually transferred to the vacuum (reflection, image forces, space charge accumulation effects are here neglected). Similarly to the analysis carried out in Section III-4, we express the current density of escaping electrons in the form

$$J_s = (-e) \int_{\sqrt{2m^*\chi/\hbar^2}}^{+\infty} dk_z \int_{-\infty}^{+\infty} dk_x \int_{-\infty}^{+\infty} dk_y \frac{2}{(2\pi)^3} \frac{1}{e^{(E_c(\mathbf{k})-\mu)/k_B T}+1}\, v_z \,, \tag{26}$$

where $v_z = \hbar k_z/2m^*$, and m^* is the effective mass of the conduction band, assumed to be parabolic. Since

$$E_c(\mathbf{k}) - \mu = E_c + \frac{\hbar^2}{2m^*}\, (k_x^2 + k_y^2 + k_z^2) - \mu \gg k_B T \,,$$

we can safely neglect the unity in the Fermi distribution function in Eq. (26), and

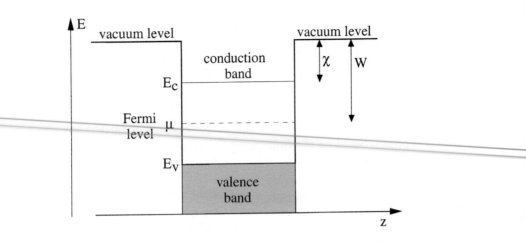

Fig. 9 Model semiconductor for the calculation of the thermionic emission. The work function W is the difference between the energy of an electron at rest in the vacuum and the chemical potential. The electronic affinity χ is the energy difference between the vacuum level and the bottom of the conduction band. The work function W equals $\chi + (E_c - \mu)$.

obtain

$$J_s = \frac{(-e)}{4\pi^3} \frac{\hbar}{m^*} \int_{\sqrt{2m^*\chi/\hbar^2}}^{+\infty} k_z dk_z \int_{-\infty}^{+\infty} dk_x \int_{-\infty}^{+\infty} dk_y \, \exp\left[\frac{\mu}{k_B T} - \frac{\hbar^2\left(k_x^2 + k_y^2 + k_z^2\right)}{2m^* k_B T}\right].$$

The integrals are all of elementary type (and already encountered in Section III-4); we thus recover the Richardson expression (25) (with m^* replacing m) as an estimation of the thermionic emission in semiconductors.

5 Non-equilibrium carrier distributions

5.1 Drift and diffusion currents

In the previous sections, we have dealt with the distribution of electrons and holes at the thermodynamic equilibrium in uniformly doped (or undoped) semiconductors; in this and next section, still in uniformly doped materials, we consider non-equilibrium carrier distributions and transport phenomena of interest in the physics of semiconductors and devices.

In non-equilibrium and non-stationary conditions (produced for instance by applied fields, absorption of electromagnetic radiation of sufficiently high frequency, carrier injection or extraction, etc.), the electron and hole concentrations depend in general on space coordinate and time. In the following, we denote by $n(\mathbf{r}, t)$ and $p(\mathbf{r}, t)$ the actual non-equilibrium electron and hole concentrations in the semiconductor; the equilibrium electron and hole concentrations (at a given temperature T) are denoted as previously by n_0 and p_0; the differences $\Delta n(\mathbf{r}, t) = n(\mathbf{r}, t) - n_0$ and $\Delta p(\mathbf{r}, t) =$

$p(\mathbf{r}, t) - p_0$, whether positive or not, are referred as the "excess carrier concentrations" of electrons and holes, respectively.

We examine first the carrier transport in a semiconductor, with a uniform concentration of carriers, in the presence of an applied electric field. We consider the simplest possible model of semiconductor with two bands only: a spherical conduction band of effective mass m_c^*, and a spherical valence band of effective mass m_v^* (see Fig. 3). A classical Drude-like picture of the dynamics and damping of carriers (or also a more rigorous approach based on the Boltzmann transport equation, see Chapter XI) gives for the electron current density

$$\mathbf{J}_n = \frac{n\, e^2\, \tau_n}{m_c^*}\, \mathbf{E} \equiv n\, e\, \mu_n\, \mathbf{E} \; , \qquad (27a)$$

where n is the density of electrons, e is the absolute value of the electronic charge, τ_n is the average interval between collisions suffered by an electron, and $\mu_n \equiv e\,\tau_n/m_c^*$ is the electron mobility, i.e. the average drift velocity of carriers divided by the applied electric field. Similarly for the hole current density we have

$$\mathbf{J}_p = \frac{p\, e^2\, \tau_p}{m_v^*}\, \mathbf{E} \equiv p\, e\, \mu_p\, \mathbf{E} \; . \qquad (27b)$$

Notice that $\mathbf{J}_n$ and $\mathbf{J}_p$ are proportional to the electric field (Ohm's law), and this is the expected behaviour for sufficiently small electric fields.

In the following, we assume that the linear relationships (27) between current density and electric field hold. However we wish to mention that deviations from Ohm's law are expected at sufficiently high electric fields in semiconductors, due to several mechanisms. In metals, the high value of the conductivity (because of the high carrier density), and the consequent joule heating, strongly limit the values of applicable electric fields without sublimation of the material; vice versa, in semiconductors high electric field effects may occur. As long as the applied electric field has low values ($\lesssim 10^3$ volt/cm), the carriers lose energy mainly by emitting low energy acoustic phonons. As the field is increased, at around 10^4 volt/cm, energy dissipation is through emission of optical phonons; in this region the drift velocity of the carriers is nearly independent from the electric field (*saturation velocity*). For still higher electric fields avalanche processes may become active. Another interesting source of deviations from Ohm's law occurs in semiconductor with secondary valleys in the conduction band structure at energies (slightly) higher than the principal minimum. In an n-type GaAs semiconductor, for instance, high frequency current oscillations (Gunn effect) are produced by the application of a constant electric field of the order of kvolt/cm. Electric fields of this order allow electrons in the lower energy band (of small mass and high mobility) to migrate into the higher valleys (of large mass and small mobility). Upon application of an electric field electrons move to higher valleys with a consequent decrease in current; vice versa the transition of the electrons to the lower valley leads to current increase; periodic current oscillations may thus result upon application of a constant potential.

We discuss now carrier transport in semiconductors in the presence both of an *electric field* and a *concentration gradient*. An intuitive classical picture that includes the concept of diffusion according to Fick's law, (or also a more rigorous approach based on the Boltzmann transport equation and discussed in Chapter XI), gives for the electron current density the expression

$$\boxed{\mathbf{J}_n = e\,n\,\mu_n\,\mathbf{E} + e\,D_n\,\nabla n}\,, \tag{28}$$

where D_n is the diffusion coefficient. The current density $\mathbf{J}_n$ is the sum of two terms: the *drift term* proportional to the field (and already discussed), and the *diffusion term* proportional to the concentration gradient. The diffusion term takes into account the fact that the carriers tend to move from high concentration regions to lower concentration regions; for electrons the diffusion term and the concentration gradient have the same direction. In metals, n is so high to make the drift current in general dominant over the diffusion current; vice versa in semiconductors the two contributions may be of comparable size.

The mobility μ_n and the diffusion coefficient D_n are not independent but are related by the Einstein relation, already discussed in Section XI-6.1, and here re-examined for non-degenerate semiconductors on a more intuitive basis. Consider in fact a semiconductor, in stationary conditions and thermal equilibrium, with a superimposed potential $\phi(\mathbf{r})$; in an open circuit arrangement we have $\mathbf{J}_n(\mathbf{r}) \equiv 0$, and Eq. (28) gives

$$0 \equiv n(\mathbf{r})\,\mu_n\,[-\nabla\phi(\mathbf{r})] + D_n\,\nabla n(\mathbf{r})\,. \tag{29}$$

On the other hand, the bottom of the conduction band depends continuously on the space variable $\mathbf{r}$ via the relation

$$E_c(\mathbf{r}) = E_c + (-e)\,\phi(\mathbf{r})\,. \tag{30a}$$

From Eq. (5a), which applies to *non-degenerate* semiconductors, also the carrier density depends on $\mathbf{r}$ in the form

$$n(\mathbf{r}) = N_c(T)\,e^{-[E_c(\mathbf{r})-\mu]/k_B T}\,, \tag{30b}$$

where μ is independent of the position, since the system is at thermal equilibrium. From the gradient of both members of Eq. (30b) and the use of Eq. (30a), we have

$$\nabla n(\mathbf{r}) = n(\mathbf{r})\,\frac{1}{k_B T}\,[-\nabla E_c(\mathbf{r})] = n(\mathbf{r})\,\frac{e}{k_B T}\,\nabla\phi(\mathbf{r})\,;$$

after comparison with Eq. (29), we obtain the Einstein relation for the diffusion coefficient

$$\boxed{e\,D_n = \mu_n\,k_B T}\,. \tag{31}$$

A similar reasoning can be done for the hole current density; we have

$$\boxed{\mathbf{J}_p = e\,p\,\mu_p\,\mathbf{E} - e\,D_p\,\nabla p}\,, \tag{32}$$

where the first term arises from the drift and the second term arises from the diffusion; notice that the diffusion term of the hole current is opposite to the direction of the hole concentration gradient. The diffusion coefficient and the mobility are linked in non-degenerate semicondutors by the Einstein relation

$$\boxed{e\,D_p = \mu_p\,k_B T}\;.\tag{33}$$

The total current density $\mathbf{J}$ is then given by the equation

$$\mathbf{J} = e\,(n\,\mu_n + p\,\mu_p)\,\mathbf{E} + e\,D_n\,\nabla n - e\,D_p\,\nabla p\;,\tag{34}$$

which is the sum of the drift and the diffusion currents of electrons and holes.

Continuity equations for carriers in the presence of electron–hole generation–recombination processes

In the theory of transport, the equation of continuity ensures that the carriers crossing a closed surface balance the change of the carriers enclosed therein. In semiconductors, we must express the continuity equation both for electrons and holes, taking into account that the electron and hole concentrations change not only because of drift and diffusion mechanisms, but also because of generation and recombination processes due to internal or external causes. We indicate by $g_n = g_n(\mathbf{r}, t)$ the number of electrons generated in the conduction band per unit time and unit volume, and $r_n = r_n(\mathbf{r}, t)$ the number of electrons subtracted from the conduction band per unit time and volume (space and time coordinates $\mathbf{r}$ and t are often omitted, for simplicity). Similarly, g_p and r_p refer to the holes generated or subtracted in the valence band.

Generation and recombination processes may occur via a number of different mechanisms, and we mention in particular the following. (i) Electron and hole concentrations can change because of transitions from occupied valence states to empty conduction states or, vice versa, from occupied conduction states to empty valence states (band-to-band transitions); in this case we have evidently $g_n = g_p$ and $r_n = r_p$. (ii) Electron and hole concentrations can change because of transitions to or from impurity levels (band-to-impurity transitions); the impurity centers of relevance (if any) are those deep within the energy gap, since at temperatures of ordinary interest all the shallow impurity levels (donors and acceptors) are thermally ionized. (iii) Electron and hole concentrations can change because of carrier injection or extraction from "external sources" (for instance states at physical surfaces, applied electrodes etc).

From the above considerations we expect that the generation–recombination rates for electrons and holes may in general be different. In the following, for simplicity, we shall confine our attention exclusively to mechanisms and regions where the *generation–recombination processes involve electron–hole pairs*, and thus the generation and recombination rates for electrons and holes are the same.

Let $n(\mathbf{r}, t)$ indicate the electron concentration in the semiconductor, and $\mathbf{J}_n$ the electron current density. In the presence of electron–hole generation–recombination processes at a rate g_{eh} and r_{eh}, the continuity equation for electrons takes the form

$$\frac{\partial n}{\partial t} = \frac{1}{e}\,\mathrm{div}\,\mathbf{J}_n + g_{\mathrm{eh}} - r_{\mathrm{eh}}\;;\tag{35a}$$

in fact, if we multiply both members of Eq. (35a) by the volume element dV and integrate over any given volume V enclosed by a surface S, we obtain a very intuitive balance between the change of the number of electrons in the volume V, and the electrons crossing the surface or being supplied by the generation–recombination processes.

In a completely similar way, we have that the time and space dependent hole concentration $p(\mathbf{r}, t)$ satisfies the continuity equation

$$\frac{\partial p}{\partial t} = -\frac{1}{e} \operatorname{div} \mathbf{J}_p + g_{\text{eh}} - r_{\text{eh}} \ . \tag{35b}$$

We can also write a continuity equation for the total current density $\mathbf{J} = \mathbf{J}_n + \mathbf{J}_p$ and the total electron and hole charge density $\rho = (-e)\, n + e\, p$. From Eqs. (35) we obtain

$$\frac{\partial}{\partial t}(p - n) = -\frac{1}{e} \operatorname{div} \mathbf{J} \ ,$$

or equivalently

$$\frac{\partial}{\partial t}(\Delta p - \Delta n) = -\frac{1}{e} \operatorname{div} \mathbf{J} \ . \tag{36}$$

The above continuity equation holds when generation–recombination processes occur in pairs; notice also that Eq. (36) is independent from the rates g_{eh} and r_{eh}.

The continuity equation (35a) for the electron concentration is more conveniently written using Eq. (28) and assuming, for simplicity, that the electric field and concentration gradients are in the x-direction. We obtain the basic equation

$$\boxed{\frac{\partial n}{\partial t} = D_n \frac{\partial^2 n}{\partial x^2} + \mu_n \frac{\partial (n\, E)}{\partial x} + g_{\text{eh}} - r_{\text{eh}}} \ . \tag{37a}$$

The interpretation of the right part of Eq. (37a) is very simple: the first term represents the change of concentration due to the diffusion, the second term the drift and diffusion in the presence of an electric field, and finally the last terms the generation and recombination processes. In a completely similar way, the continuity equation for holes reads

$$\boxed{\frac{\partial p}{\partial t} = D_p \frac{\partial^2 p}{\partial x^2} - \mu_p \frac{\partial (p\, E)}{\partial x} + g_{\text{eh}} - r_{\text{eh}}} \ . \tag{37b}$$

The two basic continuity equations (37) for electrons and holes are not independent; they are coupled through the presence of the electric field and through the presence of generation–recombination processes. The electric field is related to the total external charge density through the Poisson equation $\operatorname{div} \mathbf{E} = \rho_{\text{tot}}/\varepsilon_s$. In the case the concentration of ionized impurity centers is the same as at thermal equilibrium, we have

$$\boxed{\operatorname{div} \mathbf{E} = \frac{e}{\varepsilon_s}(\Delta p - \Delta n)} \ , \tag{38}$$

where $\Delta p = p - p_0$ is the excess hole concentration, $\Delta n = n - n_0$ is the excess electron concentration, and $\varepsilon_s = \varepsilon_r \varepsilon_0$ is the static dielectric constant of the semiconductor (we use here standard SI units; $\varepsilon_0 = 10^{-11}$ Farad/m).

Considerations on charge neutrality in uniformly doped semiconductors

Consider a homogeneously doped n-type semiconductor with equilibrium carrier concentrations n_0 and p_0. At thermal equilibrium, the *charge neutrality condition* is satisfied over all the sample. We wish now to justify that also non-equilibrium carrier concentrations must satisfy a *charge quasi-neutrality condition* throughout the sample.

Suppose that the carrier density of the semiconductor is changed, for instance by applying a voltage pulse to injecting electrodes. Let the disturbed electron concentration and hole concentration be $n = n_0 + \Delta n$ and $p = p_0 + \Delta p$, with $\Delta n \neq \Delta p$; in this situation a space-charge $e\,(\Delta p - \Delta n)$ is set up in the material, and an electric field $\mathbf{E}$ arises that satisfies the Poisson equation (38). The very important result that we now show is that the unbalance $\Delta p - \Delta n$ can not survive in the uniformly doped semiconductor: the unbalance $\Delta p - \Delta n$ quickly goes to zero (or almost to zero) because of the drift current of the majority carriers in the electric field $\mathbf{E}$, created by the unbalance itself.

Consider the time variation of $e\,(\Delta p - \Delta n)$ by means of the continuity equation (36); we have

$$\frac{\partial}{\partial t}(\Delta p - \Delta n) = -\frac{1}{e}\,\text{div}\,\mathbf{J} = -\frac{1}{e}\,\text{div}\,\sigma_0 \mathbf{E}\ ; \qquad (39)$$

in the last passage, we have neglected diffusion terms in $\mathbf{J}$ and we have indicated by σ_0 the conductivity of the medium (essentially due to the majority carriers); from Eq. (38) and Eq. (39), we obtain

$$\frac{\partial}{\partial t}(\Delta p - \Delta n) = -\frac{\sigma_0}{\varepsilon_s}(\Delta p - \Delta n)\ ;$$

this equation shows that the space-charge decreases exponentially with time constant $t_D = \varepsilon_s/\sigma_0$ (*dielectric relaxation time*).

To estimate t_D, consider for instance an n-type GaAs semiconductor with $N_D = 10^{14}\,\text{cm}^{-3}$. In the saturation regime we have $n_0 \approx N_D = 10^{20}\,\text{m}^{-3}$; the electron mobility (see Table 1) is $\mu_n = 0.85\,\text{m}^2/\text{volt}\cdot\text{sec}$; the conductivity $\sigma_0 = n_0\,e\,\mu_n$ becomes $\sigma_0 = 10^{20}\cdot 1.6\cdot 10^{-19}\cdot 0.85 \approx 10\,(\text{ohm}\cdot\text{m})^{-1}$; with $\varepsilon_s = \varepsilon_0\,\varepsilon_r = 10^{-11}\times 10 = 10^{-10}\,\text{F/m}$ we estimate $t_D = 10^{-11}$ sec.

The above considerations show that the space-charge neutrality condition $\Delta n \approx \Delta p$ is reached within the very short dielectric relaxation time t_D, because of the drift of the majority carriers: any deviation from the space-charge neutrality condition $\Delta n \approx \Delta p$ tends to be neutralized very quickly, within an interval of the order of the dielectric relaxation time. This estimate is confirmed by more detailed analyses, that show that charge quasi-neutrality holds in ordinary situations in homogeneously doped materials. In the next chapter, in the study of semiconductors with non-uniform doping (pn junction, for instance) or non-uniform energy band structure (heterostructures, for instance) we will see that space-charge regions with large built-in electric fields are formed in the non-homogeneity regions.

5.2 Generation and recombination of electron–hole pairs in semiconductors

In order to complete the description of carrier transport equations in doped semi-conductors, we add a few qualitative considerations on the generation and recombination processes of electron–hole pairs in the bulk of homogeneously doped semiconductors.

The total generation rate g_{eh} of electron–hole pairs can be split in the form

$$g_{eh} = g_{th} + g_{ext} , \tag{40}$$

where g_{th} denotes the thermal generation rate of electron–hole pairs, and g_{ext} denotes any other external generation mechanisms at work (for instance: illumination of the specimen with a beam of photons of energies higher than the energy gap; bombardment with charged particles that lose kinetic energy by creating electron–hole pairs; avalanche multiplication processes due to the presence of strong electric fields, direct tunneling of electrons from the valence band to the conduction band in strong electric fields, etc.). The thermal generation rate of electron–hole pairs $g_{th}(T)$, at temperature T, depends in general on the type of semiconductor and several features (i.e. electromagnetic modes in the medium, photon occupancy at temperature T, optical absorption coefficient, possible presence of traps or localized surface states), but is essentially independent of the concentrations n and p.

Similarly to Eq. (40), the recombination rate for electron–hole pairs can be split into a thermal contribution and possibly other external mechanisms at work. The simplest model for the thermal recombination rate $r_{th}(T)$ assumes that annihilation processes are proportional to the product of the concentrations of electrons in the conduction band and holes in the valence band, i.e.

$$r_{th} = A(T)\, n\, p . \tag{41}$$

At thermal equilibrium, we have evidently $g_{th}(T) = r_{th}(T) = A(T)\, n_0\, p_0$; thus the *net thermal generation–recombination rate* can be written as

$$g_{th} - r_{th} = A(T)\, n_0\, p_0 - A(T)\, n\, p . \tag{42}$$

The above expression becomes particularly simple and meaningful in the case of semiconductors with large difference in the density of majority and minority carriers, as shown by the following arguments.

We consider here the case of an n-type semiconductor, where $n_0 \gg p_0$. We also suppose that the deviations Δn and Δp are much smaller than n_0, i.e. $\Delta n \approx \Delta p \ll n_0$ (this is called *low disturbance or low injection regime*). In an n-type semiconductor in the low disturbance regime, we can replace n with n_0 in the right-hand side of Eq. (42), and we have

$$g_{th} - r_{th} = A(T)\, n_0\, p_0 - A(T)\, n_0\, p = A(T)\, n_0\, (p_0 - p) . \tag{43}$$

The quantity $A(T) n_0$ has the dimensions of the inverse of a time. It is thus convenient to define the *minority carrier lifetime* t_p in the form

$$t_p = \frac{1}{A(T) n_0} . \tag{44a}$$

Notice that t_p can be interpreted as the lifetime of the minority carriers in the presence of a (hostile) concentration n_0 of majority carriers. Eq. (43) can be re-written as

$$g_{\text{th}} - r_{\text{th}} = \frac{p_0}{t_p} - \frac{p}{t_p} . \tag{44b}$$

We see that the *thermal generation rate of electron–hole pairs is given by* p_0/t_p ; this is just the *equilibrium value of the minority carrier density divided by the minority carrier lifetime* t_p. Similarly the thermal recombination rate is the *actual value of minority carrier density divided by the lifetime* t_p.

We now specify Eq. (37b) in the case the only electron–hole generation–recombination mechanism at work is the thermal one, described by Eq. (44b); we obtain

$$\boxed{\frac{\partial p}{\partial t} = D_p \frac{\partial^2 p}{\partial x^2} - \mu_p E \frac{\partial p}{\partial x} - \mu_p p \frac{\partial E}{\partial x} + \frac{p_0 - p}{t_p}} . \tag{45}$$

This is the basic continuity equation for holes minority carriers in an n-type semiconductor. In the case of a homogeneous semiconductor in the absence of electric fields and carrier gradients, we see that the minority carrier concentration (and hence also the majority carrier concentration) relaxes towards the equilibrium value with time constant t_p (typical values are $t_p \approx 10^{-5}$ sec).

Similar considerations can be done for p-type semiconductors, where holes are the majority carriers and electrons the minority ones. The partner equation of (45) is

$$\boxed{\frac{\partial n}{\partial t} = D_n \frac{\partial^2 n}{\partial x^2} + \mu_n E \frac{\partial n}{\partial x} + \mu_n n \frac{\partial E}{\partial x} + \frac{n_0 - n}{t_n}} . \tag{46}$$

Eq. (46) is the *basic continuity equation for electrons minority carriers in a p-type semiconductor.*

6   Solutions of typical transport equations in uniformly doped semiconductors

Example 1. Injection of minority carriers at one end of an n-type semiconductor in steady conditions

Consider a bar of homogeneously doped n-type semiconductor and suppose that an external source (for instance light irradiation or particle bombardment) maintains an excess of holes $\Delta p(x) = \Delta p(0)$, and an excess of electrons $\Delta n(x) = \Delta n(0) = \Delta p(0)$ for any $x \leq 0$. Carriers diffuse in the $x > 0$ part of the semiconductor, and we determine the excess concentration $\Delta p(x)$ and $\Delta n(x)$ in stationary conditions; we also determine

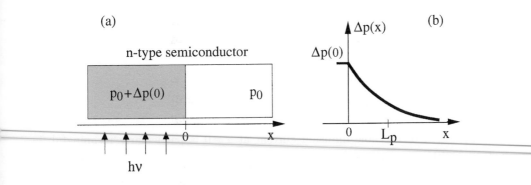

Fig. 10 (a) Schematic representation of an n-type sample subject to irradiation (b) Distribution of excess holes (and excess electrons) for $x > 0$.

the internal electric field, that automatically sets in and accompanies the diffusion process (even in the open circuit arrangement of Fig. 10).

The basic equation (45) controls the *dynamics of the minority carriers* in the region $x > 0$ (where only the thermal electron–hole generation–recombination mechanism is present). We ignore (momentarily) the internal electric field, which may accompany the diffusion processes, and also consider steady conditions; Eq. (45) simplifies in the form

$$D_p \frac{d^2 p}{dx^2} - \frac{p - p_0}{t_p} = 0 \ . \tag{47a}$$

The boundary conditions for the carrier concentration $p(x)$ at $x = 0$ and at $x = \infty$ are given by

$$p(0) \equiv p_0 + \Delta p(0) \quad \text{and} \quad p(\infty) \equiv p_0 \ . \tag{47b}$$

The solution of Eq. (47a), which fulfills the boundary conditions (47b), is

$$p(x) - p_0 = \Delta p(0) \, e^{-x/L_p} \ , \tag{48a}$$

where

$$L_p^2 = D_p \, t_p \ . \tag{48b}$$

The length L_p is called the *minority carrier diffusion length*. With $D_p \approx 10 \, \text{cm}^2/\text{sec}$, $t_p \approx 10^{-5} \, \text{sec}$ the diffusion length of minority carriers is $L_p \approx 0.01 \, \text{cm}$. We plot in Fig. 10b the behaviour of the excess of minority carriers concentration $\Delta p(x) = p(x) - p_0$ for $x > 0$. Notice that the mean square diplacement of the excess particle distribution (48a) is

$$\langle x^2 \rangle = \frac{\displaystyle\int_0^\infty x^2 \, \exp(-x/L_p) \, dx}{\displaystyle\int_0^\infty \exp(-x/L_p) \, dx} = 2 \, L_p^2 = 2 \, D_p \, t_p \ . \tag{48c}$$

The hole diffusion current in the x-direction becomes

$$J_p^{(\text{diff})}(x) = -e\, D_p\, \frac{\partial p}{\partial x} = e\, D_p\, \frac{\Delta p(0)}{L_p}\, e^{-x/L_p} \; ; \tag{49a}$$

in particular for $x = 0$, we have

$$J_p^{(\text{diff})}(0) = e\, D_p\, \frac{\Delta p(0)}{L_p} = e\, L_p\, \frac{\Delta p(0)}{t_p} \; . \tag{49b}$$

The expression of the hole diffusion current at $x = 0$ is particularly meaningful; in fact $L_p\, \Delta p(0)/t_p$ represents the number of holes that must be supplied per unit time through the plane $x = 0$ to keep approximately an excess of carriers $\Delta p(0)$ (with lifetime t_p) in the hostile region of width L_p in the n-type semiconductor.

Due to the local quasi-neutrality conditions already discussed, the excess of majority carriers is given with good approximation

$$\Delta n(x) \approx \Delta p(x) \equiv \Delta p(0)\, e^{-x/L_p} \; .$$

The electron diffusion current $J_n^{(\text{diff})}(x) = e\, D_n\, \partial n/\partial x$ in the x-direction becomes

$$J_n^{(\text{diff})}(x) = -e\, D_n\, \frac{\Delta p(0)}{L_p}\, e^{-x/L_p} = -b\, J_p^{(\text{diff})}(x) \tag{50a}$$

where $b = D_n/D_p = \mu_n/\mu_p$.

From Eq. (49a) and Eq. (50a) we see that the sum of the diffusion currents due to electrons and holes is not zero. However, the *total current* (diffusion current plus drift current of electrons and holes) obeys equation (36), and so it is a constant in steady situation, and actually vanishes in the open circuit situation we are investigating. We thus must have at any x point

$$J_{\text{tot}}(x) = J_p^{(\text{diff})}(x) + J_n^{(\text{diff})}(x) + \sigma_n\, E(x) \equiv 0 \; , \tag{50b}$$

where $\sigma_n = e\, n\, \mu_n$ is the conductivity of the electrons in the n-type material, and the much smaller conductivity of the minority carriers is neglected. From Eq. (50b) we recover for the electric field

$$E(x) = \frac{1}{\sigma_n}\, (b - 1)\, e\, D_p\, \frac{\Delta p(0)}{L_p}\, e^{-x/L_p} \; . \tag{50c}$$

This shows that also the electric field decreases exponentially in the region $x > 0$; it is also evident that for $b = 1$ the electron and hole diffusion currents are equal and opposite, and the electric field is zero. In most ordinary conductivity situations, the electric field (50c) is anyway so small that corrections to the field-free Eq. (47a) are hardly of any significance.

Example 2. Injection (or extraction) of carriers at both ends of an n-type semiconductor in steady conditions

Consider a homogeneously doped n-type semiconductor and suppose to maintain for $x \leq 0$ an excess (or a deficit) of holes $\Delta p(0)$ and for $x \geq w$ an excess (or deficit) of holes $\Delta p(w)$ (see Fig. 11).

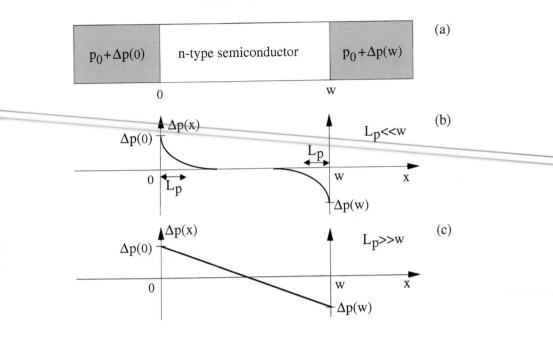

Fig. 11 (a) Schematic representation of an n-type semiconductor, with hole concentration $p_0 + \Delta p(0)$ for $x < 0$ and $p_0 + \Delta p(w)$ for $x > w$; we assume $\Delta p(0) > 0$ and $\Delta p(w) < 0$. (b) Distribution of excess holes (and excess electrons) as a function of x in the case $L_p \ll w$. (c) Distribution of excess holes (and excess electrons) as a function of x in the case $L_p \gg w$.

Similarly to Example 1, the minority carrier distribution is described by the field-free steady-state continuity equation

$$D_p \frac{d^2 p}{dx^2} - \frac{p - p_0}{\tau_p} = 0 \ . \tag{51a}$$

The boundary conditions for the carrier concentration $p(x)$ at $x = 0$ and $x = w$ are

$$p(0) \equiv p_0 + \Delta p(0) \quad \text{and} \quad p(w) \equiv p_0 + \Delta p(w) \ . \tag{51b}$$

We consider the two linearly independent solutions $\exp(\pm x/L_p)$ of the second order differential equation (51a), and determine the linear combination that fulfills the boundary conditions (51b). We obtain

$$\Delta p(x) = \Delta p(0) \frac{\sinh (w - x)/L_p}{\sinh w/L_p} + \Delta p(w) \frac{\sinh x/L_p}{\sinh w/L_p} \quad 0 \le x \le w \ . \tag{52a}$$

The hole diffusion current $J_p^{(\mathrm{diff})}(x) = -e\, D_p \, \partial \Delta p(x)/\partial x$ associated to the above carrier distribution is

$$J_p^{(\mathrm{diff})}(x) = e\, D_p \frac{\Delta p(0)}{L_p} \frac{\cosh\left[(w - x)/L_p\right]}{\sinh (w/L_p)} - e\, D_p \frac{\Delta p(w)}{L_p} \frac{\cosh (x/L_p)}{\sinh (w/L_p)} \ . \tag{52b}$$

In particular, at the points $x = 0$ and $x = w$ we have

$$J_p^{(\text{diff})}(0) = e\,D_p \frac{\Delta p(0)}{L_p} \frac{1}{\tanh\left(w/L_p\right)} - e\,D_p \frac{\Delta p(w)}{L_p} \frac{1}{\sinh\left(w/L_p\right)} \qquad (52c)$$

$$J_p^{(\text{diff})}(w) = e\,D_p \frac{\Delta p(0)}{L_p} \frac{1}{\sinh\left(w/L_p\right)} - e\,D_p \frac{\Delta p(w)}{L_p} \frac{1}{\tanh\left(w/L_p\right)} . \qquad (52d)$$

We discuss the concentration of minority carriers and diffusion currents, given by Eqs. (52), in the two limiting cases $w \gg L_p$ and $w \ll L_p$. In the case the width w is much larger than the diffusion length, Eqs. (52) simplify, and we have in particular

$$\Delta p(x) = \Delta p(0)\,e^{-x/L_p} + \Delta p(w)\,e^{-(w-x)/L_p} \qquad 0 \le x \le w \qquad w \gg L_p$$

and

$$J_p^{(\text{diff})}(0) = e\,D_p \frac{\Delta p(0)}{L_p} , \qquad J_p^{(\text{diff})}(w) = -e\,D_p \frac{\Delta p(w)}{L_p} . \qquad (53)$$

We thus see that the concentration of minority carriers, as well as diffusion currents penetrating in the semiconductors from the two ends, are completely decoupled (as shown in Fig. 11b).

A much more interesting situation occurs in the case $L_p \gg w$. With appropriate series expansion of the terms at the right member of Eqs. (52), we obtain

$$\Delta p(x) = \Delta p(0)\frac{w - x}{w} + \Delta p(w)\frac{x}{w} + \ldots = \Delta p(0) + [\Delta p(w) - \Delta p(0)]\frac{x}{w} + \ldots$$

and

$$J_p^{(\text{diff})}(0) = J_p^{(\text{diff})}(w) = e\,D_p \frac{\Delta p(0) - \Delta p(w)}{w} . \qquad (54)$$

From expressions (54) and (53), we see that the diffusion current is greatly increased when $w \ll L_p$.

Due to the local quasi-neutrality condition the excess majority carrier concentration is given approximately by $\Delta n(x) \approx \Delta p(x)$; from the requirement that the total current must be zero, the (small) electric field in the sample can be obtained, with considerations similar to those worked out in Example 1.

Example 3. Injection of a narrow pulse of minority carriers and measurement of drift mobility (Haynes–Shockley experiment)

Consider an n-type semiconductor and suppose that a narrow pulse of holes (and electrons) is produced by a source in the plane $x = 0$ at the time $t = 0$ (for instance by irradiation of light, or in practice by a voltage-pulse applied to an emitting contact). We wish to determine the excess hole (and electron) concentration $\Delta p(x,t)$ at successive instants $t > 0$ in the semiconductor.

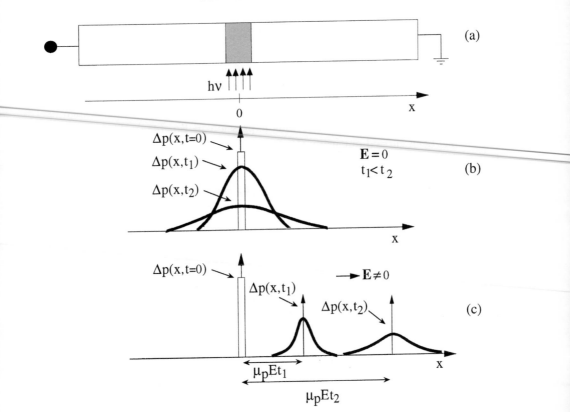

Fig. 12 (a) Schematic representation of an n-type semiconductor with a narrow pulse of holes (and electrons) optically injected at $x = 0$ and $t = 0$. (b) Distribution of excess holes (and electrons) in the absence of applied electric fields. (c) Distribution of excess holes (and electrons) in the presence of a constant and uniform electric field.

The equation that governs the dynamics of the holes minority carriers in the injection experiment is the basic equation (45); neglecting momentarily the terms containing the electric field, we have

$$\frac{\partial p}{\partial t} = D_p \frac{\partial^2 p}{\partial x^2} - \frac{p - p_0}{t_p} \ . \tag{55}$$

The solution of this equation with the boundary condition $p(0,0) = p_0 + N\,\delta(t)\,\delta(x)$ is

$$\Delta p(x,t) = \frac{N}{\sqrt{4\pi D_p t}} \exp\left(-\frac{x^2}{4 D_p t}\right) \exp\left(-\frac{t}{t_p}\right) \ ; \tag{56}$$

in fact it is easily seen by inspection that the above expression of $\Delta p(x,t)$ satisfies the differential equation (55), as well as the boundary conditions $\Delta p(0,0) = N\,\delta(t)\,\delta(x)$. The behaviour of $\Delta p(x,t)$ at different times in the semiconductor is illustrated in Fig. 12b.

It is of interest to evaluate the average square distance $\langle x^2(t) \rangle$ of the carriers at time t; we have

$$\langle x^2(t) \rangle = \int_0^\infty x^2 \, \Delta p(x,t) \, dx \left/ \int_0^\infty \Delta p(x,t) \, dx \right. = 2 \, D_p \, t \ . \tag{57}$$

The average distance $\langle x^2 \rangle$ travelled by the carriers during their lifetime is then $\langle x^2 \rangle = 2 \, D_p \, t_p$, in agreement with the previous expression in Eq. (48c). The calculation of $\langle x^2(t) \rangle$ has been carried out using the following elementary integrals of the Gaussian function

$$\int_0^\infty e^{-a \, x^2} \, dx = \frac{1}{2} \sqrt{\frac{\pi}{a}} \qquad \int_0^\infty x^2 \, e^{-a \, x^2} \, dx = \frac{1}{4a} \sqrt{\frac{\pi}{a}} \qquad a > 0 \ .$$

We discuss now the injection of a narrow pulse of holes in the presence of a uniform and constant electric field $\mathbf{E}$ (in the x-direction); instead of Eq. (55), we have to consider the partial differential equation of the form

$$\frac{\partial p}{\partial t} = D_p \, \frac{\partial^2 p}{\partial x^2} - \frac{p - p_0}{t_p} - \mu_p \, E \, \frac{\partial p}{\partial x} \ ;$$

the solution of the above equation with the boundary condition $\Delta p(0,0) = N \, \delta(t) \, \delta(x)$ is

$$\Delta p(x,t;E) = \frac{N}{\sqrt{4 \, \pi \, D_p \, t}} \exp \left[-\frac{(x - \mu_p \, E \, t)^2}{4 \, D_p \, t} \right] \exp \left(-\frac{t}{t_p} \right) \ .$$

The expression $\Delta p(x,t;E)$ has the same form as $\Delta p(x,t;E \equiv 0)$ of Eq. (56), except that x is now replaced by $x - \mu_p \, E$; thus the excess carriers (both holes and electrons) follow the electric field with drift velocity $\mu_p \, E$, as shown in Fig. 12c.

The above results are the basis for the understanding of the Haynes–Shockley experiment for measuring the drift mobility of minority carriers. The method consists in the injection of a narrow pulse of minority carriers into a filament and sweeping the pulse along the filament by means of an electric field; if L is the distance between the emitter and collector contacts, from the measurement of the transit time $t = L/\mu_p \, E$ the mobility of minority carriers can be obtained.

Further reading

F. Bassani, G. Iadonisi and B. Preziosi "Electronic Impurity Levels in Semiconductors" Rep. Progr. Phys. **37**, 1099 (1974)

P. Bhattacharya "Semiconductor Optoelectronic Devices" (Prentice-Hall, Englewood Cliffs, New Jersey 1997)

M. Jaros "Physics and Applications of Semiconductor Microstructures" (Clarendon Press, Oxford 1989)

R. S. Muller and T. I. Kamins "Device Electronics for Integrated Circuits" (Wiley, New York 1986)

S. T. Pantelides "The Electronic Structure of Impurities and Other Point Defects in Semiconductors" Rev. Mod. Phys. **50**, 797 (1978)

B. Sapoval and C. Hermann "Physics of Semiconductors" (Springer, New York 1995)

K. Seeger "Semiconductor Physics" (Springer, Berlin 1989, fourth edition)

R. A. Smith "Semiconductors" (Cambridge University Press, 1978)

S. M. Sze "Semiconductor Devices. Physics and Technology" (Wiley, New York 1985)

C. M. Wolfe, N. Holonyak and G. E. Stillman "Physical Properties of Semiconductors" (Prentice-Hall, Englewood Cliffs, New Jersey 1989)

XIV

Transport in inhomogeneous semiconductors

This chapter is the natural extension of the previous one and deals with semiconducting structures that are inhomogeneous, because of non-uniform doping or non-uniform band structure, or both. It would be impossible to describe all the electron devices that can be tailored in this way; thus we focus by necessity on simple selected models, which nevertheless should give a feeling of the art of band structure engineering.

We begin this chapter with a discussion of pn homojunctions, obtained by doping different regions of the same semiconductor with acceptor and donor impurities. We then pass to the description of devices, whose building blocks are pn homojunctions. In particular we describe the classical pnp (or npn) bipolar transistor, and the junction field effect transistor; in the former case the transistor action is achieved through the interaction of two back-to-back junctions, while in the latter it is achieved through modulation of the width of a conductivity channel. We also consider a variety of structures that can be envisaged when the flexibility in doping is accompanied also by the flexibility in the composing materials; this analysis includes semiconductor heterostructures, metal–semiconductor junctions, metal–oxide–semiconductor contacts, and their role in the physics of devices.

1 Properties of the pn junction at equilibrium

A pn junction (or pn diode) consists of a semiconducting crystal with spatially dependent concentration of shallow impurities; acceptor impurities are prevalent in the p-

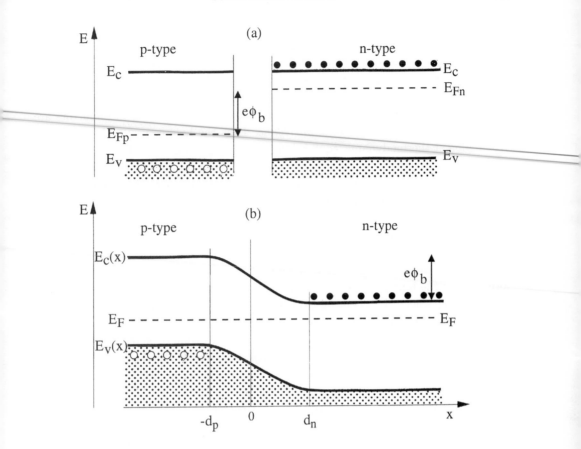

Fig. 1 (a) Energy bands and *Fermi levels* in two physically separated p-type and n-type non-degenerate semiconductors. The majority carriers (electrons in n-type and holes in p-type regions) are indicated by full dots and empty dots, respectively, while minority carriers are not indicated for simplicity (b) Energy band profile and *Fermi level* across the pn junction at equilibrium.

type region, while donor impurities are prevalent in the n-type region. The pn junction is perhaps the simplest electronic device manifactured with dishomogeneous doping of a semiconductor; the study of its physical properties is vital also for the comprehension of many other semiconducting devices.

In the following, just for the sake of simplicity, we consider an *ideal abrupt pn junction*; it consists of a semiconducting single crystal with a uniform concentration N_a of acceptor impurities (at the left, say, of the $x = 0$ plane) and a uniform concentration N_d of donor impurities (at the right of the $x = 0$ plane). In reality the transition from the p region to the n region is not discontinuous, but gradual; furthermore a small number of donor impurities may be present in the p-type region (and vice versa); these details, however, are here disregarded, because they do not change the essential aspects of the model.

Before considering the pn junction at equilibrium, we represent schematically in Fig. 1a the energy bands and the Fermi energies of the p-type semiconductor and n-type semiconductor, not yet in contact. As seen in Section XIII-3, the Fermi energy E_{Fp} for the p-type semiconductor is rather near to the top E_v of the valence band, while the Fermi energy E_{Fn} of the n-type semiconductor is rather near to the bottom E_c of the conduction band. For simplicity, we assume that both semiconductors are *non-degenerate and in the saturation regime.* Thus the hole concentration in the p-type region at thermodynamic equilibrium is $p_{p0} = N_a$; similarly, in the n-type region, $n_{n0} = N_d$ (the concentration of minority carriers in both regions is then determined by the mass-action law). In the n-type material, the majority carriers are free electrons, of density n_{n0}, embedded in the background of immobile positively ionized donors with the same density N_d; similarly, in the p-type material, the majority carriers are free holes, of density p_{p0}, embedded in the background of immobile negatively ionized acceptors with the same density N_a.

Let us now imagine that the p-type and n-type specimens are joined together. On formation of the junction, some electrons leave the n-type material and move towards the p-region, where the chemical potential is lower; the leaving electrons, once in the hostile p-region, recombine with holes. Similarly, holes from the p-type material move to the n-region and recombine with electrons. A region depleted of the majority carriers is thus formed near the junction (*depletion region*). In the depletion layer a *space-charge* is present: it is constituted by an amount of immobile non-neutralized positive donor centers in the n-side part of the semiconductor and by an equal amount of immobile non-neutralized negative acceptor centers in the p-side part. The space-charge produces a strong electric field so to oppose the flux of carriers; eventually, any net flux of carriers vanishes, when the Fermi level is spatially uniform throughout the all system (see Fig. 1b). At equilibrium, drift and diffusion currents of electrons and holes must balance exactly at any point of the junction, and we have:

$$J_n(x) = J_n^{(\text{drift})}(x) + J_n^{(\text{diff})}(x) = n\,e\,\mu_n\,F + e\,D_n\,\frac{dn}{dx} \equiv 0 \qquad (1a)$$

and

$$J_p(x) = J_p^{(\text{drift})}(x) + J_p^{(\text{diff})}(x) = p\,e\,\mu_p\,F - e\,D_p\,\frac{dp}{dx} \equiv 0 \qquad (1b)$$

(carrier gradients ∇n and ∇p, and electric field $\mathbf{F} = -\nabla\phi$, are assumed in the x-direction).

Let us indicate with $\phi(x)$ the electrostatic potential created by the internal space-charge region. As shown below, the potential $\phi(x)$ across the structure *varies slowly with respect to the lattice constant* (typical lengths of space-charge regions are in fact of the order of 10^3 Å or so). We can thus retain that, locally, the potential $\phi(x)$ rigidly shifts every energy level of the semiconductor by the x-dependent quantity $(-e)\,\phi(x)$; the x-dependence of the band structure energies is referred to as *energy band profile.* In particular, the band-edges E_c and E_v (bottom of the conduction band and top of the valence band) are given by

$$E_c(x) = E_c + (-e)\,\phi(x) \qquad E_v(x) = E_v + (-e)\,\phi(x) \ . \tag{2a}$$

Notice that the built-in electric field $F = -d\phi(x)/dx$ can be expressed as

$$F = \frac{1}{e}\frac{dE_c(x)}{dx} \quad \text{or equivalently} \quad F = \frac{1}{e}\frac{dE_v(x)}{dx} \ . \tag{2b}$$

(a similar expression holds of course for any other level). Thus a flat band profile is synonymous of vanishing electric field, while a steep band bending denotes a region of stong electric field.

We can now evaluate quantitatively the extent of the space-charge region, the potential barrier and the electric field in the pn junction. We begin to consider the height of the electrostatic *potential barrier* ϕ_b (also called *built-in potential* or *contact potential*). From Fig. 1, we see by inspection that $e\,\phi_b$ is given by the difference of the Fermi levels, before carriers transfer takes place; it holds

$$\boxed{e\,\phi_b = E_{Fn} - E_{Fp}} \ . \tag{3}$$

As discussed in Section XIII-3, the Fermi levels for n-type and p-type materials in the saturation regime are given by

$$E_{Fn} = E_c + k_B T \ln\frac{N_d}{N_c(T)} \qquad E_{Fp} = E_v + k_B T \ln\frac{N_v(T)}{N_a} \ ,$$

where $N_c(T)$ and $N_v(T)$ indicate the effective density-of-states in the conduction and valence bands, respectively. Thus

$$e\,\phi_b = E_{Fn} - E_{Fp} = E_G + k_B T \ln\frac{N_a\,N_d}{N_v(T)\,N_c(T)} \ ,$$

and the potential barrier ϕ_b is of the order of E_G/e.

In the pn junction at equilibrium, the free electron carrier concentration $n(x)$ depends on the distance x from the junction via the relation

$$n(x) = N_c(T)\,e^{-[E_c(x)-E_F]/k_B T} \tag{4a}$$

where $E_c(x)$ is the bottom of the conduction band, given by Eq. (2a). Equation (4a) shows that electrons become numerous in regions where the bottom of the conduction band and the Fermi level are close. From Eq. (4a) and Eq. (2a), we see that the carrier density $n_1(x)$ and $n_2(x)$, at any two arbitrary points of the semiconducting structure at thermodynamic equilibrium, are related by the expression

$$\boxed{n_1(x) = n_2(x)\,e^{e\,[\phi(x_1)-\phi(x_2)]/k_B T}} \ . \tag{4b}$$

This useful relation can also be obtained directly from Eq. (1a) by appropriate integration.

In a completely similar way we have for holes

$$p(x) = N_v(T)\,e^{-[E_F-E_v(x)]/k_B T} \ . \tag{5a}$$

Thus we see that holes are more numerous in regions where the top of the valence band and the Fermi level are closer. We have also

$$\boxed{p_1(x) = p_2(x)\, e^{-e\,[\phi(x_1)-\phi(x_2)]/k_B T}}\;.$$

(5b)

It is of interest to apply Eq. (4b) in the case x_1 is well inside the p-type material and x_2 is well inside the n-type material. Then $n(x_1) = n_{p0}$ and $n(x_2) = n_{n0}$, where n_{p0} and n_{n0} indicate the equilibrium concentration of electrons in the neutral bulk p-type material and in the neutral bulk n-type material, respectively. We have thus

$$n_{p0} = n_{n0}\, e^{-e\,\phi_b/k_B T}\;.$$

(6a)

We can do a completely similar reasoning for holes and obtain

$$p_{p0} = p_{n0}\, e^{e\,\phi_b/k_B T}\;,$$

(6b)

where p_{p0} and p_{n0} indicate the equilibrium concentration of holes in the p-side and in the n-side neutral parts of the semiconductor, respectively.

Space-charge region and internal electric field

We have indicated schematically in Fig. 1b the energy band profile of a pn junction at equilibrium; far from the junction, the n and p materials are neutral and have bulk equilibrium properties. For instance, well inside the bulk p-type material, the concentration of holes is constant and equal to N_a. As we approach the junction plane, the difference between the Fermi level E_F and $E_v(x)$ increases. Any increase exceeding a few $k_B T$ depresses severely the number of holes, as can be inferred from Eq. (5a); thus, on the p-side of the junction, a layer depleted of majority carriers is formed; in the depletion layer (of width d_p to be determined) the net charge density is practically $-e\,N_a$. Similarly, there is formation of a spatial region with charge density $+e\,N_d$ and extension d_n (to be determined) on the n-side of the junction. The space-charge region, shown in Fig. 2a, consists essentially of ionized donor impurities and ionized acceptor impurities in the regions of widths d_n and d_p from the junction, respectively. In the *depletion layer approximation*, the space-charge $\rho(x)$ is approximated in the form

$$\rho(x) = \begin{cases} -e\,N_a & -d_p < x < 0 \\ +e\,N_d & 0 < x < d_n \end{cases}\;.$$

(7)

The electrostatic potential $\phi(x)$ satisfies the Poisson equation

$$-\frac{d^2\phi}{dx^2} = \frac{\rho(x)}{\varepsilon}$$

(8)

($\varepsilon = \varepsilon_r\,\varepsilon_0$ is the static dielectric constant of the semiconductor; we use here SI units; $\varepsilon_0 = 10^{-11}$ Farad/m).

We integrate Eq. (8) in the region $-d_p \leq x \leq 0$ with the boundary conditions $\phi(-d_p) = 0$ and $\phi'(-d_p) = 0$. We observe in fact that for $x \leq -d_p$, we have a homogeneous semiconductor in open circuit situation with null diffusion and drift current density, and hence null electric field $-\phi'(x)$; the potential $\phi(x)$ for $x \leq -d_p$

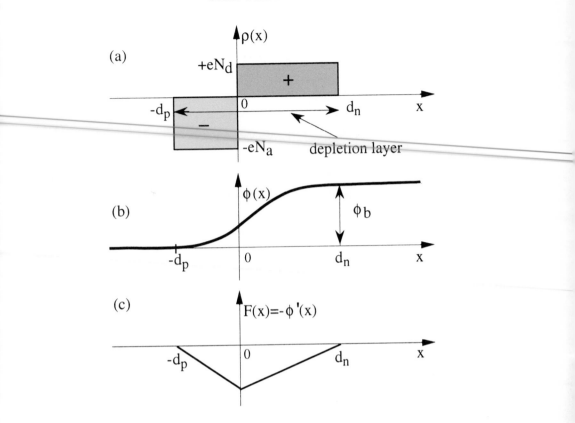

Fig. 2 Charge density $\rho(x)$ at a pn junction in the depletion layer approximation, electrostatic potential $\phi(x)$, and internal electric field $F(x)$.

is thus constant, and we choose this arbitrary constant equal to zero. Similarly we integrate Eq. (8) in the region $0 \leq x \leq d_n$ with the boundary conditions $\phi(d_n) = \phi_b$ and $\phi'(d_n) = 0$, and obtain

$$
\phi(x) = \begin{cases} \dfrac{e\,N_a}{2\varepsilon}\,(x + d_p)^2 & -d_p \leq x \leq 0 \\[2mm] \phi_b - \dfrac{e\,N_d}{2\varepsilon}\,(x - d_n)^2 & 0 \leq x \leq d_n \end{cases} . \tag{9}
$$

We now require the continuity of the potential $\phi(x)$ and of the electric field $-\phi'(x)$ at $x = 0$; the two conditions determine the extensions d_n and d_p of the depletion layers. The continuity of the electric field at $x = 0$ gives

$$
N_a\,d_p = N_d\,d_n \;, \tag{10a}
$$

a condition which is synonymous of overall charge neutrality of the system. The continuity of the potential at $x = 0$ gives

$$
\phi_b = \frac{e}{2\varepsilon}\,[N_a\,d_p^2 + N_d\,d_n^2] \;. \tag{10b}
$$

The solution of the two equations (10) gives

$$d_p = \left[\frac{N_d}{N_a} \frac{1}{N_a + N_d} \frac{2\,\varepsilon\,\phi_b}{e} \right]^{1/2} \quad \text{and} \quad d_n = \left[\frac{N_a}{N_d} \frac{1}{N_a + N_d} \frac{2\,\varepsilon\,\phi_b}{e} \right]^{1/2}. \tag{11a}$$

The total width of the depletion layer is

$$w = d_p + d_n = \left[\frac{N_a + N_d}{N_a\,N_d} \frac{2\,\varepsilon\,\phi_b}{e} \right]^{1/2}. \tag{11b}$$

As an example, let us take for instance $\phi_b = 1$ volt; $N_a = N_d = 10^{16}\,\text{cm}^{-3} = 10^{22}\,m^{-3}$; $\varepsilon = \varepsilon_r\,\varepsilon_0 = 10 \cdot 10^{-11}\,\text{F/m} = 10^{-10}\,\text{F/m}$; we have $w = 5 \cdot 10^{-7}\,\text{m} = 5000\,\text{Å}$. The average value of the electric field ϕ_b/w equals $2 \cdot 10^4$ volt/cm.

Considerations on other types of homojunctions

In the pn junctions considered until now, we have (tacitly) assumed that the acceptor and donor doping levels N_a and N_d are moderate and comparable. In particular circumstances, it is convenient to consider junctions with very high concentrations of acceptors, or donors, or both; these junctions are indicated as p^+n, pn^+ and p^+n^+ structures. These homojunctions, when treated in the depletion layer approximation, are just particular cases of the formalism considered so far. For instance the one-sided abrupt junction p^+n denotes a homojunction with $N_a \gg N_d$; in this case from Eqs. (11), we have that $d_p \ll d_n$; the width of the depletion layer becomes

$$w \cong d_n = \left[\frac{1}{N_d} \frac{2\,\varepsilon\,\phi_b}{e} \right]^{1/2}. \tag{12}$$

It is instructive to consider also the n^+n homojunctions (or the p^+p homojunctions); for our purpose, a few qualitative remarks are sufficient. In contrast to pn junctions, in n^+n junctions the electrons are the majority carriers in the whole structure; the concentration of electrons changes gradually from a very high value in the n^+-side to a high value on the n-side. Differently from the pn junctions, the n^+n junctions present an accumulation of majority carriers in the n-region rather than a depletion; thus while a pn contact is a region of high resistivity with *rectifying properties* (see next section), an n^+n contact is a region of low resistivity. Well designed n^+n junctions are often used to realize *ohmic contacts* in electronic circuits (a contact is said to be "ohmic" when the potential drop across it can be safely disregarded with respect to the potential drops in other parts of the electronic circuit).

2   Current–voltage characteristics of the pn junction

In a pn diode at equilibrium, an internal potential barrier ϕ_b is established so that no net flow of carriers occurs. In equilibrium situation, the drift and diffusion components of the current density both for electrons and for holes must cancel (see Eq. 1). When a bias potential V is applied to the diode, the effective potential barrier is modified; drift and diffusion currents are no longer balanced, and current flows across the junction.

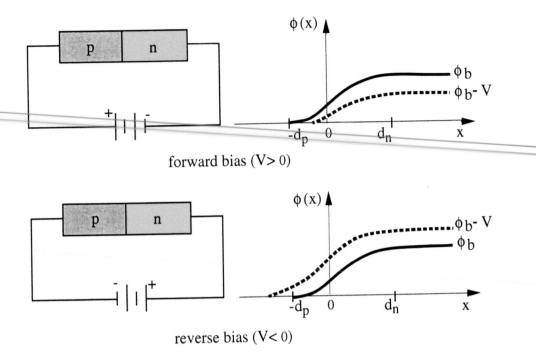

Fig. 3 Forward and reverse bias for a pn junction.

To determine the current–voltage characteristic of the device, and its rectifying properties, consider a pn junction with a bias voltage V applied to it (see Fig. 3). The external voltage V is counted positive if the potential barrier between p and n regions is decreased (forward bias) and thus the p-region tends to be positive with respect to the n-region; vice versa, the external potential V is considered negative, if the potential barrier between the regions is increased (reverse bias). In Fig. 3 we illustrate schematically a direct-biased and a reverse-biased pn diode.

We have already noticed that the space-charge region has high resistivity (compared with the bulk p and n regions) since the majority carriers are removed from it. We can thus assume that the external potential V applied to the diode (almost completely) drops across the depletion region; the total potential drop across the depletion layer in a biased diode is $\phi_b - V$ (it is assumed $V < \phi_b$). The width of the depletion layers in the presence of an external potential V can be obtained from Eqs. (11), replacing ϕ_b with $\phi_b - V$; we have

$$d_p = \left[\frac{N_d}{N_a} \frac{1}{N_a + N_d} \frac{2\varepsilon(\phi_b - V)}{e} \right]^{1/2} , \qquad d_n = \left[\frac{N_a}{N_d} \frac{1}{N_a + N_d} \frac{2\varepsilon(\phi_b - V)}{e} \right]^{1/2} ,$$

$$(13a)$$

and

$$w = d_p + d_n = \left[\frac{N_a + N_d}{N_a N_d} \frac{2\varepsilon(\phi_b - V)}{e} \right]^{1/2} ,$$

$$(13b)$$

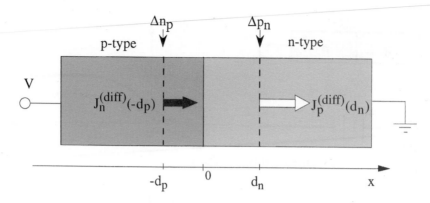

Fig. 4 Schematic representation of a pn homojunction with an applied potential V on the p-side (the n-side is grounded); Δn_p and Δp_n denote the change in the concentration of minority carriers at the boundaries between the space charge region and the neutral regions. The figure refers to the case $V > 0$, so that Δn_p and Δp_n are positive; $J_n^{(\text{diff})}(-d_p)$ denotes the electron diffusion current at $x = -d_p$, and $J_p^{(\text{diff})}(d_n)$ the hole diffusion current at $x = d_n$. (Notice that the electron current is in the opposite direction to the electron flow, while the hole current and the hole flow are in the same direction). The diode current is approximately the sum $J_n^{(\text{diff})}(-d_p) + J_p^{(\text{diff})}(d_n)$.

where $\phi_b - V > 0$. Our following considerations are restricted to forward voltages smaller than the barrier height ϕ_b, while for reverse biases we have no restriction (except for the onset of avalanche breakdown or Zener breakdown).

We can examine quantitatively the change of minority carriers just at the boundaries of the depletion layer region of the pn homojunction (see Fig. 4). We assume that in situations of quasi-equilibrium, the carrier densities and electrostatic potentials at different points are still related by equations (4b) and (5b) (strictly valid only at the thermodynamic equilibrium); with this assumption, the electron concentrations at $-d_p$ and d_n are related by

$$n(-d_p) = n(d_n)\, e^{-e\,(\phi_b - V)/k_B T} \tag{14}$$

(a similar treatment holds for holes). It is also reasonable to assume the low-injection condition, i.e. the majority carrier concentrations are changed by a negligible amount at the boundary of the neutral regions; in other words we can put $n(d_n) \approx n_{n0}$ in Eq. (14). We have

$$n(-d_p) = n_{n0}\, e^{-e\,(\phi_b - V)/k_B T} = n_{p0}\, e^{e\,V/k_B T} \ ,$$

where use has been made of Eq. (6a). *We thus see that the minority (electron) concentration at the left end of the depletion layer is controlled by the factor* $\exp(e\,V/k_B T)$. In the forward bias condition we have injection of minority carriers, while extraction occurs in the reverse bias condition. The change $\Delta n_p(-d_p)$ in electron concentration at the end of the depletion layer (in the p-side part of the semiconductor) is

$$\Delta n_p(-d_p) = n_{p0}\,(e^{e\,V/k_B T} - 1) \ . \tag{15}$$

For $x < -d_p$ the semiconductor is (approximately) neutral; the excess (or deficit) of minority carriers $\Delta n_p(-d_p)$ gives rise to a diffusion electron current (studied in detail in Section XIII-6). The *electron diffusion current* at $x = -d_p$ is given by

$$J_n^{(\mathrm{diff})}(-d_p) = e\,D_n\frac{\Delta n_p(-d_p)}{L_n} = e\,D_n\frac{n_{p0}}{L_n}\left(e^{eV/k_BT} - 1\right) .$$

The *total electron current* at the boundary $x = -d_p$ is the sum of the diffusion component and drift component; the latter is negligible, because so is the concentration of minority carriers, as well as the electric field in the quasi-neutral region $x < -d_p$. We have thus

$$J_n(-d_p) \approx J_n^{(\mathrm{diff})}(-d_p) = e\,D_n\frac{n_{p0}}{L_n}\left(e^{eV/k_BT} - 1\right) . \tag{16a}$$

Similarly, for the hole current density on the right boundary $(x = d_n)$ of the space-charge region we have

$$J_p(d_n) \approx J_p^{(\mathrm{diff})}(d_n) = e\,D_p\frac{p_{n0}}{L_p}\left(e^{eV/k_BT} - 1\right) . \tag{16b}$$

The space-charge region is so narrow that we can reasonably assume that the net balance of generation and recombination processes vanishes within it. In this case, *electron and hole currents are constant throughout the depletion layer*, and the total current through the diode is then given by the sum of the contributions (16); it follows

$$\boxed{J = J_s\left(e^{eV/k_BT} - 1\right)} , \tag{17a}$$

where

$$J_s = e\,D_p\frac{p_{n0}}{L_p} + e\,D_n\frac{n_{p0}}{L_n} = e\,L_p\frac{p_{n0}}{t_p} + e\,L_n\frac{n_{p0}}{t_n} ; \tag{17b}$$

the characteristic $J - V$ given by Eq. (17a) is called the *ideal diode equation*, and J_s is called the *reverse saturation current* of the diode. Notice that J_s is composed by the currents of the minority carriers across the depletion region. Holes are created thermally in the n-type region at a rate p_{n0}/t_p; those created within a distance L_p from the barrier region, reach the depletion layer by diffusion and are then swept by the internal electric field to the p-side of the semiconductor. Similar considerations hold for electrons thermally generated in the p-type region.

In Fig. 5 we report the ideal $I - V$ characteristic of a pn junction. The differential resistance (or impedance), defined as $R = (dI/dV)^{-1}$, is in general rather low in the direct bias configuration (typical values can be of the order of $10\,\Omega$ or so). Notice instead that the impedance is very high for reverse bias polarization, and can be of the order of $10^5\,\Omega$ or so. Notice also that, with appropriate degree of doping, it is possible to make electron current negligible with respect to hole current (or vice versa). For instance in a p^+n junction, where the p-side is strongly doped with respect to the n-side, the hole current may represent almost all of the total current.

It is of interest to give some further consideration on the differential resistance of a

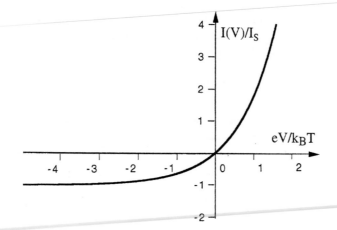

Fig. 5 Ideal $I-V$ characteristic for a pn junction with $I(V) = I_s\,[\exp(e\,V/k_BT) - 1]$.

diode in the direct bias configuration $0 < V < \phi_b$. For simplicity, we focus on a p$^+$n junction, whose $I - V$ characteristic (17) takes the simplified form

$$I(V) = A\,e\,D_p\frac{p_{n0}}{L_p}\left(e^{e\,V/k_BT} - 1\right) , \qquad (18)$$

where A is the area of the junction. Using Eq. (18) the differential resistance of the diode becomes

$$\frac{1}{R_{\mathrm{diode}}(V)} = \frac{\partial I}{\partial V} = A\,e\,D_p\frac{p_{n0}}{L_p}\,e^{e\,V/k_BT}\,\frac{e}{k_BT} .$$

It is seen by inspection that $R_{\mathrm{diode}}(V)$ decreases rapidly as V increases. In particular for $V = 0$ we have

$$\frac{1}{R_{\mathrm{diode}}(0)} = \frac{A}{L_p}\,e\,p_{n0}\mu_p , \qquad (19a)$$

where use has been made of the Einstein relation $e\,D_p = \mu_p\,k_B\,T$. For $V = \phi_b$ we have

$$\frac{1}{R_{\mathrm{diode}}(\phi_b)} = \frac{A}{L_p}\,e\,p_{p0}\mu_p , \qquad (19b)$$

where use has been made of Eq. (6b). The inverse of the second member of Eq. (19b) denotes the resistance R_s of a length L_p of semiconductor with dopant concentration $N_a = p_{p0}$. From Eq. (19b), it is seen that $R_{\mathrm{diode}}(V)$ may indeed become rather low for V approaching ϕ_b, but is always larger than the resistance R_s defined above.

The ideal rectifier equation (17) has been obtained with a number of simplifying assumptions, that we briefly summarize as follows: abrupt depletion layer, carrier densities at the boundaries in quasi-equilibrium, low injection condition, no appreciable net generation-recombination processes in the space-charge region. We do not discuss the extensions of the theory necessary to overcome these limitations.

There are other mechanisms that can modify the $I - V$ characteristic of a diode.

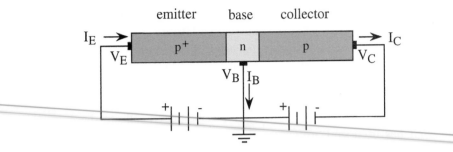

Fig. 6 Schematic representation of a pnp junction transistor in the grounded base arrangement.

When the reverse voltage exceeds a critical value, the inverse current begins to increase rapidly (breakdown effect). This can be due to direct tunneling of electrons from the valence band in the p-side semiconductor to the conduction band in the n-side, or to "avalanche" multiplication of carriers in strong electric fields. Also the forward $I - V$ characteristics of a strongly doped p^+n^+ junction may present a current component due to carriers tunneling through thin barriers, with the possibility of negative differential resistance.

Finally we should notice that the most important applications of pn junctions regards not only electronic devices, but also photonic devices. Among them, we mention light-emitting diodes or diode lasers (with conversion of electrical energy into optical energy), solar cells and photovoltaic cells (with conversion of optical energy into electric energy), photodetectors (electronic detection of photons). In the following we focus on the use of pn junctions as building blocks in transistors.

3   The bipolar junction transistor

A bipolar transistor consists of a single crystal containing two back-to-back junctions in interaction; the device relies on the behaviour of two types of carriers and is thus called bipolar. We consider the essential aspects of the pnp transistor (for the npn transitor one can follow a completely specular treatment).

A pnp junction transistor is built by two p-type regions separated by a thin n-type layer of the same material. In Fig. 6 the three regions (emitter, base, collector) are shown: the emitter is a strongly doped p^+ region, the base is a weakly doped n-type region, and the collector is a moderately doped p-region. The width of the base is much smaller than the diffusion length of minority carriers (holes) in the n-region.

We begin with a qualitative description of the operation of the pnp device, and we pass later to a semi-quantitative analysis. In the normal operating mode, the base–emitter junction is biased in the forward sense, and the base–collector junction is biased in the reverse sense. Taking the potential V_B of the base as reference zero potential, (i.e. $V_B \equiv 0$), we have $V_E > 0$ and $V_C \ll 0$. The forward bias of the emitter-base p^+n

junction determines a current due to the injection of holes in the base (the current due to the injection of the electrons in the emitter is relatively small, since the emitter is strongly doped, while the base is weakly doped). If the base width is small with respect to the diffusion length of minority carriers in the base, practically all the hole current from the emitter is collected by the collector. The *transistor* name has been coined to address this "tran(sfer)-(res)istor" action from the low-impedance forward-biased emitter–base junction to the high-impedance reverse-biased collector-base junction.

From what has been said, we expect that the collector current I_C is only slightly smaller than the emitter current I_E; it is customary to define the "current transfer" parameter α as the ratio

$$\alpha = \frac{I_C}{I_E} \; ; \tag{20a}$$

for a well designed transistor, α is very close to 1, and a typical value is $\alpha \approx 0.99$. It is also convenient to define the "current gain" parameter β as the ratio between the collector current I_C and the base current $I_B = I_E - I_C$; we have:

$$\beta = \frac{I_C}{I_B} = \frac{I_C}{I_E - I_C} = \frac{\alpha}{1 - \alpha} \; . \tag{20b}$$

We thus see that for $\alpha \approx 0.99$, the current gain factor β is typically 100. As we shall see by a more quantitative analysis, a large current gain requires a very thin base, and large diffusion length of the minority carriers of the base.

We analyse now quantitatively the currents flowing in the pnp transistor, when the emitter and the collector potentials are V_E and V_C, respectively, and the base is grounded. In Fig. 7 we give a schematic representation of the structure of the pnp transistor; we indicate also the limits of the depletion layer regions, and the change in density of minority carriers in the emitter, base and collector. We indicate by $\Delta n_E(x_E)$ the change of electron concentration at the boundary plane $x = x_E$ of the space-charge region within the p$^+$-side of the semiconductor; $\Delta p_B(x_1)$ denotes the change of hole concentration at the boundary plane $x = x_1$ of the depletion layer at the emitter–base junction; similar notations are used for $\Delta p_B(x_2)$ and $\Delta n_C(x_C)$. Using the results obtained in Section 2, we have

$$\Delta n_E(x_E) = n_{E0} \left[e^{e\, V_E/k_B T} - 1 \right] , \tag{21a}$$

$$\Delta p_B(x_1) = p_{B0} \left[e^{e\, V_E/k_B T} - 1 \right] , \qquad \Delta p_B(x_2) = p_{B0} \left[e^{e\, V_C/k_B T} - 1 \right] , \tag{21b}$$

$$\Delta n_C(x_C) = n_{C0} \left[e^{e\, V_C/k_B T} - 1 \right] \tag{21c}$$

(n_{E0}, p_{B0} and n_{C0} are the equilibrium concentrations of minority carriers in the emitter, in the base and in the collector, respectively).

The electron diffusion current, at the left boundary $x = x_E$ of the space charge region between emitter and base is

$$J_{nE}^{(\text{diff})}(x_E) = e\, D_E \frac{\Delta n_E(x_E)}{L_E} = e\, D_E \frac{n_{E0}}{L_E} \left[e^{e\, V_E/k_B T} - 1 \right] , \tag{22a}$$

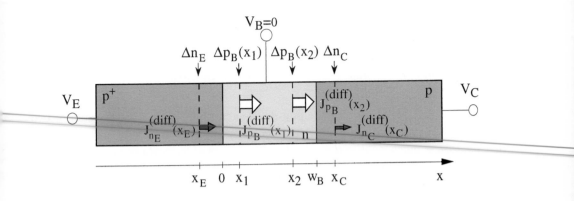

Fig. 7 Effect of applied emitter and collector potentials on the concentration of minority carriers at the borders of the space-charge regions. The electron diffusion current at $x = x_E$, the hole diffusion current at $x = x_1$ and $x = x_2$, and the electron diffusion current at $x = x_C$ are indicated (the signs of the currents correspond to $V_E > 0$ and $V_C < 0$). The emitter current is $J_E = J_{nE}^{(\text{diff})}(x_E) + J_{pB}^{(\text{diff})}(x_1)$; the collector current is $J_C = J_{pB}^{(\text{diff})}(x_2) + J_{nC}^{(\text{diff})}(x_C)$; the base current is $J_B = J_E - J_C$.

where $\Delta n_E(x_E)$ is the excess (or deficit) of electron concentration at the boundary of the neutral region, D_E is the diffusion constant of the minority carrier in the emitter, and L_E the diffusion length. The hole diffusion current $J_{pB}^{(\text{diff})}(x_1)$, in the presence of hole concentration excess $\Delta p_B(x_1)$ at $x = x_1$ and $\Delta p_B(x_2)$ at $x = x_2$ has been calculated previously (see Eqs. XIII-52); with some obvious changes of notations, we have

$$J_{pB}^{(\text{diff})}(x_1) = e\,D_B \frac{p_{B0}}{L_B} \frac{1}{\tanh(w_B/L_B)} \left[e^{e\,V_E/k_B T} - 1 \right]$$

$$- e\,D_B \frac{p_{B0}}{L_B} \frac{1}{\sinh(w_B/L_B)} \left[e^{e\,V_C/k_B T} - 1 \right] \qquad (22b)$$

where $w_B = x_2 - x_1$ is the width of the neutral region in the base (in ordinary situations, we can safely assume that the width w_B of the neutral region in the base and the width of the base are the same). The sum of the contributions (22a) and (22b) gives the emitter current density J_E; in a similar way, we can obtain the collector current density J_C. Their expressions can be written in the compact form

$$\boxed{\begin{aligned} J_E &= a_{11} \left(e^{e\,V_E/k_B T} - 1 \right) - a_{12} \left(e^{e\,V_C/k_B T} - 1 \right) \\ J_C &= a_{21} \left(e^{e\,V_E/k_B T} - 1 \right) - a_{22} \left(e^{e\,V_C/k_B T} - 1 \right) \end{aligned}} \qquad (23)$$

where

$$\begin{cases} a_{11} = e\,D_E\,(n_{E0}/L_E) + e\,D_B\,(p_{B0}/L_B)\,\coth(w_B/L_B) \\ a_{12} = a_{21} = e\,D_B\,(p_{B0}/L_B)\,[\sinh(w_B/L_B)]^{-1} \\ a_{22} = e\,D_C\,(n_{C0}/L_C) + e\,D_B\,(p_{B0}/L_B)\,\coth(w_B/L_B) \end{cases} \qquad (24)$$

In spite of the simplifications inherent Eqs. (23) and Eqs. (24), these results are useful to give a rather intuitive discussion of the transistor action. Let us consider first the particular case in which $w_B \gg L_B$; the parameters (24) of the model become

$$a_{11} = e\,D_E\,\frac{n_{E0}}{L_E} + e\,D_B\,\frac{p_{B0}}{L_B}\,, \qquad a_{22} = e\,D_C\,\frac{n_{C0}}{L_C} + e\,D_B\,\frac{p_{B0}}{L_B}\,, \qquad a_{12} = a_{21} = 0\,. \quad (25)$$

We thus see that for $w_B \gg L_B$ the pnp device is nothing more than the sum of two independent back-to-back junctions. The achievement of the transistor action requires that *the base is much smaller than the diffusion length of minority carriers in the base*; in the case $w_B \ll L_B$ the parameters (24) become

$$a_{11} = e\,D_E\,\frac{n_{E0}}{L_E} + e\,D_B\,\frac{p_{B0}}{w_B}\,, \qquad a_{22} = e\,D_C\,\frac{n_{C0}}{L_C} + e\,D_B\,\frac{p_{B0}}{w_B}\,, \qquad a_{12} = a_{21} = e\,D_B\,\frac{p_{B0}}{w_B}\,,$$

$$(26)$$

and the coupling parameter $a_{12} = a_{21}$ is now different from zero.

In order to show in a simple way the central features of Eqs. (23), let us consider the particular case of the forward mode operation, characterized by $e\,V_E \gg k_B T$ and $V_C \ll 0$. We can simplify expressions (23) in the form

$$J_E \approx a_{11}\,e^{e\,V_E/k_B T}\,, \qquad J_C \approx a_{21}\,e^{e\,V_E/k_B T}\,, \qquad (27)$$

and the current gain parameter of the transistor, defined in Eq. (20b), becomes

$$\beta = \frac{J_C}{J_E - J_C} = \frac{a_{21}}{a_{11} - a_{21}}\,. \qquad (28)$$

In the case $w_B \gg L_B$, Eqs. (25) hold and the current gain of the pnp structure vanishes. In the case $w_B \ll L_B$, Eqs. (26) hold and we obtain

$$\beta = \frac{e\,D_B\,(p_{B0}/w_B)}{e\,D_E\,(n_{E0}/L_E)} = \frac{D_B}{D_E}\,\frac{N_{aE}}{N_{dB}}\,\frac{L_E}{w_B}\,, \qquad (29)$$

where $N_{aE}\,(= n_i^2/n_{E0})$ is the acceptor impurity doping in the emitter and, similarly, $N_{dB}\,(= n_i^2/p_{B0})$ is the donor impurity doping in the base. When $\beta \gg 1$, we see that a small variation of the base current is associated with a much bigger variation of the collector current and the transistor operates as a current amplifier of gain β. From Eq. (29) it is evident that a large current gain β requires a heavy doping of the emitter region with respect to the base, and a base width w_B very small with respect to the diffusion length of holes in the base.

4   The junction field-effect transistor (JFET)

The junction field-effect transistor consists of a slice of doped semiconductor sandwiched between two reverse-biased junctions (see Fig. 8). In essence the device is formed by a conductive channel, for instance a bar of n-type semiconductor, with two (ohmic) contacts at the ends: the *source* and the *drain*. The potential of the source is taken as the reference potential ($V_S = 0$), and the potential of the drain is positive ($V_D > 0$). The two rectifying p$^+$n junctions are reverse biased by means of a

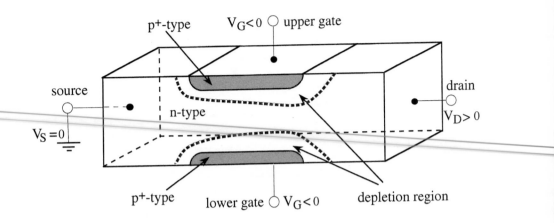

Fig. 8 Schematic illustration of the junction field-effect transistor.

negative potential applied to the *gate* ($V_G < 0$). The principle of the junction field effect transistor is just to control the width of the conductive channel in the doped semiconductor, by varying the reverse bias V_G of the junctions, and thus the size of the depletion regions.

We can give a semi-quantitative account of the $I - V$ characteristics of a junction field-effect transistor, with the the help of Fig. 9, which illustrates the central part and essential aspects of the device.

In order to calculate the current through the conductive channel, we assume for simplicity that the space-charge regions between the (upper and lower) gates and the channel are totally depleted of carriers, with abrupt boundaries. We also assume that the potential $V(x)$ across the channel varies slowly with the distance x (gradual channel approximation); in this case the width of the depleted region can be obtained adapting to the present geometry the results of the one-dimensional p^+n junction, expressed by Eq. (12). We have

$$w(x) = \sqrt{\frac{2\varepsilon \left[V(x) + \phi_b - V_G \right]}{e\, N_d}} . \tag{30}$$

For an n-channel junction field-effect transistor, the gate voltage is negative, so we denote $-V_G$ with the absolute value $|V_G|$.

It is convenient to define the *pinch-off voltage* V_P as the voltage for which the depletion width equals half-height of the device; V_P is determined by the equality

$$a \equiv \sqrt{\frac{2\,\varepsilon\, V_P}{e\, N_d}} .$$

The width $w(x)$ of the depletion region at the point x of the channel, given by Eq. (30), can be written in the form

$$w(x) = a \sqrt{\frac{V(x) + \phi_b + |V_G|}{V_P}} . \tag{31}$$

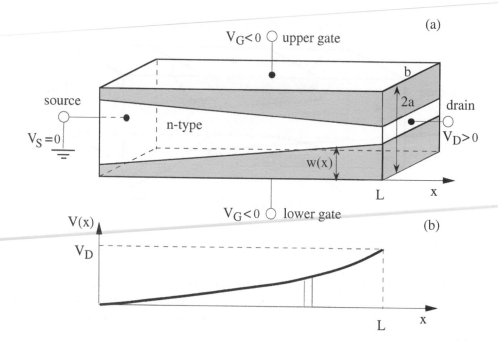

Fig. 9 (a) Model for a semi-quantitative account of the $I - V$ characteristics of a junction field-effect transistor. (b) Voltage between source and drain.

The above equation makes sense provided $w(x) \leq a$, i.e. provided the normal conducting channel exists for any x, up to the end point $x = L$, where $V(x) = V_D$. In carrying out the present analysis, one thus requires that

$$V_D + \phi_b + |V_G| \leq V_P .\tag{32}$$

When condition (32) is satisfied, the conductive channel is open and the current I_D through the device can be estimated with the following considerations. In the active part of the channel, the current density is given by the equation

$$J = n e \mu_n E = \sigma E ,$$

where the electron mobility μ_n (and the conductivity σ) is assumed to be indepedent from the strength of the electric field. The active layer in the plane perpendicular to x extends for an area equal to $[2a - 2w(x)] b$; the drain current (taken as positive in the $-x$ direction) is

$$I_D = [2a - 2w(x)] b \sigma \frac{dV(x)}{dx} .\tag{33}$$

Multiplying both members of the above equation by dx, and using expression (31), we have

$$I_D \, dx = \sigma \, 2a \, b \left[1 - \frac{V(x) + \phi_b + |V_G|}{V_P}\right] dV(x) .$$

Integration with respect to x in the range $0 \leq x \leq L$ gives

$$I_D = \sigma \frac{2ab}{L} \left[V_D - \frac{2}{3} \frac{(V_D + \phi_b + |V_G|)^{3/2} - (\phi_b + |V_G|)^{3/2}}{V_P^{1/2}} \right]. \tag{34}$$

Notice that $\sigma \, 2ab/L$ is just the inverse of the resistance R of a bar of semiconductor, with conductivity σ, and dimensions $2a$, b, and L.

Equation (34) is applicable provided the constraint (32) is satisfied, so that a portion of normal conducting channel remains open up to the drain. For small drain-to-source voltages V_D, and appropriate expansion to first order in V_D of the second member of Eq. (34), one obtains

$$I_D = \frac{1}{R} \left[1 - \sqrt{\frac{\phi_b + |V_G|}{V_P}} \right] V_D \; ;$$

the I_D–V_D characteristic is thus linear for small values of V_D.

The maximum value of V_D, for which Eq. (34) holds, is called saturation value $V_{D,\text{sat}}$. From the constraint (32), we have

$$V_{D,\text{sat}} = V_P - \phi_b - |V_G| \; ; \tag{35}$$

the saturation current, obtained inserting Eq. (35) into Eq. (34), is

$$I_{D,\text{sat}} = \frac{1}{R} \left[V_{D,\text{sat}} - \frac{2}{3} V_P + \frac{2}{3} V_P \left(1 - \frac{V_{D,\text{sat}}}{V_P} \right)^{3/2} \right]. \tag{36}$$

In the case $V_{D,\text{sat}} \ll V_P$, we see that $I_{D,\text{sat}}$ depends quadratically on $V_{D,\text{sat}}$ and takes the form

$$I_{D,\text{sat}} = \frac{1}{R} \frac{1}{2} \frac{V_{D,\text{sat}}^2}{V_P} \; . \tag{37}$$

The above analysis must be reconsidered with caution when V_D, although smaller than $V_{D,\text{sat}}$, is very close to it. In this case the conductivity channel is almost closed at the drain; from Eq. (33) we see that when $w(x) \approx a$, the flow of a finite drain current implies that the electric field $dV(x)/dx$ is becoming arbitrarily large. Thus the assumption that the mobility of the electrons is constant is no more acceptable; it is in fact well known that the average drift velocity of carriers in semiconductors is linear with the electric field for small fields, but tends to saturate for strong electric fields. Without entering a detailed analysis of transport processes in high electric fields, we can expect that for $V_D \approx V_{D,\text{sat}}$, the drain voltage loses control of the channel current and I_D remains (approximately) constant as the drain voltage is further increased (saturation region). Notice that for $V_D \geq V_{D,\text{sat}}$, the source and the drain are *completely separated by a depletion layer*, through which the saturation electron current $I_{D,\text{sat}}$ keeps flowing.

The results of Eq. (34) and the inherent discussion are presented in Fig. 10. In this figure we give the I_D–V_D characteristics of a junction field-effect transistor, in the case

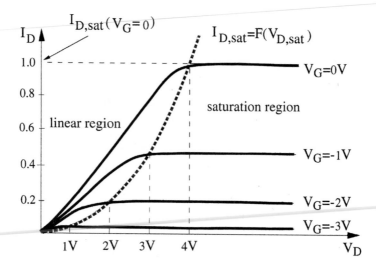

Fig. 10 Idealized $I_D - V_D$ characteristics of a junction field-effect transistor. For $V_D \geq V_{D,\text{sat}}$ the drain current remains constant. The current $I_{D,\text{sat}}(V_G = 0)$ has been taken as the unit of current. The curve $I_{D,\text{sat}} = F(V_{D,\text{sat}})$ is given by Eq. (36) of the text. The parameters of the JFET of the figure are taken to be $\phi_b = 1\,$V and $V_p = 5\,$V; $V_{D,\text{sat}}$ equals $V_p - \phi_b - |V_G|$.

$V_P = 5\,$volt and $\phi_b = 1\,$volt, for several gate voltages in the interval $-(V_P - \phi_b)$ and zero; we also indicate the locus of points $I_{D,\text{sat}}$ versus $V_{D,\text{sat}}$, given by Eq. (36). From Fig. 10, we see that for a given V_G, the drain current first increases linearly with the voltage (linear region), then gradually levels off to the saturation value (saturation region); the linear and saturation regimes are approximatively separated by the curve of Eq. (36). From the knowledge of the current–voltage characteristics of a transistor, it is (in principle) possible the design of JFET circuits.

5   Semiconductor heterojunctions

Until now we have considered homopolar junctions, formed by a single *host semiconductor* partly doped with donors and partly with acceptors. For particular applications it may be very useful to consider junctions between two different semiconductors, with different energy gaps. If the two semiconductors, although formed by different chemical elements, are similar for what concerns the nature of the crystalline binding and the lattice structure, it is possible to form ideal (or nearly ideal) "heterostructures". Molecular beam epitaxy (MBE) and metal–organic chemical–vapour deposition (MOCVD) have allowed the realization of several heterostructures for electronic and photonic devices.

The most studied heterojunction is the one between GaAs and the $\text{Al}_x\text{Ga}_{1-x}\text{As}$ alloy, where x can vary from 0 to 1; the band gap passes from $\approx 1.42\,$eV in pure gallium arsenide to $\approx 2.17\,$eV in pure aluminum arsenide, while the lattice constants of the two solids are basically the same within less than 0.1%. The energy band diagrams (referred to the vacuum level) for the two isolated semiconductors are schematically shown in Fig. 11a.

Of particular importance in the study of heterostructures, is the conduction band offset ΔE_c and the valence band offset ΔE_v, defined as

$$\Delta E_c = E_{c2} - E_{c1} \qquad \Delta E_v = E_{v1} - E_{v2} \;. \tag{38a}$$

These two quantities are not independent; in fact their sum equals the difference between the band gaps of the two semiconductors

$$\Delta E_c + \Delta E_v = E_{G2} - E_{G1} \;. \tag{38b}$$

In the case of GaAs-Al$_x$Ga$_{1-x}$As heterostructures, ΔE_c accounts for about 66% of the energy gap difference.

The conduction band offset can be expressed in the form

$$\Delta E_c = \chi_1 - \chi_2 + \delta_{\text{int}} \;, \tag{38c}$$

where χ_1 and χ_2 are the bulk electron affinities of the two solids (i.e. the energy difference between an electron at rest outside the solid and an electron at the bottom of the conduction band), and δ_{int} is a correction due to the specific microscopic properties of the interface under attention. The origin of δ_{int} is the quantum modification of wavefunctions at the interface that separates the two chemically different materials (this quantum modification is generally confined within a few lattice planes from the interface), and the charged double layer formed therein. For simplicity, we neglect the corrective term δ_{int} at the semiconductor–semiconductor interfaces (or also at the metal–semiconductor interfaces encountered in the following); often, the quantum correction δ_{int} is (tacitly) embodied with empirical or semi-empirical adjustments of the band offset.

We consider now equilibrium properties of semiconductor heterostructures, for instance between p-type GaAs and n-type Al$_x$Ga$_{1-x}$As. When the two semiconductors are brought into contact, carriers are exchanged between the two materials, until a unique Fermi level results in the whole structure. The potential barrier ϕ_b at the interface is given by

$$\boxed{e\,\phi_b = E_{F1} - E_{F2}}\;, \tag{39}$$

where E_{F1} and E_{F2} are the two Fermi levels before contact. Very much as in the pn junction, a space-charge region is formed on both sides of the interface: in the n-type material we have a layer of width d_n of non-neutralized donor impurities (with charge density $+eN_d$); in the p-type material we have a layer of width d_p of non-neutralized acceptor impurities (with charge density $-eN_a$).

The electrostatic potential $\phi(x)$ across the heterostructure at equilibrium can be easily obtained following (with trivial modifications) the same procedure already applied in Section 1 for homojunctions. The potential $\phi(x)$, due to the space-charge region at the heterostructure, is given by a straightforward modification of Eq. (9);

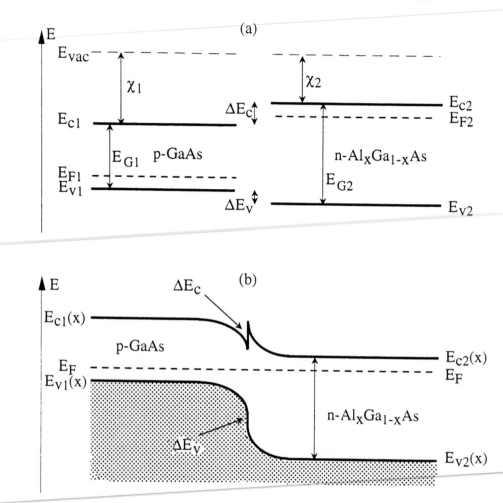

Fig. 11 (a) Energy bands and Fermi levels in two separated semiconductors: p-type GaAs and n-type $Al_xGa_{1-x}As$ (b) Energy band profile and Fermi level across the heterostructure at thermal equilibrium.

we have

$$\phi(x) = \begin{cases} (e\,N_a/2\,\varepsilon_1)\,(x + d_p)^2 & -d_p \le x \le 0 \\ \phi_b - (e\,N_d/2\,\varepsilon_2)\,(x - d_n)^2 & 0 \le x \le d_n \end{cases} \tag{40a}$$

where ε_1 and ε_2 are the dielectric constants of the two materials. The two boundary conditions that allow to determine the widths d_p and d_n of the depletion layers are

$$\phi(0^-) = \phi(0^+) \qquad \varepsilon_1\,\phi'(0^-) = \varepsilon_2\phi'(0^+) \,. \tag{40b}$$

A simple calculation gives

$$d_p = \left[\frac{N_d}{N_a} \frac{1}{\varepsilon_1\,N_a + \varepsilon_2\,N_d} \frac{2\varepsilon_1\,\varepsilon_2\,\phi_b}{e} \right]^{1/2} \qquad d_n = \left[\frac{N_a}{N_d} \frac{1}{\varepsilon_1\,N_a + \varepsilon_2\,N_d} \frac{2\varepsilon_1\,\varepsilon_2\,\phi_b}{e} \right]^{1/2}$$

$$\tag{40c}$$

(in the particular case $\varepsilon_1 = \varepsilon_2 = \varepsilon$, Eqs. (40) reduce to the corresponding expressions of Section 1).

The profiles of the bands at equilibrium can be easily obtained adding the contribution $-e\,\phi(x)$ to the band structure of the materials not yet in contact (see Fig. 11); of course the discontinuities ΔE_c and ΔE_v at the interface must be preserved. This latter fact is of particular importance: indeed by appropriate tailoring the band offsets (for instance varying the concentration x in $Al_x Ga_{1-x} As$) and appropriate choice of doping level, the conduction band of p-GaAs near the contact can be drawn close or below the Fermi level of the system. Under these circumstances, in the p-doped side of the interface, it is possible to realize a thin *inversion layer of electrons*, of fundamental importance for basic and applicative research. Notice that appropriate p-doping of the wide bandgap semiconductor and n-doping of the small bandgap semiconductor can allow the formation of an inversion layer of holes. The application of heterostructures in the physics of devices are numerous; we mention here their use for wide bandgap emitters in highly efficient bipolar transistors, the realization of heterostructure lasers, their use as window layers for detectors and photovoltaic cells (so that light can be absorbed in the active region of interest).

6   Metal–semiconductor contacts and MESFET transistor

Metal–semiconductor contacts at thermal equilibrium

In the previous sections, we have considered contacts formed by semiconductor homojunctions or heterojunctions. We now study some properties of metal–semiconductor contacts (*Schottky contacts*). It is in fact important to recognize (at least in principle) when the contact has *blocking* or *ohmic* character; this is of major interest also in view of the necessity to weld metallic contacts on semiconducting elements in electronic circuits.

The properties of metal–semiconductor junctions can be obtained using and extending the models already adopted for homojunctions and heterojunctions. Additionally, we make the basic assumption that the *junction is ideal* (i.e. microscopically free of localized states, defects, strains, charged double layers etc.), so that the basic transport properties of the contact are just determined by the difference of the Fermi levels in the two solids (before contact) and by the doping in the semiconductor. In general, however, a metal–semiconductor interface implies a strong local variation of the quantum mechanical potential, with possible disruption in local properties (these complications are not present in homojunctions, and hardly relevant in well lattice-matched heterojunctions). Thus the results of the simplified models here adopted are to be intended only as orientative; the actual realization of high performance contacts requires care and experience.

We consider specifically the contact between a metal and an n-type semiconductor (the contact between a metal and a p-type semiconductor could be done with similar procedures). We begin the discussion starting from the energy band diagrams and

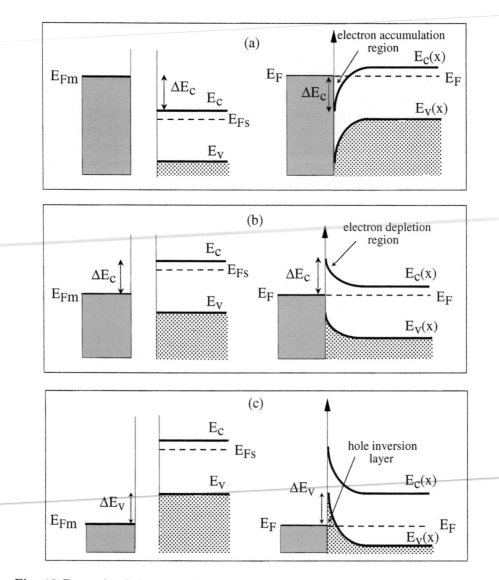

Fig. 12 Energy band diagrams of a metal and an n-doped semiconductor before contact (left panel) and after contact (right panel); the energy bands of the two materials before contact are referred to the vacuum level. Case (a) correponds to an ohmic contact. Case (b) corresponds to a blocking contact. Case (c) corresponds to the formation of an inversion layer. The energy difference between the Fermi level of the metal and the band edges of the semiconductor are indicated by ΔE_c and ΔE_v; these discontinuities are the same before and after contact.

Fermi levels E_{Fm} and E_{Fs} of the two materials, the metal and the n-type semiconductor, not yet in contact (the energy bands of the two materials are referred to the vacuum level); as usual, the semiconductor is supposed to be non-degenerate, with the Fermi level E_{Fs} within the energy gap and separated by the band edges E_c and

E_v by several $k_B T$. Fig. 12 illustrates the possible occurrence (at least in case of ideal contacts) of three different situations, depending on the relative position of the Fermi level of the metal with respect to the Fermi level and band edges of the semiconductor.

In Fig. 12a, we illustrate schematically the situation in which $E_{Fm} > E_{Fs}$ before the contact takes place. When the two solids are brought together, a unique Fermi level is established throughout the whole structure. In the n-doped semiconductor we have *accumulation of electrons near the junction*; a ready flow of electrons is possible and *the contact is said to be ohmic*. In Fig. 12b, we illustrate the situation in which initially we have $E_v < E_{Fm} < E_{Fs}$. When the contact is established, a depletion layer in the semiconductor region adjacent the junction is formed; this region of high resistance gives rise to a *blocking contact*. In Fig. 12c, we illustrate what happens (at least in principle) when $E_{Fm} < E_v < E_{Fs}$. In this case, besides the depletion layer in the n-doped semiconductor, an *inversion layer of holes* is formed at the interface (the control of the inversion layer through a metal–oxide–semiconductor structure is the subject of the next section).

We consider now in more detail the electrostatic potential and the physical properties of a blocking contact at thermal equilibrium. In Fig. 13a, we report the band diagram of the metal and the n-type semiconductor, not yet in contact. The Fermi level of the metal is taken to be in the interval $E_v < E_{Fm} < E_{Fs}$; W_m and W_s denote the work function of the metal and the semiconductor respectively (i.e. the energy difference between the vacuum level and the Fermi level of the two solids); χ_s indicates the electron affinity of the semiconductor, and $E_G = E_c - E_v$ the energy gap.

The parameters indicated in Fig. 13a are connected by some obvious relations. For the study of the band profile at the contact, an essential parameter is the built-in potential barrier ϕ_b, defined as usual by the difference of the Fermi levels (before contact)

$$\boxed{e\,\phi_b = E_{Fs} - E_{Fm}}\,. \tag{41a}$$

Furthermore, of interest is the discontinuity between the Fermi level of the metal and the bottom of the conduction band of the semiconductor (or the top of the valence band). We have

$$\Delta E_c = E_c - E_{Fm}\,, \quad \Delta E_v = E_{Fm} - E_v\,, \quad \Delta E_c + \Delta E_v = E_G\,. \tag{41b}$$

In terms of work function and electron affinity, we have

$$e\,\phi_b = W_m - W_s\,, \quad \Delta E_c = W_m - \chi_s\,. \tag{41c}$$

We now put in contact the metal with the n-type semiconductor. On formation of the junction, some electrons leave the n-type semiconductor and move in the metal, where the chemical potential is lower; eventually, any net flux of carriers vanishes when the Fermi level is spatially uniform throughout the whole structure. A depletion layer of extension d_s is formed in the semiconductor region, with space-charge density $\rho(x) = +e\,N_d$ ($0 \leq x \leq d_s$), due to fixed ionized donors. On the metal side, just at the surface there is a corresponding negative charge (the free electron density in

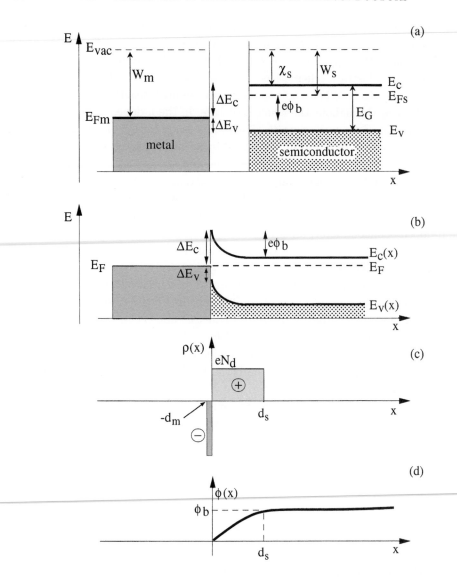

Fig. 13 (a) Energy bands of separated metal and n-type semiconductor. It is assumed that the Fermi energy of the metal is lower than the Fermi energy of the semiconductor, but still within the forbidden energy gap of the semiconductor. (b) Energy band profile for the metal and the n-type semiconductor in contact at equilibrium. The space charge region $\rho(x)$ and the internal electrostatic potential barrier $\phi(x)$ are also shown.

the metal is so high that the space charge region within the metal is confined to the first few layers next to the surface). There is thus an electric field in the space-charge region, and a corresponding barrier potential $\phi(x)$. The band structure profile, space-charge density and electrostatic potential are illustrated in Figs. 13b, c and d respectively.

The electrostatic potential $\phi(x)$ and the width of the depletion layer at the equilibrium can be evaluated from the Poisson equation

$$-\nabla^2 \phi(x) = \frac{e\,N_d}{\varepsilon} \qquad 0 \le x \le d_s \ . \tag{42a}$$

We integrate this equation with the boundary condition $\phi(d_s) = \phi_b$ and $\phi'(d_s) = 0$, and obtain

$$\phi(x) = \phi_b - \frac{e\,N_d}{2\,\varepsilon}\,(x - d_s)^2 \qquad 0 \le x \le d_s \ ; \tag{42b}$$

the boundary conditions $\phi(0) = 0$ gives for the width d_s of the depletion layer the expression

$$d_s = \left[\frac{1}{N_d} \frac{2\,\varepsilon\,\phi_b}{e} \right]^{1/2} \ . \tag{42c}$$

Non-equilibrium metal–semiconductor diode

The previous discussion refers to a metal–semiconductor junction at thermal equilibrium. If we apply a potential V to the metal side with respect to the semiconductor side, the total electrostatic potential across the contact becomes $\phi_b - V$ (Fig. 14). The potential V is counted as positive if the metal side is positively biased with respect to the semiconductor (forward bias); the potential V is counted as negative if the metal side is negatively biased with respect to the semiconductor (reverse bias). The thickness of the depletion layer, in the presence of an applied potential V, is given by

$$d_s = \left[\frac{1}{N_d} \frac{2\,\varepsilon\,(\phi_b - V)}{e} \right]^{1/2} \ . \tag{43}$$

As in a pn junction, the thickness of the barrier region decreases for forward bias, and increases for reverse bias.

We can estimate the electronic current through the circuit considering the metal and the semiconductor as two thermionic emitters facing each other. This thermionic model (tacitly) assumes that the mean free path of the carriers is longer than the space-charge region of the material; in the opposite situation that the thickness of the depletion layer is large compared with the mean free path of electrons, an appropriate drift-diffusion treatment should be considered.

Within the thermionic model, we can analyse the total current into the component flowing from the metal to the semiconductor and vice versa. The thermionic current flowing from the metal to the semiconductor is given by an expression of the type

$$J^{(m \to s)} = A^*\, e^{-\Delta E_c / k_B T} = J_0 \ , \tag{44a}$$

where ΔE_c is the barrier that an electron at the Fermi energy in the metal must overcome to pass to the semiconductor. The quantity A^* takes into account the number and the details of transmission and reflection of the electrons arriving at the metal–semiconductor surface; without being too specific, we can reasonably guess that the order of magnitude of A^* is given by the Richardson expression (see Section III-4).

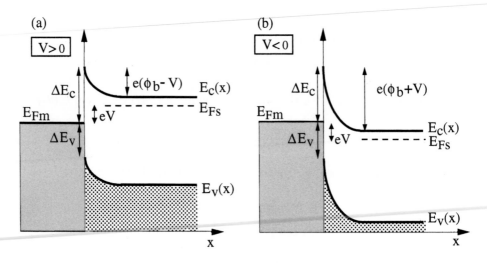

Fig. 14 Schematic energy band profile of a junction between a metal and an n-type semiconductor under forward bias (a) and reverse bias (b).

We notice that the current of Eq. (44a) is approximately independent of the applied potential V, as the barrier height ΔE_c is not changed by V.

The current that flows from the semiconductor to the metal at thermal equilibrium must equal $-J_0$ (see also Section XIII-4). With an applied potential V, we see that an electron at the Fermi energy in the (bulk) semiconductor must overcome a barrier $\Delta E_c - eV$ to pass to the metal. Thus the thermionic current flowing from the semiconductor to the metal is

$$J^{(s \rightarrow m)} = A^* e^{-(\Delta E_c - eV)/k_B T} = J_0 e^{eV/k_B T} . \tag{44b}$$

Summing up algebraically the currents (44a) and (44b), we obtain that the net thermionic emission of the metal–semiconductor diode is given by

$$\boxed{J = J_0 \left(e^{eV/k_B T} - 1 \right)} . \tag{45}$$

The current–voltage charateristic is controlled by an exponential factor, similarly to the pn junction, and gives rise to rectifying behaviour. The saturation current J_0 is given (at least approximately) by the thermionic emission above a barrier of height $\Delta E_c = W_m - \chi_s$. In ordinary situations, the thermionic contribution of the majority carriers, given by Eq. (45), is much more important than the contribution due to hole injection (or extraction) at the interface, and the metal–semiconductor diode is basically a unipolar device.

The above considerations refer to metal–semiconductor contacts with $E_v < E_{Fm} < E_{Fs}$ (see Fig. 12b). In the case that $E_{Fm} > E_{Fs}$ (see Fig. 12a), it is easily seen by inspection that an accumulation of electron carriers occurs at the interface; in close similarity with the n^+n junction, the contact is now ohmic with low resistance.

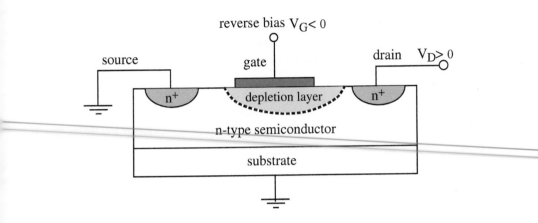

Fig. 15 Schematic illustration of the metal–semiconductor field effect transistor. The principle of operation is quite similar to the junction field effect transistor; the blocking p^+n junctions of the JFET are here replaced by the blocking metal–semiconductor contact.

Another means to realize ohmic contacts is by strong doping, so that the passage through the barrier can also occur by tunneling effect.

Metal–semiconductor field-effect transistor (MESFET)

An important application of the metal–semiconductor diode is the metal–semiconductor field-effect transistor (MESFET). The device is very similar in principle to the junction field-effect transistor and is schematically shown in Fig. 15.

The metal–semiconductor junction controls the width of the depletion layer and hence the current which flows from the source to the drain. The current–voltage characteristics of the device, for several gate voltages, can be estimated in close analogy to what done for JFET, taking into account some obvious modifications; the p^+n junctions of JFET are here replaced by a metal–semiconductor junction, and there is no lower gate in Fig. 15. We do not enter in further details, since the principles of operation of a MESFET are so similar to those presented in Section 4.

7 The metal–oxide–semiconductor structure and MOSFET transistor

The metal–oxide–semiconductor (MOS) structure

As a natural development of the metal–semiconductor structure, we study now the metal–insulator–semiconductor structure, which is obtained when the metal and the semiconductor are separated by a thin insulating layer. For silicon structures, the insulator is a layer of silicon dioxide obtained by oxidation of the semiconductor itself (the band gap of SiO_2 is $\approx 10\,eV$). The metal–oxide–semiconductor structure is denoted with the MOS acronym. We assume that interfaces between the composing material

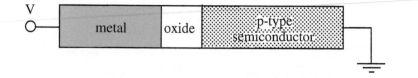

Fig. 16 Schematic representation of a biased metal–oxide–semiconductor junction; no current flows in the structure, and carriers are at thermal equilibrium both in the metal and in the semiconductor.

are appropriately technologically fabricated, with negligible concentration of surface traps or other defects.

The peculiar advantage and flexibility of the MOS structure with respect to a standard Schottky contact is easily understood (see Fig. 16). When a voltage is applied across a MOS structure, no current flows because of the presence of the insulator; carriers remain thus at the thermodynamic equilibrium both in the metal and in the semiconductor, although at two different Fermi levels; the relative position of the Fermi levels E_{Fm} and E_{Fs} in the metal and in the semiconductor can be tailored as desired by varying the sign and size of the applied potential. Accumulation or depletion of free carriers at the oxide–semiconductor junction, or even the appearance of an inversion layer, can thus be easily controlled.

We study now the carrier distribution within a MOS structure, in the case the semiconductor is p-doped (the reasoning for n-type semiconductor would follow a similar procedure). We assume for simplicity (but without loss of generality) that the Fermi level of the metal E_{Fm} and the Fermi level of the semiconductor E_{Fs}, not yet in contact, are the same (this assumption is called "ideal MOS structure"); thus when the metal and the semiconductor are connected with applied voltage $V \equiv 0$, no net flow of carriers occurs, no space-charge region is formed, and the energy bands *remain flat* (flat-band condition). In Fig. 17a we report the Fermi level and the band diagram of an "ideal" MOS with zero applied voltage. [Notice that a "non-ideal" MOS reaches the flat-band condition when biased with a potential V_{FB} such that $e\,V_{FB} = E_{Fm} - E_{Fs}$. The quantity V_{FB} is called "flat-band voltage". The results for ideal MOS in applied voltage V also hold for non-ideal MOS in applied voltage $V - V_{FB}$; thus from now on we consider only ideal MOS structures].

Suppose a potential V is applied to the metal side, while the semiconductor side is grounded (as indicated in Fig. 16). In Fig. 17 we show qualitatively the energy band profile of an ideal MOS structure under bias voltage V. For $V < 0$ we have in the semiconductor (near the junction) an accumulation of holes. If $V > 0$ we have in the semiconductor a depletion of holes. Of the highest interest is the case $V \gg 0$, which leads to an inversion layer formation (see Fig. 17d). All these situations can be realized and controlled continuously with the applied potential V. The comparison between Fig. 17 (where different situations correspond to different bias voltages) and Fig. 12 (where different situations correspond to different materials forming the junction)

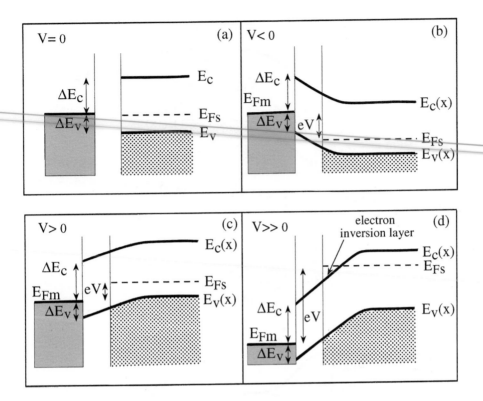

Fig. 17 Band structure of an "ideal" metal–oxide–semiconductor system and energy band profiles for: $V = 0$ (flat-band situation), $V < 0$ (hole accumulation), $V > 0$ (hole depletion), $V \gg 0$ (inverse layer formation). The energy differences between the Fermi level of the metal and the band edges of the semiconductor are indicated by ΔE_c and ΔE_v; these discontinuities are independent of the applied electric field.

makes evident the flexibility of the MOS structure, the most important component for integrated circuits.

We study now in detail the electrostatic potential and the energy band profile in the case the potential V applied to the metal is moderately positive, so that we have a depletion of holes, but not yet a strong accumulation of electrons (Fig. 18). The space-charge density $\rho(x)$ is essentially formed by the ionized acceptor impurities (of charge $-e$) in the semiconductor region $0 \le x \le d_s$ (with d_s to be determined), and a corresponding number of ionized atoms of the metal, confined in proximity of the metal surface at $x = -d_{\text{ox}}$. We have

$$\rho(x) = \begin{cases} -e\,N_a & 0 \le x \le d_s \\ Q_m\,\delta(x + d_{\text{ox}}) & Q_m = e\,N_a\,d_s \end{cases}, \tag{46}$$

where δ is the Dirac function.

The electrostatic potential within the semiconductor (of dielectric constant ε_s) is

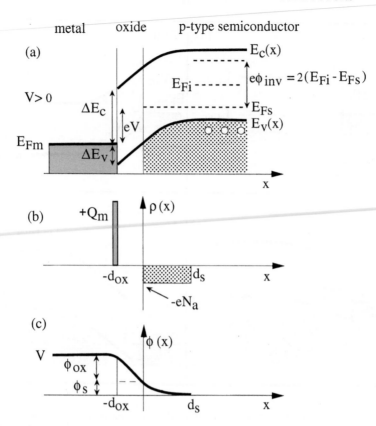

Fig. 18 Energy band profile, charge density and electrostatic potential of an ideal metal–oxide–semiconductor structure under (moderate) positive bias of the metal.

determined via the Poisson equation

$$-\frac{d^2\phi}{dx^2} = -\frac{e\,N_a}{\varepsilon_s} \qquad 0 \leq x \leq d_s \ .$$

Integrating twice the above equation with the trivial boundary conditions that the electric field and the potential are zero at $x = d_s$ we obtain

$$\phi(x) = \frac{e\,N_a\,d_s^2}{2\,\varepsilon_s}\,(1 - \frac{x}{d_s})^2 \qquad 0 \leq x \leq d_s \ . \tag{47a}$$

The potential ϕ_s at the surface of the semiconductor is thus

$$\boxed{\phi_s = \frac{e\,N_a\,d_s^2}{2\,\varepsilon_s}} \ ; \tag{47b}$$

the surface potential ϕ_s represents the potential drop across the depletion layer formed within the semiconductor.

We can now calculate the potential drop ϕ_{ox} in the oxide region. From Eq. (47a), we can obtain the electric field at the semiconductor surface: $E_{\text{surf}} = -\phi'(0) = e\,N_a\,d_s/\varepsilon_s$. In the oxide region, the electric field is given by $E_{\text{ox}} = e\,N_a\,d_s/\varepsilon_{\text{ox}}$, where ε_{ox} is the dielectric constant of the insulating material. In the oxide region the charge is zero, the electric field is constant and the potential varies linearly with the x-coordinate; the potential drop in the oxide $\phi_{\text{ox}} = E_{\text{ox}}\,d_{\text{ox}}$ is given by

$$\phi_{\text{ox}} = \frac{d_{\text{ox}}}{\varepsilon_{\text{ox}}}\,e\,N_a\,d_s\ .$$

The above expression, with the help of Eq. (47b), can be conveniently re-written in the form

$$\boxed{\phi_{\text{ox}} = \sqrt{K\,\phi_s}} \qquad (48a)$$

where

$$K = \frac{d_{\text{ox}}^2}{\varepsilon_{\text{ox}}^2}\,2\,e\,N_a\,\varepsilon_s\ ; \qquad (48b)$$

the parameter K has the dimension of a potential, and is specific of the given oxide–semiconductor junction; it depends on the width of the oxide, doping of the semiconductor, and dielectric constants of both.

The potential V applied to the metal is evidently the sum of the potential drop in the semiconductor and the potential drop ϕ_{ox} in the oxide (see Fig. 18); we have

$$\boxed{V = \phi_s + \sqrt{K\,\phi_s}}\ ; \qquad (49)$$

this equation can be inverted and gives ϕ_s (and thus d_s) as a function of the applied voltage V. Before continuing our elaboration of the properties of the MOS structure and MOSFET transistor, we notice the importance of the parameter K in the partition of the applied voltage between the semiconductor and the oxide. [Since the elaboration and design of MOS structures is slightly complicated by the presence of the parameter K, the reader interested in overall aspects is suggested to specify to the case $K = 0$ all the formulae given below. The $K = 0$ situation would apply to a fictitious ideal insulator which remains insulating for extremely narrow widths].

Onset of the strong inversion condition

When the bias potential V applied to the metal increases, it is possible to reach a situation in which the concentration of minority carriers (the electrons) at the surface of the p-type semiconductor equals the concentration of holes in the bulk p-type semiconductor; we can estimate the onset of this "strong inversion situation" with the following argument.

Let us indicate with E_{Fi} and E_{Fs} the Fermi level of the intrinsic semiconductor and the Fermi level of the p-doped semiconductor, respectively (see Fig. 18). We observe that

$$n_{p0} = n_{i0}\,e^{-(E_{Fi}-E_{Fs})/k_B T}\ , \qquad p_{p0} = p_{i0}\,e^{(E_{Fi}-E_{Fs})/k_B T}\ ;$$

since $n_{i0} = p_{i0}$ we also have

$$n_{p0}\, e^{2\,(E_{Fi}-E_{Fs})/k_B T} = p_{p0}\,. \tag{50}$$

We define as ϕ_{inv} the potential such that

$$\boxed{e\,\phi_{\mathrm{inv}} \equiv 2\,(E_{Fi} - E_{Fs})}\,. \tag{51}$$

From Eq. (50) and the definition (51), we see that *the concentration of electron carriers at the surface equals the equilibrium majority carrier concentration* p_{p0} when $\phi_s \equiv \phi_{\mathrm{inv}}$. Thus the potential ϕ_{inv}, defined by Eq. (51), can be considered as the onset of the strong inversion situation.

Let us indicate with d_s^* the length of the depletion layer in the p-type semiconductor, just at the beginning of the strong inversion condition. From Eq. (47b), we have $\phi_{\mathrm{inv}} \equiv e\,N_a\,d_s^{*2}/2\varepsilon_s$, and thus

$$d_s^* = \sqrt{\frac{2\varepsilon_s\,\phi_{\mathrm{inv}}}{e\,N_a}}\,. \tag{52a}$$

From Eq. (49), we see that the threshold potential required to achieve the onset of the strong inversion condition is

$$V_T = \phi_{\mathrm{inv}} + \sqrt{K\,\phi_{\mathrm{inv}}}\,. \tag{52b}$$

If the potential V of the metal is further increased with respect to the threshold voltage, we can assume that d_s^* remains constant and an inversion electron layer begins to appear near the surface; the negative charge Q_n per unit surface formed at the oxide–semiconductor interface is such that

$$-Q_n\,\frac{d_{\mathrm{ox}}}{\varepsilon_{\mathrm{ox}}} = V - V_T = V - \phi_{\mathrm{inv}} - \sqrt{K\,\phi_{\mathrm{inv}}}\,.$$

From the last equation, we see that the surface charge density Q_n of mobile electrons is related to the applied potential $V\,(> V_T)$ by the expression

$$\boxed{Q_n = -\frac{\varepsilon_{\mathrm{ox}}}{d_{\mathrm{ox}}}\,(V - \phi_{\mathrm{inv}} - \sqrt{K\,\phi_{\mathrm{inv}}})}\,. \tag{52c}$$

Effect of a bias between the inversion layer and the bulk semiconductor in MOS structures

We consider now the situation in which, by means of an additional electrode, a potential difference V_c is applied *between the "channel" of electrons in the surface inversion layer and the bulk semiconductor* (see Fig. 19). If the potential V_c is positive, we have a kind of reverse biased np junction between the inversion layer and the bulk semiconductor. Although there is some current betweeen the channel and the semiconductor, under reverse bias we can neglect such a current.

In the case a bias potential $V_c > 0$ is applied to the surface layer, we easily see that the expressions (52) maintain their validity provided we systematically replace ϕ_{inv}

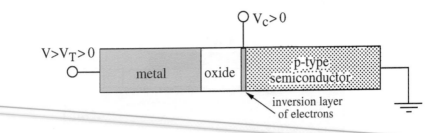

Fig. 19 Schematic representation of a biased metal–oxide–semiconductor junction, with the metal bias voltage V larger than the threshold value for the creation of the electron inversion layer; V_c is the bias potential for the channel of electrons.

by $\phi_{\rm inv} + V_c$. The maximum length of the depletion layer in the case of bias potential $V_c > 0$ becomes

$$d_s^* = \sqrt{\frac{2\varepsilon_s}{e\, N_a}\,(\phi_{\rm inv} + V_c)}\; . \tag{53a}$$

The threshold potential is now

$$V_T = \phi_{\rm inv} + V_c + \sqrt{K\,(\phi_{\rm inv} + V_c)}\; . \tag{53b}$$

When V exceeds the above threshold potential, the charge density Q_n per unit surface of mobile electrons is

$$Q_n = -\frac{\varepsilon_{\rm ox}}{d_{\rm ox}} \left[V - \phi_{\rm inv} - V_c - \sqrt{K\,(\phi_{\rm inv} + V_c)} \right] . \tag{53c}$$

The MOSFET transistor

We have now all the elements for a semi-quantitative discussion of the $I - V$ characteristics of an ideal metal–oxide–semiconductor field-effect transistor (MOSFET); a cross-section of the device is schematically given in Fig. 20. A voltage is applied between the *source* and *drain* electrodes; the source is grounded ($V_s = 0$) and the potential of the drain is positive ($V_D > 0$). Moving from source to drain, we find two pn^+ junctions, one of which biased in the forward direction, and the other biased in the reverse direction with high impedance. There is a third electrode, the *gate*, which is insulated from the semiconductor by an oxide layer and held at a positive potential V_G. The positive potential of the gate exceeds the threshold value required to convert the p-type semiconductor into an n-type material at the interface; when the n-type inversion layer is formed, electrons can flow between the two n-type source and drain regions, and the current through the channel can be controlled by the gate potential.

For the calculation of the current–voltage characteristics of a MOSFET, we begin to consider the situation in which the inversion layer channel extends from source to drain. Let us indicate with $V_c(x)$ the potential at a generical point x of the channel ($0 \le x \le L$); at the two ends of the channel we have $V_c(0) = 0$ and $V_c(L) = V_D$. If the channel potential is assumed to be a smoothly changing function of the position,

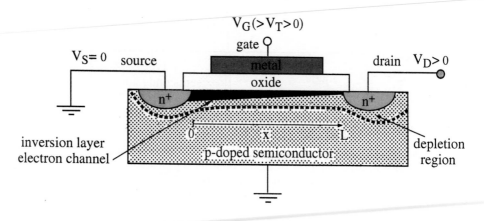

Fig. 20 Schematic representation of a MOSFET transistor.

we can use Eq. (53c) for the calculation of the electron surface charge induced at a position x along the channel. In the gradual channel approximation, we obtain

$$Q_n(x) = -\frac{\varepsilon_{\text{ox}}}{d_{\text{ox}}} \left[V_G - \phi_{\text{inv}} - V_c(x) - \sqrt{K \left(\phi_{\text{inv}} + V_c(x) \right)} \right] . \qquad (54)$$

Assuming a constant mobility in the channel, we have for the drain current (taken as positive in the $-x$ direction)

$$I_D = Q_n(x)\, W\, \mu_n\, E(x) = Q_n(x)\, W\, \mu_n\, \left(-\frac{dV_c}{dx} \right) ,$$

where μ_n is the low field electron mobility, W is the device width (orthogonal to the cross-section of Fig. 20), $-dV_c(x)/dx$ is the component of the electric field parallel to the semiconductor–oxide interface. With separation of variables and use of Eq. (54), we obtain

$$I_D\, dx = W\, \mu_n\, \frac{\varepsilon_{\text{ox}}}{d_{\text{ox}}} \left[V_G - \phi_{\text{inv}} - V_c(x) - \sqrt{K \left(\phi_{\text{inv}} + V_c(x) \right)} \right] dV_c(x) .$$

Integrating from the source to the drain, we have

$$I_D = A \left[\left(V_G - \phi_{\text{inv}} - \frac{1}{2} V_D \right) V_D - \frac{2}{3} \sqrt{K} \left(\phi_{\text{inv}} + V_D \right)^{3/2} + \frac{2}{3} \sqrt{K}\, \phi_{\text{inv}}^{3/2} \right] , \qquad (55)$$

where $A = (W\, \mu_n / L)(\varepsilon_{\text{ox}}/d_{\text{ox}})$ is a constant.

Equation (55) is applicable provided the n channel remains open up to the drain. For small drain-to-source voltages such that $V_D \ll \phi_{\text{inv}}$, Eq. (55) gives

$$I_D = A \left[V_G - \phi_{\text{inv}} - \sqrt{K\, \phi_{\text{inv}}} \right] V_D ,$$

and we have thus a linear region in the $I_D - V_D$ characteristics for small V_D.

The maximum value of the drain current occurs at the saturation potential, which is the potential for which Eq. (54) vanishes for $x = L$. We have thus

$$V_G - \phi_{\text{inv}} - V_{D,\text{sat}} - \sqrt{K\left(\phi_{\text{inv}} + V_{D,\text{sat}}\right)} \equiv 0 \ . \tag{56a}$$

Using Eq. (55), we obtain for the saturation current:

$$I_{D,\text{sat}} = A\left[\left(V_G - \phi_{\text{inv}} - \frac{1}{2}V_{D,\text{sat}}\right)V_{D,\text{sat}} - \frac{2}{3}\sqrt{K}\left(\phi_{\text{inv}} + V_{D,\text{sat}}\right)^{3/2} + \frac{2}{3}\sqrt{K}\phi_{\text{inv}}^{3/2}\right]$$

$$= A\left[\sqrt{K}\left(\phi_{\text{inv}} + V_{D,\text{sat}}\right)^{1/2} + \frac{1}{2}V_{D,\text{sat}}^2 - \frac{2}{3}\sqrt{K}\left(\phi_{\text{inv}} + V_{D,\text{sat}}\right)^{3/2} + \frac{2}{3}\sqrt{K}\phi_{\text{inv}}^{3/2}\right] \ . \tag{56b}$$

If $V_{D,\text{sat}} \ll \phi_{\text{inv}}$ we can perform a series development in Eq. (56b) and obtain

$$I_{D,\text{sat}} = A\frac{1}{2}V_{D,\text{sat}}^2 \ ; \tag{56c}$$

we thus see that, in the case $V_{D,\text{sat}} \ll \phi_{\text{inv}}$, the current $I_{D,\text{sat}}$ depends quadratically on $V_{D,\text{sat}}$ (similarly to the expression (37)).

The reader can notice the close analogy, both from a physical point of view and from the semi-quantitative estimates, with the junction field-effect transistor. The $I_D - V_D$ characteristics, for several gate voltages, have shapes similar to those of Fig. 10; the knowledge of the current–voltage characteristics of the MOSFET transistor makes (in principle) possible its design in integrated circuits.

Further reading

T. Ando, A. Fowler and F. Stern "Electronic Properties of Two Dimensional Systems" Rev. Mod. Phys. **54**, 437 (1982)

P. Bhattacharya "Semiconductor Optoelectronic Devices" (Prentice-Hall, Englewood Cliffs, New Jersey 1997)

F. Capasso "Band-Gap Engineering: from Physics and Materials to New Semiconductor Devices" Science **235**, 172 (1987)

F. Capasso and G. Margaritondo eds. "Heterojunction Band Discontinuities. Physics and Device Applications" (North-Holland, Amsterdam 1987)

R. S. Muller and T. I. Kamins "Device Electronics for Integrated Circuits" (Wiley, New York 1986)

B. Sapoval and C. Hermann "Physics of Semiconductors" (Springer, New York 1995)

K. Seeger "Semiconductor Physics" (Springer, Berlin 1989, fourth edition)

J. Singh "Semiconductor Devices: an Introduction" (McGraw Hill, New York 1994)

R. A. Smith "Semiconductors" (Cambridge University Press, 1987)

S. M. Sze "Semiconductor Devices. Physics and Technology" (Wiley, New York 1985)

C. M. Wolfe, N. Holonyak and G. E. Stillman "Physical Properties of Semiconductors" (Prentice-Hall, Englewood Cliffs, New Jersey 1989)

A. Yariv "Optical Electronics" (Saunders College Publishing, Philadelphia 1995, fourth edition)

XV
Electron gas in magnetic fields

In this chapter, we begin the study of the electronic structure of matter in the presence of magnetic fields, and we start with the treatment of the electron gas in magnetic fields. In the next chapter, we consider diamagnetism and paramagnetism of substances, such as ions, molecules or impurities in solids; Chapter XVII concerns cooperative effects and magnetic ordering; the peculiar magnetic properties of superconductors (Meissner effect, magnetic flux quantization) are contained in Chapter XVIII.

When a magnetic field is applied to an otherwise freely moving electron, the classical motion is given by a helical path along the magnetic field direction. Quantistically, the cyclotron motion perpendicular to the field is quantized into discrete Landau levels; this has important implications on the physical properties of the electron gas and in particular on the magnetic susceptibility. We will see that the modification of the electron orbital motion due to the field leads to diamagnetism (Landau diamagnetism), while the spin of the electron gives rise to a paramagnetic contribution (spin or Pauli paramagnetism). Under very special conditions (strong magnetic fields on very pure specimens at low temperatures) the quantization of Landau levels manifests itself in characteristic oscillatory effects occurring in density-of-states, thermodynamic functions and several physical properties; in particular the magnetic moment exhibits oscillations as a function of the magnetic field (de Haas–van Alphen effect), and precise information on the Fermi surface can be inferred.

Magnetic fields strongly influence the transport properties of ordinary (i.e. three-dimensional) metals or semiconductors, and we discuss some aspects of the rich phenomenology related to the classical Hall effect and magnetoresistivity. We also describe the profound modifications that may occur in the transport properties of the two-dimensional electron gas and in particular the striking quantum Hall effect. Here the quantization due to magnetic field manifests itself in well defined plateaus in the Hall resistance, which allow to measure the fundamental resistance h/e^2 with an accuracy as high as one part in ten millions.

1　Magnetization and magnetic susceptibility

Definitions and generalities

In general, upon application of an external magnetic field H, a specimen acquires a magnetization M(*magnetic moment per unit volume*). When M is proportional to H, the *magnetic susceptibility* χ is defined as

$$\chi = \frac{M}{H} \; ; \tag{1a}$$

otherwise, the magnetic susceptibility χ takes the more general definition

$$\chi = \frac{\partial M}{\partial H} \; . \tag{1b}$$

If gaussian units are used χ is dimensionless.

The static magnetic susceptibility χ can be measured for instance with the method of the Gouy balance; precision up to 10^{-10} can be obtained. We have to remember that the magnetic susceptibility is typically of the order of -10^{-6} for several diamagnetic substances, and of the order of 10^{-4} for several paramagnetic substances at ordinary temperatures. However, the range of possible values of χ in materials is extremely wide; for instance, in the case of perfect diamagnetism in superconductors, the susceptibility takes the value $\chi = -1/4\pi$, and the magnetic forces can outnumber the gravitational forces with typical levitation effects.

We consider now some definitions and properties, useful in the quantum treatment of the magnetic susceptibility of electronic systems. At $T = 0$, the *magnetization* of a homogeneous system of volume V in a *uniform magnetic field H* is defined as

$$M(T{=}0, H) = -\frac{1}{V} \frac{\partial E_0(H)}{\partial H} \; , \tag{2a}$$

where $E_0(H)$ is the energy of the ground-state of the sample in the presence of H. The magnetic susceptibility at zero temperature then becomes

$$\chi(T{=}0, H) = -\frac{1}{V} \frac{\partial^2 E_0(H)}{\partial H^2} \; . \tag{2b}$$

The magnetic susceptibility $\chi(T=0,H)$ is related to the second derivative of the ground-state energy $E_0(H)$ with respect to the external field; if the curvature of $E_0(H)$ versus H is positive, χ is negative, and vice versa. In the (rather frequent) cases that $E_0(H)$ depends quadratically on H, the magnetic susceptibility is independent from the applied field.

At a finite temperature T, if $F(T,H)$ is the free energy of the system, the magnetization is given by the thermodynamic relation

$$M(T,H) = -\frac{1}{V}\frac{\partial F(T,H)}{\partial H}. \tag{3a}$$

The meaning of this relation becomes obvious, if one considers the general expression of the free energy given in Eq. (III-A13), and here re-written for convenience

$$F = -k_B T \ln Z = -k_B T \ln \sum_n e^{-E_n(H)/k_B T},$$

where $E_n(H)$ are the energy values of the system in the presence of an external magnetic field H. Inserting this expression into Eq. (3a), we obtain

$$M(T,H) = -\frac{1}{V}\frac{\sum_n \dfrac{\partial E_n(H)}{\partial H} e^{-E_n(H)/k_B T}}{\sum_n e^{-E_n(H)/k_B T}} \; ;$$

from the above expression, it is seen that the magnetization $M(T,H)$ is just the thermal equilibrium average of the the magnetization $-(1/V)\partial E_n(H)/\partial H$ of each pure state of the system. The magnetic susceptibility at temperature T becomes

$$\chi(T,H) = -\frac{1}{V}\frac{\partial^2 F(T,H)}{\partial H^2}. \tag{3b}$$

In the particular case that $T=0$, one recovers back Eq. (2b). In the case $F(T,H)$ changes quadratically with H, the magnetic susceptibility is independent from the applied field.

In the following of this chapter, we focus on electronic systems (such as the free-electron gas) with small magnetic susceptibility ($\chi \approx \pm 10^{-6}$ or so); in these systems, in the analysis of the energy eigenvalues $E_n(H)$, we can safely assume negligible the difference between the external magnetic field H in the absence of the specimen and the actual magnetic field B in the presence of the specimen; in other words, the magnetic field to be embodied in the Hamiltonian of the electronic system can be taken to be the external magnetic field applied to the system. Before passing to the study of the magnetic susceptibility of specific quantum systems, we wish to comment briefly on the absence of magnetism from a purely classical point of view.

Absence of magnetism from purely classical arguments

From pure classical arguments, the magnetic susceptibility of any dynamical system is rigorously zero; this result, known as *Bohr–van Leeuwen theorem*, can be proved as follows. Consider a many-electron system with N spinless electrons, whose classical state is characterized by $6N$ coordinates and momenta $(q_1, q_2, \ldots q_{3N}, p_1, p_2, \ldots, p_{3N})$. The classical partition function is given by

$$Z_{\text{classical}} = \int \exp[-H_0(q_i, p_i)/k_B T] \, dq_1 \, dq_2 \ldots dq_{3N} \, dp_1 \, dp_2 \ldots dp_{3N} \ ,$$

where $H_0(q_i, p_i)$ $(i = 1, \ldots, 3N)$ is the Hamiltonian of the electron system.

In the presence of a magnetic field described by the vector potential $\mathbf{A}(\mathbf{r})$, each momentum $\mathbf{p}_i$ in the Hamiltonian H_0 must be replaced by the generalized momentum $\mathbf{P}_i = \mathbf{p}_i + (e/c)\mathbf{A}(\mathbf{r}_i)$. This is the only way by which the magnetic field enters in the expression of the Hamiltonian (the electron spin, and hence the interaction energy of the magnetic field with each electron spin, is taken as a quantum phenomenon and disregarded in the present considerations of classical nature). The terms containing the vector potential can be eliminated by a simple shift in the momentum integration. Thus, the classical partition function does not depend on the magnetic field; hence the magnetization and the magnetic susceptibility vanish.

As an example, consider the partition function of a free electron in a uniform magnetic field H, directed along the z-axis. A possible choice of the vector potential is $\mathbf{A}(\mathbf{r}) = (-H y, 0, 0)$, and in fact it is easily verified that $\operatorname{curl} \mathbf{A}(\mathbf{r}) = (0, 0, H)$. We have for the classical partition function

$$Z_{\text{classical}} = \int_V d\mathbf{r} \int_{-\infty}^{+\infty} d\mathbf{p} \, \exp \frac{-\left[(p_x - \frac{e}{c} H y)^2 + p_y^2 + p_z^2\right]}{2m \, k_B T} = V \, (2\pi m \, k_B T)^{3/2} \ ;$$

thus the classical partition function is not influenced by the presence of magnetic fields. A quantum theory is necessary from the very beginning to describe magnetic phenomena.

2 Energy levels and density-of-states of a free-electron gas in magnetic fields

In this section we consider the effects produced by a uniform magnetic field on the dynamics of a free-electron gas. From a classical point of view, the dynamics of a free charged particle in a magnetic field is composed by a circular cyclotron motion in the plane perpendicular to the magnetic field and a free motion parallel to it. The quantum picture leads to quantization of the motion of the electron perpendicularly to the magnetic field, and converts the free-particle kinetic energy $E = \hbar^2 \mathbf{k}^2/2m$ into a set of one-dimensional sub-bands, each corresponding to a given Landau level.

Before beginning with the study of magnetic fields, it is convenient to report the energies and the density-of-states of free particles in the one-, two-, and three-dimensional cases (see Section II-7). The dimension of the sample is supposed to be L_z for the one-dimensional case, $S = L_x L_y$ for the two-dimensional case, and $V = L_x L_y L_z$ for the three-dimensional case. The dispersion relations of the free-electron gas, and the corresponding density-of-states (*excluding spin degeneracy*), can be summarized as follows:

$$E(k_z) = E_0 + \frac{\hbar^2 k_z^2}{2m} \qquad\qquad D^{(1d)}(E) = L_z \frac{(2m)^{1/2}}{h} \frac{1}{\sqrt{E - E_0}} \Theta(E - E_0) \quad (4a)$$

$$E(k_x, k_y) = E_0 + \frac{\hbar^2}{2m}(k_x^2 + k_y^2) \qquad D^{(2d)}(E) = S \pi \frac{2m}{h^2} \Theta(E - E_0) \qquad\qquad (4b)$$

$$E(\mathbf{k}) = E_0 + \frac{\hbar^2}{2m}(k_x^2 + k_y^2 + k_z^2) \; D^{(3d)}(E) = V \, 2\pi \frac{(2m)^{3/2}}{h^3} \sqrt{E - E_0} \, \Theta(E - E_0) \quad (4c)$$

where $\Theta(x)$ is the step function ($\Theta(x) = 0$ for $x < 0$, $\Theta(x) = 1$ for $x > 0$). When the spin degeneracy is included, the density-of-states for both spin directions is obtained multiplying by two the results of Eqs. (4).

2.1 Energy levels of the two-dimensional electron gas in magnetic fields

Our analysis of the magnetic properties of matter starts considering a two-dimensional electron gas. As discussed in Chapter XIV, appropriate semiconductor interfaces or metal-oxide-semiconductor structures provide the technical means to realize this system: carriers can be considered as freely moving parallel to the planar structure, while in the perpendicular direction they are described by wavefunctions strongly localized in a thin layer of atomic planes next to the interface.

The (effective) Hamiltonian of a two-dimensional electron gas (in xy plane) in a uniform magnetic field (parallel to z) can be written as

$$H_0 = \frac{1}{2m}\left(p_x + \frac{e}{c}A_x\right)^2 + \frac{1}{2m}\left(p_y + \frac{e}{c}A_y\right)^2 , \qquad (5)$$

where e is the absolute value of the electron charge, the effective mass m is taken equal to the electron mass (for simplicity), and $\mathbf{A}(\mathbf{r})$ is the vector potential of the applied magnetic field. The Hamiltonian of Eq. (5) focuses on the modification of the *orbital motion* of carriers due to the magnetic field, and neglects the interaction of the spin magnetic moment with the applied field (this interaction can be worked out separately, as discussed below in Section 4); spin degeneracy is taken into account introducing a factor two, whenever appropriate.

A possible choice of the vector potential, corresponding to a uniform magnetic field of intensity H in the z-direction, is given by the so-called first Landau gauge

$$\mathbf{A}(\mathbf{r}) = (-Hy, 0, 0) ; \qquad (6)$$

it is seen by inspection that $\text{curl}\,\mathbf{A}(\mathbf{r}) = H(0, 0, 1)$. The vector potential is not uniquely determined; other frequently used gauges are the second Landau gauge $\mathbf{A}(\mathbf{r}) = (0, Hx, 0)$ and the symmetric gauge $\mathbf{A}(\mathbf{r}) = (1/2)(-Hy, Hx, 0)$.

The energies of the Hamiltonian H_0, in the absence of the external magnetic field, are given by

$$E(k_x, k_y) = \frac{\hbar^2}{2m}(k_x^2 + k_y^2) .$$

Using Eq. (4b), we see that the corresponding two-dimensional density-of-states (excluding spin degeneracy) for a sample in a rectangular box of surface S is given by

$$D(E, H=0) = \frac{2\pi m}{h^2} S \quad \text{for } E > 0 \,. \tag{7}$$

In the presence of the magnetic field, and adopting the gauge specified by Eq. (6), the Hamiltonian of Eq. (5) becomes

$$H_0 = \frac{1}{2m}\left(p_x - \frac{eH}{c}y\right)^2 + \frac{1}{2m}p_y^2 \,. \tag{8}$$

The above Hamiltonian does not contain x, and then p_x is a constant of motion with values $\hbar k_x$; the Hamiltonian of Eq. (8) can be rewritten in the form

$$H_0 = \frac{1}{2m}p_y^2 + \frac{1}{2m}\left(\hbar k_x - \frac{eH}{c}y\right)^2 = \frac{1}{2m}p_y^2 + \frac{1}{2}m\,\omega_c^2\,(y - y_0)^2 \,, \tag{9a}$$

where

$$\omega_c = \frac{eH}{mc} \tag{9b}$$

is the *cyclotron resonance frequency*, and

$$y_0 = \frac{\hbar c}{eH}k_x \tag{9c}$$

is the center of oscillations. Eq. (9a) is the familiar Hamiltonian of a displaced linear oscillator of frequency ω_c; its eigenvalues are

$$E_{nk_x} = \left(n + \frac{1}{2}\right)\hbar\omega_c \,, \tag{10}$$

where $n = 0, 1, 2, \ldots$ indicates the *Landau level*. Thus the presence of a magnetic field modifies the continuous kinetic energy $E(k_x, k_y) = (\hbar^2/2m)(k_x^2 + k_y^2)$ into discretized Landau levels, as indicated schematically in Fig. 1. The energy separation between adjacent Landau levels is $\hbar\omega_c = (e\hbar/mc)H = 2\mu_B H$; the value of the Bohr magneton is $\mu_B = (e\hbar/2mc) = 0.05788\,\text{meV}/\text{Tesla}$, and then $\hbar\omega_c/H = 0.1158\,\text{meV}/\text{Tesla}$ (1 Tesla $= 10^4$ gauss).

The eigenfunctions corresponding to the eigenvalues of Eq. (10) are

$$\phi_{nk_x}(\mathbf{r}) = \frac{1}{\sqrt{L_x}}e^{i\,k_x\,x}H_n(y - y_0) \,,$$

where L_x is the length, in the x-direction, of the rectangular box where the electron is confined and H_n are the Hermite functions for the harmonic oscillator. The wavefunctions $\phi_{nk_x}(\mathbf{r})$ in the presence of a magnetic field (in the chosen Landau gauge) are plane waves in the x-direction, multiplied by harmonic oscillator wave functions in the y-direction, with typical extension $l_n = \sqrt{2n+1}\,l_0$, where $l_0 = (\hbar c/eH)^{1/2}$ is the classical cyclotron radius of an electron with kinetic energy $\hbar\omega_c/2$. For $H = 10^4$ gauss, with $hc/e = 4.136 \times 10^{-7}$ gauss·cm^2, the magnetic length l_0 becomes $l_0 = 2.57 \times 10^{-6}$ cm $= 257$ Å.

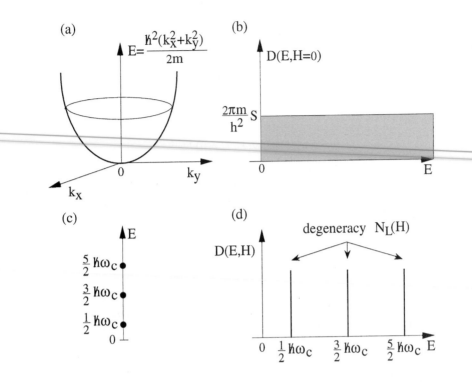

Fig. 1 Energy levels (Fig. 1a) and density-of-states excluding spin degeneracy (Fig. 1b) of a two-dimensional electron gas in the absence of a magnetic field. The energy levels and the density-of-states in the presence of a magnetic field are indicated in Fig. 1c and Fig. 1d, respectively. The degeneracy of the Landau levels is $N_L(H) = (e/hc)HS = D(E, H = 0)\hbar\omega_c$.

The energy levels of Eq. (10) do not depend on k_x; thus their degeneracy is given by the number of allowed k_x values for the system. We can determine the orbital degeneracy $N_L(H)$ of a Landau level by imposing the condition

$$0 \leq y_0 \equiv \frac{\hbar c}{eH} k_x < L_y \quad \text{or equivalently} \quad 0 \leq k_x < \frac{eH}{\hbar c} L_y \ .$$

The number N_L of allowed k_x values in the above interval gives the number of harmonic oscillators, whose origin is confined within the sample; the degeneracy N_L of each Landau level is then

$$\boxed{N_L(H) = \frac{L_x}{2\pi} \frac{eH}{\hbar c} L_y \equiv \frac{e}{hc} H S} \ . \tag{11}$$

The orbital degeneracy $N_L(H)$ is *thus proportional to the magnetic field, whose effect is that of piling up a large number of states into discrete Landau levels.*

The orbital degeneracy N_L can be written also in the effective form

$$N_L(H) = \frac{\Phi(H)}{\Phi_0} \ ,$$

where $\Phi(H) = H\,S$ is the flux of H through the sample, and

$$\Phi_0 = \frac{hc}{e} = 4.136 \times 10^{-7}\,\text{gauss} \cdot \text{cm}^2$$

provides the natural unit of flux.

The degeneracy $N_L(H)$ of each Landau level can be written also in the following alternative form

$$N_L(H) = D(E, H{=}0)\,\hbar\omega_c \,, \tag{12}$$

where $D(E, H = 0) = 2\pi m\,S/h^2$ indicates the constant two-dimensional density-of-states in the absence of the magnetic field (see Eq. 7). The quantity $D(E, H{=}0)\hbar\omega_c$ represents the number of states in the energy interval $\hbar\omega_c$ between two successive Landau levels; Eq. (12) explicitly shows that the overall number of states is unaffected by the magnetic field.

There is another instructive way to express the Landau degeneracy of Eq. (11). Let us consider a sample with N free electrons (of both spin directions) in the surface S, and surface density $n_s = N/S$. In the presence of a magnetic field H perpendicular to the sample, the energy levels are quantized in Landau levels, whose degeneracy increases with H ; thus, it is always possible to find a *threshold value H_t of the magnetic field*, at which all the available electrons fully occupy the ground Landau level (above H_t, all the available electrons are accommodated in the ground Landau level without filling it). The *threshold field H_t, for which all the available electrons fully occupy the ground Landau level, assuming no removal of spin degeneracy*, is determined by the relation

$$2\,N_L(H_t) = 2\frac{e}{hc}\,H_t\,S \equiv N \;.$$

Thus the characteristic field H_t is given by

$$H_t = \frac{1}{2}\frac{N}{S}\frac{hc}{e} = \frac{1}{2}n_s\frac{hc}{e} \;. \tag{13}$$

For n_s in the range $10^{10} - 10^{12}$ electrons/cm^2, and $hc/e = 4.136 \times 10^{-7}$ gauss $\cdot$ cm^2, we have H_t in the range $10^3 - 10^5$ gauss. The orbital degeneracy of the Landau levels can be cast in the form

$$N_L(H) = \frac{1}{2}N\frac{H}{H_t} \;. \tag{14}$$

Expressions (11), (12) and (14) for the degeneracy of the Landau levels are all equivalent and particularly significant in discussing magnetic field effects.

In a quite similar way, we can define the characteristic field H_t^* as the field for which *all the available electrons fully occupy the ground Landau level with definite spin orientation*; H_t^* is given by the expression

$$H_t^* = n_s\frac{hc}{e} \tag{15}$$

and we evidently have $H_t^* = 2\,H_t$.

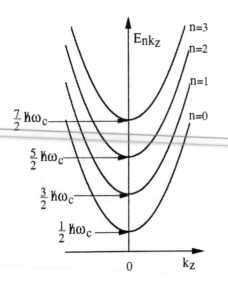

Fig. 2 Schematic representation of the energy levels for a three-dimensional electron gas in the presence of a magnetic field in the z-direction.

2.2 Energy levels of the three-dimensional electron gas in magnetic fields

A free electron, in a uniform magnetic field of vector potential $\mathbf{A}(\mathbf{r}) = (-Hy, 0, 0)$, is described by the Hamiltonian

$$H_0 = \frac{1}{2m}\left(\mathbf{p} + \frac{e}{c}\mathbf{A}\right)^2 = \frac{1}{2m}\left(p_x - \frac{eH}{c}y\right)^2 + \frac{1}{2m}p_y^2 + \frac{1}{2m}p_z^2 ; \qquad (16)$$

(the interaction of the spin magnetic moment with the applied field is neglected at this stage, and is discussed in Section 4).

The eigenvalues of the Hamiltonian (16) are straightforwardly obtained from the eigenvalues (10) taking into account the free motion parallel to the direction z of the magnetic field; we have

$$E_{nk_z} = \left(n + \frac{1}{2}\right)\hbar\omega_c + \frac{\hbar^2 k_z^2}{2m}, \qquad (17)$$

where each state of quantum numbers n and k_z has degeneracy $N_L(H) = (e/hc)\,H\,S$ (excluding spin degeneracy), and $S = L_x L_y$ is the section of the sample of volume $V = L_x L_y L_z$ perpendicularly to the z-axis. The energy levels E_{nk_z} are schematically indicated in Fig. 2; the magnetic field governs the spacing $\hbar\omega_c$ between the one-dimensional Landau sub-bands and their degeneracy $N_L(H)$.

For a three-dimensional electron gas the density-of-states for both spin directions in the absence of a magnetic field is obtained from Eq. (4c) (including a factor 2 due to spin degeneracy); we have

$$D(E, H{=}0) = A\,E^{1/2} \quad \text{for} \quad E > 0 \qquad (18a)$$

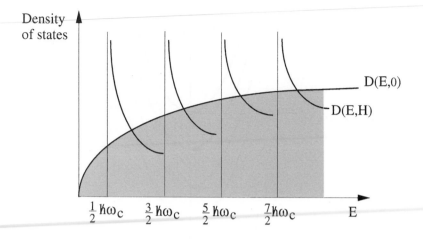

Fig. 3 Density-of-states of a three-dimensional electron gas in the presence and in the absence of a magnetic field.

where

$$A = V \, 4\pi \, (2m)^{3/2} h^{-3} \; . \tag{18b}$$

The density-of-states corresponding to the n-th sub-band of Eq. (17) can be obtained using Eq. (4a) for the one-dimensional character of the dispersion along z, and taking into account the orbital degeneracy $N_L(H) = (e/hc) \, H \, S$ of the Landau levels and the spin degeneracy; we have

$$D_n(E, H) = 2 \, L_z \frac{(2m)^{1/2}}{h} \frac{1}{\sqrt{E - (n + \frac{1}{2}) \hbar\omega_c}} N_L(H) \quad \text{for} \; \; E > \left(n + \frac{1}{2}\right) \hbar\omega_c \; .$$

We notice that $2 \, L_z \, (2m)^{1/2} \, h^{-1} \, N_L = (1/2) \, \hbar\omega_c \, A$. Summing up the contributions from every Landau sub-bands, we obtain for the total density-of-states (including spin degeneracy)

$$D(E, H) = \frac{1}{2}\hbar\omega_c \, A \sum_{n=0}^{\infty} \frac{1}{\sqrt{E - (n + \frac{1}{2}) \hbar\omega_c}} \, \Theta\left[E - \left(n + \frac{1}{2}\right) \hbar\omega_c\right] \; , \tag{19}$$

where $\Theta(x)$ is the step function, i.e. $\Theta(x) = 0$ for $x < 0$, $\Theta(x) = 1$ for $x > 0$. From Eq. (19), we see that the density of energy levels in the presence of a magnetic field becomes infinite just above the edge of any Landau level and behaves as $E^{-1/2}$ in the whereabouts (see Fig. 3). A further effect of the presence of the magnetic field is the increase ("blue shift") of $\hbar\omega_c/2$ of the energy spectrum threshold.

Until now we have considered the magnetic field effects on free electrons characterized by the dispersion relation $E(\mathbf{k}) = \hbar^2 k^2 / 2m$. In some materials (for instance simple metals, or a number of cubic semiconductors), the conduction band in the energy region of interest can be reasonably approximated with a parabolic energy band dispersion $E(\mathbf{k}) = E_c + \hbar^2 k^2/2m_c^*$, where m_c^* is the effective conduction mass and E_c is the bottom of the conduction band. In this case, upon application of a uniform magnetic field, the energy of the electrons becomes quantized in the form

$$E_{nk_z} = E_c + \left(n + \frac{1}{2}\right) \hbar\omega_c^* + \frac{\hbar^2 k_z^2}{2\,m_c^*} \, ,$$

where the cyclotron frequency is $\omega_c^* = e\,H/m_c^*c$.

The above considerations for the motion of electrons in a conduction band of effective mass m_c^* can be repeated for holes in semiconductors; in the case of a parabolic topmost valence band of effective mass m_v^*, the energy of holes becomes quantized in the form

$$E_{nk_z} = E_v - \left(n + \frac{1}{2}\right) \hbar\omega_v^* - \frac{\hbar^2 k_z^2}{2\,m_v^*}$$

$(m_v^* > 0; \omega_v^* = e\,H/m_v^*c)$. The energy levels around the fundamental gap for an isotropic two-band model semiconductor are shown in Fig 4. In small gap semiconductors, the effective masses of electrons and holes are in general rather small; this makes it easier to resolve in magneto-optic spectroscopy (or other experiments) several Landau levels, as their energy separation, at parity of magnetic field, increases for small effective masses. We wish also to mention that the effect of magnetic fields has been generalized to more complicated band structures, including the case of degenerate bands, anisotropic effects, and spin–orbit interaction.

In strong magnetic fields, the continuum of energy levels in the valence and conduction bands tend to pile up into strongly degenerate Landau sub-bands (see for instance Fig. 3 and Fig. 4); were it not for the energy dispersion in the magnetic field direction, a full discretization of the energy levels would occur. The bunching effect of the energy levels, produced by the magnetic field, is at the origin of the oscillatory behaviour of several physical properties; among them we mention the *oscillatory magneto-absorption effect* observed in semiconductors (pioneered by Zwerdling in late fifties), and the oscillatory behaviour of the magnetization observed in metals (this latter subject is discussed in next section).

As an example of magneto-absorption measurements in semiconductors, we report in Fig. 5a typical magneto-absorption curves concerning the direct transition at $\mathbf{k}=0$ of germanium from the valence bands (heavy holes) to the conduction band; with increasing values of the magnetic field, pronounced minima in the transmission curve can be detected, together with a shift to higher energies of the direct absorption edge. Fig. 5b shows the energy position of consecutive transmission minima (i.e. absorption maxima) as a function of the magnetic field; the figure shows a fan of straight lines, which extrapolate with high accuracy to the value of $0.803\,\text{eV}$ at $H = 0$; the so determined value of the direct gap of germanium exceeds by 0.14 eV the indirect gap of the material. An accurate analysis of the fine structure of the magneto-band effect provides invaluable information also on the light holes in germanium and on the detailed structure of the energy bands around the threshold of the direct transitions.

(a)   (b)

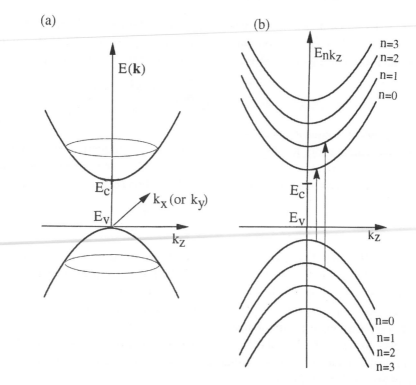

Fig. 4 (a) Schematic representation of an isotropic two-band model semiconductor. (b) Schematic representation of quantization of electron states in the conduction band and in the valence band in the presence of a uniform magnetic field in the z-direction. Arrows indicate examples of optical transitions between valence and conduction states with the same values of n and k_z.

3   Orbital magnetic susceptibility and de Haas–van Alphen effect

3.1   Orbital magnetic susceptibility of a two-dimensional electron gas

In this section we discuss the orbital magnetic susceptibility of the two-dimensional free-electron gas; the treatment is confined at $T=0$, because in this case the expression of the magnetic susceptibility, and the physical arguments are rather simple. Besides its intrinsic interest, the two-dimensional model also provides qualitative guidelines for an understanding of the three-dimensional electron gas; this is of particular value, in view of the fact that the full treatment of the magnetic susceptibility of the three-dimensional gas, given in Section 3.2 and in Appendix A, is rather demanding.

Consider a two-dimensional free-electron gas with N electrons (of both spin directions) distributed over a surface S (or, more precisely, distributed in an appropriate layer formed by a few atomic planes adjacent the surface S). In the presence of a magnetic field H perpendicular to the surface, the electronic energy levels are discretized and are given by Eq. (10); the energies of the Landau levels (measured from

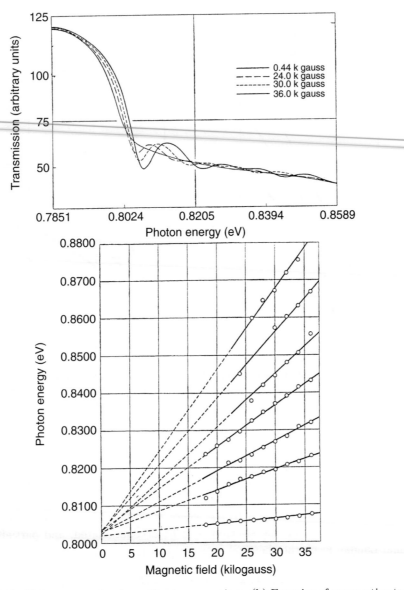

Fig. 5 (a) Oscillatory magneto-absorption in germanium. (b) Energies of consecutive transmission minima as a function of the magnetic field. The lowest curve corresponds to transitions between Landau levels with $n=0$; the successive curves correspond to transitions with higher quantum states of both conduction and valence bands [from S. Zwerdling and B. Lax, Phys. Rev. **106**, 51 (1957); copyright 1957 by the American Physical Society].

the conduction band edge) are

$$\varepsilon_n(H) = (n + \frac{1}{2})\,\hbar\omega_c = (n + \frac{1}{2})\,\frac{\hbar e}{m\,c}\,H \qquad (n = 0, 1, 2, \dots)\,.$$

In Fig. 6a we plot the energies $\varepsilon_n(H)$ $(n = 0, 1, 2, \dots)$ as a function of H, and notice the typical fan diagram of the family of curves.

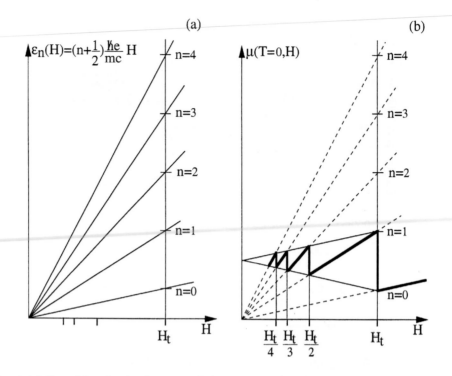

Fig. 6 (a) Fan of Landau levels versus H for several values of the quantum number n. (b) Chemical potential $\mu(T=0, H)$ versus H for a two-dimensional free-electron gas at zero temperature, showing discontinuities at $H = H_t/i$, with i integer number.

We can now study the behaviour of the chemical potential $\mu(T=0, H)$ versus H at zero temperature (for fermions, the chemical potential μ is also referred to as the Fermi energy E_F). For this purpose, we remember that *energy and degeneracy of the Landau levels are proportional to* H. The total degeneracy of the Landau levels (spin included) is $2N_L(H) = N\,H/H_t$, where H_t is the characteristic field defined in Eq. (13). If $H > H_t$ all the electrons can be accommodated, and partially fill, the ground Landau level; the chemical potential becomes

$$\mu(T=0, H) = \frac{1}{2}\hbar\omega_c = \frac{1}{2}\frac{\hbar e}{mc}H \quad \text{if } H > H_t \ .$$

In the case $H_t/2 < H < H_t$, all the electrons can be accommodated by filling completely the $n = 0$ Landau level and filling partially the $n = 1$ Landau level, and we have

$$\mu(T=0, H) = \frac{3}{2}\hbar\omega_c = \frac{3}{2}\frac{\hbar e}{mc}H \quad \text{if } \frac{H_t}{2} < H < \frac{H_t}{1} \ .$$

It can be immediately seen that whenever $H = H_t, H_t/2, H_t/3, \ldots$ the chemical potential $\mu(T=0, H)$ presents sudden jumps; between them, the chemical potential changes linearly with H, and has the behaviour shown in Fig. 6b.

We can now obtain the ground-state energy of the free-electron gas in magnetic fields at $T = 0$. If $H > H_t$ all particles are in the ground Landau level ε_0; the total energy of the system is

$$E_0(H) = \frac{1}{2}\hbar\omega_c N = \mu_B H N \, ,$$

where $\mu_B = e\hbar/2mc$ is the Bohr magneton. We notice that whenever $H = H_t, H_t/2, H_t/3, \ldots$ the Landau levels up to $n = 0, 1, 2, \ldots$ are *fully occupied* and all the other levels are *completely empty*. If H is such that

$$\frac{H_t}{i+1} < H < \frac{H_t}{i} \tag{20}$$

where i is an integer number, we have that all Landau levels ε_n with $n < i$ are *fully occupied*, the Landau level with $n \equiv i$ is *partially occupied*, and all Landau levels with $n > i$ are *completely empty*.

For H in the interval defined by Eq. (20), the ground energy of the two-dimensional electron gas becomes

$$E_0(H) = 2 N_L(H) \sum_{n=0}^{i-1}(n + \frac{1}{2}) \hbar\omega_c + [N - i\, 2\, N_L(H)] (i + \frac{1}{2}) \hbar\omega_c \, .$$

From the equality

$$\sum_{n=0}^{i-1}(n + \frac{1}{2}) = \frac{1}{2} i^2 \, ,$$

it follows

$$E_0(H) = N\, (i + \frac{1}{2}) \hbar\omega_c - 2\, N_L(H)\, (i^2 + i)\, \frac{1}{2}\, \hbar\omega_c \, .$$

Using the relations $\hbar\omega_c = 2\,\mu_B H$ and $2\, N_L(H) = N\, H/H_t$, we have

$$\boxed{E_0(H) = \mu_B H_t N \left[(2i+1)\frac{H}{H_t} - (i^2 + i) \left(\frac{H}{H_t}\right)^2 \right] \quad \text{for} \quad \frac{H_t}{i+1} < H < \frac{H_t}{i}} \, . \tag{21}$$

In Fig. 7a we plot $E_0(H)$ as a function of H; notice the oscillating behaviour of $E_0(H)$, with minima of equal value in correspondence to H_t/i, and maxima of values $\mu_B H_t N\, [(2i+1)^2/4i(i+1)]$ midway two successive minima. In Fig. 7b, we also plot $E_0(H)$ versus the variable $1/H$, to show pictorially that the oscillations of $E_0(H)$ have a constant period equal to $1/H_t$ in the inverse magnetic field; this period is called the de Haas–van Alphen period for the two-dimensional gas.

We can obtain the magnetization of the sample (magnetic moment per unit surface) at zero temperature from the definition

$$M(T{=}0, H) = -\frac{1}{S}\frac{\partial E_0(H)}{\partial H} \, .$$

We have

$$M(T{=}0, H) = \begin{cases} -\mu_B n_s & \text{if } H > H_t \\ -\mu_B n_s \left[(2i+1) - 2i(i+1)\dfrac{H}{H_t}\right] & \text{if } \dfrac{H_t}{i+1} < H < \dfrac{H_t}{i} \ (i = 1, 2, 3, ..) \end{cases}$$

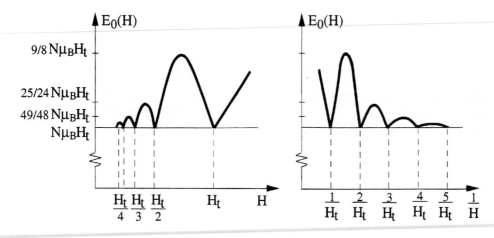

Fig. 7 Ground-state energy of a two-dimensional free-electron gas in a magnetic field as a function of H and $1/H$.

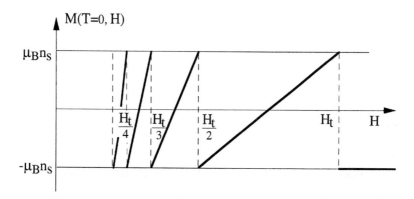

Fig. 8 Magnetic moment per unit surface at $T = 0$ of a two-dimensional free-electron gas in a magnetic field.

where $n_s = N/S$ is the number of electrons (of both spin directions) per unit surface; the behaviour of $M(T = 0, H)$ is shown in Fig. 8. It is seen that the magnetization has a sharp saw-tooth form with jumps from positive to negative values at the end of each de Haas–van Alphen period.

For the magnetic susceptibility of the sample, $\chi = (\partial M/\partial H)$, we have

$$\chi(T{=}0, H) = \begin{cases} 0 & \text{if } H > H_t \\ \dfrac{2\mu_B\, n_s}{H_t} i(i+1) & \text{if } \dfrac{H_t}{i+1} < H < \dfrac{H_t}{i} \;\; i = 1, 2, 3, \ldots \\ -\dfrac{2\mu_B\, n_s}{H_t} \delta\left(\dfrac{H}{H_t} - \dfrac{1}{i}\right) & \text{if } H = \dfrac{H_t}{i} \;\; i = 1, 2, 3, \ldots \end{cases} \tag{22}$$

The corresponding curve is shown in Fig. 9; we can see that the magnetic susceptibility has a paramagnetic contribution with a staircase aspect and a diamagnetic contribution with spikes of equal intensity for $H = H_t/i$.

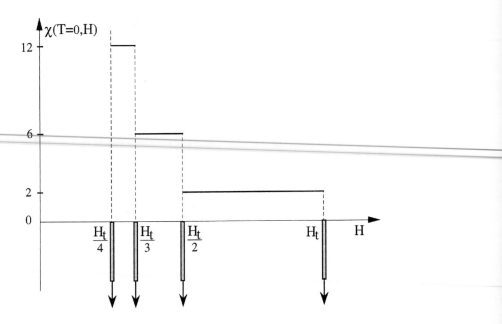

Fig. 9 Magnetic susceptibility (in units of $2\mu_B \, n_s/H_t$) of the two-dimensional free-electron gas at zero temperature.

The origin of the infinitely sharp negative spikes is related to the saw-tooth form of the magnetization. It should be noticed that if the δ-like spikes *in each interval* are replaced by a finite constant (preserving the area), the paramagnetic and diamagnetic contributions in Eq. (22) *exactly give zero net result* (and this is the classical result predicted by the Bohr–van Leeuwen theorem). In summary, the orbital susceptibility of the two-dimensional electron gas presents sudden negative jumps at the values H_t/i (i integer) and constant positive values between them. When we pass to three-dimensional structures and consider the dispersion along the k_z axis, we expect a smoothing of the discontinuities and an oscillatory behaviour of the magnetic susceptibility.

3.2   Orbital magnetic susceptibility of a three-dimensional electron gas

We begin the study of the magnetic properties of a three-dimensional free-electron gas with some qualitative considerations, and later we pass to a quantitative analysis. For a three-dimensional gas, considerations similar to the two-dimensional case hold for *any slice* $k_z = $const, since the k_z quantum number as well as the energy contribution $\hbar^2 k_z^2/2m$ are unaffected by the magnetic field in the z-direction. The qualitative difference between the three- and two-dimensional electron gas essentially arises because we have to take into account *simultaneously* all the different sections $k_z = $const. Thus, while in the two-dimensional electron gas sudden jumps in the chemical potential occur at the values $H = H_t/i$ (see Fig. 6b), in three dimensions we expect that the oscillations of the chemical potential are negligibly small; for $\hbar\omega_c \ll \mu$ we can safely assume that the chemical potential μ is constant and independent of H.

A further remarkable difference between the two- and three-dimensional case is that, in the latter situation, there is an overall increase in energy with the magnetic field and a consequent diamagnetism, called Landau diamagnetism. In the two-dimensional electron gas we have seen that whenever Landau levels are *either fully occupied or fully empty*, there is a balance between the states which have increased their energy and those which have decreased their energy, and the total energy is unaffected at $H = H_t/i$ (see Fig. 7). In the three-dimensional case this argument is still valid for any thin slice k_z=const; but since we have to consider so many k_z slices, the net effect is an increase in the energy. We can estimate this increase in energy through the relationship

$$\Delta E(H) \approx D(E_F)\frac{1}{2}\hbar\omega_c\,\frac{1}{2}\hbar\omega_c \,, \tag{23}$$

where $D(E_F)$ is the density-of-states at the Fermi energy, and $D(E_F)\hbar\omega_c/2$ is an estimate of the number of states increasing their energy by the amount $\hbar\omega_c/2$ [notice that the naive estimate (23) is not far from the exact value $D(E_F)\hbar^2\omega_c^2/24$, given in Eq. (26) below]. From Eq. (23), we see that the energy change $\Delta E(H)$ is positive and proportional to H^2; the magnetic susceptibility $\chi = -(1/V)\,\partial^2\Delta E(H)/\partial H^2$ explains qualitatively the origin of the Landau diamagnetism.

Besides considering the overall increase in energy, we can consider in detail population and depopulation effects of the Landau levels crossing the Fermi energy; as the applied magnetic field is varied, we expect oscillations in the ground-state energy $E_0(H)$ and, as a consequence, in the magnetic moment. With reference to the electrons in a thin slice around k_z=const=0, we shall expect that the period of these oscillations is such that

$$(j+\frac{1}{2})\,\hbar\omega_c = E_F$$

(with j integer) or equivalently

$$\frac{1}{H} = (j+\frac{1}{2})\frac{\hbar\,e}{m\,c\,E_F} \tag{24a}$$

where E_F is the Fermi level. Thus we expect a contribution to the magnetic susceptibility, which is *periodic in inverse magnetic field* $1/H$ with period $\hbar\,e/m\,c\,E_F$, called the de Haas–van Alphen period. The experimental determination of the de Haas-van Alphen oscillations requires the sharpness of the Fermi–Dirac distribution function, i.e. $k_BT \ll \hbar\omega_c$; it also requires the sharpness of Landau levels, i.e. $\omega_c\tau \gg 1$ (the relaxation time τ represents the mean time between successive collisions suffered by the electrons); in general, the de Haas–van Alphen effect can be detected for high magnetic fields (of the order of Tesla), in high pure samples, at a few Kelvin degrees.

It is possible to elaborate Eq. (24a), which holds for the free-electron gas and for its spherical Fermi surface, into a form suitable to more complicated Fermi surfaces.

If we write $E_F = \hbar^2 k_F^2/2m$ in Eq. (24a), we obtain for the period of oscillations

$$\Delta\left(\frac{1}{H}\right) = \frac{e}{\hbar c}\frac{2\pi}{A_{\text{extremal}}}, \tag{24b}$$

where $A_{\text{extremal}} = \pi k_F^2$ is the extremal area determined by the intersection of the Fermi surface with planes perpendicular to the magnetic field. Eq. (24b) is valid also in real metals with more complicated Fermi surfaces and allows to determine extremal cross-sectional areas for any given direction of the external magnetic field.

The quantum mechanical treatment of the orbital magnetic susceptibility for the three-dimensional electron gas is given in Appendix A. According to Eq. (A20), the free energy of the electron gas in a uniform magnetic field H at temperature T is given by

$$F(T,H) = F(T,0) + \frac{1}{24}\hbar^2\omega_c^2\, D(E_F) + \frac{1}{4\pi^2}\hbar^2\omega_c^2\, D(E_F)\left(\frac{\hbar\omega_c}{2E_F}\right)^{1/2}$$
$$\cdot\sum_{s=1}^{\infty}\frac{(-1)^s}{s^{5/2}}\frac{2\pi^2 s\, k_B T}{\hbar\omega_c}\frac{1}{\sinh(2\pi^2 s\, k_B T/\hbar\omega_c)}\cos\left(2\pi s\frac{F_0}{H} - \frac{\pi}{4}\right) \tag{25}$$

where $F(T,0)$ is the free energy in the absence of the magnetic field, $D(E_F)$ is the density-of-states (for both spin directions) at the chemical potential and the "frequency" F_0 of the oscillations is defined as $F_0 = (E_F/\hbar\omega_c)H = E_F/2\mu_B$; notice that F_0 has the dimension of a magnetic field. The presence of oscillatory terms in the right-hand side of Eq. (25) explains the oscillatory behaviour of the magnetic moment $M = -(1/V)\,\partial F/\partial H$ as a function of the magnetic field, in *the low temperature limit* $k_B T \ll \hbar\omega_c$; when the condition $k_B T \ll \hbar\omega_c$ is not satisfied, the oscillations are damped through a factor of the type $\exp(-2\pi^2 k_B T/\hbar\omega_c)$.

In the *high temperature limit* ($\hbar\omega_c \ll k_B T$) we can neglect the oscillatory terms in Eq. (25), and we have

$$F(T,H) = F(T,0) + \frac{1}{24}\hbar^2\omega_c^2\, D(E_F). \tag{26}$$

The free energy (26) increases as H^2; the magnetic susceptibility $\chi = -(1/V)\,\partial^2 F/\partial H^2$ gives rise to the Landau diamagnetism

$$\boxed{\chi_L = -\frac{1}{12}\left(\frac{e\hbar}{mc}\right)^{1/2}\frac{D(E_F)}{V}.} \tag{27}$$

For the free-electron gas, we can write

$$\chi_L = -\frac{1}{12}\frac{e^2\hbar^2}{m^2c^2}\frac{3}{2}\frac{n}{E_F} = -\frac{1}{12\pi^2}\frac{e^2}{mc^2}k_F \tag{28}$$

(use has been made of the standard relations $E_F = \hbar^2 k_F^2/2m$ and $k_F^3 = 3\pi^2 n$). With $e^2/mc^2 = r_0 = 2.82 \times 10^{-13}$ cm and $k_F \approx (1/10^{-8})\,\text{cm}^{-1}$, we have that the order of magnitude of the susceptiblity is $\chi_L \approx -10^{-6}$.

As an example of experimental de Haas–van Alphen oscillations, we report in Fig. 10

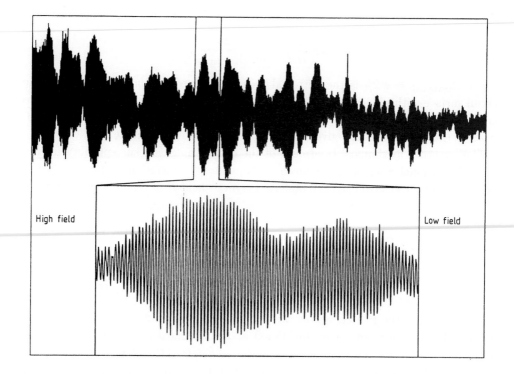

Fig. 10 Experimental de Haas–van Alphen magnetization oscillations for a sample containing randomly oriented particles of lithium dispersed in paraffin wax. The full trace covers the field range 12.9 - 7.9 T at a temperature of 20 mK. The portion of the trace between 11.1 and 10.8 T is shown expanded in the inset [from M. B. Hunt, P. H. P. Reinders and M. Springford, J. Phys. Condens. Matter **1**, 6589 (1989)].

the behaviour of the magnetization of lithium grains as a function of the external field H. In lithium metal there is one conduction electron per atom and the Brillouin zone is large enough to contain the free-electron Fermi sphere; the Fermi surface of bcc lithium looks like a distorted spherical surface, pushed (slightly) outwards in the < 110 > directions and inwards in the < 100 > directions. In the free-electron model the Fermi energy of lithium is E_F=4.78 eV, and the quantity F_0 takes the value F_0=41250 Tesla (taking μ_B=0.05788 meV/Tesla). In Fig. 10 about 2000 de Haas–van Alphen oscillations are observed over the field range 12.9 - 7.9 Tesla; the oscillations show a complex beat structure, which is related to the anisotropy of the extremal cross-sectional areas of the randomly oriented particles.

4   Spin paramagnetism of a free-electron gas

Until now, in the discussion of the magnetic field effects on the free-electron gas, we have neglected the interaction of the spin magnetic moment with the applied field;

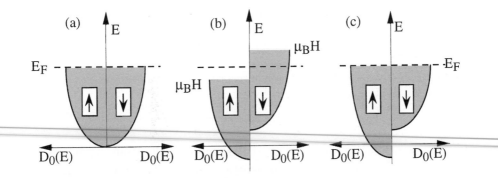

Fig. 11 Schematic representation of the origin of the Pauli spin paramagnetism of a free-electron gas; arrows indicate the direction up or down of the electron spin magnetic moment; $D_0(E)$ denotes the density-of-states for one spin direction.

when this interaction is taken into account, the Hamiltonian of an electron in a uniform magnetic field of vector potential $\mathbf{A}(\mathbf{r})$, becomes

$$H_0 = \frac{1}{2m}\left(\mathbf{p} + \frac{e}{c}\mathbf{A}\right)^2 + g_0\mu_B\,\mathbf{s}\cdot\mathbf{H}\;, \tag{29}$$

where $\mathbf{s}$ is the electron spin operator (in units $\hbar$), and $g_0 \approx 2$ is the giromagnetic factor for the electron spin. The eigenvalues of the Hamiltonian (29) are

$$E_{n\,k_z} = \left(n + \frac{1}{2}\right)\hbar\omega_c + \frac{\hbar^2 k_z^2}{2m} \pm \mu_B H;$$

they are obtained from Eq. (17) by adding the terms $\pm\mu_B H$ for spin up and spin down electrons.

The magnetic susceptibility of the free-electron gas including spin contribution, could be worked out with straightforward generalization of the calculations presented in Appendix A; in a more simple way, the effect of the spin contribution to the magnetic susceptibility can be better illustrated with the following argument.

Consider a metal, with N free electrons (of both spin directions) in the volume V; the standard occupation of states for both spin directions in the absence of magnetic field is indicated in Fig. 11a. In Fig. 11b we indicate the situation occurring in the presence of a uniform field H (while keeping the electrons in their original states); there is an energy shift by $-\mu_B H$ (or $+\mu_B H$) for electrons with magnetic moment parallel (or antiparallel) to the applied magnetic field. In Fig. 11c equilibrium is recovered, and an excess of electrons with parallel magnetic moment is established.

The number of electrons ΔN that flip their magnetic moment from $-\mu_B$ to μ_B are those contained in the energy interval $\mu_B H$ around the Fermi level; we have

$$\Delta N = \frac{1}{2}\,D(E_F)\,\mu_B H,$$

where $D(E_F)/2 = D_0(E_F)$ is the density-of-states at the Fermi level for one spin direction. The magnetic moment M per unit volume is then

$$M = \frac{1}{V} \Delta N \, 2\mu_B = \mu_B^2 \, H \, \frac{D(E_F)}{V} \; ;$$

the ratio M/H gives for the Pauli spin susceptibility the value

$$\boxed{\chi_P = \mu_B^2 \, \frac{D(E_F)}{V}} \; . \tag{30}$$

In the case of the free-electron gas, the net effect of the Pauli paramagnetic susceptibility (30) and the Landau diamagnetic susceptibility (27) is a paramagnetic behaviour; in fact $\chi_L = -(1/3)\chi_P$. In actual materials, the "effective mass" for orbital motion, as well as the "effective giromagnetic factor", can be significantly different from the free-electron values of Eq. (29); the same occurs for the orbital and spin susceptibility, and this explains why some metals may have a net diamagnetic behaviour. For a more quantitative account, correlation and exchange effects among electrons should be considered, because they may significantly influence the magnetic susceptibility.

5   Magnetoresistivity and classical Hall effect

General considerations and phenomenological aspects

Transport effects in crystals in the presence of electric fields and temperature gradients have been considered in Chapter XI. In this section, we study some aspects of transport phenomena due to the simultaneous presence of electric and magnetic fields; the possible presence of thermal gradients adds further variety to the phenomenology, but here we confine our attention to samples at uniform temperature.

The study of magnetic field effects on the transport properties of metals and semiconductors has become a well established and invaluable tool for the investigation of mobile carriers in crystals. In particular the Hall measurements, aimed at the determination of carrier concentration and charge sign, are routinely used for the characterization of materials. Also magnetoresistivity measurements, which determine the resistivity of materials in the presence of magnetic fields, offer a wide range of effects. In metals with closed Fermi surfaces (such as alkali metals), the magnetoresistivity does saturate for any crystal orientation (i.e. it approaches a constant value for sufficiently high magnetic fields, irrespective of orientation); the same occurs for n-type and p-type semiconductors. In metals with equal number of electrons and holes (such as Bi, Sb and others), the magnetoresistivity does not saturate for any crystal orientation and keeps on increasing as the magnetic field increases; the same occurs for semiconductors with equal numbers of electrons and holes. We also mention that in metals with open Fermi surfaces (such as Cu, Ag, Au and others), the magnetoresistivity saturates for most of the crystal orientations but does not saturate for others. Finally, and most importantly, for two-dimensional systems the magnetoresistance is quantized and the spectacular quantum Hall effect occurs. In this section we consider some aspects of the traditional magnetoresistivity and Hall phenomenology in three-dimensional crystals, while in the next section we consider the quantized Hall effect in two-dimensional systems.

In isotropic media the application of a (small) electric field drives a current density parallel and proportional to it, and the linear relationship holds

$$\mathbf{J} = \sigma\,\mathbf{E} \tag{31a}$$

where the conductivity σ is a scalar quantity. In the presence of a magnetic field, carriers are deflected and in general the current density is no more parallel to the electric field; the conductivity becomes a tensor even for an isotropic material. Relation (31a) has to be replaced by the more general expression

$$J_i = \sum_j \sigma_{ij}(H)\,E_j \ , \tag{31b}$$

where $\sigma_{ij}(H)\,(i,j = x,y,z)$ are the components of the *magnetoconductivity tensor* $\sigma(H)$. Similar considerations can be done for the resistivity of an isotropic medium in the presence of a magnetic field; the relationship between electric field and current density becomes

$$E_i = \sum_j \rho_{ij}(H)\,J_j \ , \tag{31c}$$

where $\rho_{ij}(H)\,(i,j = x,y,z)$ are the components of the *magnetoresistivity tensor* $\rho(H)$. The magnetoconductivity tensor and magnetoresistivity tensor are the inverse of each other, and it holds

$$\rho_{ij}(H) = \left(\frac{1}{\sigma(H)}\right)_{ij} \ . \tag{31d}$$

The transport parameters $\rho_{ij}(H)$ are often determined experimentally *using the standard geometry* in which a magnetic field $\mathbf{H}$ is applied orthogonally to a long and thin current carrying conductor and the current flows along the x-direction (see Fig. 12); the x- and y-directions are often referred to as "longitudinal" and "transverse" directions, respectively.

In the standard geometry, in which transport is in the xy plane and furthermore $J_y \equiv 0$ (in stationary conditions), the density current $\mathbf{J}$ and the electric field $\mathbf{E}$ are related by

$$\begin{pmatrix} E_x \\ E_y \end{pmatrix} = \begin{pmatrix} \rho_{xx}(H) & \rho_{xy}(H) \\ \rho_{yx}(H) & \rho_{yy}(H) \end{pmatrix} \begin{pmatrix} J_x \\ J_y \equiv 0 \end{pmatrix} \ .$$

The above matrix equation can be written explicitly in the form

$$E_x = \rho_{xx}(H)\,J_x \tag{32a}$$

$$E_y = \rho_{yx}(H)\,J_x \ . \tag{32b}$$

Thus the diagonal element $\rho_{xx}(H)$ of the magnetoresistivity tensor is measured by

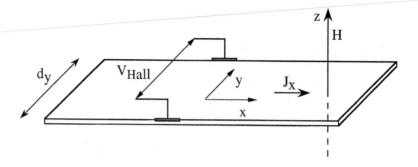

Fig. 12 Standard geometry for Hall effect and magnetoresistivity measurements. V_{Hall} is the Hall potential, and J_x is the current density in the flow direction.

the ratio between the longitudinal electric field E_x and the current density J_x in the x-direction. The off-diagonal component $\rho_{yx}(H)$ is measured by the ratio between the transverse electric field E_y and the current density J_x. It can also be inferred by inspection that $\rho_{xy}(H) = -\rho_{yx}(H)$ (as shown also in the models discussed below).

The transverse electric field E_y, also called *Hall field*, is produced by the space charges accumulated (in stationary conditions) at the borders of the conductor because of the deflection due to the magnetic field. One or the other of the off-diagonal magnetoresistivity components (i.e. $\rho_{yx}(H)$ or its opposite $\rho_{xy}(H)$) are also known as *Hall resistivity*. Most often it is convenient to report the *Hall coefficient*, defined as

$$R_{\text{Hall}}(H) = \frac{1}{H}\,\rho_{yx}(H) = \frac{1}{H}\frac{E_y}{J_x}\;. \tag{33}$$

Notice that the *Hall potential* is $V_{\text{Hall}} = E_y\,d_y$, where d_y is the transverse dimension of the sample; in the absence of magnetic field, both V_{Hall} and ρ_{yx} vanish.

We pass now to study the Hall effect and the magnetoresistivity in a few simple models. We consider first the case of a single type of carriers in a spherical band model, with a unique relaxation time. Next, we consider the case in which holes and electrons are present, both with isotropic masses. The models described below, provide an orientative picture of the transport phenomena in the presence of magnetic fields in somewhat idealized situations. We wish to remark that the description of magneto-transport effects in realistic materials is rather demanding and requires a proper account of several features (such as energy dependence of the relaxation time, deviations from spherical bands, detailed shape of the Fermi surfaces, accurate analysis of the Boltzmann transport equations). We cannot enter in these and other aspects, and we refer for more elaborated models and discussions to the classic book by R. A. Smith "Semiconductors" (Cambridge University Press, 1978)].

Model 1 Magnetoresistivity and Hall effect in an isotropic one-band model

We consider here the magnetoresistivity and the Hall effect in the case of a single type of carriers (electrons or holes) in a spherical energy band. For simplicity we use a modelistic approach to the motion of electrons (or holes); the treatment with the more rigorous Boltzmann equation would give in the present case the same results.

The classical equation of motion of an electron, in a dissipative medium, in the presence of an electric field $\mathbf{E}$ and a magnetic field $\mathbf{H}$ is

$$m^* \frac{d\mathbf{v}}{dt} = (-e)\,\mathbf{E} + \frac{(-e)}{c}\mathbf{v} \times \mathbf{H} - \frac{m^*}{\tau}\mathbf{v}\,, \tag{34a}$$

where m^* is the effective mass of the electron, and a damping term with constant relaxation time τ has been included. In stationary conditions $d\mathbf{v}/dt = 0$, and Eq. (34a) becomes

$$\mathbf{v} = -\frac{e\tau}{m^*}\mathbf{E} - \frac{e\tau}{m^*c}\mathbf{v} \times \mathbf{H}\,. \tag{34b}$$

We specify the above equation in the geometry of Fig. 12, with the electric field in the xy plane and the magnetic field $\mathbf{H} = (0, 0, H)$ in z-direction; Eq. (34b) becomes

$$\begin{cases} v_x = -\dfrac{e\tau}{m^*}E_x - \omega_c\,\tau\,v_y \\[2mm] v_y = -\dfrac{e\tau}{m^*}E_y + \omega_c\,\tau\,v_x \end{cases}, \tag{35a}$$

where $\omega_c = eH/m^*c$ is the cyclotron frequency. From Eqs. (35a) we have

$$\begin{cases} v_x = -\dfrac{e\tau}{m^*}\dfrac{1}{1 + \omega_c^2\tau^2}\,(E_x - \omega_c\,\tau\,E_y) \\[2mm] v_y = -\dfrac{e\tau}{m^*}\dfrac{1}{1 + \omega_c^2\tau^2}\,(\omega_c\,\tau\,E_x + E_y) \end{cases}. \tag{35b}$$

Thus the current density $\mathbf{J} = n(-e)\,\mathbf{v}$ (where n is the electron density) is related to the electric field via the magnetoconductivity tensor $\sigma(H)$ given by

$$\sigma(H) = \frac{n\,e^2\,\tau}{m^*}\frac{1}{1 + \omega_c^2\tau^2}\begin{pmatrix} 1 & -\omega_c\,\tau \\ \omega_c\,\tau & 1 \end{pmatrix}. \tag{36}$$

Notice that $\sigma_{xy}(H) = -\sigma_{yx}(H)$, which is a particular case of the general Onsager relations.

Eq. (36) provides the magnetoconductivity for a *spherical band with constant (i.e. energy independent) relaxation time*. From inversion of the matrix (36) we obtain the magnetoresistivity tensor

$$\rho(H) = \frac{m^*}{n\,e^2\,\tau}\begin{pmatrix} 1 & +\omega_c\,\tau \\ -\omega_c\,\tau & 1 \end{pmatrix}. \tag{37}$$

From Eq. (37), we see that the diagonal (or parallel) magnetoresistivity $\rho_{xx}(H)$, the Hall magnetoresistivity $\rho_{yx}(H)$, and the Hall coefficient have the expressions

$$\rho_{xx}(H) = \frac{m^*}{n\,e^2\,\tau}\,, \quad \rho_{yx}(H) = -\frac{H}{n\,e\,c}\,, \quad R_{\text{Hall}}(H) = -\frac{1}{n\,e\,c}\,. \tag{38}$$

Thus in the spherical one-band model with a single relaxation time, the diagonal magnetoresistivity turns out to be independent from H, and we have $\rho_{xx}(H) = \rho_{xx}(0) = m^*/(n\,e^2\,\tau)$. Even more important, the Hall coefficient is independent both from the effective mass and from the relaxation time; it depends only on the carrier concentration and charge sign. Also notice that in the case of positive holes, the off-diagonal matrix elements in Eq. (36) and Eq. (37) change sign.

The results summarized in Eqs. (38), obtained in the rather idealized one-band model, are to be taken only as orientative, and cannot be used as they stand for quantitative descriptions of realistic conductors. It is important in fact to notice that a proper account of the energy dependence of the relaxation time, or of the anisotropy of the energy bands, modify the results of Eqs. (38); in particular, a dependence of $\rho_{xx}(H)$ on H is actually always observed in experiments. For these reasons, we consider the slightly more sophisticated two-band model, representing two groups of carriers.

Model 2 Magnetoresistivity and Hall effect in an isotropic two-band model

Interesting new features appear in the study of magnetoresistivity and Hall effect within the two-band model. For simplicity we suppose that the two bands are spherical, with effective masses m_1 and m_2; we also assume that the relaxation times τ_1 and τ_2 are constant for each group of carriers. The two-band model is useful to provide insight of transport phenomena in crystals with two groups of carriers of the same type (but different masses or relaxation times) or for mixed type carriers; we consider specifically this last situation.

Consider a material with n electrons (per unit volume) of mass m_1 and relaxation time τ_1, and p holes of mass m_2 and relaxation time τ_2. The magnetoconductivity is just the sum of the contributions from each group of carriers. Using Eq. (36) for electrons, and the appropriate modified form for positive holes, we obtain

$$\sigma(H) = \begin{pmatrix} A_1 & -B_1 \\ B_1 & A_1 \end{pmatrix} + \begin{pmatrix} A_2 & B_2 \\ -B_2 & A_2 \end{pmatrix} = \begin{pmatrix} A_1 + A_2 & -B_1 + B_2 \\ B_1 - B_2 & A_1 + A_2 \end{pmatrix} \tag{39}$$

where

$$A_1 = \frac{\sigma_1}{1 + \omega_1^2 \tau_1^2} \ , \quad B_1 = \frac{\sigma_1 \omega_1 \tau_1}{1 + \omega_1^2 \tau_1^2} \ , \quad \sigma_1 = \frac{n e^2 \tau_1}{m_1} \ , \tag{40a}$$

$$A_2 = \frac{\sigma_2}{1 + \omega_2^2 \tau_2^2} \ , \quad B_2 = \frac{\sigma_2 \omega_2 \tau_2}{1 + \omega_2^2 \tau_2^2} \ , \quad \sigma_2 = \frac{p e^2 \tau_2}{m_2} \ . \tag{40b}$$

The magnetoresistivity tensor is obtained by inverting the magnetoconductivity tensor (39); we have

$$\rho(H) = \frac{1}{(A_1 + A_2)^2 + (B_1 - B_2)^2} \begin{pmatrix} A_1 + A_2 & B_1 - B_2 \\ -B_1 + B_2 & A_1 + A_2 \end{pmatrix} \ . \tag{41}$$

We consider first the parallel component $\rho_{xx}(H)$ of the magnetoresistivity tensor of Eq. (41); using expressions (40) we obtain

$$\rho_{xx}(H) = \frac{\sigma_1 + \sigma_2 + \sigma_1 \omega_2^2 \tau_2^2 + \sigma_2 \omega_1^2 \tau_1^2}{(\sigma_1 + \sigma_2)^2 + (\sigma_1 \omega_2 \tau_2 - \sigma_2 \omega_1 \tau_1)^2} \ . \tag{42}$$

It is straightforward to verify that $\rho_{xx}(H) > \rho_{xx}(0)$; thus $\rho_{xx}(H) - \rho_{xx}(0)$ is an essentially *positive quantity for any value of the magnetic field.*

For very high magnetic fields (such that $\omega_1 \tau_1 \gg 1$ and $\omega_2 \tau_2 \gg 1$), Eq. (42) shows that in general $\rho_{xx}(H \to \infty)$ is finite, and thus there is saturation of the magnetoresistivity; the only remarkable exception occurs when

$$\sigma_1 \omega_2 \tau_2 \equiv \sigma_2 \omega_1 \tau_1$$

which is equivalent to

$$n = p \ .$$

When the two groups of carriers of opposite type (electrons and holes) have the same concentration, then $\rho_{xx}(H \to \infty) = \infty$ and no saturation occurs.

We consider now the off-diagonal magnetoresistivity transport parameter $\rho_{yx}(H)$; from Eq. (41) and Eqs. (40), we have

$$\rho_{yx}(H) = \frac{-\sigma_1 \omega_1 \tau_1 \left(1 + \omega_2^2 \tau_2^2\right) + \sigma_2 \omega_2 \tau_2 \left(1 + \omega_1^2 \tau_1^2\right)}{(\sigma_1 + \sigma_2)^2 + (\sigma_1 \omega_2 \tau_2 - \sigma_2 \omega_1 \tau_1)^2} \ . \tag{43a}$$

For high magnetic fields (i.e. $\omega_i \tau_i \gg 1$), the above expression simplifies in the form

$$\rho_{yx}(H \to \infty) \approx -\frac{\omega_1 \tau_1 \, \omega_2 \tau_2}{\sigma_1 \omega_2 \tau_2 - \sigma_2 \omega_1 \tau_1} = -\frac{H}{(n - p) \, e \, c} \ . \tag{43b}$$

The Hall parameter becomes

$$R_{\text{Hall}}(H \to \infty) = -\frac{1}{(n - p) \, e \, c} \ , \tag{43c}$$

a result which is independent on relaxation time and is *governed by the difference of the density of electrons and holes.* It is easy to understand qualitatively the limiting result (43c); for high values of H, the deflection of carriers produced by the magnetic field increases. Since electrons and holes have opposite charges and move in opposite directions, they are deflected on *the same side*; thus the effective number of carriers entering in Eq. (43c) is given by the difference of the electron and hole concentrations.

6 The quantum Hall effect

In the previous section we have considered some effects of magnetic fields on transport properties in three-dimensional materials. In this section we present some aspects of transport measurements under strong magnetic fields for the two-dimensional electron gas; the observed effects have opened a new area of investigation and brought major advances in the comprehension of the two-dimensional systems.

The transport properties of *two-dimensional* conductors, when observed in high purity samples, at very low temperatures and strong magnetic fields, show striking departure from the classical behaviour; in particular the Hall resistance $\rho_{xy}(H)$ versus H exhibits flat plateaus, from which the universal constant h/e^2 can be obtained. Very

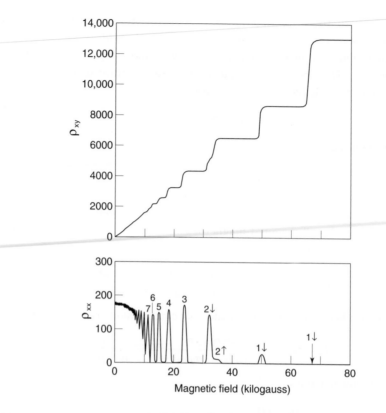

Fig. 13 Hall resistance $\rho_{xy}(H)$ and parallel component of the resistance $\rho_{xx}(H)$ (in ohm) of a two-dimensional electron gas at the GaAs-Al$_x$Ga$_{1-x}$As interface. The quantum Hall effect is already evident at about 10 kgauss. The numbers and the arrows above $\rho_{xx}(H)$ maxima refer to the Landau quantum number and the spin polarization of the levels [from M. A. Paalanen, D. C. Tsui and A. C. Gossard, Phys. Rev. B**25**, 5566 (1982); copyright 1982 by the American Physical Society].

accurate measurements performed on two-dimensional electron gases have given the value

$$\frac{h}{e^2} = 25812.806\,\Omega$$

with an error less than one part per million, even for samples of different origin [for metrological aspects see E. R. Cohen and B. N. Taylor "The 1986 Adjustment of the Fundamental Physical Constants" Rev. Mod. Phys. **59**, 1121 (1987)].

The quantum Hall effect was first reported for the two-dimensional electron gas in the inversion layer of a silicon MOSFET at $T = 1.5\,\mathrm{K}$ and $H = 18\,\mathrm{Tesla}$ by K. von Klitzing, G. Dorda and M. Pepper, Phys. Rev. Lett. **45**, 494 (1980). In the device, the density of surface electrons can be controlled and changed by varying the gate voltage; the Hall resistance shows fixed values $(1/i)(h/e^2)$ (with i integer number) at experimentally well-defined surface carriers concentrations, while the parallel component of the electric resistance is vanishingly small.

Degenerate two-dimensional electron systems can be also realized at the interface between GaAs and (n-doped) $Al_xGa_{1-x}As$; nearly ideal semiconductor heterostructures are prepared by molecular beam epitaxy techniques. As already discussed in Section XIV-5, the electrons at the interface are confined by the potential well, originated from the conduction band offset; the motion perpendicular to the interface is quantized, and, even when all the carriers are trapped in the lowest ground state, the motion parallel to the interface is still free-like. When a strong magnetic field H is applied perpendicularly to the two-dimensional electron gas, the quantum Hall effect is observed as a sequence of flat plateaus in the Hall resistance $\rho_{xy}(H)$ plotted as a function of H; in the same regions the parallel component of the electric resistance $\rho_{xx}(H)$ becomes vanishingly small, as shown in Fig. 13. The effective mass of electrons in GaAs ($m^* \approx 0.07\,m_0$) is approximately three times lighter than that of electrons in silicon inversion layers, and thus the quantum Hall regime in $GaAs$-$Al_xGa_{1-x}As$ heterostructures is reached at smaller values of magnetic fields (see Fig. 13).

For a qualitative understanding of some features of the quantum Hall effect, it is convenient to examine first the Hall effect within a classical modelistic point of view; later we introduce heuristically the effect of quantization.

Consider a two-dimensional electron system (in the xy plane) in the presence of a strong perpendicular magnetic field. Let n_s denote the *surface carrier density* (more precisely, n_s is the number of carriers, per unit surface, in the conductive layer of planes adjacent the surface). Similarly, let $\mathbf{J}^{(\text{surf})} = n_s(-e)\mathbf{v}$ denote the *surface current density* (notice that in the previous section $\mathbf{J}$ and n were bulk quantities). In the presence of a magnetic field, carriers are deflected and in general $\mathbf{J}^{(\text{surf})}$ is no more parallel to the electric field (in the xy plane); rather we have

$$J_i^{(\text{surf})} = \sum_j \sigma_{ij}(H)\,E_j \;, \tag{44a}$$

where $\sigma_{ij}(H)\,(i,j = x,y)$ are the components of the conductance tensor $\sigma(H)$. Similarly, the relation between electric field and current density is

$$E_i = \sum_j \rho_{ij}(H)\,J_j^{(\text{surf})} \;, \tag{44b}$$

where $\rho_{ij}(H)\,(i,j = x,y)$ are the components of the resistance tensor $\rho(H) = 1/\sigma(H)$.

In the standard geometry of Fig. 12, similarly to Eq. (32b), we have

$$\rho_{yx}(H) = \frac{E_y}{J_x^{(\text{surf})}} = \frac{d_y\,E_y}{d_y\,J_x^{(\text{surf})}} = \frac{V_{\text{Hall}}}{I_x} \;.$$

Thus we see that for the two-dimensional system the Hall resistance $\rho_{yx}(H)$ is given by the Hall voltage divided by the current flowing in the sample; $\rho_{yx}(H)$ can thus be measured with high accuracy, since the precise sample dimensions are not relevant and tend to drop out.

The magnetoconductance of a two-dimensional electron system in a modelistic approach with constant relaxation time, is given by Eq. (36), once n is replaced by n_s;

we have

$$\sigma(H) = \frac{n_s e^2 \tau}{m^*} \frac{1}{1 + \omega_c^2 \tau^2} \begin{pmatrix} 1 & -\omega_c \tau \\ \omega_c \tau & 1 \end{pmatrix}. \tag{45}$$

In the *ideal collisionless regime*, taking the limit $\tau \to \infty$ of expression (45), we obtain

$$\sigma(H) = \frac{n_s e c}{H} \begin{pmatrix} 0 & -1 \\ 1 & 0 \end{pmatrix}. \tag{46}$$

From Eq. (46), valid in the absence of scattering events, we see that the diagonal conductance $\sigma_{xx}(H)$ is zero and the Hall conductance $\sigma_{xy}(H) = -n_s ec/H$ is inversely proportional to H. From Eq. (44a) and Eq. (46), in the standard geometry of Fig. 12, we have $E_x = 0$ and $J_x^{(\text{surf})} = -(n_s e c/H) E_y$. Thus a current flows in the x-direction with null electric field in the x-direction; the current $J_x^{(\text{surf})}$ is driven by the transverse electric field E_y, acting in conjunction with the perpendicular magnetic field H [this is so because a charged particle under the action of an electric field and a perpendicular magnetic field, describes a cycloid perpendicular both to the electric field and the magnetic field].

From inversion of the magnetoconductance matrix (46), the magnetoresistance matrix in the ideal collisionless regime is given by

$$\rho(H) = \frac{H}{n_s e c} \begin{pmatrix} 0 & 1 \\ -1 & 0 \end{pmatrix}. \tag{47}$$

Thus, for the two-dimensional electron gas *in the absence of scattering*, the classical Hall resistance $\rho_{xy}(H) = H/n_s e c$ is linear in H, and the diagonal resistance $\rho_{xx}(H)$ is always zero.

We can reasonably argue that the condition of collisionless regime holds at (and near) those very values of the magnetic field, for which *Landau levels (with definite spin polarization) are either fully occupied or completely empty*. In fact, in this situation, electrons cannot suffer either elastic or quasi-elastic scattering events (at low temperatures), because an energy gap of the order of $\hbar\omega_c$ or $\mu_B H$ separates occupied states from empty states.

It is interesting to consider the classical Hall resistance at the magnetic field values H_t^*/i $(i = 1, 2, \dots)$, where $H_t^* = n_s h c/e$ is the characteristic field defined by Eq. (14) (H_t^* is the field for which all electrons can be accommodated and fully fill the ground Landau level, with definite spin orientation). When $H = H_i = H_t^*/i$, it holds

$$\boxed{\rho_{xy}(H_i) = \frac{H_i}{n_s e c} = \frac{h}{e^2} \frac{1}{i}}; \tag{48}$$

this means that in correspondence to the values H_t^*/i of the magnetic field, the Hall resistance is quantized at $h/e^2 i$.

The experimental findings of $\rho_{xy}(H)$, and the vanishing of $\rho_{xx}(H)$, for the values of H corresponding to Landau levels *either fully occupied or fully empty*, are thus

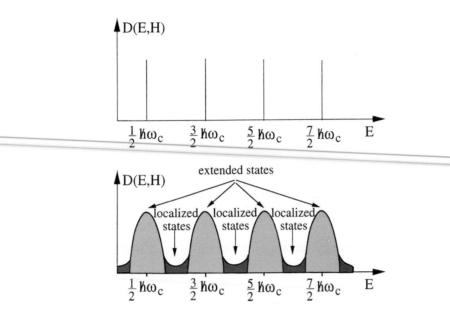

Fig. 14 Density-of-states $D(E, H)$ of the two-dimensional electron gas in a magnetic field. (a) in a pure material, the density-of-states is composed by a series of delta functions at the Landau levels; (b) in the presence of disorder, the Landau levels are broadened into bands, constituted by extended and localized states.

in agreement with the modelistic estimate of the Hall resistance in the collisionless regime; however, one must justify why the experimental value of $\rho_{xy}(H)$ presents so flat plateaus, while the modelistic collisionless model (if taken as it stands) provides $\rho_{xy}(H)$ linear in H.

A possible interpretation of the mechanism leading to the plateaus is the presence of sufficient disorder to broaden the N_L orbitally degenerate states, within each Landau level, into a *band of levels* (see Fig. 14).

If the magnetic field is sufficiently strong, the energy overlap of the broadened δ-functions is minimal. While the states near the original Landau level remain conducting, the states toward the edges become localized (and thus lost for conductivity), according to the general concepts concerning the effect of disorder. Thus Eq. (48) holds not only for $H = H_i$ but also for H around H_i, as far as the occupation of the extended states does not change (i.e. the Fermi level occurs at energies where the electronic states are localized). General considerations and sum rules are also invoked to explain why the current not being carried in the localized states is completely compensated by extra current in delocalized states; but a thorough account of the role of disorder, width of plateaus, and other features certainly need further investigations.

We have seen that in the integer quantum Hall effect, the Hall resistance is quantized at $h/e^2 i$ where i is an integer number. For an overall account of the integer Hall effect a single-electron picture is sufficient; the effect is basically linked to the energy gaps resulting from the quantization of the kinetic energy of the free electrons into discrete Landau levels. The discovery in 1980 of the integer quantum Hall effect was soon followed by the discovery of the fractional quantum Hall effect, in which the Hall resistance is quantized at $h/e^2\nu$ with ν rational fractions with odd denominators. In the original paper of D. C. Tsui, H. L. Stormer and A. C. Gossard [Phys. Rev. Lett. 48, 212 (1982)], a pronounced quantized Hall plateau of $\rho_{xy}(H)$ at $3h/e^2$ is observed when the lowest-energy spin-polarized Landau level has filling factor $\nu = 1/3$. The interpretation of the fractional Hall effect appears to be a challenging many-body problem, where an important role is played by the effect of disorder and the occurrence of gaps is linked to electron-electron interactions, as suggested in the seminal work by R. B. Laughlin, Phys. Rev. Lett. 50, 231, 1395 (1983).

Appendix A. Free energy of an electron gas in a uniform magnetic field

In this Appendix we give the mathematical evaluation of the free energy of the three-dimensional electron gas in a uniform magnetic field. We begin with the particular case of zero temperature, where the free energy is just the ground-state energy of the free-electron gas in the uniform magnetic field. We then generalize the treatment to any temperature. The analytic evaluation of the ground-state energy, and more generally of the free energy, of the electron gas in the presence of the magnetic field requires some brief digression on the Poisson sum formula, Fresnel integrals, and integrals involving trigonometric functions and Fermi–Dirac function. The mathematical details provide a better insight of the magnetic effects, previously inferred on the basis of physical arguments.

Poisson sum formula

Let $f(x)$ be an arbitrary regular function for x in the interval $[0, -\infty]$, and n be an integer number $n \geq 0$. For $n < x < n + 1$ we can write

$$f(x) = \sum_{s=-\infty}^{+\infty} a_s\, e^{-2\pi\, i\, s\, x} \quad \text{with} \quad a_s = \int_n^{n+1} f(x)\, e^{2\pi\, i\, s\, x}\, dx\ . \tag{A1}$$

From Eqs. (A1), we have in particular

$$f(n + \frac{1}{2}) = \sum_{s=-\infty}^{+\infty} (-1)^s \int_n^{n+1} f(x)\, e^{2\pi\, i\, s\, x}\, dx\ .$$

Summing up over n $(n = 0, 1, 2, \dots)$ provides the *Poisson sum formula*

$$\sum_{n=0}^{+\infty} f(n + \frac{1}{2}) = \int_0^\infty f(x)\, dx + 2 \sum_{s=1}^\infty (-1)^s \int_0^\infty f(x)\, \cos 2\pi s x\, dx\ . \tag{A2}$$

Fresnel integrals

Integrals which can be brought into the form

$$\int_0^\infty \cos(bx^2)\, dx = \int_0^\infty \sin(bx^2)\, dx = \frac{1}{2}\sqrt{\frac{\pi}{2b}} \qquad (b > 0)$$

are called *Fresnel integrals*. The little more general form

$$\int_0^\infty \cos(a \pm b\,x^2)\, dx = \frac{1}{2}\sqrt{\frac{\pi}{b}} \cos(a \pm \frac{\pi}{4}) \qquad (b > 0)\,, \qquad (A3)$$

is obtained exploiting the standard trigonometric formula

$$\cos(a \pm b\,x^2) = \cos a \, \cos bx^2 \mp \sin a \, \sin bx^2 \, .$$

Ground-state energy of the electron gas in a uniform magnetic field

The Poisson sum formula and the Fresnel integrals allow us to obtain a simple analytic expression for the ground-state energy of the three-dimensional electron gas in a magnetic field, provided $\hbar\omega_c \ll \mu$, which is the standard situation. From Eq. (19), the density-of-states in the presence of a magnetic field (including spin degeneracy) is given by

$$D(E, H) = \frac{1}{2}\hbar\omega_c\, A \sum_{n=0}^{\infty} \frac{1}{\sqrt{E - (n + \frac{1}{2})\hbar\omega_c}}\Theta\left[E - \left(n + \frac{1}{2}\right)\hbar\omega_c\right]\,, \qquad (A4)$$

where the constant A is defined by Eqs. (18), and $\Theta(x)$ denotes the step function, $\Theta(x) = 0$ for $x < 0$, $\Theta(x) = 1$ for $x > 0$. At $T = 0$ the ground-state energy of the electron system becomes

$$E_0(H) = \int_0^\mu E\, D(E, H)\, dE \, . \qquad (A5)$$

In order to calculate Eq. (A5), it is convenient to define the primitive function of the density-of-states $D(E, H)$ as

$$P_1(E, H) = \int_0^E D(E', H)\, dE' \, , \qquad (A6)$$

and similarly the next primitive as

$$P_2(E, H) = \int_0^E P_1(E', H)\, dE' \, . \qquad (A7)$$

We exploit the property that μ is independent on H in the three-dimensional electron gas. Thus

$$P_1(\mu, H) = N$$

where N is the total number of electrons (with both spin directions). An integration by parts in Eq. $(A5)$ gives

$$E_0(H) = \int_0^\mu E\, dP_1(E, H) = \mu\, P_1(\mu, H) - \int_0^\mu P_1(E, H)\, dE \ .$$

Thus the ground-state energy $E_0(H)$ takes the very manageable expression

$$\boxed{E_0(H) = N\mu - P_2(\mu, H)}\ ; \qquad (A8)$$

apart the constant term $N\mu$, the calculation of the ground-state energy at the temperature $T = 0$ simply requires the evaluation of the function $P_2(E, H)$ at the chemical potential.

From Eqs. $(A4)$, $(A6)$ and $(A7)$, the explicit expression of $P_2(E, H)$ becomes

$$P_2(E, H) = \frac{2}{3} \hbar\omega_c\, A \sum_{n=0}^{\infty} \left[E - \left(n + \frac{1}{2}\right)\hbar\omega_c \right]^{3/2} \Theta\left[E - \left(n + \frac{1}{2}\right)\hbar\omega_c \right]\ . \qquad (A9)$$

We apply the Poisson sum formula $(A2)$ to the above expression and obtain

$$P_2(E, H) = \frac{2}{3} \hbar\omega_c\, A \int_0^{E/\hbar\omega_c} (E - x\hbar\omega_c)^{3/2}\, dx$$

$$+ \frac{4}{3} \hbar\omega_c\, A \sum_{s=1}^{\infty} (-1)^s \int_0^{E/\hbar\omega_c} (E - x\hbar\omega_c)^{3/2} \cos 2\pi s x\, dx\ . \qquad (A10)$$

The first integral in the right hand side of Eq. $(A10)$ is trivial and coincides with $P_2(E, 0) \equiv (4/15)\, A\, E^{5/2}$, which is the primitive of the density-of-states in the absence of the magnetic field. The second integral can be brought to a Fresnel integral performing first the change of variables $E - x\hbar\omega_c = t^2$ and performing then two integrations by part; we have

$$\int_0^{E/\hbar\omega_c} (E - x\hbar\omega_c)^{3/2} \cos 2\pi s x\, dx = \frac{2}{\hbar\omega_c} \int_0^{\sqrt{E}} t^4 \cos\left(\frac{2\pi s}{\hbar\omega_c} E - \frac{2\pi s}{\hbar\omega_c} t^2\right) dt$$

$$= \frac{3}{2} \frac{\hbar\omega_c}{(2\pi s)^2} \sqrt{E} - \frac{3}{2} \frac{\hbar\omega_c}{(2\pi s)^2} \int_0^{\sqrt{E}} \cos\left(\frac{2\pi s}{\hbar\omega_c} E - \frac{2\pi s}{\hbar\omega_c} t^2\right) dt$$

$$= \frac{3}{8} \frac{\hbar\omega_c}{\pi^2 s^2} E^2 - \frac{3}{16} \frac{\hbar\omega_c}{\pi^2 s^2} \sqrt{\frac{\hbar\omega_c}{2s}} \cos\left(\frac{2\pi s}{\hbar\omega_c} E - \frac{\pi}{4}\right)\ ; \qquad (A11)$$

the last passage has been obtained assuming $E \gg \hbar\omega_c$ and using Eq. $(A3)$ for Fresnel integrals.

Using the relationship

$$\sum_{s=1}^{\infty} \frac{(-1)^s}{s^2} = -\frac{\pi^2}{12}\ ,$$

and inserting $(A11)$ into $(A10)$, we obtain for $P_2(E, H)$ the expression

$$P_2(E, H) = P_2(E, 0) - \frac{1}{6}\left(\frac{1}{2}\hbar\omega_c\right)^2 AE^{1/2} - \left(\frac{1}{2}\hbar\omega_c\right)^{5/2} \frac{A}{\pi^2} \sum_{s=1}^{\infty} \frac{(-1)^s}{s^{5/2}} \cos\left(\frac{2\pi s}{\hbar\omega_c}E - \frac{\pi}{4}\right).$$

$$(A12)$$

From Eq. $(A8)$, we have the following analytic result for the ground-state energy $E_0(H)$ at zero temperature

$$E_0(H) = E_0(0) + \frac{1}{24}\hbar^2\omega_c^2 D(\mu) + \frac{1}{4\pi^2}\hbar^2\omega_c^2 D(\mu)\left(\frac{\hbar\omega_c}{2\mu}\right)^{1/2} \sum_{s=1}^{\infty} \frac{(-1)^s}{s^{5/2}} \cos\left(\frac{2\pi s}{\hbar\omega_c}\mu - \frac{\pi}{4}\right)$$

$$(A13)$$

where $D(\mu) = A\mu^{1/2}$ is the density-of-states for both spin directions at the chemical potential.

Special integrals involving trigonometric functions and Fermi–Dirac function derivative

This mathematical digression is needed in order to generalize the above procedure and to evaluate the free energy at any temperature. Consider an integral of the type

$$I_s = \int_0^{\infty} \cos\left(\frac{2\pi s}{\hbar\omega_c}E - \frac{\pi}{4}\right)\left(-\frac{\partial f}{\partial E}\right) dE, \qquad (A14)$$

where $f(E) = [\exp(E - \mu)/k_BT + 1]^{-1}$ is the Fermi–Dirac function and $\omega_c = eH/mc$ is the cyclotron frequency. The integral I_s can be performed analytically, but much caution must be applied because of the rapid oscillations exhibited by the trigonometric functions; in particular the Sommerfeld expansion (see Section III-2) is not applicable here. It is easily seen that at very high temperatures and at very low temperatures we have respectively:

$$I_s = \begin{cases} 0 & \text{if } k_BT \gg \hbar\omega_c \\ \cos\left(\frac{2\pi s}{\hbar\omega_c}\mu - \frac{\pi}{4}\right) & \text{if } k_BT \ll \hbar\omega_c \end{cases}.$$

The exact analytical expression of I_s, derived below, is

$$I_s = \frac{2\pi^2 s\, k_BT}{\hbar\omega_c} \frac{1}{\sinh(2\pi^2 s\, k_BT/\hbar\omega_c)} \cos\left(\frac{2\pi s}{\hbar\omega_c}\mu - \frac{\pi}{4}\right). \qquad (A15)$$

In order to evaluate analytically the integral $(A14)$, we notice that $-\partial f/\partial E$ is strongly peaked at $E = \mu$ and we can thus replace the lower limit of integration with $-\infty$. From the identity

$$\cos\left(\frac{2\pi s}{\hbar\omega_c}E - \frac{\pi}{4}\right) \equiv \cos\left[\left(\frac{2\pi s}{\hbar\omega_c}\mu - \frac{\pi}{4}\right) + \frac{2\pi s}{\hbar\omega_c}(E - \mu)\right],$$

and from the fact that $-\partial f/\partial E$ is an even function of $E - \mu$ we have

$$I_s = \cos\left(\frac{2\pi s}{\hbar\omega_c}\mu - \frac{\pi}{4}\right) \int_{-\infty}^{+\infty} \cos\frac{2\pi s(E - \mu)}{\hbar\omega_c}\left(-\frac{\partial f}{\partial E}\right) dE .\qquad (A16)$$

We now observe the identity

$$-\frac{\partial f}{\partial E} \equiv \frac{1}{k_B T}\frac{1}{4\cosh^2[(E - \mu)/2k_B T]} ,$$

and perform the change of variable $x = (E - \mu)/2k_B T$ to obtain

$$I_s = \frac{1}{2}\cos\left(\frac{2\pi s}{\hbar\omega_c}\mu - \frac{\pi}{4}\right) \int_{-\infty}^{+\infty} \cos\frac{4\pi s k_B T x}{\hbar\omega_c}\frac{1}{\cosh^2 x} dx .\qquad (A17)$$

In Eq. $(A17)$, we can add to the cosine function under integral the odd function $i\sin(4\pi s k_B T x/\hbar\omega_c)$ and write

$$I_s = \frac{1}{2}\cos\left(\frac{2\pi s}{\hbar\omega_c}\mu - \frac{\pi}{4}\right) \int_{-\infty}^{+\infty} \exp\left(i\frac{4\pi s k_B T}{\hbar\omega_c}x\right)\frac{1}{\cosh^2 x} dx .$$

We close the integration contour on the upper part of the complex plane and notice that $\cosh z = 0$ for $z = z_n = (n + 1/2)\pi i$ ($n = 0, 1, 2, \dots$). We also remark that $\cosh z = i(z - z_n)$ and $1/\cosh^2 z = -1/(z - z_n)^2$ for $z \approx z_n$; thus the function $1/\cosh^2 z$ has poles of second order when $z = (n + 1/2)\pi i$. We obtain

$$\int_{-\infty}^{+\infty}\frac{\exp\left(i\frac{4\pi s k_B T}{\hbar\omega_c}x\right)}{\cosh^2 x} dx = \sum_{n=0}^{\infty}2\pi i\exp\left[i\frac{4\pi s k_B T}{\hbar\omega_c}\left(n+\frac{1}{2}\right)\pi i\right]i\frac{4\pi s k_B T}{\hbar\omega_c}(-1)$$

$$= \frac{8\pi^2 s k_B T}{\hbar\omega_c}\exp\left(-\frac{2\pi^2 s\,k_B T}{\hbar\omega_c}\right)\sum_{n=0}^{\infty}\exp\left(-\frac{4\pi^2 s k_B T}{\hbar\omega_c}\right)^n = \frac{4\pi^2 s k_B T}{\hbar\omega_c}\frac{1}{\sinh(2\pi^2 s k_B T/\hbar\omega_c)} ,$$

where we have performed the sum over n using the geometric series. Eq. $(A15)$ is thus proved.

Evaluation of the free energy of an electron gas in a uniform magnetic field

We have now all the ingredients for evaluating the free energy of an electron gas in a uniform magnetic field. The free energy of a system of independent fermions is given by Eq. (III-B7), here re-written for convenience in the form

$$F(T, H) = N\mu - k_B T\sum_{n}\ln\left[1 + e^{(\mu - E_n)/k_B T}\right] ,$$

where $E_n = E_n(H)$ are all the possible one-electron energy states. Indicating by $D(E, H)$ the density-of-states in the presence of the magnetic field, the free energy can be written as

$$F(T, H) = N\mu - k_B T\int_0^{\infty}\ln\left[1 + e^{(\mu - E)/k_B T}\right] D(E, H) dE .\qquad (A18)$$

In Eq. $(A18)$ we indicate $D(E, H)\, dE = dP_1(E, H)$, where $P_1(E, H)$ is the primitive $(A6)$ of the density-of-states, and then we perform a first integration by parts; then we indicate with $P_2(E, H)$ the second primitive of the density-of-states and perform a second integration by parts; we also notice that

$$\frac{d}{dE} \ln\left[1 + e^{(\mu - E)/k_B T}\right] = \frac{-1}{k_B T} \frac{1}{e^{(E - \mu)/k_B T} + 1} = \frac{-1}{k_B T} f(E) \,,$$

where $f(E)$ is the standard Fermi–Dirac function; we obtain

$$F(T, H) = N\mu - \int_0^\infty P_2(E, H) \left(-\frac{\partial f}{\partial E}\right) dE \,. \tag{A19}$$

This form of the free energy can be easily recognized as the generalization of Eq. $(A8)$; in fact $(-\partial f/\partial E)$ is strongly peaked at $E = \mu$ and becomes a δ-function at $T = 0$. Using the expression of $P_2(E, H)$ given in Eq. $(A12)$ and the special integrals $(A15)$ we obtain for the free energy

$$F(T, H) = F(T, 0) + \frac{1}{24} \hbar^2 \omega_c^2 D(\mu) + \frac{1}{4\pi^2} \hbar^2 \omega_c^2 D(\mu) \left(\frac{\hbar\omega_c}{2\mu}\right)^{1/2}$$

$$\cdot \sum_{s=1}^\infty \frac{(-1)^s}{s^{5/2}} \frac{2\pi^2 s\, k_B T}{\hbar\omega_c} \frac{1}{\sinh(2\pi^2 s\, k_B T/\hbar\omega_c)} \cos\left(\frac{2\pi s}{\hbar\omega_c}\mu - \frac{\pi}{4}\right) \,. \tag{A20}$$

In the low temperature limit $k_B T \ll \hbar\omega_c$, Eq. $(A20)$ coincides with Eq. $(A13)$; in the high temperature limit $k_B T \gg \hbar\omega_c$, the oscillating terms in Eq. (A20) can be neglected. From Eq. $(A20)$ the magnetic moment $M = -(1/V)\,\partial F/\partial H$ can be obtained.

Appendix B. Generalized orbital magnetic susceptibility of the free-electron gas

In the main text we have considered the effect a *uniform magnetic field* on the three-dimensional gas; after solving exactly the Schrödinger equation for an elecron in a uniform magnetic field, we have shown that the magnetic orbital susceptibility of the free-electron gas consists of two contributions: (i) the Landau diamagnetism (ii) the oscillating de Haas–van Alphen contribution. In this Appendix we consider the generalized magnetic susceptibility $\chi(q)$ for applied magnetic fields, characterized by a given $\mathbf{q}$ wavevector; for this analysis we have to resort to a perturbative approach. The study of the wavevector dependence of the generalized magnetic susceptibility, besides its own interest, is very instructive from several points of view: (i) The perturbative approach, when considered in the long wavelength limit, makes evident the origin of the Landau diamagnetism (the oscillating de Haas–van Alphen contribution is not contained in the perturbative treatment). (ii) More importantly, the present considerations provide some insight on the microscopic properties of those materials,

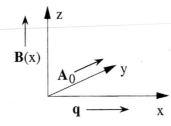

Fig. 15 Directions of the wavevector $\mathbf{q}$, vector potential amplitude $\mathbf{A}_0$, and magnetic field.

which are candidate to exhibit perfect diamagnetism (for the discussion of the perfect diamagnetism of superconductors see Chapter XVIII).

The Hamiltonian of a free electron, in the presence of a magnetic field described by a vector potential $\mathbf{A}(\mathbf{r})$, is given by

$$H = \frac{1}{2m}\left[\mathbf{p} + \frac{e}{c}\mathbf{A}(\mathbf{r})\right]^2 = \frac{\mathbf{p}^2}{2m} + \frac{e}{mc}\mathbf{A}(\mathbf{r})\cdot\mathbf{p} + \frac{e^2}{2mc^2}\mathbf{A}^2(\mathbf{r})\ . \tag{B1}$$

As usual the Coulomb gauge is being adopted, so that $\operatorname{div}\mathbf{A}(\mathbf{r})=0$, and $\mathbf{A}(\mathbf{r})$ and $\mathbf{p}$ commute. The interaction of the magnetic moment of the electron with the magnetic field is neglected, since in this Appendix we focus on the orbital magnetic susceptibility.

In the following the vector potential $\mathbf{A}(\mathbf{r})$ is taken in the form

$$\mathbf{A}(\mathbf{r}) = \mathbf{A}_0\left(e^{i\,\mathbf{q}\cdot\mathbf{r}} + e^{-i\,\mathbf{q}\cdot\mathbf{r}}\right) \tag{B2}$$

with $\mathbf{A}_0\cdot\mathbf{q}=0$, as required by the Coulomb gauge; the magnetic field $\mathbf{B}=\operatorname{curl}\mathbf{A}$ associated to the vector potential ($B2$) is

$$\mathbf{B}(\mathbf{r}) = i\,\mathbf{q}\times\mathbf{A}_0\left(e^{i\,\mathbf{q}\cdot\mathbf{r}} - e^{-i\,\mathbf{q}\cdot\mathbf{r}}\right) = -2\,\mathbf{q}\times\mathbf{A}_0\,\sin\mathbf{q}\cdot\mathbf{r}\ .$$

The directions of $\mathbf{q}$, $\mathbf{A}_0$, $\mathbf{B}$ are chosen as indicated in Fig. 15. The average in space of the square of the applied magnetic field, here denoted as B_a^2, is given by

$$B_a^2 = 2\,q^2\,A_0^2\ . \tag{B3}$$

When the magnetic susceptibility is very small, the fields $\mathbf{B}$ and $\mathbf{H}$ are approximately the same, and so the average fields B_a and H_a can be taken as equal.

Consider a free-electron gas of density $n = N/V$, with N electrons confined in a volume V. The ground state $|\Psi_0\rangle$ of the non-interacting free-electron gas is given by the Slater determinant, formed with doubly occupied plane waves with wavevectors filling the Fermi sphere of radius k_F, where $k_F^3 = 3\pi^2 n$. We have

$$|\Psi_0\rangle = A\{W_{\mathbf{k}_1}\alpha\ W_{\mathbf{k}_1}\beta\ \ldots\ W_{\mathbf{k}_{N/2}}\alpha\ W_{\mathbf{k}_{N/2}}\beta\}\ ,$$

where A is the antisymmetrization operator and $W(\mathbf{k}_i, r) = (1/\sqrt{V})\exp(i\mathbf{k}_i\cdot\mathbf{r})$ are normalized plane waves (see Section IV-7). Similarly, we shall denote by $|\Psi_n\rangle$ the single-particle excitations of the Fermi sea, with an electron in a plane wave of energy larger than the Fermi energy E_F and a hole in a plane wave below it.

In the presence of a magnetic field, the change in energy of the ground-state of the system due to the terms linear and quadratic in the vector potential $\mathbf{A}(\mathbf{r})$ in Eq. ($B1$), can be treated by perturbation theory. We remark that the expectation value of the operator $(e/mc)\,\mathbf{A}\cdot\mathbf{p}$ on the ground-state $|\Psi_0\rangle$ vanishes (as can be seen by inspection, or inferred from time-reversal symmetry); to second order in $\mathbf{A}$ we have

$$\Delta E = \Delta E_D + \Delta E_P\ ,$$

where

$$\Delta E_D = \frac{e^2}{2\,m\,c^2}\langle\Psi_0|\sum_{i=1}^{N}\mathbf{A}^2(\mathbf{r}_i)|\Psi_0\rangle \tag{$B4$}$$

and

$$\Delta E_P = \frac{e^2}{m^2\,c^2}\sum_{n}\frac{|\langle\Psi_n|\sum\limits_{i=1}^{N}\mathbf{A}(\mathbf{r}_i)\cdot\mathbf{p}_i|\Psi_0\rangle|^2}{E_0 - E_n}\ . \tag{$B5$}$$

Notice that the term ΔE_D is always positive and leads to diamagnetism, while the term ΔE_P is always negative and leads to paramagnetism.

The evaluation of the diamagnetic term ΔE_D, with $\mathbf{A}(\mathbf{r})$ given by Eq. ($B2$), is rather simple; we have in fact

$$\Delta E_D = \frac{e^2}{2mc^2}2\sum_{\mathbf{k}}f(\mathbf{k})\langle\frac{1}{\sqrt{V}}e^{i\mathbf{k}\cdot\mathbf{r}}|A_0^2(e^{i\mathbf{q}\cdot\mathbf{r}}+e^{-i\mathbf{q}\cdot\mathbf{r}})^2|\frac{1}{\sqrt{V}}e^{i\mathbf{k}\cdot\mathbf{r}}\rangle$$

$$= \frac{e^2}{2mc^2}2\sum_{\mathbf{k}}f(\mathbf{k})\,2\,A_0^2\ , \tag{$B6$}$$

where $f(\mathbf{k})$ is the Fermi–Dirac distribution function at $T = 0$, and the factor 2 in front of the sum over $\mathbf{k}$ appears because summation over spin variables has been performed. The region in k-space contributing to the sum in Eq. ($B6$) is reported in Fig. 16a; we have $2\sum f(\mathbf{k}) = N = nV$ and then

$$\boxed{\Delta E_D = \frac{n\,e^2}{m\,c^2}A_0^2\,V}\ , \tag{$B7$}$$

where n is the electronic density.

The evaluation of the paramagnetic term requires some straightforward algebraic manipulations; we have

$$\Delta E_P = \frac{e^2\,\hbar^2}{m^2\,c^2}2\sum_{\mathbf{k}}\frac{|\mathbf{A}_0\cdot\mathbf{k}|^2\,f(\mathbf{k})\,[1 - f(\mathbf{k}+\mathbf{q})]}{E_{\mathbf{k}} - E_{\mathbf{k}+\mathbf{q}}} + \text{a similar term with } \mathbf{q}\to-\mathbf{q}\ ,$$

$$\tag{$B8$}$$

where the factor 2 appears because spin summation has been performed. It is seen by inspection that the two terms in the right-hand side of Eq. ($B8$) are equal; furthermore,

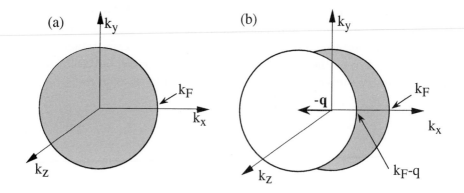

Fig. 16 Schematic representation in the **k**-space of the Fermi sphere of radius k_F, representing the ground-state of the free-electron gas. In (a) we have shadowed the region where the Fermi–Dirac distribution function $f(\mathbf{k}) = 1$, at zero temperature. In (b) we have shadowed the region where $f(\mathbf{k})\left[1 - f(\mathbf{k} + \mathbf{q})\right] = 1$.

using the standard parabolic dispersion relation $E(\mathbf{k}) = \hbar^2 k^2 / 2m$, we obtain

$$\Delta E_P = -\frac{8\,e^2}{m\,c^2} \sum_{\mathbf{k}} \frac{|\mathbf{A}_0 \cdot \mathbf{k}|^2\, f(\mathbf{k})\left[1 - f(\mathbf{k} + \mathbf{q})\right]}{(\mathbf{k} + \mathbf{q})^2 - k^2} . \tag{B9}$$

The region in **k**-space contributing to the sum in Eq. ($B9$) is schematically indicated in Fig. 16b.

The expression of ΔE_P can be easily worked out analytically (and is reported at the end of this Appendix). We have

$$\boxed{\Delta E_P = -\frac{n\,e^2}{m\,c^2}\, A_0^2\, V\, F_P(q)} \tag{B10}$$

where $F_P(q)$ is the positive and monotonically decreasing function

$$F_P(q) = \frac{3}{8}\left[\frac{5}{3} - \frac{q^2}{4\,k_F^2} + \frac{k_F}{q}\left(1 - \frac{q^2}{4\,k_F^2}\right)^2 \ln\left|\frac{2k_F + q}{2k_F - q}\right|\right]. \tag{B11}$$

Notice that

$$F_P(q) = 1 - \frac{1}{4}\frac{q^2}{k_F^2} \quad \text{for} \quad q \ll k_F ; \tag{B12}$$

we also have $F_P(q) = (4/3)\,k_F^2/q^2$ for $q \gg k_F$.

Using Eq. ($B6$) and Eq. ($B10$), we obtain for the total energy change of the ground-state the expression

$$\Delta E = \Delta E_D + \Delta E_P = \frac{n\,e^2}{m\,c^2}\, A_0^2\, V\left[1 - F_P(q)\right]. \tag{B13}$$

We can now evaluate the expression $\chi(q) = -(1/V)\,\partial^2 \Delta E/\partial H_a^2$, which gives the generalized q-dependent magnetic susceptibility; assuming $H_a \approx B_a$ and using relation $(B3)$, we have

$$\chi(q) = -\frac{1}{V} \frac{1}{2q^2} \frac{\partial^2 \Delta E}{\partial A_0^2} \, . \qquad (B14)$$

From Eq. $(B13)$ and Eq. $(B14)$ we obtain

$$\boxed{\chi(q) = -\frac{n\,e^2}{m\,c^2} \frac{1}{q^2} \left[1 - F_P(q)\right]} \, . \qquad (B15)$$

In the long wavelength limit $q \to 0$, from Eq. $(B12)$ and Eq. $(B15)$ we have

$$\chi(q = 0) = -\frac{n\,e^2}{m\,c^2} \frac{1}{4\,k_F^2} = -\frac{e^2\,k_F}{12\,\pi^2\,m\,c^2} \, , \qquad (B16)$$

and we recover the already known Landau result of Eq. (28).

Considerations on the magnetic susceptibility of superconductors

The free-electron gas with "ideal" plane-wave-type wavefunctions has theoretically infinite conductivity (were it not for impurities, lattice vibrations, and other defects) and small orbital diamagnetism, with $\chi \approx -10^{-6}$ according to Eq. $(B16)$. *The "ideal" free-electron gas is thus the prototype example of materials with perfect conductivity and very poor diamagnetism.* Similarly, ordinary metals with "ideally" Bloch-type electronic wavefunctions would exhibit perfect conductivity and poor diamagnetism.

Materials characterized by perfect conductivity and perfect diamagnetism are known as superconductors. For an introduction to superconductivity we refer to Chapter XVIII; here we only wish to speculate on which mechanism could be behind the manifestation of perfect diamagnetism; for this purpose, we focus heuristically on the perturbative treatment of the generalized magnetic susceptibility of an electron system, discussed above, for small wavevector transfer $q \to 0$.

The simplest microscopic model of a superconductor is that of a "modified electron gas", where an attractive two-body interaction mechanism is active between electrons in a small energy shell around the Fermi level. Without entering here in details, in the "modified electron gas" it is found that a (small) gap opens at the Fermi level, and the structure of the modified energy levels and wavefunctions around E_F is such that the paramagnetic contribution is suppressed, i.e. $\Delta E_P \equiv 0$ for $q \to 0$. Since the diamagnetic energy change ΔE_D is not vulnerable to the modification of energy levels and wavefunctions around the Fermi level (see Fig. 16a), we can expect that the total energy change of the "modified electron gas" in the long wavelength limit is given by

$$\Delta E = \frac{n\,e^2}{m\,c^2} A_0^2 \, V = \frac{n\,e^2}{m\,c^2} \frac{B_a^2}{2\,q^2} \, V \, , \qquad (B17)$$

where in the last passage use has been made of Eq. $(B3)$.

From Eq. $(B17)$, we can thus speculate that the penetration in the sample of a uniform magnetic field ($B_a \neq 0$ and $q = 0$) is not energetically possible, and this is the essential of the Meissner effect (for a detailed analysis see Section XVIII-6). From $B_a = H + 4\pi M = 0$, we can infer a magnetic susceptibility of

$$\chi = M/H = -1/4\pi$$

for the "modified electron gas"; such a value is about one million times larger than the magnetic susceptibility ($\chi \approx -10^{-6}$) of the ordinary electron gas. The "blurring" of electron properties even in a very small energy shell around the Fermi surface may thus produce profound consequences in the magnetic properties of the "modified electron gas".

An integral of interest

We evaluate here the expression of ΔE_P, given by Eq. ($B9$). It is seen by inspection that the sum with the term $f(\mathbf{k}) f(\mathbf{k} + \mathbf{q})$ vanishes; we can thus write Eq. ($B9$) in the form

$$\Delta E_P = -\frac{8 e^2}{m c^2} \sum_{\mathbf{k}} \frac{|\mathbf{A}_0 \cdot \mathbf{k}|^2 f(\mathbf{k})}{2 \mathbf{k} \cdot \mathbf{q} + q^2} .$$

Converting the sum over $\mathbf{k}$ into $V/(2\pi)^3$ times the integral in $d\mathbf{k}$, and choosing the geometry of Fig. 15, we have

$$\Delta E_P = -\frac{8 e^2}{m c^2} \frac{V}{(2\pi)^3} \frac{1}{q} A_0^2 \int_{\substack{\text{Fermi} \\ \text{sphere}}} \frac{k_y^2}{2 k_x + q} \, d\mathbf{k} . \tag{B18}$$

We evaluate the following integral

$$I = \int_{\substack{\text{Fermi} \\ \text{sphere}}} \frac{k_y^2}{2 k_x + q} \, d\mathbf{k} = \int_{\substack{\text{Fermi} \\ \text{sphere}}} \frac{k_y^2}{2 k_z + q} \, d\mathbf{k} .$$

Using polar coordinates with the polar axis in the z-direction, we obtain

$$I = \int \frac{k^2 \sin^2\theta \sin^2\phi}{2k \cos\theta + q} k^2 \, dk \, \sin\theta \, d\theta \, d\phi = \pi \int_0^{k_F} k^4 \, dk \int_0^\pi \frac{\sin^3\theta}{2k \cos\theta + q} \, d\theta . \tag{B19}$$

The integral in θ can be performed writing $\sin^3\theta \, d\theta \equiv -(1 - \cos^2\theta) \, d\cos\theta$ and using the indefinite integral

$$\int \frac{1 - x^2}{2 k x + q} \, dx = -\frac{x^2}{4 k} + \frac{q}{4 k^2} x + \frac{4 k^2 - q^2}{8 k^3} \ln|2 k x + q| .$$

The integral ($B19$) thus becomes

$$I = \pi \int_0^{k_F} \frac{q k^2}{2} \, dk + \pi \int_0^{k_F} \frac{1}{8} (4 k^2 - q^2) k \ln\left|\frac{2k + q}{2k - q}\right| dk$$

$$= \frac{\pi}{6} q k_F^3 + \frac{\pi}{8 \cdot 16} \int_0^{k_F} \ln\left|\frac{2k + q}{2k - q}\right| d(4 k^2 - q^2)^2$$

Performing the integration by parts, we obtain

$$I = \frac{\pi q k_F^3}{8} \left[\frac{5}{3} - \frac{q^2}{4 k_F^2} + \frac{k_F}{q} \left(1 - \frac{q^2}{4 k_F^2} \right)^2 \ln \left| \frac{2k_F + q}{2k_F - q} \right| \right] . \tag{B20}$$

From Eq. (B18) and Eq. (B20), we obtain $\Delta E_P = -(n e^2/m c^2) A_0^2 V F_P(q)$, with

$$F_P(q) = \frac{3}{8} \left[\frac{5}{3} - \frac{q^2}{4 k_F^2} + \frac{k_F}{q} \left(1 - \frac{q^2}{4 k_F^2} \right)^2 \ln \left| \frac{2k_F + q}{2k_F - q} \right| \right] .$$

Further Reading

A. G. Aronov and Yu. V. Sharvin "Magnetic Flux Effects in Disordered Conductors" Rev. Mod. Phys. **59**, 755 (1987)

T. Chakraborty and P. Pietiläinen "The Quantum Hall Effects" (Springer, Berlin 1995, 2nd edition)

S. Datta "Electronic Transport in Mesoscopic Systems" (Cambridge University Press 1995)

W. A. Harrison and M. B. Webb eds. "The Fermi Surface" (Wiley, New York 1960)

M. J. Kelly "Low Dimensional Semiconductors. Materials, Physics, Technology, Devices" (Clarendon Press, Oxford 1995)

A. H. Morrish "The Physical Principles of Magnetism" (Wiley, New York 1965)

M. Peshkin and A. Tonomura " The Aharonov–Bohm Effect" (Springer, Berlin 1989)

A. B. Pippard "Dynamics of Conduction Electrons" (Gordon and Breach, New York 1965)

R. E. Prange and S. M. Girvin (editors) "The Quantum Hall Effect" (Springer, New York 1990)

A. Shapere and F. Wilczek (editors) "Geometric Phases in Physics" (World Scientific, Singapore 1989)

M. Stone "Quantum Hall Effect" (World Scientific, Singapore 1992)

S. V. Vonsovskii "Magnetism" (Halstead, New York 1975)

R. M. White "Quantum Theory of Magnetism" (Springer, Berlin 1983)

A. H. Wilson "Theory of Metals" (Cambridge University Press 1954)

XVI

Magnetic properties of localized systems and Kondo impurities

In the previous chapter we have considered the effects of magnetic fields on the *free electron gas*. In this chapter we discuss the effects of magnetic fields on *localized electronic systems*, such as atoms, ions, molecules, and particularly magnetic impurities embedded in solids. It is assumed that the units in the solids are sufficiently far apart, that any mutual interaction among themselves can be neglected; the study of cooperative magnetic effects is postponed to the next chapter.

We begin with a preliminary study of magnetic field effects on atomic or molecular systems with closed-shell ground states and no permanent magnetic moment. These systems are most often diamagnetic, but there are a few remarkable exceptions of paramagnetism (Van Vleck paramagnetism); in both cases, the magnetic susceptibility is practically independent of temperature. We consider then the magnetic field effects in atoms, ions or impurities, whose ground states have a permanent magnetic moment. These electronic systems exhibit in general a temperature dependent paramagnetism (Curie paramagnetism), because of the balance between the orientation effect of the magnetic field on permanent magnetic dipoles and the opposite effect of temperature. The remaining part of this chapter is focused on the problem of localized magnetic impurities dissolved in normal metals, and on the rich phenomenology related to the Kondo problem.

1 Quantum mechanical treatment of magnetic susceptibility

From purely classical argument, the magnetic susceptibility of any dynamical system is zero, according to the Bohr–van Leeuwen theorem (see Section XV-1); thus a quantum mechanical treatment is necessary from the very beginning.

Consider an atomic or polynuclear system (such as a molecule, a cluster or a solid), composed by N_e electrons in interaction among themselves and with the nuclei (fixed in some configuration). The non-relativistic many-electron Hamiltonian of the system can be written as

$$H_0 = \sum_{i=1}^{N_e} \frac{\mathbf{p}_i^2}{2m} + \sum_{i=1}^{N_e} V(\mathbf{r}_i) + \frac{1}{2} \sum_{i \neq j}^{N_e} \frac{e^2}{|\mathbf{r}_i - \mathbf{r}_j|} \; . \tag{1}$$

The first term in the right-hand side of Eq. (1) is the kinetic energy of the electrons; the second term is the electronic–nuclear interaction energy, and the last one represents the electron–electron Coulomb repulsion. For simplicity in Eq. (1) the relativistic corrections, and in particular the spin–orbit interaction, are neglected; the spin–orbit operator couples together the spin and the orbital motion of the electrons, has far reaching consequences in the magnetic properties of systems with increasing atomic numbers, and will be considered later.

We consider now a uniform magnetic field $\mathbf{H}$, described in the symmetric gauge by the vector potential

$$\mathbf{A}(\mathbf{r}) = \frac{1}{2} \mathbf{H} \times \mathbf{r} \; ; \tag{2}$$

it is easily verified that $\mathrm{curl}\mathbf{A}(\mathbf{r})=\mathbf{H}$ and also $\mathrm{div}\mathbf{A}(\mathbf{r})=0$ (so that the operators $\mathbf{A}$ and $\mathbf{p}$ commute). In the presence of a uniform magnetic field, the Hamiltonian (2) of the many-electron system is modified by the following terms. (i) In the expression of the electronic kinetic energy, the momentum $\mathbf{p}_i$ is replaced by the generalized momentum $\mathbf{P}_i = \mathbf{p}_i + (e/c)\mathbf{A}(\mathbf{r}_i)$. (ii) The interaction energy of the spin magnetic moment of each electron with the magnetic field must be included; this term is of the form $-\boldsymbol{\mu}\cdot\mathbf{H} = g_0\mu_B\mathbf{s}\cdot\mathbf{H}$, where $\mathbf{s}$ is the Pauli operator (in units of $\hbar$) for half-spin particles, $\mu_B = e\hbar/2mc$ is the Bohr magneton, e is the absolute value of the electronic charge, and the giromagnetic factor can be taken as $g_0 = 2$ (free electrons without relativistic corrections).

The Hamiltonian (1), with the modifications specified above, takes the form

$$H = \sum_{i=1}^{N_e} \frac{1}{2m} \left[\mathbf{p}_i + \frac{e}{c}\mathbf{A}(\mathbf{r}_i) \right]^2 + \sum_{i=1}^{N_e} V(\mathbf{r}_i) + \frac{1}{2} \sum_{i \neq j}^{N_e} \frac{e^2}{|\mathbf{r}_i - \mathbf{r}_j|} + 2\mu_B \mathbf{H} \cdot \sum_{i=1}^{N_e} \mathbf{s}_i \; . \tag{3}$$

It is convenient to define the total spin operator $\mathbf{S}$ (in units of $\hbar$) as

$$\mathbf{S} = \sum_{i=1}^{N_e} \mathbf{s}_i \; , \tag{4a}$$

and the total angular momentum operator $\mathbf{L}$ (in units of $\hbar$) as

$$\mathbf{L} = \frac{1}{\hbar} \sum_{i=1}^{N_e} \mathbf{r}_i \times \mathbf{p}_i \; . \tag{4b}$$

In the presence of a uniform magnetic field (described in the symmetric gauge), the Hamiltonian (3) of the many-electron system becomes

$$H = H_0 + \mu_B \mathbf{H} \cdot (\mathbf{L} + 2\mathbf{S}) + \frac{e^2}{8\,m\,c^2} \sum_{i=1}^{N_e} (\mathbf{H} \times \mathbf{r}_i)^2 \; , \tag{5}$$

where H_0 is the unperturbed Hamiltonian of the system.

The first term added to H_0 in Eq. (5) is the Zeeman operator $H_Z = -\boldsymbol{\mu} \cdot \mathbf{H}$, giving the energy of the magnetic dipole $\boldsymbol{\mu} = -\mu_B(\mathbf{L} + 2\mathbf{S})$ in the magnetic field; the second term is proportional to the square of the magnetic field strength; since both terms are in general small with respect to H_0, we can treat them by perturbation theory. In adopting perturbation theory to determine the effect of a magnetic field on the ground state of H_0, we have as usual to distinguish whether the ground state is degenerate or non-degenerate.

Suppose that the ground state Ψ_0 of H_0 is *non-degenerate* (nor quasi-degenerate; i.e. Ψ_0 is well separated in energy from the excited states Ψ_n). The expectation value of the Zeeman operator on Ψ_0 is zero, since $\langle \Psi_0 | \mathbf{L} | \Psi_0 \rangle = \langle \Psi_0 | \mathbf{S} | \Psi_0 \rangle \equiv 0$; this can be seen by inspection (the non-degenerate wavefunction Ψ_0 can be taken as real), or by parity considerations of Ψ_0, $\mathbf{L}$ and $\mathbf{S}$ under time-reversal operator. Thus corrections linear in the magnetic field do not occur, and the system is said to have *no permanent magnetic moment*. Corrections quadratic in the magnetic field involve the expectation value of $(\mathbf{H} \times \mathbf{r})^2$ on the state Ψ_0 and the Zeeman operator to second order.

From perturbation theory, consistently selecting terms up to H^2, we have for the energy shift ΔE_0 induced by a magnetic field

$$\Delta E_0 = \Delta E_D + \Delta E_P \; , \tag{6a}$$

where

$$\Delta E_D = \frac{e^2}{8\,m\,c^2} \langle \Psi_0 | \sum_{i=1}^{N_e} (\mathbf{H} \times \mathbf{r}_i)^2 | \Psi_0 \rangle \tag{6b}$$

and

$$\Delta E_P = \sum_{n(\neq 0)} \frac{|\langle \Psi_n | \mu_B \mathbf{H} \cdot (\mathbf{L} + 2\mathbf{S}) | \Psi_0 \rangle|^2}{E_0 - E_n} \; . \tag{6c}$$

If we have N independent units in the volume V, the energy change of the ground state of the system is $N \Delta E_0$; from Eq. (XV-2b) the magnetic susceptibility (when thermal excitations are negligible) becomes

$$\boxed{\chi = -\frac{N}{V} \frac{\partial^2 \Delta E_0}{\partial H^2}} \; . \tag{7}$$

Thus, the *positive term* ΔE_D contributes to the magnetic susceptibility with a *diamagnetic contribution* $\chi_D = -(N/V)\,\partial^2 \Delta E_D/\partial H^2$, while the *negative term* ΔE_P gives rise to a *paramagnetic contribution* $\chi_P = -(N/V)\,\partial^2 \Delta E_P/\partial H^2$.

If the ground state of H_0 is *degenerate*, then the diagonalization of the Zeeman operator in the subspace removes the degeneracy and leads in general to energy separations proportional to the magnetic field strength. The system is said to have a *permanent magnetic moment*, and presents in general a Curie-type paramagnetism; in this case, second-order perturbations are often of minor importance, but must sometimes be considered for better quantitative results. We pass now to illustrate the above concepts in some typical situations.

2 Magnetic susceptibility of closed-shell systems

Diamagnetism of closed-shell atoms or ions

Consider a closed-shell atom or ion, and let Ψ_0 denote the non-degenerate ground-state wavefunction of the system; Ψ_0 is also eigenfunction of the total spin operator and total angular momentum operator (with the nucleus site taken as origin) with zero eigenvalue; we have

$$\mathbf{S}\,|\Psi_0\rangle = 0 \tag{8a}$$

$$\mathbf{L}\,|\Psi_0\rangle = 0 . \tag{8b}$$

Thus in Eqs. (6) the paramagnetic term is absent, and the ground-state energy shift becomes

$$\boxed{\Delta E_0 = \frac{e^2}{8\,m\,c^2}\langle \Psi_0|\sum_{i=1}^{N_e}(\mathbf{H}\times\mathbf{r}_i)^2|\Psi_0\rangle .} \tag{9}$$

Notice that, according to Eqs. (6), the determination of the energy shift of the ground state of a system in a magnetic field requires in general the knowledge both of *the ground state and of the excited states*; however, in the case of *closed-shell atoms or ions* only the ground state is needed.

From the spherical symmetry of the ground-state charge distribution of a closed-shell atom or ion it follows

$$\langle \Psi_0|\sum_{i=1}^{N_e}(\mathbf{H}\times\mathbf{r}_i)^2|\Psi_0\rangle = \frac{2}{3}H^2\langle \Psi_0|\sum_{i=1}^{N_e}r_i^2\,|\Psi_0\rangle .$$

We can thus write Eq. (9) in the form

$$\Delta E_0 = \frac{e^2}{12\,m\,c^2}H^2\langle r^2\rangle , \tag{10a}$$

where

$$\langle r^2 \rangle = \langle \Psi_0 | \sum_{i=1}^{N_e} r_i^2 | \Psi_0 \rangle \tag{10b}$$

represents the sum of the mean square of the radii of the electron orbits.

Consider now a system with N independent atoms (or ions) in the volume $V = N\Omega$. From Eq. (7) and Eqs. (10), we obtain for the diamagnetic susceptibility the Larmor expression

$$\boxed{\chi = -\frac{N}{V} \frac{e^2}{6 \, m \, c^2} \langle r^2 \rangle} \, . \tag{11}$$

The above equation can be re-written in the form

$$\chi = -\frac{a_B^3}{\Omega} \frac{e^2}{6 \, m \, c^2} \frac{1}{a_B} \frac{\langle r^2 \rangle}{a_B^2} \, ,$$

where $e^2 / m \, c^2 = r_0 = 2.82 \cdot 10^{-13}$ cm is the electron radius and $a_B = 0.529 \cdot 10^{-8}$ cm is the Bohr radius; the order of magnitude of the Larmor diamagnetic susceptibility is then $\chi = -10^{-6}$.

Magnetic susceptibility of closed-shell molecules

Consider a closed-shell (diatomic or polyatomic) molecule, and assume that the ground state Ψ_0 is non-degenerate and well separated in energy from the other excited molecular states. Since spin-up and spin-down electrons are coupled in the ground state, we have that Ψ_0 is eigenfunction of the total electron spin $\mathbf{S}$ with zero eigenvalue, and

$$\mathbf{S} \, | \Psi_0 \rangle = 0 \tag{12a}$$

similarly to Eq. (8a). We notice that the potential of a polynuclear molecule is no more central; thus in general Ψ_0 is not eigenfunction of the total electronic angular momentum $\mathbf{L}$, and $\mathbf{L} | \Psi_0 \rangle \neq 0$, differently from Eq. (8b). However, we can still write for the expectation value of the operator $\mathbf{L}$ on the non-degenerate wavefunction Ψ_0 the relation

$$\langle \Psi_0 \, | \mathbf{L} \, | \, \Psi_0 \, \rangle = 0 \, ; \tag{12b}$$

this can be seen by inspection (the non-degenerate wavefunction Ψ_0 can be taken as real), or by parity considerations of Ψ_0 and $\mathbf{L}$ under time-reversal operator.

From Eqs. (6) and Eqs. (12), one obtains for the energy shift of the ground state

$$\Delta E_0 = \frac{e^2}{8 \, m \, c^2} \langle \Psi_0 | \sum_{i=1}^{N_e} (\mathbf{H} \times \mathbf{r}_i)^2 | \Psi_0 \rangle + \sum_{n(\neq 0)} \frac{|\langle \Psi_n | \mu_B \mathbf{H} \cdot \mathbf{L} | \Psi_0 \rangle|^2}{E_0 - E_n} \, . \tag{13}$$

The first term, in the right-hand side of Eq. (13) is always positive, and gives a diamagnetic contribution to the susceptibility; the second term is always negative and gives a paramagnetic contribution; a net diamagnetic effect usually occurs. However for a number of molecules, whose lowest excitation energies $E_n - E_0$ are very small, paramagnetism may dominate over diamagnetism, and a net paramagnetic effect occurs (Van Vleck paramagnetism). The Van Vleck paramagnetism, as well as ordinary diamagnetism, is practically independent of temperature.

The computation of (13) requires the knowledge not only of the ground state of the molecule, but also of the whole excitation spectrum. Furthermore expression (13) is rigorously gauge independent only if the sum is extended over all excited states. If in actual calculations the sum over the excited states is truncated, the expression (13) is no more gauge invariant. This poses the problem of which gauge to adopt in approximate calculations, but we do not go through the theoretical and empirical answers to this problem.

3 Permanent magnetic dipoles in atoms or ions with partially filled shells

A permanent magnetic moment in atoms (or ions) can result only from incompletely filled shells. Consider a free atom (or ion) with an incomplete shell of orbital angular momentum l. Such a shell contains a total number $n_{tot} = 2(2l + 1)$ of spin–orbitals (namely 2, 6, 10, 14 in the case of s, p, d, f shells, respectively). Let $n < n_{tot}$ be the number of electrons in the shell; in the hypothetical case that the electrons do not interact, the electronic configuration would be $n_{tot}!/n!(n_{tot} - n)!$ times degenerate.

In free atoms the above degeneracy is removed (at least partially) by the electron–electron Coulomb interaction. In atoms with increasing atomic number, the role of spin–orbit interaction becomes of increasing importance, and must be properly taken into account. In the presence of an external magnetic field the Zeeman interaction has also to be considered. Of course, depending on the situation, the hierarchy of strength of the different contributions may vary, and additional terms (as for instance crystalline field effects if the ion is embedded in a crystalline matrix) may become of relevance.

In removing the degeneracy of a partially filled electronic shell, the first physical mechanism is the electron–electron Coulomb interaction. In general, this can be taken into account through the Russell–Saunders (or LS) coupling scheme: the orbital angular momentum vectors $\mathbf{l}_i$ of the electrons combine to give a resultant vector $\mathbf{L}$, and the spin angular momentum vectors $\mathbf{s}_i$ of the electrons combine to give a resultant vector $\mathbf{S}$. The energy levels so obtained are indicated with the notation ^{2S+1}L (where the value of L is indicated with the letters S, P, D, F, G, H, I for $L = 0, 1, 2, \dots, 6$, respectively). A state LS, with total orbital angular momentum L and total spin S, is degenerate $(2L + 1)(2S + 1)$ times. If we include spin–orbit interaction in the form $\lambda \mathbf{L} \cdot \mathbf{S}$, a state LS is split into multiplets of definite total angular momentum J with $J = |L - S|, \dots, |L + S|$; a multiplet is indicated with the notation $^{2S+1}L_J$.

It is well known that the degeneracy of a given multiplet is fully removed by a (weak) magnetic field, that produces $2J + 1$ magnetic levels, separated by $g\mu_B$; thus, the (effective) magnetic moment associated with a given multiplet is

$$\boldsymbol{\mu} = -g\,\mu_B\,\mathbf{J} \tag{14a}$$

where $g = g(JLS)$ is the Landé factor and μ_B is the Bohr magneton. For a given multiplet (JLS assigned), the Landé factor is expressed as

$$g = 1 + \frac{J(J+1) + S(S+1) - L(L+1)}{2J(J+1)} \tag{14b}$$

(for this and other aspects of atomic properties, see for instance L. D. Landau and E. M. Lifshitz "Quantum Mechanics" Pergamon Press, Oxford 1958, vol.3).

The determination of the states arising from a given electronic configuration is in general rather complicated; however in the study of magnetism we often need only the ground-state quantum numbers SLJ. Except for the heaviest ions, where spin–orbit coupling is rather strong and the Russell–Saunders coupling scheme may become inadequate, the lowest lying spectroscopic term can be obtained by applying Hund's rules. These give a simple prescription for obtaining the quantum numbers SLJ for the ground state of atoms (or ions) with incomplete filled shell.

The ground state for a system of equivalent electrons (electrons belonging to the same shell) is determined by the following three rules, essentially due to exchange, correlation and relativistic effects.

(i) *Hund's first rule.* The electron spins combine to give the maximum multiplicity $2S + 1$ consistent with the Pauli principle. This occurs because parallel spins imply electrons tendentially kept apart in real space, thus decreasing the Coulomb repulsion.

(ii) *Hund's second rule.* The orbital momenta combine to give the maximum L consistent with the maximum spin multiplicity of rule (i) and with the exclusion principle. This is justified by the fact that, at parity of other conditions, the electron wavefunctions of higher L are more spread in real space and electron–electron Coulomb interaction is minimized.

(iii) *Hund's third rule.* The angular momenta $\mathbf{S}$ and $\mathbf{L}$ couple antiparallel giving $J = |L - S|$ if the shell is less than half-filled; $\mathbf{S}$ and $\mathbf{L}$ couple parallel giving $J = L + S$ for more than half-filled shells. The parallel or antiparallel coupling can be justified with the electron–hole symmetry of partially occupied or partially empty shells.

As an example, consider the ion Fe^{2+}, whose external electron configuration is $3d^6$. The lowest spectroscopic term is obtained from the Hund's rules, as schematically shown in Fig. 1. We indicate the five one-electron d-orbitals with $l_z = 2, 1, 0, -1, -2$, and accommodate the electrons following the Hund's rules. To obtain the highest value of S_Z, we place as many electrons as possible with spin up. Since there are five d orbitals, we can place 5 electrons with spin up and the sixth electron with spin down in the $l_z = 2$ level. We have thus $S = S_Z^{max} = 2$ and $L = L_Z^{max} = 2$. Since the shell is more than half-filled, the value of the total angular momentum is $J = L + S = 4$; the ground state of Fe^{2+} is thus $^5D_4$.

As another example consider the ion Cr^{3+}, whose external electron configuration is $3d^3$. Then $S = 1/2 + 1/2 + 1/2 = 3/2$ and $L = 2 + 1 + 0 = 3$; since the shell is less than half-filled we have $J = |L - S| = 3/2$. The ground state is thus $^4F_{3/2}$. With the Hund's rules, the quantum numbers of the ground states of atoms or ions can be easily worked out.

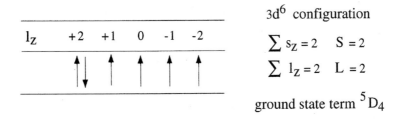

$$\sum s_z = 2 \quad S = 2$$

$$\sum l_z = 2 \quad L = 2$$

ground state term $^5D_4$

Fig. 1 Application of the Hund's rules for the ground state of Fe^{2+} (configuration $3d^6$).

4   Paramagnetism of localized magnetic moments

Consider an atom (or ion) whose ground state has angular momentum J and magnetic moment $\boldsymbol{\mu} = -g\,\mu_B\,\mathbf{J}$. The interaction energy of the magnetic moment with an applied magnetic field can be described by the Zeeman operator

$$H_Z = -\boldsymbol{\mu}\cdot\mathbf{H} = g\,\mu_B\,\mathbf{J}\cdot\mathbf{H}\ . \tag{15}$$

In the magnetic field, the ground state of the atom splits into $2J+1$ Zeeman sublevels separated by the energy $g\,\mu_B H$.

We suppose that, at temperatures of practical interest, only the $2J+1$ lowest states can be thermally excited. In this case the thermodynamical properties can be calculated starting from the partition function

$$Z = \sum_{m=-J}^{+J} e^{-m\,g\,\mu_B H/k_B T}\ . \tag{16}$$

We confine first our attention to the case $J = 1/2$. With two terms only in the sum (16), the partition function becomes

$$Z = e^{g\,\mu_B H\,J/k_B T} + e^{-g\,\mu_B H\,J/k_B T} \quad (J = 1/2)\ .$$

The thermal average $\langle\mu_z\rangle$ at a given temperature is

$$\langle\mu_z\rangle = k_B T \frac{\partial \ln Z}{\partial H} \equiv g\,\mu_B J \frac{e^{g\,\mu_B H\,J/k_B T} - e^{-g\,\mu_B H\,J/k_B T}}{e^{g\,\mu_B H\,J/k_B T} + e^{-g\,\mu_B H\,J/k_B T}} = g\,\mu_B J \tanh \frac{g\,\mu_B H\,J}{k_B T}\ .$$

The average magnetization $M = (N/V)\langle\mu_z\rangle$ in the field direction of a paramagnetic gas with N (independent) atoms in the volume V takes the expression

$$\boxed{M = \frac{N}{V} g\,\mu_B J \tanh \frac{g\,\mu_B H\,J}{k_B T}} \quad (J = 1/2)\ . \tag{17}$$

In most ordinary conditions it holds $\mu_B H \ll k_B T$ (*paramagnetic region*); in this case the spacing of Zeeman sublevels is much smaller than the thermal energy $k_B T$, and $\tanh x$ can be approximated with x in Eq. (17). Remembering that $\mu^2 = g^2 \mu_B^2 J(J+1) =$

$(3/4)g^2\mu_{\rm B}^2$ for $J = 1/2$, we obtain

$$\boxed{\chi = \frac{M}{H} = \frac{N}{V}\frac{\mu^2}{3k_{\rm B}T}} \; ; \tag{18}$$

the dependence $\chi \propto 1/T$ of the magnetic susceptibility on the temperature is known as Curie law. From measurements of χ as a function of $1/T$ one can infer the experimental value of μ. The Curie law maintains its validity down to a few Kelvin degrees in typical magnetic fields.

We can estimate the order of magnitude of the paramagnetic susceptibility (18). For $\mu = \mu_{\rm B} = e\hbar/2mc$, $T = T_0 = 273.15\,{\rm K}$, and $V = N\Omega$, expression (18) becomes

$$\chi = \frac{1}{\Omega}\left(\frac{e\hbar}{2mc}\right)^2\frac{1}{3k_{\rm B}T_0} = \frac{a_{\rm B}^3}{\Omega}\frac{e^2}{6mc^2}\frac{1}{a_{\rm B}}\frac{\hbar^2}{2ma_{\rm B}^2}\frac{1}{k_{\rm B}T_0} \; ;$$

with $e^2/mc^2 = 2.82 \cdot 10^{-13}\,{\rm cm}$, $a_{\rm B} = 0.529 \cdot 10^{-8}\,{\rm cm}$, $\hbar^2/2ma_{\rm B}^2 = 1\,{\rm Ryd.} = 13.606\,{\rm eV}$, $k_{\rm B}T_0 = 0.0235\,{\rm eV}$, and $\Omega \approx 100\,a_{\rm B}^3$, we obtain $\chi \approx 10^{-4}$. Equation (18) thus gives as order of magnitude $\chi \approx 10^{-4}(\mu^2/\mu_{\rm B}^2)(T_0/T)$.

At very low temperatures and very high magnetic fields one can have situations in which the thermal energy is much smaller than the spacing of the Zeeman sublevels (we remember that $\mu_{\rm B} = 0.05788\,{\rm meV/Tesla}$ and $k_{\rm B} = 0.0862\,{\rm meV/K}$). When $k_{\rm B}T \ll \mu_{\rm B}H$, $\tanh x \approx 1$ and Eq. (17) gives

$$M = \frac{N}{V}g\,\mu_{\rm B}J \; ; \tag{19}$$

this saturation value denotes that all magnetic moments are lined up along **H**.

Although we have considered specifically the case $J = 1/2$, the cases with higher values of J lead to quite similar physical results, and simply require a little more algebra. The partition function for a generic J is

$$Z = \sum_{m=-J}^{+J} e^{-m\,g\,\mu_{\rm B}H/k_{\rm B}T} = \frac{\sinh[(2J+1)g\,\mu_{\rm B}H/2k_{\rm B}T]}{\sinh[g\,\mu_{\rm B}H/2k_{\rm B}T]} \; . \tag{20}$$

In fact, defining $\alpha = g\,\mu_{\rm B}H/k_{\rm B}T$, the sum over m in Eq. (20) can be recognized as the geometrical series of $2J+1$ terms with first term $\exp(\alpha J)$ and ratio $\exp(-\alpha)$; it follows

$$Z = e^{\alpha J}\frac{1-e^{-\alpha(2J+1)}}{1-e^{-\alpha}} = \frac{e^{\alpha(2J+1)/2}-e^{-\alpha(2J+1)/2}}{e^{\alpha/2}-e^{-\alpha/2}} = \frac{\sinh[(2J+1)\alpha/2]}{\sinh(\alpha/2)} \; .$$

Using Eq. (20) we have for the thermal average of the magnetization in the direction of the applied field

$$\langle\mu_z\rangle = k_{\rm B}T\frac{\partial\ln Z}{\partial H} \equiv g\,\mu_{\rm B}J\,B_J\left(\frac{g\,\mu_{\rm B}H\,J}{k_{\rm B}T}\right) \; ,$$

where

$$B_J(x) = \frac{2J+1}{2J}\coth\frac{(2J+1)x}{2J} - \frac{1}{2J}\coth\frac{x}{2J}$$

Table 1 Ground states of rare earth ions with partially filled f shell, and effective Bohr magneton numbers.

ion	electronic configuration	ground state term	calculated $p = g[J(J+1)]^{1/2}$	measured p
La^{3+}	$4f^0 5s^2 5p^6$	$^1S_0$	0.	diamagnetic
Ce^{3+}	$4f^1 5s^2 5p^6$	$^2F_{5/2}$	2.54	2.4
Pr^{3+}	$4f^2 5s^2 5p^6$	$^3H_4$	3.58	3.5
Nd^{3+}	$4f^3 5s^2 5p^6$	$^4I_{9/2}$	3.62	3.5
Pm^{3+}	$4f^4 5s^2 5p^6$	$^5I_4$	2.68	—
Sm^{3+}	$4f^5 5s^2 5p^6$	$^6H_{5/2}$	0.84	1.5
Eu^{3+}	$4f^6 5s^2 5p^6$	$^7F_0$	0.	3.4
Gd^{3+}	$4f^7 5s^2 5p^6$	$^8S_{7/2}$	7.94	8.0
Tb^{3+}	$4f^8 5s^2 5p^6$	$^7F_6$	9.72	9.5
Dy^{3+}	$4f^9 5s^2 5p^6$	$^6H_{15/2}$	10.65	10.6
Ho^{3+}	$4f^{10} 5s^2 5p^6$	$^5I_8$	10.61	10.4
Er^{3+}	$4f^{11} 5s^2 5p^6$	$^4I_{15/2}$	9.58	9.5
Tm^{3+}	$4f^{12} 5s^2 5p^6$	$^3H_6$	7.56	7.3
Yb^{3+}	$4f^{13} 5s^2 5p^6$	$^2F_{7/2}$	4.54	4.5
Lu^{3+}	$4f^{14} 5s^2 5p^6$	$^1S_0$	0.	diamagnetic

is called *Brillouin function*. The average magnetization in the direction of H for a paramagnetic gas, with N independent atoms of spin J in the volume V, becomes

$$M = \frac{N}{V} g\,\mu_{\mathrm{B}} J\, B_J \left(\frac{g\,\mu_{\mathrm{B}} H\, J}{k_{\mathrm{B}} T} \right). \tag{21}$$

The Curie law $\chi = M/H \propto 1/T$ follows from the limit $B_J(x) \to x(J+1)/3J$ for small x. The saturation region result (18) follows from the limit $B_J(x) \to 1$ for large x.

Paramagnetism of rare earth ions

Paramagnetism of atoms or ions requires the existence of partially filled electronic shells, and we consider here the rare earth group (incomplete $4f$ shell) and later the iron group (incomplete $3d$ shell). Other groups are the palladium group (incomplete $4d$ shell), the platinum group (incomplete $5d$ shell) and the uranium group or actinides (incomplete $5f$, $6d$ shells).

In Table 1 we report the ground states $^{2S+1}L_J$ of trivalent lanthanum group ions, obtained following the Hund's rules; we give also the calculated *effective number of Bohr magnetons* $p = g[J(J+1)]^{1/2}$. Finally, the values of p derived from measured values of χ and from the Curie law are also given for comparison; the experimental values refer to impurities in insulating crystals, and not to the free ion situation (see for further details the book of J. H. Van Vleck, or the book of S. Chikazumi, cited at the end of this chapter). A comparison is meaningful only if the crystalline potential has negligible effect on f-electrons; this is indeed a reasonable assumption because the paramagnetic electrons are in well localized inner shells and are screened by the outer-

Table 2 Ground states of ions with partially filled d shell, and effective Bohr magneton numbers.

ion	electronic configuration	ground-state term	$g[J(J+1)]^{1/2}$	$2[S(S+1)]^{1/2}$	p measured values
Ti^{3+} V^{4+}	$3d^1$	$^2D_{3/2}$	1.55	1.73	1.8
V^{3+}	$3d^2$	$^3F_2$	1.63	2.83	2.8
Cr^{3+} V^{2+}	$3d^3$	$^4F_{3/2}$	0.77	3.87	3.7 3.8
Mn^{3+} Cr^{2+}	$3d^4$	$^5D_0$	0.	4.90	5.0 4.8
Fe^{3+} Mn^{4+}	$3d^5$	$^6S_{5/2}$	5.92	5.92	5.9 5.9
Fe^{2+}	$3d^6$	$^5D_4$	6.70	4.90	5.4
Co^{2+}	$3d^7$	$^4F_{9/2}$	7.54	3.87	4.8
Ni^{2+}	$3d^8$	$^3F_4$	5.59	2.83	3.2
Cu^{2+}	$3d^9$	$^2D_{5/2}$	3.55	1.73	1.9

The "p calculated values" spanning header covers the two columns $g[J(J+1)]^{1/2}$ and $2[S(S+1)]^{1/2}$.

most $5s^2 5p^6$ electrons [however, crystalline electric field effects may become important at low temperatures, as in the case of praseodymium and its magnetic excitations; see for instance J. Jensen and A. R. Mackintosh "Rare Earth Magnetism" (Clarendon, Oxford 1991)]. It is seen from Table 1 that there is a good overall agreement between calculated and measured values of the effective number of Bohr magnetons, the only exceptions being Sm^{3+} and mainly Eu^{3+}; the discrepancy has been explained noticing that in this case excited levels mix to some extent with the ground state in the presence of a magnetic field, and the assumption of considering only the ground-state term becomes clearly inadequate.

Paramagnetism of iron group ions

We give in Table 2 the measured and computed effective number of Bohr magnetons for some iron group ions; the measured values of p, derived from the experimental susceptibility and from the Curie law, refer to impurities in insulating crystals (see for further details the book of J. H. Van Vleck or the book of S. Chikazumi, cited at the end of this chapter). We see that in this case the effective number of Bohr magnetons cannot be estimated by the relation $p = g[J(J+1)]^{1/2}$; rather it is reasonably well described by the relation $p = 2[S(S+1)]^{1/2}$. Thus there is evidence that the LS coupling is broken because of the crystal field effects, and that the orbital angular momentum L appears to be "quenched".

The origin of the quenching of the angular momentum, and the difference with respect to the rare earth ions, can be understood qualitatively with the following arguments. (i) The LS coupling in rare earth ions is larger because of the higher localization of f shells. (ii) The paramagnetic electrons are in an inner shell in rare earth ions, while they are in the external shell in transition group ions. Thus the crystalline potential is negligible on f states, because of screening of outermost electrons, while it is very important for d electrons with a consequent quenching of the orbital angular momentum $\mathbf{L}$, as discussed below.

Quenching of the orbital angular momentum: an example

We can understand qualitatively the tendency of the crystal field (or other similar perturbations) to quench the orbital angular momentum, considering the elementary example of an atom (or ion) with one electron in an incomplete d shell. When a magnetic field is applied to the free atom, the coupling of the orbital magnetic moment $-\mu_{\mathrm{B}}\boldsymbol{\ell}$ of the electron under consideration with the external field is described by the Zeeman operator

$$H_Z = \mu_{\mathrm{B}}\boldsymbol{\ell}\cdot\mathbf{H}\ . \tag{22}$$

The Zeeman operator, diagonalized within the fivefold degenerate wavefunctions of the d level, produces the splitting into five levels of energies $\pm 2\mu_{\mathrm{B}}H, \pm\mu_{\mathrm{B}}H, 0$.

Suppose now that the atom is put in a cubic environment; the crystal field separates the fivefold degeneracy of the d orbital into a triplet and a doublet, whose wavefunctions are

$$\begin{cases} \psi_1 \equiv d_{xz} = \sqrt{\dfrac{15}{\pi}}\dfrac{xz}{r^2} = \dfrac{1}{\sqrt{2}}(-Y_{21} + Y_{2-1}) \\[2ex] \psi_2 \equiv d_{xy} = \sqrt{\dfrac{15}{4\pi}}\dfrac{xy}{r^2} = \dfrac{-i}{\sqrt{2}}(Y_{22} - Y_{2-2}) \\[2ex] \psi_3 \equiv d_{yz} = \sqrt{\dfrac{15}{4\pi}}\dfrac{yz}{r^2} = \dfrac{i}{\sqrt{2}}(Y_{21} + Y_{2-1}) \end{cases} \tag{23a}$$

and

$$\begin{cases} \psi_4 \equiv d_{3z^2-r^2} = \sqrt{\dfrac{5}{16\pi}}\dfrac{3z^2 - r^2}{r^2} = Y_{20} \\[2ex] \psi_5 \equiv d_{x^2-y^2} = \sqrt{\dfrac{15}{16\pi}}\dfrac{x^2 - y^2}{r^2} = \dfrac{1}{\sqrt{2}}(Y_{22} + Y_{2-2}) \end{cases} \tag{23b}$$

(Y_{2m} with $m = -2, -1, \ldots, +2$ are the spherical harmonics of order 2). If we express the Zeeman operator (22) on the above basis (and take the magnetic field in the z direction), we obtain for the matrix elements $M_{ij} = \langle\psi_i|\mu_{\mathrm{B}}\, l_z\, H|\psi_j\rangle$ the expression

$$M = \mu_{\mathrm{B}}H \begin{pmatrix} 0 & 0 & -i & 0 & 0 \\ 0 & 0 & 0 & 0 & 2i \\ i & 0 & 0 & 0 & 0 \\ 0 & 0 & 0 & 0 & 0 \\ 0 & -2i & 0 & 0 & 0 \end{pmatrix}. \tag{24}$$

It is immediate to verify that the eigenvalues of the above 5×5 matrix are $\pm 2\mu_{\mathrm{B}}H, \pm\mu_{\mathrm{B}}H, 0$ (as obviously expected). However if we add a cubic crystal field contribution, this produces a decoupling between the upper left 3×3 corner and the lower right 2×2 corner of the matrix given in Eq. (24). We have that the states of the doublet (with both null eigenvalues) behave as if a *total quenching* of the orbital momentum had occured, while the states of the triplet behave as if a *partial quenching* of the angular momentum had occurred from $l = 2$ to $l = 1$. Although the treatment of realistic situations may be quite complicated, from the above simplified model the tendency of "suppression" of the orbital angular momentum by crystal field effects can be inferred. Also notice that if the degeneracy of the manifold is totally removed by the crystal field of a low symmetry environment, then the quenching of the orbital angular momentum would be complete since the diagonal matrix elements of $\boldsymbol{\ell}$ are all vanishing.

5   Localized magnetic states in normal metals

The embedding of a magnetic atom in a normal metal leads to a variety of interesting effects. The foreign magnetic impurity, coupled to the (easily polarizable) Fermi sea of the host material, may lose or not its magnetic properties. For instance, iron group elements when dissolved in non-magnetic metals can lose their magnetic moments; on the contrary rare earth elements generally maintain their magnetic moments. Even more important for the entailed consequences, is the fact that *a preserved magnetic moment coupled to the Fermi sea can lead to a correlated non-magnetic ground state of the whole system*, constituted by the magnetic impurity plus Fermi sea. Very peculiar properties then occur at temperatures sufficiently low that thermal excitations are negligible. In this section we study the problem of preservation or disappearance of magnetic moment for impurities embedded in normal metals, while in the next section we consider the occurrence of a correlated non-magnetic ground state.

The presence of a foreign magnetic atom in a normal metal gives rise to a complex situation, that can be analysed following the Anderson reasoning on local magnetic moment preservation or loss. The system under consideration consists of a potentially magnetic impurity, embedded in a normal metal; in its essential aspects, the system can be described by means of a few phenomenological parameters through the following Anderson model

$$ H = \sum_{\mathbf{k}\sigma} E_{\mathbf{k}} n_{\mathbf{k}\sigma} + E_d(n_{d\uparrow} + n_{d\downarrow}) + \frac{1}{\sqrt{N}} \sum_{\mathbf{k}\sigma} \left[V_{\mathbf{k}d} c_{\mathbf{k}\sigma}^\dagger c_{d\sigma} + V_{\mathbf{k}d}^* c_{d\sigma}^\dagger c_{\mathbf{k}\sigma} \right] + U n_{d\uparrow} n_{d\downarrow} \ . \quad (25) $$

The first term in the right-hand side of Eq. (25) represents the conduction band of the host normal metal (for instance an s-like conduction band); $E_{\mathbf{k}}$ is the energy of the conduction band electronic state of wavevector $\mathbf{k}$ (and spin σ); $c_{\mathbf{k}\sigma}^\dagger$ and $c_{\mathbf{k}\sigma}$ are the corresponding creation and annihilation operators; $n_{\mathbf{k}\sigma} = c_{\mathbf{k}\sigma}^\dagger c_{\mathbf{k}\sigma}$ is the electron number operator. The second term in Eq. (25) represents a single localized non-degenerate impurity orbital of energy E_d (the possible degeneracy of the impurity orbital does not change the essential aspects of the treatment and is thus neglected); $n_{d\sigma} = c_{d\sigma}^\dagger c_{d\sigma}$ is the number operator for the impurity orbital. The third term in Eq. (25), called sd mixing, represents the hybridization energy between the localized state and the con-

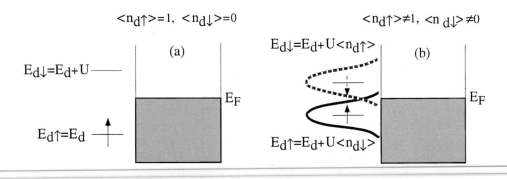

$\langle n_{d\uparrow}\rangle = 1,\ \langle n_{d\downarrow}\rangle = 0$

$\langle n_{d\uparrow}\rangle \neq 1,\ \langle n_{d\downarrow}\rangle \neq 0$

(a)

(b)

$E_{d\downarrow} = E_d + U \langle n_{d\uparrow}\rangle$

$E_{d\downarrow} = E_d + U$ ——

E_F

E_F

$E_{d\uparrow} = E_d$

$E_{d\uparrow} = E_d + U \langle n_{d\downarrow}\rangle$

Fig. 2 Schematic representation of the Anderson model. In Fig. (a) the hybridization of the localized level with the Fermi sea is neglected; it is assumed that $E_d < E_F$ and $E_d + U > E_F$, so that $\langle n_{d\uparrow}\rangle = 1$ and $\langle n_{d\downarrow}\rangle = 0$ (or vice versa); then $E_{d\uparrow} = E_d + U\langle n_{d\downarrow}\rangle = E_d$, while $E_{d\downarrow} = E_d + U$. In Fig. (b) the hybridization is switched on, and it tends to equalize spin-up and spin-down occupancies and energies.

duction band wavefunctions of the metal; the interaction energy between the Bloch wavefunction of vector $\mathbf{k}$ and the localized d-orbital is proportional to $1/\sqrt{N}$ (where N is the number of unit cells of the metal) and is denoted as $V_{\mathbf{k}d}/\sqrt{N}$; for simplicity, in the following $V_{\mathbf{k}d}$ is taken to be independent from $\mathbf{k}$ and is indicated as V_{sd}. The last term in Eq. (25) represents the Coulomb repulsive energy ($U > 0$) between electrons of opposite spins in the localized d orbital. It is evident that the Anderson Hamiltonian (25) contains two competing terms: the Coulomb repulsion between opposite spins tends to create unbalance of spin-up and spin-down occupancy of the localized state, while the hybridization tends to equalize them. This is indicated schematically in Fig. 2.

The basic difficulty toward a rigorous solution of the Hamiltonian (25) is the presence of the correlation term $U n_{d\uparrow} n_{d\downarrow} = U\, c_{d\uparrow}^\dagger c_{d\uparrow}\, c_{d\downarrow}^\dagger c_{d\downarrow}$, product of four creation or annihilation operators. In order to find a workable (although approximate) way to handle it, we first drop it altogether and then we attempt some approximate account. We consider thus the Hamiltonian H_0 obtained from Eq. (25) in the particular case $U \equiv 0$; we have

$$H_0 = \sum_{\mathbf{k}\sigma} E_{\mathbf{k}} n_{\mathbf{k}\sigma} + E_d\,(n_{d\uparrow} + n_{d\downarrow}) + \frac{1}{\sqrt{N}} \sum_{\mathbf{k}\sigma} \left[V_{sd}\, c_{\mathbf{k}\sigma}^\dagger c_{d\sigma} + V_{sd}^*\, c_{d\sigma}^\dagger c_{\mathbf{k}\sigma} \right]. \quad (26)$$

The Hamiltonian (26) is an ordinary one-electron Hamiltonian, representing a localized state interacting with a continuum. It is evident that spin-up and spin-down occupancy of the localized d state are equal, and no net magnetic moment exists. Furthermore, following the same concepts and techniques adopted in the case of the Fano effect (see Section XII-6), we have that the localized state E_d develops into a resonance, because of the hybridization with the band continuum.

The shape of the resonance can be easily worked out from the diagonal matrix element of the Green's function $G^0(E) = (E - H_0)^{-1}$ on the localized state $c_{d\sigma}^\dagger |0\rangle$; by means of the renormalization approach (see Section V-8.4) we immediately obtain

$$G_{dd}^0(E) = \cfrac{1}{E - E_d - |V_{sd}|^2 \cfrac{1}{N} \sum_{\mathbf{k}} \cfrac{1}{E - E_{\mathbf{k}}}} \tag{27}$$

where, as usual, the energy E is defined with the addition of a small imaginary part $i\varepsilon$ and the limit $\varepsilon \to 0^+$ is understood.

The sum over $\mathbf{k}$ appearing at the denominator of Eq. (27) can be split into a real part and an imaginary part; the real part represents an energy shift of the resonant level (and can be supposed to be embodied in E_d itself); for the imaginary part, in the limit $\varepsilon \to 0^+$, we have

$$\text{Im} \sum_{\mathbf{k}} \frac{1}{E + i\varepsilon - E_{\mathbf{k}}} = -\sum_{\mathbf{k}} \frac{\varepsilon}{(E - E_{\mathbf{k}})^2 + \varepsilon^2} = -\pi \sum_{\mathbf{k}} \delta(E - E_{\mathbf{k}}) = -\pi D_0(E)$$

where $D_0(E)$ denotes the density-of-states of the conduction band of the unperturbed metal (for one spin direction). In the following, for simplicity, we also assume that $D_0(E) \approx D_0(E_F)$ in the energy range of interest around the Fermi energy.

With the above approximations, the Green's function (27) can be written as

$$G_{dd}^0(E) = \frac{1}{E - E_d + i\Gamma} \, , \tag{28a}$$

where

$$\Gamma = \pi |V_{sd}|^2 \frac{1}{N} D_0(E_F) = \pi |V_{sd}|^2 n_0(E_F) \, ; \tag{28b}$$

the quantity $n_0(E_F) = D_0(E_F)/N$ denotes the electronic density-of-states at the Fermi level for one spin direction and per unit cell. The order of magnitude of $n_0(E_F)$ is $n_0(E_F) \approx 1/W_F$, where W_F is the energy width of the part of the conduction band (from its bottom to the Fermi energy) occupied by electrons. The parameter Γ is thus determined by the hybridization energy $|V_{sd}|$ times the dimensionless ratio $|V_{sd}| n_0(E_F) \approx |V_{sd}|/W_F$. The *local density-of-states at the impurity site* is

$$n_d^0(E) = -\frac{1}{\pi} \text{Im} \, G_{dd}^0(E) = \frac{1}{\pi} \frac{\Gamma}{(E - E_d)^2 + \Gamma^2} \, ; \tag{28c}$$

the local density-of-states has thus a Lorentzian shape with half-width parameter Γ.

The integrated density-of-states up to the Fermi energy gives the average number of electrons at the impurity site with given spin direction, and can be easily calculated using the indefinite integral

$$\int \frac{b}{(x - a)^2 + b^2} \, dx = \arctan \frac{x - a}{b} \, .$$

We have

$$\langle n_d^0 \rangle = \int_{-\infty}^{E_F} n_d^0(E) \, dE = \frac{1}{2} + \frac{1}{\pi} \arctan \frac{E_F - E_d}{\Gamma} \, . \tag{29}$$

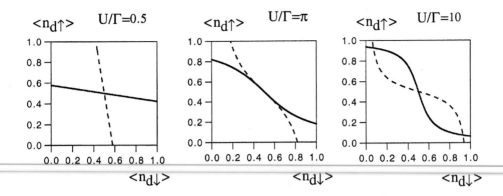

Fig. 3 Graphical solution of the self-consistent equations (31) for different values of the dimensionless ratio U/Γ.

We have now to consider (at least in some approximate form) the correlation term $U\,n_{d\uparrow}\,n_{d\downarrow}$ so far neglected. We use the unrestricted Hartree–Fock approximation to replace the operator $U\,n_{d\uparrow}\,n_{d\downarrow}$ with $U\,n_{d\uparrow}\,\langle n_{d\downarrow}\rangle$ for spin-up electrons and with $U\,\langle n_{d\uparrow}\rangle\,n_{d\downarrow}$ for spin-down electrons. Following the same arguments used above for obtaining Eq. (29), we arrive at the two coupled equations

$$
\begin{aligned}
\langle n_{d\uparrow}\rangle &= \tfrac{1}{2} + \tfrac{1}{\pi}\arctan\frac{E_F - E_d - U\langle n_{d\downarrow}\rangle}{\Gamma} \\
\langle n_{d\downarrow}\rangle &= \tfrac{1}{2} + \tfrac{1}{\pi}\arctan\frac{E_F - E_d - U\langle n_{d\uparrow}\rangle}{\Gamma}
\end{aligned}
\tag{30}
$$

These two equations must be solved self-consistently; it is seen by inspection that one can have solutions with $\langle n_{d\uparrow}\rangle \equiv \langle n_{d\downarrow}\rangle$ corresponding to non-magnetic behaviour, and solutions $\langle n_{d\uparrow}\rangle \neq \langle n_{d\downarrow}\rangle$ corresponding to magnetic behaviour.

Consider, for instance, the case in which the energy of the d state is located at $U/2$ below the Fermi energy $(E_F - E_d = \tfrac{1}{2}U)$. In this case Eqs. (30) take the simplified form

$$
\begin{cases}
\langle n_{d\uparrow}\rangle = \tfrac{1}{2} + \tfrac{1}{\pi}\arctan\left[\tfrac{U}{\Gamma}(\tfrac{1}{2} - \langle n_{d\downarrow}\rangle)\right] \\
\langle n_{d\downarrow}\rangle = \tfrac{1}{2} + \tfrac{1}{\pi}\arctan\left[\tfrac{U}{\Gamma}(\tfrac{1}{2} - \langle n_{d\uparrow}\rangle)\right]
\end{cases}
\tag{31}
$$

We can solve graphically the self-consistent equations (31), as illustrated in Fig. 3. We notice that when $\Gamma \gg U$ we have *only* non-magnetic solutions of type $\langle n_{d\uparrow}\rangle \equiv \langle n_{d\downarrow}\rangle = 1/2$; if $\Gamma \ll U$ we have also the magnetic solution $\langle n_{d\uparrow}\rangle \approx 1$, $\langle n_{d\downarrow}\rangle \approx 0$ (or vice versa). The transition between the magnetic regime and the non-magnetic regime occurs when $\pi\Gamma = U$. With similar procedures, we can analyse the occurrence or not of magnetic and non-magnetic solutions of Eqs. (30) at any values of the dimensionless parameters $\pi\Gamma/U$ and $(E_F - E_d)/U$, and obtain the phase diagram reported in Fig. 4.

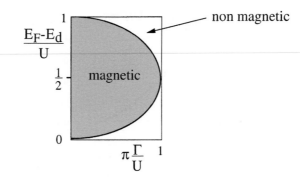

Fig. 4 Phase diagram corresponding to the Hartree–Fock solution of the Anderson Hamiltonian for a magnetic impurity in a normal metal [from P. W. Anderson, Phys. Rev. B**124**, 41 (1961)].

As qualitative application of what has been said so far, we summarize in Table 3 the existence or non-existence of localized magnetic moments for transition element impurities dissolved in various host metals; the existence of a localized magnetic moment can be inferred experimentally from a Curie-type temperature-dependent contribution to the magnetic susceptibility. From Table 3, it can be seen, for instance, that the Cr impurity remains magnetic when inserted in metals such as Au, Cu, Ag, and is non-magnetic when inserted in Al, because of the higher value of the density-of-states at the Fermi level in the latter case. Similarly, it can be seen for instance that Cr, Mn and Fe impurities remain magnetic in Au, while Ti and Ni impurities are non-magnetic in Au; this trend can be understood considering that the orbital energy E_d decreases passing from Ti to Ni; at the beginning of the series (when E_d is higher or near the Fermi level of gold) and at the end of the series (when E_d is well below the Fermi level of gold) no magnetic moment is possible. It can also be noticed that, for iron group impurities, magnetism may or may not occur (estimates of U and Γ parameters are $U \approx 10\,\mathrm{eV}$ and $\Gamma \approx 2 - 5\,\mathrm{eV}$). For rare earth impurities ($U \approx 15\,\mathrm{eV}$ and $\Gamma \approx 1\,\mathrm{eV}$ or less) the magnetic regime is expected to occur very frequently.

6   Dilute magnetic alloys and the resistance minimum phenomenon

6.1   Some phenomenological aspects

In the previous section, we have seen that a magnetic impurity, when inserted in an ordinary metal, may lose or preserve its magnetic moment; which one of the two cases is likely to occur, has been discussed on the basis of the phenomenological Anderson model. In this section, we consider some transport properties of normal metals containing a small concentration of impurity atoms, which remain magnetic (*dilute magnetic alloys*). In experiments, typical concentrations are less than few hundred magnetic atoms per million host atoms (at this dilution, interaction between impuri-

Table 3 Existence or non-existence of magnetic moments for transition element impurities in various host metals [from A. J. Heeger, Solid State Physics **23**, 283 (1969)].

	Au	Cu	Ag	Al
Ti	no	–	–	no
V	?	–	–	no
Cr	yes	yes	yes	no
Mn	yes	yes	yes	?
Fe	yes	yes	–	no
Co	?	?	–	no
Ni	no	no	–	no

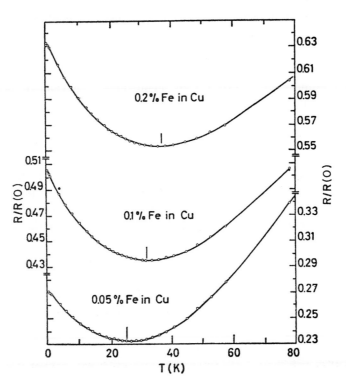

Fig. 5 Resistance at low temperatures for Cu with 0.05% , 0.1% and 0.2% of Fe impurities [from J. P. Franck, F. D. Manchester and D. L. Martin, Proc. Roy. Soc. (London) A**263**, 494 (1961)].

ties can be neglected); typical host materials are copper, silver, gold, magnesium, zinc; typical magnetic impurities are chromium, manganese, iron, cobalt, nickel, vanadium, titanium.

It is a well established fact that several thermal, electrical and magnetic properties of ordinary metals in the presence of dilute magnetic impurities appear to be "anomalous", with respect to the properties observed in the presence of ordinary (non-magnetic) impurities. For instance, the resistivity of an ordinary metal containing non-magnetic impurities is well described at low temperatures by a law of the type

$$\rho(T) = \rho_0[1 + AT^5 + \dots] , \tag{32}$$

where ρ_0 is the so-called residual resitivity due to the impurities, and the term with the fifth power of temperature is due to lattice vibrational effects. Thus the resistivity (and hence the resistance, which is proportional to the resistivity in a given chosen sample geometry) decreases monotonically with decreasing temperature and becomes constant at very low temperatures. On the contrary, the resistivity of a normal metal in the presence of dilute magnetic impurities exhibits a rather shallow minimum (Kondo minimum) at very low temperatures T_K, typically of the order of a few kelvin degrees; the resistivity eventually saturates (i.e. reaches a constant value) as the temperature is further decreased. In Fig. 5 we report the "resistance minimum phenomenon" for dilute alloys of iron in copper, as an example.

As the temperature is decreased also the magnetic moment of the impurity, embedded in the ordinary metal, exhibits an "anomalous" behaviour; in particular the magnetic susceptibility ceases to follow the Curie law, and rather saturates to a constant value at $T = 0$, thus indicating that the magnetic moment of the impurity is fully compensated and quenched by the conduction electrons of the host metal at very low temperatures. This is also confirmed by an extra contribution to the specific heat, corresponding to an entropy change of the order of $k_B \ln 2$, for each magnetic impurity.

Resistivity, magnetic susceptibility, specific heat and several other properties such as magnetoresistance, thermoelectric effects, appear to be anomalous in ordinary metals with dilute magnetic impurities as the temperature is decreased in the neighbourhood or below the temperature T_K. All these effects and phenomenology are collectively referred to as the *Kondo effect* and the essentially new physical situation that occurs at low temperatures is referred to as the *Kondo problem*. In essence this problem consists in investigating the quantum properties of a system made by a single localized magnetic impurity, interacting with a degenerate free electron gas. It is not our purpose to consider this system in all the details; rather we focus on semi-quantitative models, that can at least provide some guidelines in the complicated and challenging Kondo phenomenology.

6.2   The resistance minimum phenomenon

An impurity, magnetic or non-magnetic, embedded in a normal metal interacts with the conduction electrons of the metal, thus giving a contribution to the resistivity. The reason why in dilute magnetic alloys the resistivity presents a broad minimum, while for non-magnetic alloys the resistivity drops monotonically to the residual resistivity, is related to the different kind of perturbation produced by impurities with or without magnetic moments. Such an effect is rather subtle, as the essential difference between the scattering amplitudes of spin-dependent and spin-independent interaction terms appears only at second (and higher orders) of perturbation theory, while at first order of perturbation theory there is no qualitative difference between them.

Let us consider the simplest possible interaction Hamiltonian for a normal impurity, and then for a magnetic impurity, embedded in an ordinary metal. For a normal impurity (located at the point $\mathbf{r} = 0$), the perturbation acting on the free-electron gas can be described in the one-electron picture by an impurity potential $V_{\text{imp}}(\mathbf{r})$, which extends in the space to a distance of the order of k_F^{-1} from the origin.

For a magnetic impurity of spin $\mathbf{S}$ (localized at the origin), it seems reasonable to assume a coupling with the conduction electrons of the form

$$H_K = -J\Omega\, \mathbf{S} \cdot \mathbf{s}\, \delta(\mathbf{r}) , \tag{33}$$

where $\mathbf{s}$ is the spin of the electron in the conduction band, Ω is the volume of the unit cell, and J is a phenomenological parameter with the dimension of an energy [in the case Ω is omitted in Eq. (33), J becomes a phenomenological parameter with the dimension of energy times volume]. The contact type form (33) proposed by Kondo, is justified by more detailed microscopic models, which also show that in general $J < 0$ (antiferromagnetic spin–spin interaction); thus the interaction (33) shows the tendency of screening of the localized magnetic moments by the conduction band electron spins. Kondo first showed that the scattering by a spin-dependent potential of the type (33) could explain the resistivity minimum which is observed in ordinary metals containing magnetic impurities. An elementary justification of the phenomenological spin-dependent interaction (33) is considered below in Section 6.3.

First-order calculation of scattering amplitude from a magnetic impurity

We remember that the spin of an electron is $1/2$ (in units $\hbar$) and the spin operators are represented by the matrices

$$s_x = \frac{1}{2}\begin{pmatrix} 0 & 1 \\ 1 & 0 \end{pmatrix} , \quad s_y = \frac{1}{2}\begin{pmatrix} 0 & -i \\ i & 0 \end{pmatrix} , \quad s_z = \frac{1}{2}\begin{pmatrix} 1 & 0 \\ 0 & -1 \end{pmatrix} .$$

We indicate with $\alpha = \begin{pmatrix} 1 \\ 0 \end{pmatrix}$ and $\beta = \begin{pmatrix} 0 \\ 1 \end{pmatrix}$ the standard spin-up and spin-down wavefunctions of the operator s_z; with straightforward calculations we have

$$
\begin{aligned}
\langle \alpha | \mathbf{s} \cdot \mathbf{S} | \alpha \rangle &= \tfrac{1}{2} S_z , & \langle \beta | \mathbf{s} \cdot \mathbf{S} | \alpha \rangle &= \tfrac{1}{2}(S_x + i\, S_y) = \tfrac{1}{2} S_+ , \\
\langle \beta | \mathbf{s} \cdot \mathbf{S} | \beta \rangle &= -\tfrac{1}{2} S_z , & \langle \alpha | \mathbf{s} \cdot \mathbf{S} | \beta \rangle &= \tfrac{1}{2}(S_x - i\, S_y) = \tfrac{1}{2} S_- .
\end{aligned}
\tag{34}
$$

With the help of Eqs. (34), we can now calculate the matrix elements and the scattering amplitudes of interest.

At first-order of perturbation theory the spin-dependent interaction operator H_K gives an *energy-independent scattering amplitude* (and hence an energy independent scattering cross-section), and nothing basically new appears with respect to an ordinary impurity potential. Consider for instance the scattering amplitude for an electron of the Fermi sea from an initial state $\mathbf{k}_i \uparrow$ to the final state $\mathbf{k}_f \uparrow$ (as usual, plane waves are normalized to one on the volume $V = N\Omega$ of the crystal); we have

$$t^{(1)}_{\mathbf{k}_f \alpha \leftarrow \mathbf{k}_i \alpha} = \langle \mathbf{k}_f \alpha | H_K | \mathbf{k}_i \alpha \rangle = \langle \mathbf{k}_f \alpha | - J\Omega\, \mathbf{S} \cdot \mathbf{s}\, \delta(\mathbf{r}) | \mathbf{k}_i \alpha \rangle$$

$$= \frac{1}{N\Omega} \int e^{-i\,\mathbf{k}_f \cdot \mathbf{r}}\, \delta(\mathbf{r})\, e^{-i\,\mathbf{k}_i \cdot \mathbf{r}}\, d\mathbf{r} (-J\Omega)\langle \alpha | \mathbf{S} \cdot \mathbf{s} | \alpha \rangle = -\frac{J}{N}\frac{1}{2}S_z \ . \tag{35}$$

Similar expressions can be obtained for scattering amplitude between spin–orbitals with different spins. Scattering amplitudes and hence transition probabilities (obtained with appropriate averages over the initial magnetic impurity states) are anyway energy independent to first-order perturbation theory.

Second-order calculation of scattering amplitude from a magnetic impurity

To second order of perturbation theory, interesting new features appear. Quite generally we can write for the scattering amplitude

$$t^{(2)}_{\mathbf{k}_f \alpha \leftarrow \mathbf{k}_i \alpha} = \sum_{\mathbf{q}\sigma} \frac{\langle \mathbf{k}_f \alpha | H_K | \mathbf{q}\,\sigma \rangle \langle \mathbf{q}\,\sigma | H_K | \mathbf{k}_i \alpha \rangle}{E_{\mathbf{k}_i} - E_{\mathbf{q}}}(1 - f_{\mathbf{q}})$$

$$+(-1)\sum_{\mathbf{q}\sigma} \frac{\langle \mathbf{q}\,\sigma | H_K | \mathbf{k}_i \alpha \rangle \langle \mathbf{k}_f \alpha | H_K | \mathbf{q}\,\sigma \rangle}{E_{\mathbf{q}} - E_{\mathbf{k}_f}} f_{\mathbf{q}} \tag{36}$$

where $f_{\mathbf{q}} = 1$ for $E_{\mathbf{q}} < E_F$ and $f_{\mathbf{q}} = 0$ for $E_{\mathbf{q}} > E_F$; $1 - f_{\mathbf{q}}$ and $f_{\mathbf{q}}$ take into account the Pauli exclusion principle in the intermediate state; the intermediate state must be empty in direct processes (first term in the right-hand side of Eq. 36) and full in the exchange processes (second term in the right-hand side of Eq. 36). The two kinds of processes are schematically indicated in Fig. 6.

In elastic scattering $E_{\mathbf{k}_i} = E_{\mathbf{k}_f}$, and Eq. (36) becomes

$$t^{(2)}_{\mathbf{k}_f \alpha \leftarrow \mathbf{k}_i \alpha} = \sum_{\mathbf{q}\sigma} \frac{1}{E_{\mathbf{k}_i} - E_{\mathbf{q}}} \{ \langle \mathbf{k}_f \alpha | H_K | \mathbf{q}\,\sigma \rangle \langle \mathbf{q}\,\sigma | H_K | \mathbf{k}_i \alpha \rangle (1 - f_{\mathbf{q}})$$

$$+ \langle \mathbf{q}\,\sigma | H_K | \mathbf{k}_i \alpha \rangle \langle \mathbf{k}_f \alpha | H_K | \mathbf{q}\,\sigma \rangle f_{\mathbf{q}} \} \ . \tag{37}$$

For an "ordinary" (spin-independent) perturbation, the product of two matrix elements does not depend on their order; in the case in Eq. (37) the order is inessential, the terms with $f_{\mathbf{q}}$ cancel out exactly, and what is left represents in general a (small) correction to the first-order amplitude (for an "ordinary" scattering potential, the Pauli principle in the intermediate states appears as non operative, and can thus be neglected altogether). In the present case of the spin-dependent operator $H_K = -J\Omega\,\mathbf{S} \cdot \mathbf{s}\,\delta(\mathbf{r})$, the order is vital and this gives an essentially new effect. Using Eqs. (34), we can perform explicitly the sum over spin variables in Eq. (37) and obtain

$$t^{(2)}_{\mathbf{k}_f \alpha \leftarrow \mathbf{k}_i \alpha} = \left(\frac{J}{N}\right)^2 \frac{1}{4} \sum_{\mathbf{q}} \frac{1}{E_{\mathbf{k}_i} - E_{\mathbf{q}}} \{(S_z^2 + S_- S_+)(1 - f_{\mathbf{q}}) + (S_z^2 + S_+ S_-)\, f_{\mathbf{q}} \} \ .$$

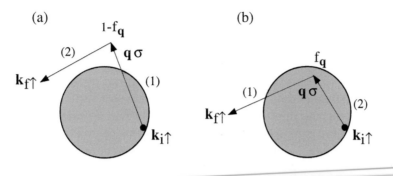

Fig. 6 (a) Direct process in second-order Born approximation for scattering of an electron from the initial state $\mathbf{k}_i \uparrow$ to the final state $\mathbf{k}_f \uparrow$ via an empty intermediate state $\mathbf{q}\sigma$. (b) Exchange process contributing to second-order scattering amplitude: the filled state $\mathbf{q}\sigma$ is first scattered to the final state $\mathbf{k}_f \uparrow$ and is then occupied by the electron in the state $\mathbf{k}_i \uparrow$. In Fig. (a) and Fig. (b) the initial states of the Fermi gas can be represented by the same Slater determinant; it is then easily seen that the two final Slater determinants, at the end of the processes of Fig. (a) and Fig. (b), have a pair of electronic functions exchanged; hence the factor (-1) in Eq. (36).

From standard commutation rules, we have

$$S_+ S_- = S_- S_+ + 2 S_z \; ,$$

and thus

$$t^{(2)}_{\mathbf{k}_f \alpha \leftarrow \mathbf{k}_i \alpha} = \left(\frac{J}{N} \right)^2 \frac{1}{4} \sum_{\mathbf{q}} \frac{1}{E_{\mathbf{k}_i} - E_{\mathbf{q}}} \left(S_z^2 + S_- S_+ + 2 S_z f_{\mathbf{q}} \right) \; . \tag{38}$$

In order to estimate the sum over $\mathbf{q}$, suppose that the conduction band extends in the energy interval $(E_F - \Delta_c, E_F + \Delta_c)$, and here the density-of-states (per one spin direction) is constant, say $D_0(E_F)$; Eq. (38) becomes

$$t^{(2)}_{\mathbf{k}_f \alpha \leftarrow \mathbf{k}_i \alpha} = \left(\frac{J}{N} \right)^2 \frac{1}{4} \left(S_z^2 + S_- S_+ \right) D_0(E_F) \int_{E_F - \Delta_c}^{E_F + \Delta_c} \frac{1}{E_{\mathbf{k}_i} - E} \, dE$$

$$+ \left(\frac{J}{N} \right)^2 \frac{1}{2} S_z \, D_0(E_F) \int_{E_F - \Delta_c}^{E_F} \frac{1}{E_{\mathbf{k}_i} - E} \, dE \; . \tag{39}$$

The integrations in Eq. (39) are straightforward, and it can be seen that for $E_{\mathbf{k}_i} \approx E_F$ the first term in the right-hand side of Eq. (39) becomes negligibly small, while the second term in the right hand side of Eq. (39) gives

$$t^{(2)}_{\mathbf{k}_f \alpha \leftarrow \mathbf{k}_i \alpha} = \left(\frac{J}{N} \right)^2 \frac{1}{2} S_z \, D_0(E_F) \ln \frac{\Delta_c}{|E_{\mathbf{k}_i} - E_F|} \; ; \tag{40}$$

thus $t^{(2)}$ is strongly energy dependent, and actually diverges when $E_{\mathbf{k}_i} = E_F$.

At a given temperature T, we must average $t^{(2)}$ for $E_{\mathbf{k}_i}$ in the thermal interval $k_B T$ around E_F; putting $E_{\mathbf{k}_i} - E_F \approx k_B T$ in Eq. (40), and using Eq. (35), we obtain for the scattering amplitude up to second order

$$t = t^{(1)} + t^{(2)} = -\frac{J}{N}\frac{1}{2}S_z\left[1 - J\,n_0(E_F)\ln\frac{\Delta_c}{k_BT}\right]\,, \tag{41}$$

where $n_0(E_F) = D_0(E_F)/N$ denotes the density-of-states at the Fermi level for one spin direction and per unit cell. The resistivity is related (after performing appropriate averages on initial spin directions and integrals on scattering angles) to the modulus square of the above expression; the temperature dependence is all embodied in the expression $[1 - J\,n_0(E_F)\ln(\Delta_c/k_BT)]^2$, and the resistivity takes thus the form

$$\rho(T) = A\left[1 - J\,n_0(E_F)\ln\frac{\Delta_c}{k_BT}\right]\,, \tag{42}$$

where A denotes a constant, and it is assumed $J\,n_0(E_F) \approx J/W_F \ll 1$ (W_F is the energy width of the part of the conduction band occupied by electrons).

From Eq. (42) we see that if $J < 0$, the contribution to the resistivity of the magnetic impurities increases logarithmically as the temperature approaches absolute zero. Since the contribution to the resistivity due to phonons decreases with decreasing temperature, we expect that the total resistivity goes through a minimum. Of course a more complete theory of the Kondo effect must take into account several other features, not embodied in the highly simplified model here considered; in particular, the increase in resistance on the low temperature side must eventually saturate, rather than diverge logarithmically. The strong energy dependence of the scattering relaxation time is also at the origin of the "giant" thermoelectric effect observed in some materials containing magnetic impurities.

6.3 Microscopic origin of the Kondo interaction: a molecular model

In the previous section we have seen that a phenomenological spin-dependent Hamiltonian of the type $H_K = -J\,\Omega\,\mathbf{S}\cdot\mathbf{s}\,\delta(\mathbf{r})$ can explain the resistance minimum phenomenon of dilute magnetic alloys. Here we consider the simplest justification of the contact term H_K coupling the spin of the magnetic impurity and the spins of the conduction electrons.

Before considering a more realistic approach for the localized magnetic moment in interaction with the Fermi sea (see Section 7), we make here the drastic assumption to mimic the Fermi sea with just *one extended orbital* ψ_k (of energy ε_k), which can be doubly occupied by electrons; the magnetic impurity is represented with a *single localized orbital* ψ_f (of energy ε_f), where double occupation is avoided because of Coulomb correlation effects. We consider thus the *two-orbital molecule* described by the Anderson-type Hamiltonian

$$H_m = \varepsilon_k\left[c_{k\uparrow}^{\dagger}c_{k\uparrow} + c_{k\downarrow}^{\dagger}c_{k\downarrow}\right] + \varepsilon_f\left[c_{f\uparrow}^{\dagger}c_{f\uparrow} + c_{f\downarrow}^{\dagger}c_{f\downarrow}\right]$$

$$+ V_{sf}\left[c_{k\uparrow}^{\dagger}c_{f\uparrow} + c_{k\downarrow}^{\dagger}c_{f\downarrow}\right] + V_{sf}^{*}\left[c_{f\uparrow}^{\dagger}c_{k\uparrow} + c_{f\downarrow}^{\dagger}c_{k\downarrow}\right] + Uc_{f\uparrow}^{\dagger}c_{f\uparrow}c_{f\downarrow}^{\dagger}c_{f\downarrow}\,. \tag{43}$$

The physical meaning of the various terms can be easily understood from analogy with Eq. (25). The "molecular model" described by the Hamiltonian H_m, is in fact the simplest version of the Anderson phenomenological Hamiltonian (25), with just one itinerant orbital ψ_k; the parameter U represents the Coulomb repulsive energy between electrons of opposite spin in the localized f orbital.

We consider now two electrons in the molecule described by the Hamiltonian (43). In the particular case $U = 0$, i.e. in the independent electron approximation, the Hamiltonian (43) trivially describes two orbitals, of energy ε_k and ε_f, interacting via the off-diagonal matrix element V_{sf}. We are thus left with the determinantal equation

$$\begin{vmatrix} \varepsilon_k - E & V_{sf} \\ V_{sf}^* & \varepsilon_f - E \end{vmatrix} = 0 \,,$$

which produces a bonding molecular state and an antibonding molecular state, of energies E_b and E_a respectively. We assume that for the hybridization parameter V_{sf} it holds $|V_{sf}| \ll \varepsilon_k - \varepsilon_f = \Delta$; in this case $E_b = \varepsilon_f - |V_{sf}|^2/\Delta$ and $E_a = \varepsilon_k + |V_{sf}|^2/\Delta$. The two electrons of the molecule can be put with antiparallel spins in the bonding orbital (singlet state $S=0$ with energy $2E_b = 2\varepsilon_f - 2|V_{sf}|^2/\Delta$), or in the antibonding orbital ($S=0$ and $2E_a = 2\varepsilon_k + 2|V_{sf}|^2/\Delta$), or one in the bonding orbital and the other in the antibonding orbital ($S=0, 1$, with energy $E_a + E_b = \varepsilon_k + \varepsilon_f$). The energy levels of the two-electron system, when $U=0$, are reported in Fig. 7a. It can be noticed that the ground state of the molecule is a singlet and the lowest excitation energy equals $\Delta + 2|V_{sf}|^2/\Delta \approx \Delta$; this excitation corresponds to the transfer of an electron from the bonding state to the antibonding state and can be pictured as a "charge fluctuation".

We consider now two electrons on the molecule described by the Hamiltonian (43) in the case correlation effects are important, and actually U is so large to prevent double occupancy of the impurity orbital. For finite values of U, the number of possible molecular states is 6; but in the limit $U \to +\infty$ we can confine our considerations to five states only (double occupation of the f state is avoided), as indicated in Fig. 8. [The one-impurity Anderson model and generalizations are discussed for instance by P. Fulde, J. Phys. F **18**, 601 (1988). For a variational study of the two-impurity Anderson model we refer to L. C. Andreani and H. Beck, Phys. Rev. B**48**, 7322 (1993) and references quoted therein].

From the five states listed in Fig. 8, we can form one triplet and two singlet states. The basis wavefunctions for the singlets are

$$\Psi_1^{(S=0)} = c_{k\uparrow}^\dagger c_{k\downarrow}^\dagger |0\rangle \quad \text{and} \quad \Psi_2^{(S=0)} = \frac{1}{\sqrt{2}} \left[c_{k\uparrow}^\dagger c_{f\downarrow}^\dagger - c_{k\downarrow}^\dagger c_{f\uparrow}^\dagger \right] |0\rangle \,.$$

The matrix elements of H_m on the states $\Psi_1^{(S=0)}$ and $\Psi_2^{(S=0)}$ are the following

$$H_{11} = 2\varepsilon_k \,, \quad H_{22} = \varepsilon_k + \varepsilon_f \,, \quad H_{12} = \sqrt{2} V_{sf} \,, \quad H_{21} = \sqrt{2} V_{sf}^* \,.$$

The determinantal equation $\|H_{ij} - E\,\delta_{ij}\| = 0$ gives the two eigenvalues

$$E = \varepsilon_k + \frac{1}{2} \left[\varepsilon_k + \varepsilon_f \pm \sqrt{(\varepsilon_k - \varepsilon_f)^2 + 8|V_{sf}|^2} \right] \,.$$

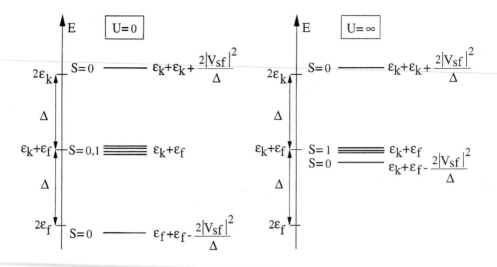

Fig. 7 Energy levels of the two-electron two-orbital molecule, with one extended orbital of energy ε_k and one localized orbital of energy ε_f, in the case the Coulomb repulsive energy U between electrons of opposite spin on the localized orbital is zero or infinity. The hybridization parameter $|V_{sf}|$ is assumed much smaller than $\Delta = \varepsilon_k - \varepsilon_f$; typical values are $\Delta \approx 1\,\mathrm{eV}$, $|V_{sf}| \approx 10$–$100\,\mathrm{meV}$, $|V_{sf}|^2/\Delta \approx 1\,\mathrm{meV}$.

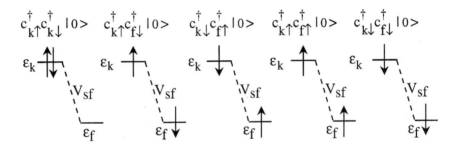

Fig. 8 Possible molecular states for two electrons in a two-orbital molecule (double occupancy of the localized orbital ψ_f is not possible in the $U=\infty$ limit considered in the figure). The extended orbital ψ_k (of energy ε_k) hybridizes with the localized orbital ψ_f (of energy ε_f) via the matrix element V_{sf}, leading to the formation of the singlet molecular ground state.

Suppose that the hybridization parameter is such that $|V_{sf}| \ll \varepsilon_k - \varepsilon_f \equiv \Delta$; by a series development we obtain

$$E_{\mathrm{ground}}^{(S=0)} = \varepsilon_k + \varepsilon_f - \frac{2|V_{sf}|^2}{\Delta} \, , \qquad E_{\mathrm{excited}}^{(S=0)} = \varepsilon_k + \varepsilon_k + \frac{2|V_{sf}|^2}{\Delta} \, . \qquad (44a)$$

For the triplet state, the wavefunctions for $S_z = 1, 0, -1$ are, respectively,

$$c_{k\uparrow}^{\dagger} c_{f\uparrow}^{\dagger} |0\rangle \, , \qquad \frac{1}{\sqrt{2}} \left[c_{k\uparrow}^{\dagger} c_{f\downarrow}^{\dagger} + c_{k\downarrow}^{\dagger} c_{f\uparrow}^{\dagger} \right] |0\rangle \, , \qquad c_{k\downarrow}^{\dagger} c_{f\downarrow}^{\dagger} |0\rangle \, ,$$

and the energy of the triplet state is

$$E^{(S=1)} = \varepsilon_k + \varepsilon_f .\qquad(44b)$$

The energy levels of the correlated ($U = \infty$) molecular model, constructed by two orbitals and two electrons, are indicated in Fig. 7b. It can be noticed that the ground state of the molecule is a singlet; *the singlet state is lower in energy than the triplet state*, as a result of the fact that virtual excitations are possible in the former case, by means of the hybridization matrix element V_{sf}. It can be noticed that the lowest excitation energy is rather weak and equals $2|V_{sf}|^2/\Delta$; this excitation corresponds to a spin-flip of one of the two electrons bound in the ground singlet state and can be pictured as a "spin fluctuation". Typical values of interest of the model parameters are $\Delta \approx 1\,\text{eV}$ and $|V_{sf}| \approx 10\text{--}100\,\text{meV}$. The molecular model, whatever rough it may appear, contains in germs what is an essential aspect of the Kondo phenomenology, namely: a localized magnetic impurity hybridizes and forms a weakly bound singlet ground state with the surrounding extended orbital, and thus appears to lose its magnetic moment (for temperatures $k_B T$ smaller than $2|V_{sf}|^2/\Delta$). The quenching of the magnetic moment of sufficiently low temperatures is discussd in further detail in Section 7.

Another remarkable feature of the correlated molecular model is to provide a justification of the spin–spin coupling Hamiltonian, assumed heuristically in Eq. (33). In fact, it is possible to express the energy difference between the triplet excited state and the singlet ground state in such a way to put in evidence the effective coupling between the spins $\mathbf{s}_1$ and $\mathbf{s}_2$ of the two electrons. As far as we limit our attention to the ground singlet state $\Psi_{\text{ground}}^{(S=0)}$ and the triplet state $\Psi^{(S=1)}$, we can observe that the many-body Hamiltonian H_m (given by Eq. 43) is equivalent to the much more manageable spin Hamiltonian

$$H = -2\,J\,\mathbf{s}_1 \cdot \mathbf{s}_2 .\qquad(45)$$

In fact, from

$$(\mathbf{s}_1 + \mathbf{s}_2)^2 = \mathbf{s}_1^2 + \mathbf{s}_2^2 + 2\,\mathbf{s}_1 \cdot \mathbf{s}_2$$

we have

$$2\,\mathbf{s}_1 \cdot \mathbf{s}_2 = (\mathbf{s}_1 + \mathbf{s}_2)^2 - \mathbf{s}_1^2 - \mathbf{s}_2^2 .$$

We notice that $\mathbf{s}_1^2 = \mathbf{s}_2^2 = s(s+1) = 3/4$; furthermore, $(\mathbf{s}_1 + \mathbf{s}_2)^2$ equals zero for singlet states and equals two for triplet states; then we have

$$\mathbf{s}_1 \cdot \mathbf{s}_2 = \begin{cases} -\dfrac{3}{4} & \text{for singlet states} \\[2mm] \dfrac{1}{4} & \text{for triplet states} \end{cases} .\qquad(46)$$

Thus the many-body Hamiltonian (43) is "equivalent" (at least in an appropriate subspace of wavefunctions) to a spin–spin Hamiltonian of the type $-2\,J\,\mathbf{s}_1 \cdot \mathbf{s}_2$ with antiferromagnetic interaction ($J = -|V_{sf}|^2/\Delta < 0$). Such equivalence, justified here in the extremely simplified one-impurity Anderson model of Eq. (43), is corroborated by the general formal relation between the Anderson and the Kondo Hamiltonians [see J. R. Schrieffer and P. A. Wolff, Phys. Rev. B**149**, 491 (1966)].

7   Magnetic impurity in normal metals at very low temperatures

Singlet ground state of a magnetic impurity embedded in an ordinary electron gas

In the previous section we have considered some properties of the "electron gas plus magnetic impurity system" on the basis of a rather intuitive model: the preserved magnetic impurity is described as a localized spin $\mathbf{S}$, weakly coupled through the contact interaction term $H_K = -J\Omega\,\mathbf{S}\cdot\mathbf{s}\,\delta(\mathbf{r})$ with the conduction electron spin density at the impurity. In this section we study the mechanism by which the *impurity forms a weakly bound singlet ground state with the surrounding extended orbitals*; as a consequence, at sufficiently low temperatures, the "electron gas plus magnetic impurity system" appears to collapse into a *singlet bound state*, constituted by the localized spin dressed and quenched by a cloud of conduction spin polarization. This quenching effect already appears in germs in the very idealized molecular model of Section 6.3. We consider now a more realistic (although still simplified) microscopic model for a single magnetic impurity embedded in a sea of conduction electrons. The magnetic impurity is schematized with a single localized orbital, of energy ε_f well below the Fermi energy, which can accept no more than one electron.

The model that we are considering can be described by the Anderson Hamiltonian of the type (25), here rewritten for convenience in slightly different notations

$$H = \sum_{\mathbf{k}} \varepsilon_{\mathbf{k}} \left[c_{\mathbf{k}\uparrow}^{\dagger} c_{\mathbf{k}\uparrow} + c_{\mathbf{k}\downarrow}^{\dagger} c_{\mathbf{k}\downarrow} \right] + \varepsilon_f \left[c_{f\uparrow}^{\dagger} c_{f\uparrow} + c_{f\downarrow}^{\dagger} c_{f\downarrow} \right]$$

$$+ \frac{1}{\sqrt{N}} \sum_{\mathbf{k}\sigma} \left[V_{\mathbf{k}f} c_{\mathbf{k}\sigma}^{\dagger} c_{f\sigma} + V_{\mathbf{k}f}^{*} c_{f\sigma}^{\dagger} c_{\mathbf{k}\sigma} \right] + U\, n_{f\uparrow}\, n_{f\downarrow} \ . \qquad (47)$$

In Eq. (47), $c_{\mathbf{k}\sigma}^{\dagger}$ and $c_{\mathbf{k}\sigma}$ are the creation and annihilation operators corresponding to the conduction band wavefunctions of the metal, of wavevector $\mathbf{k}$, spin σ and energy $\varepsilon_{\mathbf{k}}$ (for convenience we use the Fermi energy as the reference zero of energy). The operators $c_{f\sigma}^{\dagger}$ and $c_{f\sigma}$ are the creation and annihilation operators corresponding to the single impurity orbital, of energy ε_f (measured with respect to the Fermi energy). The Coulomb repulsion between the f-electrons is assumed to be so high ($U \to \infty$) to allow occupancy not higher than one of the f orbital. The hybridization matrix elements $V_{\mathbf{k}f}$ between plane waves and the f orbital are considered constant and denoted by V_{sf}. The model, with its essential phenomenological parameters, is indicated schematically in Fig. 9.

We consider the impurity in the magnetic regime (described in Section 5), and we suppose specifically that the parameter Γ, given by Eq. (28b) is such that

$$\Gamma = \pi\, |V_{sf}|^2\, n_0(E_F) \ll |\varepsilon_f| \ ; \qquad (48)$$

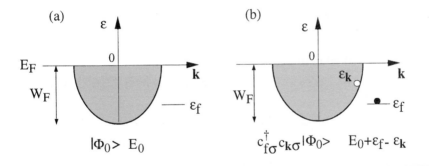

Fig. 9 (a) Schematic representation of the model Hamiltonian describing a magnetic impurity embedded in a sea of conduction electrons; the magnetic impurity is schematized with a single impurity orbital, which can accept no more than one electron. The ground-state wavefunction and energy of the filled Fermi sea are denoted by $|\Phi_0\rangle$ and E_0, respectively; W_F is the energy width of the conduction band occupied by electrons, and the Fermi level E_F is taken to be zero. (b) Schematic representation of the basis states interacting with the filled Fermi sea $|\Phi_0\rangle$, and leading to the formation of the singlet state in the Kondo problem.

the above condition specifies that the half-width Γ of the f-orbital is negligible with respect to $|\varepsilon_f|$, which represents the difference between the Fermi energy and the f-orbital energy. In this case, according to the unrestricted Hartree–Fock treatment of Section 5, a local magnetic moment is preserved with $\langle n_{f\uparrow}\rangle \approx 1$ and $\langle n_{f\downarrow}\rangle \approx 0$ (or vice versa). However, the unrestricted Hartree–Fock treatment is only approximate; if one considers the problem from a closer point of view, it is seen that the impurity magnetic moment (at low temperatures) is quenched by the spin polarization of the conduction electrons, and the magnetic impurity and the electron gas are bound together into a singlet ground state.

The properties of the one-impurity Anderson Hamiltonian (47), and related versions, have been investigated in the literature with a number of techniques, including variational methods, renormalization group calculations, Bethe Ansatz (see, for instance A. C. Hewson "The Kondo Problem to Heavy Fermions", Cambridge University Press, 1993). Here, to keep technicities at the minimum, we adopt the simplest variational representation for the ground state of the Anderson model (47).

To describe the singlet ground state $|\Psi\rangle$ of the whole system, formed by the magnetic impurity in interaction with the Fermi sea, consider the subspace spanned by the singlet state $|\Phi_0\rangle$ (which is the filled Fermi sea of the conduction electrons with total energy E_0) and the singlet states $c^\dagger_{f\sigma}c_{k\sigma}|\Phi_0\rangle$ (of energy $E_0 + \varepsilon_f - \varepsilon_k$), in which an electron from an occupied conduction state is transferred to the localized impurity state. In this subspace, $|\Psi\rangle$ can be expressed in the form

$$|\Psi\rangle = A\,|\Phi_0\rangle + \sum_{\mathbf{k}} a(\mathbf{k})\left[c^\dagger_{f\uparrow}c_{\mathbf{k}\uparrow} + c^\dagger_{f\downarrow}c_{\mathbf{k}\downarrow}\right]|\Phi_0\rangle \ . \tag{49}$$

Notice that basis states with two or more holes below the Fermi energy and one or more electrons above it are not included in the expansion (49); this simplification is justified only in the case of sufficiently small hybridization parameter V_{sf}.

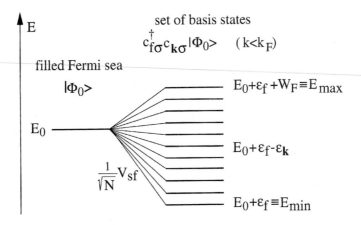

Fig. 10 Representation of the Anderson Hamiltonian on the basis of states formed by $|\Phi_0\rangle$ and $c_{f\sigma}^\dagger c_{k\sigma}|\Phi_0\rangle$ with $k < k_F$, and connected by the hybridization interaction $V_{sf}/\sqrt{N}$.

The Anderson Hamiltonian (47), on the basis of the states $|\Phi_0\rangle$ and $c_{f\sigma}^\dagger c_{k\sigma}|\Phi_0\rangle$ (with $k < k_F$, and $\sigma = \alpha$ or β) takes the form schematically indicated in Fig. 10. It is immediate to obtain the Green's function on the state $|\Phi_0\rangle$ with the renormalization method (see Section V-8.4); it is given by

$$G_{00}(E) = \cfrac{1}{E - E_0 - 2|V_{sf}|^2 \cfrac{1}{N} \sum_{k}^{(occ)} \cfrac{1}{E - [E_0 + \varepsilon_f - \varepsilon_k]}} \,, \tag{50a}$$

where the factor 2 takes into account spin degeneracy and the sum over $\mathbf{k}$ extends to the occupied states of the Fermi sea. The poles of the Green's function occur at the energies E such that

$$E - E_0 - 2|V_{sf}|^2 \frac{1}{N} \sum_{k}^{(occ)} \frac{1}{E - E_0 - \varepsilon_f + \varepsilon_k} = 0 \,, \tag{50b}$$

and the equation can be conveniently solved with graphical methods.

For this purpose, let us indicate by ε the quantity

$$\varepsilon = E - E_0 - \varepsilon_f \,;$$

then Eq. (50b) becomes

$$\varepsilon + \varepsilon_f = 2|V_{sf}|^2 \frac{1}{N} \sum_{k}^{(occ)} \frac{1}{\varepsilon + \varepsilon_k} \,. \tag{51}$$

The standard graphical solution of Eq. (51) is shown in Fig. 11. The solution of Eq. (51) with negative ε individuates a peculiar collective state (as already encountered in a number of problems, different from the physical point of view, but rather similar in the underlying mathematical aspects).

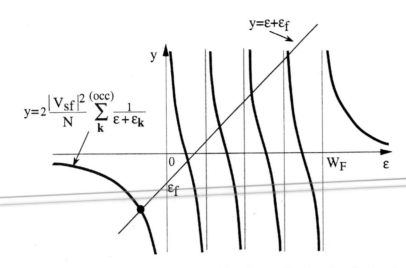

Fig. 11 Graphical solution of the eigenvalue problem for the Kondo effect (notice that $\varepsilon_f < 0$ and also $0 \leq -\varepsilon_{\mathbf{k}} \leq W_F$).

The numerical solution for the ground state, i.e. the collective state, of Eq. (51) is also immediate. We notice that for negative ε, the quantities $\varepsilon + \varepsilon_{\mathbf{k}}$ appearing in Eq. (51) are all different from zero since $-W_F \leq \varepsilon_{\mathbf{k}} \leq 0$. Thus we have

$$\sum_{\mathbf{k}}^{(occ)} \frac{1}{\varepsilon + \varepsilon_{\mathbf{k}}} = D_0(E_F) \int_{-W_F}^0 \frac{1}{\varepsilon + x} \, dx = D_0(E_F) \ln \left| \frac{\varepsilon}{\varepsilon - W_F} \right| , \tag{52}$$

where as usual we have changed the sum over $\mathbf{k}$ into a continuous integral over the energy; to avoid inessential details, we have supposed that the density-of-states of the conduction band (for one spin direction) is constant in the energy interval of interest around the Fermi level and equal to $D_0(E_F)$.

Inserting Eq. (52) into Eq. (51), and assuming $|\varepsilon| \ll W_F$ and $|\varepsilon| \ll |\varepsilon_f|$, we obtain

$$\varepsilon_f = 2|V_{sf}|^2 n_0(E_F) \ln \frac{|\varepsilon|}{W_F} , \tag{53}$$

where $n_0(E_F) = D_0(E_F)/N$ denotes the density-of-states at the Fermi level for one spin direction and per unit cell. The energy gain of the singlet ground state due to the hybridization is thus

$$\boxed{|\varepsilon| = W_F \exp \left[-\frac{|\varepsilon_f|}{2|V_{sf}|^2 n_0(E_F)} \right] .} \tag{54}$$

It is convenient to define the Kondo temperature, so that

$$k_B T_K = |\varepsilon| \ ;$$

for temperatures T much smaller than T_K, thermal excitations are negligible, and the system is essentially described by the non-magnetic singlet ground state; for temperatures comparable or higher than T_K the standard local moment regime occurs.

We are now in the position to understand the finite, although large, paramagnetic susceptibility of the magnetic impurity for $T < T_K$. In the presence of a magnetic field H we suppose that the energy ε_f of the f state is split into $\varepsilon_f \pm \mu_B H$. We can proceed with a treatment similar to that used above, and we obtain that Eq. (53) is now replaced by

$$\varepsilon_f = |V_{sf}|^2 \, n_0(E_F) \left[\ln \frac{|\varepsilon(H) + \mu_B H|}{W_F} + \ln \frac{|\varepsilon(H) - \mu_B H|}{W_F} \right] , \tag{55}$$

where $\varepsilon(H)$ indicates the binding energy of the ground state of the system in the presence of the magnetic field. With $\mu_B H$ smaller than $|\varepsilon(H)|$, Eq. (55) reads

$$\varepsilon_f = |V_{sf}|^2 \, n_0(E_F) \, \ln \frac{\varepsilon^2(H) - \mu_B^2 H^2}{W_F^2} . \tag{56}$$

Comparison of Eq. (53) and Eq. (56) gives

$$\varepsilon^2(H) - \mu_B^2 H^2 = |\varepsilon|^2 .$$

We have thus

$$|\varepsilon(H)| = \sqrt{|\varepsilon|^2 + \mu_B^2 H^2} = |\varepsilon| + \frac{1}{2} \frac{\mu_B^2 H^2}{|\varepsilon|} .$$

Since $\varepsilon(H)$ is negative, and $|\varepsilon| = k_B T_K$, the above equation can be written as

$$\varepsilon(H) = \varepsilon(H\!=\!0) - \frac{1}{2} \frac{\mu_B^2 H^2}{k_B T_K} .$$

If we have a number $N^{(\mathrm{imp})}$ of isolated (i.e. non-interacting) impurities in the volume V, differentiating twice with respect to H yields the contribution $\chi^{(\mathrm{imp})}$ to the magnetic susceptibility:

$$\chi^{(\mathrm{imp})} = -\frac{N^{(\mathrm{imp})}}{V} \frac{\partial^2 \varepsilon(H)}{\partial H^2} = +\frac{N^{(\mathrm{imp})}}{V} \frac{\mu_B^2}{k_B T_K} . \tag{57a}$$

It is interesting to compare the above expression with the Pauli paramagnetism of the Fermi gas $\chi_P = (1/V) D(E_F) \mu_B^2$ of Eq. (XV-30); for free electrons, we have

$$\chi_P = \frac{1}{V} \frac{3}{2} \frac{N}{E_F} \mu_B^2 = \frac{3}{2} \frac{N}{V} \frac{\mu_B^2}{k_B T_F} ; \tag{57b}$$

thus the smaller Kondo temperature appearing in Eq. (57a) explains the giant contribution to the susceptibility from every magnetic impurity.

Before closing, a few more comments are worthwhile on the fundamental result (54), after re-writing it in a slightly different form. We express $|\varepsilon|=k_B T_K$ and $W_F=k_B T_F$, and also put $J_{eff}=2J=-2|V_{sf}|^2/|\varepsilon_f|$ in evident analogy with the molecular model of Section 6.3. From Eq. (54), we obtain that the Kondo temperature T_K and the Fermi temperature T_F are linked by the equation

$$\boxed{T_K = T_F \exp[-1/|J_{eff}|n_0(E_F)] .} \tag{58}$$

This relation is analogous in some ways to the Cooper pair binding energy encountered in superconductivity (see Section XVIII-2). It can be noticed that the exponential in the second member of Eq. (58) is *not* an analytic function of $|J_{eff}|$ for $|J_{eff}| \to 0$, and thus cannot be expanded in powers of $|J_{eff}|$; this explains the difficulties encountered attacking the Kondo problem with ordinary perturbation theory. Expression (58) was earlier inferred in the pioneer papers of Abrikosov and Suhl who applied the many-body formalism to the treatment of the Kondo Hamiltonian $-J_{eff}\Omega \, \mathbf{S} \cdot \mathbf{s} \, \delta(\mathbf{r})$; for this the Kondo temperature T_K is also referred to in the literature as the Abrikosov–Suhl temperature.

Considerations on heavy fermion systems

Although this chapter concerns dilute magnetism, we conclude it with some qualitative remarks on the challenging problem of heavy fermion metals. In general these systems are compounds containing lanthanides (such as cerium and other rare earth elements with incomplete $4f$ shell) or actinides (such as uranium or neptunium with incomplete $5f$ shell).

At temperatures higher than a given critical temperature T^* (of the order of a few kelvin), the properties of the heavy fermion metal can be described as consisting of an ordinary metal plus a set of independent localized spins; the f electrons in incomplete shells give rise to a local magnetic moment $-g\mu_B \mathbf{J}$, and the magnetic susceptibility obeys the Curie law

$$\chi = \frac{N_f}{V} \frac{g^2 J(J+1) \mu_B^2}{3k_B T} \; ,$$

where N_f is the number of atoms with incomplete f shells in the crystal of volume V.

At $T \leq T^*$ the hybridization of the localized f electrons with the itinerant conduction electrons makes its effects felt and measurable; hybridization gives rise to a Kondo type quenching of the local moments, with a paramagnetic susceptibility which is two or three orders of magnitude higher than that of normal metals. This fact can be interpreted as if the f electrons occupy a narrow band of width $k_B T^*$ (instead of $k_B T_F$), and have a very large effective mass. The interpretation is corroborated by the fact that for $T \ll T^*$ the specific heat behaves as

$$C_V = \gamma T \; ,$$

the Sommerfeld constant γ being two or three orders of magnitude higher than that of normal metals. The "Wilson ratio" between the giant magnetic susceptibility and the giant specific heat remains of the same order as in ordinary metals.

Another interesting effect concerns the resistivity of heavy fermion metals. For $T > T^*$ the scattering of Kondo atoms is incoherent; for $T < T^*$ the f electrons are in coherent Bloch states and the resistivity begins still to decrease; at very low temperatures some heavy fermion metals become superconductors. The extension of the single-impurity Kondo problem to a lattice of Kondo impurities, with possible interplay of magnetism and superconductivity, presents a number of conceptual and technical difficulties, and we refer to the literature for information on this very active field of research.

Further Reading

A. Abragam and B. Bleany "Electron Paramagnetic Resonance of Transition Ions" (Clarendon, Oxford 1970)

N. Andrei, K. Furuya and J. H. Lowenstein "Solution of the Kondo Problem" Rev. Mod. Phys. **55**, 331 (1983)

I. B. Bersuker "Electronic Structure and Properties of Transition Metal Compounds. Introduction to the Theory" (Wiley, New York 1996)

S. Chikazumi "Physics of Ferromagnetism" (Clarendon Press, Oxford 1997)

P. Fulde, J. Keller and G. Zwicknagl "Theory of Heavy Fermion Systems" Solid State Physics **41**, 1 (1988) (edited by H. Ehrenreich and D. Turnbull, Academic Press)

A. J. Heeger "Localized Moments and Nonmoments in Metals: the Kondo Effect" Solid State Physics **23**, 283 (1969)

A. C. Hewson "The Kondo Problem to Heavy Fermions" (Cambridge University Press, 1993)

J. Jensen and A. R. Mackintosh "Rare Earth Magnetism" (Clarendon, Oxford 1991)

J. Kondo "Theory of Dilute Magnetic Alloys" Solid State Physics **23**, 183 (1969)

R. Kubo and T. Nagamiya eds. "Solid State Physics" (McGraw Hill, New York 1969)

A. H. Morrish "The Physical Principles of Magnetism" (Wiley, New York 1965)

G. T. Rado and H. Suhl eds. "Magnetism" Vol.V: Magnetic Properties of Alloys (Academic Press, New York 1973)

J. H. Van Vleck "The Theory of Electric and Magnetic Susceptibilities" (Oxford University Press 1952)

R. M. White "Quantum Theory of Magnetism" (Springer, Berlin 1983)

K. G. Wilson "The Renormalization Group: Critical Phenomena and the Kondo Problem" Rev. Mod. Phys. **47**, 773 (1975)

XVII
Magnetic ordering in crystals

In the previous two chapters, among various topics on magnetism, we have discussed the occurrence of permanent magnetic moments in crystals and their paramagnetic susceptibility. In particular we have considered the situation of well localized electronic wavefunctions with formation of localized magnetic moments (typified by the Curie paramagnetism), as well as the alternative situation of itinerant electronic wavefunctions (typified by the Pauli paramagnetism). It is well known that many paramagnetic materials, below a critical temperature, present *magnetic order* even in the absence of applied magnetic fields. The most familiar order is the ferromagnetic one, with localized moments lined up in the same direction so that a spontaneous magnetization is apparent; but several other types of magnetic ordering (antiferromagnetic, ferrimagnetic, helical etc.) are possible.

The occurrence of magnetic ordering in crystals implies some coupling mechanism between localized or delocalized magnetic moments. Among the microscopic models of coupling between the localized moments, we focus on the Heisenberg spin Hamiltonian, because of its formal simplicity and the individuation of the electrostatic origin of the effective spin-spin interaction. In paramagnetic crystals characterized by itinerant electronic wavefunctions (band paramagnetism), the occurrence of magnetic ordering suggests the importance of correlation effects among the independent particle wavefunctions; the simplest microscopic model of magnetism derived from electron band structure, and the only one we briefly describe, is the Stoner–Hubbard model of itinerant magnetism.

The phase transition to magnetic ordering is first studied within mean field theories; this permits a justification of the Weiss molecular field assumptions, and a reasonably

Table 1 Ferromagnetic Curie point (in K) and saturation magnetization (in gauss) for some materials (above T_c, the rare earth dysprosium makes a transition to helices).

Material	T_c	$M(T=0)$
Fe	1043	1752
Co	1394	1446
Ni	630	510
Gd	293	1980
Dy	85	3000

simple description of some aspects of the wide and rich phenomenology accompanying magnetic ordering and phase transitions. We also consider aspects beyond mean field theories. In particular, at low temperatures, we study with the spin-wave theory the elementary excitations (magnons) of the Heisenberg ferromagnets; we also examine in the neighbourhood of the crytical temperature some relevant features of Ising models with the renormalization group theory.

The vastity of the field of magnetism makes unavoidable drastic abridgements and simplifications; in particular we confine our attention to some elementary aspects of bulk magnetism in periodic crystals. Other interesting subjects such as, for instance, surface magnetism, magnetism of multilayers and superlattices, cluster magnetism, domain structure of magnets have been omitted. Although the topics we cover are so limited and simplified, they should give orientative guidelines in the subject of magnetic ordering in matter.

1   Ferromagnetism and the Weiss molecular field

Phenomenological aspects of ferromagnetism

We begin the study of cooperative effects in magnetic materials, by considering a system in which the microscopic magnetic moments have the same magnitude and tend to line up in the same direction. A ferromagnetic specimen, at sufficiently low temperatures, exhibits spontaneous magnetization. The spontaneous magnetization $M(T)$ depends on the temperature and vanishes above a *critical temperature T_c, called ferromagnetic Curie temperature*. The critical temperature T_c and *the saturation magnetization $M(T=0)$* for some substances are reported in Table 1 [for further data see for instance F. Keffer, Handbuch der Physik, Vol.18 part 2 (Springer, Berlin 1966); for Fe and Ni see also T. Tanaka and K. Miyatani, J. Appl. Phys. **82**, 5658 (1997)]. Among the elements only a few are ferromagnetic; there is instead a relatively large number of ferromagnetic alloys and oxides.

For an estimate of the order of magnitude of the saturation magnetization, consider for simplicity a crystal with a localized magnetic moment $\mu = \mu_B = e\hbar/2mc$ in every unit cell of volume Ω; assuming parallel alignment of all the microscopic magnetic moments at zero temperature, the saturation magnetization becomes

$$M(0) = \frac{\mu_B}{\Omega} = \frac{a_B^3}{\Omega} \frac{\mu_B}{a_B^3} \ . \tag{1a}$$

We notice that

$$\frac{\mu_B^2}{a_B^3} = \left(\frac{e\,\hbar}{2\,m\,c} \right)^2 \frac{1}{(\hbar^2/me^2)^3} = \frac{1}{4} mc^2 \alpha^4 \ ,$$

where $\alpha = e^2/\hbar c$ is the fine structure constant. Using $\alpha^{-1} = 137.036$, $mc^2 = 0.511\,\text{MeV}$, and for the Bohr magneton $\mu_B = 0.05788$ meV/Tesla, we obtain

$$\frac{\mu_B^2}{a_B^3} = 0.363\,\text{meV} = 4.21\,\text{K} \quad \text{and} \quad \frac{\mu_B}{a_B^3} = 62716 \text{ gauss} \ . \tag{1b}$$

From Eq. (1a) and Eq. (1b), taking $\Omega \approx 100\,a_B^3$, it can be seen that the expected order of magnitude of the saturation magnetization is 1000 gauss, in agreement with the observed values of Table 1.

The spontaneous magnetization of a ferromagnetic crystal can be explained only by some interaction mechanism, which favours a parallel alignment of microscopic magnetic moments. From the order of magnitude of the critical temperatures given in Table 1, we notice immediately that the *magnetic dipolar coupling cannot be the origin of such interaction*. In fact the order of magnitude of the dipolar interaction energy between two magnetic dipoles with $\mu = \mu_B$ at distance a is

$$E_{\text{dip}} \approx \frac{\mu_B^2}{a^3} = \frac{a_B^3}{a^3} \frac{\mu_B^2}{a_B^3} \ ; \tag{1c}$$

taking $a \approx 4\text{-}5\,a_B$, $a^3 \approx 100\,a_B^3$, and using Eq. (1b), we have $E_{\text{dip}} \approx 0.1\,\text{K}$. Thus random thermal fluctuations would destroy alignment of magnetic moments at very low temperatures, of the order of tenths of kelvin degrees, much smaller than the observed critical temperatures, which are of the order of thousand kelvin degrees. The mechanism of spin alignment must be explained by an interaction among spins, which is larger than the dipolar magnetic interaction by a factor 10^4 or so.

The Weiss molecular field

Consider a crystal of volume V, formed by N equal magnetic units (atoms or ions), each of angular momentum J and magnetic moment $\mu = g\,\mu_B\,J(J+1)$, localized at the sites of a Bravais lattice. In the previous chapter, we have discussed the magnetization of the sample in an applied field, under the assumption that the microscopic moments are independent. For instance, from Eq. (XVI-17), the magnetization of a paramagnetic substance, with N units with angular momentum $J = 1/2$ and giromagnetic factor $g = 2$, is given by

$$M = \frac{N}{V} \mu_B \tanh \frac{\mu_B\,H}{k_B\,T} \ , \tag{2a}$$

where H is the applied magnetic field. In a paramagnetic substance, composed by independent magnetic dipoles, M is proportional to H for small H, and no spontaneous magnetization can occur. With the approximation $\tanh x \approx x$ for small x in Eq. (2a), one finds $M = (N/V)\,\mu_B^2\,H/k_B\,T$; it follows that the magnetic susceptibility $\chi = M/H$ obeys the *Curie law*

$$\chi = \frac{C}{T} \tag{2b}$$

with the *Curie constant C* given by

$$C = \frac{N}{V}\frac{\mu_B^2}{k_B} . \tag{2c}$$

In the case of paramagnetic substances with $J \neq 1/2$, expressions similar to Eqs. (2) hold with the hyperbolic tangent function replaced by the appropriate Brillouin function (and other minor changes if $g \neq 2$). For simplicity in this section we only consider paramagnetic substances with $J = 1/2$ and $g = 2$.

The first phenomenological mechanism leading to magnetism was proposed by Weiss in 1907. It is based on the assumption that the effective magnetic field acting on a given dipole is given by

$$H_{\mathrm{eff}} = H + \lambda M , \tag{3}$$

where H is the external magnetic field, λ is an appropriate constant, M is the magnetization and λM provides the cooperative effect. Originally the Weiss constant λ was considered as a phenomenological constant; the interpretation of λ in terms of microscopic quantum models appears later with the works of Heisenberg.

If we make the Weiss assumption that the effective field acting on a given dipole is $H + \lambda M$, we obtain for a ferromagnetic substance (with $J = 1/2$ and $g = 2$) the basic equation

$$\boxed{M = \frac{N}{V}\,\mu_B\,\tanh\frac{\mu_B\,(H + \lambda M)}{k_B\,T}} . \tag{4}$$

It is evident that the case $\lambda = 0$ corresponds to ordinary paramagnetism, while λM with $\lambda > 0$ describes a ferromagnetic cooperative effect. The rest of this section is devoted to extract explicitly the implications of the basic equation (4), when λ is a positive constant.

We begin to look for spontaneous magnetization, putting $H = 0$ in Eq. (4); we obtain

$$\boxed{M = \frac{N}{V}\,\mu_B\,\tanh\frac{\mu_B\,\lambda M}{k_B\,T}} . \tag{5}$$

To solve this equation graphically it is convenient to introduce the dimensionless variable

$$x = \frac{\mu_B\,\lambda M}{k_B\,T} , \tag{6a}$$

and to define the quantity T_c as

$$T_c = \frac{N}{V}\frac{\mu_B^2}{k_B}\,\lambda ; \tag{6b}$$

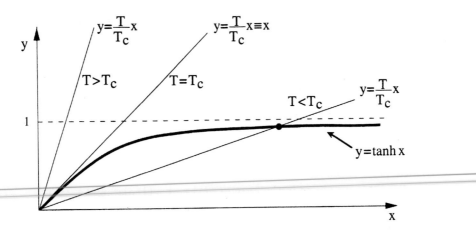

Fig. 1 Graphical solution of the mean field equation $(T/T_c)\, x = \tanh x$, for the spontaneous magnetization of a ferromagnetic crystal.

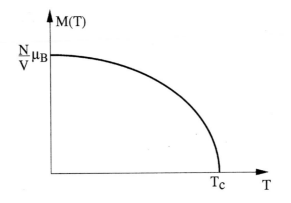

Fig. 2 Spontaneous magnetization as function of temperature.

the identification of the quantity T_c with the critical temperature will become apparent soon. Eq. (5) thus takes the form

$$\frac{T}{T_c} x = \tanh x \ . \tag{7}$$

The solution of this equation is obtained plotting separately the first member (a family of straight lines) and the second member as functions of x, and looking for the intersections between the curves; the graphical solution is shown in Fig. 1.

With the help of Fig. 1, it is immediate to note that for $T \geq T_c$ we have only the trivial solution $x = 0$ and thus $M(T) = 0$; however, below the critical temperature $T < T_c$ we have also a non-trivial solution with non-zero $M(T)$. The spontaneous magnetization $M(T)$ as a function of temperature can be obtained graphically, and its typical behaviour is indicated in Fig. 2.

It is straightforward to obtain analytically the behaviour of the magnetization in the limit of very low temperatures $T \ll T_c$ or near the critical point $T \approx T_c$.

Low temperature limit. Near $T \approx 0$ we have $x \to \infty$; from Eq. (7) we obtain $(T/T_c)\, x = 1 - 2\, \exp(-2\, x) \approx 1 - 2\, \exp(-2\, T_c/T)$. Using Eqs. (6), we obtain

$$M(T) \approx \frac{N}{V}\, \mu_B \left[1 - 2\, e^{-2\, T_c/T} \right] \; ; \tag{8}$$

thus, near zero temperature, $M(T)$ deviates from its saturation value $M(0) = (N/V)\mu_B$ by an exponential term.

Critical point region. Near the critical temperature the magnetization is small and so is x; Eq. (7) is simplified in the form $(T/T_c)\, x = x - (1/3)x^3$ and $x = \sqrt{3}(1 - T/T_c)^{1/2}$; the magnetization becomes

$$M(T) = \sqrt{3}\, \frac{N}{V}\, \mu_B\, \frac{T}{T_c} \left(1 - \frac{T}{T_c} \right)^{1/2} .$$

It follows

$$M(T) \approx (T_c - T)^{1/2} \tag{9a}$$

for $T \to T_c$ and $T < T_c$.

At the critical temperature we can also consider how the magnetization depends on the applied external field. From Eq. (4) and the first two terms of the series development of the hyperbolic tangent function, $\tanh x = x - (1/3)\, x^3$, we have

$$M = \frac{N}{V}\, \mu_B^2\, \frac{H + \lambda M}{k_B\, T_c} - \frac{1}{3} \frac{N}{V}\, \mu_B^4 \left(\frac{H + \lambda M}{k_B\, T_c} \right)^3 .$$

Using Eq. (6b) and straightforward manipulations, we obtain

$$M \approx H^{1/3} . \tag{9b}$$

Another quantity of interest is the specific heat at zero-field. The internal energy of the ferromagnet, contributed by the spontaneous magnetization, is $-\frac{1}{2} H_{\text{eff}}\, M = -\frac{1}{2}\, \lambda M^2$; using Eq. (9a), it is seen that the specific heat, which is the derivative of the internal energy with respect to the temperature, is zero above the critical temperature and nearly constant just below it.

We consider now the magnetization of the ferromagnetic material for $T > T_c$, in the presence of an external magnetic field H. In the limit of sufficiently small H, also M is small, and we can approximate $\tanh x \approx x$ in Eq. (4). For $T > T_c$ we have thus

$$M = \frac{N}{V} \mu_B^2\, (H + \lambda M)\, \frac{1}{k_B\, T} .$$

Solving the above equation for M, we obtain for the magnetic susceptibility $\chi = M/H$ the *Curie–Weiss law*

$$\chi = \frac{N}{V}\, \frac{\mu_B^2}{k_B\, (T - T_c)} = \frac{C}{T - T_c} . \tag{9c}$$

Table 2 Definition of critical exponents for some quantities in magnetic systems. The values of the critical exponents in the mean field theory are given. The values for the three-dimensional and for the two-dimensional Ising model are also reported. It can be noticed that the critical exponents satisfy the relations $\alpha + 2\beta + \gamma = 2$; $\alpha + \beta(1+\delta) = 2$; $\gamma = (2-\eta)\nu$; $\gamma = \beta(\delta - 1)$; $\nu d = 2 - \alpha$ (d=dimensionality).

		mean field approximation	three-dimensional Ising	two-dimensional Ising
zero-field specific heat	$C \approx |T - T_c|^{-\alpha}$	$\alpha = 0$ (discontinuity)	$\alpha = 0.10$	$\alpha = 0$ (logarithm)
zero-field magnetization ($T < T_c$)	$M \approx (T_c - T)^\beta$	$\beta = 1/2$	$\beta = 0.33$	$\beta = 1/8$
zero-field isothermal susceptibility	$\chi \approx |T - T_c|^{-\gamma}$	$\gamma = 1$	$\gamma = 1.24$	$\gamma = 7/4$
magnetization at $T = T_c$	$M \approx H^{1/\delta}$	$\delta = 3$	$\delta = 4.8$	$\delta = 15$
correlation length	$\xi \approx |T - T_c|^{-\nu}$	$\nu = 1/2$	$\nu = 0.63$	$\nu = 1$
pair correlation function at $T = T_c$	$\Gamma(r) \approx \dfrac{1}{r^{d-2+\eta}}$	$\eta = 0$ ($d = 4$)	$\eta = 0.04$ ($d = 3$)	$\eta = 1/4$ ($d = 2$)

Thus, in the Weiss molecular field theory, χ diverges as $(T - T_c)^{-1}$ as the temperature decreases toward the critical point. The tendency to parallel alignment of spins, enhances the magnetic susceptibility of the systems and leads to a phase transition below T_c.

Near the critical temperature, the Weiss molecular field theory shows that several significant quantities, such as zero field magnetization, magnetization at critical temperature, zero field isothermal susceptibility, zero field specific heat, follow simple power laws with appropriate exponents. The definition and the values of some critical exponents of more common use are summarized for convenience in Table 2; for further details and relationships among critical point exponents see for instance the book of J. M. Yeomans (1992); see also P. Butera and M. Comi, Phys. Rev. B**58**, 11552 (1998) and references quoted therein.

Inadequacies of the Weiss molecular field

The Weiss approach accounts for some basic features of ferromagnetic behaviour (with the advantage of so little mathematical labour!). However the Weiss molecular field, which is essentially a mean field theory, presents several inadequacies, that are sum-

Table 3 Measured critical exponents for several materials; the data are taken from the paper by L. P. Kadanoff et al., Rev. Mod. Phys. **39**, 395 (1967), and from T. Tanaka and K. Miyatani, J. Appl. Phys. **82**, 5658 (1997).

Material	critical exponent β	critical exponent γ
Fe	0.35	1.27
Co		1.21
Ni	0.38	1.27
Gd		1.33
EuS	0.33	

marized below; concepts and theories developed to overcome these inadequacies are the subject of the rest of this chapter.

Origin of the Weiss molecular field. The more basic problem raised by the Weiss molecular field is the microscopic origin of the coupling between magnetic dipoles. We can estimate the order of magnitude of the Weiss parameter λ in terms of T_c; from Eq. (6b) we have

$$\lambda = \frac{k_B T_c}{(N/V)\,\mu_B^2} \ .$$

To estimate the value of λ, we take $N/V \approx 1/(5\,a_B^3) \approx (1/100)\,a_B^{-3}$; $\mu_B^2/a_B^3 \approx 1\,\mathrm{meV}$ (see Eq. 1b), $T_c = 1000\,\mathrm{K}$ and $k_B\,T_c \approx 100\,\mathrm{meV}$; we obtain $\lambda \approx 10^4 - 10^5$.

The phenomenological values of λ in the range $10^4 - 10^5$ *exclude* that the molecular field originates from magnetic dipole–dipole interactions; these dipolar interactions, once treated "mutatis mutandis" according to the Lorentz cavity field of Section IX-7.3, would give $H_{\mathrm{eff}} = H + \lambda M$ with $\lambda \approx 4\pi/3$ (or so), a value thousand times smaller than the phenomenological values of λ. As we shall see in Section 2, the microscopic origin of the molecular field is the much stronger effective interaction of electrostatic origin, related to the electron–electron Coulomb interaction and the Pauli exclusion principle. The dipolar interactions, on the other hand, have a leading role in the tendency of a ferromagnetic specimen to break into "domains" of macroscopic size, with magnetization vectors oriented in different directions, so to minimize the magnetic dipole–dipole interaction energy. In the following, we will neglect this and other consequences of the dipolar interactions among spins; rather we will focus on the effective coupling of electrostatic origin among the localized magnetic moments. This effective coupling can lead not only to ferromagnetism but also to other types of magnetic ordering (as discussed in Section 3).

Low temperature region. Near zero temperature the mean field theory predicts that spontaneous magnetization deviates from saturation by an exponential term. In contrast, experiments show a $T^{3/2}$ dependence; a more accurate analysis in terms of the concepts of spin waves and magnons is needed (see Section 4).

Critical point region. Near the critical temperature, the mean field theory shows that several significant quantities (such as magnetization and susceptibility) follow power law forms. However the measured exponents of some ferromagnetic materials are rather different from the mean field result, although only minor differences seem to occur among the different materials, as seen from Table 3; thus near the critical temperature a more detailed analysis is necessary. Another evident limit of the mean field theory is the fact that it is unable to correctly determine the effect of dimensionality on the phase transitions. The analysis of critical exponents in second order phase transitions is of particular importance, and some aspects are discussed in Section 5 and Section 6.

Itinerant band magnetism. The description of ferromagnetism, starting from a picture of magnetic moments localized at the lattice sites, may become inadequate for transition metals and other materials, in which the interplay between band formation and magnetization is relevant; some aspects are considered in Section 7.

2 Microscopic origin of the coupling between localized magnetic moments

The Heisenberg model for the coupling among localized spins

The localized picture of magnetism assumes that microscopic magnetic moments are set at the lattice sites and interact cooperatively among themselves. A popular form of interaction among spins is represented by the Heisenberg model

$$H = -\sum_{m \neq n} J_{mn} \mathbf{S}_m \cdot \mathbf{S}_n , \tag{10}$$

where $\mathbf{S}_m$ denotes the total angular momentum of the atom (or ion) at the mth site, and $-2 J_{mn} \mathbf{S}_m \cdot \mathbf{S}_n$ ($m \neq n$) denotes the contribution to the energy from a pair of atoms (or ions) in the mth and nth sites, respectively; to avoid double counting of interactions, the factor 2 in front of J_{mn} is dropped in the summation in Eq. (10). The parameters J_{mn} are referred to as the exchange parameters (even if, in almost all situations, they cannot be identified with bielectronic exchange integrals). Among the major merits of the Heisenberg model, we can mention the entailed electrostatic origin of the spin–spin coupling and its formal simplicity.

Often spin–spin interactions are relevant only for nearest neighbour sites. In the case of spins lying on a Bravais lattice, with only nearest neighbour interactions and same parameter J, the Heisenberg Hamiltonian can be recast in the form

$$H = -J \sum_{\mathbf{t}_m \mathbf{t}_i^{(I)}} \mathbf{S}_m \cdot \mathbf{S}_{m+i} = -J \sum_{\mathbf{t}_m \mathbf{t}_i^{(I)}} (S_m^z S_{m+i}^z + S_m^x S_{m+i}^x + S_m^y S_{m+i}^y) , \tag{11}$$

where $\mathbf{t}_m$ ($m = 1, 2, \ldots, N$) denote the N translations of the crystal, and $\mathbf{t}_i^{(I)}$ ($i = 1, 2, \ldots, n_c$) denote the n_c translation vectors connecting a given site of the Bravais lattice with its first neighbours (n_c is the coordination mumber).

An obvious limitation of the model Hamiltonian of Eq. (11) is that an interaction of the form $-J \mathbf{S}_m \cdot \mathbf{S}_n$ between electron spins is isotropic and does not account for preferential orientations of the magnetic moments in the crystal (crystal axes could be felt through the spin–orbit interaction). A more realistic model of some magnets with localized moments is the anisotropic XY model, defined as

$$H_{XY} = -J_z \sum_{\mathbf{t}_m \, \mathbf{t}_i^{(I)}} S_m^z S_{m+i}^z - J_\perp \sum_{\mathbf{t}_m \, \mathbf{t}_i^{(I)}} (S_m^x S_{m+i}^x + S_m^y S_{m+i}^y) \, . \tag{12}$$

For $J_\perp = J_z = J$ we recover the isotropic Heisenberg model of Eq. (11). For $J_\perp = 0$ we obtain the Ising model

$$H_{\text{Ising}} = -J \sum_{\mathbf{t}_m \, \mathbf{t}_i^{(I)}} S_m^z S_{m+i}^z \, . \tag{13}$$

The spin Hamiltonians (10), (11), (12) and (13) are the prototypes of several other related models, of theoretical or experimental interest [see for instance J. Yeomans "The Theory and Applications of Axial Ising Models" in Solid State Physics **41**, 151 (1991)].

At first sight it could appear surprising that the (non-relativistic) Hamiltonian of a many-electron system (which depends only on *space variables and not on the spin variables*) can be mimicked by an *effective spin–spin Hamiltonian*, as the one of Eq. (10). The basic reason is that real space and spin space of a many-electron system are always interconnected quantum mechanically (regardless of the fact that the many-body Hamiltonian acts only on the space variables), because of the use of antisymmetrized total wavefunctions. Alignment or not of spins in the *spin space* imposes requirements on the wavefunctions *in real space*, where the electrostatic interactions are at work. Under appropriate circumstamces, an effective spin–spin Hamiltonian may result, whose ultimate origin are just the electrostatic interactions in real space.

Because of such electrostatic origin, there is no surprise that the interaction energy J can assume values of the order of $10 \approx 100$ meV (or so), i.e. several orders of magnitude higher than the magnetic dipolar interaction of Eq. (1). The sign of the coupling can be either positive or negative, or even oscillatory with spin distance. In the case $J > 0$ the spins tend to line up in the same direction and the *coupling is said to be ferromagnetic*; in the case $J < 0$ the spins tend to line up in opposite directions and *the coupling is said antiferromagnetic*.

In order to give some plausibility arguments on the microscopic origin and range of applicability of the Heisenberg Hamiltonian, we consider qualitatively the *simplest atomic and molecular systems with two electrons*, and we show under which constraints an effective Hamiltonian of the type "scalar product of spins" can be recovered, with coupling that can be either ferromagnetic or antiferromagnetic, depending on physical situations. The hope is to infer hints generalizable to actual many-electron systems; this could appear a cavalier undertaking, but the essential reasoning privileges the consequences of the *Pauli exclusion principle* rather than other aspects such as the number of particles.

Elementary examples of two-electron systems with ferromagnetic or antiferromagnetic spin–spin interaction

We begin our qualitative analysis of prototype two-electron systems by considering situations in which an effective coupling of antiferromagnetic type $-2\,J\,\mathbf{s}_1\cdot\mathbf{s}_2$ $(J<0)$ between the electronic spins $\mathbf{s}_1$ and $\mathbf{s}_2$ appears to be at work.

In Section XVI-6.3, in the discussion of the Kondo effect, we have considered in detail the energy levels of an idealized molecule with two electrons and two orbitals, one of them of extended nature (with the possibility of double occupation by electrons of opposite spin) and the other of localized nature (and so large Coulomb repulsive energy between electrons of opposite spin to prevent double occupancy). The hybridization between the two orbitals makes the singlet state (with total spin $S=0$) lower in energy than the triplet state (with total spin $S=1$); in Section XVI-6.3, it has also been shown that the energy difference and multiplicity of the ground state and first excited molecular state can be mimicked by the spin–spin Hamiltonian $-2\,J\,\mathbf{s}_1\cdot\mathbf{s}_2$ with antiferromagnetic interaction.

A much more familiar two-electron system in which the singlet state is lower in energy than the triplet state is represented by the hydrogen molecule. In the independent particle approximation (or Hartree–Fock approximation), the ground state molecular orbital of the H_2 molecule is non-degenerate, and the ground state of the molecule is obtained accommodating two electrons of either spin in it. In the ground molecular orbital there is accumulation of the bonding charge in the region intermediate between the nuclei, where the electron cloud enjoys attractive interaction from both nuclei. If we try to change the spin direction of one of the electrons, we must by necessity transfer an electron from the ground molecular orbital to a higher energy molecular orbital, with a consequent penalty in energy. We can interpret the energy difference between the singlet and the triplet state as if an antiferromagnetic interaction $-2\,J\,\mathbf{s}_1\cdot\mathbf{s}_2$ with $J<0$ between the electron spins were operative.

We discuss now examples of two-electron systems with ferromagnetic spin–spin interaction. Consider the case of two electrons in *incomplete shells* of ions or atoms (for instance in the configurations of the type np^2 or nd^2 or nf^2). It is well-known that the two external (or "optical") electrons line up their spins in the ground state, in agreement with Hund's first rule; this means that the triplet state is lower in energy than the singlet state, and an effective coupling of ferromagnetic type $-2\,J\,\mathbf{s}_1\cdot\mathbf{s}_2$ (with $J>0$) appears to be at work.

To estimate the singlet–triplet separation avoiding inessential details, we consider the system with *two electrons* and *two degenerate orbitals*, whose orthonormal wave-functions are denoted by ϕ_1 and ϕ_2. Accommodation of the two electrons on the four available spin–orbitals ($\phi_1\,\alpha$, $\phi_1\,\beta$, $\phi_2\,\alpha$, $\phi_2\,\beta$) gives rise to six possible determinantal states. For simplicity, we assume that double occupation of the same orbital is avoided (because of the strong Coulomb repulsive energy between electrons of opposite spin in localized orbitals). We are thus left with four determinantal states (illustrated in Fig. 3); from them, we can form the singlet and triplet states given respectively

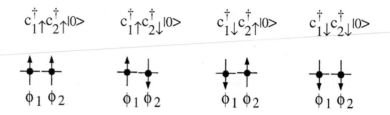

Fig. 3 The four possible two-electron states of a system with two orbitals and two electrons (double occupation of the same orbital is avoided).

by

$$\Psi_S = \frac{1}{\sqrt{2}}\left[\phi_1(\mathbf{r}_1)\,\phi_2(\mathbf{r}_2) + \phi_2(\mathbf{r}_1)\,\phi_1(\mathbf{r}_2)\right]\frac{1}{\sqrt{2}}\left[\alpha(1)\,\beta(2) - \beta(1)\,\alpha(2)\right] \qquad (14a)$$

and

$$\Psi_T = \frac{1}{\sqrt{2}}[\phi_1(\mathbf{r}_1)\phi_2(\mathbf{r}_2) - \phi_2(\mathbf{r}_1)\phi_1(\mathbf{r}_2)]\begin{cases} \alpha(1)\alpha(2) \\ \dfrac{1}{\sqrt{2}}[\alpha(1)\beta(2) - \beta(1)\alpha(2)] \\ \beta(1)\beta(2) \end{cases} . \qquad (14b)$$

We now evaluate the energy difference between the singlet and the triplet states. The Hamiltonian of the two-electron system has the general form

$$H = H_0(\mathbf{r}_1) + H_0(\mathbf{r}_2) + \frac{e^2}{|\mathbf{r}_1 - \mathbf{r}_2|} \,,$$

where $H_0(\mathbf{r}) = (\mathbf{p}^2/2m) + V(\mathbf{r})$ is a one-electron operator and e^2/r_{12} is the electron–electron Coulomb repulsion (the effect of core electrons is assumed to be embodied someway in the model potential $V(\mathbf{r})$). The one-electron operator needs not be further specified, since the energy splitting of the singlet and triplet states (14) is only due to the bielectronic operator e^2/r_{12}. A straightforward calculation gives for the difference

$$E_S - E_T = \langle\Psi_S|H\Psi_S\rangle - \langle\Psi_T|H\Psi_T\rangle = 2J \,, \qquad (15a)$$

where

$$J = \langle\phi_1\phi_2|\frac{e^2}{r_{12}}|\phi_2\phi_1\rangle \equiv \int \phi_1^*(\mathbf{r}_1)\phi_2^*(\mathbf{r}_2)\,\frac{e^2}{|\mathbf{r}_1 - \mathbf{r}_2|}\,\phi_2(\mathbf{r}_1)\phi_1(\mathbf{r}_2)\,d\mathbf{r}_1\,d\mathbf{r}_2 \,. \qquad (15b)$$

The bielectronic exchange integral J, defined by Eq. (15b), represents the self-energy of the (complex) charge distribtion $\phi_1^*\phi_2$ and is thus a *positive definite quantity* (see Appendix IV-A); it follows that the ground state of the two-electron system is the (triplet) magnetic state. This is so because the Coulomb repulsion between electrons is lowered in the triplet states with respect to the singlet state; notice in fact that $\Psi_T \equiv 0$ if $\mathbf{r}_1 \equiv \mathbf{r}_2$. The direct exchange mechanism mimics a ferromagnetic interaction between the spins of the two external electrons.

Another significant example of a system with two ("optical") electrons with ferromagnetic like interaction is provided by the O_2 molecule; in the independent particle approximation, the (antibonding) π molecular orbitals are twofold degenerate, corresponding to the two possible orientations perpendicular to the axis of the molecule; on them, we can accommodate the two external electrons of the molecule, with their spins lined up to decrease the Coulomb repulsion.

Other microscopic mechanisms leading to spin–spin coupling and Ruderman–Kittel–Kasuya–Yosida interaction

The highly idealized two-electron systems discussed so far justify, at least qualitatively, the spin–spin interaction models in particular physical situations. In actual materials with many electrons the justification of the validity (when possible at all!) of the Hamiltonian of type (10) requires extreme caution. It is not unfrequent that computational models of the exchange parameters J_{mn} in real materials are at times unsuccessful even in predicting the correct signs of J_{mn}; for this reason the interactions J_{mn} are often considered as disposable or semiempirical parameters.

In certain crystals, a spin–spin coupling may result due to a *superexchange interaction* with the following mechanism: two atoms with magnetic moments interact with the polarization produced on a third non-magnetic atom (or ion). An interesting case is that of transition-metal fluorides MnF_2, FeF_2, CoF_2; these materials are antiferromagnetic at low temperature and the coupling between cations is mediated through the halogen wavefunctions. This kind of superexchange (often antiferromagnetic) characterizes several magnetic insulators [for a thorough discussion of magnetic exchange interactions in insulators and semiconductors see P. W. Anderson, Solid State Physics **14**, 99 (1963)].

In some metals, such as rare earth metals, an effective interaction between localized spins may occur through the polarization of the free electron gas, where the localized spins are embedded. The magnetic moments of the incomplete f shells of rare earth atoms are so well localized that any direct interaction between f-electron wavefunctions on different sites are safely negligible. The f electrons, however, are coupled to the conduction electrons, induce a spin polarization, and an effective coupling between localized f-electron spins can occur. This effective interaction is known as the Ruderman–Kittel–Kasuya–Yosida (RKKY) interaction; it is oscillatory with the distance between the localized spins, and thus can give rise both to ferromagnetic and antiferromagnetic kind of interaction.

It is worthwhile to analyse in some detail the RKKY coupling mechanism. Consider a magnetic impurity of spin $\mathbf{S}$, located at the origin, and embedded in a pure bulk metal; we assume that the interaction between the localized spin $\mathbf{S}$ and the spin operator $\mathbf{s}$ for an electron in the conduction band can be described in the contact form

$$H_{\text{int}} = -J\,\Omega\,\mathbf{S} \cdot \mathbf{s}\,\delta(\mathbf{r})\ , \tag{16}$$

where J is a phenomenological constant with the dimension of an energy, and Ω is the volume of the unit cell of the crystal [it is irrelevant to include or not Ω in Eq. (16); if Ω is omitted, then J becomes a phenomenological constant with the dimension of energy times volume]. The conduction band wavefunctions of the metal are assumed to be plane waves of the form $W_{\mathbf{k}\sigma}(\mathbf{r}) = (1/\sqrt{V}) \exp(i\,\mathbf{k} \cdot \mathbf{r}) |\sigma\rangle$, normalized to one on the volume V of the crystal.

In the Born approximation, at the lowest order in the perturbation H_{int}, the conduction state $W_{\mathbf{k}\uparrow}$ is scattered into the state $\widetilde{W}_{\mathbf{k}\uparrow}$ given by

$$\widetilde{W}_{\mathbf{k}\uparrow}(\mathbf{r}) = W_{\mathbf{k}\uparrow}(\mathbf{r}) + \int g(\mathbf{r} - \mathbf{r}_0, E)\,(-J)\,\Omega\,\mathbf{S} \cdot \mathbf{s}\,\delta(\mathbf{r}_0)\,W_{\mathbf{k}\uparrow}(\mathbf{r}_0)\,d\mathbf{r}_0 \;, \tag{17a}$$

where $g(\mathbf{r} - \mathbf{r}_0, E)$ is the free particle Green's function

$$g(\mathbf{r} - \mathbf{r}_0, E) = -\frac{2m}{\hbar^2}\,\frac{e^{i\,k\,|\mathbf{r}-\mathbf{r}_0|}}{4\pi\,|\mathbf{r} - \mathbf{r}_0|} \qquad k = \frac{\sqrt{2\,m\,E}}{\hbar} \qquad (E > 0)\;. \tag{17b}$$

The expression of the free particle Green's function has been obtained in Section V-7.1, and is provided (in atomic units) by Eq. (V-55).

From Eqs. (17) we obtain

$$\widetilde{W}_{\mathbf{k}\uparrow}(\mathbf{r}) = \frac{1}{\sqrt{V}}\,e^{i\,\mathbf{k}\cdot\mathbf{r}}|\uparrow\rangle + \frac{1}{\sqrt{V}}\,\frac{2m\,J\,\Omega}{\hbar^2}\,\frac{e^{i\,k\,r}}{4\pi\,r}\,\mathbf{S} \cdot \mathbf{s}|\uparrow\rangle\;. \tag{18}$$

The density of electrons with spin up corresponding to the wave function $\widetilde{W}_{\mathbf{k}\uparrow}(\mathbf{r})$ is

$$n_{\mathbf{k}\uparrow}(\mathbf{r}) = \left| \frac{1}{\sqrt{V}}\,e^{i\,\mathbf{k}\cdot\mathbf{r}} + \frac{1}{\sqrt{V}}\,\frac{2m\,J\,\Omega}{\hbar^2}\,\frac{e^{i\,k\,r}}{4\pi\,r}\,S_z\,\frac{1}{2} \right|^2$$

$$= \frac{1}{V} + \frac{1}{V}\left[\frac{m\,J\,\Omega}{\hbar^2}\,S_z\,\frac{e^{i\,k\,r}}{4\pi\,r}\,e^{-i\,\mathbf{k}\cdot\mathbf{r}} + c.c. \right] + O(J^2)\;,$$

where c.c. indicates the complex conjugate of the previous term; similarly $n_{\mathbf{k}\downarrow}(\mathbf{r}) = 0$ to $O(J^2)$. We sum now the above expression over the $\mathbf{k}$ vectors within the Fermi sphere, and as usual we transform the sum over $\mathbf{k}$ into $V/(2\pi)^3$ times the integral in $d\mathbf{k}$; we obtain for the spin-up electron density

$$n_\uparrow(\mathbf{r}) = n_0 + \frac{m\,J\,\Omega}{\hbar^2}\,S_z\,\frac{1}{(2\pi)^3} \int_{\substack{\text{Fermi} \\ \text{sphere}}} \left[\frac{e^{i\,k\,r}}{4\pi\,r}\,e^{-i\,\mathbf{k}\cdot\mathbf{r}} + c.c. \right] d\mathbf{k}\;, \tag{19}$$

where n_0 is the density of electrons for one spin direction.

With the help of the integral on the angular variables

$$\int e^{-i\,\mathbf{k}\cdot\mathbf{r}}\,\sin\theta\,d\theta\,d\phi = \int e^{+i\,\mathbf{k}\cdot\mathbf{r}}\,\sin\theta\,d\theta\,d\phi = \frac{4\pi}{k\,r}\,\sin k\,r\;,$$

expression (19) becomes

$$n_\uparrow(\mathbf{r}) = n_0 + \frac{m\,J\,\Omega}{\hbar^2}\,S_z\,\frac{1}{(2\pi)^3}\,\frac{1}{r^2} \int_0^{k_F} 2\,\cos kr \cdot \sin kr \cdot k \cdot dk\;.$$

Using $2\cos kr\sin kr = \sin 2kr$, we can easily perform the integral in dk and obtain

$$n_\uparrow(\mathbf{r}) = n_0 - \frac{4\,m\,J\,\Omega}{\hbar^2}\,\frac{k_F^4}{(2\pi)^3}\,F(2\,k_F\,r)\,S_z\;, \tag{20a}$$

where

$$F(x) = \frac{x\cos x - \sin x}{x^4}\;.$$

Notice that $F(x) \approx -1/6x$ for small x, and $F(x) \approx \cos x/x^3$ for large x.

A completely similar calculation can be performed for the spin-down electron density; we obtain

$$n_\downarrow(\mathbf{r}) = n_0 + \frac{4\,m\,J\,\Omega}{\hbar^2}\,\frac{k_F^4}{(2\pi)^3}\,F(2\,k_F\,r)\,S_z\;. \tag{20b}$$

From Eqs. (20), we see that there is *no net charge polarization* (or electric field gradient); in fact

$$n_\uparrow(\mathbf{r}) + n_\downarrow(\mathbf{r}) \equiv 2\,n_0\;;$$

however *there is a spin polarization*

$$\boxed{n_\uparrow(\mathbf{r}) - n_\downarrow(\mathbf{r}) = -\frac{8\,m\,J\,\Omega}{\hbar^2}\,\frac{k_F^4}{(2\pi)^3}\,F(2\,k_F\,r)\,S_z}\;. \tag{21}$$

Since $F(x) \approx \cos x/x^3$ for large x, we see that the spin polarization is of long-range nature, oscillates with distance and falls off with the inverse cube power law r^{-3} at large distances; notice the analogies between the spin polarization oscillations produced by a magnetic impurity and the charge density oscillations (Friedel oscillations, Section VII-5) produced by an ordinary impurity in a metal.

We can now obtain the indirect exchange coupling between two spins $\mathbf{S}_1$ and $\mathbf{S}_2$, localized at $\mathbf{r} = 0$ and $\mathbf{R}$ respectively, and interacting with the Fermi sea electrons via a contact coupling. The first spin $\mathbf{S}_1$ creates a spin polarization; in fact, according to Eq. (18), the plane wave functions are modified in the form

$$\widetilde{W}_{\mathbf{k}\,\sigma}(\mathbf{r}) = \frac{1}{\sqrt{V}}\,e^{i\,\mathbf{k}\cdot\mathbf{r}}|\sigma\rangle + \frac{1}{\sqrt{V}}\,\frac{2\,m\,J\,\Omega}{\hbar^2}\,\frac{e^{i\,k\,r}}{4\pi\,r}\,\mathbf{S}_1\cdot\mathbf{s}|\sigma\rangle\;;$$

the spin $\mathbf{S}_2$ at $\mathbf{R}$ feels this spin polarization via contact interaction with an energy

$$E(R) = \sum_{\mathbf{k}\,\sigma}^{(\text{occ})} \langle \widetilde{W}_{\mathbf{k}\,\sigma}| - J\,\Omega\,\mathbf{S}_2\cdot\mathbf{s}\,\delta(\mathbf{r} - \mathbf{R})|\widetilde{W}_{\mathbf{k}\,\sigma}\rangle$$

$$= -\frac{2\,m\,J^2\,\Omega^2}{\hbar^2}\,\frac{1}{2}\,\mathbf{S}_1\cdot\mathbf{S}_2\,\frac{1}{V}\sum_{\mathbf{k}}^{(\text{occ})}\left[\frac{e^{i\,k\,R}}{4\pi\,R}\,e^{-i\,\mathbf{k}\cdot\mathbf{R}} + c.c.\right]\;. \tag{22}$$

In Eq. (22) the isotropic scalar product $(1/2)\,\mathbf{S}_1\cdot\mathbf{S}_2$ appears because of the general property

$$\langle\uparrow |\mathbf{S}_1\cdot\mathbf{s}\;\mathbf{S}_2\cdot\mathbf{s}|\uparrow\rangle + \langle\downarrow |\mathbf{S}_1\cdot\mathbf{s}\;\mathbf{S}_2\cdot\mathbf{s}|\downarrow\rangle = \frac{1}{2}\,\mathbf{S}_1\cdot\mathbf{S}_2\;;$$

the above identity can be easily worked out by inspection, using the matrix elements of Eq. (XVI-34), or more formally using general properties of spin operators.

The integration appearing in Eq. (22) has been already done just above, and gives

$$E(R) = \frac{4\,m\,J^2\,\Omega^2}{\hbar^2}\,\frac{k_F^4}{(2\pi)^3}\,F(2\,k_F\,R)\,\mathbf{S}_1\cdot\mathbf{S}_2 \ . \tag{23a}$$

The above expression can also be written in the form

$$E(R) = \frac{2\,J^2}{E_F}\,\frac{(\Omega\,k_F^3)^2}{(2\pi)^3}\,F(2\,k_F\,R)\,\mathbf{S}_1\cdot\mathbf{S}_2 \ . \tag{23b}$$

The Ruderman–Kittel–Kasuya–Yosida interaction (23) is ferromagnetic-like for small R; for large R it can be either ferromagnetic or antiferromagnetic depending on the distance of the two spins. The oscillatory behaviour of $F(2\,k_F\,R)$ is responsible for the great variety of magnetic behaviour occurring in rare earth metals and compounds, including ferro- and antiferromagnetism, incommensurate order, spiral structures etc. [see for instance J. Jensen and A. R. Mackintosh "Rare Earth Magnetism" (Clarendon, Oxford 1991)].

Microscopic interpretation of the Weiss molecular field and mean field theory

We are now in the position to understand the microscopic origin of the Weiss molecular field. Consider a magnetic crystal composed by N spins $\mathbf{S}_i$ (and magnetic moment $\mu_0\,\mathbf{S}_i$) localized on the sites of a Bravais lattice, and interacting via a spin–spin Hamiltonian. In the presence of an external magnetic field H in the z-direction, the Hamiltonian of the system of spins can be written as

$$H_0 = -\sum_{m\neq n} J_{mn}\,\mathbf{S}_m\cdot\mathbf{S}_n - \mu_0\,H\,\sum_m S_m^z \ , \tag{24}$$

where the first term is the Heisenberg spin–spin Hamiltonian, and the second term is the Zeeman energy. We wish now to establish a relationship between the exchange parameters and the Weiss constant λ.

For simplicity we assume that the spin Hamiltonian couples only nearest neighbour spins, with exchange energy J. We focus our attention on one particular site, say the site at the origin, and isolate from H_0 the part $H(\mathbf{S}_0)$ containing the spin $\mathbf{S}_0$. We have

$$H(\mathbf{S}_0) = -2\,J\,\sum_{n=1}^{n_c}\mathbf{S}_0\cdot\mathbf{S}_n - \mu_0\,H\,S_0^z$$

$$= -2\,J\,\sum_{n=1}^{n_c}[S_0^z\,S_n^z + S_0^x\,S_n^x + S_0^y\,S_n^y] - \mu_0\,H\,S_0^z \ , \tag{25}$$

where n_c is the coordination number, i.e. the number of neighbours interacting with the chosen spin at the origin.

We replace the spin operators S_n^z, S_n^x, S_n^y of the neighbouring sites by their average values $\langle S_n^z\rangle$, $\langle S_n^x\rangle$, $\langle S_n^y\rangle$. Assuming the magnetization along the z-axis, we have

$$M = \frac{N}{V}\,\mu_0\,\langle S_n^z\rangle \ , \qquad \langle S_n^x\rangle = \langle S_n^y\rangle = 0 \ ,$$

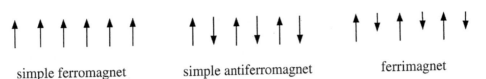

simple ferromagnet   simple antiferromagnet   ferrimagnet

Fig. 4 Schematic illustration of ferromagnetic and antiferromagnetic order. The ferrimagnetic order, in which spins of different values alternate, is also shown.

and Eq. (25) becomes

$$H(\mathbf{S}_0) = -2\,J\,n_c\,\frac{M}{(N/V)\,\mu_0}\,S_0^z - \mu_0\,H\,S_0^z\,. \tag{26}$$

Comparison of Eq. (26) and Eq. (3) gives for the Weiss constant

$$\lambda = \frac{2\,J\,n_c}{(N/V)\,\mu_0^2}\,.$$

The Weiss constant can be estimated taking $N/V \approx 1/(5\,a_B^3) \approx (1/100)\,a_B^{-3}$; $\mu_0^2/a_B^3 \approx$ $\mu_B^2/a_B^3 \approx 1\,\mathrm{meV}$ (see Eq. 1b), and $J \approx 10 - 100\,\mathrm{meV}$; we obtain $\lambda \approx 10^4 - 10^5$. In conclusion, the spin–spin model Hamiltonian can be taken as a reasonable microscopic starting point for the justification of the Weiss field in several magnetic crystals.

3   Antiferromagnetism in the mean field approximation

In the Heisenberg theory of ferromagnetism the coupling energy J between nearest neighbour spins is positive and parallel alignment of local spins is favoured. In the case the coupling energy J is negative, antiparallel orientation of neighbour spins is preferred, and a tendency of spins to order in a structure where up and down spins alternate is expected (Fig. 4).

The discussion of the magnetization of an antiferromagnetic material in the mean field approximation can be done in close analogy with the ferromagnetic case. It is convenient to consider the *two sublattice model* in which the complete lattice of atoms is divided into two identical sublattices, A and B, such that all nearest neighbours of an A site are B sites and vice versa. For instance, in the case of a simple cubic lattice, we can assign the site at the origin to sublattice A, the six nearest neighbours to sublattice B, the nearest neighbours of B to sublattice A etc.; the two interpenetrating sublattices so obtained are fcc lattices displaced as in the well-known NaCl structure. As a second example, in the case the complete lattice is bcc, the two composing and interpenetrating sublattices are both simple cubic, one formed with corner sites and one formed with body centred sites of the initial bcc lattice. Notice also that in the case of a fcc lattice, no simple partition is possible in which all nearest neighbours of A sites are B sites, and vice versa.

The Weiss molecular field approach has been generalized by Néel to the treatment of antiferromagnetism, by assuming that the effective field acting on sites A (or B)

contains a cooperative contribution, proportional and opposite to the magnetization on sublattice B (or A). In addition to the nearest neighbour AB (or BA) antiferromagnetic interaction, we also assume an antiferromagnetic interaction for second nearest neighbour AA and BB (this is not strictly necessary, but otherwise the model would be somewhat more restrictive).

At the absolute temperature $T=0$, each of the two sublattices has maximum magnetization. Increasing the temperature the magnetization on each of the two sublattices decreases; however, differently from ferromagnetism the net magnetization of an antiferromagnet is zero at any temperature. Above a critical temperature T_N, called the *Néel temperature*, the magnetization on each sublattice vanishes, and the magnetic susceptibility decreases with a law of the type $\chi \propto 1/(T + \Theta)$, with Θ of the order of T_N. We show that all these facts can be accounted for within the mean field theory.

In the two sublattice model, the effective fields at A and B sites are given by

$$\mathbf{H}_a^{(\text{eff})} = \mathbf{H} - \lambda_1 \mathbf{M}_b - \lambda_2 \mathbf{M}_a \tag{27a}$$

$$\mathbf{H}_b^{(\text{eff})} = \mathbf{H} - \lambda_1 \mathbf{M}_a - \lambda_2 \mathbf{M}_b \tag{27b}$$

where $\mathbf{H}$ is the external field, $\mathbf{M}_a$ and $\mathbf{M}_b$ are the magnetization of A and B sublattices, λ_1 and λ_2 are *positive phenomenological constants* (λ_2 could also be zero, or even negative; but in any case we expect and assume that $|\lambda_2| < \lambda_1$). With the effective fields provided by Eqs. (27), we can now obtain the magnetization of the two sublattices. We suppose that the local spins have $S = 1/2$ and $g = 2$. We indicate with z the direction of the applied field $\mathbf{H}$, and with M_a and M_b the z-components of the sublattice magnetization. We have for the magnetization

$$M_a = \frac{1}{2} \frac{N}{V} \mu_B \tanh \frac{\mu_B (H - \lambda_1 M_b - \lambda_2 M_a)}{k_B T} \tag{28a}$$

$$M_b = \frac{1}{2} \frac{N}{V} \mu_B \tanh \frac{\mu_B (H - \lambda_1 M_a - \lambda_2 M_b)}{k_B T} \tag{28b}$$

where N is the total number of sites and V is the volume of the crystal. We consider now the high temperature region ($T > T_N$) and the low temperature region ($T < T_N$).

Temperature region above the Néel temperature ($T > T_N$). In the temperature region above the Néel temperature, we expect that both M_a and M_b are parallel to the applied field H and vanish when the applied field decreases to zero. For reasonably small applied fields, we can use the approximation $\tanh x \approx x$ in Eqs. (28); we obtain

$$M_a = \frac{1}{2} \frac{C}{T} (H - \lambda_1 M_b - \lambda_2 M_a) \tag{29a}$$

$$M_b = \frac{1}{2} \frac{C}{T} (H - \lambda_1 M_a - \lambda_2 M_b) \tag{29b}$$

Table 4 Néel temperature T_N and Θ temperature (in K) for some antiferromagnetic materials [for further data see for instance F. Keffer, Handbuch der Physik, Vol. 18 part 2 (Springer, Berlin 1966)].

Material	T_N	Θ	Material	T_N	Θ
MnO	122	610	MnF$_2$	67	80
FeO	198	507	FeF$_2$	78	117
CoO	291	330	CoF$_2$	38	50
NiO	600	2000	KMnF$_3$	88	158

where C denotes the constant

$$C = \frac{N}{V} \frac{\mu_B^2}{k_B} \ .$$

Summing up the two relations expressed by Eqs. (29), we have

$$M = M_a + M_b = \frac{C}{T} H - \frac{\Theta}{T} M$$

with

$$\Theta = C \frac{\lambda_1 + \lambda_2}{2} \ . \tag{30}$$

The magnetic susceptibility of the antiferromagnet thus becomes

$$\boxed{\chi = \frac{M}{H} = \frac{C}{T + \Theta}} \ .$$

The susceptibility is reduced with respect to the non-interacting case ($\lambda_1 = \lambda_2 = 0 = \Theta$) because the magnetic moments tend to arrange in an antiparallel way.

Temperature region below the Néel temperature ($T \leq T_N$). In the temperature region below the Néel temperature, we expect a spontaneous magnetization on each sublattice, even in the absence of an external magnetic field. We furthermore expect that $M_a = -M_b$. Eq. (28a) for the spontaneous magnetization of sublattice A becomes

$$M_a = \frac{1}{2} \frac{N}{V} \mu_B \tanh \frac{\mu_B (\lambda_1 - \lambda_2) M_a}{k_B T} \ , \tag{31}$$

and a similar equation holds for M_b. Eq. (31) for the spontaneous magnetization in each sublattice has the same structure as Eq. (5); it follows in particular that the critical Néel temperature T_N is given by

$$T_N = C \frac{\lambda_1 - \lambda_2}{2} \ . \tag{32}$$

From Eq. (30) and Eq. (32) we obtain

$$\frac{T_N}{\Theta} = \frac{\lambda_1 - \lambda_2}{\lambda_1 + \lambda_2} \ .$$

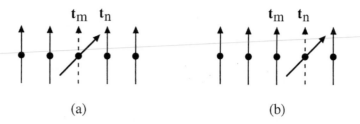

t_m t_n t_m t_n

(a)   (b)

Fig. 5 Schematic representation of trial excited states of the spin Hamiltonian. In (a) there is a spin deviation at the site t_m , and in (b) there is a spin deviation at the adjacent site t_n. In the spin-wave theory, the operators which couple spin deviations on different sites produce a wave-like Bloch state.

In the case the antiferromagnetic AA and BB interactions are negligible (i.e. $\lambda_2 = 0$) we have $T_N = \Theta$. With λ_2 positive, we have $T_N < \Theta$, as it is experimentally observed in a number of materials (see Table 4).

We do not dwell further on the analysis and application of the mean field theory to cooperatively interacting localized magnetic moments. The mean field theory, with the support of appropriate microscopic spin–spin Hamiltonians, can describe numerous other types of magnetic order present in the nature. Besides ferromagnetism and antiferromagnetism, we can mention ferrimagnetism, helimagnetism, modulated structures in the presence of competing interactions, and we refer to the literature for further information.

4   Spin waves and magnons in ferromagnetic crystals

So far, in the discussion of magnetic ordering, we have considered the Heisenberg Hamiltonian for a system of spins coupled by exchange interaction, and we have applied to it the Weiss (or Néel) molecular field approximation. To improve the understanding of magnetism, one should know more about the eigenvalues and eigenstates of the spin Hamiltonian. For this purpose there are two important techniques: (i) the spin–wave theory, which is particularly appropriate for magnetic systems in the low-temperature limit;   (ii) the renormalization group formalism, which focuses on the physics of phase transitions. In this section we consider some aspects of the spin–wave theory.

It has been shown originally by Bloch that the low-lying energy states of a periodic system of localized spins, coupled by exchange interaction, are wavelike. The energy of a spin wave is quantized, and the unit of energy is called *magnon*. Spin waves have been studied for all types of magnetic ordering (ferromagnetic, antiferromagnetic, ferrimagnetic, helical etc.). For simplicity, we confine our attention to ferromagnetic materials (see Fig. 5); for them the treatment of wavelike excitations is rather intuitive, compared with the much more demanding cases of antiferromagnetism and other types of magnetic ordering.

Elementary properties of spin operators

Before discussing the basic aspects of the spin–wave theory, it is useful to remind some elementary properties of spin operators. Consider the three component operators S^x, S^y, S^z of a given spin operator $\mathbf{S}$. The commutation rules are $[S^x, S^y] = i\, S^z$ and cyclic permutations of the indices x, y, z. The operators S^x, S^y, S^z are not independent, but are connected by the identity

$$\mathbf{S} \cdot \mathbf{S} = S_x^2 + S_y^2 + S_z^2 = S(S+1) \ .$$

It is convenient to define the raising operator S^+ and the lowering operator S^- as

$$S^+ = S^x + i\, S^y \ , \quad S^- = S^x - i\, S^y \ . \tag{33}$$

The operator S^z is connected to the raising and lowering operators by the commutation relation

$$S^z = \frac{1}{2}[S^+, S^-] \ .$$

The effect of S^+ or S^- on an eigenstate of S^z with eigenvalue M is

$$S^+ |M\rangle = \sqrt{(S-M)(S+1+M)}\,|M+1\rangle \tag{34a}$$

$$S^- |M\rangle = \sqrt{(S+M)(S+1-M)}\,|M-1\rangle \ , \tag{34b}$$

where $-S \leq M \leq +S$.

It is convenient to define the spin deviation operator $\hat{n} = S - S^z$, which is diagonal on the representation where S^z is diagonal, and whose eigenvalues are integer numbers ranging from 0 to $2S$; $\hat{n}$ represents the spin deviations from the maximum value of spin S. Thus the states $|S\rangle$, $|S-1\rangle$, ... $|-S\rangle$ correspond to spin deviations $|0\rangle$, $|1\rangle$, ... $|2S\rangle$, respectively. We now rewrite expressions (34) in the newly relabelled basis and we have

$$S^+ |n\rangle = \sqrt{n}\,\sqrt{2S-(n-1)}\,|n-1\rangle \tag{35a}$$

$$S^- |n\rangle = \sqrt{2S-n}\,\sqrt{n+1}\,|n+1\rangle \ , \tag{35b}$$

where the restriction $0 \leq n \leq 2S$ holds, and the arguments of the square root in Eqs. (35) are always non-negative.

In formal analogy with the properties of annihilation and creation operators for the harmonic oscillator (Appendix IX-A), it is convenient to introduce the boson operators a and $a^\dagger$, which applied to the spin deviations $|n\rangle$ give

$$a|n\rangle = \sqrt{n}\,|n-1\rangle \ , \quad a^\dagger |n\rangle = \sqrt{n+1}\,|n+1\rangle \ , \quad a^\dagger a\,|n\rangle = n\,|n\rangle \ .$$

From the above relations, and Eqs. (35) and Eqs. (33), the operators S^+, S^-, S^z can be expressed in the form

$$\left\{ \begin{array}{l} S^+ = \sqrt{(2S - a^\dagger a)}\,a \\[4pt] S^- = a^\dagger \sqrt{(2S - a^\dagger a)} \\[4pt] S^z = S - a^\dagger a \end{array} \right. \tag{36}$$

The last of Eqs. (36) is easily verified using the standard commutation rule for boson operators $[a, a^\dagger] = 1$; we have in fact

$$S^z = \frac{1}{2}(S^+ S^- - S^- S^+) = \frac{1}{2}\left[\sqrt{(2S - a^\dagger a)}\, aa^\dagger \sqrt{(2S - a^\dagger a)} - a^\dagger(2S - a^\dagger a)\, a\right]$$

$$= \frac{1}{2}\left[(2S - a^\dagger a) + (2S - a^\dagger a)a^\dagger a - 2S\, a^\dagger a + a^\dagger a^\dagger aa\right] = S - a^\dagger a .$$

Expressions (36) are known as the Holstein–Primakoff transformations.

Elementary excitations in ferromagnetic crystals

Consider an ideal ferromagnetic crystal, of volume V, formed by N spins $\mathbf{S}_n$ localized at the points $\mathbf{t}_n$ of a simple Bravais lattice. We assume a ferromagnetic interaction among nearest neighbour spins $(J > 0)$, and consider the isotropic Heisenberg Hamiltonian

$$H = -J \sum_{\mathbf{t}_m\, \mathbf{t}_i^{(I)}} \mathbf{S}_m \cdot \mathbf{S}_{m+i} , \tag{37a}$$

where the sum over $\mathbf{t}_m$ $(m = 1, 2, \dots, N)$ runs over all translation vectors of the crystal, and the sum over $\mathbf{t}_i^{(I)}$ $(i = 1, 2, \dots, n_c)$ runs over the n_c translational vectors connecting a lattice site of a Bravais lattice with its nearest neighbour ones. We can also write

$$H = -J \sum_{\mathbf{t}_m\, \mathbf{t}_i^{(I)}} (S_m^z S_{m+i}^z + \frac{1}{2} S_m^+ S_{m+i}^- + \frac{1}{2} S_m^- S_{m+i}^+) , \tag{37b}$$

where the scalar product between spin operators has been expressed with the help of the raising and lowering spin operators, defined by Eqs. (33).

The ground state $|0\rangle$ of the ferromagnetic crystal is that in which all spins are lined up in a given direction (say the z direction), and can be written as

$$|0\rangle = |S\rangle_{\mathbf{t}_1} \cdot |S\rangle_{\mathbf{t}_2} \cdot \dots \cdot |S\rangle_{\mathbf{t}_N} .$$

Applying the operator H to the state $|0\rangle$ one has

$$H |0\rangle = -J S^2 N n_c |0\rangle ;$$

thus the state $|0\rangle$ is an eigenstate of H with energy $E_0 = -J S^2 N n_c$, where n_c is the coordination number. The fact that we can individuate explicitly the ground state of a ferromagnet makes things much easier. We can now try to individuate the excited states. For this purpose it is convenient to perform on the periodic Hamiltonian (37) appropriate canonical transformations, suggested by the Bloch theorem.

We use the Holstein–Primakoff relations (36) in the simplified form

$$\begin{cases} S_m^+ = \sqrt{(2S - a_m^\dagger a_m)}\, a_m \approx \sqrt{2S}\, a_m \\[2mm] S_m^- = a_m^\dagger \sqrt{(2S - a_m^\dagger a_m)} \approx \sqrt{2S}\, a_m^\dagger \\[2mm] S_m^z = S - a_m^\dagger a_m \end{cases} \tag{38}$$

The approximation made for S_m^+ and S_m^- is justified in the low temperature limit; in this limit, the number of excitations is small, the thermal average $\langle a_m^\dagger a_m \rangle$ is expected to be of the order $O(1/N)$ and can be safely neglected with respect to $2S$. Inserting Eqs. (38) into Eq. (37b), we obtain

$$H = -J \sum_{\mathbf{t}_m \, \mathbf{t}_i^{(I)}} \left[(S - a_m^\dagger a_m)(S - a_{m+i}^\dagger a_{m+i}) + S a_m a_{m+i}^\dagger + S a_m^\dagger a_{m+i} \right] ; \qquad (39a)$$

in line with the approximations made in Eqs. (38) in the low temperature limit, we neglect in the Hamiltonian (39a) any term containing products of four operators, and we have

$$H = E_0 + 2 J S n_c \sum_{\mathbf{t}_m} a_m^\dagger a_m - 2 J S \sum_{\mathbf{t}_m \, \mathbf{t}_i^{(I)}} a_m^\dagger a_{m+i} , \qquad (39b)$$

where $E_0 = -J S^2 N n_c$ is the ground-state energy of the ferromagnet.

We now perform the standard canonical transformations suggested by the Bloch theorem

$$a_m = \frac{1}{\sqrt{N}} \sum_{\mathbf{q}} e^{i \mathbf{q} \cdot \mathbf{t}_m} a_{\mathbf{q}} \quad \text{and} \quad a_m^\dagger = \frac{1}{\sqrt{N}} \sum_{\mathbf{q}} e^{-i \mathbf{q} \cdot \mathbf{t}_m} a_{\mathbf{q}}^\dagger , \qquad (40a)$$

or conversely

$$a_{\mathbf{q}} = \frac{1}{\sqrt{N}} \sum_{\mathbf{t}_m} e^{-i \mathbf{q} \cdot \mathbf{t}_m} a_m \quad \text{and} \quad a_{\mathbf{q}}^\dagger = \frac{1}{\sqrt{N}} \sum_{\mathbf{t}_m} e^{i \mathbf{q} \cdot \mathbf{t}_m} a_m^\dagger , \qquad (40b)$$

where the allowed N vectors $\mathbf{q}$ are defined in the first Brillouin zone. We have

$$\left[a_{\mathbf{q}} , a_{\mathbf{q}'}^\dagger \right] = \delta_{\mathbf{q},\mathbf{q}'}$$

and all other commutators are zero.

Using the relations (40a), the Hamiltonian (39b) can be recast in the form

$$H = E_0 + 2 J S n_c \sum_{\mathbf{q}} [1 - \gamma(\mathbf{q})] \, a_{\mathbf{q}}^\dagger a_{\mathbf{q}} , \qquad (41a)$$

where

$$\gamma(\mathbf{q}) = \frac{1}{n_c} \sum_{\mathbf{t}_i^{(I)}} e^{i \mathbf{q} \cdot \mathbf{t}_i^{(I)}} \qquad (41b)$$

and $\mathbf{t}_i^{(I)}$ ($i = 1, 2, \dots, n_c$) indicate the translation vectors connecting a site with its first neighbours. We thus obtain an Hamiltonian which is equivalent to a set of harmonic oscillators with frequency

$$\hbar \omega(\mathbf{q}) = 2 J S n_c [1 - \gamma(\mathbf{q})] . \qquad (41c)$$

The quanta of energy $\hbar \omega(\mathbf{q})$ are called *magnons*.

For a simple cubic lattice, for instance, the dispersion relation (41c) reads

$$\hbar \omega(\mathbf{q}) = 2 J S [6 - 2 \cos q_x a - 2 \cos q_y a - 2 \cos q_z a] .$$

For small values of q ($q \ll 2\pi/a$) we have the quadratic expression

$$\hbar\omega(\mathbf{q}) = 2\,J\,S\,a^2\,q^2 = D\,q^2 \tag{42}$$

with $D = 2\,J\,S\,a^2$. A quadratic expression holds in general for cubic crystals.

The energy–wavevector dispersion relation of the spin waves can be fruitfully studied by means of inelastic neutron scattering experiments. From measurements of the energy difference of the ingoing and outgoing neutron energies and wavevectors, the energies and the corresponding wavevectors of the spin wave can be obtained, and the spin-wave relationship $\hbar\omega(\mathbf{q})$ versus $\mathbf{q}$ can be worked out.

Low temperature magnetization in ferromagnetic crystals

We can now show how the spontaneous magnetization varies with temperature in a ferromagnet (at least in the low temperature limit, when the interaction among spin waves can be neglected and there is a small number of excitations). The average number of quasi particles, or magnons, in a state of wavevector $\mathbf{q}$ is given by the Bose–Einstein statistics

$$\langle n_{\mathbf{q}} \rangle = \frac{1}{e^{\hbar\omega(\mathbf{q})/k_B T} - 1} \ .$$

The spontaneous magnetization at $T=0$ of a ferromagnetic crystal of volume V and N spins $\mathbf{S}$, of magnetic moment $\mu_0\,\mathbf{S}$, is $M(0) = (N/V)\,\mu_0\,S$; the spontaneous magnetization (at sufficiently low temperatures, where the spin–wave theory holds) is given by the expression

$$M(T) = M(0) - \mu_0\,S\,\frac{1}{V}\,\sum_{\mathbf{q}} \langle n_{\mathbf{q}} \rangle \ .$$

Converting, as usual, the sum over $\mathbf{q}$ into an integral over the Brillouin zone times $V/(2\pi)^3$, one finds

$$M(0) - M(T) = \mu_0\,S\,\frac{1}{(2\pi)^3}\,\int_{\text{B.Z.}} \frac{d\mathbf{q}}{\exp\left(\hbar\omega(\mathbf{q})/k_B T - 1\right)} \ . \tag{43a}$$

At small temperatures, only the occupation numbers of magnons of small q are significant; we thus use the quadratic form $\hbar\omega(\mathbf{q}) = D\,q^2$ (for cubic lattices) and extend the integral over the whole reciprocal space (with negligible error). We obtain

$$M(0) - M(T) = \mu_0\,S\,\frac{1}{(2\pi)^3}\,\int \frac{d\mathbf{q}}{\exp\left(D\,q^2/k_B T - 1\right)} \ . \tag{43b}$$

It is convenient to introduce polar coordinates in the reciprocal space, and operate the change of variable $D\,q^2/k_B T = x$; this gives

$$M(0) - M(T) = \mu_0\,S\,\frac{1}{4\pi^2}\,\left(\frac{k_B T}{D}\right)^{3/2}\,\int_0^\infty \frac{x^{1/2}}{e^x - 1}\,dx \ . \tag{43c}$$

Thus $M(0) - M(T)$ is proportional to $T^{3/2}$ a famous law due to Bloch, confirmed by experimental results. Notice that the mean field theory fails to give the correct power law $T^{3/2}$ of Eq. (43c), and rather gives the exponential behaviour $\exp(-2T_c/T)$ of Eq. (8).

It is of interest to notice that the integral in Eq. (43b), at small wavevectors and finite temperature, takes the form $\int (1/q^2)\,d\mathbf{q}$; this integral converges at small wavevectors only if $d\mathbf{q}$ is three-dimensional. Thus if one calculates the spontaneous magnetization of the two- and the one-dimensional ferromagnets in analogy to the three-dimensional case, one finds an integral which diverges at small wavevectors. From this divergence at small q, it can be inferred that no spontaneous magnetization occurs in the one- and two-dimensional isotropic Heisenberg models, a conclusion corroborated by more sophisticated treatments.

Considerations on spin waves in antiferromagnetic crystals

The extension of the spin–wave formalism to antiferromagnetic crystals requires particular care and appropriate implementations, as it becomes apparent from the following considerations. The simplest model of an antiferromagnetic crystal is that of two interpenetrating sublattices, say A and B, with spin up and spin down, respectively.

By analogy with the ferromagnetic state, one could attempt to write the ground state of the antiferromagnet in the form

$$|S\rangle_{1A} \cdot |-S\rangle_{1B} \cdot \ldots \cdot |S\rangle_{mA} \cdot \ldots \cdot |-S\rangle_{nB} \cdot \ldots \;;$$

however such a state interacts ("resonates") with states in which the spin on a site of sublattice A is decreased by 1 and the spin on a neighbour site of sublattice B is increased by 1. Thus for antiferromagnets we cannot say easily what the ground state is; in fact the ground state is not simply of the type $|\uparrow\downarrow\uparrow\downarrow\uparrow\downarrow \ldots\rangle$, but must be expressed as a superposition of all spin configurations with total spin component S_z equal to zero (resonant ground-state picture). The quantum fluctuations of the spins make the classical picture and the quantum picture rather different; actually, even in three-dimensional antiferromagnets at $T=0$, the full magnetization of a given sublattice is never achieved. We do not dwell on this and other aspects of the spin–wave theory in antiferromagnetism and other magnetic structures; for a thorough analysis, based on the Holstein–Primakoff transformations, we refer to the literature [see for instance W. Jones and N. H. March "Theoretical Solid State Physics" (Wiley-Interscience, New York 1973)].

5 The Ising model with the transfer matrix method

In the previous sections, we have discussed some aspects of magnetism essentially on the basis of the Heisenberg exchange Hamiltonian, of the type of Eq. (10). In particular we have considered the mean field theory and we have seen that its inadequacies in the low temperature limit can be (partially) settled with the spin–wave theory. The mean field theory presents serious limitations also in the study of the magnetic properties near the critical temperature; in particular, the mean field theory is unable to determine correctly the effect of dimensionality on the phase transitions and on the critical exponents. In order to study some elementary aspects of phase transitions at criticality, we consider the Ising approximation to the Heisenberg Hamiltonian; the Ising model, in fact, permits a trivial simplification of the eigenvalue problem and a consequent easier attention to the physics of phase transitions.

Transfer matrix method for the one-dimensional Ising model

The one-dimensional Ising model, with nearest neighbour spin interactions and in the presence of an external magnetic field, can be described as follows. We introduce on each site of a linear chain the spin variable σ_i, which can take only the values ± 1. The interaction energy between nearest neighbour spins and the Zeeman interaction energy are described by the two terms of the following Hamiltonian

$$H = -J \sum_{i=1}^{N} \sigma_i \sigma_{i+1} - h \sum_{i=1}^{N} \sigma_i , \tag{44}$$

where the sum extends over a (very large) number N of sites, and periodic boundary conditions are adopted, so that $\sigma_{N+1} \equiv \sigma_1$ (in the thermodynamic limit $N \to \infty$ the choice of boundary conditions becomes irrelevant). We assume $J > 0$, which favours a parallel alignment of spins; h is the strength of the applied external magnetic field times the magnetic moment on each site (taken as unity for simplicity).

In the case of a one-dimensional Ising model, we can determine exactly the partition function and the thermodynamic functions, and then show that no spontaneous magnetization and no phase transition occur for a one-dimensional chain.

The partition function of the one-dimensional Ising chain is given by

$$Z = \sum_{\{\sigma\}} e^{\beta J (\sigma_1 \sigma_2 + \sigma_2 \sigma_3 + \sigma_3 \sigma_4 + \dots) + \beta h (\sigma_1 + \sigma_2 + \sigma_3 + \dots)} , \tag{45a}$$

where $\beta = 1/k_B T$, and the sum over $\{\sigma\}$ runs on all the 2^N spin configurations, obtained by assigning the values $\sigma_i = \pm 1$ to each of the N spins. The partition function (45a) can also be written in the form

$$Z = \sum_{\{\sigma\}} \left[e^{\beta J \sigma_1 \sigma_2 + \beta h (\sigma_1 + \sigma_2)/2} \right] \left[e^{\beta J \sigma_2 \sigma_3 + \beta h (\sigma_2 + \sigma_3)/2} \right] \dots$$

$$= \sum_{\{\sigma\}} P(\sigma_1, \sigma_2) P(\sigma_2, \sigma_3) \dots P(\sigma_N, \sigma_1) , \tag{45b}$$

where the function $P(\sigma_i, \sigma_j)$ is defined as

$$P(\sigma_i, \sigma_j) = e^{\beta J \sigma_i \sigma_j + \beta h (\sigma_i + \sigma_j)/2} .$$

It is convenient to collect the four possible values of the function $P(\sigma_i, \sigma_j)$ in the form of a 2×2 *transfer matrix* P defined as

$$P = \begin{pmatrix} e^{\beta J + \beta h} & e^{-\beta J} \\ e^{-\beta J} & e^{\beta J - \beta h} \end{pmatrix} .$$

The partition function (45b) thus takes the compact form

$$\boxed{Z = \text{Tr } P^N} , \tag{46}$$

and the trace of the matrix P^N can be explicitly calculated as follows.

The two eigenvalues of the P matrix are real, positive and given by

$$\lambda_\pm = e^{\beta J} \left[\cosh \beta h \pm \sqrt{\sinh^2 \beta h + e^{-4\beta J}} \right] ;$$

in the particular case of vanishing applied magnetic field, we have $\lambda_+(h = 0) = 2 \cosh \beta J$ and $\lambda_-(h{=}0) = 2 \sinh \beta J$.

The partition function (46) becomes

$$Z = \lambda_+^N + \lambda_-^N \approx \lambda_+^N ;$$

since N is very large, the relevant contribution is given only by the larger of the two eigenvalues of P. For the free energy $F = -k_B T \ln Z$, we have

$$F = -N k_B T \ln \lambda_+ = -N J - N k_B T \ln \left[\cosh \beta h + \sqrt{\sinh^2 \beta h + e^{-4\beta J}} \right] .$$

The thermodynamic function $F(h, T)$ is a continuous function of h and T, without singularities; no phase transition is thus expected. In particular, it is easily verified that the spontaneous magnetization, given by $-\partial F / \partial h$ in the limit $h \rightarrow 0$, vanishes at any temperature; thus a linear chain cannot be ferromagnetic.

We consider now the mean energy of the linear chain of interacting spins, and its heat capacity, in the absence of a magnetic field. We have

$$Z(h{=}0, T) = [2 \cosh \beta J]^N \tag{47a}$$

and

$$\langle E \rangle = k_B T^2 \frac{\partial \ln Z}{\partial T} = -N J \tanh \beta J . \tag{47b}$$

The heat capacity becomes

$$C_V = \frac{\partial \langle E \rangle}{\partial T} = N k_B \left(\frac{J}{k_B T} \right)^2 \frac{1}{\cosh^2 (J/k_B T)} , \tag{47c}$$

and the behaviour of C_V is shown in Fig. 6.

From the behaviour of C_V, and from the fact that the spontaneous magnetization vanishes at any temperature, we can infer the following. (i) In a linear chain there is no long-range order (otherwise a spontaneous magnetization would occur). (ii) In a linear chain, at low temperatures $k_B T \ll k_B T_J = J$, there is a short-range order; this short-range order is relaxed gradually as the temperature T increases at or above the temperature T_J.

We can understand qualitatively why it is so, noticing that in a linear chain the long-range order is broken simply by a change of spin orientation at a single site, as

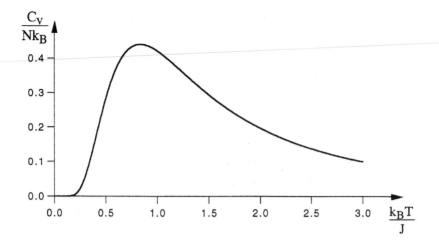

Fig. 6 Specific heat of a linear chain of spins coupled by exchange interaction in the Ising model.

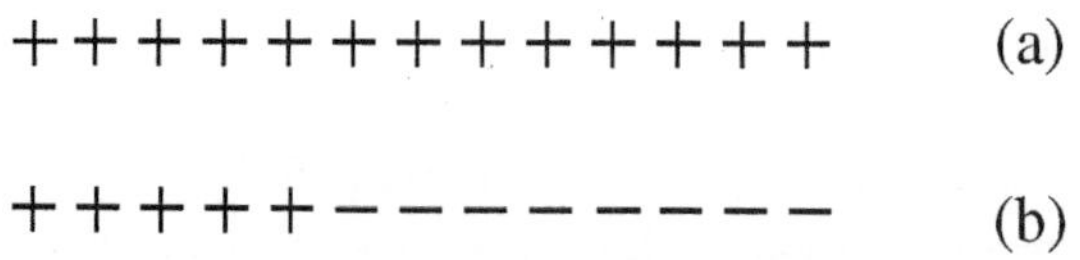

$$+ + + + + + + + + + + +$$ (a)

$$+ + + + + - - - - - - - -$$ (b)

Fig. 7 (a) Schematic representation of a linear Ising chain with all spins parallel to each other. (b) Suppression of long-range order in a linear chain, by a single break.

schematized in Fig. 7. The free energy of a linear chain of spins, with a single break, is given by

$$F = E_0 + 2\,J - k_B T \ln N \ ,$$

where E_0 is the ground-state energy of a linear chain of fully ordered spins, $2J$ is the energy required to create a single break, $S = k_B \ln N$ is the entropy due to the fact that the break can occur at each of the N sites of the lattice. For any finite temperature T and for N sufficiently large, we can have a decrease of free energy (and thus a more stable situation) breaking the long-range order of the linear chain, rather than leaving it perfect.

Considerations on the Ising model for two-dimensional crystals

Consider a two-dimensional lattice and two connected regions of opposite spins separated by a borderline involving N sites (see Fig. 8). The free energy for the situation of Fig. 8b is approximately

$$F \approx E_0 + 2\,N\,J - k_B T \ln 3^N \ ,$$

where E_0 is the energy of the fully ordered lattice of spins, $2\,N\,J$ is (approximately) the energy required to create N breaks, and $S = k_B \ln 3^N$ is an estimate of the entropy;

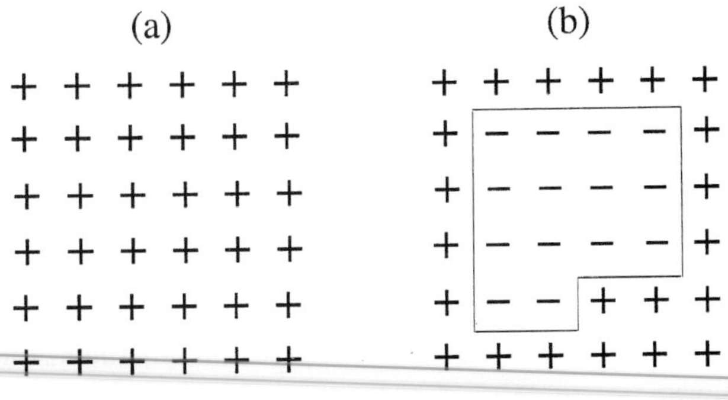

Fig. 8 Break of long-range order in a two-dimensional system.

in fact at each point of the borderline we have approximately 3 choices of keeping on with the border line. Thus if

$$k_B T < \frac{2 J}{\ln 3}$$

the ordered state is stable.

The Ising model has been solved *exactly* by Onsager for a bidimensional square lattice, using appropriate generalizations of the transfer matrix concepts (see for instance K. Huang "Statistical Mechanics", second edition, Wiley, New York 1987, Chapter 15). The Onsager solution constitutes a remarkable piece of theoretical work; at the end of a tour-de-force treatment Onsager has shown that the system is ferromagnetic for temperatures lower than the critical temperature given by

$$k_B T_c = \frac{2 J}{\ln(1 + \sqrt{2})} \; ,$$

or equivalently

$$\frac{J}{k_B T_c} = \frac{1}{2} \ln(1 + \sqrt{2}) = 0.441 \; . \tag{48}$$

Finally, for three-dimensional crystals, the topological situation further favours long-range order.

6   The Ising model with the renormalization group theory

The basic idea of the renormalization group theory to describe systems at the criticality is to define a transformation on the initial Hamiltonian so that the number of sites of the system (or more generally the number of degrees of freedom of the system) is decreased, say from N to $N/2$, the scale of lengths is increased by $2^{1/d}$ (where d is the dimensionality), and at the same time the partition function is not changed. It is apparent that the same transformation can be applied to the transformed system itself; the aim is to find the *fixed points* of this mapping procedure, since systems at criticality are invariant under scale change.

It is not our purpose to illustrate in full generality the renormalization group formalism and its procedures; we choose instead to focus on one- and two-dimensional Ising models, and to illustrate for these systems, where technicalities can be kept at an elementary level, the wealth of concepts and implications of the physics of phase transitions. We should also notice that the renormalization–decimation procedure, described in Section V-8.4 for the electronic properties of crystals, is aimed at keeping invariant the Green's function of the system (rather than the partition function); although quite different in technical aspects and purpose, the renormalization-decimation procedure of Section V-8.4 owes much conceptually to the Wilson renormalization group theory of phase transitions.

Renormalization group theory for the one-dimensional Ising model

The one-dimensional Ising model offers the simplest opportunity to define a renormalization group theory transformation. We have already seen, using the transfer matrix method, that the one-dimensional Ising model does not present a phase transition; it is however instructive to recover this same conclusion from the point of view of renormalization group theory; the key point in fact is that the renormalization theory can be easily applied to Ising models in higher dimensions, differently from the already discussed transfer matrix method or other procedures.

The Hamiltonian of the zero-field one-dimensional Ising model, with nearest neighbour interactions only, is given by

$$H = -J \sum_{i=1}^{N} \sigma_i \, \sigma_{i+1} \, , \tag{49a}$$

where $J > 0$, the variables σ_i can assume the values ± 1, and the sum extends over a (very large) number N of sites. Together with the Hamiltonian (49a), it is convenient to consider the *reduced Hamiltonian* $\overline{H} = H/k_B T$ defined as

$$\overline{H} = -K \sum_{i=1}^{N} \sigma_i \, \sigma_{i+1} \, , \tag{49b}$$

where $K = J/k_B T$ is the only coupling parameter of the present model. Notice that for $T \to 0$, the coupling parameter K goes to infinity; the limit $K = \infty$ implies perfect alignment of spins, and is called *ferromagnetic point*. Notice also that for $T \to \infty$ the coupling parameter K vanishes; the value $K = 0$ is called *paramagnetic point*.

The partition function corresponding to the Hamiltonian (49a), or to the reduced Hamiltonian (49b), is

$$Z(K, N) = \sum_{\{\sigma\}} e^{K \, (\sigma_1 \, \sigma_2 + \sigma_2 \, \sigma_3 + \sigma_3 \, \sigma_4 + \sigma_4 \, \sigma_5 + \dots)}$$

where the sum must be performed over the 2^N possible configurations $\{\sigma\}$; a configuration is specified by the sequence of values ± 1 given to the variables $\sigma_1, \sigma_2, \dots, \sigma_N$.

We can also write

$$Z(K, N) = \sum_{\{\sigma\}} \left[e^{K\,\sigma_1\,\sigma_2}\, e^{K\,\sigma_2\,\sigma_3} \right] \left[e^{K\,\sigma_3\,\sigma_4}\, e^{K\,\sigma_4\,\sigma_5} \right] \cdots \tag{50}$$

where we have collected in square brackets the terms containing the variables $\sigma_2, \sigma_4, \ldots$ (as our intention is to eliminate all even site variables).

Performing the sum over σ_2 in Eq. (50) one obtains

$$Z(K, N) = \sum_{\{\sigma(\neq \sigma_2)\}} \left[e^{K\,\sigma_1 + K\,\sigma_3} + e^{-K\,\sigma_1 - K\,\sigma_3} \right] \left[e^{K\,\sigma_3\,\sigma_4}\, e^{K\,\sigma_4\,\sigma_5} \right] \cdots$$

We now show that the term within the first square parenthesis can be written in the exponential form $\exp(a + b\,\sigma_1\,\sigma_3) = A \exp(b\,\sigma_1\,\sigma_3)$ or more explicitly

$$\left[e^{K\,\sigma_1 + K\,\sigma_3} + e^{-K\,\sigma_1 - K\,\sigma_3} \right] \equiv A(K)\, e^{b(K)\,\sigma_1\,\sigma_3} \ . \tag{51a}$$

In fact, the requirement that Eq. (51a) holds for each of the four possibilities $\sigma_1, \sigma_3 = \pm 1$ gives

$$\begin{cases} \sigma_1 = \sigma_3 = +1 \ (\text{or} \ -1) & 2\cosh 2K = A(K)\, e^{b(K)} \\[2mm] \sigma_1 = -\sigma_3 = +1 \ (\text{or} \ -1) & 2 = A(K)\, e^{-b(K)} \end{cases} \ .$$

From the above two equations we have

$$A(K) = 2\sqrt{\cosh 2K} \quad \text{and} \quad b(K) = \frac{1}{2}\ln\cosh 2K \equiv K^{(1)} \ . \tag{51b}$$

The elimination of all even sites in a single stroke allows us to express the partition function (50) in the form

$$Z(K, N) = [A(K)]^{N/2} \cdot Z(K^{(1)}, \frac{N}{2}) \ , \tag{52}$$

where $Z(K^{(1)}, N/2)$ is the partition function corresponding to the renormalized Hamiltonian

$$\overline{H}^{(1)} = -K^{(1)} \sum_{i=1}^{N/2} \sigma_i\,\sigma_{i+1} \ . \tag{53}$$

In the renormalized Hamiltonian $\overline{H}^{(1)}$, the number of preserved sites is half of the original number, the length scale is now $a^{(1)} = 2a$, and the coupling constant is $K^{(1)} = (1/2)\ln\cosh 2K$; the renormalization procedure is schematically shown in Fig. 9.

The transformed Hamiltonian (53) in the present case has the same form as the original Hamiltonian (49), and is just ready as it stands for iterative procedures. The recursion transformation of the coupling constant is thus

$$\boxed{K^{(1)} = \frac{1}{2}\ln\cosh 2K} \ , \tag{54a}$$

and in general

$$\boxed{K^{(i+1)} = \frac{1}{2}\ln\cosh 2K^{(i)}} \ , \tag{54b}$$

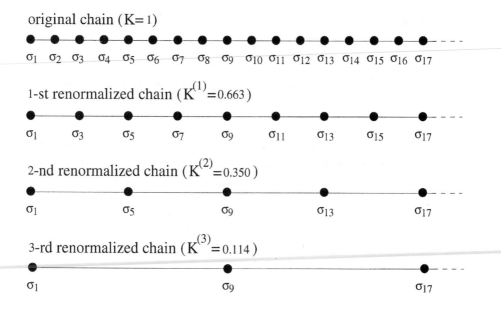

Fig. 9 Schematic indication of the renormalization procedure for the one-dimensional Ising chain of spins with coupling constant K (taken for instance equal to 1). The sum over spin variables $\sigma_2, \sigma_4, \sigma_6, \ldots$ is performed and this leaves a new chain that involves only the spin variables $\sigma_1, \sigma_3, \sigma_5, \ldots$ interacting with coupling constant $K^{(1)} = (1/2) \ln \cosh 2K$. The procedure is iterated and this leaves a new chain that involves only the spin variables $\sigma_1, \sigma_5, \sigma_9, \ldots$ interacting with coupling constant $K^{(2)} = (1/2) \ln \cosh 2K^{(1)}$. At the 20-th iteration, for instance, $2^{20} \equiv 1\cdot048\cdot576$, and the preserved sites are $\sigma_1, \sigma_{1\cdot048\cdot577}, \sigma_{2\cdot097\cdot153}, \ldots$ and the coupling constant is $K^{(20)} < 10^{-6}$.

(with $i = 1, 2, \ldots$). The fixed points of the mapping procedure (54) are defined as those points for which

$$\boxed{K^* = \frac{1}{2} \ln \cosh 2K^*} \, . \tag{55}$$

The fixed points of a mapping procedure, besides the trivial values zero and infinity of the coupling parameters, may include non-trivial values. A system is said to be sub-critical or super-critical if the mapping procedure makes it flow to the trivial fixed points zero or infinity (called paramagnetic and ferromagnetic fixed points, respectively). A system is critical if the mapping procedure makes it flow to a non-trivial fixed point. The non-trivial fixed points of a mapping procedure are the fingerprints of phase transitions and also determine the critical exponents (as shown below).

We can obtain the fixed points of the mapping procedure (54) either by direct solution of Eq. (55) or by the "flow diagram" graphical procedure. Suppose we specify some initial value of K; then we apply Eqs. (54) recursively. Since it is easily verified that $K^{(1)}(K) < K$ for any K, we find $K^{(i)} \to 0$ (for large values of i). The flow diagram of Eqs. (54) is indicated in Fig. 10. For the one-dimensional Ising problem, the flow diagram indicates that the only fixed points of the whole procedure are the trivial

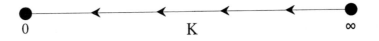

Fig. 10 Flow diagram of the recursive transformation $K^{(i+1)} = (1/2) \ln \cosh 2K^{(i)}$ for the coupling parameter of the one-dimensional Ising model. Arrows indicate the flow direction; the two fixed points are at $K^* = 0$ and $K^* = \infty$.

fixed points $K^* = 0$ (i.e. the paramagnetic one) and $K^* = \infty$ (i.e. the ferromagnetic one). The mapping procedure (54) does not have non-trivial fixed points, and thus no phase transition exists for the one-dimensional Ising model. The coupling constant $K^{(i)}$, for increasing values of i, moves towards zero whatever is the chosen initial value of K, and the one-dimensional Ising model is always a sub-critical system.

We can now obtain a recursive relation between the reduced free energy per spin $\overline{f}(K)$ and $\overline{f}(K^{(1)})$, defined as

$$\overline{f}(K) = -\frac{1}{N} \ln Z(K, N) \quad \text{and} \quad \overline{f}(K^{(1)}) = -\frac{1}{N/2} \ln Z\left(K^{(1)}, \frac{N}{2}\right) . \tag{56}$$

From Eqs. (56) and Eq. (52), we obtain

$$\overline{f}(K^{(1)}) = 2\,\overline{f}(K) + \ln A(K) , \tag{57a}$$

and in general

$$\overline{f}(K^{(i+1)}) = 2\,\overline{f}(K^{(i)}) + \ln A(K^{(i)}) \tag{57b}$$

(with $i = 1, 2, \dots$). In particular for $K \to 0$ we have $\ln A(K \approx 0) = \ln 2$ and $\overline{f}(K^{(1)}) = \overline{f}(K \approx 0) = -\ln 2$; this is what expected from $Z(K \approx 0, N) \approx 2^N$. Similarly, for $K \to \infty$ we have $\ln A(K \to \infty) \approx K$ and $\overline{f}(K^{(1)}) = \overline{f}(K \to \infty) = -K$ this is in agreement with what expected for $Z(K \to \infty, N) \approx \exp(K\,N)$. The recursion relations (54) and (57) do not predict a phase transition, since the free energy is obtained as the result of a finite number of operations involving regular functions, and is thus a regular function itself.

Renormalization group theory for the two-dimensional Ising model

Consider a square lattice, formed by N spins with only nearest neighbour interactions $J > 0$, and schematically indicated in Fig. 11. The reduced zero-field Ising Hamiltonian of the square lattice is

$$\overline{H} = -K \sum_{\langle i\,j \rangle} \sigma_i \, \sigma_j , \tag{58}$$

where $K = J/k_B T$, and $\langle i\,j \rangle$ denotes distinct nearest neighbour pairs. The partition function corresponding to the Hamiltonian (58) can be written as

$$Z(K, N) = \sum_{\{\sigma\}} e^{K(\sigma_0 \sigma_1 + \sigma_0 \sigma_2 + \sigma_0 \sigma_3 + \sigma_0 \sigma_4)} \, e^{K(\sigma_5 \sigma_1 + \sigma_5 \sigma_4 + \sigma_5 \sigma_9 + \sigma_5 \sigma_{10})} \dots \tag{59}$$

where the first few sites near the origin have been labelled following Fig. 11.

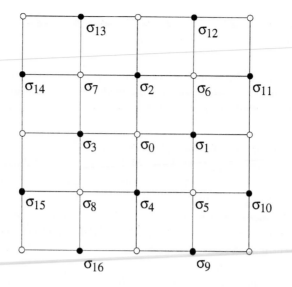

Fig. 11 Two-dimensional square Ising lattice with nearest neighbour spin interactions; labels of sites are introduced for the application of the renormalization procedure.

We perform the sum over σ_0 in Eq. (59); the partition function, just after the "elimination" of the site σ_0, can be written as

$$Z(K, N) = \sum_{\{\sigma(\neq\sigma_0)\}} \left[e^{K(\sigma_1+\sigma_2+\sigma_3+\sigma_4)} + e^{-K(\sigma_1+\sigma_2+\sigma_3+\sigma_4)} \right] \cdots$$

We can easily check that for all sixteen possible values of $\sigma_1, \sigma_2, \sigma_3, \sigma_4 = \pm 1$ we can write

$$\left[e^{K(\sigma_1+\sigma_2+\sigma_3+\sigma_4)} + e^{-K(\sigma_1+\sigma_2+\sigma_3+\sigma_4)} \right]$$

$$= A(K)\, e^{b(K)[\sigma_1\sigma_2+\sigma_2\sigma_3+\sigma_3\sigma_4+\sigma_4\sigma_1+\sigma_1\sigma_3+\sigma_2\sigma_4]+c(K)\sigma_1\sigma_2\sigma_3\sigma_4} .$$

We have in fact

$$\begin{cases} \sigma_1 = \sigma_2 = \sigma_3 = \sigma_4 = +1 \text{ (or } -1) & 2\cosh 4K = A(K)\, e^{6\,b(K)+c(K)} \\ \sigma_1 = \sigma_2 = \sigma_3 = -\sigma_4 = +1 \text{ (or } -1) & 2\cosh 2K = A(K)\, e^{-c(K)} \\ \sigma_1 = \sigma_2 = -\sigma_3 = -\sigma_4 = +1 \text{ (or } -1) & 2 = A(K)\, e^{-2\,b(K)+c(K)} \end{cases}.$$

We thus obtain the explicit expression

$$A(K) = 2\cosh^{1/2} 2K \,\cosh^{1/8} 4K$$

$$b(K) = \frac{1}{8}\ln\cosh 4K$$

$$c(K) = \frac{1}{8}\ln\cosh 4K - \frac{1}{2}\ln\cosh 2K .$$

The elimination in a single stroke of all the "white sites" in Fig. 11 allows us to express the partition function (59) in the form

$$Z(K, N) = [A(K)]^{N/2} \cdot Z\left(\{K^{(1)}\}, \frac{N}{2}\right) , \tag{60}$$

where $Z(\{K^{(1)}\}, N/2)$ is the partition function corresponding to the renormalized Hamiltonian

$$\overline{H}^{(1)} = -2b(K) \sum_{\langle i\,j \rangle} \sigma_i \sigma_j - b(K) \sum_{[i\,j]} \sigma_i \sigma_j - c(K) \sum_{\text{square}} \sigma_i \sigma_j \sigma_k \sigma_l \tag{61}$$

(the factor 2 appears because each nearest neighbour coupling is generated by two different sums on adjacent white sites). In the reduced Hamiltonian $\overline{H}^{(1)}$, $\langle i\,j \rangle$ denotes nearest neighbour pairs (of preserved sites) interacting with coupling parameter $K^{(1)}_{(nn)} = 2\,b(K)$, $[i\,j]$ denotes next nearest neighbour pairs interacting with parameter $K^{(1)}_{(nnn)} = b(K)$, and the last term indicates spins on squares interacting with $K^{(1)}_{(sq)} = c(K)$. The number of preserved sites is half of the original number of sites, and the length scale is now $a^{(1)} = \sqrt{2}\,a$; $\{K^{(1)}\}$ is a shorthand notation for $K^{(1)}_{(nn)}$, $K^{(1)}_{(nnn)}$, $K^{(1)}_{(sq)}$.

The transformed Hamiltonian $\overline{H}^{(1)}$ does not have the same form as the original one of Eq. (58), because of the presence of the sum over second nearest neighbours and the sum over the sites of the square. We have thus to make some appropriate manipulations on the form of $\overline{H}^{(1)}$, before proceeding to a new renormalization, otherwise interactions of rapidly increasing complexity would be generated. This step is a key point of renormalization procedure, and can be solved (at least approximately) with the help of physical intuition and appropriate technical procedures. We consider here simple possible approximations of the renormalized Hamiltonian of Eq. (61), with cutoff at nearest neighbour and then at next nearest neighbour interactions. Needless to say, countless more accurate solutions are available in the literature, but our selection here is motivated mainly by sake of technical simplicity, rather than sophistication.

Approximate solution of the two-dimensional Ising model with cutoff at nearest neighbour interaction

A rough approximation for the solution of the two-dimensional Ising model would be just to drop in Eq. (61) all the "offending terms" (i.e. the sum over second nearest neighbours and the sum over the sites on the square). The renormalization procedure in this case would give $K^{(1)} = (1/4) \ln \cosh 4K$; since $K^{(1)}(K)$ is always smaller than K, the flow diagram would be very similar to what already obtained for the one-dimensional model and no phase transition would occur. Thus the simple dropping the offending terms is a too drastic cure and the phase transition is just suppressed.

From the above discussion, we see that the interactions beyond nearest neighbour spins appearing in $\overline{H}^{(1)}$ must be taken into account some way; among the various

$$0 \qquad\qquad\qquad K_C{=}0.507 \qquad\qquad\qquad \infty$$

Fig. 12 Flow diagram for the recursion relation $K^{(i+1)} = (3/8)\ln\cosh 4K^{(i)}$ in the approximate solution of the two-dimensional Ising model.

possible procedures, we follow here a very instructive suggestion given by H. J. Maris and L. P. Kadanoff, Am. J. Phys. **46**, 652 (1978); this paper also contains references for more sophisticated elaborations. It seems reasonable to mimic somehow the aligning effect of second nearest neighbours on the reference spin, before dropping them. Since the number of second and first neighbours in the square lattice is the same, we make the approximation $K^{(1)} = K^{(1)}_{(nn)} + K^{(1)}_{(nnn)} = 3\,b(K)$; we obtain

$$K^{(1)} = \frac{3}{8}\ln\cosh 4K \;, \tag{62a}$$

and in general

$$K^{(i+1)} = \frac{3}{8}\ln\cosh 4K^{(i)} \;, \tag{62b}$$

(with $i = 1, 2, \ldots$).

Suppose we specify some initial value of K; then we apply Eq. (62a) and iterate it recursively. In the limit of large number of recursions, we find three types of behaviours. (i) For $0 < K < K_c \equiv 0.507$ we find that $K^{(i)} \to 0$; this is evident for small K considering the series development of Eq. (62a), that gives $K^{(1)} = 3\,K^2$ (thus $K^{(1)}(K)$ is systematically smaller than K for small K). (ii) For $K > K_c$ we find $K^{(i)} \to \infty$; this is evident for large values of K considering the series development of Eq. (62a), that gives $K^{(1)} = (3/2)\,K$ (thus $K^{(1)}(K)$ is systematically larger than K for large K). (iii) There is just a unique value $K = K_c$ for which the effective coupling remains unchanged.

The fixed points of the mapping procedure (62) satisfy the equation

$$K^* = \frac{3}{8}\ln\cosh 4K^* \;;$$

the solutions are the trivial values $K^* = 0$, $K^* = \infty$, and the non-trivial value $K^* = 0.507 = K_c$. The flow diagram of Eqs. (62) has a non-trivial fixed point as shown in Fig. 12. If K starts just at the right of K_c, it increases to infinity; if K starts just at the left of K_c, it decreases to zero; at the very value of K_c, the effective coupling between any two preserved nearest neighbour spins (no matter how far apart in real space they are) remains K_c; this fixed point is thus associated with the phase transition. The exact solution of the two-dimensional Ising problem provided by Onsager gives $K_c^{(\text{exact})} = 0.441$ (it is indeed a rewarding feature that, with almost no mathematical or numerical labour, we have arrived so near to the exact result!).

We can now estimate the critical exponents. Consider the system near criticality, i.e.

$K \approx K_c$; the reduced free energy per spin must contain a non-analytic contribution near the critical point, that we indicate by $\overline{f}_s(K)$. Let us now perform a renormalization transformation, where the number of spins is reduced from N to $N/2$; taking the logarithm of both members of Eq. (60) (and from the observation that $A(K)$ is in any case a regular function of K), we have that the singular part of the reduced free energy satisfies the relation

$$\overline{f}_s(K^{(1)}) = 2\,\overline{f}_s(K) \ . \tag{63a}$$

As suggested by exactly soluble models, we assume a non-analyticity of $\overline{f}_s$ with power-type law

$$\overline{f}_s(K) = c\,|K - K_c|^p \ , \tag{63b}$$

where c is a constant. We can obtain the value of p by requiring that Eq. (63a) is satisfied; we have

$$|K^{(1)} - K_c|^p \equiv 2\,|K - K_c|^p \ . \tag{63c}$$

The above equation is easily solved by "linearizing" the transformation function $K^{(1)}(K)$ near the critical point K_c. We have

$$K^{(1)}(K) = K_c + (K - K_c)\,\lambda_c \ , \tag{64a}$$

where

$$\lambda_c = \left(\frac{dK^{(1)}}{dK}\right)_{K=K_c} = \frac{3}{2}\tanh 4K_c = 1.449 \ .$$

Notice that λ_c, which is larger than one, is a *relevant parameter*; unless $K \equiv K_c$ the system flows away from criticality, as evident from Eq. (64a). Inserting Eq. (64a) into Eq. (63c), we obtain

$$\lambda_c^p = 2 \quad \text{and} \quad p = \frac{\ln 2}{\ln \lambda_c} = 1.869 \ . \tag{64b}$$

The specific heat, which is related to the second derivative of Eq. (63b), has the behaviour

$$C_V \approx |K - K_c|^{p-2} \approx |T - T_c|^{p-2} \ ;$$

the critical exponent α (defined in Table 2) becomes

$$\alpha = 2 - p = 2 - \frac{\ln 2}{\ln \lambda_c} = 0.131 \ . \tag{65a}$$

This exponent is just nearly equal to zero and should be compared with the Onsager exact result, which provides a logarithmic divergence in the specific heat, i.e. $\alpha = 0$.

We can do a similar reasoning to obtain the critical exponent for the correlation length, which is an estimate of the spatial extension of the spin–spin correlation function. Let us indicate with $\xi(K)$ the correlation length for the system near criticality, i.e. for $K \approx K_c$; let us now perform a renormalization transformation where the number of spins is reduced from N to $N/2$; the new lattice retains its original topology with a change of scale of $\sqrt{2}$. Lengths, that are now measured in terms of the new lattice parameter, are reduced by a factor $1/\sqrt{2}$; thus we have

$$\xi(K^{(1)}) = \frac{1}{\sqrt{2}} \xi(K) \ .$$

As suggested by exactly soluble models, we assume that the correlation length obeys a power type law $\xi(K) \approx |K - K_c|^{-\nu}$; by substitution into the above equation and linearization, we obtain

$$\nu = \frac{\ln \sqrt{2}}{\ln \lambda_c} = 0.935 \ , \tag{65b}$$

a value that compares favourably with the exact Onsager result $\nu = 1$. Notice that the critical exponents (65) satisfy the equality $\nu d = 2 - \alpha$ $(d = 2)$, which is one of the relations mentioned in Table 2.

Approximate solution of the two-dimensional Ising model with cutoff at second nearest neighbours

We now consider another instructive approximation for the solution of the two-dimensional Ising model, following the guidelines given by K. G. Wilson, Rev. Mod. Phys. **47**, 773 (1975). The reduced zero-field Ising Hamiltonian for a square lattice of spins can be written as

$$\overline{H} = -K \sum_{\langle i\,j \rangle} \sigma_i \sigma_j - L \sum_{[i\,j]} \sigma_i \sigma_j \ , \tag{66}$$

where $\langle i\,j \rangle$ denotes nearest neighbour pairs interacting with coupling parameter K, and $[i\,j]$ denotes next nearest neighbour pairs interacting with coupling parameter L. The Ising lattice is schematically indicated in Fig. 14; in the particular case $K \neq 0$ and $L = 0$, we recover the Hamiltonian of Eq. (58) and the Ising lattice of Fig. 11.

Similarly to the renormalization procedure already applied to Fig. 11, we eliminate all white sites of Fig. 13 (under the basic simplifying assumption that second nearest neighbour coupling acting between white sites is neglected). After the renormalization, making appropriate use of the coupling parameters of Eq. (61) and neglecting four spin coupling, we obtain

$$\begin{cases} K^{(1)} = \frac{1}{4} \ln \cosh 4K + L \\ L^{(1)} = \frac{1}{8} \ln \cosh 4K \end{cases} . \tag{67}$$

The fixed points of the transformations (67) are given by the solutions of the two equations

$$\begin{cases} K^* = \frac{1}{4} \ln \cosh 4K^* + L^* \\ L^* = \frac{1}{8} \ln \cosh 4K^* \end{cases} .$$

We have a ferromagnetic fixed point at $(K^*, L^*) = (\infty, \infty)$, a paramagnetic fixed point at $(K^*, L^*) = (0, 0)$, and a non-trivial fixed point at $(K^*, L^*) = (0.507, 0.169)$.

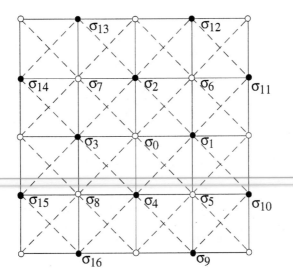

Fig. 13 Two-dimensional square Ising lattice with nearest neighbour spin interactions (full lines) and next nearest neighbour spin interactions (dotted lines); labels of sites are introduced for the application of the renormalization procedure.

Suppose we specify a system with some initial values of K and L in the KL plane. Then we apply the transformations (67) and iterate them recursively. The region of the KL plane for which the iteration flows to the fixed point $(K^*, L^*) = (0, 0)$ defines the region of *sub-critical systems*; the region of the KL plane for which the iteration flows to the fixed point $(K^*, L^*) = (\infty, \infty)$ describes *super-critical systems*. The line in the KL plane for which the iteration flows to the non-trivial fixed point $(K^*, L^*) = (0.507, 0.169)$ describes the *critical systems*; the critical line and the non-critical regions are indicated in Fig. 14. In particular, if $0 < K < K_c \equiv 0.632$ (and $L = 0$) then $K^{(i)} \to 0$ and $L^{(i)} \to 0$. If instead $K_c < K < \infty$ (and $L = 0$) then $K^{(i)} \to \infty$ and $L^{(i)} \to \infty$. Finally if $K = K_c$ (and $L = 0$) the iteration flows to the (non-trivial) fixed point.

We have already noticed that the correlation length $\xi(K, L)$ for the square lattice, after a renormalization transformation, scales in the form

$$\xi(K^{(1)}, L^{(1)}) = \frac{1}{\sqrt{2}} \xi(K, L) \,.$$

If the original system is non-critical, $\xi(K, L)$ is finite; the new system has a smaller correlation length $\xi(K^{(1)}, L^{(1)})$ and thus it is further away from criticality. If $\xi(K, L) = \infty$, the same occurs for the transformed correlation length $\xi(K^{(1)}, L^{(1)})$ and more generally for $\xi(K^{(i)}, L^{(i)})$. Since the different systems belonging to the critical line are taken over by the renormalization transformations to the non-trivial fixed point, this can be used to calculate the critical exponents; it follows that all the systems on the critical line have the same critical exponents (this fact is called "universality").

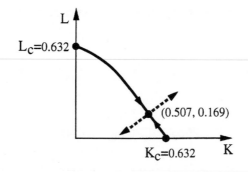

Fig. 14 Critical line representing the systems which are taken over by the renormalization transformations to the non-trivial fixed point $(K^*, L^*) = (0.507, 0.169)$. The critical line is the borderline between sub-critical and super-critical systems, which are taken over to the fixed points $(K^*, L^*) = (0,0)$ and $(K^*, L^*) = (\infty, \infty)$, respectively.

Critical exponents can be obtained after linearization of the transformation functions (67) around the non-trivial fixed point (it is a fundamental tenet of Wilson formalism that all transformation functions are regular functions; hence linearization is always possible). We have

$$\delta K^{(1)} = \tanh 4K^* \, \delta K + \delta L \quad \text{and} \quad \delta L^{(1)} = \frac{1}{2} \tanh 4K^* \, \delta K \, .$$

In matrix form we can write

$$\begin{pmatrix} \delta K^{(1)} \\ \delta L^{(1)} \end{pmatrix} = \begin{pmatrix} \tanh 4K^* & 1 \\ \frac{1}{2} \tanh 4K^* & 0 \end{pmatrix} \begin{pmatrix} \delta K \\ \delta L \end{pmatrix} . \tag{68a}$$

The eigenvalues of the above transformation matrix are determined by the secular equation

$$\begin{vmatrix} \tanh 4K^* - \lambda & 1 \\ \frac{1}{2} \tanh 4K^* & -\lambda \end{vmatrix} = 0 \, . \tag{68b}$$

With $K^* = 0.507$, we obtain the eigenvalues $\lambda_1 = 1.329$ (relevant eigenvalue) and $\lambda_2 = -0.363$ (irrelevant eigenvalue). The critical exponent ν is thus given by $\nu = \ln \sqrt{2} / \ln \lambda_1 = 1.218$ (the exact result provided by Onsager is $\nu = 1$); the comparison is reasonable, in spite of the drastic simplifications made to justify the transformations (67). We conclude here this brief presentation of some elementary aspects of phase transitions in the two-dimensional Ising model and refer to the literature for an analysis of other topological structures and more accurate solutions [see for instance F. Ravndal "Scaling and Renormalization Groups", Nordita, Copenhagen 1976].

7 The Stoner–Hubbard itinerant electron model for magnetism

The physical picture of magnetism, behind Heisenberg-type Hamiltonians, is that of localized spins at the lattice sites, cooperatively interacting among themselves. This phenomenological picture seems suitable for magnetic insulators and also for rare earth metals, whose f bands are quite narrow; the localized spin picture may become inadequate for transition metals with unfilled d bands, where the electrons participating in the magnetic state are itinerant. It would be desirable for these situations a physical picture of magnetism that starts from electrons described by Bloch functions, and then takes proper account of relevant correlation effects; the prototype model for the investigation of itinerant magnetism is constituted by the Stoner–Hubbard Hamiltonian, which embodies electronic correlations at a reasonably intuitive level.

Consider for simplicity a metal with a simple Bravais lattice (with N unit cells and volume $V = N\Omega$ and described by a single conduction band, partially filled by the available electrons; in the one-electron approximation the conduction electrons are described by the Hamiltonian

$$H_0 = \sum_{\mathbf{k}} E(\mathbf{k}) \left[c_{\mathbf{k}\uparrow}^{\dagger} c_{\mathbf{k}\uparrow} + c_{\mathbf{k}\downarrow}^{\dagger} c_{\mathbf{k}\downarrow} \right]$$

where $E(\mathbf{k})$ are the conduction band energies, $c_{\mathbf{k}\sigma}^{\dagger}$ and $c_{\mathbf{k}\sigma}$ are creation and annihilation operators for electrons of spin σ in the conduction band state of vector $\mathbf{k}$. It is well-known that the one-electron approximation fails to take apart electrons of opposite spin. We thus introduce by hand a term that represents the Coulomb repulsion between electrons with opposite spins on the same site, and arrive at the Stoner–Hubbard model Hamiltonian of the type

$$H = \sum_{\mathbf{k}} E(\mathbf{k}) \left[c_{\mathbf{k}\uparrow}^{\dagger} c_{\mathbf{k}\uparrow} + c_{\mathbf{k}\downarrow}^{\dagger} c_{\mathbf{k}\downarrow} \right] + U \sum_{\mathbf{t}_m} c_{m\uparrow}^{\dagger} c_{m\uparrow} c_{m\downarrow}^{\dagger} c_{m\downarrow} , \tag{69}$$

where U is a phenomenological positive parameter that describes the Coulomb repulsion when two electrons of opposite spins are on the same site, and $c_{m\sigma}^{\dagger}$ and $c_{m\sigma}$ are creation and annihilation operators for electrons with spin σ in the lattice site $\mathbf{t}_m$. The site operators $c_{m\sigma}^{\dagger}$ (and $c_{m\sigma}$) are expressed as linear combinations of the band operators $c_{\mathbf{k}\sigma}^{\dagger}$ (and $c_{\mathbf{k}\sigma}$) with coefficients given by the standard Bloch phase factors.

The Hamiltonian (69) can lead to macroscopic magnetic effects. Consider, for simplicity, a material with N electrons to be accommodated in the conduction band with $2N$ states (the factor 2 takes into account spin degeneracy). If U is negligible with respect to the bandwidth, the ground state is non-magnetic with the lowest lying $N/2$ orbital Bloch states doubly occupied. In the opposite case in which the band dispersion is negligible, i.e. $E(\mathbf{k})$ is nearly independent of $\mathbf{k}$, then the ground state is magnetic with all states with spin-up occupied and all states with spin-down empty (or vice versa).

To give a semiquantitative analysis of intermediate situations, let us calculate the energy levels and the magnetic susceptibility of an electron system described by the Stoner–Hubbard Hamiltonian (69). The many-body electron Hamiltonian (69) is quite complicated due to the correlation term; we adopt here the so-called "unrestricted Hartree–Fock approximation" and we replace $c^\dagger_{m\uparrow} c_{m\uparrow}$ (or $c^\dagger_{m\downarrow} c_{m\downarrow}$) with their expectation value on the ground state (we suppose to operate at $T = 0$). We indicate by $N_\uparrow$ and $N_\downarrow$ the total number of conduction electrons with spin-up and spin-down, respectively, and by $n_\uparrow = N_\uparrow/N$ and $n_\downarrow = N_\downarrow/N$ the average number of electrons per site. We also consider an applied magnetic field H in the z direction and the Zeeman energy $-\boldsymbol{\mu} \cdot \mathbf{H} = g_0 \mu_B \mathbf{s} \cdot \mathbf{H}$ (where $g_0 \approx 2$ and $\mu_B = e\hbar/2mc$ is the Bohr magneton). The energy dispersion curve for electrons with spin up becomes

$$E_{\mathbf{k}\uparrow} = E(\mathbf{k}) + U n_\downarrow + \mu_B H \; ; \tag{70a}$$

similarly for electrons with spin down we have

$$E_{\mathbf{k}\downarrow} = E(\mathbf{k}) + U n_\uparrow - \mu_B H \; . \tag{70b}$$

Let us indicate with $D(E)$ the density-of-states, for both spin directions, corresponding to the dispersion curve $E = E(\mathbf{k})$. For the total number of electrons with spin-down we have

$$N_\downarrow = \int f(E + U n_\uparrow - \mu_B H) \frac{1}{2} D(E)\, dE = \int_{E_0}^{E_F - U n_\uparrow + \mu_B H} \frac{1}{2} D(E)\, dE \; , \tag{71}$$

where $f(E)$ is the Fermi–Dirac distribution function (at $T = 0$), E_F is the Fermi energy, and E_0 is the bottom of the conduction band. Using Eq. (71) for $N_\downarrow$ and a similar expression for $N_\uparrow$ we have

$$N_\downarrow - N_\uparrow = \int_{E_0}^{E_F - U n_\uparrow + \mu_B H} \frac{1}{2} D(E)\, dE - \int_{E_0}^{E_F - U n_\downarrow - \mu_B H} \frac{1}{2} D(E)\, dE$$

$$= [U(n_\downarrow - n_\uparrow) + 2\mu_B H] \frac{1}{2} D(E_F) \; ,$$

where $D(E)$ has been approximated by $D(E_F)$ around the Fermi energy; we thus obtain

$$N_\downarrow - N_\uparrow = \frac{\mu_B H\, D(E_F)}{1 - \frac{1}{2} \dfrac{D(E_F)}{N} U} \; .$$

The magnetization is given by $M = \mu_B (N_\downarrow - N_\uparrow)/V$, and the magnetic susceptibility becomes

$$\boxed{\chi = \frac{1}{V} \frac{\mu_B^2\, D(E_F)}{1 - \frac{1}{2} \dfrac{D(E_F)}{N} U}} \; . \tag{72}$$

In the case $U = 0$, Eq. (72) regains the Pauli magnetic susceptibility of independent particles (see Eq. XV-30); if the (essentially positive) quantity U is different from zero

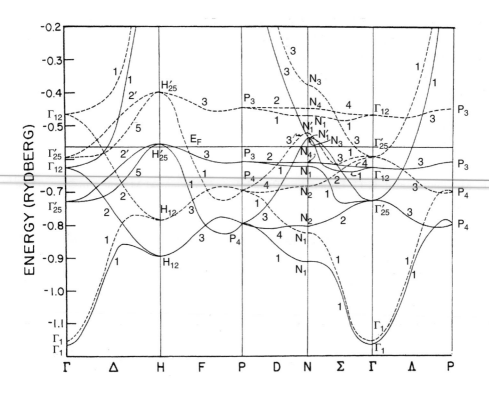

Fig. 15 Energy bands in ferromagnetic iron along some lines of high symmetry; solid lines are majority-spin states, dashed are minority-spin states [from J. Callaway and C. S. Wang, Phys. Rev. B16, 2095 (1977); copyright 1977 by the American Physical Society].

we see that the magnetic susceptibility is increased. In the case

$$\frac{1}{2}\frac{D(E_F)}{N}U = 1 \tag{73}$$

the magnetic susceptibility diverges, and we expect a transition to ferromagnetism; relation (73) represents the Stoner criterion for onset of band magnetism.

Consider Eqs. (70) in the case the material is ferromagnetic (and $H = 0$); in the simplified picture we are considering, the spin-up and spin-down energy bands are split in energy by the amount $\Delta = U(n_\downarrow - n_\uparrow)$; the "exchange splitting" Δ is thus proportional to the magnetization and is independent from $\mathbf{k}$. In actual materials, majority and minority spin bands are not split rigidly over the whole Brillouin zone; nevertheless the trend for bands with opposite spins to have a constant splitting in the energy region around the Fermi level is confirmed. In Fig. 15 we give as an example the band structure of ferromagnetic iron.

The understanding of the band structure of ferromagnetic materials has received a great benefit from the extension of the density functional formalism (see Section IV-7) to spin polarized crystals. The "spin-density functional" formalism is based on the generalized theorem of Hohenberg and Kohn, which states that the ground-state energy of an inhomogeneous electron gas is a functional of the electron density and spin density. Ground-state properties, such as the magnetic moment, can be calculated with great accuracy. Photoemission spectroscopy, especially in angle and spin-resolved version, has been of great value for the investigations of the details of energy bands of spin polarized crystals.

Further Reading

P. M. Chaikin and T. C. Lubensky "Principles of Condensed Matter Physics" (Cambridge University Press, Cambridge 1995)

S. Chikazumi "Physics of Ferromagnetism" (Clarendon Press, Oxford 1997)

C. Domb and M. S. Green eds "Phase Transitions and Critical Phenomena" vols. I-VIII (Academic Press, London 1972-1983)

P. Fulde "Electron Correlation in Molecules and Solids" (Springer, Berlin 1995, third edition)

J. Jensen and A. R. Mackintosh "Rare Earth Magnetism" (Clarendon, Oxford 1991)

W. Jones and N. H. March "Theoretical Solid State Physics" (Wiley-Interscience, New York 1973)

C. Kittel "Quantum Theory of Solids" (Wiley, New York 1967)

D. H. Martin "Magnetism in Solids" (London Cliffe Books, London 1967)

D. C. Mattis "The Theory of Magnetism" vols. I and II (Springer, Berlin 1981)

A. H. Morrish "The Physical Principles of Magnetism" (Wiley, New York 1965)

H. E. Stanley "Phase Transitions and Critical Phenomena" (Oxford University Press, Oxford 1971)

R. M. White "Quantum Theory of Magnetism" (Springer, Berlin 1983)

J. M. Yeomans "Statistical Mechanics of Phase Transitions" (Clarendon Press, Oxford 1992)

H. M. Ziman "The Principles of the Theory of Solids" (Cambridge University Press, Cambridge 1972)

XVIII

Superconductivity

The history of superconductivity is full of fascinating surprises and challenging developments. The milestone work of Kamerlingh Onnes in 1911 on the electrical resistivity of mercury has opened a new world to the physical investigation, and the discovery of Bednorz and Müller in 1986 of the superconductivity in barium-doped lanthanum cuprate has given a novel impetus to the subject.

The origin of superconductivity is linked to the possible occurrence of a (small) effective attractive interaction between conduction electrons (or valence holes in p-type conductors) and the consequent formation of electron pairs (or hole pairs), at sufficiently low temperatures. The mechanism of pairing is at the origin of perfect conductivity, perfect diamagnetism, anomalous specific heat and thermodynamical properties, magnetic flux quantization, coherent tunneling and several other effects. Empirical laws and semi-empirical models have accompanied the accumulation of the wide and rich phenomenology of superconductors. Eventually, the fundamental work of Bardeen, Cooper and Schrieffer (1957) has transformed an endless list of peculiar effects and conjectures into a logically consistent theoretical framework.

The subject of superconductivity is quite extensive, and by necessity in this chapter we can focus only on some relevant experimental and theoretical aspects. A special attention is devoted to the quantum microscopic theory of superconductivity of Bardeen, Cooper and Schrieffer; without its concepts no serious discussion would be possible at all. The necessary formalism is presented in a self-contained way, at an accessible technical level; with this tool, the reader is in a position to appreciate a most remarkable piece of theoretical work and to follow the intimate relation among the various phenomenological aspects. For the novice to the subject of superconductivity, we suggest the preliminary reading of those sections (or pieces thereof) of descriptive nature; in a subsequent reading, the framework that links together the phenomenological facts can be better appreciated.

1   Some phenomenolgical aspects of superconductors

Old and new superconducting materials

Superconductivity was discovered in 1911 by Kamerlingh Onnes, soon after helium had been liquefied by him. In studying the electrical resistance of mercury at low temperatures, Kamerlingh Onnes found that, at about 4.2 K and in a range of 0.01 K, the electrical resistance sharply dropped by several orders of magnitude to non-measurable values; freezing the metal below the critical temperature apparently led to a new resistanceless state, referred to as the superconducting state. The resistanceless state is actually a state of zero resistivity ρ (and not a state of very low resistivity); when a current is started, for instance by magnetic induction, in a closed superconducting ring which is maintained at temperature well below T_c, it circulates for years without any detectable decay. [The non-dissipative current flow in the superconducting ring, besides of course zero resistance, also entails that the magnetic flux threading the ring is quantized, as discussed in Section 7.2].

Since 1911, superconductivity has been found in more than 25 metallic elements and in more than one thousand alloys. The element with the highest transition temperature is niobium with $T_c = 9.25$ K. From 1972 to 1986 the alloy Nb_3Ge kept the record of highest critical temperature with $T_c = 23.3$ K. The run to the reachment of materials with the highest critical temperature has been sharply accelerated from 1986 after the epochal discovery of Bednorz and Müller of the superconducting properties of La-based cuprates, with T_c in excess of 30 K [J. G. Bednorz and K. A. Müller, Z. Phys. B64, 189 (1986)]. Soon, other families of superconducting cuprate oxides were discovered, and the achievements of materials with critical temperatures above 77 K (the boiling point of liquid nitrogen) passed suddenly from dream to reality [M. K. Wu et al., Phys. Rev. Lett. 58, 908 (1987)].

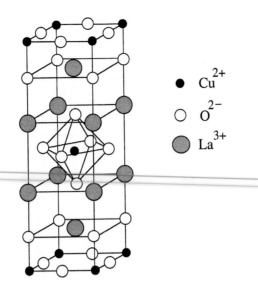

Fig. 1 Crystal structure of La_2CuO_4.

The copper oxide superconductors are characterized by crystal structures with one (or more) sheets of CuO_2, separated by insulating block layers. The first discovered high-T_c compounds belong to the family of La-based superconductors, and are obtained with appropriate doping of the insulating parent compound La_2CuO_4, whose body-centered-tetragonal structure is shown in Fig. 1. The CuO_2 planes are $\approx 6.6\,\text{Å}$ apart and separated by two LaO planes; each copper ion is sorrounded by an elongated octahedron of oxygens, with Cu-O distance $\approx 1.9\,\text{Å}$ within the plane and $\approx 2.4\,\text{Å}$ perpendicular to it.

The configurations of the outer electrons of the atoms forming the La_2CuO_4 compound are La: $5d^1\,6s^2$; O: $2s^2\,2p^4$; Cu: $3d^{10}\,4s^1$. In the crystal, lanthanum loses three electrons and takes the closed-shell configuration La^{3+}; oxygen completes the $2p$ shell and becomes O^{2-}; to conserve charge neutrality, the copper atom loses the $4s$ electron and one d electron, and takes the ionic configuration Cu^{2+}: $3d^9$. There is thus a hole in the d shell of the Cu^{2+} ion and, in an independent-electron model, the material should be a metal; in reality, hopping of holes between copper sites is prevented by the strong Coulomb repulsion, at work whenever two holes are on the same site; this correlation mechanism produces what is called a *Mott–Hubbard insulator*, since the Cu^{2+} $3d$ holes remain localized around their parent atoms. The magnetic moments of the Cu^{2+} ions in different unit cells are coupled antiferromagnetically, because of the super-exchange interaction through oxygens, typical of oxides (see Section XVII-2); so La_2CuO_4 is an antiferromagnetic insulator, with Néel critical temperature of about 320 K.

Upon doping, La^{3+} are randomly replaced by Sr^{2+} or Ba^{2+} ions; thus fewer electrons are donated to the CuO_2 planes, and a number of mobile holes are produced

Table 1 Critical temperature of some selected superconductors, and zero-temperature critical field. For elemental materials, the critical field $H_c(0)$ is given in gauss; for the compounds, which are type-II superconductors, the upper critical field $H_{c2}(0)$ is given in Tesla (1 Tesla $= 10^4$ gauss) [the data for metallic elements and binary compounds of V and Nb are taken from G. Burns "High-Temperature Superconductivity" (Academic Press, New York 1992); the data for the other compounds are taken from D. R. Harshman and A. P. Mills, Phys. Rev. B**45**, 10684 (1992); a more extensive list of data can be found in the mentioned references].

Metallic elements	T_c(K)	$H_c(0)$(gauss)
Al	1.17	105
Sn	3.72	305
Pb	7.19	803
Hg	4.15	411
Nb	9.25	2060
V	5.40	1410

Binary compounds	T_c(K)	$H_{c2}(0)$(Tesla)
V_3Ga	16.5	27
V_3Si	17.1	25
Nb_3Al	20.3	34
Nb_3Ge	23.3	38

Other compounds	T_c(K)	$H_{c2}(0)$(Tesla)
UPt_3 (heavy fermion)	0.53	2.1
$PbMo_6S_8$ (Chevrel phase)	12	55
κ-[BEDT-TTF]$_2$Cu[NCS]$_2$ (organic phase)	10.5	≈ 10
Rb_2CsC_{60} (fullerene)	31.3	≈ 30

Cuprate oxides	T_c(K)	$H_{c2}(0)$(Tesla)
$La_{2-x}Sr_xCuO_4$ ($x \approx 0.15$)	39	≈ 45
$YBa_2Cu_3O_7$	92	≈ 140
$Bi_2Sr_2CaCu_2O_8$	89	≈ 107
$Tl_2Ba_2Ca_2Cu_3O_{10}$	125	≈ 75

(essentially) in the $2p$ orbitals of oxygen (hybridized with $3d_{x^2-y^2}$ orbitals of copper). We can thus crudely schematize $La_{2-x}Sr_xCuO_4$ by means of a localized picture for the Cu^{2+} states and an itinerant picture of the oxygen $2p$ holes; a superconducting phase is found for $0.05 < x < 0.30$, with the optimum value at $x \approx 0.15$.

Soon after the discovery of lanthanum-based superconductors, other families of superconducting cuprate oxides have been synthetized. In spite of their apparent complexity, the basic structure of the cuprates can be described as an alternating sequence of electronically active metallic CuO_2 layers and other block layers, which act as "charge reservoirs" and provide the carriers (electrons or holes) to the active layers. A particular attention has been given to the yttrium-based superconductor $YBa_2Cu_3O_{6+x}$ (also called YBCO); the increase of the oxygen contents, when passing from $YBa_2Cu_3O_6$ to $YBa_2Cu_3O_7$, increases the holes on the conducting planes. Another family of cuprate superconductors is constituted by $Nd_{2-x}Ce_xCuO_4$; as Nd^{3+} is replaced by Ce^{4+}, the CuO_2 planes get the excess of electrons, and a superconducting phase is found with an optimum doping at $x \approx 0.15$. Several other families have been investigated; in particular, Hg-bearing compounds appear promising materials in the race for higher critical temperatures.

In 1991 fullerene has also entered the arena of challenging superconductor materials. Solid fullerene is constituted by molecules C_{60} (see Fig. II-12) arranged in a fcc lattice. The molecules C_{60} have a closed-shell ground state, and are weakly bound by van der Waal forces in the solid. When doped with alkali metals, excess electrons are accommodated in the lower unoccupied molecular orbital and solid C_{60} becomes conductive. An explosion of interest in fullerene has followed the discovery that the potassium-doped K_3C_{60} becomes superconductor with critical temperature at 18 K [A. F. Hebard et al., Nature **350**, 600 (1991)]; this critical temperature is striking high for a molecular superconductor, especially when compared with the critical temperature of 0.55 K in potassium intercalated graphite.

Cuprate oxides compounds, together with fullerene and other materials of lower critical temperature (such as heavy fermions, organic crystals), have contributed to expand the interest in the fundamental and applicative aspects of superconductivity; in Table 1 we report the critical temperature of some materials (just for orientative purpose and without any attempt of completeness or record update).

General considerations and phenomenological aspects

Superconductivity is related to the presence of an effective attractive interaction among conduction electrons (in n-type conductors) or valence holes (in p-type conductors) [from now on, for brevity, we use the terminology appropriate to the case that metallic conductivity is due to electron carriers; however, the general considerations are applicable also in the specular case that metallic conductivity is due to hole carriers]. In the so-called "conventional superconductivity" it is generally accepted that the attractive electron–electron interaction among electrons near the Fermi energy is mediated by the phonon field; in the more recent "non-conventional superconductors" other pairing mechanisms could play a role; in either cases the Bardeen–Cooper–Schrieffer theory provides the appropriate general framework to interpret experimental properties.

Above the critical temperature, the effect of the attractive interaction between electrons is in general irrelevant (apart causing superconducting fluctuations and related phenomena). Below the critical temperature, the effect of the attractive interaction leads to the formation of highly-correlated pairs of electrons (Cooper pairs) in an energy shell around the Fermi surface. As we shall see, many physical properties of this *correlated electron gas* are dramatically different from the corresponding properties of the *normal electron gas* of non-interacting electrons; in particular the pairing mechanism entails not only perfect conductance, but also perfect diamagnetism, anomalous specific heat and transport properties, energy gap in quasiparticle spectrum, dissipationless tunneling through non-conducting layers, and many other effects. Before

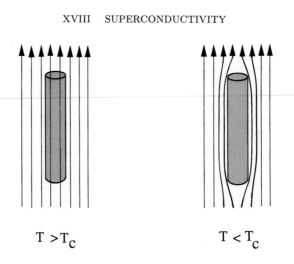

$T > T_c$ $T < T_c$

Fig. 2 Exclusion of a weak external magnetic field from the interior of a superconductor.

discussing these effects in the light of the microscopic quantum theory, we need to present a few phenomenological facts on the magnetic properties of superconductors, because of their particular role in the historical development of superconductivity.

Besides resistanceless, a distinctive feature of superconductors is perfect diamagnetism: a magnetic field (smaller than a critical value) cannot penetrate into the interior of a superconductor, regardless of its connectivity. This effect was first observed by Meissner and Ochsenfeld in 1933, and is usually referred as *Meissner effect*. In Fig. 2 we show schematically the Meissner–Ochsenfeld effect; the magnetic field is expelled from the bulk of a sample, that is cooled below the critical temperature. Contrary to the superconductor, a normal conductor would keep the magnetic field in which it is embedded, when it becomes ideally conducting (we discuss in detail this point in Section 6). Thus *perfect conductance does not automatically imply perfect diamagnetism; it is the pairing mechanism, with the resulting gap in the quasiparticle excitation spectrum and wavefunction modifications, that links in a unique destiny perfect conductance and perfect diamagnetism in superconducting materials.*

The absence of magnetic field in the interior of a superconductor has its origin in the "diamagnetic screening currents", which flow in a very thin surface layer of the order of 10^{-5} cm thick. It is worth noticing that even if it is well known that are these self-generated lossless "supercurrents" to make $\mathbf{B} = 0$ in the bulk superconductor, it is generally convenient to accept the diamagnetic description of the superconductor, attributing to it an internal magnetization $\mathbf{M}$ (magnetic moment per unit volume). Since

$$\mathbf{B} = \mathbf{H} + 4\pi\mathbf{M}$$

vanishes *inside* the superconductor, we have that the static magnetic susceptibility of a superconductor is $\chi = M/H = -1/4\pi$. In Chapter XV, in treating the orbital diamagnetism of a normal metal, we found $\chi \approx -10^{-6}$; the magnetic susceptibility of a superconductor is thus about *one million times larger* than the susceptibility of

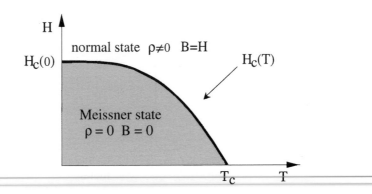

H

normal state $\rho \neq 0$ B=H

$H_C(0)$

$H_C(T)$

Meissner state
$\rho = 0$ B $= 0$

T_C T

Fig. 3 Schematic phase diagram illustrating normal and superconducting regions of a type-I superconductor.

a normal metal (even if the normal metal were ideally conducting); a superconducting material is thus also a superdiamagnetic material. The energy per unit volume, associated with the presence of the screening currents holding the field out of the superconducting sample, is

$$-\int_0^H M\,dH = \frac{H^2}{8\pi} \ .$$ (1)

The strong diamagnetic behaviour of superconductors is responsible of peculiar phenomena such as magnetic levitation. It is now a routine classroom experiment to show disks of high-T_c materials floating freely above magnets (or vice versa), when the samples are cooled below the critical temperature; with high-T_c materials, experiments can be easily performed at liquid nitrogen temperature [see for instance E. H. Brandt, Am. J. Phys. **58**, 43 (1990) and references quoted therein].

Sufficiently small magnetic fields are fully expelled by a superconductor; on the other hand sufficiently strong magnetic fields destroy the superconducting state. The modality of conversion to the normal state depends both on the intrinsic properties of the specimen and also on the geometry of the sample (the geometry determines appropriate demagnetization factors). In order to put in evidence genuine microscopic properties of the sample, we consider *exclusively* the structure of long, thin, cylindrical rods with axis parallel to the applied magnetic fields (in this standard geometry, demagnetization factors are zero). The behaviour of superconductors in magnetic fields divides them into two classes; correspondingly they are called type-I (or soft) superconductors and type-II (or hard) superconductors.

Type-I superconductors. The behaviour of type-I superconductors, at a given temperature T and in a uniform external magnetic field H, can be described as follows. If H is smaller than a critical value $H_c(T)$, the superconductor completely expels the magnetic flux from its interior (Meissner effect); as the external field is increased above the critical value $H_c(T)$, the *entire specimen reverts from the superconducting to the normal state*.

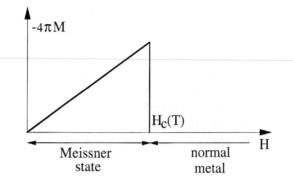

Fig. 4 Magnetization versus applied field for type-I superconductors.

The equilibrium curve $H_c(T)$ divides the $H - T$ plane in the "normal" region and in the "superconducting" (or "Meissner") region, as schematically shown in Fig. 3. In a number of superconducting materials of type-I, the variation with the temperature has approximately the parabolic form

$$H_c(T) = H_c(0) \left(1 - \frac{T^2}{T_c^2} \right) \tag{2}$$

where $H_c(0)$ is the critical field extrapolated at $T = 0$; for $T \to T_c$ from below, we have the linear behaviour $H_c(T) \propto T_c - T$.

A plot of the magnetization M versus the applied magnetic field H is shown in Fig. 4. For $H < H_c(T)$ we have $B = H + 4\pi M \equiv 0$ and thus $-4\pi M = H$. For $H > H_c(T)$ we have a normal metal; for the present considerations, the magnetic susceptibility of the normal metal (see Chapter XV) can be safely disregarded and hence $B = H$. Most of the elemental superconducting metals are type-I superconductors (exceptions include niobium and vanadium, which are type-II superconductors).

Type-II superconductors. The behaviour of type-II superconductors, at a given temperature T and in a uniform external magnetic field H, is the following. If H is smaller than a critical value $H_{c1}(T)$, the flux is completely expelled from the sample (Meissner state). As the external magnetic field is increased above the lower critical value $H_{c1}(T)$ and below an upper critical value $H_{c2}(T)$, the flux partially penetrates the sample and subdivides into flux-bearing regions, arranged into a regular triangular lattice; each tube of flux is called *filament* or *vortex*, and the sample is said to be in a *mixed state* or *vortex state*. As the external magnetic field increases, the density of vortices increases; finally, above $H_{c2}(T)$, the field penetrates uniformly and the whole material returns to the normal state.

The equilibrium curves $H_{c1}(T)$ and $H_{c2}(T)$ divide the H-T plane in normal, mixed, and Meissner regions, as schematically shown in Fig. 5. Notice that the upper critical field $H_{c2}(T)$ may be as high as several tens of Tesla; thus the mixed state can sustain high currents in high magnetic fields, still in a zero resistivity situation (provided

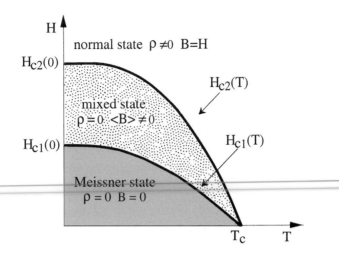

Fig. 5 Schematic phase diagram illustrating normal, mixed and Meissner regions of a type-II superconductor (the zero resistivity of the mixed state occurs if flux lines are "pinned" by appropriate material defects); in the mixed state, $\langle B \rangle$ denotes the average magnetic field in the superconductor.

appropriate material inhomogeneities, called "pinning centers", are present to hamper the motion of the flux tubes); for this reason type-II superconductors are of major interest for superconducting high field magnets and other technological applications. Also for the fields $H_{c1}(T)$ and $H_{c2}(T)$ approximate parabolic behaviour of type (2) holds.

A plot of the magnetization M versus the applied magnetic field is shown in Fig. 6. Below $H_{c1}(T)$ the superconductor shows perfect diamagnetism within all the sample; above $H_{c2}(T)$ the magnetization is negligible; between these two critical fields, the region of mixed state appears, in which normal and superconducting regions coexist in the sample. In Fig. 6 it is also illustrated the construction of the *thermodynamic critical field* $H_c(T)$: the low field magnetization is extrapolated linearly to the field $H_c(T)$ so to preseve the total area under the magnetization curve. In a type-II superconductor, we can thus distinguish three critical fields $H_{c1}(T) < H_c(T) < H_{c2}(T)$; a vortex state exists in the range between $H_{c1}(T)$ and $H_{c2}(T)$. In a type-I superconductor, the lower critical field, the thermodynamic critical field, and the upper critical field (which destroys superconductivity on the whole sample) coincide and are all denoted as $H_c(T)$.

In the mixed state the magnetic flux penetrates the sample under the form of thin filaments, or vortices, carrying a quantum of magnetic flux. A quantized flux vortex is essentially formed by a core of normal metal (carrying a magnetic field) surrounded by a region of superconducting material (carrying the screening supercurrents which reduce the magnetic field to zero within the penetration depth). Each vortex carries a quantum of magnetic flux equal to

$$\Phi_0 = \frac{h\,c}{2\,e} = 2.0679 \times 10^{-7} \text{ gauss} \cdot \text{cm}^2 \, . \tag{3}$$

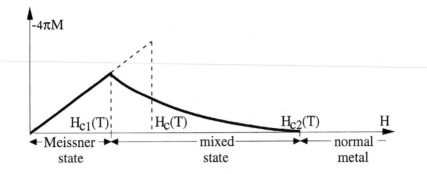

Fig. 6 Magnetization versus applied field H for a type-II superconductor. The equivalent area construction of the thermodynamic field $H_c(T)$ is also illustrated.

As we will see in Section 7, the classification of superconductors into two classes is determined by the sign of the surface energy of a superconductor–normal metal interface: positive in type-I superconductors and negative in type-II superconductors.

The phenomenology concerning superconductors is particular rich; it is not limited to perfect conductivity and magnetic properties, but includes a long list of peculiar effects. At this stage, however, it is convenient to describe the fundamental aspects of the microscopic quantum theory of superconductivity; with this tool in hands, we will continue more fruitfully the analysis of the experimental facts of superconductivity.

2   The Cooper pair idea

The occurrence of an effective attractive interaction among conduction electrons of a metal, whatever striking it might appear at first sight, is the key point of the microscopic interpretation of superconductivity. A possible mechanism leading to mutual attraction is the indirect electron–electron interaction via phonons, suggested by Fröhlich. In reasonably simple metals, we are accustomed to describe the conduction electrons as a system of independent fermions, weakly interacting with lattice vibrations. Because of this coupling, one electron interacts with the lattice and polarizes it, and another electron interacts with the polarized lattice. This (second-order) indirect electron–electron interaction can lead to an effective attractive interaction, as discussed in more detail in Appendix A. This coupling mechanism is operative among electrons lying near the Fermi energy E_F in an energy shell of the order of $\hbar\omega_D$, where ω_D is the Debye phonon frequency of the material.

Other mediator systems that could lead to effective pairing of carriers have been suggested in the literature, especially in connection with high-temperature cuprates. Proposed microscopic mechanisms include modified phonon coupling (with attention to anharmonic vibrational modes or Jahn–Teller active modes), role of strong anisotropy, coupling via magnetic excitations, or via bipolarons, or excitons, and other exotic mechanisms [see for instance the review article by E. Dagotto, Rev. Mod. Phys. **66**,

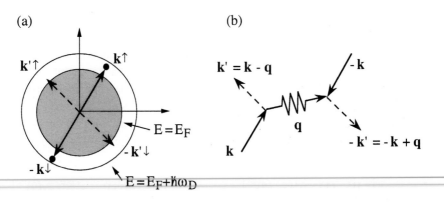

Fig. 7 (a) Schematic representation of a *single* Cooper pair, added to the ground-state of a free-electron gas. Two "extra electrons" in the pair state $(\mathbf{k}\uparrow, -\mathbf{k}\downarrow)$ scatter freely to the pair states $(\mathbf{k}'\uparrow, -\mathbf{k}'\downarrow)$, in the energy region $E_F < E_{\mathbf{k}}$, $E'_{\mathbf{k}} < E_F + \hbar\omega_D$, where the phonon-mediated attractive interaction is operative, and form a bound Cooper pair. (b) Schematic representation of the scattering of two electrons with wavevectors $(\mathbf{k}, -\mathbf{k})$ into the state $(\mathbf{k}', -\mathbf{k}')$ via the emission and subsequent absorption of a phonon of momentum $\hbar\mathbf{q}$.

763 (1994)]. We do not wish to enter here in the details and justifications of them: rather, *regardless of the origin of the net attractive interaction, we wish to analyse its consequences.*

A significant step towards the microscopic theory is due to Cooper demonstration that the normal Fermi sea becomes unstable, if a whatever small attractive interaction is operative among electrons [L. N. Cooper, Phys. Rev. **104**, 1189 (1956)]. This can be understood along the following lines.

The ordinary ground state of a free-electron gas at $T = 0$ is obtained filling with electrons all the states with $k < k_F$, where k_F is the Fermi wavevector. Consider two "extra" electrons added to the normal Fermi sea, as schematically indicated in Fig. 7. Let the two electrons interact with each other via a (small) *attractive* two-body potential $U(\mathbf{r}_1, \mathbf{r}_2)$ and feel the other electrons of the Fermi sea only through the Pauli exclusion principle, i.e. they cannot occupy the states of the filled Fermi sea. The Schrödinger equation for these two "extra" electrons can be written as

$$\left[\frac{\mathbf{p}_1^2}{2m} + \frac{\mathbf{p}_2^2}{2m} + U(\mathbf{r}_1, \mathbf{r}_2)\right]\psi(\mathbf{r}_1\sigma_1, \mathbf{r}_2\sigma_2) = E\,\psi(\mathbf{r}_1\sigma_1, \mathbf{r}_2\sigma_2)\,, \tag{4}$$

where $\mathbf{r}_1, \mathbf{r}_2, \sigma_1, \sigma_2$ are the space and spin coordinates of the two electrons.

In order to solve Eq. (4) we start from a complete set of spin-orbitals formed by the product of the plane waves $W_{\mathbf{k}}(\mathbf{r}) = (1/\sqrt{V})\exp(i\,\mathbf{k}\cdot\mathbf{r})$ by the spin functions α and β (the plane waves are normalized to one in the volume $V = N\Omega$ of the crystal, constituted by N unit cells of volume Ω). From any two spin-orbitals $\mathbf{k}_1\,\sigma_1$ and $\mathbf{k}_2\,\sigma_2$, of momentum $\hbar\mathbf{k}_1$ and $\hbar\mathbf{k}_2$ and spin σ_1 and σ_2, we can form the corresponding two-particle Slater determinant, whose total momentum is $\hbar\mathbf{K} = \hbar\mathbf{k}_1 + \hbar\mathbf{k}_2$. In a homogeneous system, the two-body interaction potential entering Eq. (4) has the form $U(\mathbf{r}_1, \mathbf{r}_2) = U(\mathbf{r}_1 - \mathbf{r}_2)$, and the mixing of two determinantal states can occur only if the total momentum is conserved, i.e. $\hbar\mathbf{k}_1 + \hbar\mathbf{k}_2 = \hbar\mathbf{k}_1' + \hbar\mathbf{k}_2' = \hbar\mathbf{K}$. The total momentum $\hbar\mathbf{K}$, that describes the motion of the center of mass of the two electron system, is thus a constant of motion. In the present discussion we are interested in the lowest possible energy of the pair; thus we neglect the motion of the center of mass by taking $\mathbf{K} = 0$, and consider only couples of spin-orbitals of wavevector $\mathbf{k}$ and $-\mathbf{k}$.

We write now explicitly the four Slater determinants formed with spin-orbitals $\mathbf{k}$ and $-\mathbf{k}$ and spin functions α and β. For instance, the determinantal state formed with the spin-orbitals $W_{\mathbf{k}\alpha}$ and $W_{-\mathbf{k}\alpha}$ has the expression

$$\psi_1 = \frac{1}{\sqrt{2}} \begin{vmatrix} W_{\mathbf{k}\alpha}(1) & W_{\mathbf{k}\alpha}(2) \\ W_{-\mathbf{k}\alpha}(1) & W_{-\mathbf{k}\alpha}(2) \end{vmatrix} = \frac{1}{\sqrt{2}} \frac{1}{V} \begin{vmatrix} e^{i\,\mathbf{k}\cdot\mathbf{r}_1}\,\alpha(1) & e^{i\,\mathbf{k}\cdot\mathbf{r}_2}\,\alpha(2) \\ e^{-i\,\mathbf{k}\cdot\mathbf{r}_1}\,\alpha(1) & e^{-i\,\mathbf{k}\cdot\mathbf{r}_2}\,\alpha(2) \end{vmatrix} .$$

We have thus

$$\psi_1 = A\{W_{\mathbf{k}\alpha}, W_{-\mathbf{k}\alpha}\} = \frac{1}{V}\frac{1}{\sqrt{2}} \left[e^{i\,\mathbf{k}\cdot(\mathbf{r}_1-\mathbf{r}_2)} - e^{-i\,\mathbf{k}\cdot(\mathbf{r}_1-\mathbf{r}_2)} \right] \alpha(1)\,\alpha(2) ,$$

and similarly

$$\psi_2 = A\{W_{\mathbf{k}\alpha}, W_{-\mathbf{k}\beta}\} = \frac{1}{V}\frac{1}{\sqrt{2}} \left[e^{i\mathbf{k}\cdot(\mathbf{r}_1-\mathbf{r}_2)}\alpha(1)\beta(2) - e^{-i\mathbf{k}\cdot(\mathbf{r}_1-\mathbf{r}_2)}\beta(1)\alpha(2) \right]$$

$$\psi_3 = A\{W_{\mathbf{k}\beta}, W_{-\mathbf{k}\alpha}\} = \frac{1}{V}\frac{1}{\sqrt{2}} \left[e^{i\mathbf{k}\cdot(\mathbf{r}_1-\mathbf{r}_2)}\beta(1)\,\alpha(2) - e^{-i\mathbf{k}\cdot(\mathbf{r}_1-\mathbf{r}_2)}\alpha(1)\,\beta(2) \right]$$

$$\psi_4 = A\{W_{\mathbf{k}\beta}, W_{-\mathbf{k}\beta}\} = \frac{1}{V}\frac{1}{\sqrt{2}} \left[e^{i\mathbf{k}\cdot(\mathbf{r}_1-\mathbf{r}_2)} - e^{-i\mathbf{k}\cdot(\mathbf{r}_1-\mathbf{r}_2)} \right] \beta(1)\beta(2) .$$

From the above four Slater determinants, we obtain one singlet state and one triplet state. The singlet state $\psi^{(S=0)}$, also denoted as $(\mathbf{k}\uparrow, -\mathbf{k}\downarrow)$, is given by

$$\psi^{(S=0)} = \frac{1}{V}\frac{1}{\sqrt{2}} \left[e^{i\mathbf{k}\cdot(\mathbf{r}_1-\mathbf{r}_2)} + e^{-i\mathbf{k}\cdot(\mathbf{r}_1-\mathbf{r}_2)} \right] \frac{1}{\sqrt{2}} \left[\alpha(1)\,\beta(2) - \beta(1)\,\alpha(2) \right] . \tag{5a}$$

The three partner functions of the triplet state are

$$\psi^{(S=1)} = \frac{1}{V}\frac{1}{\sqrt{2}} \left[e^{i\mathbf{k}\cdot(\mathbf{r}_1-\mathbf{r}_2)} - e^{-i\mathbf{k}\cdot(\mathbf{r}_1-\mathbf{r}_2)} \right] \begin{cases} \alpha(1)\,\alpha(2) \\ \frac{1}{\sqrt{2}}\left[\alpha(1)\,\beta(2) + \beta(1)\,\alpha(2)\right] \\ \beta(1)\,\beta(2) \end{cases} . \tag{5b}$$

The singlet spin function appearing in Eq. (5a) is antisymmetric for exchange of the particles 1 and 2; thus the orbital part in Eq. (5a) is symmetric in real space. The triplet spin functions appearing in Eq. (5b) are symmetric for exchange of the particles 1 and 2; thus the orbital part in Eq. (5b) is antisymmetric in real space.

The most general pair wavefunction of total spin zero and total momentum zero can be expressed as a linear combination of basis functions (5a) in the form

$$\psi(\mathbf{r}_1\sigma_1, \mathbf{r}_2\sigma_2) = \sum_{\mathbf{k}} g(\mathbf{k}) \frac{1}{V} e^{i\,\mathbf{k}\cdot(\mathbf{r}_1-\mathbf{r}_2)} \frac{1}{\sqrt{2}} [\alpha(1)\,\beta(2) - \alpha(2)\,\beta(1)] \qquad (6a)$$

with $g(\mathbf{k}) = g(-\mathbf{k})$, i.e. $g(\mathbf{k})$ even function of $\mathbf{k}$. Similarly, the most general pair wavefunction of total spin $S = 1$, definite spin component (say $S_z = 1$), and total momentum zero, has an expansion of type

$$\psi(\mathbf{r}_1\sigma_1, \mathbf{r}_2\sigma_2) = \sum_{\mathbf{k}} g(\mathbf{k}) \frac{1}{V} e^{i\,\mathbf{k}\cdot(\mathbf{r}_1-\mathbf{r}_2)} \alpha(1)\,\alpha(2) \qquad (6b)$$

with $g(\mathbf{k}) = -g(-\mathbf{k})$, i.e. $g(\mathbf{k})$ odd function of $\mathbf{k}$.

We take into account the exclusion principle between the extra pair and the "passive" free-electron gas by requiring

$$g(\mathbf{k}) = 0 \quad \text{for} \quad E_{\mathbf{k}} < E_F \ ,$$

where $E_{\mathbf{k}} = \hbar^2 k^2/2m$ is the unperturbed free electron energy. Moreover, from the considerations on phonon-mediated electron–electron coupling of Appendix A, it is assumed that $U(\mathbf{r}_1 - \mathbf{r}_2)$ is attractive only among electrons in an energy shell of the order of $\hbar\omega_D$ around the Fermi level E_F; thus we take

$$g(\mathbf{k}) = 0 \quad \text{for} \quad E_{\mathbf{k}} > E_F + \hbar\omega_D \ .$$

The Debye energy $\hbar\omega_D$ is much smaller than the Fermi energy E_F (measured from the bottom of the conduction band) in most ordinary metals; thus we can assume that the condition $\hbar\omega_D \ll E_F$ is verified.

The matrix elements of the potential $U(\mathbf{r}_1 - \mathbf{r}_2)$ between an initial state with two electrons with wavevectors $(\mathbf{k}', -\mathbf{k}')$ and a final state with two electrons with wavevectors $(\mathbf{k}, -\mathbf{k})$ is

$$U_{\mathbf{k}\mathbf{k}'} = \int\int \frac{1}{V} e^{-i\,\mathbf{k}\cdot(\mathbf{r}_1-\mathbf{r}_2)} U(\mathbf{r}_1 - \mathbf{r}_2) \frac{1}{V} e^{i\,\mathbf{k}'\cdot(\mathbf{r}_1-\mathbf{r}_2)} d\mathbf{r}_1 \, d\mathbf{r}_2$$

$$= \frac{1}{V} \int e^{-i\,(\mathbf{k}-\mathbf{k}')\cdot\mathbf{r}} U(\mathbf{r}) \, d\mathbf{r} \ ,$$

where $\mathbf{r} = \mathbf{r}_1 - \mathbf{r}_2$; notice that $U_{\mathbf{k}\mathbf{k}'}$ is inversely proportional to the volume $V = N\Omega$ of the sample.

We now transform the Schrödinger equation (4) into an integral equation for the coefficients $g(\mathbf{k})$ following the by now familiar procedure adopted in several other physical problems (excitons, plasmons, impurities, etc.). We insert expressions (6) (with $\mathbf{k}$ relabelled as $\mathbf{k}'$) into Eq. (4), premultiply both members of the resulting equation by $(1/V) \exp[-i\,\mathbf{k}\cdot(\mathbf{r}_1 - \mathbf{r}_2)]$ (times the singlet or triplet spin function), integrate over $\mathbf{r}_1$ and $\mathbf{r}_2$ and sum over spin variables, and obtain

$$(2\,E_{\mathbf{k}} - E)\,g(\mathbf{k}) + \sum_{\mathbf{k}'} U_{\mathbf{k}\mathbf{k}'}\,g(\mathbf{k}') = 0 \quad E_F < E_{\mathbf{k}}, E_{\mathbf{k}'} < E_F + \hbar\omega_D \ ; \qquad (7a)$$

with $\mathbf{k}$ and $\mathbf{k'}$ thought of as continuous variables, Eq. (7a) represents an integral equation, with kernel $U_{\mathbf{k}\mathbf{k'}}$, for the function $g(\mathbf{k})$ of the expasion coefficients.

To solve the integral equation (7a), we consider the particularly significant and simple case that the scattering amplitudes $U_{\mathbf{k}\mathbf{k'}}$ can be taken as constant in the (small) energy shell of width $\hbar\omega_D$ around the Fermi level. It is convenient to express $U_{\mathbf{k}\mathbf{k'}}$ in the form $U_{\mathbf{k}\mathbf{k'}} = -U_0/N$, where U_0 is a *positive energy*, independent of the number N of unit cells of the crystal. Eq. (7a) then becomes

$$(2\,E_{\mathbf{k}} - E)\,g(\mathbf{k}) - U_0\,\frac{1}{N}\,\sum_{\mathbf{k'}} g(\mathbf{k'}) = 0 \quad E_F < E_{\mathbf{k}}, E_{\mathbf{k'}} < E_F + \hbar\omega_D \ . \quad (7b)$$

It is seen by inspection that the integral equation (7b) does not modify the energies of the triplet states of the electron pair; in fact, for triplet states $g(\mathbf{k}) = -g(-\mathbf{k})$, thus $\sum g(\mathbf{k}) \equiv 0$ and no effect of U_0 occurs. On the contrary, for singlet states the integral equation (7b) presents a bound state with energy lower than twice the Fermi energy (this hints at a possible instability of the Fermi sea); it can also be noticed that in the case of a spherical conduction band also $g(\mathbf{k})$ is a spherically symmetric function, and the *relative motion of the paired electrons is then described by an s-wave*.

To determine the energy E_{pair} of the Cooper pair in the bound singlet state ($E_{\text{pair}} < 2\,E_F$), let us put $E = E_{\text{pair}}$ in Eq. (7b), divide it by $(2\,E_{\mathbf{k}} - E_{\text{pair}})$ (which is always a non-vanishing and positive quantity), and sum up over $\mathbf{k}$. We obtain the integral compatibility equation

$$1 = U_0\,\frac{1}{N}\,\sum_{\mathbf{k}} \frac{1}{2\,E_{\mathbf{k}} - E_{\text{pair}}} \quad E_F < E_{\mathbf{k}} < E_F + \hbar\omega_D \ .$$

The sum over the discrete wavevectors $\mathbf{k}$ can be converted into an energy integral in the usual form

$$1 = U_0\,\frac{1}{N}\,\int_{E_F}^{E_F + \hbar\omega_D} D_0(E)\,\frac{1}{2\,E - E_{\text{pair}}}\,dE \ , \quad (7c)$$

where $D_0(E)$ is the electronic density-of-states for one spin direction of the metal under consideration.

In Eq. (7c), in the small energy range of width $\hbar\omega_D$ around E_F, we can neglect the energy dependence of $D_0(E)$; we denote by $n_0(E_F) \equiv D_0(E_F)/N$ the density-of-states for one spin direction per unit cell at the Fermi energy, and obtain

$$1 = U_0\,n_0(E_F)\,\int_{E_F}^{E_F + \hbar\omega_D} \frac{1}{2\,E - E_{\text{pair}}}\,dE = \frac{1}{2}\,U_0\,n_0(E_F)\,\ln\frac{2\,E_F + 2\,\hbar\omega_D - E_{\text{pair}}}{2\,E_F - E_{\text{pair}}} \ .$$

It follows

$$\frac{2\,E_F - E_{\text{pair}}}{2\,E_F + 2\,\hbar\omega_D - E_{\text{pair}}} = e^{-2/U_0\,n_0(E_F)} \ ;$$

indicating by $\Delta_b = 2\,E_F - E_{\text{pair}}$ the binding energy of the two electron pair, one obtains

$$\Delta_b = \hbar\omega_D\,\frac{e^{-1/U_0 n_0(E_F)}}{\sinh\left[\,1/U_0 n_0(E_F)\,\right]} \ .$$

In the *weak coupling limit*, i.e. when $U_0 \, n_0(E_F) \ll 1$ and then $\Delta_b \ll \hbar\omega_D$, we have

$$\Delta_b = 2 \, \hbar\omega_D \exp\left[-2/U_0 \, n_0(E_F)\right] . \qquad (8)$$

The model considered until now, however simplified it might appear, is nevertheless useful to illustrate several key points. From Eq. (8), it can be seen that, no matter small U_0 is, the free electron gas is unstable ($\Delta_b > 0$) and electrons are expected to group in "singlet s-wave Cooper pairs". We also notice that the binding energy Δ_b in Eq. (8) is not an analytic function of U_0 for $U_0 \to 0$ and thus cannot be expanded in powers of U_0; conventional perturbation theory expressed in powers of the coupling constant is not valid due to the circumstance that the effect of U_0 is very significant near the Fermi energy, no matter small U_0 might be.

Another interesting message that can be inferred from Eq. (8) is that superconductivity is more likely to occur in materials with high values of U_0 (this entails high phonon contribution to electron scattering and electron resistivity) than in materials with low values of U_0. For instance a poor conductor such as lead becomes superconductor at 7.19 K, while excellent conductors (such as copper, silver, gold, sodium, potassium) do not appear to become superconductors even at the lowest temperatures. We notice that the singlet mechanism of coupling $(\mathbf{k}\uparrow, -\mathbf{k}\downarrow)$ tends to be broken, for instance, by the presence of magnetic impurities; in fact, it is well established experimentally that even a small amount of magnetic impurities may severely depress superconductivity.

A comparison with the integral equations we have encountered in the electronic structure calculations of crystals (for instance impurity, excitons and plasmons) shows that, in the case of the Cooper pair treatment, it is the *finite density-of-states* at the Fermi level responsible of the pair binding. In the Cooper model no threshold for the potential strength exists, differently for instance from what happens in the Slater–Koster model for impurities in three-dimensional crystals: in that case we are faced with a density-of-states of the form $n(E) \propto (E - E_G)^{1/2}$; a bound state of an impurity in the Slater–Koster model exists only for sufficiently *strong* attractive interaction.

Before concluding this section, it is instructive to describe the single Cooper pair by the second quantization formalism (see Appendix IV-B). The treatment of the single "two-electron" pair with the many-body apparatus represents a useful algebraic check, before considering the (somewhat) more demanding many-body treatment of superconductivity; it should allow the not yet expert reader to visualize better physical facts beyond the formalism.

The normal ground-state $|\Psi_N\rangle$ of a free-electron metal can be represented as a Slater determinant, formed with spin-orbitals (product of plane waves and spin functions) with wavevectors $\mathbf{k}$ up to the Fermi wavevector k_F. In second quantization formalism, we can write $|\Psi_N\rangle$ in the form

$$|\Psi_N\rangle = \prod_{\mathbf{k}}^{k<k_F} (c_{\mathbf{k}\uparrow}^\dagger c_{-\mathbf{k}\downarrow}^\dagger) |0\rangle , \qquad (9)$$

where $|0\rangle$ is the vacuum state, $c_{\mathbf{k}\uparrow}^\dagger$ creates an electron with momentum $\hbar\mathbf{k}$ and spin up (and obvious meaning for $c^\dagger_{-\mathbf{k}\downarrow}$). In $|\Psi_N\rangle$ all the electrons are independent, the order of the creation operators is irrelevant, and the choice to create electrons in pairs $(\mathbf{k}\uparrow, -\mathbf{k}\downarrow)$ has been done for convenience, in view of the forthcoming elaborations.

Let us consider now the wavefunction (6a) of the single Cooper pair; in the second quantization form it can be written in the compact form

$$|\Psi_{\text{Cooper pair}}\rangle = \sum_{\mathbf{k}} g(\mathbf{k})\, c^\dagger_{\mathbf{k}\uparrow} c^\dagger_{-\mathbf{k}\downarrow} |\Psi_N\rangle . \tag{10a}$$

Eq. (10a) shows by inspection that the two extra electrons added to the normal metal ground state have total spin equal to zero and total momentum equal to zero. The Hamiltonian of the system of interacting fermions in the second quantized form becomes

$$H = \sum_{\mathbf{k}} E_{\mathbf{k}}(c^\dagger_{\mathbf{k}\uparrow} c_{\mathbf{k}\uparrow} + c^\dagger_{-\mathbf{k}\downarrow} c_{-\mathbf{k}\downarrow}) + \sum_{\mathbf{k}\mathbf{k}'} U_{\mathbf{k}\mathbf{k}'}\, c^\dagger_{\mathbf{k}\uparrow} c^\dagger_{-\mathbf{k}\downarrow} c_{-\mathbf{k}'\downarrow} c_{\mathbf{k}'\uparrow} . \tag{10b}$$

The first term in H is the kinetic energy of a set of independent fermions of wavevectors $\mathbf{k}$ and energies $E_{\mathbf{k}} = \hbar^2 k^2/2m$; the second term is the pairing electron–electron interaction. The creation and annihilation operators obey the standard anticommutation rules

$$\{c_{\mathbf{k}\sigma}, c_{\mathbf{k}'\sigma'}\} = \{c^\dagger_{\mathbf{k}\sigma}, c^\dagger_{\mathbf{k}'\sigma'}\} = 0 , \qquad \{c_{\mathbf{k}\sigma}, c^\dagger_{\mathbf{k}'\sigma'}\} = \delta_{\mathbf{k}\mathbf{k}'}\, \delta_{\sigma\sigma'} . \tag{11}$$

From Eqs. (10) and Eq. (11) we can reobtain all the results of the isolated Cooper pair, and in particular the compatibility equation (7).

The next step of the theory is to consider not only a single pair added to the Fermi sea, but to allow the possibility of pairing for all the electrons of the sea, on the same footing. This is done by the approach outlined below.

3   Ground state for a superconductor in the BCS theory at zero temperature

3.1   Variational determination of the ground-state wavefunction

The Cooper model (1956) is only a rough (although for some aspects illuminating) starting point for a theory of superconductivity; it opens the way to the major breakthrough (1957) in the microscopic theory of superconductivity by Bardeen, Cooper and Schrieffer (BCS). In the Cooper treatment we have seen that the normal electron gas in the presence of a (net) attractive electron–electron interaction is unstable, even with respect to the formation of a single Cooper pair. A realistic theory of superconductivity must be able to describe a *cooperative condensation process* in which many pairs of electrons of the normal Fermi sea are formed so to minimize the total energy of the system (at $T = 0$).

These demanding physical and formal requirements can be adequately described by the following BCS variational form of the ground-state wavefunction for superconductors

$$|\Psi_S\rangle = \prod_{\mathbf{k}}(u_{\mathbf{k}} + v_{\mathbf{k}}c^{\dagger}_{\mathbf{k}\uparrow}c^{\dagger}_{-\mathbf{k}\downarrow})|0\rangle \quad, \tag{12}$$

where $|0\rangle$ is the vacuum state, $u_{\mathbf{k}}$ and $v_{\mathbf{k}}$ are real and even functions of $\mathbf{k}$, chosen in such a way to minimize the ground-state energy. Under the constraint $u^2_{\mathbf{k}}+v^2_{\mathbf{k}}=1$ (which assures the normalization to unity of Ψ_S), $v_{\mathbf{k}}$ represents the probability amplitude that the pair $(\mathbf{k}\uparrow, -\mathbf{k}\downarrow)$ is occupied and $u_{\mathbf{k}}$ represents the probability amplitude that the pair $(\mathbf{k}\uparrow, -\mathbf{k}\downarrow)$ is unoccupied. In the superconducting ground state described by Ψ_S, electrons are involved only as pairs; in fact by carrying out explicitly the products in Eq. (12) we see that there is a term without Cooper pairs and terms with one, two, and so on Cooper pairs. Notice also that the wavefunction (12) of the superconductor is reduced to the wavefunction (9) of the normal metal in the particular case in which $u_{\mathbf{k}}$ and $v_{\mathbf{k}}$ are given by

$$\begin{cases} u_{\mathbf{k}} = 0 \quad v_{\mathbf{k}} = 1 \quad \text{for } k < k_F \\ u_{\mathbf{k}} = 1 \quad v_{\mathbf{k}} = 0 \quad \text{for } k > k_F \end{cases} . \tag{13}$$

An aspect of the trial wavefunction (12) is that the total number of electrons is not well defined. Actually, it is not convenient to be tied up with a fixed number of electrons; in a manner which is usual in such cases, we proceed to minimize the expectation value $\langle\Psi_S|H|\Psi_S\rangle$ under the constraint $\langle\Psi_S|N_{op}|\Psi_S\rangle = N$, where

$$N_{op} = \sum_{\mathbf{k}}(c^{\dagger}_{\mathbf{k}\uparrow}c_{\mathbf{k}\uparrow} + c^{\dagger}_{-\mathbf{k}\downarrow}c_{-\mathbf{k}\downarrow})$$

is the particle number operator and N is the number of electrons in the actual metal. The standard method to treat minimization plus one (or more) appropriate constraint consists in introducing one (or more) Lagrange multiplier μ, minimizing without restraints the quantity $\langle\Psi_S|H - \mu N_{op}|\Psi_S\rangle$ and finally determining μ through the condition $\langle\Psi_S|N_{op}|\Psi_S\rangle = N$; it turns out that μ is the Fermi energy, and it is the same for the normal or superconducting state.

In essence we have thus arrived at the minimization of the quantity

$$W_S = \langle\Psi_S|H_{BCS}|\Psi_S\rangle , \tag{14a}$$

where $H_{BCS} \equiv H - \mu N_{op}$ is the so-called Bardeen–Cooper–Schrieffer Hamiltonian given by

$$H_{BCS} = \sum_{\mathbf{k}}\varepsilon_{\mathbf{k}}(c^{\dagger}_{\mathbf{k}\uparrow}c_{\mathbf{k}\uparrow} + c^{\dagger}_{-\mathbf{k}\downarrow}c_{-\mathbf{k}\downarrow}) + \sum_{\mathbf{k}\mathbf{k}'}U_{\mathbf{k}\mathbf{k}'}c^{\dagger}_{\mathbf{k}\uparrow}c^{\dagger}_{-\mathbf{k}\downarrow}c_{-\mathbf{k}'\downarrow}c_{\mathbf{k}'\uparrow} , \tag{14b}$$

and $\varepsilon_{\mathbf{k}} = E_{\mathbf{k}} - \mu = \hbar^2\mathbf{k}^2/2m - \mu$ denotes the single particle energy measured with respect to the Fermi energy.

The expectation value (14a) can be easily evaluated keeping in mind the practical recipe of using the anticommutation rules (11) for systematically shifting the fermion annihilation operators to the right (or creation operators to the left) until they arrive to operate on the vacuum state (giving zero). Here are some matrix elements of interest. We first note that

$$\langle 0|(u_{\mathbf{k}} + v_{\mathbf{k}} c_{-\mathbf{k}\downarrow} c_{\mathbf{k}\uparrow})(u_{\mathbf{k}} + v_{\mathbf{k}} c_{\mathbf{k}\uparrow}^\dagger c_{-\mathbf{k}\downarrow}^\dagger)|0\rangle = u_{\mathbf{k}}^2 + v_{\mathbf{k}}^2 \; ; \tag{15a}$$

the normalization of the wavefunction $|\Psi_S\rangle$ is given by

$$\langle \Psi_S|\Psi_S\rangle = \prod_{\mathbf{k}} (u_{\mathbf{k}}^2 + v_{\mathbf{k}}^2) = 1 \quad \text{if} \quad u_{\mathbf{k}}^2 + v_{\mathbf{k}}^2 = 1 \quad \text{for every } \mathbf{k} \; . \tag{15b}$$

Similarly for the number operator we have

$$\langle \Psi_S|c_{\mathbf{k}\uparrow}^\dagger c_{\mathbf{k}\uparrow}|\Psi_S\rangle = \langle 0|(u_{\mathbf{k}} + v_{\mathbf{k}} c_{-\mathbf{k}\downarrow} c_{\mathbf{k}\uparrow}) c_{\mathbf{k}\uparrow}^\dagger c_{\mathbf{k}\uparrow} (u_{\mathbf{k}} + v_{\mathbf{k}} c_{\mathbf{k}\uparrow}^\dagger c_{-\mathbf{k}\downarrow}^\dagger)|0\rangle = v_{\mathbf{k}}^2 \; . \tag{15c}$$

We have also

$$\langle \Psi_S|c_{\mathbf{k}\uparrow}^\dagger c_{-\mathbf{k}\downarrow}^\dagger|\Psi_S\rangle = \langle \Psi_S|c_{-\mathbf{k}\downarrow} c_{\mathbf{k}\uparrow}|\Psi_S\rangle = u_{\mathbf{k}} v_{\mathbf{k}} \; . \tag{15d}$$

Using the matrix elements (15) we obtain for the superconductor ground-state energy the following expression

$$W_S = \langle \Psi_S|H_{\mathrm{BCS}}|\Psi_S\rangle = 2 \sum_{\mathbf{k}} \varepsilon_{\mathbf{k}} v_{\mathbf{k}}^2 + \sum_{\mathbf{k}\mathbf{k}'} U_{\mathbf{k}\mathbf{k}'} u_{\mathbf{k}} v_{\mathbf{k}} u_{\mathbf{k}'} v_{\mathbf{k}'} \; ; \tag{16}$$

at this stage our problem has become a standard algebraic minimization problem of W_S as a function of the $u_{\mathbf{k}}$ and $v_{\mathbf{k}}$ (with $\mathbf{k}$ in the reciprocal space).

In order to minimize W_S under the constraint $u_{\mathbf{k}}^2 + v_{\mathbf{k}}^2 = 1$, let us represent $u_{\mathbf{k}}$ and $v_{\mathbf{k}}$ in polar form

$$\begin{cases} u_{\mathbf{k}} = \cos\theta_{\mathbf{k}} \\ v_{\mathbf{k}} = \sin\theta_{\mathbf{k}} \end{cases} .$$

The superconductor ground-state energy (16) becomes

$$W_S = 2 \sum_{\mathbf{k}} \varepsilon_{\mathbf{k}} \sin^2\theta_{\mathbf{k}} + \frac{1}{4} \sum_{\mathbf{k}\mathbf{k}'} U_{\mathbf{k}\mathbf{k}'} \sin 2\theta_{\mathbf{k}} \sin 2\theta_{\mathbf{k}'} \; .$$

The condition $\partial W_S/\partial\theta_{\mathbf{k}} = 0$ gives

$$2\,\varepsilon_{\mathbf{k}} \sin 2\theta_{\mathbf{k}} + \sum_{\mathbf{k}'} U_{\mathbf{k}\mathbf{k}'} \cos 2\theta_{\mathbf{k}} \sin 2\theta_{\mathbf{k}'} = 0 \; ;$$

we re-write this minimization condition in terms of $u_{\mathbf{k}}$ and $v_{\mathbf{k}}$ and obtain

$$2\,\varepsilon_{\mathbf{k}} u_{\mathbf{k}} v_{\mathbf{k}} + \sum_{\mathbf{k}'} U_{\mathbf{k}\mathbf{k}'} (u_{\mathbf{k}}^2 - v_{\mathbf{k}}^2) u_{\mathbf{k}'} v_{\mathbf{k}'} = 0 \tag{17}$$

for every $\mathbf{k}$.

To solve this set of self-consistent equations for $u_{\mathbf{k}}$ and $v_{\mathbf{k}}$, let us define the *energy gap parameters* $\Delta_{\mathbf{k}}$ (the origin of the name will be clear below)

$$\Delta_{\mathbf{k}} = -\sum_{\mathbf{k}'} U_{\mathbf{k}\mathbf{k}'} u_{\mathbf{k}'} v_{\mathbf{k}'} \; . \tag{18}$$

The variational condition (17) thus becomes

$$2\,\varepsilon_{\mathbf{k}}\,u_{\mathbf{k}}\,v_{\mathbf{k}} - \Delta_{\mathbf{k}}\,(u_{\mathbf{k}}^2 - v_{\mathbf{k}}^2) = 0 \qquad . \tag{19}$$

The solution of Eq. (19), together with the normalization condition $u_{\mathbf{k}}^2 + v_{\mathbf{k}}^2 = 1$ gives

$$u_{\mathbf{k}}^2 = \frac{1}{2}\left[1 + \frac{\varepsilon_{\mathbf{k}}}{\sqrt{\varepsilon_{\mathbf{k}}^2 + \Delta_{\mathbf{k}}^2}}\right] \quad \text{and} \quad v_{\mathbf{k}}^2 = \frac{1}{2}\left[1 - \frac{\varepsilon_{\mathbf{k}}}{\sqrt{\varepsilon_{\mathbf{k}}^2 + \Delta_{\mathbf{k}}^2}}\right] . \tag{20}$$

The choice of sign in Eq. (20) has been determined by the requirement that, for vanishing $U_{\mathbf{k}\mathbf{k}'}$ interaction and thus for vanishing $\Delta_{\mathbf{k}}$, the standard normal ground state described by Eqs. (13) is obtained. If we use Eqs. (20) to calculate $u_{\mathbf{k}'}\,v_{\mathbf{k}'}$, and insert the result into Eq. (18) we see that the energy gap parameters are determined by the self-consistent equations

$$\Delta_{\mathbf{k}} = -\frac{1}{2}\sum_{\mathbf{k}'} U_{\mathbf{k}\mathbf{k}'} \frac{\Delta_{\mathbf{k}'}}{\sqrt{\varepsilon_{\mathbf{k}'}^2 + \Delta_{\mathbf{k}'}^2}} \tag{21}$$

for every $\mathbf{k}$.

A trivial solution of equation (21) is $\Delta_{\mathbf{k}} = 0$ for every $\mathbf{k}$; this corresponds to the normal metal state with electrons filling the Fermi sphere up to k_F. In general the integral equation (21) is not easy to be solved. For simplicity, and in analogy with the procedure adopted in the treatment of a single s-wave Cooper pair, we assume that the matrix elements $U_{\mathbf{k}\mathbf{k}'}$ can be considered constant in an appropriate energy shell around the Fermi level; in the so-called *average potential approximation*, we assume

$$U_{\mathbf{k}\mathbf{k}'} = \begin{cases} -U_0/N & \text{if } |\varepsilon_{\mathbf{k}}|, |\varepsilon_{\mathbf{k}'}| < \hbar\omega_D \\ 0 & \text{otherwise} \end{cases} \tag{22}$$

where $U_0 > 0$ is a *positive constant*, independent of the states $\mathbf{k}$ and $\mathbf{k}'$ in the shell $|\varepsilon_{\mathbf{k}}|, |\varepsilon_{\mathbf{k}'}| < \hbar\omega_D$, and also independent of the number N of unit cells of the crystal. From the average potential approximation of Eq. (22), and from Eq. (18), it follows that

$$\Delta_{\mathbf{k}} = \begin{cases} \Delta_0 & \text{if } |\varepsilon_{\mathbf{k}}| < \hbar\omega_D \\ 0 & \text{otherwise} \end{cases} .$$

The integral equation (21) simplifies as

$$1 = \frac{1}{2} U_0 \frac{1}{N} \sum_{\mathbf{k}'} \frac{1}{\sqrt{\varepsilon_{\mathbf{k}'}^2 + \Delta_0^2}} \quad \text{with} \quad -\hbar\omega_D < \varepsilon_{\mathbf{k}'} < \hbar\omega_D .$$

As usual the sum over $\mathbf{k}'$ can be converted into an integral; assuming the electron density-of-states as constant in the small energy shell of interest around the Fermi level, one obtains

$$1 = \frac{1}{2} U_0\, n_0(E_F) \int_{-\hbar\omega_D}^{\hbar\omega_D} \frac{d\varepsilon}{\sqrt{\varepsilon^2 + \Delta_0^2}} \; ; \tag{23a}$$

as before, $n_0(E_F) = D_0(E_F)/N$ denotes the density-of-states for one spin direction per unit cell at the Fermi energy.

The integral in Eq. (23a) can be carried out considering that the primitive of $1/\sqrt{x^2 + a^2}$ is $\operatorname{arcsinh}(x/a)$; Eq. (23a) becomes

$$1 = U_0 \, n_0(E_F) \operatorname{arcsinh} \frac{\hbar \omega_D}{\Delta_0} \; ;$$

it follows

$$\Delta_0 = \frac{\hbar \omega_D}{\sinh\left[\, 1/U_0 n_0(E_F)\,\right]} \; .$$

In the weak coupling limit, when $U_0 n_0(E_F)$ is rather small with respect to unity, and $\exp(-1/U_0 n_0) \ll 1$, one obtains

$$\boxed{\Delta_0 = 2 \, \hbar \omega_D \exp[-1/U_0 \, n_0(E_F)]} \qquad (23b)$$

[the similarity in form of Eq. (23b) with Eq. (8) is fortuitous, as the weak coupling limit tends to produce exponentials]. For typical conventional superconductors, one has $E_F \approx 1 \, \text{eV}$ (measured from the bottom of the conduction band) and $U_0 \approx 0.1 - 0.5 \, \text{eV}$; the dimensionless coupling parameter $U_0 n_0(E_F) \approx U_0/E_F$ is in the range 0.1-0.5; thus $\Delta_0 \approx 1 \, \text{meV}$ is in general a small fraction ($\approx 0.1 - 0.01$) of $\hbar \omega_D$.

Until now, we have confined our discussion of the BCS theory of superconductivity within the "*weak coupling limit*" (i.e. $\Delta_0 \ll \hbar \omega_D$), and "*s-wave symmetry of the pairing state*"; also in the following we keep our considerations and guidelines within these approximations. We wish to mention that the BCS theory has been generalized by various authors to describe strong-coupling situations [see W. L. MacMillan, Phys. Rev. **167**, 331 (1978) and in particular the works of Eliashberg and other authors, quoted therein]. We also notice that an active debate concerns the possibility that in some unconventional superconductors the matrix elements $U_{\mathbf{k}\mathbf{k}'}$, differently from Eq. (22), are strongly dependent on the directions $\mathbf{k}$ and $\mathbf{k}'$, and not necessarily negative everywhere on the Fermi surface; in these circumstances, pairing states with non-zero angular momentum might be preferred, and we refer to the literature for a discussion of aspects of "non s-wave superconductivity" [see for instance P. W. Anderson and P. Morel, Phys. Rev. B**23**, 1911 (1961); see also B. E. C. Koltenbah and R. Joynt, Rep. Progr. Phys. **60**, 23 (1997) and references quoted therein].

3.2 Ground-state energy and isotopic effect

The *condensation energy of a superconductor* is defined as the energy difference between the superconductor ground-state energy W_S and the normal ground-state energy W_N; its expression is

$$W_S - W_N = 2 \sum_{\mathbf{k}} \varepsilon_{\mathbf{k}} \, v_{\mathbf{k}}^2 + \sum_{\mathbf{k}\mathbf{k}'} U_{\mathbf{k}\mathbf{k}'} \, u_{\mathbf{k}} \, v_{\mathbf{k}} \, u_{\mathbf{k}'} \, v_{\mathbf{k}'} - 2 \sum_{\mathbf{k}}^{k < k_F} \varepsilon_{\mathbf{k}} \; . \qquad (24)$$

Using Eq. (18), and then Eqs. (20), we obtain

$$W_S - W_N = \sum_{\mathbf{k}} \left[2\varepsilon_{\mathbf{k}} v_{\mathbf{k}}^2 - \Delta_{\mathbf{k}} u_{\mathbf{k}} v_{\mathbf{k}} \right] - \sum_{\mathbf{k}}^{k<k_F} 2\varepsilon_{\mathbf{k}}$$

$$= \sum_{\mathbf{k}} \left[\varepsilon_{\mathbf{k}} - \frac{2\varepsilon_{\mathbf{k}}^2 + \Delta_{\mathbf{k}}^2}{2\sqrt{\varepsilon_{\mathbf{k}}^2 + \Delta_{\mathbf{k}}^2}} \right] - \sum_{\mathbf{k}}^{k<k_F} 2\varepsilon_{\mathbf{k}} \ ;$$

the above expression is now evaluated in the case of average potential approximation and weak binding limit.

In the case of average potential approximation expressed by Eq. (22), $\Delta_{\mathbf{k}}$ is different from zero only for $|\varepsilon_{\mathbf{k}}| < \hbar\omega_D$, and we have

$$W_S - W_N = \sum_{\mathbf{k}}^{|\varepsilon_{\mathbf{k}}|<\hbar\omega_D} \left[\varepsilon_{\mathbf{k}} - \frac{2\,\varepsilon_{\mathbf{k}}^2 + \Delta_0^2}{2\,\sqrt{\varepsilon_{\mathbf{k}}^2 + \Delta_0^2}} \right] - \sum_{\mathbf{k}}^{-\hbar\omega_D<\varepsilon_{\mathbf{k}}<0} 2\,\varepsilon_{\mathbf{k}} \ .$$

As usual, we convert the sum over $\mathbf{k}$ into an energy integral. We remember that the primitive of $(2x^2 + a^2)/\sqrt{x^2 + a^2}$ is $x\sqrt{x^2 + a^2}$ and obtain

$$W_S - W_N = D_0(E_F) \int_{-\hbar\omega_D}^{\hbar\omega_D} \left(\varepsilon - \frac{2\,\varepsilon^2 + \Delta_0^2}{2\,\sqrt{\varepsilon^2 + \Delta_0^2}} \right) d\varepsilon - D_0(E_F) \int_{-\hbar\omega_D}^{0} 2\varepsilon\, d\varepsilon$$

$$= D_0(E_F) \left[\hbar^2\omega_D^2 - \hbar\omega_D \cdot \sqrt{\hbar^2\omega_D^2 + \Delta_0^2} \right] \ .$$

In the weak binding limit $\Delta_0 \ll \hbar\omega_D$, with a series development of the square root, we obtain for the condensation energy the expression

$$\boxed{W_S - W_N = -\frac{1}{2} D_0(E_F) \Delta_0^2} \ . \qquad (25)$$

The condensation energy (25) of the superconductor can be interpreted as originated by the electrons $D_0(E_F) \Delta_0$, the ones in the energy shell Δ_0 around the Fermi energy, which decrease their energy by about Δ_0 because of the pairing mechanism (see Fig. 8).

We can link the *thermodynamic magnetic field* $H_c(0)$ of the superconductor with the condensation energy per unit volume, via the relation

$$\frac{1}{V}(W_N - W_S) = \frac{H_c^2(0)}{8\pi} = \frac{1}{2} D_0(E_F) \Delta_0^2 \frac{1}{V} \ . \qquad (26a)$$

We express the crystal volume as $V = N\Omega$, and obtain

$$H_c^2(0) = 4\pi\, n_0(E_F) \Delta_0^2 \frac{1}{\Omega} \ ,$$

where $n_0(E_F) = D_0(E_F)/N$. To estimate the value of $H_c(0)$, we write it in the form

$$H_c^2(0) = 4\pi\, n_0(E_F) \Delta_0^2 \frac{a_B^3}{\Omega} \frac{\mu_B^2}{a_B^3} \frac{1}{\mu_B^2} \ ,$$

where a_B is the Bohr radius and μ_B is the Bohr magneton; then we take $E_F \approx$

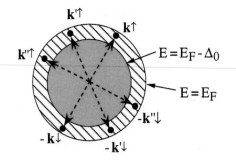

Fig. 8 Schematic representation of the origin of the condensation energy in the BCS ground state. The electron pairs $(\mathbf{k}\uparrow, -\mathbf{k}\downarrow)$ in the blurred region of $\mathbf{k}$ space, within an energy shell Δ_0 around the Fermi level, contribute to the condensation energy of the superconductor (notice that, typically, $\Delta_0 \approx 1\text{meV}$ and $E_F \approx 1\text{eV}$). The dashed lines are a remind of the binding energy of a Cooper pair.

$1\,\text{eV}$, $n_0(E_F) \approx 1/E_F = 10^{-3}\,\text{states/meV}$, $\Delta_0 \approx 1\,\text{meV}$, $a_B^3/\Omega = 1/100$, $\mu_B^2/a_B^3 = 0.363\,\text{meV}$ (see Eq. XVII-1b), $\mu_B = 5.788 \cdot 10^{-6}\text{meV/gauss}$, and estimate $H_c(0) \approx 10^3\,\text{gauss}$. Values of $\Delta_0 \approx 1\,\text{meV}$ thus imply values of $H_c(0)$ of the order of several hundred or few thousand gauss.

It is interesting to notice that Eq. (26a) also contains the so-called *isotopic effect*, which historically has been the hint of the phonon induced electron–electron coupling mechanism considered in the BCS theory. Compare in fact metals built with atoms of different isotopic masses M. It is reasonable to assume that the relevant effect of isotopic substitution is just to change the Debye frequency $\hbar\omega_D$, leaving essentially unchanged all the other parameters concerning the electronic part of the problem. In the Debye model for simple lattices $\hbar\omega_D$ is proportional to $1/\sqrt{M}$ (similarly to what expected for a simple oscillator). Using Eq. (23b) and Eq. (26a) we thus expect that

$$H_c(0) \cdot M^\alpha = \text{const} \tag{26b}$$

with $\alpha = 0.5$ (at least in the weak coupling limit). Experimental measurements show that the above relation is well verified for mercury; other non transition metals also have $\alpha = 0.5$. The experimental situation is however much richer; for instance for Mo, Re and Os, α is lower than 0.5. For the conventional superconductors Ru and Zr (as well as for several high-T_c superconductors) α vanishes to zero. It should be remembered that the theoretical model we are presenting contains by necessity several simplifying assumptions, and the actual behaviour of real materials must be discussed and analysed case by case.

3.3   Momentum distribution and coherence length

The probability of finding an electron in a state with momentum $\hbar\mathbf{k}$ and spin σ in the superconductor is given in terms of the single particle number operator by

$$\langle \Psi_S | c_{\mathbf{k}\sigma}^\dagger c_{\mathbf{k}\sigma} | \Psi_S \rangle = v_{\mathbf{k}}^2 = \frac{1}{2}\left[1 - \frac{\varepsilon_{\mathbf{k}}}{\sqrt{\varepsilon_{\mathbf{k}}^2 + \Delta_{\mathbf{k}}^2}}\right]; \tag{27a}$$

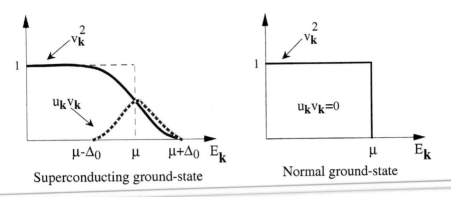

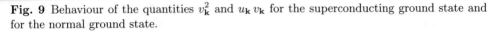

Fig. 9 Behaviour of the quantities $v_{\mathbf{k}}^2$ and $u_{\mathbf{k}} v_{\mathbf{k}}$ for the superconducting ground state and for the normal ground state.

in the normal gas, the same probability is given by

$$\langle \Psi_N | c_{\mathbf{k}\sigma}^\dagger \, c_{\mathbf{k}\sigma} | \Psi_N \rangle = \begin{cases} 1 & \text{for } k < k_F \\ 0 & \text{for } k > k_F \end{cases} . \tag{27b}$$

The two situations are schematically indicated in Fig. 9. Notice also that $u_{\mathbf{k}}^2 = 1 - v_{\mathbf{k}}^2$.

We can do a similar analysis for the pair operator $c_{\mathbf{k}\uparrow}^\dagger \, c_{-\mathbf{k}\downarrow}^\dagger$ (or $c_{-\mathbf{k}\downarrow} \, c_{\mathbf{k}\uparrow}$). In the superconducting ground state we have

$$\langle \Psi_S | c_{\mathbf{k}\uparrow}^\dagger \, c_{-\mathbf{k}\downarrow}^\dagger | \Psi_S \rangle = u_{\mathbf{k}} v_{\mathbf{k}} = \frac{1}{2} \frac{\Delta_{\mathbf{k}}}{\sqrt{\varepsilon_{\mathbf{k}}^2 + \Delta_{\mathbf{k}}^2}} . \tag{28}$$

The above quantity is always zero in the normal state, but it is different from zero in a small shell around k_F in the superconductor state; some of the most typical effects of superconductors (for instance Meissner effect and Josephson tunneling) are related to this fact.

The region in reciprocal space where $u_{\mathbf{k}} v_{\mathbf{k}}$ is different from zero has an energy width Δ_0 around the Fermi energy. From the free-electron dispersion law $E(\mathbf{k}) = \hbar^2 k^2 / 2m$, we have the relation $\delta E = (\hbar^2 k/m) \, \delta k$ between the width δk in reciprocal space and the energy width δE. With $\delta E = \Delta_0$ and $k = k_F$, we obtain $\Delta_0 = (\hbar^2 k_F/m) \, \delta k = \hbar v_F \, \delta k$. From the uncertainty principle, we have that in real space the spatial extent $\xi_0 \approx 1/\delta k$ is given by

$$\xi_0 = \frac{1}{\pi} \frac{\hbar v_F}{\Delta_0} = \frac{1}{\pi} \frac{\hbar^2 k_F}{m \Delta_0} \tag{29}$$

(the numerical factor π requires a more detailed analysis). The length ξ_0 is the so-called *BCS coherence length*, and represents the average distance in real space between the two electrons of the Cooper pair. The pair coherence length is proportional to the Fermi velocity and inversely proportional to the binding energy of the Cooper pair. Typical values of ξ_0 range from thousand Å in conventional superconductors, to some tens Å in high-T_c superconductors (where, furthermore, ξ_0 is anisotropic).

In general there is a large number of Cooper pairs within the spatial coherence length ξ_0. In fact the density of Cooper pairs is of the order of $\approx n \cdot (\Delta_0/E_F)$, where $n \equiv 1/[(4/3)\pi r_s^3 a_B^3]$ is the electron density of the metal, and Δ_0/E_F is the fraction of electrons that condensate in the ground superconducting state with an energy gain of $\approx 2\Delta_0$; we thus have that the average distance between the centers of mass of the pairs is related to the average distance $r_s a_B$ between electrons by the expression $d \approx r_s a_B (E_F/\Delta_0)^{1/3}$. For conventional superconductors, $E_F \approx 1\,\mathrm{eV}$, $\Delta_0 \approx 1\,\mathrm{meV}$, $d \approx 10\,r_s a_B \approx 10\,\text{Å}$. With ξ_0 of the order of $1000\,\text{Å}$, there are million Cooper pairs within the coherence length; the individual pairs overlap strongly in space and the binding energy $2\Delta_0$ of any pair depends cooperatively on the presence of all the other pairs.

4   Excited states of superconductors at zero temperature

4.1   The Bogoliubov canonical transformation

Until now we have studied the ground state of a superconductor *at zero temperature*. We could now study the excited states of a superconductor system (still remaining at $T = 0$) starting from the ground-state wavefunction (12), applying to it creation or annihilation operators and elaborating the trial excited states so obtained; this was the procedure originally followed by BCS. An equivalent although much more convenient procedure, based on the canonical transformations of Bogoliubov and Valatin, allows one to obtain in a single stroke the excitation spectrum of the superconductor [N. N. Bogoliubov, Nuovo Cimento **7**, 794 (1958); J. G. Valatin, Nuovo Cimento **7**, 843 (1958)].

Let us consider the superconductor BCS Hamiltonian of Eq. (14b), here re-written for convenience

$$H_{\mathrm{BCS}} = \sum_{\mathbf{k}\sigma} \varepsilon_{\mathbf{k}} c_{\mathbf{k}\sigma}^{\dagger} c_{\mathbf{k}\sigma} + \sum_{\mathbf{k}\mathbf{k}'} U_{\mathbf{k}\mathbf{k}'} c_{\mathbf{k}\uparrow}^{\dagger} c_{-\mathbf{k}\downarrow}^{\dagger} c_{-\mathbf{k}'\downarrow} c_{\mathbf{k}'\uparrow} . \tag{30}$$

We then consider the ground-state wavefunction in the form

$$|\Psi_S\rangle = \prod_{\mathbf{k}} (u_{\mathbf{k}} + v_{\mathbf{k}} c_{\mathbf{k}\uparrow}^{\dagger} c_{-\mathbf{k}\downarrow}^{\dagger}) |0\rangle , \tag{31}$$

where $u_{\mathbf{k}}$ and $v_{\mathbf{k}}$ are not yet specified real quantities, satisfying $u_{\mathbf{k}}^2 + v_{\mathbf{k}}^2 = 1$ to assure the normalization of $|\Psi_S\rangle$. In the variational BCS procedure, $u_{\mathbf{k}}$ and $v_{\mathbf{k}}$ are determined so to minimize the energy of the superconductor ground state. In the present procedure, $u_{\mathbf{k}}$ and $v_{\mathbf{k}}$ are chosen as coefficients of a canonical transformation that gives diagonal form to the Hamiltonian (30) (once appropriately simplified).

We remark that in the operator H_{BCS}, the source of difficulties is constituted by the presence of products of four fermion operators. With the aim to find an appropriate simplification, and in line with a routine procedure of mean field theory, we express the product of any two creation operators $c_{\mathbf{k}\uparrow}^{\dagger} c_{-\mathbf{k}\downarrow}^{\dagger}$ in the form

$$c_{\mathbf{k}\uparrow}^{\dagger} c_{-\mathbf{k}\downarrow}^{\dagger} = a_{\mathbf{k}} + (c_{\mathbf{k}\uparrow}^{\dagger} c_{-\mathbf{k}\downarrow}^{\dagger} - a_{\mathbf{k}}) , \tag{32}$$

where $a_{\mathbf{k}}$ is the *expectation value on the ground state*

$$a_{\mathbf{k}} = \langle \Psi_S | c_{\mathbf{k}\uparrow}^{\dagger} c_{-\mathbf{k}\downarrow}^{\dagger} | \Psi_S \rangle , \tag{33}$$

and $(c_{\mathbf{k}\uparrow}^{\dagger} c_{-\mathbf{k}\downarrow}^{\dagger} - a_{\mathbf{k}})$ is called the *fluctuation operator*. A similar split is performed for the product of any two annihilation operators $c_{-\mathbf{q}\downarrow} c_{\mathbf{q}\uparrow}$. The split in Eq. (32) is useful whenever the fluctuation operators are small, so that it appears plausible to retain in the Hamiltonian (30) only terms up to first order in the fluctuations; in such a situation the resulting simplified model Hamiltonian can be easily diagonalized and the excitation spectrum of the superconductor explicitly worked out.

Let us follow in detail the outlined procedure. The superconductor Hamiltonian (30), using the operatorial identity (32) (and the hermitian conjugate of it), takes the form

$$H_{\mathrm{BCS}} = \sum_{\mathbf{k}\sigma} \varepsilon_{\mathbf{k}} c_{\mathbf{k}\sigma}^{\dagger} c_{\mathbf{k}\sigma} + \sum_{\mathbf{k}\mathbf{k}'} U_{\mathbf{k}\mathbf{k}'} \left[a_{\mathbf{k}} + (c_{\mathbf{k}\uparrow}^{\dagger} c_{-\mathbf{k}\downarrow}^{\dagger} - a_{\mathbf{k}}) \right] \left[a_{\mathbf{k}'} + (c_{-\mathbf{k}'\downarrow} c_{\mathbf{k}'\uparrow} - a_{\mathbf{k}'}) \right]$$

(for simplicity we consider all $a_{\mathbf{k}}$ real, in close analogy to a similar assumption for the amplitudes $u_{\mathbf{k}}$ and $v_{\mathbf{k}}$ of the variational method). Neglecting the presumably small terms that are second order in the fluctuations, we arrive at the simplified Bogoliubov Hamiltonian H_B given by

$$H_B = \sum_{\mathbf{k}\sigma} \varepsilon_{\mathbf{k}} c_{\mathbf{k}\sigma}^{\dagger} c_{\mathbf{k}\sigma} + \sum_{\mathbf{k}\mathbf{k}'} U_{\mathbf{k}\mathbf{k}'} \left[a_{\mathbf{k}} c_{-\mathbf{k}'\downarrow} c_{\mathbf{k}'\uparrow} + a_{\mathbf{k}'} c_{\mathbf{k}\uparrow}^{\dagger} c_{-\mathbf{k}\downarrow}^{\dagger} - a_{\mathbf{k}} a_{\mathbf{k}'} \right] .$$

If we define the parameters

$$\Delta_{\mathbf{k}} = - \sum_{\mathbf{k}'} U_{\mathbf{k}\mathbf{k}'} a_{\mathbf{k}'} , \tag{34}$$

we can write the model Hamiltonian H_B in the form

$$H_B = \sum_{\mathbf{k}\sigma} \varepsilon_{\mathbf{k}} c_{\mathbf{k}\sigma}^{\dagger} c_{\mathbf{k}\sigma} - \sum_{\mathbf{k}} \Delta_{\mathbf{k}} \left[c_{\mathbf{k}\uparrow}^{\dagger} c_{-\mathbf{k}\downarrow}^{\dagger} + c_{-\mathbf{k}\downarrow} c_{\mathbf{k}\uparrow} \right] + \sum_{\mathbf{k}} \Delta_{\mathbf{k}} a_{\mathbf{k}} . \tag{35}$$

The model Hamiltonian (35) can be diagonalized with a suitable canonical transformation. Consider in fact the linear transformations

$$
\begin{array}{ll}
c_{\mathbf{k}\uparrow} = u_{\mathbf{k}} \gamma_{\mathbf{k}\uparrow} + v_{\mathbf{k}} \gamma_{-\mathbf{k}\downarrow}^{\dagger} & c_{-\mathbf{k}\downarrow} = u_{\mathbf{k}} \gamma_{-\mathbf{k}\downarrow} - v_{\mathbf{k}} \gamma_{\mathbf{k}\uparrow}^{\dagger} \\
c_{\mathbf{k}\uparrow}^{\dagger} = u_{\mathbf{k}} \gamma_{\mathbf{k}\uparrow}^{\dagger} + v_{\mathbf{k}} \gamma_{-\mathbf{k}\downarrow} & c_{-\mathbf{k}\downarrow}^{\dagger} = u_{\mathbf{k}} \gamma_{-\mathbf{k}\downarrow}^{\dagger} - v_{\mathbf{k}} \gamma_{\mathbf{k}\uparrow}
\end{array}
\tag{36}
$$

while the inverse transformations are

$$
\begin{array}{ll}
\gamma_{\mathbf{k}\uparrow} = u_{\mathbf{k}} c_{\mathbf{k}\uparrow} - v_{\mathbf{k}} c_{-\mathbf{k}\downarrow}^{\dagger} & \gamma_{-\mathbf{k}\downarrow} = u_{\mathbf{k}} c_{-\mathbf{k}\downarrow} + v_{\mathbf{k}} c_{\mathbf{k}\uparrow}^{\dagger} \\
\gamma_{\mathbf{k}\uparrow}^{\dagger} = u_{\mathbf{k}} c_{\mathbf{k}\uparrow}^{\dagger} - v_{\mathbf{k}} c_{-\mathbf{k}\downarrow} & \gamma_{-\mathbf{k}\downarrow}^{\dagger} = u_{\mathbf{k}} c_{-\mathbf{k}\downarrow}^{\dagger} + v_{\mathbf{k}} c_{\mathbf{k}\uparrow}
\end{array}
\tag{37}
$$

The quantities $u_{\mathbf{k}}$ and $v_{\mathbf{k}}$ are real and even functions of $\mathbf{k}$, and satisfy the constraint $u_{\mathbf{k}}^2 + v_{\mathbf{k}}^2 = 1$. The linear transformations in Eqs. (36) and Eqs. (37) are canonical, since the fermion operators $\gamma_{\mathbf{k}\sigma}$ and $\gamma_{\mathbf{k}\sigma}^{\dagger}$ satisfy the same standard anticommutation rules as those between $c_{\mathbf{k}\sigma}$ and $c_{\mathbf{k}\sigma}^{\dagger}$, given by Eqs. (11). In the particular case of an *ordinary free-electron metal* (where Eqs. 13 apply), the operators $\gamma_{k\sigma}$ and $\gamma_{k\sigma}^{\dagger}$ represent annihilation and creation operators of *free-electrons and free-holes, of wavevector* $\mathbf{k}$ *and spin* σ. In the general case of a superconductor, we will see that the operators $\gamma_{k\sigma}$ and $\gamma_{k\sigma}^{\dagger}$ describe the annihilation and creation of *quasiparticles*, of wavevector $\mathbf{k}$ and spin σ, of the correlated system, i.e. of the interacting electron gas where an appropriate attractive two-body coupling is active among electrons.

We now express the model Hamiltonian (35) in terms of γ operators. For instance, using Eqs. (36) and performing some straightforward commutation of fermion operators, we have

$$c_{\mathbf{k}\uparrow}^{\dagger} c_{\mathbf{k}\uparrow} + c_{-\mathbf{k}\downarrow}^{\dagger} c_{-\mathbf{k}\downarrow} = (u_{\mathbf{k}}^2 - v_{\mathbf{k}}^2)\left[\gamma_{\mathbf{k}\uparrow}^{\dagger}\gamma_{\mathbf{k}\uparrow} + \gamma_{-\mathbf{k}\downarrow}^{\dagger}\gamma_{-\mathbf{k}\downarrow}\right]$$
$$+ 2u_{\mathbf{k}}v_{\mathbf{k}}\left[\gamma_{\mathbf{k}\uparrow}^{\dagger}\gamma_{-\mathbf{k}\downarrow}^{\dagger} + \gamma_{-\mathbf{k}\downarrow}\gamma_{\mathbf{k}\uparrow}\right] + 2v_{\mathbf{k}}^2 .$$

Similarly we obtain

$$c_{\mathbf{k}\uparrow}^{\dagger} c_{-\mathbf{k}\downarrow}^{\dagger} + c_{-\mathbf{k}\downarrow} c_{\mathbf{k}\uparrow} = -2u_{\mathbf{k}}v_{\mathbf{k}}\left[\gamma_{\mathbf{k}\uparrow}^{\dagger}\gamma_{\mathbf{k}\uparrow} + \gamma_{-\mathbf{k}\downarrow}^{\dagger}\gamma_{-\mathbf{k}\downarrow}\right]$$
$$+ (u_{\mathbf{k}}^2 - v_{\mathbf{k}}^2)\left[\gamma_{\mathbf{k}\uparrow}^{\dagger}\gamma_{-\mathbf{k}\downarrow}^{\dagger} + \gamma_{-\mathbf{k}\downarrow}\gamma_{\mathbf{k}\uparrow}\right] + 2u_{\mathbf{k}}v_{\mathbf{k}} .$$

The model Hamiltonian (35) thus becomes

$$H_B = \sum_{\mathbf{k}}\left[\varepsilon_{\mathbf{k}}(u_{\mathbf{k}}^2 - v_{\mathbf{k}}^2) + 2\Delta_{\mathbf{k}}u_{\mathbf{k}}v_{\mathbf{k}}\right]\left[\gamma_{\mathbf{k}\uparrow}^{\dagger}\gamma_{\mathbf{k}\uparrow} + \gamma_{-\mathbf{k}\downarrow}^{\dagger}\gamma_{-\mathbf{k}\downarrow}\right]$$
$$+ \sum_{\mathbf{k}}\left[2\varepsilon_{\mathbf{k}}u_{\mathbf{k}}v_{\mathbf{k}} - \Delta_{\mathbf{k}}(u_{\mathbf{k}}^2 - v_{\mathbf{k}}^2)\right]\left[\gamma_{\mathbf{k}\uparrow}^{\dagger}\gamma_{-\mathbf{k}\downarrow}^{\dagger} + \gamma_{-\mathbf{k}\downarrow}\gamma_{\mathbf{k}\uparrow}\right]$$
$$+ \sum_{\mathbf{k}}\left[2\varepsilon_{\mathbf{k}}v_{\mathbf{k}}^2 - 2\Delta_{\mathbf{k}}u_{\mathbf{k}}v_{\mathbf{k}} + \Delta_{\mathbf{k}}a_{\mathbf{k}}\right] . \tag{38}$$

A direct inspection of the model Hamiltonian (38) shows that it contains the diagonal particle number operators of the type $\gamma^{\dagger}\gamma$, but also the undesired terms of the type $\gamma^{\dagger}\gamma^{\dagger}$ and $\gamma\gamma$. At this stage we choose $u_{\mathbf{k}}$ and $v_{\mathbf{k}}$ so that the coefficient of $\gamma^{\dagger}\gamma^{\dagger}$ and $\gamma\gamma$ are zero and the model Hamiltonian (38) becomes diagonal in the particle number operators; for this aim we require that

$$\boxed{2\varepsilon_{\mathbf{k}}u_{\mathbf{k}}v_{\mathbf{k}} - \Delta_{\mathbf{k}}(u_{\mathbf{k}}^2 - v_{\mathbf{k}}^2) = 0} . \tag{39}$$

Thus we see that equation (39), which expresses the requirement of diagonalization of the model Hamiltonian (38), exactly coincides with the variational condition (19), which expresses energy minimization in the BCS theory. The solution of Eq. (39), together with the normalization condition $u_{\mathbf{k}}^2 + v_{\mathbf{k}}^2 = 1$, gives for $u_{\mathbf{k}}$ and $v_{\mathbf{k}}$ the explicit expressions already reported in Eqs. (20).

Further comparison of the present procedure with the variational procedure, makes transparent the full equivalence of the variational BCS approach and the Bogoliubov approach, based on canonical transformations. Notice in particular that the constant term in the second member of Eq. (38) coincides with the expression of the supercon-

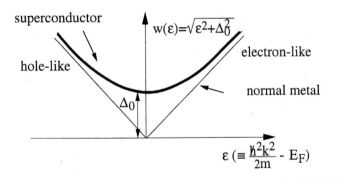

superconductor

$w(\varepsilon) = \sqrt{\varepsilon^2 + \Delta_0^2}$

electron-like

hole-like

normal metal

Δ_0

$\varepsilon \ \left(\equiv \dfrac{\hbar^2 k^2}{2m} - E_F \right)$

Fig. 10 Quasiparticle energy spectrum for the normal metal and for the superconductor at zero temperature.

ductor ground-state W_S, given by Eq. (16). Using Eqs. (20) for $u_{\mathbf{k}}$ and $v_{\mathbf{k}}$, we easily verify that the quasiparticle energies in Eq. (38) become

$$\varepsilon_{\mathbf{k}} \left(u_{\mathbf{k}}^2 - v_{\mathbf{k}}^2 \right) + 2\,\Delta_{\mathbf{k}}\, u_{\mathbf{k}}\, v_{\mathbf{k}} = \sqrt{\varepsilon_{\mathbf{k}}^2 + \Delta_{\mathbf{k}}^2} \ .$$

The model Hamiltonian (38) can be recast in the form

$$H_B = \sum_{\mathbf{k}} \sqrt{\varepsilon_{\mathbf{k}}^2 + \Delta_{\mathbf{k}}^2} \left[\gamma_{\mathbf{k}\uparrow}^{\dagger} \gamma_{\mathbf{k}\uparrow} + \gamma_{-\mathbf{k}\downarrow}^{\dagger} \gamma_{-\mathbf{k}\downarrow} \right] + W_S \quad , \tag{40}$$

where $w_{\mathbf{k}} = \sqrt{\varepsilon_{\mathbf{k}}^2 + \Delta_{\mathbf{k}}^2}$ are the energies, above the ground state, of the *quasiparticles* created by the Fermi operators $\gamma_{\mathbf{k}\uparrow}^{\dagger}$ and $\gamma_{-\mathbf{k}\downarrow}^{\dagger}$.

In order to further clarify the meaning of the Hamiltonian (40) let us consider the *average gap approximation*. In this case, the energy gap parameters $\Delta_{\mathbf{k}}$ are constant and equal to Δ_0 (in the energy shell $\pm\hbar\omega_D$ around the Fermi level). The quasiparticles excitations in the superconductor become

$$w_{\mathbf{k}} = \sqrt{\varepsilon_{\mathbf{k}}^2 + \Delta_0^2} \quad \text{(superconductor)} \ . \tag{41a}$$

In the limiting case of vanishing electron–electron interaction, i.e. for the normal metal, we would have $\Delta_0 = 0$ and hence

$$w_{\mathbf{k}} = |\varepsilon_{\mathbf{k}}| \quad \text{(normal metal)} \ . \tag{41b}$$

In Fig. 10 we show the energies of the elementary excitations in a normal metal and in a superconductor. The quasiparticle spectrum of the superconductor exhibits thus an energy gap given by Δ_0. In the superconductor, there are no electron-like states with energy in the interval $[E_F, E_F + \Delta_0]$, and no hole-like states in the energy interval $[E_F - \Delta_0, E_F]$.

The density of quasiparticle states in the superconductor can be obtained from the dispersion relation (41a) as follows. The constant energy surfaces in the reciprocal space are spheres, and the number of states $D_S(w)\, dw$ in the energy interval $[w, w+dw]$

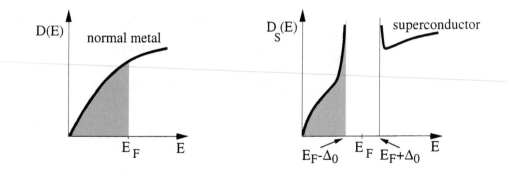

Fig. 11 Schematic density-of-states for quasiparticles in a normal metal and in a superconductor; for sake of clarity of the figure, the gap $2\,\Delta_0$ around E_F has been greatly magnified.

are thus

$$D_S(w)\,dw = \frac{2\,V}{(2\pi)^3}\,4\,\pi\,k^2\,dk = \frac{V\,k_F^2}{\pi^2}\,dk \;, \tag{42a}$$

where V is the volume of the sample, and k has been replaced by k_F, since we consider energy intervals very close to the Fermi surface. From the dispersion relation (41a), we have

$$\frac{dw}{dk} = \frac{\varepsilon}{\sqrt{\varepsilon^2 + \Delta_0^2}}\,\frac{d\varepsilon}{dk} = \frac{\sqrt{w^2 - \Delta_0^2}}{w}\,\frac{\hbar^2\,k_F}{m} \;, \tag{42b}$$

where $d\varepsilon/dk$ has been calculated at the Fermi energy. If we insert Eq. (42b) into Eq. (42a), we obtain

$$D_S(w) = \frac{m\,k_F}{\pi^2\,\hbar^2}\,V\,\frac{w}{\sqrt{w^2 - \Delta_0^2}} = D(E_F)\,\frac{w}{\sqrt{w^2 - \Delta_0^2}} \;, \tag{42c}$$

where $D(E_F) = (m\,k_F/\pi^2\,\hbar^2)\,V$ denotes the density-of-states of the normal metal at the Fermi energy; in fact we have for free electrons

$$D(E_F) = \frac{3}{2}\,\frac{N}{E_F} = \frac{3}{2}\,\frac{1}{E_F}\,n\,V = \frac{3}{2}\,\frac{1}{\hbar^2\,k_F^2/2\,m}\,\frac{k_F^2}{3\,\pi^2}\,V = \frac{m\,k_F}{\pi^2\,\hbar^2}\,V \;.$$

It is convenient at this stage to measure the energy of quasiparticles on a fixed scale (instead of referring them to the Fermi energy). We obtain for the density of quasiparticles in the superconductor the expression

$$D_S(E) = D(E_F)\,\frac{|E - E_F|}{\sqrt{(E - E_F)^2 - \Delta_0^2}} \qquad \text{for} \quad |E - E_F| > \Delta_0 \;. \tag{42d}$$

From Eq. (42d) we see that the density of quasiparticle states in the superconductor presents a singularity at the energies $E_F \pm \Delta_0$; the schematic behaviour of the density-of-states in a normal metal and in a superconductor is shown in Fig. 11.

From the above considerations on quasiparticle excitations, we also infer that breaking a pair, giving rise to two quasiparticles, requires at least an energy $2\,\Delta_0$; the quantity $2\Delta_0$ can be interpreted as the binding energy of any one pair, due to the cooperative presence of many other pairs (all in the same quantum state of zero total spin and zero total momentum) in the superconductor ground state.

4.2 Persistent currents in superconductors

The existence of an energy gap at the Fermi level in the excitation spectrum of a superconductor has profound effects on the transport properties, that are sensitive to the modifications introduced at and near the Fermi level. These include, for instance, perfect conductance, diamagnetic properties, tunneling properties, specific heat, infrared photon absorption, ultrasonic attenuation, transport coefficients; here we discuss the zero electrical resistance of a superconductor.

Until now we have considered the ground state of a superconductor by pairing electrons $(\mathbf{k}\uparrow, -\mathbf{k}\downarrow)$ in states of total spin $S = 0$ and total momentum $\mathbf{K} = \mathbf{k} - \mathbf{k} = 0$. A current carrying state of a superconductor is obtained by pairing the electrons $(\mathbf{k}+\mathbf{Q}\uparrow; -\mathbf{k}+\mathbf{Q}\downarrow)$, and following for the rest the whole BCS microscopic formalism. In fact the total momentum $\hbar\mathbf{K}$, with $\mathbf{K} = (\mathbf{k}+\mathbf{Q}) + (-\mathbf{k}+\mathbf{Q}) = 2\mathbf{Q}$, of the Cooper pairs is a constant of motion, and the kinetic energy associated with the center of mass motion of a Cooper pair is $E_{\text{kin}} = \hbar^2 (2\,Q)^2/4\,m$. For small values of $\mathbf{Q}$ (the ones of practical interest in transport), the kinetic energy E_{kin} is small with respect to the binding energy $2\Delta_0$ of any one pair, and is neglected for simplicity.

Suppose we attempt to degrade the current of the drifting superconducting state by changing the momentum of a number of drifting Cooper pairs $(\mathbf{k}+\mathbf{Q}\uparrow; -\mathbf{k}+\mathbf{Q}\downarrow)$. Due to the cooperative origin of the binding energy of a pair, changing the momentum of a single pair with respect to the common value $2\mathbf{Q}$, requires a cost in energy equal to the binding energy $2\Delta_0$. As soon as the number of pairs with random momenta increases, the energy penalty to sustain the situation becomes prohibitively large: any scattering process, that occasionally breaks or restore Cooper pairs, tends to *restore* the situation in which pairs have the same common momentum. The only process that could degrade current is the *simultaneous scattering of a huge number of drifting Cooper pairs* into a new coherent state, where the scattered pairs have the same final momentum so that the pairing energy gain is at work. It is not easy to imagine how such a scattering event could occur, and therefore our arguments justify the persistent currents in superconductors.

4.3 Electron tunneling into superconductors

A central feature of the BCS theory is the presence of an energy gap in the electron density-of-states of superconductors. A most direct evidence of the gap and of the electron structure of the superconductors is provided by the electron tunneling experiments, initiated by I. Giaever, Phys. Rev. Lett. **5**, 147, 464 (1960).

Consider first a junction constituted by two normal metals, separated by a thin insulating film (typically $10 \approx 50\,\text{Å}$). It is well known that, if a potential difference is applied across the junction, a current flows because of the capability of electrons to

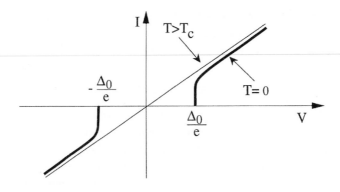

Fig. 12 Schematic representation of the $I - V$ characteristics of a normal metal–insulator–superconductor junction at zero temperature and above the critical temperature; Δ_0 is the energy gap parameter of the superconductor. At finite temperature $0 < T < T_c$ the presence of thermally excited electrons contribute to smear out and shift the strongly nonlinear $T = 0$ current–voltage characteristics.

penetrate a thin barrier. For low fields, the tunneling current is proportional to the applied voltage V. In fact, in "quasi-equilibrium" conditions, the Fermi levels of the two metals are shifted by eV. The density-of-states in the two metals, as well as tunneling probabilities, are practically independent of energy in the few millielectronvolts of interest around E_F, and this leads to the ohmic behaviour of the junction.

Let us now examine the electron tunneling (*Giaever tunneling*) across a junction formed by normal metal–insulator–superconductor (NIS junction). In the ordinary metal (at zero temperature) all states below the Fermi energy E_F are filled, while all states above E_F are empty, with zero gap between occupied and empty states. In the superconductor the quasiparticle energies differ from the Fermi energy at least by the energy gap Δ_0. When a bias potential V is applied to a NIS junction, one-particle states are not available in the superconducting material for accepting or supplying electrons, unless the bias voltage exceeds Δ_0/e. When $V > \Delta_0/e$, the $I - V$ characteristics and in particular the differential conductance $G = dI/dV$ is related to the density-of-states of quasiparticles in the superconductor (since the density-of-states in the metal can be taken as constant in the few millielectronvolts of interest). Finally when $V \gg \Delta_0/e$ the ohmic behaviour of the junction is recovered. These essential features of the $I - V$ characteristics of the NIS junction are schematically indicated in Fig. 12.

Let us now consider the tunneling between two equal superconductors, of energy gap parameter Δ_0, separated by a thin insulating barrier (SIS junction). At $T = 0$ we find that there is no quasiparticle tunneling until the bias voltage V exceeds $2\Delta_0/e$; in the case the two superconductors are different, the threshold voltage is $(\Delta_1 + \Delta_2)/e$. At the threshold voltage, we expect a discontinuous jump of the current, because of the singularity in the density-of-states of quasiparticles in the two superconductors (see Eq. 42); eventually for higher bias voltages the ohmic behaviour is recovered. All these features of the $I - V$ characteristics of the SIS tunnel junction can be clearly seen in the specific example reported in Fig. 13.

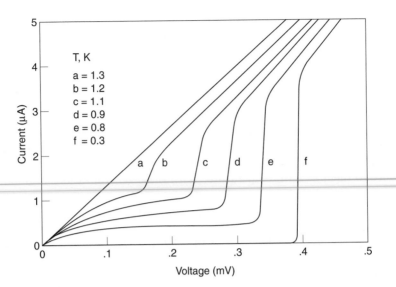

Fig. 13 Typical current–voltage characteristics of an Al-Al$_2$O$_3$-Al superconductor–insulator–superconductor junction (only quasiparticle tunneling current is considered). The onset of the sharp break in the curves for $T < T_c$ occurs when eV $= 2\Delta(T)$; the critical temperature is $T_c = 1.26$ K, and $2\Delta_0/e = 0.379$ mV [from D. N. Langenberg, D. J. Scalapino and B. N. Taylor, Proc. IEEE **54**, 560 (1966); copyright 1996 IEEE].

It is important to notice that in the case of SIS junctions, besides the quasiparticle tunneling current discussed so far, we can have also a supercurrent tunneling due to Cooper pairs transfer between the two superconductors; this current, called Josephson current, can be observed in SIS junctions with extremely thin insulating layers (10–15 Å); in this situation, the coupling between the two superconductors is sufficiently strong that a definite phase relationship between pairs on opposite sides of the insulating barrier can be maintained (and controlled by electromagnetic fields). Some aspects of the Josephson current will be discussed in Section 8.

5 Treatment of superconductors at finite temperature and heat capacity

Self-consistency at finite temperature

At finite temperature the quasiparticle states of the superconductor are thermally excited and a number of Cooper pairs are broken; this process, in turns, is accompanied by a decrease of the energy gap in quasiparticle excitations and eventually leads to the transition to the normal state.

In the BCS variational method, the treatment of superconductors at finite temperature can be done by minimizing the free energy at temperature T. Alternatively, and equivalently, we can extend at finite temperature the approach of Section 4.1; following a procedure familiar in mean field theory of phase transitions, at finite temperature we express the product $c^\dagger_{\mathbf{k}\uparrow} c^\dagger_{-\mathbf{k}\downarrow}$ in the form

$$c^\dagger_{\mathbf{k}\uparrow} c^\dagger_{-\mathbf{k}\downarrow} = a_{\mathbf{k}} + (c^\dagger_{\mathbf{k}\uparrow} c^\dagger_{-\mathbf{k}\downarrow} - a_{\mathbf{k}}) , \tag{43}$$

where $a_{\mathbf{k}}$ is the *thermal average expectation value*

$$a_{\mathbf{k}} = \langle c^\dagger_{\mathbf{k}\uparrow} c^\dagger_{-\mathbf{k}\downarrow} \rangle_T \tag{44}$$

and $(c^\dagger_{\mathbf{k}\uparrow} c^\dagger_{-\mathbf{k}\downarrow} - a_{\mathbf{k}})$ is the *fluctuation operator from the thermal average*. Eq. (43) and Eq. (44) are the trivial generalization at finite temperature of Eq. (32) and Eq. (33), and reduce to them at zero temperature. With these generalizations in mind, we now follow step by step the whole procedure of Section 4.1.

At finite temperature, the generalization of Eq. (34) takes the form

$$\Delta_{\mathbf{k}} = - \sum_{\mathbf{k}'} U_{\mathbf{k}\mathbf{k}'} \langle c_{-\mathbf{k}'\downarrow} c_{\mathbf{k}'\uparrow} \rangle_T . \tag{45}$$

It is convenient to express the product operator $c_{-\mathbf{k}\downarrow} c_{\mathbf{k}\uparrow}$ in terms of the γ operators by means of the canonical transformations (36); we have

$$c_{-\mathbf{k}\downarrow} c_{\mathbf{k}\uparrow} = (u_{\mathbf{k}} \gamma_{-\mathbf{k}\downarrow} - v_{\mathbf{k}} \gamma^\dagger_{\mathbf{k}\uparrow}) (u_{\mathbf{k}} \gamma_{\mathbf{k}\uparrow} + v_{\mathbf{k}} \gamma^\dagger_{-\mathbf{k}\downarrow}) .$$

The thermal average of the above operators gives

$$\langle c_{-\mathbf{k}\downarrow} c_{\mathbf{k}\uparrow} \rangle_T = \langle u_{\mathbf{k}} v_{\mathbf{k}} \gamma_{-\mathbf{k}\downarrow} \gamma^\dagger_{-\mathbf{k}\downarrow} - u_{\mathbf{k}} v_{\mathbf{k}} \gamma^\dagger_{\mathbf{k}\uparrow} \gamma_{\mathbf{k}\uparrow} \rangle_T$$

$$= u_{\mathbf{k}} v_{\mathbf{k}} \langle 1 - \gamma^\dagger_{-\mathbf{k}\downarrow} \gamma_{-\mathbf{k}\downarrow} - \gamma^\dagger_{\mathbf{k}\uparrow} \gamma_{\mathbf{k}\uparrow} \rangle_T = u_{\mathbf{k}} v_{\mathbf{k}} [1 - 2 f(w_{\mathbf{k}})] ,$$

where $w_{\mathbf{k}} = \sqrt{\varepsilon^2_{\mathbf{k}} + \Delta^2_{\mathbf{k}}}$ are the quasiparticle energies, and the Fermi–Dirac function $f(E) = 1/[\exp(\beta E) + 1]$ gives their excitation probability at thermal equilibrium.

At finite temperature the self-consistent equation (45) becomes

$$\Delta_{\mathbf{k}} = - \sum_{\mathbf{k}'} U_{\mathbf{k}\mathbf{k}'} u_{\mathbf{k}'} v_{\mathbf{k}'} [1 - 2 f(w_{\mathbf{k}'})] .$$

Using Eqs. (20) and the identity

$$1 - 2 f(E) \equiv \tanh \frac{\beta E}{2} ,$$

we obtain the self-consistent equation

$$\boxed{\Delta_{\mathbf{k}} = - \frac{1}{2} \sum_{\mathbf{k}'} U_{\mathbf{k}\mathbf{k}'} \frac{\Delta_{\mathbf{k}'}}{\sqrt{\varepsilon^2_{\mathbf{k}'} + \Delta^2_{\mathbf{k}'}}} \tanh \frac{\beta \sqrt{\varepsilon^2_{\mathbf{k}'} + \Delta^2_{\mathbf{k}'}}}{2}} . \tag{46}$$

This is the desired generalization of Eq. (21), and reduces to it at zero temperature.

In the average potential approximation, equation (46) simplifies in the form

$$1 = \frac{1}{2} U_0 \, n_0(E_F) \int_{-\hbar\omega_D}^{\hbar\omega_D} \frac{d\varepsilon}{\sqrt{\varepsilon^2 + \Delta^2}} \tanh \frac{\beta \sqrt{\varepsilon^2 + \Delta^2}}{2} , \tag{47}$$

where $n_0(E_F) = D_0(E_F)/N$ denotes the density-of-states for one spin direction and per unit cell. Eq. (47) provides an implicit relation for the temperature dependence of the gap parameter $\Delta(T)$; for $T \to 0$ we have $\tanh(\beta \sqrt{\varepsilon^2 + \Delta^2}/2) \to 1$ and we recover the value Δ_0 given by Eq. (23b). As T increases, $\tanh(\beta \sqrt{\varepsilon^2 + \Delta^2}/2)$ is always less than 1 and the integral in the second member of Eq. (47) can preserve its constant value only if Δ decreases. Equation (47) thus puts in evidence that in the superconducting state a cooperative decrease of the energy gap parameter occurs as the temperature increases, leading eventually to the disappearance of the energy gap and transition to the normal state.

From Eq. (47) we see that the energy gap disappears at the critical temperature T_c implicitly determined by the relation

$$1 = U_0 \, n_0(E_F) \int_0^{\hbar \omega_D} \frac{1}{\varepsilon} \tanh \frac{\varepsilon}{2 \, k_B \, T_c} \, d\varepsilon \ ,$$

or equivalently, introducing the dimensionless variable $x = \varepsilon/k_B \, T_c$,

$$\int_0^{\hbar \omega_D / k_B \, T_c} \frac{1}{x} \tanh \frac{x}{2} \, dx = \frac{1}{U_0 \, n_0(E_F)} \ .$$

For large ratios $\hbar \omega_D / k_B \, T_c$ the integral has the value $\ln(1.13 \, \hbar \omega_D / k_B \, T_c)$, where 1.13 is an approximation for $2 \exp(\gamma)/\pi$ with γ Euler constant. Then, in the weak coupling limit $U_0 \, n_0(E_F) \ll 1$ and $\hbar \omega_D / k_B \, T_c \gg 1$, we have

$$k_B \, T_c = 1.13 \, \hbar \omega_D \, e^{-1/U_0 \, n_0(E_F)} \ .$$

Inserting this result into Eq. (23b) we have

$$\boxed{\Delta(0) = 1.76 \, k_B \, T_c} \quad . \tag{48a}$$

The behaviour of $\Delta(T)$ as a function of T as obtained by numerical integration of Eq. (47) is shown in Fig. 14. In particular for $T \to T_c$ (and $T < T_c$) the form of $\Delta(T)$ is found to be

$$\Delta(T) = 3.06 \, k_B \, T_c \left(1 - \frac{T}{T_c} \right)^{1/2} , \tag{48b}$$

and the exponent $1/2$ is a characteristic feature of the mean field theories.

Electronic heat capacity for a superconductor

We can now discuss the electronic heat capacity of a superconductor. The transition to the superconducting state is accompanied by a quite drastic change of the electronic contribution to the heat capacity; the characteristic behaviour is indicated in Fig. 15. There is a sharp jump of the heat capacity of the superconductor at the critical temperature. For $T > T_c$ the electronic heat capacity of the normal material is linear with temperature (as studied in Section III-3). For $T \approx 0$ the heat capacity decays exponentially to zero; the exponential behaviour at low temperature is of the form expected when an energy gap exists in the energy spectrum of quasiparticles.

To calculate the electronic heat capacity of a superconductor, we start from the

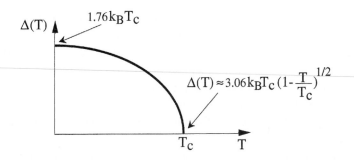

Fig. 14 Behaviour of the energy gap parameter $\Delta(T)$ for a superconductor in the BCS theory and weak coupling limit.

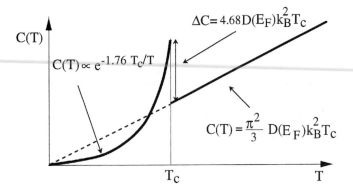

Fig. 15 Characteristic behaviour of the heat capacity of a normal metal and a superconductor, above and below the critical temperature.

general expression of the entropy $S(T)$ of a system of fermionic quasiparticles (see Eq. III-B8); in the present case we have

$$S(T) = -2\,k_B \sum_{\mathbf{k}} [f_{\mathbf{k}} \ln f_{\mathbf{k}} + (1 - f_{\mathbf{k}}) \ln(1 - f_{\mathbf{k}})] \ , \qquad (49a)$$

where the factor 2 takes into account the two spin orientations, $f_{\mathbf{k}}$ is the Fermi–Dirac distribution function

$$f_{\mathbf{k}} = \frac{1}{\exp(\beta\,w_{\mathbf{k}}) + 1} \qquad \text{and} \qquad w_{\mathbf{k}} = \sqrt{\varepsilon_{\mathbf{k}}^2 + \Delta^2(T)} \ .$$

The heat capacity is given by the expression

$$C(T) = T\,\frac{dS(T)}{dT} \ , \qquad (49b)$$

according to Eq. (III-23b).

We notice that

$$\frac{dS(T)}{dT} = \sum_{\mathbf{k}} \frac{\partial S}{\partial f_{\mathbf{k}}}\,\frac{\partial f_{\mathbf{k}}}{\partial T} \ .$$

We have

$$\frac{\partial S}{\partial f_{\mathbf{k}}} = -2\,k_B\,\ln\frac{f_{\mathbf{k}}}{1 - f_{\mathbf{k}}} = 2\,\frac{1}{T}\,\sqrt{\varepsilon_{\mathbf{k}}^2 + \Delta^2(T)}$$

and

$$\frac{\partial f_{\mathbf{k}}}{\partial T} = \frac{1}{k_B\,T^2}\,\frac{\exp(\beta\,w_{\mathbf{k}})}{[\exp(\beta\,w_{\mathbf{k}}) + 1]^2}\,\left[\sqrt{\varepsilon_{\mathbf{k}}^2 + \Delta^2(T)} - T\,\frac{d}{dT}\,\sqrt{\varepsilon_{\mathbf{k}}^2 + \Delta^2(T)}\right]\,.$$

Eq. (49b) thus gives

$$C(T) = \frac{2}{k_B T^2}\,\sum_{\mathbf{k}}\,\frac{\exp(\beta w_{\mathbf{k}})}{[\exp(\beta w_{\mathbf{k}}) + 1]^2}\,\left[\varepsilon_{\mathbf{k}}^2 + \Delta^2(T) - \frac{T}{2}\,\frac{d}{dT}\Delta^2(T)\right]\,. \tag{50a}$$

As usual, the sum over $\mathbf{k}$ in the reciprocal space can be replaced by the integral in the energy variable ε multiplied by the density-of-states $D_0(E_F)$ (the density-of-states is taken as constant in the small energy shell of interest around the Fermi level). Extending to infinity the limits of integration one obtains

$$C(T) = \frac{2\,D_0(E_F)}{k_B\,T^2}\,\int_{-\infty}^{\infty}\,\frac{\exp\left(\beta\sqrt{\varepsilon^2 + \Delta^2(T)}\right)}{\left[\exp\left(\beta\sqrt{\varepsilon^2 + \Delta^2(T)}\right) + 1\right]^2}\,\left[\varepsilon^2 + \Delta^2(T) - \frac{T}{2}\,\frac{d}{dT}\Delta^2(T)\right]\,d\varepsilon\,.$$

$$\tag{50b}$$

From Eq. (50b) the behaviour of the heat capacity as a function of temperature can be worked out. In particular for $T > T_c$ we have $\Delta(T) = 0$ and Eq. (50b) becomes

$$C(T) = \frac{4}{k_B\,T^2}\,D_0(E_F)\,\int_0^{\infty}\,\frac{e^{\beta\varepsilon}}{(e^{\beta\varepsilon} + 1)^2}\,\varepsilon^2\,d\varepsilon$$

$$= 4\,k_B^2\,T\,D_0(E_F)\,\int_0^{\infty}\,\frac{e^x}{(e^x + 1)^2}\,x^2\,dx = \frac{\pi^2}{3}\,D(E_F)\,k_B^2\,T\,,$$

which is the standard result for a normal metal (see Eq. III-25).

For $T = T_c$ a discontinuity occurs in the heat capacity because of the term involving $d\Delta^2(T)/dT$ which equals $-9.36\,k_B^2\,T_c$ (from Eq. 48b). From Eq. (50b) the value of this discontinuity is

$$\Delta C(T_c) = 9.36\,k_B^2\,T_c\,D_0(E_F)\,2\,\int_0^{\infty}\,\frac{e^x}{(e^x + 1)^2}\,dx = 4.68\,D(E_F)\,k_B^2\,T_c\,.$$

At the critical temperature, the heat capacity jump $C_s - C_n$ from the superconductor to the normal metal, divided by the normal metal heat capacity, gives the parameter free expression

$$\frac{C_s - C_n}{C_n} = \frac{4.68 \cdot 3}{\pi^2} = 1.42\,.$$

Finally, for $T \ll T_c$, Eq. (50b) exhibits an exponential decay for the superconductor heat capacity of the form $C(T) \approx \exp(-\beta\,\Delta(0)) = \exp(-1.76\,T_c/T)$.

6   Diamagnetism of superconductors and Meissner effect

6.1   The phenomenological London model

Some phenomenological aspects concerning the behaviour of superconductors in the presence of magnetic field have been presented in Section 1. We have seen that a distinctive feature of superconductivity is the Meissner effect: a (weak) magnetic field cannot penetrate in the bulk of the superconductor. The first phenomenological model to explain the Meissner effect was given by the brothers F. and H. London; it is instructive to report from their model the aspects useful to clarify the distinctive transport equations governing *ordinary conductors*, *perfect conductors* and *superconductors* (here, phenomenologically, by *superconductor* it is meant a *perfect conductor which also exhibits the Meissner effect*).

Consider first an *ordinary conductor* with n electrons per unit volume. In an ordinary conductor, the electrons can be scattered by impurities, phonons, and other defects; the carriers can be characterized with some relaxation time τ, and a resistivity $\rho = m/ne^2\tau$. In the presence of an electric field $\mathbf{E}$, the current density $\mathbf{J}$ is given by the Ohm law $\mathbf{E} = \rho\mathbf{J}$.

Consider now a *perfect conductor* with n electrons per unit volume; by perfect we mean that the electrons are not scattered by impurities, phonons or other defects. In the presence of an electric field $\mathbf{E}$, the motion of each freely moving electron is described by the (ballistic) dynamic equation

$$(-e)\,\mathbf{E} = m\,\frac{d\mathbf{v}}{dt}\ .$$

The current density $\mathbf{J} = n\,(-e)\,\mathbf{v}$ is then related to the electric field by the equation

$$\boxed{\ \mathbf{E} = \frac{m}{n\,e^2}\,\frac{\partial\mathbf{J}}{\partial t}\ }\ . \tag{51}$$

This is known as the *first London equation* and it is written on account of *perfect conductivity*. Notice that in the case of a standard conductor, with a (finite) relaxation time τ, Eq. (51) is replaced by $\mathbf{E} = (m/ne^2\tau)\mathbf{J} \equiv \rho\mathbf{J}$ (Ohm's law); thus the first London equation is simply a modification of the Ohm law needed to describe the dynamics of collisionless electrons.

Let us insert the first London equation (51) into the Maxwell equation $\operatorname{curl}\mathbf{E} = -(1/c)\,\partial\mathbf{B}/\partial t$; we obtain

$$\frac{\partial}{\partial t}\left[\operatorname{curl}\mathbf{J} + \frac{n\,e^2}{m\,c}\,\mathbf{B}\right] = 0\ . \tag{52}$$

This is a general equation for any ideal perfect conductor, but does not account "per se" for the Meissner effect; for instance, Eq. (52) is compatible with a static situation characterized by a uniform and constant magnetic field $\mathbf{B} \neq 0$ in the bulk of the sample, and current density $\mathbf{J} = 0$ (since $\operatorname{curl}\mathbf{B} = 4\pi\,\mathbf{J}/c \equiv 0$).

It has been pointed out by the London brothers that, if the quantity in square brackets in Eq. (52) is not only time-independent (as stated by Eq. 52), but actually *vanishes identically*, then the ideal perfect conductor also exhibits the Meissner effect, i.e. the property of excluding fields and currents from its interior. The equation

$$\operatorname{curl} \mathbf{J} = -\frac{n\,e^2}{m\,c}\,\mathbf{B} \tag{53}$$

is called the *second London equation*, and appeared originally in the literature as a conjecture of the London brothers for superconductors. According to Eq. (53), in any superconducting region where $\mathbf{J}$ vanishes also $\mathbf{B}$ vanishes.

The first London equation (51) implies, in particular, that a perfect conductor in stationary conditions cannot sustain an electric field. It is easily seen that the second London equation (53) implies, in particular, that a superconductor in stationary conditions cannot sustain a magnetic field in its interior, except for a thin surface layer whose depth can be calculated as follows. We combine Eq. (53) with the Maxwell equation (in stationary conditions)

$$\operatorname{curl} \mathbf{B} = \frac{4\pi}{c}\,\mathbf{J}\;; \tag{54}$$

we also exploit the operator identity $\operatorname{curl}\operatorname{curl} = \operatorname{grad}\operatorname{div} - \nabla^2$, and the relationships $\operatorname{div}\mathbf{B} = 0$ and $\operatorname{div}\mathbf{J} = 0$ (in stationary situations). We obtain

$$\nabla^2\,\mathbf{B}(\mathbf{r}) = \frac{1}{\lambda_L^2}\,\mathbf{B}(\mathbf{r}) \tag{55a}$$

$$\nabla^2\,\mathbf{J}(\mathbf{r}) = \frac{1}{\lambda_L^2}\,\mathbf{J}(\mathbf{r}) \tag{55b}$$

where the *London penetration length* λ_L is given by

$$\lambda_L = \frac{c}{\omega_p} = \sqrt{\frac{m\,c^2}{4\pi\,n\,e^2}} \tag{56}$$

and ω_p is the plasma frequency corresponding to the electron density n. For electron concentrations typical of metals, the magnetic penetration length λ_L is of the order of $10^2 \approx 10^3\,\text{Å}$.

We solve now the above equations (55) for a semi-infinite slab of superconductor with the geometry indicated in Fig. 16. We consider two simple and particularly significant situations.

(i) The magnetic field $\mathbf{B}$ is parallel to the z-axis and homogeneous in the xy plane. This means that $\mathbf{B}$ can be written in the form $\mathbf{B} = (0, 0, B(z))$. The equation $\operatorname{div}\mathbf{B} = 0$ gives $\partial B(z)/\partial z = 0$, and thus $B(z)$ is constant. Equation (55a) gives $B(z) = 0$ for $z > 0$ and shows that it is impossible to have a magnetic field normal to the superconductor surface.

(ii) The magnetic field $\mathbf{B}$ is parallel to the x-axis and homogeneous in the xy plane (see Fig. 16). This means that $\mathbf{B}$ can be written in the form $\mathbf{B} = (B(z), 0, 0)$. The

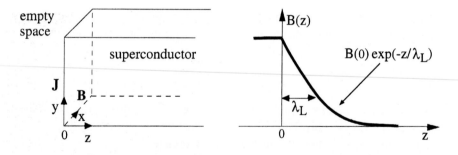

Fig. 16 Semi-infinite slab geometry for illustration of screening currents and penetration depth of the magnetic field parallel to the surface.

equation div $\mathbf{B} = 0$ is now automatically satisfied. From Eq. (55a) we have

$$\frac{\partial^2 B(z)}{\partial z^2} = \frac{1}{\lambda_L^2} B(z) ,$$

whose solution is

$$B(z) = B(0)\, e^{-z/\lambda_L} \quad (z > 0) .$$

Thus, except for a narrow surface region of width of the order of λ_L, the magnetic field parallel to the surface cannot penetrate into a superconductor (Meissner effect).

The London gauge for superconductors

Before concluding this section, we wish to consider some other consequences entailed by the fundamental equation (53). For simplicity we assume stationary situation. Equation (53) can be elaborated in a more convenient form, by expressing the magnetic induction $\mathbf{B}(\mathbf{r})$ in terms of the vector potential $\mathbf{A}(\mathbf{r})$ associated to it through the relation

$$\mathbf{B}(\mathbf{r}) = \operatorname{curl}\mathbf{A}(\mathbf{r}) . \tag{57}$$

It is well known that the vector potential $\mathbf{A}(\mathbf{r})$ is defined within gauge transformations of the type $\mathbf{A}'(\mathbf{r}) = \mathbf{A}(\mathbf{r}) + \nabla\phi(\mathbf{r})$, where $\phi(\mathbf{r})$ is any regular and single-valued function. We now insert Eq. (57) into Eq. (53), and we obtain

$$\operatorname{curl}\left[\mathbf{J}(\mathbf{r}) + \frac{n\,e^2}{m\,c}\,\mathbf{A}(\mathbf{r})\right] = 0 \quad (\mathbf{r}\ \text{within the superconductor}) . \tag{58}$$

The vector $\mathbf{J}(\mathbf{r}) + (n\,e^2/m\,c)\,\mathbf{A}(\mathbf{r})$ is irrotational within the volume of the superconductor and can thus be expressed as $-\nabla\chi(\mathbf{r})$; notice that $\chi(\mathbf{r})$ is a regular single-valued function in the case of a simply-connected geometry, while it is a multiply-valued function in the case of a multiply-connected geometry (as illustrated in Fig. 17). We can thus write

$$\boxed{\mathbf{J}(\mathbf{r}) = -\frac{n\,e^2}{m\,c}\,\mathbf{A}(\mathbf{r}) - \nabla\chi(\mathbf{r})} \ . \tag{59}$$

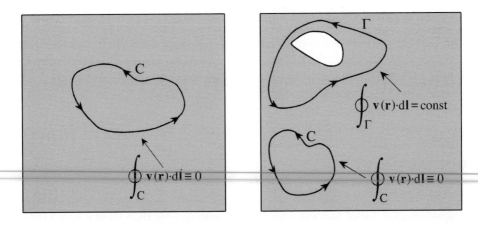

Fig. 17 Schematic illustration of a simply-connected domain and multiply-connected domain wherein it is defined an irrotational vector field, i.e. a field $\mathbf{v}(\mathbf{r})$ such that curl $\mathbf{v}(\mathbf{r}) \equiv 0$. In Fig. (17a), we have $\oint_C \mathbf{v}(\mathbf{r}) \cdot d\mathbf{l} \equiv 0$ for any closed path C within the connected domain. In Fig. (17b), the circuitation $\oint_C \mathbf{v}(\mathbf{r}) \cdot d\mathbf{l}$ vanishes for any circuit C that does not enclose the hole, while $\oint_\Gamma \mathbf{v}(\mathbf{r}) \cdot d\mathbf{l} \equiv$ const (in general different from zero) for any closed circuit Γ, that encloses the hole. In either cases, we can construct the function $\chi(\mathbf{r}) = \int_{P_0}^{P(\mathbf{r})} \mathbf{v}(\mathbf{r}') \cdot d\mathbf{l}$ and write $\mathbf{v}(\mathbf{r}) = -\nabla \chi(\mathbf{r})$; in the simply-connected geometry $\oint_C \mathbf{v}(\mathbf{r}) \cdot d\mathbf{l} \equiv 0$ and $\chi(\mathbf{r})$ is an ordinary single-valued function; in the multiply-connected geometry $\oint_C \mathbf{v}(\mathbf{r}) \cdot d\mathbf{l} \neq 0$ (in general) and $\chi(\mathbf{r})$ is a multiply-valued function.

In the case of a simply-connected geometry, $\chi(\mathbf{r})$ is a regular one-valued function, and we can exploit the arbitrariness in $\mathbf{A}(\mathbf{r})$ to embody $\nabla \chi(\mathbf{r})$ into $\mathbf{A}(\mathbf{r})$ itself; thus we arrive at the London equation

$$\boxed{\mathbf{J}(\mathbf{r}) = -\frac{n\,e^2}{m\,c}\,\mathbf{A}(\mathbf{r})} \quad . \tag{60}$$

Eq. (60) shows that it is always possible, for *connected superconductors*, to construct a gauge in which $\mathbf{J}(\mathbf{r})$ and $\mathbf{A}(\mathbf{r})$ are proportional (*this gauge is called London gauge*).

Notice that Eq. (60) is not gauge-invariant and holds exclusively in the London gauge; this particular choice of gauge is uniquely specified by the following requirements (i) curl $\mathbf{A}(\mathbf{r}) = \mathbf{B}(\mathbf{r})$, (ii) div $\mathbf{A}(\mathbf{r}) = 0$ (so that div $\mathbf{J} = 0$ in stationary conditions), (iii) the normal component of $\mathbf{A}$ at the surface is assigned (and equal to the normal component of $\mathbf{J}$ times $-(m\,c/n\,e^2)$. [It is evident that if one goes from one gauge $\mathbf{A}(\mathbf{r})$ to another gauge $\mathbf{A}'(\mathbf{r})$, the vector field $\mathbf{v}(\mathbf{r}) = \mathbf{A}(\mathbf{r}) - \mathbf{A}'(\mathbf{r})$ must satisfy (j) curl $\mathbf{v}(\mathbf{r}) = 0$, (jj) div $\mathbf{v}(\mathbf{r}) = 0$, (jjj) normal component of $\mathbf{v}(\mathbf{r})$ at the surface equal to zero. Because of (jj) and (jjj), the force lines of $\mathbf{v}(\mathbf{r})$ are closed within the connected region; then (j) entails that $\mathbf{v}(\mathbf{r})$ vanishes identically, so that $\mathbf{A}(\mathbf{r}) \equiv \mathbf{A}'(\mathbf{r})$ is uniquely determined]. In the case of multiply-connected geometry, the generalized equation (59) must be considered.

6.2   Pippard electrodynamics and effective magnetic penetration depth

In the London model, the electrodynamics of a (simple connected) superconductors is determined by the equation $\mathbf{J}(\mathbf{r}) = -(n\,e^2/m\,c)\,\mathbf{A}(\mathbf{r})$. The London electrodynamics is a *local one*: the current density at some point depends on the vector potential at the same point. The penetration depth of fields tangential to the surface occurs in a length given by the London length λ_L.

The microscopic justification of Eq. (60) stands in the "rigidity" of the superconducting ground-state wavefunction under a perturbation corresponding to a slowly varying vector potential $\mathbf{A}(\mathbf{r})$; in fact the mixing of the ground-state of the superconductor with its excited states tends to be suppressed because of the combined effect of the presence of an energy gap and of the structure of the matrix elements of the perturbation. A detailed analysis based on the microscopic BCS theory shows that the London equation (60) holds in the case $\mathbf{A}(\mathbf{r})$ is slowly varying on the length scale given by the coherence length ξ_0 of the Cooper pair. In the case $\mathbf{A}(\mathbf{r})$ varies rapidly in space (i.e. its Fourier components with $q > 1/\xi_0$ cannot be neglected), the microscopic BCS theory well justifies the non-local phenomenological equation introduced by Pippard

$$\mathbf{J}(\mathbf{r}) = -\frac{n\,e^2}{m\,c}\,\frac{3}{4\,\pi\,\xi_0}\,\int\frac{[\mathbf{A}(\mathbf{r}_1)\cdot(\mathbf{r}-\mathbf{r}_1)]\,(\mathbf{r}-\mathbf{r}_1)}{(\mathbf{r}-\mathbf{r}_1)^4}\,e^{-|\mathbf{r}-\mathbf{r}_1|/\xi_0}\,d\mathbf{r}_1\;. \tag{61}$$

In Eq. (61) n is the total density of conduction electrons (at $T = 0$ all conduction electrons participate to the screening supercurrents); the characteristic length ξ_0 represents the coherence length of the Cooper pair and is given by Eq. (29). The Pippard equation (61) shows that the current density is given by a weighted integral of the vector potential over a distance of the order of the correlation length (the energy gap Δ_0 at the Fermi level is responsible for this *non-local* behaviour).

In the case $\mathbf{A}(\mathbf{r})$ is slowly varying on the length scale ξ_0, we can take $\mathbf{A}(\mathbf{r})$ out of the integral in the Pippard equation (61). We obtain

$$\mathbf{J}(\mathbf{r}) = -\frac{n\,e^2}{m\,c}\,\mathbf{A}(\mathbf{r})\,\frac{3}{4\,\pi\,\xi_0}\,\int\frac{x_1^2}{r_1^4}\,e^{-r_1/\xi_0}\,d\mathbf{r}_1 = -\frac{n\,e^2}{m\,c}\,\mathbf{A}(\mathbf{r})\;.$$

We recover thus the London equation (60) and the penetration depth λ_L is given by Eq. (56). Superconductors for which $\lambda_L \gg \xi_0$ are well described by the local London electrodynamics, and are called *London superconductors*.

We consider now superconductors for which $\lambda_L \ll \xi_0$ (*Pippard superconductors*). In this case the Pippard equation (61) can be integrated (with some labour) for the slab geometry to obtain the effective penetration depth λ_{eff} of fields and currents. We can however give an estimation of the penetration depth with the following argument by Pippard. When the penetration depth $\lambda_{\mathrm{eff}} \ll \xi_0$ and one attempts to extract $\mathbf{A}(\mathbf{r})$ from the integral in Eq. (61), one recovers a modified London equation of the type

$$\mathbf{J}(\mathbf{r}) \approx -\frac{n\,e^2}{m\,c}\,\frac{\lambda_{\mathrm{eff}}}{\xi_0}\,\mathbf{A}(\mathbf{r})\;. \tag{62a}$$

The reduction factor $\lambda_{\mathrm{eff}}/\xi_0$ appears because $\mathbf{A}(\mathbf{r})$ is different from zero only in a

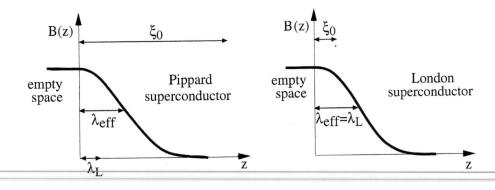

Fig. 18 Qualitative distinction between Pippard and London superconductors; $B(z)$ denotes the magnetic field parallel to the surface, in the geometry indicated in Fig. 16

surface layer of thickness λ_{eff}; thus the integration in the direction normal to the surface must be confined to a length of the order λ_{eff} (instead of extending to a depth of order ξ_0). The modified local relationship (62a), when combined with the Maxwell equation (54), gives for the effective penetration depth λ_{eff} the equation

$$\lambda_{\text{eff}}^2 = \frac{m c^2}{4 \pi n e^2} \frac{\xi_0}{\lambda_{\text{eff}}} = \lambda_L^2 \frac{\xi_0}{\lambda_{\text{eff}}} . \tag{62b}$$

It follows

$$\lambda_{\text{eff}} \approx \lambda_L \left(\frac{\xi_0}{\lambda_L} \right)^{1/3} \tag{62c}$$

and the actual penetration depth λ_{eff} exceeds λ_L by a factor of order $(\xi_0/\lambda_L)^{1/3}$. The qualitative distinction between Pippard and London superconductors is schematically shown in Fig. 18.

Until now we have considered ideal superconductors (*clean superconductors*), in which the ordinary mean free path l_e of electrons in the normal metal is understood to be much larger than the coherence length of the superconductor. The Pippard relation (61), which is strictly valid for bulk and clean superconductors has been extended to account for the presence of impurities. We expect that the weighting factor $\exp(-|\mathbf{r} - \mathbf{r}_1|/\xi_0)$ is depressed by the quantity $\exp(-|\mathbf{r} - \mathbf{r}_1|/l_e)$. We can thus guess that the Pippard equation for *dirty superconductors* takes the form

$$\mathbf{J}(\mathbf{r}) = -\frac{n e^2}{m c} \frac{3}{4 \pi \xi_0} \int \frac{[\mathbf{A}(\mathbf{r}_1) \cdot (\mathbf{r} - \mathbf{r}_1)] (\mathbf{r} - \mathbf{r}_1)}{(\mathbf{r} - \mathbf{r}_1)^4} e^{-|\mathbf{r} - \mathbf{r}_1|/\xi_P} d\mathbf{r}_1 , \tag{63a}$$

where the Pippard correlation length is defined as

$$\frac{1}{\xi_p} = \frac{1}{\xi_0} + \frac{1}{l_e} . \tag{63b}$$

Let us consider the extremely dirty limit $l_e \ll \xi_0$; in this case $\xi_p \approx l_e$. In the case $\mathbf{A}(\mathbf{r})$ is slowly varying on the scale of l_e, one can extract $\mathbf{A}(\mathbf{r})$ from the integral in

Eq. (63a) and obtains

$$\mathbf{J}(\mathbf{r}) = -\frac{n e^2}{m c} \mathbf{A}(\mathbf{r}) \frac{3}{4 \pi \xi_0} \int \frac{x_1^2}{r_1^4} e^{-r_1/l_e} d\mathbf{r}_1 = -\frac{n e^2}{m c} \frac{l_e}{\xi_0} \mathbf{A}(\mathbf{r}) . \tag{63c}$$

The penetration depth corresponding to the Eq. (63c) is

$$\lambda_{\text{eff}} \approx \lambda_L \sqrt{\frac{\xi_0}{l_e}} .$$

Thus we see that in dirty superconductors we can greatly increase the penetration depth (leaving almost unchanged all the other parameters of the sample), and even transform Pippard superconductors into London superconductors. Alloying is the most usual technique to increase the penetration depth; even a small amount of non-magnetic impurities may strongly influence the electron mean free path.

For what concerns the temperature dependence of the penetration depth, in a number of situations it can be described by the semi-empirical form

$$\lambda_{\text{eff}}(T) = \lambda_{\text{eff}}(0) \frac{1}{\sqrt{1 - (T/T_c)^4}} ; \tag{64}$$

from it, we expect that $\lambda_{\text{eff}}(T) \rightarrow (T_c - T)^{-1/2}$ for T approaching T_c from below.

7   Macroscopic quantum phenomena

7.1   Order parameter in superconductors and Ginzburg–Landau theory

The BCS microscopic theory of superconductivity is very illuminating but also rather demanding in the formal and technical aspects, in spite of the fact that the systems considered up to now are bulk and clean homogeneous materials. In several problems of remarkable theoretical and technological interest, one has to consider non homogeneous systems with boundary effects, impurity effects, space-varying Cooper pairs density and magnetic fields, etc. Important physical effects near the normal metal-superconductor phase transition can be well described and understood within the Ginzburg–Landau phenomenological theory of second-order phase transitions. This approach is an ingenious body of assumptions, which are based on the phenomenology of superconductors and some findings of the microscopic theory, avoiding however inessential details.

Let us specify some elements of the theory with reference to the transition to the superconducting phase. For the description of the superconducting phase, Ginzburg and Landau introduce a macroscopic complex quantity, the *order parameter* ψ, which characterizes the ordering which is reached passing from the disordered phase (for $T > T_c$) to the ordered one (for $T < T_c$). It is assumed that the complex order parameter $\psi(\mathbf{r})$ at a given temperature T in the superconducting phase is related to the local number of Cooper pairs by the relation

$$n_{\text{pairs}}(\mathbf{r}) = |\psi(\mathbf{r})|^2 .$$

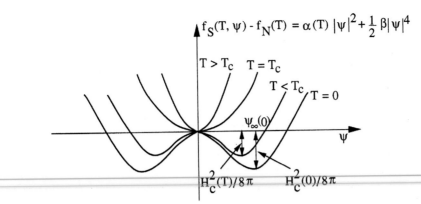

Fig. 19 Schematic behaviour of the Ginzburg–Landau free-energy change (in the absence of fields and gradients) for a typical second-order phase transition.

The basic assumption of the Ginzburg–Landau theory is that the superconductor free energy density $f_S(T, \psi)$, in the temperature region around the transition temperature T_c, is a functional of the order parameter of the form

$$f_S(T, \psi) = f_N(T) + \alpha(T)\,|\psi|^2 + \frac{1}{2}\,\beta(T)\,|\psi|^4 + \frac{1}{2m^*}\left|\left(\mathbf{p} - \frac{e^*}{c}\mathbf{A}\right)\psi(\mathbf{r})\right|^2, \qquad (65)$$

where $f_N(T)$ is the free energy density in the normal phase ($\psi = 0$). This form, suggested by physical intuition, can be made plausible by the following considerations.

Consider first a bulk superconductor in the absence of fields ($\mathbf{A}(\mathbf{r}) = 0$) and gradients ($\psi(\mathbf{r})$=const, and thus $\mathbf{p}\,\psi(\mathbf{r}) = -i\hbar\nabla\psi(\mathbf{r}) = 0$). Eq. (65) becomes in this case

$$f_S(T, \psi) - f_N(T) = \alpha(T)\,|\psi|^2 + \frac{1}{2}\,\beta(T)\,|\psi|^4 . \qquad (66)$$

This can be recognized as an expansion of the free energy change up to second order in $\psi\psi^*$, and the neglecting of higher order terms seems reasonable in the vicinity of the critical temperature, where $\psi\psi^*$ becomes small. The two phenomenological quantities α and β must satisfy $\beta(T) > 0$ and $\alpha(T) > 0$ for $T > T_c$, and $\beta(T) > 0$ and $\alpha(T) < 0$ for $T < T_c$; in fact only with these constraints, the free energy change of Eq. (66) admits a negative minimum, at a finite value $|\psi|$, for $T < T_c$ (see Fig. 19). The simplest possible temperature dependence (corroborated also by experimental evidence) is just $\beta(T)$ positive (and leading term independent on T for $T \approx T_c$) and $\alpha(T)$ proportional to $T - T_c$ for $T \approx T_c$; namely

$$\begin{cases} \alpha(T) = \alpha_0\,(T - T_c) & \text{with } \alpha_0 > 0 \quad \text{for } T \approx T_c \\ \beta(T) = \beta_0 & \text{with } \beta_0 > 0 \quad \text{for } T \approx T_c \end{cases} . \qquad (67)$$

According to Eq. (67), in a temperature range reasonably near to T_c, we consider $\alpha(T)$ linear in $T - T_c$, and disregard the temperature dependence of β.

The last term in the second member of Eq. (65) is written as if $\psi(\mathbf{r})$ represents a true quantum mechanical wavefunction for a particle of charge e^* and mass m^*. From the comparison with the BCS microscopic theory, we identify the charge e^* of the Cooper pair with $-2\,e$ (e is the absolute value of the electronic charge); ignoring effective mass effects, we also identify $m^* = 2m$.

At this stage we can go back to Eq. (65) and minimize $f_S(T, \psi)$ with respect to arbitrary variations of ψ (or ψ^*); the variational calculation shows that the order parameter function ψ satisfies the Ginzburg–Landau equation

$$\frac{1}{2\,m^*} \left(\mathbf{p} - \frac{e^*}{c} \mathbf{A} \right)^2 \psi(\mathbf{r}) + \beta \, |\psi(\mathbf{r})|^2 \, \psi(\mathbf{r}) = -\alpha(T) \, \psi(\mathbf{r}) \qquad . \tag{68}$$

The Ginzburg–Landau equation is complemented by the expression of the supercurrent density in terms of $\psi(\mathbf{r})$ (in the way usual in quantum mechanics)

$$\mathbf{J}_S(\mathbf{r}) = \frac{-i\,\hbar\,e^*}{2\,m^*} \left[\psi^*(\mathbf{r}) \, \nabla \psi(\mathbf{r}) - \psi(\mathbf{r}) \, \nabla \psi^*(\mathbf{r}) \right] - \frac{e^{*2}}{m^*\,c} \, |\psi(\mathbf{r})|^2 \, \mathbf{A}(\mathbf{r}) \qquad . \tag{69}$$

Equation (69) assumes a particularly significant form if we write the complex macroscopic wavefunction $\psi(\mathbf{r})$ in the polar form

$$\psi(\mathbf{r}) = |\psi(\mathbf{r})| \, e^{i\,\theta(\mathbf{r})} \,, \tag{70a}$$

where $|\psi(\mathbf{r})|$ is the modulus and $\theta(\mathbf{r})$ the phase of the wavefunction. Using Eq. (70a), the Ginzburg–Landau equation (69) for the current density becomes

$$\mathbf{J}_S(\mathbf{r}) = |\psi(\mathbf{r})|^2 \left[\frac{e^*\,\hbar}{m^*} \, \nabla \theta(\mathbf{r}) - \frac{e^{*2}}{m^*\,c} \, \mathbf{A}(\mathbf{r}) \right] \,. \tag{70b}$$

Intuitively, the order parameter $\psi(\mathbf{r})$ can be interpreted as the wavefunction which microscopically describes the motion of the center of mass of a Cooper pair (and of all the other Cooper pairs in the superconductor due to their condensation into the same "two-electron" quantum state): the magnitude $|\psi(\mathbf{r})|^2$ describes the local density of Cooper pairs, while the phase $\theta(\mathbf{r})$ describes locally the motion of the center of mass (see Eq. 70b when fields are negligible).

We now examine more closely the structure of the Ginzburg–Landau equations, in order to link the phenomenological quantities α and β to empirically known quantities, such as the thermodynamic critical field $H_c(T)$ and the effective penetration depth $\lambda_{\text{eff}}(T)$; a common procedure for isotropic systems is the following.

Consider Eq. (68) in the bulk superconductor, in the absence of fields and gradients, i.e. $\mathbf{A}(\mathbf{r}) = 0$ and $\psi(\mathbf{r}) = \text{const} = \psi_\infty$; we see that the value ψ_∞ of the order parameter in the bulk superconductor is given by

$$|\psi_\infty|^2 = -\frac{\alpha(T)}{\beta} \tag{71}$$

when $T < T_c$ (while $\psi_\infty = 0$ for $T > T_c$). From Eq. (66) the free energy difference becomes

$$f_S(T, \psi_\infty) - f_N(T) = -\frac{\alpha^2(T)}{2\,\beta}$$

for $T < T_c$. The first member of this equation can be identified with the condensation energy per unit volume $-H_c^2(T)/8\pi$. From the empirical knowledge of the thermodynamic critical field $H_c(T)$ we obtain a first useful relationship

$$\frac{H_c^2(T)}{8\pi} = \frac{\alpha^2(T)}{2\beta} \ . \tag{72}$$

From the temperature dependence of $\alpha(T)$, described in Eq. (67), we have that $H_c(T) \to T_c - T$ for T approaching T_c from below.

Consider now a superconductor and assume that $\psi(\mathbf{r}) = \psi_\infty$ not only in the bulk but everywhere up to the surface. Eq. (69) becomes in this case

$$\mathbf{J}_S(\mathbf{r}) = -\frac{e^{*2}}{m^* c} |\psi_\infty|^2 \mathbf{A}(\mathbf{r}) \ .$$

This equation, in conjunction with the Maxwell equation curl $\mathbf{B} = (4\pi/c)\mathbf{J}_S(\mathbf{r})$, allows to determine the temperature-dependent effective penetration depth

$$\lambda_{\text{eff}}(T) = \sqrt{\frac{m^* c^2}{4\pi e^{*2} |\psi_\infty|^2}} \equiv \sqrt{\frac{m^* c^2 \beta}{4\pi e^{*2} |\alpha(T)|}} \ . \tag{73}$$

From the temperature dependence of $\alpha(T)$, described in Eq. (67), we have that $\lambda_{\text{eff}}(T)$ diverges as $(T_c - T)^{-1/2}$ for T approaching T_c from below.

The Ginzburg–Landau theory, besides the expression of the penetration depth, contains another characteristic length, which represents the typical distance over which spatial changes of ψ can occur. Consider, for example, a semi-infinite superconductor in the $z > 0$ halfplane. For this one-dimensional model, the Ginzburg–Landau equation (68), in the absence of magnetic fields, gives

$$-\frac{\hbar^2}{2m^*} \frac{d^2\psi(z)}{dz^2} + \beta |\psi(z)|^2 \psi(z) = -\alpha(T)\psi(z) \ .$$

For $T < T_c$, we divide both members of the above equation by $|\alpha(T)|$ and obtain

$$-\frac{\hbar^2}{2m^* |\alpha(T)|} \frac{d^2\psi(z)}{dz^2} + \frac{|\psi(z)|^2}{|\psi_\infty|^2} \psi(z) = \psi(z) \ . \tag{74}$$

The coefficient of the first term of Eq. (74) allows to define naturally a new scale length $\xi_{GL}(T)$ given by

$$\xi_{GL}^2(T) = \frac{\hbar^2}{2m^* |\alpha(T)|} \ . \tag{75}$$

This length, referred to as the Ginzburg–Landau coherence length, represents the space scale on which the order parameter $\psi(\mathbf{r})$ varies when one introduces some inhomogeneity (for instance a surface). To prove this statement, we remark that the solution of Eq. (74), with the boundary conditions $\psi(0) = 0$ and $\psi(\infty) = \psi_\infty$, is given by

$$\psi(z) = |\psi_\infty| \tanh \frac{z}{\sqrt{2}\,\xi_{GL}(T)} \ .$$

Thus $\xi_{GL}(T)$ governs the way in which $\psi(z)$ approaches its bulk value ψ_∞ within the superconductor.

In superconductors, it is important to compare the Ginzburg–Landau spatial coherence length $\xi_{GL}(T)$ with the penetration depth $\lambda_{\text{eff}}(T)$; both lengths diverge as $(T_c - T)^{-1/2}$ for T approaching T_c from below. At a boundary between a normal region and a superconducting region, the coherence length governs the region where the condensation energy is partial, while the effective penetration depth governs the region where the Meissner effect is partial. It is thus useful to define the dimensionless and temperature independent parameter

$$\kappa = \frac{\lambda_{\text{eff}}(T)}{\xi_{GL}(T)} = \frac{1}{\sqrt{2}\,\pi}\,\frac{m^*\,c}{|e^*|\,\hbar}\,\sqrt{\beta}\ . \tag{76}$$

As we shall see in Section 7.3, the parameter κ determines whether the material is a type-I or a type-II superconductor. The Ginzburg–Landau equations (68) and (69) have been particularly successful (at times beyond hope) in the description of numerous phenomena, including the basic distinction between type-I and type-II superconductors, boundary formation, thin films, impurity effects etc. In the following we focus on some general aspects implied by the Ginzburg–Landau equations.

7.2   Magnetic flux quantization

Consider a multiply-connected superconductor, for instance a superconducting ring, as illustrated in Fig. 20. The magnetic flux, threading a loop lying in the bulk superconductor, cannot have arbitrary values; rather it has to be an integer multiple of the flux quantum $\Phi_0 = h\,c/2\,e$. The origin of the magnetic flux quantization can be seen with the following arguments.

Consider the Ginzburg–Landau equation (70b) for the current density, here rewritten in the form

$$\mathbf{J}_S(\mathbf{r}) = \frac{e^*}{m^*}\,|\psi(\mathbf{r})|^2\left[\hbar\,\nabla\theta(\mathbf{r}) - \frac{e^*}{c}\,\mathbf{A}(\mathbf{r})\right]\ . \tag{77}$$

Let Γ be a loop completely inside the superconductor and deep enough from the surfaces of the superconducting ring that magnetic field and screening current have dropped to zero (the circuit Γ must avoid surface regions of the order of the penetration depth, but for the rest is arbitrary). *Inside the superconductor in its Meissner state,* we have $\mathbf{J}_S(\mathbf{r}) = 0$; as a consequence in any point $\mathbf{r}$ of the circuit Γ we have

$$\nabla\theta(\mathbf{r}) = -\frac{2\,e}{\hbar\,c}\,\mathbf{A}(\mathbf{r})\ , \tag{78}$$

where e^* has been set equal to $-2\,e$ for a Cooper pair.

The line integral of $\nabla\theta(\mathbf{r})$ around the closed circuit Γ within the superconductor gives

$$\oint \nabla\theta(\mathbf{r})\cdot d\mathbf{l} = -\frac{2\,e}{\hbar\,c}\oint \mathbf{A}(\mathbf{r})\cdot d\mathbf{l} = -\frac{2\,e}{\hbar\,c}\iint \text{curl}\,\mathbf{A}(\mathbf{r})\cdot d\mathbf{S}\ ,$$

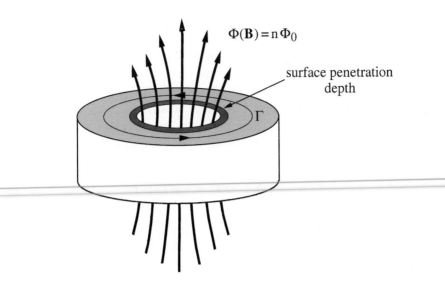

Fig. 20 Quantization of the magnetic flux through a multiply-connected superconductor with a ring shape.

where the last passage exploits the Stokes theorem. Thus we have

$$\Delta\theta = -\frac{2e}{\hbar c}\Phi_\Gamma(\mathbf{B}) , \qquad (79a)$$

where $\Phi_\Gamma(\mathbf{B})$ is the flux of $\mathbf{B}$ through a surface that embraces the line Γ, and $\Delta\theta$ gives the variation of the phase $\theta(\mathbf{r}) = \arg\psi(\mathbf{r})$ after going around Γ once. Since $\psi(\mathbf{r})$ is a single-valued function we have that $\Delta\theta$ must be a multiple of 2π; this condition, inserted into Eq. (79a) gives the flux quantization result

$$\Phi_\Gamma(\mathbf{B}) = \frac{hc}{2e}n = n\Phi_0 , \qquad (79b)$$

where n is an integer and $\Phi_0 = hc/2e$ is the elemental quantum of flux. Notice that in the case the superconductor is simply connected, then n equals zero; but if the superconductor is multiply connected, or, equivalently if it contains normal regions, then n may be an integer number different from zero.

7.3 Type-I and type-II superconductors

One of the most important achievements of the Ginzburg–Landau theory is the natural classification between type-I and type-II superconductors. As discussed in Section 1, type-I superconductors are characterized by the fact that the entire superconducting specimen reverts to the normal state, when an external magnetic field higher than the critical value $H_c(T)$ is applied to the sample (in the standard geometry of a long, thin cylinder shape). In type-II superconductors, for applied fields in the range $H_{c1}(T)$ and $H_{c2}(T)$, the flux begins to penetrate partially in the superconductors; this mixed state reverts back to the ordinary metal state for applied magnetic fields higher than $H_{c2}(T)$. The Ginzburg–Landau theory allows to understand easily the occurrence of type-I and type-II behaviour, the upper and lower critical fields, and the nature of the mixed state.

Let us consider the Ginzburg–Landau equation (68) for a superconductor of type-II, at temperature T, with a uniform applied magnetic field H extremely close to $H_{c2}(T)$. Since we are extremely close to the superconductor–metal transition, the density of superelectrons $|\psi(\mathbf{r})|^2 \to 0$. We can thus linearize the Ginzburg–Landau equation (68) and write

$$\frac{1}{2\,m^*} \left(\mathbf{p} - \frac{e^*}{c}\,\mathbf{A} \right)^2 \psi(\mathbf{r}) = -\alpha(T)\,\psi(\mathbf{r}) \ . \tag{80}$$

The operator in the first member of Eq. (80) is just the Hamiltonian of a free particle of mass m^* and charge e^* in a uniform magnetic field H; its lowest eigenvalue is

$$E_0 = \frac{1}{2}\,\hbar\,\omega_c = \frac{1}{2}\,\hbar\,\frac{|e^*|\,H}{m^*\,c}$$

(as seen in Section XV-2). Uniform magnetic field and beginning of superconductivity are compatible at the critical field $H_{c2}(T)$ such that:

$$\frac{1}{2}\,\hbar\,\frac{|e^*|\,H_{c2}(T)}{m^*\,c} = |\alpha(T)| \ .$$

We have thus for the upper critical field the expression

$$H_{c2}(T) = \frac{2\,m^*\,c}{|e^*|\,\hbar}\,|\alpha(T)| \equiv \sqrt{2}\,\kappa\,H_c(T) \ , \tag{81}$$

where $H_c(T)$ is the thermodynamic critical field, and the last passage in Eq. (81) can be verified by inspection from Eq. (72) and Eq. (76).

Thus we see that for $\kappa > 1/\sqrt{2}$ the critical field $H_{c2}(T) > H_c(T)$; the superconducting state appears at and below $H_{c2}(T)$. Thus the discrimination between type-I and type-II superconductors is just the value $\kappa = 1/\sqrt{2}$ of the Ginzburg–Landau parameter. Most of the clean superconducting elemental metals are type-I superconductors; in contrast, most alloys are type-II superconductors; the high-T_c cuprates are extreme type-II superconductors with very large values $\kappa \approx 100$.

We can now give a few considerations on the mixed state that characterizes type-II superconductors. In type-II superconductors, starting from an applied magnetic field just higher than $H_{c1}(T)$, the flux begins to penetrate partially in the superconductor. It seems plausible to speculate that it enters in the form of cylindrical flux tubes, each carrying a quantum of flux $\Phi_0 = h\,c/2\,e$. In fact, considering an isolated flux tube, and a surrounding circuit Γ well *inside* the superconductor region, the magnetic flux quantization concept of Section 7.2 restricts the threading flux to integer values of the flux quantum Φ_0. A more detailed analysis shows that the single quantum flux formation is favoured. The single-valuedness of the order parameter also imposes that $|\psi(\mathbf{r})|$ must be equal to zero along the axis of the tube. Thus we expect that a vortex can be approximately visualized as a "core" of normal metal completely surrounded by the superconducting material; the radius of the core is expected to be of the order of the Ginzburg–Landau correlation length $\xi_{GL}(T)$, and the radius from the axis where screening supercurrents flow is of the order of $\lambda_{\mathrm{eff}}(T)$. These qualitative considerations are corroborated by detailed calculations.

We can also estimate the lower critical field $H_{c1}(T)$. The estimate is done by requiring that the flux within a cylinder of radius $\lambda_{\mathrm{eff}}(T)$ is just a flux quantum

$$\pi \lambda_{\mathrm{eff}}^2(T) \, H_{c1}(T) \approx \Phi_0 = \frac{h\,c}{2\,e} \; .$$

Thus we expect approximately

$$H_{c1}(T) \approx \frac{h\,c}{|e^*|} \frac{1}{\pi \lambda_{\mathrm{eff}}^2(T)} = 2\sqrt{2} \, \frac{H_c(T)}{\kappa} \; , \tag{82}$$

where the last passage in Eq. (82) can be verified by inspection from Eq. (72), Eq. (73) and Eq. (76). We thus see that $H_c(T)$ is approximately the geometrical mean between $H_{c1}(T)$ and $H_{c2}(T)$. Many other aspects of vortices, regular array of flux tubes, flow and pinning in the presence of a current density and inhomogeneities could be studied, but we confine ourselves to the brief outline given so far.

8 Cooper pair tunneling between superconductors and Josephson effects

In Section 4.3, we have considered tunneling of "normal electrons" (or "quasiparticles") from a superconducting SIS junction, composed by the superconducting films (of the same material) separated by a very thin insulating layer. The highly nonlinear current–voltage $I - V$ characteristic at zero temperature is schematically indicated in Fig. 21; the onset of quasiparticle tunneling occurs for $V = 2\Delta_0/e$, where Δ_0 is the energy gap parameter of the superconductor.

For superconducting tunnel junctions with extremely thin insulating layers (10-15 Å), the electron pair correlations extend through the insulating barrier. In this situation, it has been predicted by Josephson that "paired electrons" can tunnel without dissipation from one superconductor to the other superconductor on the opposite side of the insulating layer [B. D. Josephson, Phys. Letters 1, 251 (1962)]. The direct supercurrent of pairs, for currents less than a certain critical value I_J, flows with *zero voltage drop across the junction (dc Josephson effect)*, as illustrated in Fig. 21. The width of the insulating barrier of the junction limits the maximum supercurrent that can flow across the junction, but introduces no resistance in the flow. Josephson also predicted that, in the case a *constant finite voltage* V is established across the junction, an alternating supercurrent $I_J \sin(\omega_J t + \phi_0)$ flows with frequency $\omega_J = 2\,e\,V/\hbar$ *(ac Josephson effect)*.

We discuss here only the most elementary aspects of this subject, remarkable both for its fundamental aspects and technological applications. Consider two superconductors separated by a thin insulating barrier of width b, in the geometry indicated in Fig. 22. In the case the insulating barrier is infinitely thick, the superconductor on the

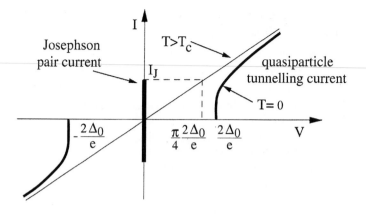

Fig. 21 Schematic representation of the direct current $I - V$ characteristics (at $T = 0$) of a superconductor–insulator–superconductor junction displaying the Josephson current. For $V \equiv 0$ a direct supercurrent can flow (up to a maximum value I_J). For $0 < V < 2\Delta_0/e$ an alternate supercurrent flows with frequency $\omega = 2eV/\hbar$, and no direct supercurrent is observed. For $V > 2\Delta_0/e$ the quasiparticle tunneling current is reported.

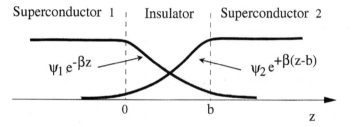

Fig. 22 Schematic representation of a junction between two superconductors separated by a thin insulating barrier of width b. The behaviour of the order parameter wavefunction for the superconductor–insulator–superconductor junction is indicated.

left side would be characterized by the order parameter ψ_1, that can be written as

$$\psi_1 = |\psi_1| \, e^{i\,\theta_1} \,, \tag{83a}$$

with $|\psi_1|$ and θ_1 uniform on the whole volume of superconductor 1. Similarly the superconductor on the right side would be characterized by the order parameter

$$\psi_2 = |\psi_2| \, e^{i\,\theta_2} \,, \tag{83b}$$

with $|\psi_2|$ and θ_2 space independent on the volume of superconductor 2. When the two superconductors are separated by a thin insulating barrier, we can expect that the superconducting order parameters ψ_1 and ψ_2 decay within the insulating region; we can guess that the order parameter $\psi(z)$ within the barrier can be expressed in the form

$$\psi(z) = \psi_1 \, e^{-\beta z} + \psi_2 \, e^{+\beta\,(z-b)} \,, \tag{83c}$$

where β characterizes the damping within the barrier.

We insert Eq. (83c) into the Ginzburg–Landau equation (69) (assuming negligible the magnetic fields) and obtain

$$J_S(z) = \frac{-i\hbar e^*}{2m^*}\left[\psi^*(z)\frac{d}{dz}\psi(z) - \psi(z)\frac{d}{dz}\psi^*(z)\right] = \frac{-i\hbar e^*}{m^*}\beta e^{-\beta b}(\psi_1^*\psi_2 - \psi_1\psi_2^*)$$

$$= \frac{2\,\hbar\,e^*}{m^*}\,\beta\,e^{-\beta b}\,|\psi_1|\,|\psi_2|\,\sin(\theta_2 - \theta_1)\ .$$

We thus see that the Josephson supercurrent, that flows between two superconductors separated by an insulating barrier, is related to the phase difference $\gamma = \theta_2 - \theta_1$ of the order parameters in the two superconductors by the relation

$$\boxed{I = I_J \sin\gamma}\ ,\tag{84a}$$

where I_J depends on the geometrical and physical properties of the junction. Eq. (84a) has been derived on the basis of a heuristic approach. A detailed analysis based on the microscopic BCS theory provides not only Eq. (84a), but also the maximum current $I = I_J$, which can flow across the junction without dissipation. The critical value I_J of an ideal SIS tunnel junction is given by the current that would flow applying a voltage equal to $(\pi/4)(2\,\Delta_0/e)$ to the normal junction, in the same geometrical conditions. Notice finally that our treatment assumes that the magnetic field flux that threads the junction is negligible; in general, the dependence of I_J on the magnetic field (*diffractive pattern*) should be appropriately considered.

Consider now a superconducting junction biased with a (constant or time dependent) voltage V. If a potential difference is established between the two superconductors, the relative energy difference between Cooper pairs belonging to different superconductors is $2\,e\,V$. In perfect analogy with the quantum mechanical rate of change of phases of ordinary eigenfunctions (energy divided by $\hbar$), we expect for the time variation of the relative phase

$$\boxed{\frac{d\gamma}{dt} = \frac{2\,e\,V}{\hbar}}\ .\tag{84b}$$

It is now easy to see that the above equations (84) describe both the dc Josephson effect as well as the ac Josephson effect.

If the potential V across the SIS junction is zero, we see from Eq. (84b) that γ is constant; from Eq. (84a) we have that any supercurrent with intensity ranging from $-I_J$ to $+I_J$ can flow through the junction (the actual value is determined by the external circuit); this is the origin of the zero resistance spike at $V = 0$ in the $I - V$ characteristics for a Josephson junction shown in Fig. 21. A schematic illustration of the dc Josephson effect is reported in Fig. 23.

Suppose that a *constant potential* V $(0 < V < 2\,\Delta_0/e)$ is established across the SIS junction (for simplicity we assume that an "ideal" voltage source maintains a constant voltage V across the SIS junction; this independently from the normal current, that

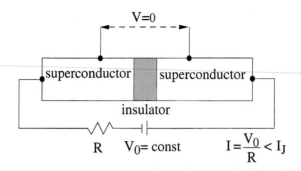

Fig. 23 Schematic illustration of the dc Josephson effect; a direct supercurrent (up to a maximum value I_J) flows without dissipation through the insulating layer.

could be flowing through the circuit). Integration of Eq. (84b) gives

$$\gamma(t) = \frac{2\,e\,V}{\hbar}t + \gamma_0 \ .$$

The phase difference is a linear function of time and the Josephson current is

$$I = I_J \sin\left(\frac{2\,e\,V}{\hbar}t + \gamma_0\right) \ .$$

This pair current is oscillatory with angular frequency $\omega_J \equiv 2\,\pi\,\nu_J = (2\,e/\hbar)\,V$. The frequency ν_J can be expressed as

$$\nu_J = \frac{2\,e}{\hbar}\,V = 483.6\,MHz \cdot \frac{V}{\mu\text{volt}} \ ,$$

where V is measured in microvolt.

As suggested by Josephson, experimental evidence of the ac current can be conveniently obtained by applying across the junction both a constant voltage V and a radiofrequency or microwave voltage, say $V_r \cos\omega_r\,t$. When the combined dc and ac voltage $V_{\text{tot}} = V + V_r \cos\omega_r\,t$ is applied, from Eq. (84b) we have

$$\frac{d\gamma}{dt} = \frac{2\,e\,V}{\hbar} + \frac{2\,e\,V_r}{\hbar}\,\cos\omega_r\,t \ . \tag{85a}$$

By integration, we obtain

$$\gamma(t) = \frac{2\,e\,V}{\hbar}t + \frac{2\,e\,V_r}{\hbar\,\omega_r}\,\sin\omega_r\,t + \gamma_0 \ , \tag{85b}$$

where γ_0 is an appropriate integration constant. The Josephson current is thus

$$I(t) = I_J \sin\left[\frac{2\,e\,V}{\hbar}t + \gamma_0 + \frac{2\,e\,V_r}{\hbar\,\omega_r}\,\sin\omega_r\,t\right] \ . \tag{85c}$$

This is a frequency modulated current. To analyse it, we use the following well-known mathematical expressions

$$\cos(a \sin x) = \sum_{n=-\infty}^{+\infty} J_n(a) \cos nx , \qquad \sin(a \sin x) = \sum_{n=-\infty}^{+\infty} J_n(a) \sin nx ,$$

and hence

$$\sin(b + a \sin x) = \sum_{n=-\infty}^{+\infty} J_n(a) \sin(b + nx) ,$$

where J_n are Bessel functions of first kind of order n. Eq. (85c) becomes

$$I(t) = I_J \sum_{n=-\infty}^{+\infty} J_n\left(\frac{2 e V_r}{\hbar \omega_r}\right) \sin\left[\frac{2 e V}{\hbar} t + \gamma_0 + n \omega_r t\right] .$$

We see from this expression that whenever V satisfies the relation

$$2 e V \equiv n \hbar \omega_r \qquad (n = 0, 1, 2, \ldots) , \tag{86}$$

a *zero frequency supercurrent* is obtained (this is called *inverse ac effect*). Thus well defined vertical spikes, known as Shapiro spikes, are expected to occur in the dc current voltage characteristics, whenever Eq. (86) is satisfied. The heights of the Shapiro spikes are $I_J J_n (2 e V_r / \hbar \omega_r)$; in particular for the zero voltage spike, the critical dc Josephson current becomes

$$I_c = I_J J_0 \left(\frac{2 e V_r}{\hbar \omega_r}\right) ,$$

and its dependence on the amplitude of the radiofrequency voltage can be used to measure it. The Shapiro spikes are so well defined that they are used for accurate determination of the ratio $2 e/\hbar$. We do not pursue further the current–voltage characteristics of superconductor junctions; we can mention however their use as fast switches, and the interest for lossless computer elements.

Before concluding, we also briefly mention the superconducting quantum interference devices (SQUID). Consider a superconducting circuit with two Josephson junctions in parallel, as schematized in Fig. 24. For simplicity we assume that both junctions have the same critical current I_J, that the flux threading the junction is negligible and the self-inductance of the circuit is negligible (these and other refinements could be incorporated in a more careful analysis).

Let us indicate with γ_A and γ_B the phase differences across the junctions A and B respectively. We have

$$\gamma_A = \theta(A_2) - \theta(A_1) \quad \text{and} \quad \gamma_B = \theta(B_2) - \theta(B_1) ,$$

where A_1 and A_2 are points just at the left and at the right of the junction A (B_1 and B_2 of junction B), and θ is the phase of the superconducting wavefunction. The total current flowing in the circuit is

$$I = I_J (\sin \gamma_A + \sin \gamma_B) = 2 I_J \sin \frac{\gamma_A + \gamma_B}{2} \cos \frac{\gamma_A - \gamma_B}{2} . \tag{87}$$

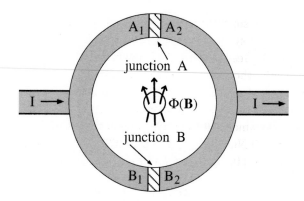

Fig. 24 Schematic diagram of a dc superconducting quantum interference device for detecting an enclosed magnetic field of flux $\Phi(\mathbf{B})$.

In the interior of a superconductor in its Meissner state $\mathbf{J}_S = 0$; then from expression (70b) we have

$$\nabla\theta(\mathbf{r}) = -\frac{2e}{\hbar c}\,\mathbf{A}(\mathbf{r})\ .$$

Let us note that

$$\gamma_A - \gamma_B = [\theta(B_1) - \theta(A_1)] + [\theta(A_2) - \theta(B_2)]$$

$$= -\frac{2e}{\hbar c}\int_{A_1}^{B_1}\mathbf{A}(\mathbf{r})\cdot d\mathbf{l} - \frac{2e}{\hbar c}\int_{B_2}^{A_2}\mathbf{A}(\mathbf{r})\cdot d\mathbf{l} = -\frac{2e}{\hbar c}\oint\mathbf{A}(\mathbf{r})\cdot d\mathbf{l}$$

$$= -\frac{2e}{\hbar c}\,\Phi(\mathbf{B}) = -2\pi\,\frac{\Phi(\mathbf{B})}{\Phi_0}\ .$$

Eq. (87) then gives for the current the expression

$$I = 2\,I_J\cos\pi\,\frac{\Phi(\mathbf{B})}{\Phi_0}\,\sin\frac{\gamma_A + \gamma_B}{2}\ .$$

The above expression takes its maximum value

$$I_{\max} = 2\,I_J\,\Big|\cos\pi\,\frac{\Phi(\mathbf{B})}{\Phi_0}\Big|\ . \tag{88}$$

We see that $I_{\max}$ changes from maxima to minima for a change of flux as small as $\Phi_0/2$; very small magnetic fields can thus be measured with high precision.

The sensitive superconducting quantum interference devices offer the opportunity of a number of interesting applications. We mention, for instance, clinical applications and in particular magneto-encephalography and magneto-cardiography; these techniques use an appropriate array of SQUIDs to measure the tiny magnetic fields produced by the electric currents involved in the activity of the brain or of the heart [see for instance M. Hämäläinen, R. Hari, R. J. Ilmoniemi, J. Knuutila, and O. V. Lounasmaa, Rev. Mod. Phys. **65**, 413 (1993) and references quoted therein].

We also mention that sophisticated SQUID experiments have been proposed for the determination of the symmetry of the order parameter in non-conventional superconductors by V. B. Geshkenbein, A. I. Larkin and A. Barone, Phys. Rev. B**36**, 235 (1987). In principle, the intrinsic relative phase of the order parameters between two different directions of a given superconductor can be inferred from SQUID interference experiments. These investigations have opened a wide debate concerning the s-wave versus d-wave symmetry of the order parameter in high T_c superconductors [see for instance A. Barone, Il Nuovo Cimento **16**D, 1635 (1994), and D. J. Van Harlingen, Rev. Mod. Phys. **67**, 515 (1995), and references quoted therein].

APPENDIX A. The phonon-induced electron–electron interaction

The origin of the phenomena related to the superconductivity is a small attractive interaction between electrons; in the conventional superconductivity it is generally accepted that this indirect interaction occurs via the phonon field (Fröhlich attractive mechanism). While the quantitative analysis of effective interactions constitute a formidable problem, from a qualitative point of view simple models can be considered to evidentiate some basic aspects; in particular it is not difficult to show that one should expect an attractive interactions for electrons lying near the Fermi surface in an energy shell of the order of $\hbar\omega_D$, ω_D being the Debye frequency. To make this result plausible, we exploit a well-known standard canonical transformation, and then we apply it to the indirect interaction via the phonon field.

A canonical transformation

Let us consider a Hamiltonian H; we can always perform a canonical transformation of the type

$$\widetilde{H} = e^{-S} H e^{S} = \left(1 - S + \frac{1}{2!} S^2 - \dots\right) H \left(1 + S + \frac{1}{2!} S^2 + \dots\right)$$

$$= H + [H, S] + \frac{1}{2!} [[H, S], S] + \frac{1}{3!} [[[H, S], S], S] + \dots \qquad (A1)$$

with S any arbitrary operator. It is evident that the eigenvalues of $\widetilde{H}$ and H coincide.

Suppose that the operator H is split in the form

$$H = H_0 + H_1 . \qquad (A2)$$

Then Eq. ($A1$) reads

$$\widetilde{H} = H_0 + H_1 + [H_0, S] + [H_1, S] + \frac{1}{2} [[H_0, S], S] + \dots$$

We can exploit the arbitrariness of S to satisfy the relation

$$H_1 + [H_0, S] = 0 . \qquad (A3)$$

With such a choice, $\widetilde{H}$ becomes

$$\widetilde{H} = H_0 + \frac{1}{2}[H_1, S] + \dots$$

If we imagine to replace H_1 with λH_1 (with λ eventually put equal to 1) we see that the terms neglected in $\widetilde{H}$ are at least of order λ^3. Thus the eigenvalues of H and the eigenvalues of $\widetilde{H}$, where

$$\widetilde{H} = H_0 + H_{\text{indirect}}$$

$$H_{\text{indirect}} = \frac{1}{2}[H_1, S],$$

are equal to order λ^2 in the coupling parameter λ.

From Eq. $(A3)$, we see that the operator S in a representation in which H_0 is diagonal is

$$\langle n|S|m\rangle = \frac{\langle n|H_1|m\rangle}{E_m - E_n}$$

(to avoid singularities we assume that the diagonal matrix elements of H_1 in a representation in which H_0 is diagonal are zero; if not so, these terms could be included in H_0). The expression of the matrix elements of H_{indirect} on the eigenstates of H_0 is

$$\langle f|H_{\text{indirect}}|i\rangle = \frac{1}{2}\langle f|H_1 S - S H_1|i\rangle = \frac{1}{2}\sum_\alpha [\langle f|H_1|\alpha\rangle\langle\alpha|S|i\rangle - \langle f|S|\alpha\rangle\langle\alpha|H_1|i\rangle]$$

$$= \frac{1}{2}\sum_\alpha \langle f|H_1|\alpha\rangle\langle\alpha|H_1|i\rangle \left[\frac{1}{E_i - E_\alpha} + \frac{1}{E_f - E_\alpha}\right]. \qquad (A4)$$

The meaning of H_{indirect} can be further clarified if we note that $\langle i|H_{\text{indirect}}|i\rangle$ is the standard second order correction given by perturbation theory. It can be noticed that the matrix elements $\langle f|H_{\text{indirect}}|i\rangle$ can also be obtained by the renormalization procedure of Section V-8.4 (simply disregarding the coupling of the intermediate states $|\alpha\rangle$ with any other state but $|i\rangle$ and $|f\rangle$, and putting the energy E equal to E_i and E_f).

Indirect electron–electron interaction

The mechanism leading to the indirect electron–electron interaction via phonons is schematically represented in Fig. 25: one electron of wavevector $\mathbf{k}$ "emits" a phonon of wavevector $\mathbf{q}$ and is scattered into the state $\mathbf{k} - \mathbf{q}$; the phonon of wavevector $\mathbf{q}$ is immediately "absorbed" by another electron of wavevector $\mathbf{k}'$ that scatters into the state $\mathbf{k}' + \mathbf{q}$. The quantum mechanical analysis of this process provides a net attractive interaction between pairs of electrons, whose energy difference is smaller than the phonon energy $\hbar\omega_\mathbf{q}$. To show this, we can apply the canonical transformation considered before to obtain the effective electron–electron interaction.

We consider final and initial electron states, coupled with the vacuum phonon state (the indirect interaction we obtain is, however, independent from the phonon occupa-

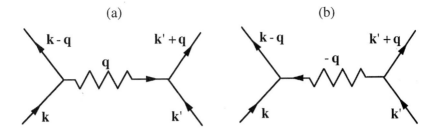

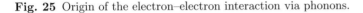

Fig. 25 Origin of the electron–electron interaction via phonons.

tion numbers). With reference to Fig. 25a, the states of interest and their unperturbed energy (neglecting electron–phonon interaction) are:

$$|i\rangle = |\mathbf{k}, \mathbf{k}'; 0\rangle \qquad E_i = E_{\mathbf{k}} + E_{\mathbf{k}'}$$
$$|\alpha\rangle = |\mathbf{k} - \mathbf{q}, \mathbf{k}'; 1\rangle \qquad E_\alpha = E_{\mathbf{k}-\mathbf{q}} + E_{\mathbf{k}'} + \hbar\omega_{\mathbf{q}} \qquad (A5)$$
$$|f\rangle = |\mathbf{k} - \mathbf{q}, \mathbf{k}' + \mathbf{q}; 0\rangle \qquad E_f = E_{\mathbf{k}-\mathbf{q}} + E_{\mathbf{k}'+\mathbf{q}} \ .$$

Let us consider the electron–phonon interaction term of the type

$$H_1 = M_{\mathbf{q}} \, e^{-i\,\mathbf{q}\cdot\mathbf{r}} \, a_{\mathbf{q}}^\dagger + M_{\mathbf{q}}^* \, e^{+i\,\mathbf{q}\cdot\mathbf{r}} \, a_{\mathbf{q}} \ . \qquad (A6)$$

The matrix element for the scattering of an electron from the state $\mathbf{k}$ into the state $\mathbf{k} - \mathbf{q}$ with creation of a phonon $\mathbf{q}$ on the vacuum phonon state is $M_{\mathbf{q}} \langle 1|a_{\mathbf{q}}^\dagger|0\rangle = M_{\mathbf{q}}$; similarly the scattering amplitude from the state $\mathbf{k}'$ into the state $\mathbf{k}'+\mathbf{q}$ with absorption of a phonon $\mathbf{q}$ is $M_{\mathbf{q}}^* \langle 0|a_{\mathbf{q}}|1\rangle = M_{\mathbf{q}}^*$. Applying Eq. $(A4)$ with the states $(A5)$ we obtain

$$\langle f|H_{\text{indirect}}|i\rangle = \frac{1}{2}|M_{\mathbf{q}}|^2 \left[\frac{1}{E_{\mathbf{k}} - E_{\mathbf{k}-\mathbf{q}} - \hbar\omega_{\mathbf{q}}} + \frac{1}{E_{\mathbf{k}'+\mathbf{q}} - E_{\mathbf{k}'} - \hbar\omega_{\mathbf{q}}} \right] , \qquad (A7)$$

By assuming $\hbar\omega_{\mathbf{q}} \approx \hbar\omega_D$, it is seen that in the energy shell of the order of $\hbar\omega_D$ around the Fermi energy this virtual process leads to an attractive interaction. It is customary to add to the above indirect matrix element $(A7)$ the term calculated from Fig. 25b; the sum gives

$$\langle f|H_{\text{indirect}}|i\rangle = |M_{\mathbf{q}}|^2 \left[\frac{\hbar\omega_{\mathbf{q}}}{(E_{\mathbf{k}} - E_{\mathbf{k}-\mathbf{q}})^2 - \hbar^2\omega_{\mathbf{q}}^2} + \frac{\hbar\omega_{\mathbf{q}}}{(E_{\mathbf{k}'+\mathbf{q}} - E_{\mathbf{k}'})^2 - \hbar^2\omega_{\mathbf{q}}^2} \right] . \qquad (A8)$$

The indirect interaction $(A7)$ or $(A8)$ has been calculated assuming initial and final electron states coupled to the vacuum phonon state. It is easily seen that the indirect interaction is independent from the phonon state considered; the key point is just the *non-commutation* between the phonon creation and annihilation operators appearing in Eq. $(A6)$. [In the hypotetical case of "ordinary scattering" with commuting operators, no indirect interaction would occur; the role played by non-commutativity of creation and annihilation phonon operators is formally similar to the role played by non-commutativity of raising and lowering spin operators in the Kondo problem of Section XVI-6].

Further Reading

A. A. Abrikosov "Fundaments of the Theory of Metals" (North-Holland, Amsterdam 1988)

M. Acquarone (editor) "High Temperature Superconductivity" (World Scientific, Singapore 1996)

J. Bardeen, L. N. Cooper and J. R. Schrieffer "Theory of Superconductivity" Phys. Rev. **108**, 1175 (1957)

A. Barone and G. Paternò "Physics and Applications of the Josephson Effect" (Wiley, New York 1982)

J. G. Bednorz and K. A. Müller "Possible High- Superconductivity in the Ba-La-Cu-O System" Z. Phys. B**64**, 189 (1986); "Perovskite-Type Oxides. The New Approach to High-T_c Superconductivity" Rev. Mod. Phys. **60**, 585 (1988).

G. Burns "High-Temperature Superconductivity" (Academic Press, New York 1992)

M. Cyrot and D. Pavuna "Introduction to Superconductivity and High-T_c Materials" (World Scientific, Singapore 1992)

P. G. de Gennes "Superconductivity of Metals and Alloys" (Benjamin, New York 1966; Addison-Wesley, Reading, Massachusetts 1989)

R. P. Huebener "Magnetic Flux Structure in Superconductors" (Springer, Berlin 1979)

B. D. Josephson "Possible New Effects in Superconducting Tunneling" Phys. Letters **1**, 251 (1962)

R. D. Parks (editor) "Superconductivity" Vol. 1 and 2 (Dekker, New York 1969).

J. C. Phillips "Physics of High-T_c Superconductors" (Academic Press, New York 1989)

C. P. Poole, H. A. Farach and R. J. Creswick "Superconductivity" (Academic Press, San Diego 1995)

C. N. R. Rao (editor) "Chemistry of High T_c Superconductors" (World Scientific, Singapore 1991)

G. Rickayzen "Theory of Superconductivity" (Wiley, New York 1965)

A. C. Rose-Innes and E. H. Rhoderick "Introduction to Superconductivity" (Pergamon Press, Oxford 1978).

J. R. Schrieffer "Theory of Superconductivity" (Benjamin, New York 1983)

D. R. Tilley and J. Tilley "Superfluidity and Superconductivity" (Adam Hilger, Bristol 1990)

M. Tinkham "Introduction to Superconductivity" (McGraw-Hill, New York, 2nd edition 1996)

J. R. Waldram "Superconductivity of Metals and Cuprates" (Institute of Physics, Bristol 1996)

Subject index